Intro Stats

SECOND EDITION

Intro Stats

SECOND EDITION

Richard D. De Veaux

Williams College

Paul F. Velleman

Cornell University

David E. Bock

Cornell University

Boston San Francisco New York
London Toronto Sydney Tokyo Singapore Madrid
Mexico City Munich Paris Cape Town Hong Kong Montreal

Publisher	Greg Tobin
Executive Editor	Deirdre Lynch
Senior Project Editor	Rachel S. Reeve
Editorial Assistant	Sara Oliver
Managing Editor	Ron Hampton
Senior Production Supervisor	Jeffrey Holcomb
Cover and Interior Design	Barbara T. Atkinson
Marketing Manager	Phyllis Hubbard
Marketing Assistant	Celena Carr
Photo Research	Beth Anderson and Naomi Kornhauser
Associate Media Producer	Sara Anderson
Senior Technical Art Specialist	Joe Vetere
Senior Manufacturing Buyer	Evelyn Beaton
Senior Media Buyer	Ginny Michaud
Project Manager	Bonnie Boehme, Nesbitt Graphics, Inc.
Production Coordinator	Marilyn Dwyer, Nesbitt Graphics, Inc.
Artist	Lisa Travis, Nesbitt Graphics, Inc.
Composition	Nesbitt Graphics, Inc.
Cover Image	© Gandee Vasan / Getty Images. Goldfish jumping from a crowded bowl, landing in an empty bowl (composite).

For permission to use copyrighted material, grateful acknowledgment is made to the copyright holders listed in Appendix C, which is hereby made part of this copyright page.

Many of the designations used by manufacturers and sellers to distinguish their products are claimed as trademarks. Where those designations appear in this book, and Addison-Wesley was aware of a trademark claim, the designations have been printed in initial caps or all caps.

If you purchased this book within the United States or Canada, you should be aware that it has been wrongfully imported without the approval of the Publisher or Author.

De Veaux, Richard D.
   Intro stats / Richard D. De Veaux, Paul F. Velleman, David E. Bock.-- 2nd ed.
      p. cm.
   Includes index.
   ISBN 0-321-31520-0
   1. Statistics--Textbooks. I. Velleman, Paul F., 1949- II. Bock, David E. III. Title.

QA276.12.D4 2006
519.5--dc22

2 3 4 5 6 7 8 9 10-QWT-08 07 06

*To Sylvia, who has helped me in more ways than she'll ever know,
and to Nicholas, Scyrine, Frederick, and Alexandra,
who make me so proud in everything that they are and do*
—Dick

To my sons, David and Zev, from whom I've learned so much
—Paul

*To Greg and Becca, great fun as kids and great friends as adults,
and especially to my wife and best friend, Joanna, for her
understanding, encouragement, and love*
—Dave

Meet the Authors

Richard D. De Veaux is an internationally known educator and lecturer. He has taught at the Wharton School and the Princeton University School of Engineering, where he won a "Lifetime Award for Dedication and Excellence in Teaching." Since 1994, he has been Professor of Statistics at Williams College. Dick has won both the Wilcoxon and Shewell awards from the American Society for Quality. He is a fellow of the American Statistical Association. Dick is also well known in industry, where for the past 20 years he has consulted for such companies as Hewlett-Packard, Alcoa, DuPont, Pillsbury, General Electric, and Chemical Bank. He has also sometimes been called the "Official Statistician for the Grateful Dead."

Dick holds degrees from Princeton University in Civil Engineering (B.S.E.) and Mathematics (A.B.) and from Stanford University in Dance Education (M.A.) and Statistics (Ph.D.) where he studied with Persi Diaconis. His research focuses on the analysis of large data sets and data mining in science and industry.

In his spare time he is an avid cyclist and swimmer. He also is the founder and bass for the "Diminished Faculty," an a cappella Doo-Wop quartet at Williams College. Dick is the father of four children.

Paul F. Velleman has an international reputation for innovative Statistics education. He is the author and designer of the multimedia statistics CD-ROM *ActivStats*, for which he was awarded the EDUCOM Medal for innovative uses of computers in teaching statistics, and the ICTCM Award for Innovation in Using Technology in College Mathematics. He also developed the award-winning statistics program, Data Desk, and the Internet site Data And Story Library (DASL) (http://dasl.datadesk.com), which provides data sets for teaching Statistics. Paul co-authored (with David Hoaglin) *ABCs of Exploratory Data Analysis*.

Paul has taught Statistics at Cornell University since 1975. He holds an A.B. from Dartmouth College in Mathematics and Social Science, and M.S. and Ph.D. degrees in Statistics from Princeton University, where he studied with John Tukey. His research often deals with statistical graphics and data analysis methods.

Paul is a Fellow of the American Statistical Association and of the American Association for the Advancement of Science.

Out of class, Paul sings baritone in a barbershop quartet and tries to keep up with his younger son on figure skates. He is the father of two boys.

David E. Bock taught mathematics at Ithaca High School for 35 years. He has taught Statistics at Ithaca High School, Tompkins-Cortland Community College, Ithaca College, and Cornell University. Dave has won numerous teaching awards, including the MAA's Edyth May Sliffe Award for Distinguished High School Mathematics Teaching (twice), Cornell University's Outstanding Educator Award (three times), and has been a finalist for New York State Teacher of the Year.

Dave holds degrees from the University at Albany in Mathematics (B.A.) and Statistics/Education (M.S.)

Dave has been a reader for the AP Statistics exam, serves as a Statistics consultant to the College Board, and leads workshops and institutes for AP Statistics teachers. He is currently K–12 Education and Outreach Coordinator and a senior lecturer for the Mathematics Department at Cornell University.

Dave relaxes by biking and hiking. He and his wife have enjoyed many days camping across Canada and through the Rockies. They have a son, a daughter, and twin granddaughters.

CONTENTS

*Indicates an optional chapter.

*Indicates an optional chapter.

PREFACE

About This Edition

We've been thrilled with the feedback we've received from instructors working with the first edition of *Intro Stats.* We've been even more pleased by the feedback from students. Our goal was to create a book that students would find clear and accessible—a book that they would actually read. We tried to share our excitement about doing Statistics with them by emphasizing in every chapter that Statistics is about understanding the world. They have told us that we are on the right track. Both students and instructors appreciated our *Think, Show* and *Tell* rubric, our *What Can Go Wrong?* sections, the sidebar stories, and even the occasional irreverent footnote.

The feedback from instructors asked for more. You suggested adding some breaks along the way where students could check their understanding as they read each chapter. You suggested that while pairing the exercises (so one with an answer in the back was followed by one without) helped students, we could help still more by improving the grading of the exercises. And, you said that a brief summary of concepts would be a useful addition to our list of the key terms and skills at the end of each chapter. We've tried to respond to these suggestions by adding a *Just Checking* section once or twice a chapter, expanding and sorting the exercises, and adding *What Have We Learned?* to the chapter summaries.

We made other improvements, too. We've streamlined some of the text and added clarifications where necessary, all the while keeping the full flavor and readability of the first edition. We hope that this second edition of *Intro Stats* will help students in all disciplines to understand and appreciate the exciting field of Statistics even more.

Our Goal: Read This Book!

The best text in the world is of little value if students don't read it. Here are some of the ways we have tried to make *Intro Stats,* Second Edition more approachable:

- *Readability.* You'll see immediately that this book doesn't read like other Statistics texts. The style is both colloquial and informative, engaging students to actually read the book to see what it says.

- *Humor.* We know that humor is the best way to promote learning. You will find quips and wry comments throughout the narrative, in margin notes, and in footnotes. If we sometimes cross the line and seem to trivialize something important, we apologize.

- *Informality.* Our informal diction doesn't mean that the subject matter is covered lightly or informally. We have tried to be precise and, wherever possible, to offer deeper explanations and justifications than those found in most introductory texts.

- *Focused lessons.* The chapters are shorter than in most other texts, to make it easier to focus on one topic at a time.

- *The need to read.* Students who plan just to skim the book may find our presentation a bit frustrating. The important concepts, definitions, and sample solutions don't sit in little boxes. This is a book that needs to be read, so we've tried to make the reading experience enjoyable.

New Features

In the second edition, we have added several new features:

- *Just Checking.* Even though we've written the book to be easy to read, there are sophisticated concepts that students need to understand. So, about once or twice a chapter, we ask students to pause and think about what they've just read. To help them check their understanding, we ask some thought questions to ensure that they understand the material. These questions are designed to be a quick check; most involve very little calculation. We've put the answers at the end of the chapter so students can easily check themselves.

- *ActivStats Pointers.* The CD-ROM bound in new copies of the book includes the *ActivStats* multimedia materials. Many students choose to look at these first, before reading the chapter or attending a class on each subject. Others may wish to just pick out simulations, interactive exercises, and animated explanations for particular topics. We've included occasional pointers to *ActivStats* activities that parallel the discussions in the book. *ActivStats* matches the book chapter by chapter, so these pointers just name the activity. The chapter number is the same as in the book.

- *What Have We Learned?* We've expanded our chapter-ending summaries into a more complete overview called *What Have We Learned?* This section highlights the new concepts, defines the new terms introduced in the chapter, and lists the skills that the student should have acquired. Students can think of this as a study guide. If they understand the concepts in the summary, know the terms, and have the skills, they're probably ready for the exam.

- *Instructor's Guide.* We wrote our Instructor's Resource Guide for the first edition to help instructors prepare for class. It contained chapter-by-chapter comments on the major concepts, tips on presenting topics (and what to avoid), teaching examples, and a list of resources. The Guide received such a positive response that we decided to print that material at the front of every chapter in the Instructor's Edition. We hope you'll find it useful whether you're a seasoned veteran or just embarking on teaching a data-oriented Statistics course for the first time.

- *Exercises.* In this edition, we've improved the exercises two ways. First, we've expanded them, adding 6 to 10 new exercises in every chapter. And we've improved their arrangement, too. Generally the exercises start with a straightforward application of the ideas of the chapter in small bites. Next, the exercises tackle larger problems, but are broken into several parts to help guide the student through the logic of a complete analysis. Finally, there are exercises that ask the student to synthesize and incorporate their ideas with less guidance. Exercises in a book like this are almost never solved with a single number. A complete solution needs to discuss the problem and justify the method (*Think*), to show a worked solution or computer output (*Show*), and to draw a reasoned conclusion that responds to the initial motivation for the exercise (*Tell*). The topics selected reflect those that we all encounter in everyday life, so the importance of having a clearly constructed conclusion is clear. Data for marked exercises, ⊤ , are available on the CD-ROM and Web site, formatted for various technologies.

Continuing Features

We have written *Intro Stats,* Second Edition with several basic features and themes in mind:

- *Think, Show, Tell.* The mantra of *Think, Show,* and *Tell* is repeated in every chapter. It emphasizes the importance of thinking about a Statistics question (What do we know? What do we hope to learn? Are the assumptions and conditions satisfied?) and reporting our findings (the *Tell* step). The *Show* step contains the mechanics of calculating results, and we try to convey our belief that it is only one part of the process. This rubric is highlighted in the Step-by-Step examples that guide the students through the process of analyzing the problem with the general explanation on the left and the worked-out problem on the right.

- *What Can Go Wrong?* The most common mistakes for the new user of statistical methods usually involve misusing a method, not miscalculating a statistic. The slander that "you can prove anything with Statistics" arises from these misuses. We acknowledge these mistakes with a *What Can Go Wrong?* section found near the end of the chapters. One of our goals is to arm students with the tools to detect statistical errors and to offer practice in debunking misuses of statistics, whether intentional or not. In this spirit, our exercises include some that probe the understanding of such failures.

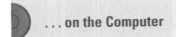

- *By Hand.* Even though we encourage the use of technology to calculate statistical quantities, we realize the pedagogical benefits of occasionally doing a calculation by hand. The *By Hand* boxes break apart the calculation of many simpler formulas to help the student through the calculation of a worked example.

- *Reality Check.* We regularly remind students that Statistics is about understanding the world with data. If our results make no sense, they are probably wrong, no matter how carefully we think we did the calculations. So we ask students to stop for a reality check before interpreting their result.

- *Notation Alert.* Throughout this book we emphasize the importance of clear communication, and proper notation is part of the vocabulary of Statistics. Students who know that in Algebra n can stand for any variable may be surprised to learn that in Statistics n is always and only the sample size. We've found that it helps students when we are clear about the letters and symbols statisticians use to mean very specific things, so we've included notation alerts whenever we introduce a special notation that students will see again.

- *Connections.* The book presents material in 27 short, focused chapters, but they tell one overall story. Each chapter has a *Connections* section to link key terms and concepts with previous discussions and to point out the continuing themes. The *Connections* help to fit newly learned concepts into a growing understanding of Statistics. The best way to retain new learning is to connect it with old understanding.

- *. . . on the Computer.* In the real world, Statistics is practiced with computers. We prefer not to choose a particular Statistics program. Instead, at the end of each chapter, we summarize what students can find in the most common packages, often with an annotated example. We then offer specific guidance for several of the most common packages (Data Desk®, Excel®, JMP®, Minitab®, SPSS®, TI-83/84 Plus®,[1] and TI-89®) to help students get started with the software of their choice. Computer output appearing in the book and in exercises is often generic, resembling all of the common packages to some degree.

Coverage

Textbooks are often defined more by what they choose not to cover than by what they do cover. We've been guided in the choice and order of topics by several fundamental principles. First, we have tried to ensure that each new topic fits into the growing structure of understanding that we hope students will build. We know that learning a new subject such as Statistics requires putting together many new ideas, and we have tried to facilitate that process. That goal has led us to cover some topics in a different order than is found in other texts.

As one example, we introduce inference by looking at a confidence interval for a proportion rather than the more traditional approach that starts with the one-sample mean. Everyone has seen opinion polls. Most people understand that pollsters use a sample of voters to try to predict the preferences of the entire population, and that the result is an estimate with a margin of error. Showing how to

---

[1] For brevity, we will write TI-83 / 84 Plus for the TI-83 Plus and/or TI-84 Plus. Keystrokes and output remain the same for the TI-83 Plus and the TI-84 Plus, so instructions and examples serve for both calculators.

construct and interpret a confidence interval in this context introduces key concepts of inference in a familiar setting. We next examine hypothesis tests for a proportion. The mechanics use formulas and a test statistic that is now familiar, so it is easier to focus on the logic and structure of a formal hypothesis test. After hypothesis tests, we examine inference about the difference between two proportions. We build on the understanding of confidence intervals and hypothesis tests as we introduce the added complexity of comparing two groups. When we then turn our attention to the more complicated topic of inference for means, the only new issue is the t-distribution. You'll find explanations for these and other such decisions in the Instructor's Guide commentaries.

We have been guided in our choice of topics and emphasis by the recommendations of the joint ASA/MAA Statistics Focus Group (Cobb, 1992), which included "Emphasize statistical thinking" and "More data and concepts: less theory, fewer recipes." They urge using the computer to automate calculations and graphics.

We have worked to provide materials to help each class, in its own way, follow the guidelines of the GAISE (Guidelines for Assessment and Instruction in Statistics Education) project sponsored by the American Statistical Association. That report urges (among other detailed suggestions) that Statistics education should:

1. Emphasize Statistical literacy and develop Statistical thinking,
2. Use real data,
3. Stress conceptual understanding rather than mere knowledge of procedures,
4. Foster active learning,
5. Use technology for developing concepts and analyzing data, and
6. Make assessment a part of the learning process.

We also have been guided by the syllabus of the Advanced Placement Statistics course. Although this book is not specifically intended for the AP course (another volume by the same authors aims there), we agree with the wisdom of those who designed that course in their selection of topics and their emphasis on Statistics as a practical discipline.

Nevertheless, you may find that we cover some topics in greater depth than you wish to. In most cases, we invite you to select the topics you prefer, but we encourage you to try our presentation order.

A Few Topics Deserve Special Mention

Chapter 10 on data re-expression may seem intimidating to some students because they'll find a logarithm and square root there. But when technology is available, these are simply buttons on the calculator or functions in the statistics package or spreadsheet. For the purposes of statistical analysis, that is all that the student needs to know about them. Re-expression is so useful, and its application is so straightforward, that we encourage you to consider including this chapter, even though we understand that it is not part of the usual introductory syllabus.

Chapters 12 and 13 on sample surveys and experiments, respectively, provide greater depth on these topics than is often found in introductory books. Feel free to cover only the earlier sections and leave the rest to students to read on their own if they wish. Many will be interested to learn the concepts behind the surveys they read in the papers and the experiments that lead to approving pharmaceuticals. But we do not expect students to design their own surveys or experiments, and few final exams will cover these topics in depth.

We offer you a choice of depths in covering probability. In particular, the chapters on random variables (Chapter 16) and on probability models (Chapter 17) are optional. Students who have the necessary background will benefit from them, but nothing in subsequent chapters depends on them.

The CD-ROM includes two additional chapters covering Analysis of Variance (Chapter 28) and Multiple Regression (Chapter 29). These topics point the way to further study in Statistics.

Mathematics

Mathematics traditionally appears in Statistics texts in several roles:

1. It can provide a concise, clear statement of important concepts.
2. It can describe calculations to be performed with data.
3. It can embody proofs of fundamental results.

Of these, we emphasize the first. Mathematics can make discussions of Statistics concepts, probability, and inference clear and concise. We have not shied away from using math where we felt that it helped to clarify without intimidating. We have tried to be sensitive to those who are discouraged by equations by also providing verbal descriptions and numerical examples.

This book is not concerned with proving theorems about statistics. Some of these theorems are quite interesting, and many are important. Often, though, their proofs are not enlightening to introductory Statistics students, and can distract the audience from the concepts we want them to understand.

Nor do we concentrate on calculations. Although statistics calculations are generally straightforward, they are also usually tedious. And, more to the point, they are often unnecessary. Today, virtually all statistics are calculated with technology, so there is little need for students to spend time summing squared deviations by hand. The equations we use have been selected for their focus on understanding concepts and methods. Often these equations are unsuited to hand calculation. When that happens, we've provided more calculation-friendly versions, often in a marginal note that lays out the calculation for those who find following the process a better way to learn about the result.

Technology and Data

To experience the real world of Statistics, it's best to explore real data sets using modern technology.

- *Technology.* We assume that you are using some form of technology in your Statistics course. That could be a calculator, a spreadsheet, a statistics package, or some combination of these. We also assume that you'll put little emphasis on calculating answers by hand, even though we often show how. However, this is not a technology-heavy book. Rather than adopt any particular software, we discuss generic computer output. It is more helpful to see the features that are common to statistics packages rather than relying on a particular format. We also offer guidance to help students get started on seven common software platforms: Excel, Minitab, Data Desk, JMP, SPSS, TI-83/84 Plus, and TI-89 graphing calculators. This guidance is found at the end of each chapter, specific to the methods learned in that chapter.

- *Data.* Because we use technology for computing, we don't limit ourselves to small, artificial data sets. In addition to including some small data sets, we have built examples and exercises on real data with moderate numbers of cases—usually more than you would want to enter by hand into a program or calculator. Machine-readable versions of the data are included on the CD-ROM as well as on the book's Web site, **http://www.aw-bc.com/dvb.** Once the data set is located, you can copy the data to paste into the software of your choice.

On the CD

- *ActivStats®.* Every new copy of the book comes with *ActivStats*, a multimedia program designed to support learning chapter by chapter with the book. *ActivStats* extends the book with videos of real-world stories, worked examples, animated expositions of each of the major Statistics topics, and tools for performing simulations, for performing hypothesis tests, and for visualizing confidence intervals. You can use *ActivStats* as a convenient way to work with standard distributions, for simulation exercises in understanding probability and inference, and for training in statistics software. In addition, the expositions can serve as an electronic "teaching assistant" who has infinite patience for reviewing topics. Exercises from the book are integrated into *ActivStats* so that the associated data sets are easily available to students in the selected statistics software. *ActivStats* provides hundreds of additional exercises and data sets as well. The *ActivStats* CD-ROM also includes a copy of Data Desk, a full-function stand-alone statistics package. (You may also bundle a special version of *ActivStats* that includes specific instruction for Excel, JMP, Minitab, or SPSS.)

- *Data.* Data for marked exercises, **T**, are available on the CD-ROM (and Web site), formatted for Minitab, Excel, JMP, SPSS, and the TI as well as being provided as text files. The *ActivStats* program also includes a data set browser that lets you search through all of the data available in the book (and a few hundred others as well) by title or by the topics they can be used to teach.

- *DDXL.* An Excel add-in, DDXL, is included on the CD-ROM and is designed to enhance the capabilities of Excel's statistics programs and graphics resources.

- *Additional Chapters.* The CD-ROM also includes two additional chapters covering Analysis of Variance (Chapter 28) and Multiple Regression (Chapter 29). These topics point the way to further study in Statistics.

Overview of Supplements

Supplements for the Instructor

Instructor's Edition contains answers to all exercises as well as a chapter-by-chapter Instructor's Guide which includes extra examples, teaching suggestions, and commentary. (ISBN 0-321-28724-X)

Printed Test Bank and Resource Guide complements and expands on the commentary and teaching suggestions included in the Instructor's Edition. It includes Web links and lists of other resources, as well as chapter quizzes, unit tests, and suggestions for projects. (ISBN 0-321-28723-1)

Instructor's Solutions Manual, by William Craine, contains detailed solutions to all of the exercises. (ISBN 0-321-28721-5)

TestGen® enables instructors to build, edit, print, and administer tests using a computerized bank of questions developed to cover all the objectives of the text. TestGen is algorithmically based, allowing instructors to create multiple but equivalent versions of the same question or test with the click of a button. Instructors can also modify test bank questions or add new questions by using the built-in question editor, which allows users to create graphs, import graphics, and insert math notation, variable numbers, or text. Tests can be printed or administered online via the Internet or another network. TestGen comes packaged with QuizMaster, which allows students to take tests on a local area network. The software is available on a dual platform Windows/Macintosh CD-ROM. (ISBN 0-321-28720-7)

MathXL® **for Statistics** is a powerful online homework, tutorial, and assessment system that accompanies this textbook. With MathXL for Statistics, instructors can create, edit, and assign online homework and tests using algorithmically generated exercises correlated at the objective level to your textbook. All student work is tracked in MathXL's online gradebook. Students can take chapter tests in MathXL and receive personalized study plans based on their test results. The study plan diagnoses weaknesses and links students directly to tutorial exercises for the objectives they need to study and retest. Students can also access supplemental activities directly from selected exercises.

MathXL for Statistics is available to qualified adopters. For more information, visit our Web site at **www.mathxl.com,** or contact your Addison-Wesley sales representative for a product demonstration.

MyMathLab for Statistics is a series of text-specific, easily customizable online courses for Addison-Wesley textbooks in mathematics and statistics, and for this book in particular. MyMathLab is powered by CourseCompass™—Pearson Education's online teaching and learning environment—and by MathXL—our online homework, tutorial, and assessment system. MyMathLab gives you the tools you need to deliver all or a portion of your Statistics course online, whether your students are in a lab setting or working from home. MyMathLab provides a rich and flexible set of course materials, featuring free-response exercises that are algorithmically generated for unlimited practice and mastery. Students can also use online tools, such as animations and a multimedia textbook, to independently improve their understanding and performance. Instructors can use MyMathLab's homework and test managers to select and assign online exercises correlated directly to the textbook, and you can import TestGen tests into MyMathLab for added flexibility. MyMathLab's online gradebook—designed specifically for mathematics and statistics—automatically tracks students' homework and test results and gives the instructor control over how to calculate final grades.

MyMathLab is available to qualified adopters. For more information, visit our Web site at **www.mymathlab.com,** or contact your Addison-Wesley sales representative for a product demonstration.

Supplements for the Student

***Intro Stats* Second Edition** for-sale student edition. (ISBN 0-321-28671-5)

Student's Solutions Manual, by William Craine, provides detailed, worked-out solutions to odd-numbered exercises. (ISBN 0-321-28719-3)

TI-83/84 Plus and TI-89 Manual, by Patricia Humphrey (Georgia Southern University), is organized to follow the sequence of topics in the text, and it is an easy-to-follow, step-by-step guide on how to use the TI-83/84 Plus and TI-89 graphing calculators. It provides worked-out examples to help students fully understand and use the graphing calculator. (ISBN 0-321-28725-8)

ActivStats (Mac and PC) provides an introductory Statistics course on CD-ROM using the full potential of multimedia. *ActivStats* integrates video, simulation, animation, narration, text, pictures, interactive experiments, and Web access into a rich learning environment. The course offers practice with real data via links to the Data Desk statistics package included on the disk. Using *ActivStats*, students develop a sound understanding of statistical concepts and methods. The CD-ROM follows a course scope and sequence consistent with the ASA/MAA guidelines for teaching introductory Statistics and the Advanced Placement Statistics course. *ActivStats* is designed to closely match this text. It also can be used as a source of laboratory activities, a personal study guide, or the core of a self-paced or distance learning course. Add instructor extensions to each page to tailor the material to particular course needs. The Data Desk version of *ActivStats* is included in every new copy of this book. *ActivStats* is also available separately in versions for Excel (Mac and PC, ISBN 0-321-30375-X), JMP (Mac and PC, ISBN 0-321-30374-1), Minitab (PC, ISBN 0-321-30373-3), and SPSS (PC, ISBN 0-321-30372-5). Contact your Addison-Wesley sales representative for special bundle pricing with the textbook.

Addison-Wesley Math and Statistics Tutor Center is staffed by qualified mathematics and statistics instructors who provide students with tutoring on examples and odd-numbered exercises from the textbook. Tutoring is available via toll-free telephone, toll-free fax, e-mail, and the Internet. Interactive, Web-based technology allows tutors and students to view and work through problems together in real time over the Internet. The AW Tutor Center is available to qualified adopters. For more information, please visit our Web site at **www.aw-bc.com/tutorcenter** or call us at 1-888-777-0463.

MathXL for Statistics. See the description under Instructor Supplements, or visit our Web site at **www.mathxl.com** for more information.

MyMathLab. See the description under Instructor Supplements, or visit our Web site at **www.mymathlab.com** for more information.

Web Site (www.aw-bc.com/dvb) provides additional resources for instructors and students.

Acknowledgments

Many people have contributed to this book in both of its editions. This edition would have never seen the light of day without the assistance of the incredible team at Addison-Wesley. Our Executive Editor, Deirdre Lynch, was central to the genesis, development, and realization of the book from day one. Rachel Reeve, Senior Project Editor, kept us on task as much as humanly possible with much needed humor and grace. Jeff Holcomb, Senior Production Supervisor, kept the cogs from getting into the wheels where they often wanted to wander. Marketing Manager Phyllis Hubbard made sure the word got out. Vanessa Hayes, Editorial Project Assistant, Sara Oliver, Editorial Assistant, and Celena Carr, Marketing Assistant, were essential in managing all of the behind-the-scenes work that needed to be done. Sara Anderson, Associate Media Producer, put together a top-notch media package for this book. Barbara Atkinson, Senior Designer, is responsible for the wonderful way the book looks. Evelyn Beaton, Senior Manufacturing Buyer, and Ginny Michaud, Manufacturing Buyer, worked miracles to get this book and CD in your hands, and Greg Tobin, VP/Publisher, was supportive and good-humored throughout all aspects of the project.

A special thanks goes out to Nesbitt Graphics, the compositor, for the wonderful work they did on this book, and in particular to Bonnie Boehme, the project manager, for her close attention to detail and her amazing editorial skills.

We'd also like to thank our accuracy checkers whose monumental task was to make sure we said what we thought we were saying. They are Amy Fisher, Miami University, Middletown; Jackie Miller, The Ohio State University; and Kim Robinson, Clayton College and State University.

We extend our sincere thanks for the suggestions and contributions made by the following reviewers of this edition:

Sanjib Basu, *Northern Illinois University*
Steven Bogart, *Shoreline Community College*
Ann Cannon, *Cornell College*
Rick Denman, *Southwestern University*
Jeffrey Eldridge, *Edmonds Community College*
Karen Estes, *St. Petersburg Junior College*
Richard Friary
Jonathan Graham, *University of Montana*
Nancy Heckman, *University of British Columbia*
James Helreich, *Marist College*
Susan Herring, *Sonoma State University*
Patricia Humphrey, *Georgia Southern University*

Debra Ingram, *Arkansas State University*
Rebecka Jornsten, *Rutgers University*
Michael Lichter, *State University of New York–Buffalo*
Pamela Lockwood, *Western Texas A & M University*
Wei-Yin Loh, *University of Wisconsin–Madison*
Catherine Matos, *Clayton College & State University*
Elaine McDonald, *Sonoma State University*
Gina Reed, *Gainesville College*
Kim Robinson, *Clayton College & State University*
Chamont Wang, *The College of New Jersey*
Edward Welsh, *Westfield State College*

We extend our sincere thanks for the suggestions and contributions made by the following reviewers, focus group participants, and class-testers of the previous edition:

Jon Angellotti, *Cornell University*
James Bearden, *SUNY Geneseo*
Peter Blaskiewicz, *McLennan Community College*
Steven Bogart, *Shoreline Community College*
Dana Calland, *Maysville Community College*
Grace Cascio-Houston, *Louisiana State University–Eunice*
Smiley Cheng, *University of Manitoba*
Crista Lynn Coles, *Elon University*
Jon Cryer, *University of Iowa*
Carolyn Cuff, *Westminster College*
Nasser Dastrange, *Buena Vista University*
Mary Ellen Davis, *Georgia Perimeter College*
Jody DeVoe, *Valencia Community College–East Campus*
David Elesh, *Temple University*
Karen Estes, *St. Petersburg Junior College*
Russell Euler, *Northwest Missouri State University*
Amy Fisher, *Miami University, Middletown*
William Fox, *Francis Marion University*
John Gabrosek, *Grand Valley State University*
Jinadasa Gamage, *Illinois State University*
James Gehrmann, *California State University–Sacramento*
Paramjit Gill, *Okanagan University College*
Martha Goshaw, *Seminole Community College*
Kimberly Goyette, *Temple University*
Robert Gould, *University of California, Los Angeles*
Ken Grace, *Anoka–Ramsey Community College*
David Graves, *Elmira College*
Richard Greene, *Temple University*
Scott Greene, *University of Oklahoma*
Josephine Hamer, *Western Connecticut State University*
Mary Hartz, *Mohawk Valley Community College*
Robert Hollister, *Jacksonville University*
Patricia Humphrey, *Georgia Southern University*
Debra L. Hydorn, *Mary Washington College*
Coleen Jacobson, *Elmira College*
Lloyd R. Jaisingh, *Moorehead State University*

Mohammed Kazemi, *University of North Carolina–Charlotte*
John Khoury, *Brevard Community College–Melbourne*
Catherine Kong, *Carson–Newman College*
Christopher Lacke, *Rowan University*
James Lang, *Valencia Community College–East Campus*
Elaine McDonald, *Sonoma State University*
Amy McElroy, *San Diego State University*
Josiah (Si) Meyer, *Elmira College*
Donald Miller, *St. Mary's College*
Jackie Miller, *The Ohio State University*
Panagis Moschopoulos, *University of Texas at El Paso*
Weston I. Nathanson, *California State University*
Sondra Perdue, *University of Washington–Tacoma*
William Peterson, *Middlebury College*
Kimberley Polly, *Parkland College*
Anne Puciloski, *Stonehill College*
Shane Redmond, *Southeastern Louisiana University*
Jerry Reiter, *Duke University*
Mary Richardson, *Grand Valley State University*
Scott Richter, *Western Kentucky University*
William Roberts
Kim Robinson, *Clayton College & State University*
Richard Rogers, *University of Massachusetts–Amherst*
Gerald E. Rubin, *Marshall University*
Edith Seier, *East Tennessee State University*
Nagambal Shah, *Spelman College*
Therese Shelton, *Southwestern University*
Sounny Slitine, *Palo Alto College*
Jeffrey Stuart, *Pacific Lutheran University*
Sharon Testone, *Onondaga Community College*
Theresa Vecchiarelli, *Nassau Community College*
Anita Wah, *Chabot College*
John Walker, *California Polytechnic State University–SLO*
Janit M. Winter-Becker, *Penn State Berks Lehigh Valley College*
Kenny Ye, *SUNY at Stony Brook*

INDEX OF APPLICATIONS

Note: IE = In-Text Example, E = Exercise, JC = Just Checking, and SBS = Step-By-Step.

Intro Stats

SECOND EDITION

PART I

Exploring and Understanding Data

Stats Starts Here[1]

Statistics gets no respect. People say things like "you can prove anything with Statistics." People will write off a claim based on data as "just a statistical trick." And Statistics courses don't have the reputation of being students' first choice for a fun elective.

But Statistics *is* fun. That's probably not what you heard on the street, but it's true. Statistics is about how to think clearly with data. A little practice thinking statistically is all it takes to start seeing the world more clearly and accurately.

So, What Is (Are?) Statistics?

Q: What is Statistics?
A: Statistics is a way of reasoning, along with a collection of tools and methods, designed to help us understand the world.
Q: What are statistics?
A: Statistics (plural) are particular calculations made from data.
Q: So what is data?
A: You mean, "what *are* data?" Data is the plural form. The singular is datum.
Q: OK, OK, so what are data?
A: Data are values along with their context.

It seems every time we turn around, someone is collecting data on us, from every purchase we make in the grocery store, to every click of our mouse as we surf the Web. The United Parcel Service (UPS) tracks every package it ships from one place to another around the world and stores these records in a giant database. You can access part of it if you send or receive a UPS package. The database is about 17 terabytes big—about the same size as a database that contained every book in the Library of Congress would be. (But, we suspect, not *quite* as interesting.) What can anyone hope to do with all these data?

Statistics plays a role in making sense of the complex world in which we live today. Statisticians assess the risk of genetically engineered foods or of a new drug being considered by the Food and Drug Administration (FDA). They predict the number of new cases of AIDS by regions of the country or the number of customers likely to respond to a sale at the market. And statisticians help scientists and social scientists understand how unemployment is related to environmental controls, whether enriched early education affects later performance of school

---

[1] This chapter might have been called "Introduction," but nobody reads the introduction, and we wanted you to read this. We feel safe admitting this here, in the footnote, because nobody reads footnotes either.

The ads say, "Don't drink and drive; you don't want to be a statistic." But you can't be a statistic.

We say: "Don't be a datum."

children, and whether vitamin C really prevents illness. Whenever there are data and a need for understanding the world, you need Statistics.

So our objectives in this book are to help you develop the insights to think clearly about the questions, use the tools to show what the data are saying, and acquire the skills to tell clearly what it all means.

FRAZZ reprinted by permission of United Feature Syndicate, Inc.

Statistics in a Word

Statistics is about variation. Data vary because we don't see everything and because even what we do see and measure, we measure imperfectly.

So, in a very basic way, Statistics is about the real, imperfect world in which we live.

It can be fun, and sometimes useful, to summarize a discipline in only a few words. So,

Economics is about . . . *Money (and why it is good).*

Psychology: *Why we think what we think (we think).*

Biology: *Life.*

Anthropology: *Who?*

History: *What, where, and when?*

Philosophy: *Why?*

Engineering: *How?*

Accounting: *How much?*

In such a caricature, Statistics is about . . . ***Variation.***

Data vary. People are different. We can't see everything, let alone measure it all. And even what we do measure, we measure imperfectly. So the data we wind up looking at and basing our decisions on provide, at best, an imperfect picture of the world. This fact lies at the heart of what Statistics is all about. How to make sense of it is a central challenge of Statistics.

So, How Will This Book Help?

A fair question. Most likely, this book will not turn out to be quite what you expected.

What's different?

Close your eyes and open the book to a page at random. Is there a graph or table on that page? Do that again, say, 10 times. We'll bet you saw data displayed in many ways, even near the back of the book and in the exercises.

We can understand everything we do with data by making pictures. This book leads you through the entire process of thinking about a problem, finding and showing results, and telling others about what you have discovered. At each of these steps we display data for better understanding and insight.

You looked at only a few randomly selected pages to get an impression of the entire book. We'll see soon that doing so was sound Statistics practice and reasoning.

Next, pick a chapter and read the first two sentences. (Go ahead; we'll wait.)

We'll bet you didn't see anything about Statistics. Why? Because the best way to understand Statistics is to see it at work. In this book, chapters usually start by presenting a story and posing questions. That's when Statistics really gets down to work.

There are three simple steps to doing Statistics right: *think, show,* and *tell:*

Think

Show

Tell

Think first. Know where you're headed and why. It will save you a lot of work.

Show is what most folks think Statistics is about. The *mechanics* of calculating statistics and making displays is important, but not the most important part of Statistics.

Tell what you've learned. Until you've explained your results so that someone else can understand your conclusions, the job is not done.

Step-By-Step

Each chapter applies new concepts in a worked example called a **Step-By-Step.** These examples model the way statisticians attack and solve problems. They illustrate how to think about the problem, what to show, and how to tell what it all means. These step-by-step examples will show you how to produce the kind of solutions instructors hope to see.

just checking

Sometimes, in the middle of the chapter, we've put a section called **Just Checking . . . .** There you'll find a few short questions you can answer without much calculation—a quick way to check to see if you've understood the basic ideas in the chapter. You'll find the answers at the end of the chapter's exercises.

A S   This is a good time to open *ActivStats* and view the first chapter. Unlike the book, *ActivStats* can show itself off, so you can start playing with it right away. Give it a whirl.

From time to time you'll see an icon like this in the margin to signal that the *ActivStats* multimedia materials on the CD-ROM in the back of the book have an activity that you might find helpful at this point. Typically, we've flagged simulations and interactive activities because they're the most fun and will probably help you see how things work best. The chapters in *ActivStats* are the same as those in the text—just look for the named activity in the corresponding chapter.

"Get your facts first, and then you can distort them as much as you please. (Facts are stubborn, but statistics are more pliable.)"
—Mark Twain

What Can Go Wrong?

One of the interesting challenges of Statistics is that, unlike some math and science courses, there can be more than one right answer. This is why two statisticians can testify honestly on opposite sides of a court case. And it's why some people think that you can prove anything with statistics. But that's not true. People make mistakes using statistics, sometimes on purpose in order to mislead others. Most of the unintentional mistakes people make, though, are avoidable. We're not talking about arithmetic. More often the mistakes come from using a method in the wrong situation or misinterpreting the results. Each chapter has a section called **What Can Go Wrong?** to help you avoid some of the most common mistakes.

Some exercises are marked like this. You'll find the data for these exercises **T** on the CD in the back of the book, and you may wish to use technology as you work the exercise.

by hand

. . . on the Computer

A S **Introduction to (Your Statistics Package).** *ActivStats* launches your statistics package automatically. Check it out with this activity to make sure it's all working.

You'll find all sorts of stuff in margin notes, such as stories and quotations. For example:

"Computers are useless. They can only give you answers."
—*Pablo Picasso*

While Picasso underestimated the value of good statistics software, he did know that creating a solution requires more than just *Showing* an answer—it means you have to *Think* and *Tell*, too!

"Far too many scientists have only a shaky grasp of the statistical techniques they are using. They employ them as an amateur chef employs a cookbook, believing the recipes will work without understanding why. A more cordon bleu attitude . . . might lead to fewer statistical soufflés failing to rise."
—The Economist, *June 3, 2004,* **"Sloppy stats shame science."**

Although we'll show you all the formulas you need to understand the calculations, you will most often use a calculator or computer to perform the mechanics of a statistics problem. The easiest way to calculate statistics with a computer is with a specialized program called a "statistics package." There are a number of statistics packages available, and they differ widely in the details of how to use them and in how they present their results. But they all work from the same basic information and find the same results. Rather than adopt one package for this book, we present generic output and point out common features that you should look for. The **. . . on the Computer** section of most chapters (just before the exercises) holds this information. We also give a table of instructions to get you started on any of several commonly used packages.

● **Time out.** From time to time we'll take time out to discuss an interesting or important side issue. We indicate these by setting them apart like this.[2] ●

At the end of each chapter, you'll see a brief summary of the important concepts you've covered in a section called **What Have We Learned?** That section includes a list of the **Terms** and a summary of the important **Skills** you've acquired in the chapter. You won't be able to learn the material from these summaries, but you can use them to check your knowledge of the important ideas in the chapter. If you have the skills, know the terms, and understand the concepts, you should be well prepared for the exam—and ready to use Statistics!

Beware: No one can learn Statistics just by reading or listening. The only way to learn it is to do it. So, of course, at the end of each chapter (except this one) you'll find **Exercises** designed to help you learn to use the Statistics you've just read about.

We've often paired up the exercises, grouping similar ones together. So, if you're having trouble doing an exercise, you may find a similar one either just before or just after it. You'll find answers to the odd-numbered exercises at the back of the book. But these are only "answers" and not complete "solutions." Huh? What's the difference? The answers are sketches of the complete solutions. For most problems, your solution should follow the model of the Step-by-Step examples. If your calculations match the numerical parts of the "answer," and your argument contains the elements shown in the answer, you're on the right track. Your complete solution should explain the context, show your reasoning and calculations, and state your conclusions. Don't fret too much if your numbers don't match the printed answers to every decimal place. Statistics is more about getting the reasoning correct—pay more attention to how you interpret a result than what the digit in the third decimal place was.

In the real world, there's no chapter just before the question. So in addition to the problems at the ends of chapters we've also collected a variety of problems at the end of each part of the text to make it more like the real world. This should help you to see whether you can sort out which methods to use when. If you can do that successfully, then you'll know you understand Statistics.

---

[2] Or in a footnote.

*Optional Sections and Chapters

Some sections and chapters of this book are marked with an asterisk(*). These are optional in the sense that subsequent material does not depend on them directly. We hope you'll read them anyway, as you did this section.

Onward!

It's only fair to warn you: You can't get there by just picking out the highlighted sentences and the summaries. This book is different. It's not about memorizing definitions and learning equations. It's deeper than that. And much more fun. But . . .

You have to read the book![3]

---

[3] So, turn the page.

CHAPTER 2

Data

Many years ago, most stores in small towns knew their customers personally. If you walked into the hobby shop, the owner might tell you about a new bridge that had come in for your Lionel train set. The tailor knew your dad's size, and the hairdresser knew how your mom liked her hair. There are still some stores like that around today, but we're increasingly likely to shop at large stores, by phone, or on the Internet. Even so, when you phone an 800 number to buy new running shoes, customer service representatives may call you by your first name, or ask about the socks you bought 6 weeks ago. Or the company may send an e-mail in October offering new head warmers for winter running. This company has millions of customers, and you called without identifying yourself. How did the sales rep know who you are, where you live, and what you had bought?

The answer to all these questions is data. Collecting data on their customers, transactions, and sales enables companies to know where their inventory is and what their customers prefer. These data can help them predict what their customers may buy in the future and how much of each item to stock. The store can use the data and what they learn from the data to improve customer service, mimicking the kind of personal attention a shopper had 50 years ago.

Amazon.com opened for business in July 1995, billing itself as "Earth's Biggest Bookstore." By 1997 they had a catalog of more than 2.5 million book titles and had sold books to more than 1.5 million customers in 150 countries. In 2004 they had more than 41 million active customers in over 200 countries and were ranked the 74th most valuable brand by *Business Week*. They have expanded into selling a wide selection of merchandise, from $400,000 necklaces[2] to yak cheese from Tibet to the largest book in the world (see picture at left).

---

[1] Of course, if nit-picking authors of Statistics texts ran Amazon, they'd probably insist on "Data *are* king."
[2] Please get credit card approval before purchasing online.

7

Amazon is constantly monitoring and evolving their Web site to best serve their customers and maximize their sales performance. To make changes to the site, they experiment, collecting data and analyzing what works best. As Ronny Kohavi, director of Data Mining and Personalization, says, "Data trumps intuition. Instead of using our intuition, we experiment on the live site and let our customers tell us what works for them."

But What *Are* Data?

We bet you thought you knew this instinctively. Think about it for a minute. What exactly *do* we mean by "data"?

Do data have to be numbers? The amount of your last purchase in dollars is numerical data, but some data record names or other labels. The names in Amazon.com's database are data, but not numerical.

Sometimes, data can have values that look like numerical values but are just numerals serving as labels. This can be confusing. For example, the ASIN (Amazon Standard Item Number) of a book, like 0201709104, may have a numerical value, but it's really just another name for *Intro Stats*.

Data values, no matter what kind, are useless without their context. Newspaper journalists know that the lead paragraph of a good story should establish the "Five W's": *Who, What, When, Where,* and (if possible) *Why.* Often, we add *How* to the list as well. Answering these questions can provide the **context** for data values. The answers to the first two questions are essential. If you can't answer *Who* and *What* you don't have **data,** and you don't have any useful information.

> **The W's:**
>
> **WHO**
>
> **WHAT**
>   and in what units
>
> **WHEN**
>
> **WHERE**
>
> **WHY**
>
> **HOW**

Data Tables

Here are some data Amazon might collect:

B000001OAA	10.99	Chris G.	902	Boston	15.98	Kansas	Illinois
Canada	Samuel P.	Orange County	N	B000068ZVQ	Bad Blood	Nashville	Katherine H.
Garbage	16.99	Ohio	N	Chicago	N	11.99	Massachusetts
312	Monique D.	Y	413	B00000I5Y6	440	B000002BK9	Let Go

A S **What Is (Are) Data?** Do you really know what's data and what's just numbers? The *Activ-Stats* discussion is a good place to start.

Try to guess what they represent. Why is that hard? Because these data have no *context.* It's impossible to know what they're about or what they refer to without knowing the W's. We can make the context clear if we organize the values into a **data table** such as this one.

Name	Ship to State/Country	Price	Area Code	Previous CD Purchase	Gift?	ASIN	Artist
Katharine H.	Ohio	10.99	440	Nashville	N	B00000I5Y6	Kansas
Samuel P.	Illinois	16.99	312	Orange County	Y	B000002BK9	Boston
Chris G.	Massachusetts	15.98	413	Bad Blood	N	B000068ZVQ	Chicago
Monique D.	Canada	11.99	902	Let Go	N	B000001OAA	Garbage

Now we can see that these are four purchase records, relating to CD orders from Amazon. The column titles tell *What* has been recorded. The rows tell us *Who*. The other W's might have to come from the company's database administrator.[3]

Who

In general, the rows of a data table correspond to individual **cases** about whom (or about which—if they're not people) we record some characteristics. These cases go by different names, depending on the situation. Individuals who answer a survey are referred to as **respondents**. People on whom we experiment are **subjects** or (in an attempt to acknowledge the importance of their role in the experiment) **participants,** but animals, plants, Web sites, and other inanimate subjects are often just called **experimental units.** In a database, rows are called **records**—in this example, **purchase records.** Perhaps the most generic term is **cases.** In the table, the cases are the individual CD orders.

Sometimes people just refer to data values as **observations,** without being clear about the *Who.* Be sure you know the *Who* of the data, or you may not know what the data say.

What and Why

It is wise to be careful. The *What* and *Why* of area codes are not as simple as they may first seem. When area codes were first introduced, AT&T was still the source of all telephone equipment, and phones had dials.

 To reduce wear and tear on the dials, the area codes with the lowest digits (for which the dial would have to spin least) were assigned to the most populous regions—those with the most phone numbers and thus the area codes most likely to be dialed. New York City was assigned 212, Chicago 312, and Los Angeles 213, but rural upstate New York was given 607, Joliet was 815, and San Diego 619. For that reason, at one time, the numerical value of an area code could be used to guess something about the population of its region. Since the advent of push-button phones, area codes have finally become just categories.

The characteristics recorded about each individual are called **variables.** These are usually shown as the columns of a data table, and they should have a name that identifies *What* has been measured. Variables may seem simple, but to really understand your variables, you must *Think* about what you want to know.

---

[3] In database management, this kind of information is called "metadata."

The International System of Units links together all systems of weights and measures by international agreement. There are seven base units from which all other physical units are derived:

- Distance   Meter
- Mass   Kilogram
- Time   Second
- Electric current   Ampere
- Temperature   Kelvin
- Amount of substance   Mole
- Intensity of light   Candela

A S   **Measuring Variables.**
There are many ways to measure data. See some of them in action here.

One tradition that hangs on in some quarters is to name variables with cryptic abbreviations written in uppercase letters. This can be traced back to the 1960s when the very first statistics computer programs were controlled with instructions punched on cards. The earliest punch card equipment used only uppercase letters, and the earliest statistics programs limited variable names to six or eight characters, so variables were called things like PRSRF3. Modern programs do not have such restrictive limits, so there is no reason for variable names that you wouldn't use in an ordinary sentence.

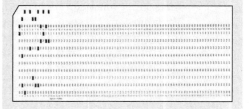

Although area codes are numbers, do we use them that way? Is 610 twice 305? Of course it is, but is that the question? Why would we want to know whether Allentown, PA (area code 610), is twice Key West, FL (305)? Variables play different roles, and you can't tell a variable's role just by looking at it.

Some variables just tell us what group or category each individual belongs to. Are you male or female? Pierced or not? . . . What kinds of things can we learn about variables like these? A natural start is to *count* how many cases belong in each category. (Are you listening to music while reading this? We could count the number of students in the class who were and the number who weren't.) We'll look for ways to compare and contrast the sizes of such categories.

Some variables have measurement **units.** Units tell how each value has been measured. But, more importantly, units such as yen, cubits, carats, angstroms, nanoseconds, miles per hour, or degrees Celsius tell us the *scale* of measurement. The units tell us how much of something we have or how far apart two values are. Without units, the values of a measured variable have no meaning. It does little good to be promised a raise of 5000 a year if you don't know whether it will be paid in euros, dollars, yen, or Estonian kroon.

What kinds of things can we learn about measured variables? We can do a lot more than just counting categories. We can look for patterns and trends. (How much did you pay for your last movie ticket? What is the range of ticket prices available in your town? How has the price of a ticket changed over the past 20 years?)

When a variable names categories and answers questions about how cases fall into those categories, we call it a **categorical variable.**[4] When a measured variable with units answers questions about the quantity of what is measured, we call it a **quantitative variable.** These types can help us decide what to do with a variable, but they are really more about what we hope to learn from a variable than about the variable itself. It's the questions we ask a variable (the *Why* of our analysis) that shape how we think about it and how we treat it.

Some variables can answer questions only about categories. If the values of a variable are words rather than numbers, it's a good bet that it is categorical. But some variables can answer both kinds of questions. Amazon could ask for your *age* in years. That seems quantitative, and would be if they want to know the average age of those customers who visit their site after 3 a.m. But suppose they want to decide which CD to offer you in a special deal—one by Raffi, Blink182, Carly Simon, or Montovani—and need to be sure they have adequate supplies on hand to meet the demand. Then thinking of your age in one of the categories child, teen, adult, or senior might be more useful. If it isn't clear whether a variable is categorical or quantitative, think about *Why* you are looking at it and what you want it to tell you.

A typical course evaluation survey asks, "How valuable do you think this course will be to you?" 1 = Worthless; 2 = Slightly; 3 = Middling; 4 = Reasonably; 5 = Invaluable. Is *educational value* categorical or quantitative? Once again, we'll look to the *Why*. A teacher might just count

---
[4] You may also see them called *qualitative*.

A S **Variables.** You're going to use your statistics package, so you'll want it to have data to work with. Several activities at the end of the first page of this *ActivStats* chapter will show you how to get data into your package and begin working with it.

the number of students who gave each response for her course, treating *educational value* as a categorical variable. When she wants to see whether the course is improving, she might treat the responses as the *amount* of perceived value—in effect, treating the variable as quantitative. But what are the units? There is certainly an *order* of perceived worth; higher numbers indicate higher perceived worth. A course that averages 4.5 seems more valuable than one that averages 2, but we should be careful about treating *educational value* as purely quantitative. To treat it as quantitative, she'll have to imagine that it has "educational value units" or some similar arbitrary construction. Because there are no natural units, she should be cautious. Variables like this that report order without natural units are often called **ordinal variables**. But saying "that's an ordinal variable" doesn't get you off the hook. You still must look to the *Why* of your study to decide whether to treat it as categorical or quantitative.

Counts Count

A S **The Circle Game.** Be your own Experimental Unit. With the computer you can experiment on yourself, then save the data. Go on to the subsequent activities to check your understanding of the variables, and use your statistics package to take a first look.

In Statistics, we often count things. When Amazon considers a special offer of free shipping to customers, they might first analyze how purchases are shipped. They'd probably start by counting the number of purchases shipped by ground transportation, by second-day air, and by overnight air. Counting is a natural way to summarize the categorical variable *shipping method*. So every time we see counts, does that mean the variable is categorical? Actually, no.

We also use counts to measure the amounts of things. How many songs are on your digital music player? How many books did you buy this semester for classes? To measure these quantities, we'd naturally count. The variables (*songs, books*) would be quantitative, and we'd consider the units to be "number of . . . ," or generically, just "counts" for short.

So we use counts in two different ways. When we count the cases in each category of a categorical variable, the category labels are the *What* and the individuals counted are the *Who* of our data. The counts themselves are not the data, but are something we summarize about the data. Amazon counts the number of purchases in each category of the categorical variable *shipping method*. For this purpose (the *Why*), the *What* is shipping method and the *Who* is purchases.

Shipping Method	Number of Purchases
Ground	20,345
Second-day	7,890
Overnight	5,432

Other times our focus is on the amount of something, which we measure by counting. Amazon might track the growth in the number of teenage customers each month to forecast CD sales (the *Why*). Now the *What* is *teens*, the *Who* is *months*, and the units are *number of teenage customers. Teen* was a category when we looked at the categorical variable *age*. But now it is a quantitative variable in its own right whose amount is measured by counting the number of customers.

Month	Number of Teenage Customers
January	123,456
February	234,567
March	345,678
April	456,789
May	. . .
. . .	. . .

Identifying Identifiers

What's your student ID number? It is numerical, but is it a quantitative variable? No, it doesn't have units. Is it categorical? Yes, but it is a special kind. Look at how many categories there are and at how many individuals are in each. There are as many categories as individuals, and only one individual in each category. While it's easy to count the totals for each category, it's not very interesting. Amazon wants to know who you are when you sign in again, and doesn't want to confuse you with some other customer. So they assign you a unique identifier.

Identifier variables themselves don't tell us anything useful about the categories because we know there is exactly one individual in each. However, they are crucial in this age of large data sets. They make it possible to combine data from different sources, to protect confidentiality, and to provide unique labels. It is easy to see that *identifier variables* are just categorical variables with exactly one individual in each category. The variables *UPS Tracking Number, Social Security Number,* and Amazon's *ASIN* are all examples of identifier variables.

You'll want to recognize when a variable is playing the role of an identifier so you won't be tempted to analyze it. There's probably a list of unique ID numbers for students in class (so you'll get your own grade confidentially), but you might worry about the professor who keeps track of the average of these numbers from class to class. Even though this semester's average happens to be higher than last's, it doesn't mean that the students are better.

Just because a variable has one case per category doesn't limit it to being an identifier. Even these variables can play different roles, depending on the question we ask of them. For their annual reports, Amazon refers to its database and looks at the variables *sales* and *year.* When analysts ask how many books Amazon sold in 1995, what role does *year* play? There's only one row for 1995, and *year* identifies it, so it is an identifier. But analysts also track sales growth over time. In this role, *year* measures time. Now it's a quantitative variable with units of . . . years. Again, the difference is in the *Why.*

Where, When, and How

A S **Three W's.** Do you know who's *Who* and what's *What?* Check it out. *Where* and *When?* Here and now, with *ActivStats!*

We must know *Who, What,* and *Why* to analyze data. Without knowing these three we don't have enough to start. Of course, we'd always like to know more. The more we know, the more we'll understand.

A S **Data Terminology.**
ActivStats includes self-tests like this one, much like the Just Checking sections of the book—but you get to play with them, not just read and answer. Try this one out to see how they work. (Usually we won't reference them here, but look for one whenever you'd like to check your understanding or review some materials.)

If possible, we'd like to know the **When** and **Where** of data as well. Values recorded in 1803 may mean something different than similar values recorded last year. Values measured in Tanzania may differ in meaning from similar measurements made in Mexico.

How the data are collected can make the difference between insight and nonsense. As we'll see later, data that come from a voluntary survey on the Internet are almost always worthless. One primary concern of Statistics, discussed in Part III, is the design of sound methods for collecting data.

Throughout this book, whenever we introduce data, we will provide a marginal note listing the W's (and H) of the data. It's a habit we recommend. The first step of any data analysis is to know why you are examining the data (what you want to know), whom each row of your data table refers to, and what the variables (the columns of the table) record. These are the *Why*, the *Who*, and the *What*. Identifying them is a key part of the *Think* step of any analysis. Make sure you know all three before you proceed to *Show* or *Tell* anything about the data.

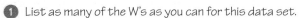

In the summer of 2003 Tour de France, Lance Armstrong averaged 40.94 kilometers per hour (km/h) for the entire course, making it the fastest Tour de France in its 100-year history. In 2004 he made history again by winning the race for an unprecedented sixth time. You can find data on all the Tour de France races in the data set Tour_de_France on the CD. Here are the first six and last six lines of the data set. Keep in mind that the entire data set has nearly 100 entries.

① List as many of the W's as you can for this data set.

② Classify each variable as categorical or quantitative; if quantitative, identify the units.

Year	Winner	Country of origin	Total time (h/min/s)	Avg. speed (km/h)	Stages	Total distance ridden (km)	Starting riders	Finishing riders
1903	Maurice Garin	France	94.33.00	25.3	6	2428	60	21
1904	Henri Cornet	France	96.05.00	24.3	6	2388	88	23
1905	Louis Trousselier	France	112.18.09	27.3	11	2975	60	24
1906	Rene Pottier	France	185.47.26	24.5	13	4637	82	14
1907	Lucien Petit-Breton	France	156.22.30	28.5	14	4488	93	33
1908	Lucien Petit-Breton	France	156.09.31	28.7	14	4488	114	36
...								
1999	Lance Armstrong	USA	91.32.16	40.30	20	3687	180	141
2000	Lance Armstrong	USA	92.33.08	39.56	21	3662	180	128
2001	Lance Armstrong	USA	86.17.28	40.02	20	3453	189	144
2002	Lance Armstrong	USA	82.05.12	39.93	20	3278	189	153
2003	Lance Armstrong	USA	83.41.12	40.94	20	3427	189	147
2004	Lance Armstrong	USA	83.36.02	40.53	20	3391	188	147

What Can Go Wrong?	• ***Don't label a variable as categorical or quantitative without thinking about the question you want it to answer.*** The same variable can sometimes take on different roles.
	• ***Just because your variable's values are numbers, don't assume that it's quantitative.*** Categories are often given numerical labels. Don't let that fool you into thinking they have quantitative meaning. Look at the context.
	• ***Always be skeptical.*** One reason to analyze data is to discover the truth. Even when you are told a context for the data, it may turn out that the truth is a bit (or even a lot) different. The context colors our interpretation of the data, so those who want to influence what you think may slant the context. A survey that seems to be about all students may in fact report just the opinions of those who visited a fan Web site. The question that respondents answered may be slanted.

What have we learned?

We've learned that data are information in a context.
- The W's help nail down the context: *Who, What, Why, Where, When,* and *hoW.*
- We must know at least the *Who, What,* and *Why* to be able to say anything useful based on the data. The *Who* are the *cases.* The *What* are the *variables.* A variable gives information about each of the cases. The *Why* helps us decide which way to treat the variables.

We treat variables in two basic ways, as *categorical* or *quantitative.*
- Categorical variables identify a category for each case. Usually we think about the counts of cases that fall in each category. (An exception is an identifier variable that just names each case.)
- Quantitative variables record measurements or amounts of something; they must have *units.*
- Sometimes we treat a variable as categorical or quantitative depending on what we want to learn from it, which means some variables can't be pigeon-holed as one type or the other. That's an early hint that in Statistics we can't always pin things down precisely.

TERMS

Context	The context ideally tells *Who* was measured, *What* was measured, *How* the data were collected, *Where* the data were collected, and *When* and *Why* the study was performed.
Data	Systematically recorded information, whether numbers or labels, together with its context.
Data table	An arrangement of data in which each row represents a case and each column represents a variable.
Case	A case is an individual about whom or which we have data.
Variable	A variable holds information about the same characteristic for many cases.
Categorical variable	A variable that names categories (whether with words or numerals) is called categorical.
Quantitative variable	A variable in which the numbers act as numerical values is called quantitative. Quantitative variables always have units.
Units	A quantity or amount adopted as a standard of measurement, such as dollars, hours, or grams.

S K I L L S   When you complete this lesson you should:

Think

- Be able to identify the *Who, What, When, Where, Why,* and *How* of data, or recognize when some of this information has not been provided.
- Be able to identify the cases and variables in any data set.
- Be able to classify a variable as categorical or quantitative depending on its use.
- For any quantitative variable be able to identify the units in which the variable has been measured (or note that they have not been provided).

Tell

- Be able to describe a variable in terms of its *Who, What, When, Where, Why,* and *How* (and be prepared to remark when that information is not provided).

Data on the Computer

A S **Examine the Data.** Take a look at your own data from your circle game experiment (p. 11). Here's a chance to get comfortable with your statistics package as you find out about the experiment test results. Several activities on this *Activ-Stats* page provide additional help and information.

Most often we find statistics on a computer using a program, or *package,* designed for that purpose. There are many different statistics packages, but they all do essentially the same things. If you understand what the computer needs to know to do what you want and what it needs to show you in return, you can figure out the specific details of most packages pretty easily.

For example, to get your data into a computer statistics package, you need to tell the computer:

- Where to find the data. This usually means directing the computer to a file stored on your computer's disk or to data on a database. Or it might just mean that you have copied the data from a spreadsheet program or Internet site, and it is currently on your computer's clipboard. Usually, the data should be in the form of a data table. Most computer statistics packages prefer the *delimiter* that marks the division between elements of a data table to be a *tab* character and the delimiter that marks the end of a case to be a *return* character.
- Where to put the data. (Usually this is handled automatically.)
- What to call the variables. Some data tables have variable names as the first row of the data, and often statistics packages can take the variable names from the first row automatically.

EXERCISES

For each description of data, identify the W's, name the variables, specify for each variable whether its use indicates it should be treated as categorical or quantitative, and, for any quantitative variable, identify the units in which it was measured (or note that they were not provided).

1. The news. Find a newspaper or magazine article in which some data are reported. For the data discussed in the article, answer the questions above. Include a copy of the article with your report.

2. Investments. According to an article in *Fortune* (Dec. 28, 1992), 401(k) plans permit employees to shift part of their before-tax salaries into investments such as mutual funds. Employers typically match 50% of the employees' contribution up to about 6% of salary. One company, concerned with what it believed was a low employee participation rate in its 401(k) plan, sampled 30 other companies with similar plans and asked for their 401(k) participation rates.

3. Oil spills. Owing to several major ocean oil spills by tank vessels, Congress passed the 1990 Oil Pollution Act, which requires all tankers to be designed with thicker hulls. Further improvements in the structural design of a tank vessel have been proposed since then, each with the objective of reducing the likelihood of an oil spill and decreasing the amount of outflow in the event of a hull puncture. To aid in this development, *Marine Technology* (Jan. 1995) reported on the spillage amount and cause of puncture for 50 recent major oil spills from tankers and carriers.

4. Oscars. *Ages of Oscar-Winning Best Actors and Actresses* by Richard Brown and Gretchen Davis gives the ages of actors and actresses at the time they won Oscars. We might use these data to see whether actors and actresses are likely to win Oscars at about the same age or not.

5. Weighing bears. Because of the difficulty of weighing a bear in the woods, researchers caught and measured 54 bears, recording their weight, neck size, length, and sex. They hoped to find a way to estimate weight from the other, more easily determined quantities.

6. Molten iron. The Cleveland Casting Plant is a large, highly automated producer of gray and nodular iron automotive castings for Ford Motor Company. According to an article in *Quality Engineering* (7 [1995]), Cleveland Casting is interested in keeping the pouring temperature of the molten iron (in degrees Fahrenheit) close to the specified value of 2550 degrees. Cleveland Casting measured the pouring temperature for 10 randomly selected crankshafts.

7. Arby's menu. A listing posted by the Arby's restaurant chain gives, for each of the sandwiches it sells, the type of meat in the sandwich, the number of calories, and the serving size in ounces. The data might be used to assess the nutritional value of the different sandwiches.

8. Firefighters. A study was conducted to compare the abilities of men and women to perform the strenuous tasks required of a shipboard firefighter (*Human Factors* 24 [1982]). The study reports the pulling force (in newtons) that a firefighter was able to exert in pulling the starter cord of a P-250 water pump. The study also gives the weight and, of course, the gender of the firefighters.

9. Babies. Medical researchers at a large city hospital investigating the impact of prenatal care on newborn health collected data from 882 births during 1998–2000. They kept track of the mother's age, the number of weeks the pregnancy lasted, the type of birth (cesarean, induced, natural), the level of prenatal care the mother had (none, minimal, adequate), the birth weight and sex of the baby, and whether the baby exhibited health problems (none, minor, major).

10. Flowers. In a study appearing in the journal *Science*, a research team reports that plants in southern England are flowering earlier in the spring. Records of the first flowering dates for 385 species over a period of 47 years indicate that flowering has advanced an average of 15 days per decade, an indication of climate warming according to the authors.

11. Fitness. Are physically fit people less likely to die of cancer? An article in the May 2002 issue of *Medicine and Science in Sports and Exercise* reported results of a study that followed 25,892 men aged 30 to 87 for 10 years. The most physically fit men had a 55% lower risk of death from cancer than the least fit group.

12. Schools. The State Education Department requires local school districts to keep these records on all students: age, race or ethnicity, days absent, current grade level, standardized test scores in reading and mathematics, and any disabilities or special educational needs the student may have.

13. Herbal medicine. Scientists at a major pharmaceutical firm conducted an experiment to study the effectiveness of an herbal compound to treat the common cold. They exposed each patient to a cold virus, then gave them either the herbal compound or a sugar solution known to have no effect on colds. Several days later they assessed each patient's condition using a cold severity scale ranging 0–5. They found no evidence of the benefits of the compound.

14. Tracking sales. A start-up company is building a database of customers and sales information. For each customer it records name, ID number, region of the country (1 = East, 2 = South, 3 = Midwest, 4 = West), date of last purchase, amount of purchase, and item purchased.

15. Cars. A survey of autos parked in student and staff lots at a large university recorded the make, country of origin, type of vehicle (car, van, SUV, etc.), and age.

16. Vineyards. Business analysts hoping to provide information helpful to grape growers compiled these data about vineyards: size (acres), number of years in existence, state, varieties of grapes grown, average case price, gross sales, and percent profit.

17. Streams. As research for an ecology class, students at a college in upstate New York collect data on streams each year. They record a number of biological, chemical, and physical variables, including the stream name, the substrate of the stream (limestone, shale, or mixed), the acidity of the water (pH), the temperature (°C), and the BCI (a numerical measure of biological diversity).

18. Age and party. The Gallup Poll conducted a representative telephone survey of 1180 American voters during the first quarter of 1999. Among the reported results were the voter's region (Northeast, South, etc.), age, party affiliation, and whether or not the person had voted in the 1998 midterm Congressional election.

19. Air travel. The Federal Aviation Administration (FAA) monitors airlines for safety and customer service. For each flight the carrier must report the type of aircraft, number of passengers, whether or not the flights departed and arrived on schedule, and any mechanical problems.

20. Fuel economy. The Environmental Protection Agency (EPA) tracks fuel economy of automobiles. Among the data they collect are the manufacturer (Ford, Toyota, etc.), vehicle type (car, SUV, etc.), weight, horsepower, and gas mileage (mpg) for city and highway driving.

21. Refrigerators. In 2002, *Consumer Reports* published an article evaluating refrigerators. It listed 41 models, giving the brand, cost, size (cu ft), type (such as top-freezer), estimated annual energy cost, an overall rating (good, excellent, etc.), and the repair history for that brand (percentage requiring repairs over the past 5 years).

22. Lotto. A study of state-sponsored Lotto games in the United States (*Chance*, Winter 1998) listed the names of the states and whether or not the state had Lotto. For states that did, the study indicated the number of numbers in the lottery, the number of matches required to win, and the probability of holding a winning ticket.

23. Sleep. In the Spring 2001 issue of *Chance* magazine, a psychology professor reported on data he had collected about his sleep patterns. He kept daily records of the number of hours of sleep he got, whether or not he suffered from "early awakening," whether or not he watched TV in the morning and in the evening, the number of hours he spent standing during the day, and his mood (happy/sad, on a scale from 10–90).

24. Indy. The 2.5-mile Indianapolis Motor Speedway has been the home to a race on Memorial Day nearly every year since 1911. Even during the first race there were controversies. Ralph Mulford was given the checkered flag first but took three extra laps just to make sure he'd completed 500 miles. When he finished, another driver, Ray Harroun, was being presented with the winner's trophy, and Mulford's protests were ignored. Harroun averaged 74.6 mph for the 500 miles. In 2003 the winner, Gil de Ferran, averaged 153.6 mph.

Here are the data for the first few and four recent Indianapolis 500 races. Included also are the pole winners (the winners of the trial the day before, when each driver drives alone to determine the position on race day).

Year	Winner	Pole Position	Average Speed (mph)	Pole Winner	Average Pole Speed (mph)
1911	Ray Harroun	28	74.602	Lewis Strang	.
1912	Joe Dawson	7	78.719	Gil Anderson	.
1913	Jules Goux	7	75.933	Caleb Bragg	.
1914	Rene Thomas	15	82.474	Jean Chassagne	.
1915	Ralph DePalma	2	89.840	Howard Wilcox	98.90
1916	Dario Resta	4	84.001	John Aitken	96.69
1919	Howdy Wilcox	2	88.050	Rene Thomas	104.78
1920	Gaston Chevrolet	6	88.618	Ralph DePalma	99.15
. . .					
2000	Juan Montoya	2	167.607	Greg Ray	223.471
2001	Helio Castroneves	11	153.601	Scott Sharp	226.037
2002	Helio Castroneves	13	166.499	Bruno Junqueira	231.342
2003	Gil de Ferran	10	156.291	Helio Castroneves	231.725

Ⓣ 25. Horse race. The Kentucky Derby is a horse race that has been run every year since 1875 at Churchill Downs, Louisville, Kentucky. The race started as a 1.5-mile race, but in 1896 it was shortened to 1.25 miles because experts felt that 3-year-old horses shouldn't run such a long race that early in the season (it has been run in May every year but one—1901—when it took place on April 29). Here are the data for the first few and a few recent races.

Date	Winner	Margin (lengths)	Jockey	Winner's Payoff ($)	Duration (min:sec)	Track Condition
May 17, 1875	Aristides	2	O. Lewis	2850	2:37.75	Fast
May 15, 1876	Vagrant	2	B. Swim	2950	2:38.25	Fast
May 22, 1877	Baden-Baden	2	W. Walker	3300	2:38.00	Fast
May 21, 1878	Day Star	1	J. Carter	4050	2:37.25	Dusty
May 20, 1879	Lord Murphy	1	C. Shauer	3550	2:37.00	Fast
. . . . . .						
May 5, 2001	Monarchos	4 3/4	J. Chavez	812000	1:59.97	Fast
May 4, 2002	War Emblem	4	V. Espinoza	1875000	2:01.13	Fast
May 3, 2003	Funny Cide	1 3/4	J. Santos	800200	2:01.19	Fast
May 1, 2004	Smarty Jones	2 3/4	S. Elliott	854800	2:04.06	Sloppy

26. **Stat students.** An online survey of students in a large Statistics class asked them to report their height, shoe size, sex, which degree program they were in, and their birth order (1 = only child or first born). The data were used for classroom illustrations.

---

just checking
Answers

1. *Who—Tour de France races; what—year, winner, country of origin, total time, average speed, stages, total distance ridden, starting riders, finishing riders; how—official statistics at race; where—France (for the most part); when—1903 to 2004; why—to see progress in speeds of cycling racing?*

2.

Variable: Year	*Type—Quantitative or Categorical*	*Units—Years*
Variable: Winner	*Type—Categorical*	
Variable: Country of origin	*Type—Categorical*	
Variable: Total time	*Type—Quantitative*	*Units—Hours/minutes/seconds*
Variable: Average speed	*Type—Quantitative*	*Units—Kilometers per hour*
Variable: Stages	*Type—Quantitative*	*Units—Counts (stages)*
Variable: Total distance	*Type—Quantitative*	*Units—Kilometers*
Variable: Starting riders	*Type—Quantitative*	*Units—Counts (riders)*
Variable: Finishing riders	*Type—Quantitative*	*Units—Counts (riders)*

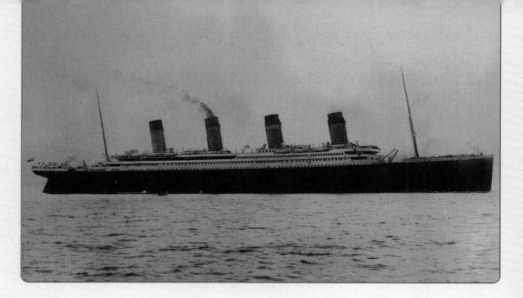

CHAPTER 3

Displaying and Describing Categorical Data

What happened on the *Titanic* at 11:40 on the night of April 14, 1912, is well known. Frederick Fleet's cry of "Iceberg, right ahead" and the three accompanying pulls of the crow's nest bell signaled the beginning of a nightmare that has become legend. By 2:15 a.m. the *Titanic*, thought by many to be unsinkable, had sunk, leaving more than 1500 passengers and crewmembers on board to meet their icy fate.

Here are some data about the passengers and crew aboard the *Titanic*. Each case (row) of the data table represents a person on board the ship. The variables are whether or not the person *Survived* (Dead or Alive), the person's *Age* (Adult or Child), *Sex* (Male or Female), and ticket *Class* (First, Second, Third, or Crew).

The problem with a data table like this—and in fact with all data tables—is that you can't *see* what's going on. And seeing is just what we want to do. We need ways to show the data so that we can see patterns, relationships, trends, and exceptions.

A S **The Incident** tells the story. Look on page 3 of this chapter in *ActivStats*. The activity includes rare film footage.

Survived	Age	Sex	Class
Dead	Adult	Male	Third
Dead	Adult	Male	Crew
Dead	Adult	Male	Third
Dead	Adult	Male	Crew
Dead	Adult	Male	Crew
Dead	Adult	Male	Crew
Alive	Adult	Female	First
Dead	Adult	Male	Third
Dead	Adult	Male	Crew

Part of a data table showing four variables for nine people aboard the *Titanic*. **Table 3.1**

The Three Rules of Data Analysis

So, what should we do with data like these? There are three things you should always do first with data:

1. **Make a picture.** A display of your data will reveal things you are not likely to see in a table of numbers and will help you to *think* clearly about the patterns and relationships that may be hiding in your data.

2. **Make a picture.** A well-designed display will *show* the important features and patterns in your data. A picture will also show you the things you did not expect to see: the extraordinary (possibly wrong) data values or unexpected patterns.

3. **Make a picture.** The best way to *tell* others about your data is with a well-chosen picture.

These are the three rules of data analysis. There are pictures of data throughout the book, and new kinds keep showing up. These days, technology makes drawing pictures of data easy, so there is no reason not to follow the three rules.

A Picture to Tell a Story

Florence Nightingale, a founder of modern nursing, was also well versed in statistics. To argue forcefully for better hospital conditions for soldiers, she invented this display, which showed that in the Crimean War, far more soldiers died of illness and infection than of battle wounds. Her campaign succeeded in improving hospital conditions and nursing for soldiers.
Figure 3.1

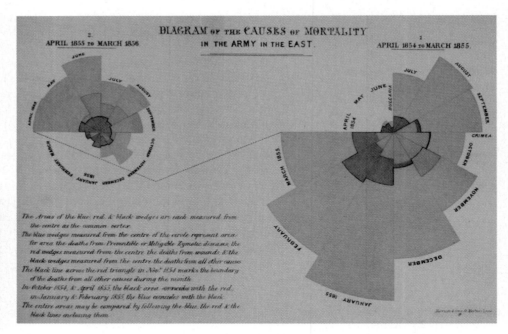

Frequency Tables: Making Piles

A S **Categorical Data.** What color is your hair? Even data on something as simple as hair color can be surprising, as you'll see here.

In order to make a picture of data, the first thing we have to do is to make piles. Making piles is the beginning of understanding about data. We pile together things that seem to go together, so we can see how the cases distribute across different categories. For categorical data, piling is easy. We just count the number of cases corresponding to each category and pile them up.

Class	Count
First	325
Second	285
Third	706
Crew	885

A frequency table of the *Titanic* passengers. **Table 3.2**

Class	%
First	14.77
Second	12.95
Third	32.08
Crew	40.21

The same data as a relative frequency table. **Table 3.3**

One way to put all 2201 people on the *Titanic* into piles is by ticket *Class,* counting up how many had each kind of ticket. We can organize these counts into a **frequency table,** which records the totals and the category names.

Even when we have thousands of cases, a variable like ticket *Class,* with only a few categories, has a frequency table that's easy to read. A frequency table with dozens or hundreds of categories would be much harder to read. We use the names of the categories to label each row in the frequency table. For ticket *Class,* these are "First," "Second," "Third," and "Crew."

Counts are useful, but sometimes we want to know the fraction or **proportion** of the data in each category, so we divide the counts by the total number of cases. Usually we multiply by 100 to express these proportions as **percentages.** A relative frequency table displays the *percentages,* rather than the counts, of the values in each category. Both types of tables show how the cases are distributed across the categories. In this way, they describe the **distribution** of a categorical variable because they name the possible categories and tell how frequently each occurs.

The Area Principle

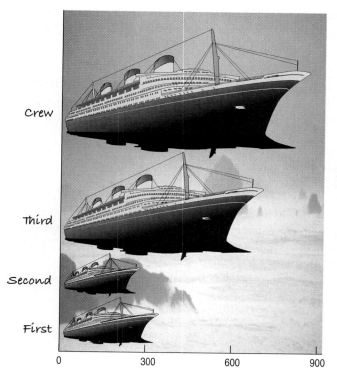

How many people were in each class on the *Titanic?* From this display it looks as though the service must have been great, since most aboard were crew members. Although the *length* of each ship here corresponds to the correct number, the impression is all wrong. In fact, only about 40% were crew. **Figure 3.2**

Now that we have the frequency table, we're ready to follow the three rules of data analysis and make a picture of the data. But a bad picture can distort our understanding rather than help it. Here's a graph of the *Titanic* data. What impression do you get about who was aboard the ship?

It sure looks like most of the people on the *Titanic* were crew members, with a few passengers along for the ride. That doesn't seem right. What's wrong? The lengths of the ships *do* match the totals in the table. (You can check the scale at the bottom.) However, experience and psychological tests show that our eyes tend to be more impressed by the *area* than by other aspects of each ship image. So, even though the *length* of each ship matches up with one of the totals, it's the associated *area* in the image that we notice. Since there were about 3 times as many crew as second-class passengers, the ship depicting the number of crew is about 3 times longer than the ship depicting second-class passengers, but it occupies about 9 times the area. As you can see from the frequency table (Table 3.2), that just isn't a correct impression.

The best data displays observe a fundamental principle of graphing data called the **area principle.** The area principle says that the area occupied by a part of the graph should correspond to the magnitude of the value it represents. Violations of the area principle are a common way to lie (or, since most mistakes are unintentional, we should say err) with Statistics.

Bar Charts

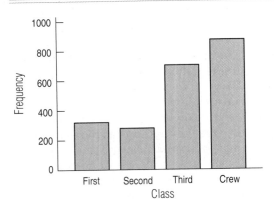

People on the *Titanic* by Ticket Class
With the area principle satisfied, we can see the true distribution more clearly.   **Figure 3.3**

A S   **Bar Charts.** Watch bar charts grow from data, then go on to the next *ActivStats* activity and use your statistics package to create some bar charts for yourself.

> For some reason, some computer programs give the name "bar chart" to any graph that uses bars. And others use different names according to whether the bars are horizontal or vertical. Don't be misled. "Bar chart" is the term for a *display of counts of a categorical variable* with bars.

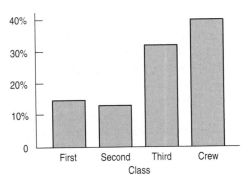

The relative frequency bar chart looks the same as the bar chart (Figure 3.3), but shows the proportion of people in each category rather than the counts.   **Figure 3.4**

Here's a chart that obeys the area principle. It's not as visually entertaining as the ships, but it does give an *accurate* visual impression of the distribution. The height of each bar shows the count for its category. The bars are the same width, so their heights determine their areas, and the areas are proportional to the counts in each class. Now it's easy to see that the majority of people on board were *not* crew, as the ships picture led us to believe. We can also see that there were about 3 times as many crew as second-class passengers. And there were more than twice as many third-class passengers as either first- or second-class passengers, something you may have missed in the frequency table. Bar charts make these kinds of comparisons easy and natural.

A **bar chart** displays the distribution of a categorical variable, showing the counts for each category next to each other for easy comparison. Bar charts should have small spaces between the bars to indicate that these are freestanding bars that could be rearranged into any order. The bars are lined up along a common base.

Usually they stick up like this 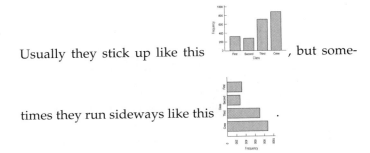, but some-

times they run sideways like this .

If we really want to draw attention to the relative *proportion* of passengers falling into each of these classes, we could replace the counts with percentages and use a **relative frequency bar chart**.

Pie Charts

Another common display that shows how a whole group breaks into several categories is a pie chart. **Pie charts** show the whole group of cases as a circle. They slice the circle into pieces whose size is proportional to the fraction of the whole in each category.

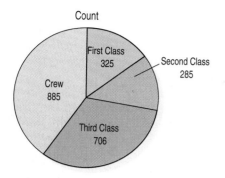

The number of *Titanic* passengers in each class.   **Figure 3.5**

Pie charts give a quick impression of how a whole group is partitioned into smaller groups. Because we're used to cutting up pies into 2, 4, or 8 pieces, pie charts are good for seeing relative frequencies near 1/2, 1/4, or 1/8. For example, you may be able to tell that the pink slice, representing the second-class passengers, is very close to 1/8 of the total. It's harder to see that there were about twice as many third-class as first-class passengers. Which category had the most passengers? Were there more crew or more third-class passengers? Comparisons such as these are easier in a bar chart.

● **Think before you draw** Our first rule of data analysis is *Make a picture.* But what kind of picture? Obviously we don't have a lot of options—yet. There's more to Statistics than pie charts and bar charts, and knowing when to use each type of graph is a critical first step in data analysis. That decision depends in part on what type of data you have.

We always have to check that the data are appropriate for whatever method of analysis we choose. Before you make a bar chart or a pie chart, always check the **Categorical Data Condition:** The data are counts or percentages of individuals in categories.

If you want to make a relative frequency bar chart or a pie chart, you'll need to also make sure that the categories don't overlap, so no individual is counted twice. If the categories do overlap, you can still make a bar chart, but the percentages won't add up to 100%. For the *Titanic* data, either kind of display is appropriate because the categories don't overlap.

Throughout this course, you'll see that doing Statistics right means selecting the proper methods. That means you have to *Think* about the situation at hand. An important first step, then, is to check that the type of analysis you plan is appropriate. Our Categorical Data Condition is just the first of many such checks. ●

Contingency Tables: Children and First-Class Ticket Holders First?

We know how many tickets of each class were sold on the *Titanic,* and we know that only about 32% of all those aboard the *Titanic* survived. After looking at the distribution of each variable by itself, it's natural and more interesting to ask how they relate. Was there a relationship between the kind of ticket a passenger held and the passenger's chances of making it into the lifeboat? To answer this question, we need to look at the two categorical variables *Class* and *Survival* together.

To look at two categorical variables together, we often arrange the counts in a two-way table. Here is a two-way table of those aboard the *Titanic,* classified according to class of ticket and whether they survived or didn't. Because the table shows how the individuals are distributed along each variable, contingent on the value of the other variable, such a table is called a **contingency table.**

A S **Children at Risk.** There's still more to the *Titanic* data. Here's a table that reveals some other aspects. And the subsequent activity shows how to make such tables on the computer.

		Class				
		First	Second	Third	Crew	**Total**
Survival	**Alive**	203	118	178	212	**711**
	Dead	122	167	528	673	**1490**
	Total	**325**	**285**	**706**	**885**	**2201**

Contingency table of ticket *Class* and *Survival*. The bottom line of "Totals" is the same as the previous frequency table.   **Table 3.4**

A bell-shaped artifact from the *Titanic.*

The margins of the table, both on the right and at the bottom, give totals. The bottom line of the table is just the frequency distribution of ticket *Class*. The right column of the table is the frequency distribution of the variable *Survival*. When presented like this, in the margins of a contingency table, the frequency distribution of one of the variables is called its **marginal distribution.**

Each **cell** of the table gives the count for a combination of values of the two variables. If you look down the column for second-class passengers to the first cell, you can see that 118 second-class passengers survived. Looking at the third-class passengers, you can see that more third-class passengers (178) survived. Were second-class passengers more likely to survive? Questions like this are more naturally addressed using percentages. The 118 survivors in second class were 41.4% of the total 285 second-class passengers, while the 178 surviving third-class passengers were only 25.2% of that class's total.

We know that 118 second-class passengers survived. We could display this number as a percentage, but as a percentage of what? The total number of passengers? (118 is 5.4% of the total, 2201.) The number of second-class passengers? (118 is 41.4% of the 285 second-class passengers.) The number of survivors? (118 is 16.6% of the 711 survivors.) All of these are possibilities, and all are potentially useful or interesting. You'll probably wind up calculating (or letting your technology calculate) lots of percentages. Most statistics programs offer a choice of total percent, row percent, or column percent for contingency tables. Here are the counts and all three percentages displayed as they might be by a computer package:

			Class				
			First	Second	Third	Crew	Total
Survival	Alive	**Count**	**203**	**118**	**178**	**212**	**711**
		% of Row	28.6%	16.6%	25.0%	29.8%	100%
		% of Column	62.5%	41.4%	25.2%	24.0%	32.3%
		% of Table	9.2%	5.4%	8.1%	9.6%	32.3%
	Dead	**Count**	**122**	**167**	**528**	**673**	**1490**
		% of Row	8.2%	11.2%	35.4%	45.2%	100%
		% of Column	37.5%	58.6%	74.8%	76.0%	67.7%
		% of Table	5.6%	7.6%	24.0%	30.6%	67.7%
	Total	**Count**	**325**	**285**	**706**	**885**	**2201**
		% of Row	14.8%	12.9%	32.1%	40.2%	100%
		% of Column	100%	100%	100%	100%	100%
		% of Table	14.8%	12.9%	32.1%	40.2%	100%

Another contingency table of ticket *Class*. This time we see not only the counts for each combination of *Class* and *Survival* (in bold) but the percentages these counts represent. For each count, there are three choices for the percentage: by row, by column, and by table total. There's probably too much information here for this table to be useful. **Table 3.5**

Each cell of this table gives the count, row percent, column percent, and table percent, in that order. This is an example of why contingency tables can look so confusing. There's too much information to sort through at one glance. While it's fine to consider all these choices, it's probably better to look at them one at a time. If we're interested in comparing the survival rates for the different passenger classes, we'd look at the percentage that survived and died within each class, using the column totals as the denominator for the percentages. These are called the column percentages, and show us what are called *conditional distributions*.

| | | Class | | | | |
		First	Second	Third	Crew	Total
Alive	Count	203	118	178	212	711
	% of Column	62.5%	41.4%	25.2%	24.0%	32.3%
Dead	Count	122	167	528	673	1490
	% of Column	37.5%	58.6%	74.8%	76.0%	67.7%
Total	Count	325	285	706	885	2201

Survival labels the Alive/Dead/Total rows.

A contingency table of *Class* by *Survival* with only counts and column percentages. Notice how much easier this table is to read than the previous one. Of course, two other similar tables could be made for row percentages and table percentages.   **Table 3.6**

● **Percent of what?**   The English language can be tricky when we talk about percentages. If you're asked, "What percent *of the survivors* were in second class?" it's pretty clear. We're restricting the *Who* in the question to the survivors, so we should look at the number of second-class passengers among all the survivors: 118 out of 711; 118/711, or 16.6%. But if you're asked, "What percent were second-class passengers who survived?" you have a different question. Be careful; here, the *Who* is everyone on board, so 2201 should be the denominator, and the answer is 118/2201 or 5.4%. And if you're asked, "What percent of the second-class passengers survived?" you have a third question. Now the *Who* is the second-class passengers, so the denominator is the 285 second-class passengers, and the answer is 118/285 or 41.4%.

Always be sure to ask "percent of what?" That will help you to know the *Who* and whether we want *row, column,* or *table* percentages. ●

Conditional Distributions

Did the chance of surviving the *Titanic* sinking depend on ticket class? From the marginal distribution, we know that the overall chance of surviving was 711/2201 or 32.3%. Now we can ask whether the distribution of survivors is roughly constant across the four classes.

When we think about the relationship between two variables, we have a choice: We can ask whether the survival rates are the same across the four ticket classes, or we can ask whether the ticket classes were distributed the same way among survivors as among nonsurvivors. Both ways of comparing distributions are interesting.

We might start by restricting our attention to the survivors. Then we're redefining the *Who* of the study. The *Who* we're interested in now is only the survivors. Their numbers are in the first row of the contingency table, so we can focus on them by splitting that row off by itself. When we do this, the natural percentages to look at are the row percentages. These give the percent of survivors holding each class of ticket. The distribution we've created is called a **conditional distribution,** since it shows the distribution of one variable for just the individuals who satisfy some condition on another variable.

	Class				
	First	**Second**	**Third**	**Crew**	**Total**
Alive	203	118	178	212	711
	28.6%	16.6%	25.0%	29.8%	100%

The *conditional distribution* of ticket *Class*, conditional on having survived.   **Table 3.7**

Now we do the same thing for the nonsurvivors. The numbers for the nonsurvivors are found in the following row:

	Class				
	First	**Second**	**Third**	**Crew**	**Total**
Dead	122	167	528	673	1490
	8.2%	11.2%	35.4%	45.2%	100%

The *conditional distribution* of ticket *Class,* conditional on having perished.   **Table 3.8**

Let's compare these conditional distributions. Among survivors, 28.6% held a first-class ticket compared to only 8.2% among those who died. That looks like a big difference. The pie charts look different, too.

Pie charts of the distribution of *Class* for the survivors and nonsurvivors separately. Do the distributions appear to be the same? We're primarily concerned with percentages here, so pie charts are a good choice.   **Figure 3.6**

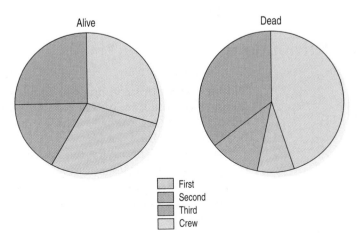

The nonsurvivors are mostly crew and third-class passengers. The survivors, on the other hand, are more uniformly split up across all four classes. If the percentages of ticket class had been about the same across the two survival groups, we would have said that survival was independent of class. But it's not. The differences we see between the two conditional distributions suggest that survival may have depended on ticket class.

It is interesting to know that *Class* and *Survival* are associated. That's an important part of the *Titanic* story. We know they're associated because we can see that the distribution of ticket classes differs between survivors and those who died. And we know how important this is because the margins show us the actual numbers of people involved.

Variables can be associated in many ways and to different degrees. The best way to tell whether two variables are associated is to ask whether they are *not*.[1] In a contingency table, when the distribution of *one* variable is the same for all categories of another, we say that the variables are **independent.** We'll see a way to check for independence formally later in the book. For now, we'll just compare the distributions.

A Statistics class reports the following data on *Sex* and *Eye color* for students in the class:

		Eye Color			
		Blue	Brown	Green/Hazel/Other	Total
Sex	Males	6	20	6	32
	Females	4	16	12	32
	Total	10	36	18	64

1. What percent of females are brown-eyed?

2. What percent of brown-eyed students are female?

3. What percent of students are brown-eyed females?

4. What's the distribution of *Eye color*?

5. What's the conditional distribution of *Eye color* for the males?

6. Compare the percent who are female among the blue-eyed students to the percent of all students who are female.

7. Does it seem that *Eye color* and *Sex* are independent? Explain.

Segmented Bar Charts

We could display the same information by dividing up bars rather than circles. The resulting **segmented bar chart** treats each bar as the "whole" and divides it proportionally into segments corresponding to the percentage in each group. We can clearly see that the distributions of ticket *Class* are different, indicating again that survival was not independent of ticket *Class*.

---

[1] This kind of "backwards" reasoning shows up surprisingly often in science—and in Statistics. We'll see it again.

A segmented bar chart for *Class* by *Survival*

Notice that although the totals for survivors and nonsurvivors are quite different, the bars are the same height because we have converted the numbers to *percentages*. Compare this display with the side-by-side pie charts of the same data in Figure 3.6.   **Figure 3.7**

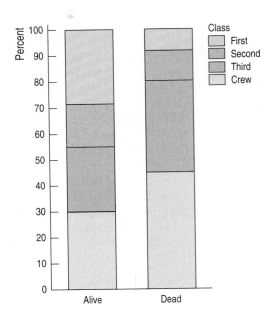

Examining Contingency Tables Step-By-Step

Medical researchers followed 6272 Swedish men for 30 years to see if there was any association between the amount of fish in their diet and prostate cancer ("Fatty Fish Consumption and Risk of Prostate Cancer," *Lancet*, June 2001). Their results are summarized in this table.

We asked for a picture of a "man eating fish." This is what we got.

	Prostate Cancer	
	No	**Yes**
Never/seldom	110	14
Small part of diet	2420	201
Moderate part	2769	209
Large part	507	42

Fish Consumption

Table 3.9

Is there an association between fish consumption and prostate cancer?

Think

Plan Be sure to state what the problem is about.

Variables Identify the variables and report the W's.

I want to know if there is an association between fish consumption and prostate cancer.

The individuals are 6272 Swedish men followed by medical researchers for 30 years. The variables record their fish consumption and whether or not they were diagnosed with prostate cancer.

Be sure to check the appropriate condition.

✔ **Categorical Data Condition:** I have counts for both fish consumption and cancer diagnosis. The categories of diet do not overlap, and the diagnoses do not overlap. It's okay to draw pie charts or bar charts.

Show

Mechanics It's a good idea to check the marginal distributions first before looking at the two variables together.

		Prostate Cancer		
		No	**Yes**	**Total**
Fish Consumption	**Never/Seldom**	110	14	**124 (2.0%)**
	Small part of diet	2420	201	**2621 (41.8%)**
	Moderate part	2769	209	**2978 (47.5%)**
	Large part	507	42	**549 (8.8%)**
	Total	**5806 (92.6%)**	**466 (7.4%)**	**6272 (100%)**

Two categories of the diet are quite small, with only 2.0% Never/Seldom eating fish and 8.8% in the "Large part" category. Overall 7.4% of the men in this study had prostate cancer.

Then, make appropriate displays to see whether there is a difference in the relative proportions.

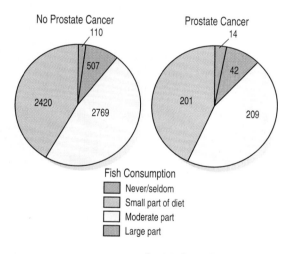

Both pie charts and bar charts can be used to compare conditional distributions.

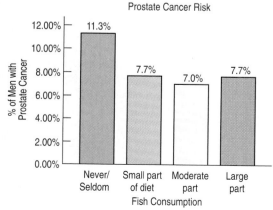

Tell

Conclusion Interpret the patterns in the table and displays in context. If you can, discuss possible real-world consequences. Be careful not to overstate what you see. The results may not generalize to other situations.

Most of the men (89.3%) in the study ate fish either as a moderate or small part of their diet. There is a 7.4% rate of prostate cancer overall for these men. From the pie charts it's hard to see a difference in cancer rates among the groups. But in the bar chart, it looks like the cancer rate for those who ate fish never/seldom may be somewhat higher.

Even though the cancer rate for those who never ate fish appears to be higher, more study would probably be needed before we would make recommendations that men change their diets.[2]

---

[2] The original study actually used pairs of twins, which enabled the researchers to discern that the risk of cancer for those who never ate fish actually *was* substantially greater. Using pairs is a special way of gathering data. We'll discuss such study design issues and how to analyze the data in later chapters.

What Can Go Wrong?

- ***Don't violate the area principle.*** This is probably the most common mistake in a graphical display. It is often made in the cause of artistic presentation. Here, for example, are two displays of the pie chart of the *Titanic* passengers by class:

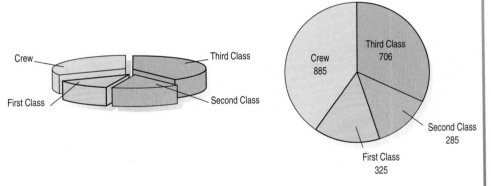

The one on the left looks pretty, doesn't it? But showing the pie on a slant violates the area principle and makes it much more difficult to compare fractions of the whole made up of each class—the principal feature that a pie chart ought to show.

- ***Keep it honest.*** Here's a pie chart that displays data on the percentage of high-school students who engage in specified dangerous behaviors as reported by the Centers for Disease Control. What's wrong with this plot?

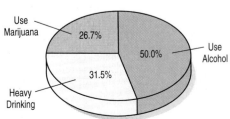

Try adding up the percentages. Or look at the 50% slice. Does it look right? Then think: What are these percentages of? Is there a "whole" that has been sliced up? In a pie chart, the proportions shown by each slice of the pie must add up to 100% and each individual must fall into only one category. Of course, showing the pie on a slant makes it even harder to detect the error.

Here's another. This bar chart shows the number of airline passengers searched by security screening.

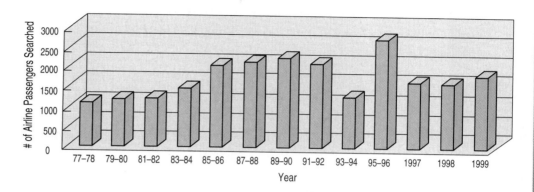

Looks like things didn't change much in the final years of the 20th century—until you read the bar labels and see that the last three bars represent single years, while all the others are for *pairs* of years. Of course, the false depth makes it harder to see the problem.

- **Don't confuse similar-sounding percentages.** These percentages sound similar but are different:

	Class				
	First	Second	Third	Crew	Total
Alive	203	118	178	212	**711**
Dead	122	167	528	673	**1490**
Total	**325**	**285**	**706**	**885**	**2201**

  (Survival)

  - The percentage of the passengers who were both in first class and survived: This would be 203/2201, or 9.4%.
  - The percentage of the first-class passengers who survived: This is 203/325, or 62.5%.
  - The percentage of the survivors who were in first class: This is 203/711, or 28.6%.

      In each instance, pay attention to the *Who* implicitly defined by the phrase. Often there is a restriction to a smaller group (all aboard the *Titanic,* those in first class, and those who survived, respectively) before a percentage is found. Your discussion of results must make these differences clear.
- **Don't forget to look at the variables separately, too.** When you make a contingency table or display a conditional distribution, be sure you also examine the marginal distributions. It's important to know how many cases are in each category.
- **Be sure to use enough individuals.** When you consider percentages, take care that they are based on a large enough number of individuals. Take care not to make a report such as this one:

    We found that 66.67% of the rats improved their performance with training. The other rat died.
- **Don't overstate your case.** Independence is an important concept, but it is rare for two variables to be *entirely* independent. We can't conclude that one variable has no effect whatsoever on another. Usually all we know is that little effect was observed in our study. Other studies of other groups under other circumstances could find different results.

Simpson's Paradox

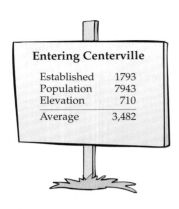

Entering Centerville

Established	1793
Population	7943
Elevation	710
Average	3,482

- ***Don't use unfair or silly averages.*** Sometimes averages can be misleading. Sometimes they just don't make sense at all. Be careful when averaging different variables that the quantities you're averaging are comparable. The Centerville sign says it all.

When using averages of proportions across several different groups, it's important to make sure that the groups really are comparable.

It's easy to make up an example showing that averaging across very different values or groups can give absurd results. Here's how that might work. Suppose there are two pilots, Moe and Jill. Moe argues that he's the better pilot of the two, since he managed to land 83% of his last 120 flights on time compared with Jill's 78%. But let's look at the data a little more closely. Here are the results for each of their last 120 flights, broken down by the time of day they flew:

		Time of Day		
		Day	**Night**	**Overall**
Pilot	**Moe**	90 out of 100 90%	10 out of 20 50%	100 out of 120 83%
	Jill	19 out of 20 95%	75 out of 100 75%	94 out of 120 78%

On-time flights by *Time of Day* and *Pilot.* Look at the percentages within each *Time of Day* category. Who has a better on-time record during the day? At night? Who is better overall?   **Table 3.10**

One famous example of Simpson's paradox arose during an investigation of admission rates for men and women at the University of California at Berkeley's graduate schools. As reported in an article in *Science*, about 45% of male applicants were admitted, but only about 30% of female applicants got in. It looked like a clear case of discrimination. However, when the data were broken down by school (Engineering, Law, Medicine, etc.) it turned out that within each school, the women were admitted at nearly the same or, in some cases, much *higher* rates than the men. How could this be? Women applied in large numbers to schools with very low admission rates (Law and Medicine, for example, admitted fewer than 10%). Men tended to apply to Engineering and Science. Those schools have admission rates above 50%. When the *average* was taken, the women had a much lower *overall* rate, but the average didn't really make sense.

Look at the day and nighttime flights separately. For day flights, Jill had a 95% on-time rate, and Moe only a 90% rate. At night, Jill was on time 75% of the time, and Moe only 50%. So Moe is better "overall," but Jill is better both during the day and at night. How can this be?

What's going on here is a problem known as **Simpson's paradox,** named for the statistician who discovered it in the 1960s. It comes up rarely in real life, but there have been several well-publicized cases of it. As we can see from the pilot example, the problem is *unfair averaging* over different groups. Jill has mostly night flights, which are more difficult, so her *overall average* is heavily influenced by her nighttime average. Moe, on the other hand, benefits from flying mostly during the day, with its higher on-time percentage. With their very different patterns of flying conditions, taking an overall average is misleading. It's not a fair comparison.

The moral of Simpson's paradox is to be careful when you average across different levels of a second variable. It's always better to compare percentages or other averages *within* each level of the other variable. The overall average may be misleading.

CONNECTIONS

All of the methods of this chapter work with *categorical variables.* You must know the *Who* of the data to know who is counted in each category and the *What* of the variable to know where the categories come from.

What have we learned?

We've learned that we can summarize categorical data by counting the number of cases in each category, sometimes expressing the resulting distribution as percents. We can display the distribution in a bar chart or a pie chart. When we want to see how two categorical variables are related, we put the counts (and/or percentages) in a two-way table called a contingency table.

- We look at the marginal distribution of each variable (found in the margins of the table).
- We also look at the conditional distribution of a variable within each category of the other variable.
- We can display these conditional and marginal distributions using bar charts or pie charts.
- If the conditional distributions of one variable are (roughly) the same for every category of the other, the variables are independent.

TERMS

Frequency table
Relative Frequency table
A frequency table lists the categories in a categorical variable and gives the count or percentage of observations of each category.

Distribution
The distribution of a variable gives
- the possible values of the variable and
- the relative frequency of each value.

Area principle
In a statistical display, each data value should be represented by the same amount of area.

Bar chart
Bar charts show a bar representing the count of each category in a categorical variable.

Pie chart
Pie charts show how a "whole" divides into categories by showing a wedge of a circle whose area corresponds to the proportion in each category.

Contingency table
A contingency table displays counts and, sometimes, percentages of individuals falling into named categories on two or more variables. The table categorizes the individuals on all variables at once, to reveal possible patterns in one variable that may be contingent on the category of the other.

Marginal distribution
In a contingency table, the distribution of either variable alone is called the marginal distribution. The counts or percentages are the totals found in the margins (last row or column) of the table.

Conditional distribution
The distribution of a variable restricting the *Who* to consider only a smaller group of individuals is called a conditional distribution.

Independence
Variables are said to be independent if the conditional distribution of one variable is the same for each category of the other. We'll show how to check for independence in a later chapter.

Simpson's paradox
When averages are taken across different groups, they can appear to contradict the overall averages. This is known as "Simpson's paradox."

S K I L L S   *When you complete this lesson you should:*

Think

- Be able to recognize when a variable is categorical and choose an appropriate display for it.
- Understand how to examine the association between categorical variables by comparing conditional and marginal percentages.

Show

- Be able to summarize the distribution of a categorical variable with a frequency table.
- Be able to display the distribution of a categorical variable with a bar chart or pie chart.
- Know how to make and examine a contingency table.
- Know how to make and examine displays of the conditional distributions of one variable for two or more groups.

Tell

- Be able to describe the distribution of a categorical variable in terms of its possible values and relative frequencies.
- Know how to describe any anomalies or extraordinary features revealed by the display of a variable.
- Be able to describe and discuss patterns found in a contingency table and associated displays of conditional distributions.

Displaying Categorical Data on the Computer

Although every package makes a slightly different bar chart, they all have similar features:

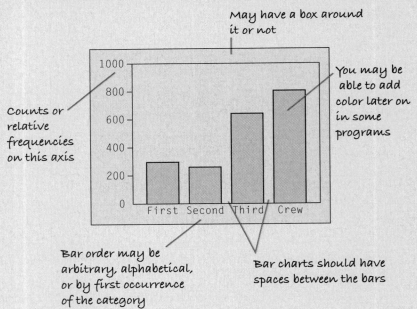

Sometimes the count or a percentage is printed above or on top of each bar to give some additional information. You may find that your statistics package sorts category names in annoying orders by default. For example, many packages sort categories alphabetically or by the order the categories are seen in the data set. Often, neither of these is the best choice.

DATA DESK

To make a bar chart or pie chart, select the variable.
In the **Plot** menu choose **Bar Chart** or **Pie Chart**.
To make a frequency table, in the **Calc** menu choose **Frequency Table.**

Comments

These commands treat the data as categorical even if they are numerals. If you select a quantitative variable by mistake, you'll see an error message warning of too many categories.

EXCEL

First make a pivot table (Excel's name for a frequency table).
From the **Data** menu choose **Pivot Table** and **Pivot Chart Report.**
When you reach the Layout window, drag your variable to the row area and drag your variable again to the data area. This tells Excel to count the occurrences of each category.
Once you have an Excel pivot table, you can construct bar charts and pie charts.
Click inside the Pivot Table.

Click the Pivot Table Chart Wizard button. Excel creates a bar chart.
A longer path leads to a pie chart; see your Excel documentation.

Comments

Excel uses the pivot table to specify the category names and find counts within each category. If you already have that information, you can proceed directly to the Chart Wizard.

JMP

JMP makes a bar chart and frequency table together.
From the **Analyze** menu, choose **Distribution.**
In the Distribution dialog, drag the name of the variable into the empty variable window beside the label "Y, Columns"; click **OK.**
To make a pie chart, choose **Chart** from the **Graph** menu.
In the Chart dialog, select the variable name from the Columns list, click on the button labeled "Statistics," and select "N" from the drop-down menu.

Click the "X, Level" button to assign the same variable name to the X-axis.

Under Options, click on the top button—labeled "Vertical"—and select "Pie" from the drop-down menu. Click **OK.**

MINITAB

To make a bar chart choose **Bar Chart** from the **Graph** menu.
Select "Counts of unique values" in the first menu, and select "Simple" for the type of graph. Click **OK.**

In the Chart dialog, enter the name of the variable that you wish to display in the box labeled "Categorical variables." Click **OK.**

SPSS

To make a bar chart choose **Bar** from the **Graphs** menu.
From the Bar Charts Dialog, choose **Simple** and indicate the nature of the data.
In the Define Simple Bar dialog, select the variable name from the source list and click on the **Category Axis** arrow.

Click the **OK** button to create the display.

A similar path makes a pie chart by choosing **Pie** rather than Bar from the **Graphs** menu.

TI-83/84 Plus

The TI-83 won't do displays for categorical variables.

TI-89

The TI-89 won't do displays for categorical variables.

EXERCISES

1. Graphs in the news. Find a bar graph of categorical data from a newspaper or magazine.
a) Is the graph clearly labeled?
b) Does it violate the area principle?
c) Does the accompanying article tell the W's of the variable?
d) Do you think the article correctly interprets the data? Explain.

2. Graphs in the news II. Find a pie chart of categorical data from a newspaper or magazine.
a) Is the graph clearly labeled?
b) Does it violate the area principle?
c) Does the accompanying article tell the W's of the variable?
d) Do you think the article correctly interprets the data? Explain.

3. Tables in the news. Find a frequency table of categorical data from a newspaper or magazine.
a) Is it clearly labeled?
b) Does it display percentages or counts?
c) Does the accompanying article tell the W's of the variable?
d) Do you think the article correctly interprets the data? Explain.

4. Tables in the news II. Find a contingency table of categorical data from a newspaper or magazine.
a) Is it clearly labeled?
b) Does it display percentages or counts?
c) Does the accompanying article tell the W's of the variables?
d) Do you think the article correctly interprets the data? Explain.

5. Magnet schools. An article in the Winter 2003 issue of *Chance* magazine reported on the Houston Independent School District's magnet schools programs. Of the 1755 qualified applicants, 931 were accepted, 298 were waitlisted, and 526 were turned away for lack of space. Find the relative frequency distribution of the decisions made, and write a sentence describing it.

6. Magnet schools, again. The *Chance* article about the Houston magnet schools program described in Exercise 5 also indicated that 517 applicants were Black or Hispanic, 292 Asian, and 946 White. Summarize the relative frequency distribution of ethnicity with a sentence or two (in the proper context, of course).

7. Causes of death. The Centers for Disease Control lists causes of death in the United States during 1999.

Cause of Death	Percent
Heart disease	30.3
Cancer	23.0
Circulatory diseases and stroke	8.4
Respiratory diseases	7.9
Accidents	4.1

a) Is it reasonable to conclude that heart or respiratory diseases were the cause of approximately 38% of U.S. deaths in 1999?
b) What percent of deaths were from causes not listed here?
c) Create an appropriate display for these data.

8. Education. In a December 2000 report, the U.S. Census Bureau listed the levels of educational attainment for Americans over 65. Create an appropriate display for these data, and write a sentence or two that might appear in a newspaper article about the report.

Education Level	Count (thousands)
No high school diploma	9,945
HS graduate, but no college	11,701
Some college, no degree	4,481
2-year degree	1,390
4-year degree	3,133
Master's degree	1,213
Ph.D. or professional degree	757

9. Ghosts. A May 2001 Gallup Poll found that many Americans believe in ghosts and other supernatural phenomena. The poll was based on telephone responses from 1012 randomly selected adults. The table shows the percentages of people who expressed belief in various phenomena.

Phenomenon	Percent Expressing Belief
Psychic healing	54
ESP	50
Ghosts	38
Astrology	28
Channeling	15

a) Is it reasonable to conclude that 66% of those polled expressed belief in either ghosts or astrology?
b) Can you tell what percent of people did not believe in any of these phenomena? Explain.
c) Create an appropriate display for these data.

10. Illegal guns. A study by the U.S. Bureau of Alcohol, Tobacco, and Firearms (BATF) (*USA Today*, 22 June 2000) surveyed 1530 investigations by the U.S. BATF into illegal gun trafficking from July 1996 through December 1998. The study reports the portion of cases that were the result of each of five gun trafficking violations:

    46% Straw purchase (legal gun buyer acting on behalf of an illegal buyer)
    21% Unlicensed sellers
    14% Gun shows and flea markets
    14% Stolen from federally licensed dealers
    10% Stolen from residences

a) State the W's for this study to the extent the story gives them.
b) What do you notice about the percentages listed? What does that probably mean?
c) Make a bar chart to display these results, and label it correctly.
d) Write a brief report on what they say about illegal gun trafficking. Note any possible problems with the data.
e) The study also noted that although corrupt licensed dealers accounted for 133 of the 1530 investigations, they were linked to 40,365 of the 84,128 firearms involved in those investigations. How does this new information about the *Who* of the study affect your conclusions?

11. Oil spills. To improve the structural design of oil tankers with the objective of reducing the likelihood of an oil spill and decreasing the amount of outflow in the event of a hull puncture, a study (*Marine Technology*, Jan. 1995) reported the spillage amount and cause of puncture for 50 recent major oil spills from tankers and carriers. Here are displays. Write a brief report interpreting what the displays show. Is a pie chart an appropriate display for these data? Why or why not?

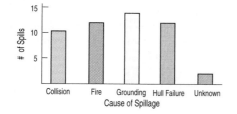

Cause of Spillage

Cause of Spillage

12. Winter Olympics 2002. 25 countries won medals in the 2002 Winter Olympics. The table lists them, along with the total number of medals each won:

Country	Medals	Country	Medals
Germany	35	Croatia	4
USA	34	Korea	4
Norway	24	Bulgaria	3
Canada	17	Estonia	3
Austria	16	Great Britain	3
Russia	16	Australia	2
Italy	12	Czech Republic	2
France	11	Japan	2
Switzerland	11	Poland	2
China	8	Spain	2
Netherlands	8	Belarus	1
Finland	7	Slovenia	1
Sweden	6		

a) Try to make a display of these data. What problems do you encounter?
b) Can you find a way to organize the data so that the graph is more successful?

13. Teens and technology. The Gallup Organization, in conjunction with CNN, *USA Today,* and the National Science Foundation, conducted a national survey of 744 children in grades 7 through 12—mostly surveying students in the "teenage" years of 13 to 17. Telephone interviews were conducted March 20–27, 1997, from Gallup interviewing centers throughout the country. The focus of the survey was on students' familiarity with and use of modern technology, with special attention given to use of computers and the Internet. The teenagers were asked if they used each of the following technologies on a daily basis and if the technology was critically important to own. For each question, the percentage of those responding "Yes" is given. Gallup dubbed the difference between the two percentages the "Importance Gap." Here are the results:

Technology	Use daily	Critically important to own	Importance gap
Computer	44%	77%	33
Telephone answering machine	46%	62%	16
VCR	39%	51%	12
Calculator	67%	71%	4
Stereo/CDs/audio	85%	69%	−16
Video games	46%	18%	−28

Consider the following graphical display showing the percentages of teens that use each of the technologies on a daily basis.

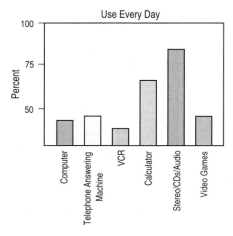

a) How much larger is the proportion of teens who use a calculator daily than the corresponding proportion for answering machines?
b) Is that the impression given by the display? Explain.
c) How would you improve this display?
d) Make an appropriate display for the Importance Gap. (*Hint:* Make the *y*-axis of your chart span the range from −30 to +35.)
e) Write a few sentences describing what you have learned about teens' attitudes toward technology.

14. Teens and technology II. Here's a display of the percentages of students who use the various technologies daily. List the errors in this display.

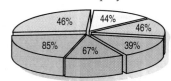

T 15. Auditing reform. In the wake of the Enron Corporation scandal, the Gallup Organization asked 1001 American adults what kind of changes, if any, are needed in the way major corporations are audited. Here is a display of the results.

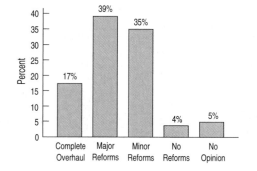

a) Make a pie chart of the same data.
b) Which chart works better to summarize the data? Why?
c) Summarize the findings of the poll in a few sentences that might appear in a newspaper article.

16. Modalities. A survey of athletic trainers (Nadler, Scott F., Prybicien, Michael, Malanga, Gerard A., Sicher, Dan. "Complications from Therapeutic Modalities: Results of a National Survey of Athletic Trainers." *Archives Physical Medical Rehabilitation* 84 [June 2003]) asked what modalities (treatment methods such as ice, whirlpool, ultrasound, or exercise) they commonly use to treat injuries. Respondents were each asked to list three modalities. The article included the following figure reporting the modalities used.

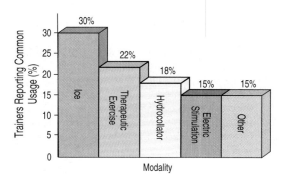

a) What problems do you see with the graph?
b) Consider the percentages for the named modalities. Do you see anything odd about them?

17. Complications. The survey of athletic trainers reported in Exercise 16 summarizes the complications most commonly associated with cryotherapy (ice). The trainers were asked to report the four most common side effects they had seen with any treatment modality. Of those identifying cryotherapy, 86 respondents reported allergic reactions, 23 reported burns, 16 reported pain intolerance, and 6 reported frostbite.
a) Make an appropriate display of these data.
b) Specify the *Who* for these data. Would the data provide the most useful information about the risks of cryotherapy?

18. Politics. Students in an Intro Stats course were asked to describe their politics as "Liberal," "Moderate," or "Conservative." Here are the results:

		Politics			
		L	**M**	**C**	**Total**
Sex	**Female**	35	36	6	77
	Male	50	44	21	115
	Total	85	80	27	192

a) What percent of the class is male?

b) What percent of the class considers themselves to be "Conservative"?

c) What percent of the males in the class consider themselves to be "Conservative"?

d) What percent of all students in the class are males who consider themselves to be "Conservative"?

e) What percent of all females in the class are "Liberals"?

f) What percent of all males in the class are "Liberals"?

g) Do *politics* and *sex* appear to be independent?

T 19. Magnet schools revisited. The *Chance* magazine article described in Exercise 5 further examined the impact of an applicant's ethnicity on the likelihood of admission to the Houston Independent School District's magnet schools programs. Those data are summarized in the table below.

	Admission Decision			
Ethnicity		**Accepted**	**Wait-listed**	**Turned away**	**Total**
Black/Hispanic	485	0	32	517
Asian	110	49	133	292
White	336	251	359	946
Total	**931**	**300**	**524**	**1755**

a) What percent of all applicants were Asian?

b) What percent of the students accepted were Asian?

c) What percent of Asians were accepted?

d) What percent of all students were accepted?

20. More politics. Look once more at the table summarizing the political views and genders of Intro Stats students in Exercise 18.

a) Produce a graphical display of the marginal distribution of politics in the class.

b) Produce a graphical display of the conditional distribution of politics among the males.

c) Produce a graphical display of the conditional distribution of politics among the females.

d) Produce a graphical display comparing the conditional distributions of sex among the three categories of politics.

e) Comment briefly on what you see from the display in d.

21. Back to school. Examine the table about ethnicity and acceptance for the Houston Independent School District's magnet schools program, shown in Exercise 19. Does it appear that the admissions decisions are made independent of the applicant's ethnicity? Explain.

T 22. Cars. A survey of autos parked in student and staff lots at a large university classified the brands by country of origin, as seen in the table.

		Driver	
		Student	**Staff**
Origin	American	107	105
	European	33	12
	Asian	55	47

a) What percent of all the cars surveyed were foreign?

b) What percent of the American cars were owned by students?

c) What percent of the students owned American cars?

d) What is the marginal distribution of origin?

e) What are the conditional distributions of origin by driver classification?

f) Do you think that origin of the car is independent of the type of driver? Explain.

T 23. Class of 2000. Prior to graduation, a high-school class of 2000 was surveyed about their plans. The table below displays the results for white and minority students. (The "Minority" group included African-American, Asian, Hispanic, and Native American students.)

		White	**Minority**
Plans	4-year college	198	44
	2-year college	36	6
	Military	4	1
	Employment	14	3
	Other	16	3

a) What percent of the graduates are white?

b) What percent of the graduates are planning to attend a 2-year college?

c) What percent of the graduates are white and planning to attend a 2-year college?

d) What percent of the white graduates are planning to attend a 2-year college?

e) What percent of the graduates planning to attend a 2-year college are white?

f) Create a graph comparing the plans of white and minority students.

g) Do you see any important differences in the post-graduation plans of white and minority students? Write a brief summary of what these data show, including comparisons of conditional distributions.

T 24. After high school. The table below compares what Ithaca High School students did after graduation in 1959, 1970, and 1980.

	1959	1970	1980
Continuing education	197	388	320
Employed	103	137	98
In the military	20	18	18
Other	13	58	45

(Left axis label: **What Graduates Did**)

a) What percent of all these graduates joined the military?
b) What percent of these students graduated in 1970?
c) What percent of the 1970 graduates joined the military?
d) Of the students in these surveys who joined the military, what percent graduated in 1970?
e) What is the marginal distribution of postgraduation activities?
f) What is the conditional distribution of postgraduation activities among the class of 1959?
g) Does this study present any evidence that postgraduation plans have changed over this 21-year period? Write a brief description of these data. Include an appropriate graph.

T 25. Canadian languages. Statistics Canada provides, on its Web site, the following data on the Canadian population (in thousands). (Zeros indicate counts below 500.)

(Left axis label: **Province**)

	English only	French only	Both	Neither	Total
Newfoundland	525	0	21	1	547
Prince Edward Island	118	0	15	0	133
Nova Scotia	813	1	84	1	900
New Brunswick	418	73	238	0	730
Quebec	359	3,952	2,661	74	7,045
Ontario	9,116	47	1,235	245	10,643
Manitoba	984	1	103	12	1,100
Saskatchewan	921	0	51	5	977
Alberta	2,455	2	179	34	2,669
British Columbia	3,342	2	249	97	3,690
Yukon Territory	27	0	3	0	31
Northwest Territories	36	0	3	1	39
Nunavut	20	0	1	4	25
Total	19,134	4,078	4,843	474	28,529

a) What percent of Canadian citizens speak only English?
b) What percent of Canadian citizens speak French?
c) What percent of Quebec residents speak French?
d) What percent of French-speaking Canadians live in Quebec?
e) Do you think that language knowledge and province of residence are independent for Canadians? Explain.

T 26. Tattoos. A study by the University of Texas Southwestern Medical Center examined 626 people to see if there was an increased risk of contracting hepatitis C associated with having a tattoo. If the subject had a tattoo, researchers asked whether it had been done in a commercial tattoo parlor or elsewhere. Write a brief description of the association between tattooing and hepatitis C, including an appropriate graphical display.

	Tattoo done in commercial parlor	Tattoo done elsewhere	No tattoo
Has hepatitis C	17	8	18
No hepatitis C	35	53	495

T 27. Weather forecasts. Just how accurate are the weather forecasts we hear every day? The table below compares the daily forecast with a city's actual weather for a year.

	Actual Weather	
	Rain	No rain
Forecast Rain	27	63
Forecast No rain	7	268

a) On what percent of days did it actually rain?
b) On what percent of days was rain predicted?
c) What percent of the time was the forecast correct?
d) Do you see evidence of an association between the type of weather and the ability of forecasters to make an accurate prediction? Write a brief explanation, including an appropriate graph.

T 28. Federal prisons. The table below shows the number of federal prison inmates serving sentences for various

(Left axis label: **Type of Offense**)

	Year 1990	Year 1998
Violent (murder, robbery, etc.)	10	13
Property (burglary, fraud, etc.)	8	9
Drugs	30	63
Public order (immigration, weapons, etc.)	9	22
Other	1	2

types of offenses in 1990 and 1998. Counts given are in thousands of prisoners.

a) Write a brief description of these data, in the proper context, highlighting any important changes you see in the prison population.

b) Do these data indicate that there was an increase in drug use in the United States during the 1990s? Explain.

T 29. Working parents. In July 1991 and again in April 2001 the Gallup Poll asked random samples of 1015 adults about their opinions on working parents. The table summarizes responses to this question:

"Considering the needs of both parents and children, which of the following do you see as the ideal family in today's society?"

Based upon these results, do you think there was a change in people's attitudes during the 10 years between these polls? Explain.

		Year	
		1991	**2001**
	Both work full time	142	131
	One works full time, other part time	274	244
Response	One works, other works at home	152	173
	One works, other stays home for kids	396	416
	No opinion	51	51

T 30. Twins. In 2000 the *Journal of the American Medical Association (JAMA)* published a study that examined pregnancies that resulted in the birth of twins. Births were classified as preterm with intervention (induced labor or cesarean), preterm without procedures, or term/postterm. Researchers also classified the pregnancies by the level of prenatal medical care the mother received (inadequate, adequate, or intensive). The data, from the years 1995–1997, are summarized in the table below. Figures are in thousands of births. (*JAMA* 284 [2000]:335–341)

Twin Births 1995–1997 (in thousands)

	Preterm (induced or cesarean)	Preterm (without procedures)	Term or post-term	Total
Intensive	18	15	28	61
Adequate	46	43	65	154
Inadequate	12	13	38	63
Total	76	71	131	278

(Level of Prenatal Care)

a) What percent of these mothers did not receive adequate medical care during their pregnancies?

b) What percent of all twin births were preterm?

c) Among the mothers who did not receive adequate medical care, what percent of the twin births were preterm?

d) Create an appropriate graph comparing the outcomes of these pregnancies by the level of medical care the mother received.

e) Write a few sentences describing the association between these two variables.

T 31. Blood pressure. A company held a blood pressure screening clinic for its employees. The results are summarized in the table below by age group and blood pressure level.

		Age		
		Under 30	**30–49**	**Over 50**
Blood Pressure	**Low**	27	37	31
	Normal	48	91	93
	High	23	51	73

a) Find the marginal distribution of blood pressure level.

b) Find the conditional distribution of blood pressure level within each age group.

c) Compare these distributions with a segmented bar graph.

d) Write a brief description of the association between age and blood pressure among these employees.

e) Does this prove that people's blood pressure increases as they age? Explain.

T 32. Obesity and exercise. In the year 2000, the Centers for Disease Control (CDC) estimated that 19.8% of Americans over 15 years old were obese. The CDC conducts a survey on obesity and various behaviors. Here is a table on self-reported exercise classified by body mass index (BMI).

	Body Mass Index		
	Normal (%)	**Overweight (%)**	**Obese (%)**
Inactive	23.8	26.0	35.6
Irregularly active	27.8	28.7	28.1
Regular, not intense	31.6	31.1	27.2
Regular, intense	16.8	14.2	9.1

(Physical Activity)

a) Are these percentages column percentages, row percentages, or table percentages?

b) Use graphical displays to show different percentages of physical activities for the three BMI groups.

c) Do these data prove that lack of exercise causes obesity? Explain.

33. Family planning. A 1954 study of 1438 pregnant women examined the association between the woman's education level and the occurrence of unplanned pregnancies, producing these data:

	Education Level		
	<3 yr HS	3+ yr HS	Some college
Number of Pregnancies	591	608	239
% Unplanned	66.2%	55.4%	42.7%

Does this indicate that more schooling taught young women better family planning? What other explanations for these data can you think of? (*Fertility Planning and Fertility Rates by Socio-Economic Status,* Social and Psychological Factors Affecting Fertility, 1954)

T 34. Pet ownership. The U.S. Census Bureau reports the number of households owning various types of pets. Specifically, they keep track of dogs, cats, birds, and horses.
a) Do you think the income distributions of the households who own these different animals would be roughly the same? Why or why not?
b) The table shows the percentages of income levels for each type of animal owned. Are these row percentages, column percentages, or table percentages?
c) Do the data support your initial guess? Explain.

INCOME DISTRIBUTION OF HOUSEHOLDS OWNING PETS (PERCENT)

Income	Pet			
	Dog	Cat	Bird	Horse
Under $12,500	14	15	16	9
$12,500 to $24,999	20	20	21	21
$25,000 to $39,999	24	23	24	25
$40,000 to $59,999	22	22	21	22
$60,000 and over	20	20	18	23
Total	100	100	100	100

T 35. Worldwide toy sales. Around the world, toys are sold through different channels. For example, in some parts of the world toys are sold primarily through large toy store chains, while in other countries, department stores sell more toys. The table below shows the percentages by region of the distribution of toys sold through various channels in 1999.
a) What percent of all toys sold through department stores are sold in Europe?
b) What percent of all toys worldwide are sold through catalogs?
c) Compare the distribution of channels for Europe to North America.
d) Summarize the distribution of toy sales by channel in a few sentences.

REGIONAL TOY SALES BY CHANNEL OF DISTRIBUTION 1999

	Type of Outlet							
	Toy chains	General merchandise	Toy, hobby, and game retailers	Department stores	Food, drug, and misc. outlets	Catalog sales	E-Tailers	Total
World (Total)	17773.75	24883.25	10109.50	5987.60	12086.15	2843.80	710.95	74395.00
North America	6571.11	12516.40	1251.64	938.73	7822.75	1877.46	312.91	31291.00
Europe	5105.40	5105.40	2552.70	2042.16	1701.80	510.54	0.00	17018.00
Asia	4294.75	4122.96	4981.91	2233.27	1546.11	0.00	0.00	17179.00
Latin and South America	440.80	1129.55	523.45	440.80	220.40	0.00	0.00	2755.00
Oceania	218.55	728.50	291.40	87.42	101.99	29.14	0.00	1457.00
Other	1143.14	1280.44	508.40	245.22	693.10	426.66	398.04	4695.00

T 36. Drivers' licenses. The following table shows the number of licensed U.S. drivers by age and by gender.

Age	Male drivers (number)	Female drivers (number)	Total
19 and under	5,029,498	4,714,021	9,743,519
20–24	8,158,599	7,807,078	15,965,677
25–29	8,988,142	8,597,716	17,585,858
30–34	9,767,476	9,387,297	19,154,773
35–39	10,621,910	10,437,549	21,059,459
40–44	10,576,976	10,516,251	21,093,227
45–49	9,578,268	9,575,363	19,153,631
50–54	8,448,424	8,419,527	16,867,951
55–59	6,394,207	6,366,285	12,760,492
60–64	4,970,258	4,944,370	9,914,628
65–69	4,182,933	4,202,950	8,385,883
70–74	3,644,990	3,822,570	7,467,560
75–79	2,820,136	3,091,013	5,911,149
80–84	1,656,789	1,854,278	3,511,067
85 and over	957,463	1,092,687	2,050,150
Total	95,796,069	94,828,955	190,625,024

a) What percent of total drivers are under 20?
b) What percent of total drivers are male?
c) Write a few sentences comparing the number of male and female licensed drivers in each age group.
d) Do a driver's age and gender appear to be independent? Explain?

37. Hospitals. Most patients who undergo surgery make routine recoveries and are discharged as planned. Others suffer excessive bleeding, infection, or other postsurgical complications and have their discharges from the hospital delayed. Suppose your city has a large hospital and a small hospital, each performing major and minor surgeries. You collect data to see how many surgical patients have their discharges delayed by postsurgical complications, and find the results shown in the following table.

Procedure	Discharge Delayed	
	Large hospital	Small hospital
Major surgery	120 of 800	10 of 50
Minor surgery	10 of 200	20 of 250

a) Overall, for what percent of patients was discharge delayed?
b) Were the percentages different for major and minor surgery?
c) Overall, what were the discharge delay rates at each hospital?

d) What were the delay rates at each hospital for each kind of surgery?
e) The small hospital advertises that it has a lower rate of postsurgical complications. Do you agree?
f) Explain, in your own words, why this confusion occurs.

38. Delivery service. A company must decide which of two delivery services they will contract with. During a recent trial period they shipped numerous packages with each service, and have kept track of how often deliveries did not arrive on time. Here are the data:

Delivery service	Type of service	Number of deliveries	Number of late packages
Pack Rats	Regular	400	12
	Overnight	100	16
Boxes R Us	Regular	100	2
	Overnight	400	28

a) Compare the two services' overall percentage of late deliveries.
b) Based on the results in part a, the company has decided to hire Pack Rats. Do you agree they deliver on time more often? Why or why not? Be specific.
c) The results here are an instance of what phenomenon?

39. Graduate admissions. A 1975 article in the magazine *Science* examined the graduate admissions process at Berkeley for evidence of gender bias. The table below shows the number of applicants accepted to each of four graduate programs.

Program	Males accepted (of applicants)	Females accepted (of applicants)
1	511 of 825	89 of 108
2	352 of 560	17 of 25
3	137 of 407	132 of 375
4	22 of 373	24 of 341
Total	1022 of 2165	262 of 849

a) What percent of total applicants were admitted?
b) Overall, were a higher percentage of males or females admitted?
c) Compare the percentage of males and females admitted in each program.
d) Which of the comparisons you made do you consider to be the most valid? Why?

Just Checking

Just Checking

Answers

1. 50.0%

2. 44.4%

3. 25.0%

4. 15.6% Blue, 56.3% Brown, 28.1% Green/Hazel/Other

5. 18.8% Blue, 62.5% Brown, 18.8% Green/Hazel/Other

6. 40% of the blue-eyed students are female, while 50% of all students are female.

7. Since blue-eyed students appear less likely to be female, it seems that they may not be independent. (But the numbers are small.)

CHAPTER 4

Displaying Quantitative Data

WHO	Months
WHAT	Changes in Enron's stock price
UNITS	Dollars
HOW	Difference of closing price on first day of each month minus the first day of previous month
WHEN	1997 through 2001
WHERE	New York Stock Exchange

Enron Corporation was once one of the world's biggest corporations. From its humble beginnings as an interstate natural gas supply company in 1985, it grew steadily throughout the 1990s, diversifying into nearly every form of energy transaction and eventually dominating the energy trading business.

Its stock price followed this spectacular growth. In 1985 Enron stock sold for about $5 a share, but at the end of 2000, Enron stock closed at a 52-week high of $89.75 and the company's stock was worth more than $6 billion. Less than a year later it hit a low of $0.25 a share, having lost more than 99% of its value. Many employees who had taken advantage of generous stock plans lost retirement packages worth hundreds of thousands of dollars. Just how volatile was Enron stock? And were there hints of trouble that might have been seen?

Rather than look at the stock prices themselves, let's look at how much they changed from month to month. For example, on February 3 (the first trading day of the month) of 1997, Enron stock sold for $0.75 less than it had on January 2 (the first trading day of that month). Table 4.1 gives the monthly changes in stock price (in dollars) for the 5 years leading up to the company's failure.

	Jan.	Feb.	Mar.	Apr.	May	June	July	Aug.	Sept.	Oct.	Nov.	Dec.
1997	−$1.44	−0.75	−0.69	−0.88	0.12	0.75	0.81	−1.75	0.69	−0.22	−0.16	0.34
1998	0.78	0.62	2.44	−0.28	2.22	−0.50	2.06	−0.88	−4.50	4.12	1.16	−0.50
1999	3.28	3.34	−1.22	0.47	5.62	−1.59	4.31	1.47	−0.72	−0.38	−3.25	0.03
2000	5.72	21.06	4.50	4.56	−1.25	−1.19	−3.12	8.00	9.31	1.12	−3.19	−17.75
2001	14.38	−1.08	−10.11	−12.11	5.84	−9.37	−4.74	−2.69	−10.61	−5.85	−17.16	−11.59

Monthly stock price change in dollars of Enron stock for the period January 1997 to December 2001. Negative amounts indicate that the stock *lost* value.   **Table 4.1**

It's hard to tell very much from the data. Don't try too hard. Tables with lots of numbers are hard to understand by just reading them. You might get a rough idea of how much the stock changed from month to month, but even that would be approximate. "Looks like it's usually less than 10 dollars in either direction," you might say.

A S **Histograms.** Watch a histogram grow from the data seen in the previous activity's video. Play with the bin width to see how it changes the look of the histogram—can't do that on paper!

Instead, let's follow the first rule of data analysis and make a picture. What kind of picture should we make? It can't be a bar chart or a pie chart. Those are for categorical variables and the values here are in dollars. We'll want to treat price change as a *quantitative* variable, not a categorical one.

Histograms: Displaying the Distribution of Price Changes

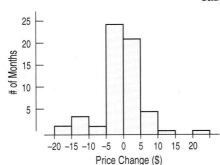

Monthly price changes of Enron stock
This histogram displays the distribution of price changes by showing how many months had price changes in each of the intervals (bins). **Figure 4.1**

How can we display a quantitative variable? With a categorical variable, life was easy. When there were only a few categories, we could make one pile for each. With quantitative variables, life is different. Usually we can't list all the individual values; there are just too many of them. So instead, we slice up the entire span of values covered by the quantitative variable into equal-width piles called **bins.** Then we count the number of values that fall into each bin. The bins and the counts in each bin give the **distribution** of the quantitative variable.

We can display the bin counts in a display called a histogram. Like a bar chart, a **histogram** plots the bin counts as the heights of bars. For the Enron data, the cases are months, so the height of each bar shows the number of months that have price changes falling into that bin. Here, the *Who* are months and the *What* are price changes. The histogram displays the distribution of price changes by showing the number of *months* that have price changes in each of the bins.

Does the distribution of *Price Change* look as you expected? It's often a good idea to *imagine* what the distribution might look like before you ask a computer or calculator to make a picture for you. That way you'll be ready if the picture shows something unexpected, and less likely to let bad data or wrong data fool you.

The first bar, which counts losses between $15 and $20, has only two months in it. We can see that although they vary, most of the monthly price changes are smaller than $5 in either direction. Only in a very few months were the changes larger than $10 in either direction. There appear to be about as many positive as negative price changes.

Determining bins by hand

The first column of the Enron data shows monthly price changes of −$1.44, $0.78, $3.28, $5.72, and $14.38. Each observation becomes a count in one of the bins of the histogram. In Figure 4.1 we've made each bin $5 wide. So, the first data point (−$1.44) goes into the bin from −$5 to $0. The next one, $0.78, goes into the bin from $0 to $5 as does the third value. $5.72 is bigger than 5 and less than 10, so it goes into the $5 to $10 bin, and $14.38 goes into the $10 to $15 bin. Continuing this for the entire data set, we find four data values between −$15 and −$10. The histogram shows this as a bar of height 4 for that bin. We also see that there are 5 times as many prices between $0 and $5, something that we couldn't easily see from the table.

Where do values on the borders of the bins go? The choice is up to you, but you have to be consistent. The standard rule is to put values at the edge of the bins into the next higher bin, placing $10.00 in the $10 to $15 bin rather than in the $5 to $10 bin.

Even if you use technology to make the histogram, you may have a choice of how many bins to use. Unless the data set is very large, you'll probably want to have between 5 and 20 bins, so you'll have to choose the width of the bin with that in mind. You might also want to choose the bins so that the border numbers come out "nicely." Always be sure that you use the *same* bin width for all the bins so that the display preserves the area principle.

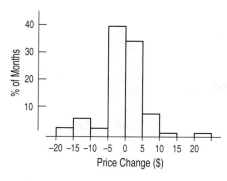

A relative frequency histogram is faithful to the area principle by displaying the percentage of cases in each bin instead of the counts. **Figure 4.2**

A bar chart has spaces between the bars because the categories could appear in any order. But in a histogram, there are no gaps because the bins slice up *all the values* of a quantitative variable. Both kinds of display satisfy the area principle because the area covered by each bar corresponds to the count of the cases falling in the range covered by that bar. Sometimes it is useful to make a **relative frequency histogram,** replacing the counts on the vertical axis with the *percentage* of the total number of cases falling in each bin. Of course, the shape of the histogram is exactly the same; only the labels are different.

Stem-and-Leaf Displays

A S **Stem-and-Leaf.** As you might expect of something called "stem-and-leaf," these displays grow as you consider each data value. The best way to see how to make them and how to understand them is to watch the growth in action in this animation.

Histograms provide an easy-to-understand summary of the distribution of a quantitative variable, but they don't show the data values themselves. Here's a histogram of the pulse rates of 24 women, taken by a researcher at a health clinic:

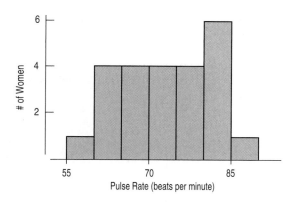

The pulse rates of 24 women at a health clinic.

The Stem-and-Leaf display was devised by John W. Tukey, one of the greatest statisticians of the 20th century. It is called a "Stemplot" in some texts and computer programs, but we prefer Tukey's original name for it. After all, he invented it, and an inventor should have the right to name his own invention.

The story seems pretty clear. We can see the entire span of the data and can easily see what a typical pulse rate might be. But is that all there is to these data?

A stem-and-leaf display is like a histogram, but it shows the individual values. It's also easier to make by hand. Here's a stem-and-leaf display of the same data:

```
8 | 8
8 | 000044
7 | 6666
7 | 2222
6 | 8888
6 | 0444
5 | 6
```

Turn the stem-and-leaf on its side (or turn your head to the right) and squint at it. It should look roughly like the histogram of the same data. Does it? Well, it's backwards because now the higher values are on the left, but other than that it has the same shape.[1]

What does the line that says **8 | 8** at the top of the display mean? It stands for a pulse of 88 beats per minute. We've taken part of the number (the "tens" place) and made that the "stem." Then we've sliced off the trailing digit (the "ones" place) and turned it into a "leaf." The next line down is **8 | 000044.** That shows that there were four pulse rates of 80 and two pulse rates of 84 bpm.

To make a stem-and-leaf display, we cut each data value into leading digits (which become the "stem") and trailing digits (the "leaves"). Then we use the stems to label the *bins*. For the pulse rate data, we chose the first digit as the stem and displayed each stem on two lines (one with digits 0–4 and the other with 5–9). We could have used only one line for each stem. It would have used only 4 bins, but it looks a bit crowded:

```
8 | 0000448
7 | 22226666
6 | 04448888
5 | 6
```

By splitting the stems in our original plot, we got a better look at the distribution. Sometimes, we may want to go further. If we decide that splitting the stem in two lines is not enough, we could try splitting it into 5, putting the digits 8 and 9 on the top line, 6 and 7 on the next, etc. This is too many bins for these data, but we'd have something like this (for the pulses from 60 to 69):

```
6* | 8888
6S |
6F | 444
6T |
60 | 0
```

Here we've used letters to show which digits will appear on each line as leaves: O for 0 and One, T for Two and Three, F for Four and Five, S for Six and Seven and * (oops, we were doing so well…) for 8 and 9.

If we had had more digits reported (if the pulses had been reported to the nearest tenth bpm, like 56.2, 57.8, etc.) we would still use only *one* digit for the leaf, rounding the data values to one decimal place after the stem before splitting them into stem and leaves.[2] Or, we might have decided to use each value 56, 57, 58 as the stem and use the number after the decimal place as the leaf, but we would have needed a lot of values for that to make sense.

---

[1] You could make the stem-and-leaf with the higher values on the bottom, if that's useful. Usually, though, higher on top makes sense.

[2] It is equally good, and perhaps a bit easier when we work by hand, to truncate, just using the next digit regardless of the third digit. The shape of the stem-and-leaf display will not change, and you will still be able to identify any individual data value easily.

Does a stem-and-leaf display satisfy the area principle? It does as long as each digit takes up the same amount of space. When doing this by hand, be careful to write thin numerals like "1" and fat ones like "3" to take up the *same amount* of horizontal space. That way, each bar's length will be proportional to the number of observations that fall into its bin.

Stem-and-leaf displays contain all the information found in a histogram and, when carefully drawn, satisfy the area principle and show the distribution. In addition, stem-and-leaf displays preserve the individual data values. Few data displays can do this as effectively. Unlike a histogram, stem-and-leaf displays also show the digits in the bins, so they can reveal unexpected patterns in the data. Did you notice anything about the pulse rates? How many seconds do you think the nurse waited while counting beats? It is clear at a glance that all the values are even. But look a bit closer. Every pulse rate is divisible by 4, something we couldn't possibly tell from the histogram. The nurse probably waited only 15 seconds and multiplied by 4 rather than counting for a whole minute.

Unlike most other displays discussed in this book, stem-and-leaf displays are great pencil-and-paper constructions. Consider using them whenever you have a quantitative variable. They are well suited to moderate amounts of data—say, between ten and a few hundred values. For larger data sets, histograms do a better job.

A S **Dotplots.** In *ActivStats* dotplots you can click on points to see their values, and even drag them around. We'll be seeing this again, so now's a great time to get acquainted.

Dotplots

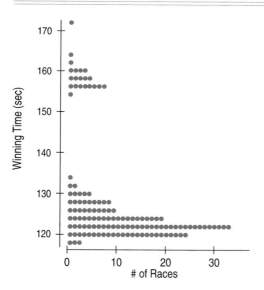

A dotplot of Kentucky Derby winning times plots each race as its own dot, showing the bimodal distribution. **Figure 4.3**

A **dotplot** is a simple display. It just places a dot along an axis for each case in the data. It's like a stem-and-leaf display, but with dots instead of digits for all the leaves. Dotplots are a great way to display a small data set (especially if you forget how to write the digits from 0 to 9). Here's a dotplot of the time (in seconds) that the winning horse took to win the Kentucky Derby in each race between the first Derby in 1875 and the 2004 Derby.

Dotplots show basic facts about the distribution. We can find the slowest and quickest races by finding times for the topmost and bottommost dots. It's also clear that there are two clusters of points, one just below 160 seconds and the other at about 122 seconds. Something strange happened to the Derby times. Once we know to look for it, we can find out that in 1896 the distance of the Derby race was changed from 1.5 miles to the current 1.25 miles. That explains the two clusters of winning times.

Some dotplots stretch out horizontally, with the counts on the vertical axis, like a histogram. Others, such as the one shown here, run vertically, like a stem-and-leaf display. Some dotplots place points next to each other when they would otherwise overlap. Others just place them on top of one another. Newspapers sometimes offer dotplots with the dots made up of little pictures.

Think Before You Draw, Again

Suddenly, we face a lot more options when it's time to invoke our first rule of data analysis and Make a picture. You'll need to *Think* carefully to decide which type of graph to make. In the previous chapter you learned to check the Categorical Data Condition before making a pie chart or a bar chart. Now, before making a stem-and-leaf display, a histogram, or a dotplot, you need to check the

> **Quantitative Data Condition:** The data are values of a quantitative variable whose units are known.

Although a bar chart and a histogram may look somewhat similar, they're not the same display. You can't display categorical data in a histogram or quantitative data in a bar chart. Always check the condition that confirms what type of data you have before proceeding with your display.

Shape, Center, and Spread

> The **mode** is sometimes defined as the single value that appears most often. That definition is fine for categorical variables because all we need to do is count the number of cases for each category. For quantitative variables, the mode is more ambiguous. What's the mode of the Enron data? No price change occurred more than twice, but two months had drops of $0.50. Should that be the mode? Probably not. It makes more sense to use the word "mode" in the more general sense of peak in a histogram, rather than as a single summary value.

Step back from a histogram or stem-and-leaf display. What can you say about the distribution? When you describe a distribution, you should always tell about three things: its **shape, center,** and **spread.**

What Is the Shape of the Distribution?

1. *Does the histogram have a single, central hump or several separated bumps?* These humps are called **modes.**[3] The Enron stock price changes have a single mode at just about $0. A histogram with one main peak, such as the price changes, is dubbed **unimodal;** histograms with two peaks are **bimodal,** and those with three or more are called **multimodal.**[4] For example, here's a bimodal histogram.

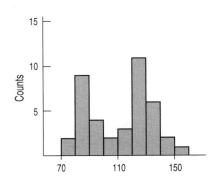

A bimodal histogram has two apparent peaks.   **Figure 4.4**

> You've heard of pie à la mode. Is there a connection between pie and the mode of a distribution? Actually, there is! The mode of a distribution is a *popular* value near which a lot of the data values gather. And "à la mode" means "in style"— *not* with ice cream. That just happened to be a *popular* way to have pie in Paris around 1900.

A histogram that doesn't appear to have any mode and in which all the bars are approximately the same height is called **uniform.**

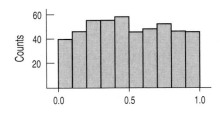

In a uniform histogram, the bars are all about the same height. The histogram doesn't appear to have a mode.
Figure 4.5

---

[3] Well, technically, it's the value on the horizontal axis of the histogram that is the mode, but anyone asked to point to the mode would point to the hump.

[4] Apparently, statisticians don't like to count past two.

2. *Is the histogram **symmetric**?* Can you fold it along a vertical line through the middle and have the edges match pretty closely, or are more of the values on one side?

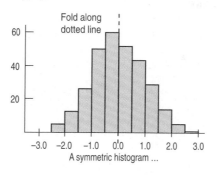
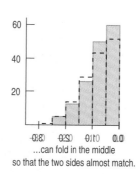

A symmetric histogram ...

...can fold in the middle so that the two sides almost match.

Figure 4.6

The (usually) thinner ends of a distribution are called the **tails.** If one tail stretches out farther than the other, the histogram is said to be **skewed** to the side of the longer tail.

A S **Attributes of Distribution Shape.** This awkwardly named activity and the others on this *ActivStats* page are easy to work with and quite informative. They show off aspects of distribution shape through animation and examples, then let you make and interpret some histograms with your statistics package.

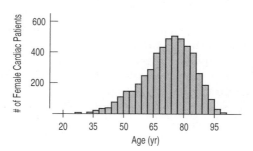
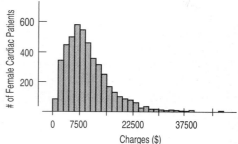

Two skewed histograms showing data on two variables for all female heart attack patients in New York state in one year. The blue one (age in years) is skewed to the left. The purple one (charges in $) is skewed to the right.   **Figure 4.7**

3. *Do any unusual features stick out?* Often such features tell us something interesting or exciting about the data. You should always mention any stragglers, or **outliers,** that stand off away from the body of the distribution. If you're collecting data on nose lengths and Pinocchio is in the group, you'd probably notice him, and you'd certainly want to mention it.

Outliers can affect almost every method we discuss in this course. So we'll always be on the lookout for them. An outlier can be the most informative part of your data. Or it might just be an error. But don't throw it away without comment. Treat it specially and discuss it when you tell about your data. Or find the error and fix it, if you can. Be sure to look for outliers. Always.

(In the next chapter you'll learn a handy rule of thumb for deciding when a point might be considered an outlier.)

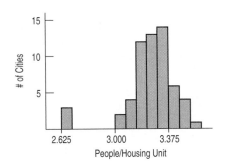

A histogram with outliers. There are three cities in the leftmost bar.
Figure 4.8

Are there any **gaps** in the distribution? The Kentucky Derby data that we saw in the dotplot on page 49 has a large gap between two groups of times, one near 120 seconds and one near 160. Gaps help us see multiple modes, and encourage us to notice when the data may come from different sources or contain more than one group.

● **Does some of this seem vague? It is!** How you characterize a distribution is often a judgment call, and it's always relative. Is that point way out on the right really an outlier, or is it just an indication of a long tail to the distribution? Generally, we start by looking at the main body of the data. If a point is separated from it by only a small gap, then it's not that unusual and probably not an outlier. If the main body seems roughly symmetric, then more distant stragglers are best regarded as outliers. If the main body of the data is skewed, then the long tail that continues that skewness is part of the overall pattern, so points would need to be farther away to be called outliers.

It may surprise you (especially if you thought this was going to be a math class) that many of the most important concepts in Statistics aren't as precise as in mathematics. When we summarize data, we want to emphasize overall features. We don't want to focus on all the *details* of the data set we're looking at, so some of the statistical concepts are deliberately left vague.

Whether a histogram is symmetric or skewed, whether it has one or more modes, whether a point is far enough from the body of the plot to be an outlier—these are all somewhat vague concepts. You may be uncomfortable with this at first, because you're used to finding a correct, precise answer. In Statistics there may be more than one right answer. That means you're entitled to your own (informed) opinion, provided you can justify it.

Keep an eye out for vague concepts in Statistics. We'll point out when concepts are vague and help you to navigate through them so you can understand the story told by the data. In fact, here come two more. . . . ●

Where Is the Center of the Distribution?

If you had to pick a single number to describe all the data, what would you pick? The center is an easy description of a typical value and a concise summary of the whole batch of numbers. When a histogram is unimodal and symmetric, it's easy to find its center. It's right in the middle. (Where else would you look?) The center of the Enron price changes is about $0. That tells us that over the period we've examined, the stock went down about as often as it went up.

For distributions with other shapes, the situation isn't as clear. If the histogram is skewed, defining the center is more of a challenge. And if the histogram has more than one mode, the center might not even be a useful concept. You might be looking at two different groups thrown together, so it's probably a good idea to find out why you don't have a single mode.

The next chapter discusses some ways to locate centers numerically. For now we'll just eyeball a picture of the distribution and give a rough idea of where the center seems to be.

How Spread Out Is the Distribution?

The center gives a typical value, but not everyone is typical. Variation matters. Statistics is about variation, but how can we describe it? We can look to see whether all the values of the distribution are tightly clustered around the center or spread out. Because distributions that vary a lot around the center are harder to predict or model, we often prefer distributions with less variability. Would you rather invest in a stock whose price gyrates wildly or one that grows steadily?[5]

You're not finished describing a distribution until you discuss its spread. Don't worry. We'll have a lot more to say about spread in the next two chapters.

> Why do banks favor a single line that feeds several teller windows rather than separate lines for each teller? The average waiting time is the same. But the time you can expect to wait is less variable when there is a single line, and people prefer consistency.

It's often a good idea to think about what the distribution of a data set might look like before we collect the data. What do you *think* the distribution of the following data sets will look like? Be sure to think in terms of shape, center, and spread.

1. Number of miles run by Saturday morning joggers at a park.

2. Hours spent by U.S. adults watching football on Thanksgiving Day.

3. Amount of winnings of all people playing a particular state's lottery last week.

4. Ages of the faculty at your school.

5. Last digit of phone numbers on your campus.

Displaying Quantitative Data Step-By-Step

WHO	CEOs
WHAT	Annual base salary in dollars
HOW	Survey from *Forbes* magazine
WHEN	1994
WHERE	United States
WHY	Inspiration?

With the current state of the economy,[6] more attention has been paid to the compensation of Chief Executive Officers (CEOs) of major companies. Let's look at the CEO salaries of the 800 largest corporations in 1994.

---

[5] Actually, that's not an easy question to answer. It depends on how much of a risk taker you happen to be.

[6] In fact, we don't know what state the economy is in as you read this, but it really doesn't matter. People always want to know how much CEOs make and will use the economy as an excuse.

Think

Plan   State what you want to find out.

Variables   Identify the variables and report the W's.

Be sure to check the appropriate condition.

I want to learn about CEO pay. I have the CEO's base salary (in $) for the 800 largest corporations reported by *Forbes* magazine in 1994.

✔ **Quantitative Data Condition:** The salaries are quantitative, with units in dollars. A histogram is an appropriate display for examining the distribution.

Show

Mechanics   We almost always make histograms with a computer or graphing calculator.

REALITY CHECK   It is always a good idea to think about what we expected to see and to check whether the histogram is close to what we expected. Could CEOs really earn this much? Yes, six- to seven-figure salaries are about what we might expect.

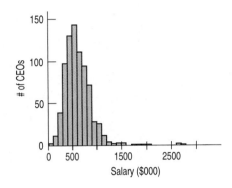

The main body of the distribution is unimodal and nearly symmetric,[7] centered around $500,000, with slightly more than half of CEOs earning salaries higher than that. But there are some high outliers. Some CEOs' salaries are higher than what is typical for most CEOs of large corporations. Even though the vast majority of CEOs have salaries below $1,000,000 a year, there are a few with salaries between $2,500,000 and $3,000,000 a year.

Tell

Conclusion   Describe the shape, center, and spread of the distribution. Be sure to report on the symmetry, number of modes, and any gaps or outliers. Remember to interpret all results in context.

Comparing Distributions

Up to now, we've looked at one distribution at a time. But this route can take us only so far. While it may be interesting to know the distribution of CEO salaries, it might be more interesting to know how salaries in high-tech industries compare with those in manufacturing. The fact is that most interesting results about data involve making comparisons or modeling relationships.

For example, many common diseases show different patterns in women and men. In the past couple of decades, researchers have been more careful to collect

---

[7] Do you think it's skewed? Well, maybe. Here, if we cut off the tail, what's left behind is pretty symmetric, so we saw this as a symmetric distribution with outliers. But you're entitled to your own opinion on such vague concepts—just be prepared to defend your choice.

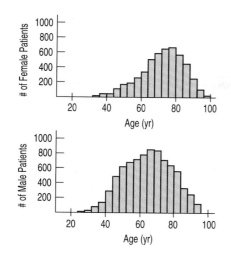

The distributions of ages for female and male heart attack patients differ in interesting ways. **Figure 4.9**

WHO	Heart attack patients
WHAT	Age (years)
WHEN	1993
WHERE	NY state

data on women. What kinds of questions might we ask about these data? Here's one: Do men and women tend to get heart attacks at different ages? That's the kind of question that's well suited to investigating with a graph. Here are two histograms of the ages of every heart attack patient hospitalized during 1993 in New York State, one for women and one for men.

What can we tell from these histograms? We'll start, as usual, by looking at shape, center, and spread.

Notice first that the *shapes* are different. The men's distribution is nearly symmetric, while the women's is clearly skewed to the left. Young women are much less likely to have heart attacks than young men. The *center* of the men's distribution appears to be in the low 60s, but for women, it's closer to 70. So, men tend to have heart attacks at an earlier age. The women's ages are highly clustered between 60 and 85 years, unlike for the men.

In general, the age at which a man might have a heart attack is less predictable than for a woman because the men's ages are more *spread* out than the women's.

Timeplots: Order, Please!

The Enron price changes were collected over time; the changes are for *consecutive* months from January 1997 to December 2001. When data are collected in a specific order like that, you should check to see if they have a pattern when plotted in that order. We can array the dots of a dotplot across the page in the data's order. Often, connecting successive points with a line helps to show the pattern.

The Enron price changes show relatively stable stock prices until the middle of 1999. Then the stock price starts to oscillate wildly. (Remember that these are the *changes* in price, not the prices themselves.) Curiously, few stock analysts seem to have been concerned about this behavior at the time. Perhaps they weren't looking at a **timeplot** such as this one.

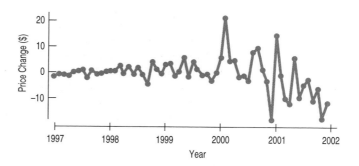

Monthly Enron stock price changes stretched out over time in a timeplot. Now it's easy to see that after being relatively stable, the stock price became somewhat volatile in 1998 and then even more so starting in 2000. **Figure 4.10**

*Re-expressing Skewed Data to Improve Symmetry[8]

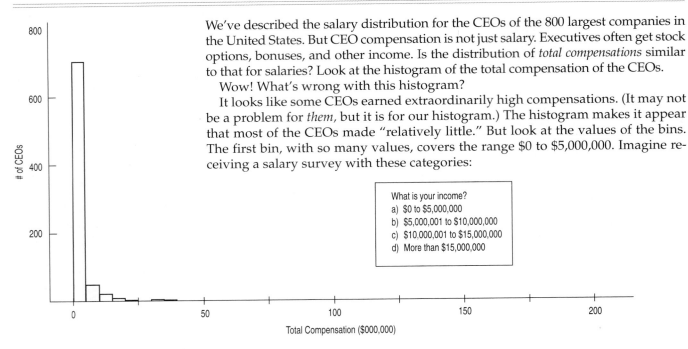

We've described the salary distribution for the CEOs of the 800 largest companies in the United States. But CEO compensation is not just salary. Executives often get stock options, bonuses, and other income. Is the distribution of *total compensations* similar to that for salaries? Look at the histogram of the total compensation of the CEOs.

Wow! What's wrong with this histogram?

It looks like some CEOs earned extraordinarily high compensations. (It may not be a problem for *them*, but it is for our histogram.) The histogram makes it appear that most of the CEOs made "relatively little." But look at the values of the bins. The first bin, with so many values, covers the range $0 to $5,000,000. Imagine receiving a salary survey with these categories:

> What is your income?
> a) $0 to $5,000,000
> b) $5,000,001 to $10,000,000
> c) $10,000,001 to $15,000,000
> d) More than $15,000,000

The total compensation for CEOs of the 800 largest companies is extremely skewed and includes some extraordinarily large values. A few earned nearly $200 million (though the bars are too short to show up clearly). **Figure 4.11**

As ridiculous as that first category appears, more than 10% of the CEOs had compensation too large to fit in that bin. In fact, 16 of them received more than $15,000,000. The reason that the histogram seems to leave so much of the area blank is that these observations are spread all along the range from about $15,000,000 to $200,000,000. After $20,000,000 there are so few for each bin that it's very hard to see the tiny bars. What we *can* see from this histogram is that this distribution is *very* skewed to the right.

Total compensation for CEOs consists of their base salaries, bonuses, and extra compensation, usually in the form of stock or stock options. Data that add together several variables, such as the compensation data, can easily have skewed distributions. It's often a good idea to separate the component variables and examine them individually to understand the data.

Skewed distributions are hard to summarize. It's hard to know what we mean by the "center" of a skewed distribution, so it's hard to pick a typical value to summarize the distribution. What would you say was a typical CEO total compensation?

One way to make a skewed distribution more symmetric is to **re-express,** or **transform,** the data by applying a simple function. For example, we could take the square root or logarithm of each data value. Variables that have a distribution that is skewed to the right often benefit from a re-expression by logarithms or square roots. Those skewed to the left may benefit from squaring the data values.

---

[8] This section is optional. It's not required for other parts of the book. But we recommend that you at least look at these ideas.

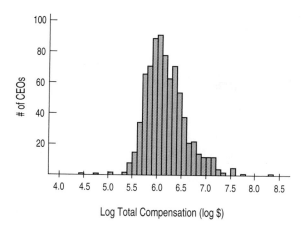

Log Total Compensation (log $)

Taking logs makes the histogram of CEO total compensation nearly symmetric. The center of the distribution is around 6. How much money is that? It's 6 + 1 = "seven figures"—or $1,000,000. **Figure 4.12**

The histogram of the logs of the total CEO compensations is much more nearly symmetric, so we can see that a typical *log compensation* is about 6.0, which corresponds to $1,000,000. Notice that nearly all the values are between 5.0 and 7.0—in other words, between six and eight figures. That's $100,000 to $10,000,000 a year, but who's counting? Because computers and calculators are available to do the calculating, you should consider re-expression as a helpful tool whenever you have skewed data.

● **Dealing with logarithms** You have probably learned about logs in math courses and seen them in psychology or science classes. In this book, we use them only for making data behave better. Base 10 logs are the easiest to understand. You can think of them as roughly one less than the number of digits you need to write the number. So 100, which is the smallest number to require 3 digits, has a $\log_{10}$ of 2. And 1000 has a $\log_{10}$ of 3. The $\log_{10}$ of 500 has to be between 2 and 3, but you'd need a calculator to find that it's about 2.7. A salary of "six figures" has a $\log_{10}$ between 5 and 6. Logs are incredibly useful for making skewed data more symmetric. But don't worry. Nobody does logs without technology and neither should you. Often, remaking a histogram or other display using the logs of the data is as easy as pushing another button. ●

What Can Go Wrong?

A data display should tell a story about the data. To do that it must speak in a clear language, making plain what variable is displayed, what any axis shows, and what the values of the data are. And it must be consistent in those decisions.

A display of quantitative data can go wrong in many ways. The most common failures arise from only a few basic errors:

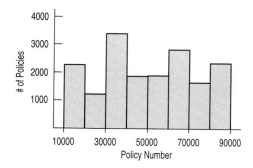

Policy Number

It's not appropriate to display these data with a histogram. **Figure 4.13**

- ● **Don't make a histogram of a categorical variable.** Just because the variable contains numbers doesn't mean it's quantitative. Here's a histogram of the insurance policy numbers of some workers. It's not very informative because the policy numbers are categorical. A histogram or stem-and-leaf display of a categorical variable makes no sense. A bar chart or pie chart may do better.
- ● **Don't look for shape, center, and spread of a bar chart.** A bar chart showing the sizes of the piles displays the distribution of a categorical variable, but the bars could be arranged in any order left to right. Concepts like symmetry, center, and spread make sense only for quantitative variables.
- ● **Don't use bars in every display—save them for histograms and bar charts.** In a bar chart, the bars indicate how many cases of a categorical variable are piled in each category. Bars in a histogram indicate the number of cases piled in each interval of a quantitative variable. In both bar charts and histograms the bars represent many data values. Some people create other displays that use bars to represent individual data values. Beware: Such graphs are neither bar charts nor histograms. For example, a student was asked to make a histogram from data showing the number of juvenile

bald eagles seen each week during the winter of 2003–2004 at a site in Rock Island, IL. Instead, he made this plot:

This isn't a histogram or a bar chart. It's an ill-conceived graph that uses bars to represent individual data values (number of eagles sighted) week by week.
Figure 4.14

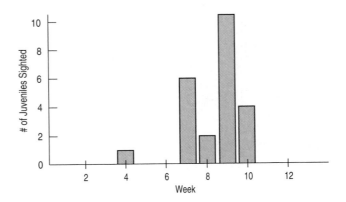

Look carefully. The student used a bar for each data value. His graph shows number of eagles week by week. That's a badly drawn timeplot, not a histogram. The result is confusing, and it wastes ink.[9] A histogram would have a tall bar above "0," to show us that there were many weeks when no eagles were sighted. Use bars wisely, to represent piles of data. Here's the histogram:

A histogram of the eagle sighting data shows the number of weeks in which different counts of eagles occurred. This display shows the distribution of juvenile eagle sightings. **Figure 4.15**

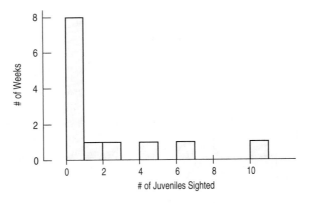

- **Choose a bin width appropriate to the data.** Computer programs usually do a pretty good job of choosing histogram bin widths. Often, there's an easy way to adjust the width, sometimes interactively. Here is the Enron price change histogram with two other choices for the bin size.

---

[9] Edward Tufte, in his book *The Visual Display of Quantitative Information*, proposes that graphs should have a high data-to-ink ratio. That is, we shouldn't waste a lot of ink to display a single number when a dot would do the job.

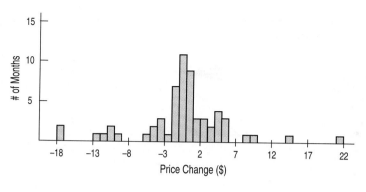

Changing the bin width changes how the histogram looks. The Enron stock price changes look different with these two choices. Neither shows the distribution well. One has too many bins, the other too few.   **Figure 4.16**

- *Avoid inconsistent scales.* Parts of displays should be mutually consistent—no fair changing scales in the middle or plotting two variables on different scales but on the same display. When comparing two groups, be sure to compare them on the same scale.
- *Label clearly.* Variables should be identified clearly and axes labeled so a reader knows what the plot displays.

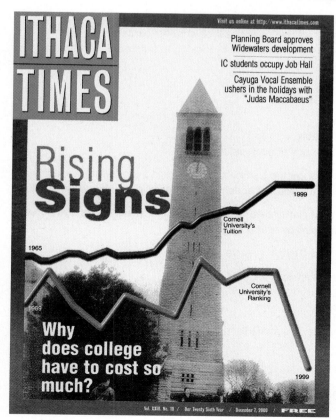

Here's a remarkable example of a plot gone wrong. It illustrated a news story about rising college costs. It uses timeplots, but it gives a misleading impression. First think about the story you're being told by this display. Then try to figure out what has gone wrong.

What's wrong? Just about everything.

- The horizontal scales are inconsistent. Both lines show trends over time, but exactly for what years? The tuition sequence starts in 1965, but rankings are graphed from 1989. Plotting them on the same (invisible) scale makes it seem that they're for the same years.
- The vertical axis isn't labeled. That hides the fact that it's inconsistent. Does it graph dollars (of tuition) or ranking (of Cornell University)?

This display violates three of the rules. And it's even worse than that. It violates a rule that we didn't even bother to mention.

- The two inconsistent scales for the vertical axis don't point in the same direction! The line for Cornell's rank shows that it has "plummeted" from 15th place to 6th place in academic rank. Most of us think that's an *improvement*, but that's not the message of this graph.

CONNECTIONS

Distributions of quantitative variables, like those for categorical variables, show the possible values and their relative frequencies. A histogram shows the distribution of values in a quantitative variable with adjacent bars. Don't confuse them with bar charts, which display categorical variables. For categorical data, the mode is the category with the biggest count. For quantitative data, modes are peaks in the histogram.

The shape of the distribution of a quantitative variable is an important concept in most of the subsequent chapters. We will be especially interested in distributions that are unimodal and symmetric. And we will continue to check for outliers because we will need to deal with them specially.

What have we learned?

We've learned how to make a picture for quantitative data to help us see the story the data have to *Tell*.

- We can display the distribution of quantitative data with a *histogram,* a *stem-and-leaf* display, or a *dotplot*.
- We *Tell* what we see about the distribution by talking about *shape, center, spread,* and any *unusual features*.
- When we want to compare the distributions of a quantitative variable for two different groups, we can look at their histograms, dotplots, or stem-and-leaf displays. Using two displays drawn on the same scale, we compare their shapes, centers, and spreads.
- We can see the trend in a quantitative variable by looking at a *timeplot* of data that have been collected over time.

TERMS

Distribution
The distribution of a variable gives
- the possible values of the variable.
- the frequency or relative frequency of each value.

Histogram (relative frequency histogram)
A histogram uses adjacent bars to show the distribution of values in a quantitative variable. Each bar represents the frequency (or relative frequency) of values falling in an interval of values.

Stem-and-leaf display
A stem-and-leaf display shows quantitative data values in a way that sketches the distribution of the data. It's best described in detail by example.

Dotplot
A dotplot graphs a dot for each case against a single axis.

Shape
To describe the shape of a distribution, look for
- single vs. multiple modes.
- symmetry vs. skewness.

Center
A value that attempts the impossible by summarizing the entire distribution with a single number, a "typical" value.

Spread
A numerical summary of how tightly the values are clustered around the "center."

Mode
A hump or local high point in the shape of the distribution of a variable is called a "mode." The apparent location of modes can change as the scale of a histogram is changed.

Unimodal	Having one mode. This is a useful term for describing the shape of a histogram when it's generally mound-shaped. Distributions with two modes are called **bimodal.** Those with more than two are **multimodal.**
Uniform	A distribution that's roughly flat is said to be uniform.
Symmetric	A distribution is symmetric if the two halves on either side of the center look approximately like mirror images of each other.
Tails	The tails of a distribution are the parts that typically trail off on either side. Distributions can be characterized as having long tails (if they straggle off for some distance) or short tails (if they don't).
Skewed	A distribution is skewed if it's not symmetric and one tail stretches out farther than the other.
Outliers	Outliers are extreme values that don't appear to belong with the rest of the data. They may be unusual values that deserve further investigation, or just mistakes; there's no obvious way to tell. Don't delete outliers automatically — you have to think about them. Outliers can affect many statistical analyses, so you should always be alert for them.
Timeplot	A timeplot displays data that change over time. Often, successive values are connected with lines to show trends more clearly.

SKILLS

Think

When you complete this lesson you should:

- Be able to identify an appropriate display for any quantitative variable.
- Be able to guess the shape of the distribution of a variable by knowing something about the data.

Show

- Know how to display the distribution of a quantitative variable with a stem-and-leaf display (by hand for smaller data sets), a dotplot, or a histogram (made by computer for larger data sets).
- Know how to make a timeplot of data that may vary over time.

Tell

- Be able to describe the distribution of a quantitative variable in terms of its shape, center, and spread.
- Be able to describe any anomalies or extraordinary features revealed by the display of a variable.
- Know how to compare the distributions of two or more groups by comparing their shapes, centers, and spreads.
- Know how to describe patterns over time shown in a timeplot.
- Be able to discuss any outliers in the data, noting how they deviate from the overall pattern of the data.

Displaying Quantitative Data on the Computer

Almost any program that displays data can make a histogram, but some will do a better job of determining where the bars should start and how they should partition the span of the data.

The vertical scale may be counts or proportions. Sometimes it isn't clear which. But the shape of the histogram is the same either way.

Most packages choose the number of bars for you automatically. Often you can adjust that choice.

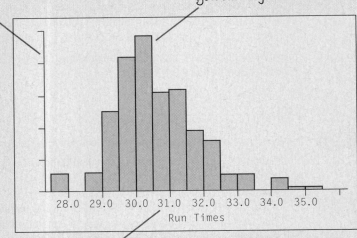

The axis should be clearly labeled so you can tell what "pile" each bar represents. You should be able to tell the lower and upper bounds of each bar.

DATA DESK

- Select the variable to display.
- In the **Plot** menu choose **Histogram.**

EXCEL

Excel cannot make histograms or dotplots without a third-party add-in.

Comments

Excel's Data Analysis add-in does offer something called a histogram, but it just makes a crude frequency table, and the Chart Wizard cannot then create a statistically appropriate histogram. The DDXL add-in provided on our CD-Rom adds these and other capabilities to Excel.

JMP

- Choose **Distribution** from the **Analyze** menu.
- In the **Distribution** dialog, drag the name of the variable that you wish to analyze into the empty window beside the label "**Y, Columns.**"
- Click **OK.**

MINITAB

- Choose **Histogram** from the **Graph** menu.
- Select "Simple" for the type of graph and click **OK.**
- Enter the name of the quantitative variable you wish to display in the box labeled "Graph variables." Click **OK.**

SPSS

- Choose **Interactive** from the **Graphs** menu.
- From the **Interactive Graphs** submenu, choose **Histogram.**
- In the **Create Histogram** dialog, drag a variable from the source list into the target. SPSS creates a variable called Counts and places it in the vertical axis target automatically to

indicate making a frequency histogram. You can specify a relative frequency histogram with the "Percentage" variable name instead.
- Click **OK.**

TI-83/84 Plus

- Turn a **Stat Plot** on.
- Choose the **histogram icon** and specify the List where the data are stored.
- **Zoom** Stat, then adjust the **Window** appropriately.

Comments

If the data are stored as a frequency table (say, with data values in L1 and frequencies in L2) set up the Plot with Xlist: L1 and Freq: L2.

TI-89

- Select F2 (**Plots**), then 1: **Plot Setup**, select a plot and press F1 to define it.
- Select plot type 4: **Histogram.** Use VAR-LINK to select the data list.
- Enter a number for the histogram bucket (bar) width.
- Press ENTER to complete the plot definition. Press F5 to display the histogram.
- Press ◆F2 to adjust the window appropriately, then press ◆F3 (**Graph**).

Comments

If the data are stored as a frequency table (say, with data values in list1 and frequencies in list2) change Use Freq and Categories to YES and use VAR-LINK to select list2 as the frequency variable on the plot definition screen.

EXERCISES

1. Statistics in print. Find a histogram that shows the distribution of a variable in a newspaper or magazine article.
   a) Does the article identify the W's?
   b) Discuss whether the display is appropriate for the data.
   c) Discuss what the display reveals about the variable and its distribution.
   d) Does the article accurately describe and interpret the data? Explain.

2. Not a histogram. Find a graph other than a histogram that shows the distribution of a quantitative variable in a newspaper or magazine article.
   a) Does the article identify the W's?
   b) Discuss whether the display is appropriate for the data.
   c) Discuss what the display reveals about the variable and its distribution.
   d) Does the article accurately describe and interpret the data? Explain.

3. Thinking about shape. Would you expect distributions of these variables to be uniform, unimodal, or bimodal? Symmetric or skewed? Explain why.
a) The number of speeding tickets each student in the senior class of a college has ever had.
b) Players' scores (number of strokes) at the U.S. Open golf tournament in a given year.
c) Weights of female babies born in a particular hospital over the course of a year.
d) The length of the average hair on the heads of students in a large class.

4. More shapes. Would you expect distributions of these variables to be uniform, unimodal, or bimodal? Symmetric or skewed? Explain why.
a) Ages of people at a Little League game.
b) Number of siblings of people in your class.
c) Pulse rates of college-age males.
d) Number of times each face of a die shows in 100 tosses.

5. Heart attack stays. The histogram shows the lengths of hospital stays (in days) for all the female patients admitted to hospitals in New York in 1993 with a primary diagnosis of acute myocardial infarction (heart attack). Write a few sentences describing this distribution (shape, center, spread, unusual features).

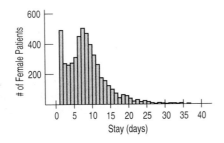

T 6. E-mails. A university teacher saved every e-mail received from students in a large Introductory Statistics class during an entire term. He then counted, for each student who had sent him at least one e-mail, how many e-mails each student had sent. Based on the histogram below, describe the distribution of e-mails.

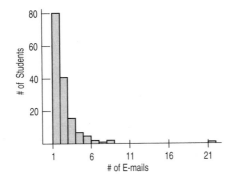

7. Sugar in cereals. The histogram displays the sugar content (as a percent of weight) of 49 brands of breakfast cereals.

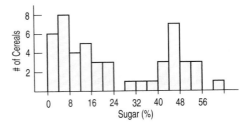

a) Describe this distribution.
b) What do you think might account for this shape?

8. Singers. The display shows the heights of some of the singers in a chorus, collected so that the singers could be positioned on stage with shorter ones in front and taller ones in back.

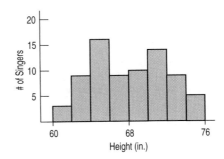

a) Describe the distribution.
b) Can you account for the features you see here?

9. Wineries. The histogram shows the sizes (in acres) of 36 wineries in the Finger Lakes region of New York.

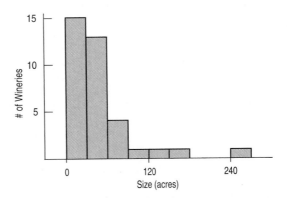

a) Approximately what percentage of these wineries are under 60 acres?
b) Write a brief description of this distribution (shape, center, spread, unusual features).

10. Run times. One of the authors collected the times (in minutes) it took him to run 4 miles on various courses during the period 1986 to 1997. Here is a histogram of the times.

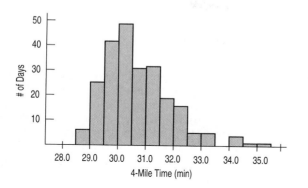

Describe the distribution and summarize the important features. What is it about running that might account for the shape you see?

11. Gasoline. In June 2004, 16 gas stations in Ithaca, NY, posted these prices for a gallon of regular gasoline.

2.029	2.119	2.259	2.049
2.079	2.089	2.079	2.039
2.069	2.269	2.099	2.129
2.169	2.189	2.039	2.079

a) Make a stem-and-leaf display of these gas prices. Use split stems; for example, use two 2.1 stems, one for prices between $2.10 and $2.149, the other for prices $2.15 to $2.199.
b) Describe the shape, center, and spread of this distribution.
c) What unusual feature do you see?

12. The Great One. During his 20 seasons in the NHL, Wayne Gretzky scored 50% more points than anyone who ever played professional hockey. He accomplished this amazing feat while playing in 280 fewer games than Gordie Howe, the previous record holder. Here are the number of games Gretzky played during each season:

79, 80, 80, 80, 74, 80, 80, 79, 64, 78, 73, 78, 74, 45, 81, 48, 80, 82, 82, 70

a) Create a stem-and-leaf display for these data using split stems.
b) Describe the shape of the distribution.
c) Describe the center and spread of this distribution.
d) What unusual feature do you see? What might explain this?

T 13. Home runs. The stem-and-leaf display shows the number of home runs hit by Mark McGwire during the 1986–2001 seasons. Describe the distribution, mentioning its shape and any unusual features.

T 14. Bird species. The Cornell Lab of Ornithology holds an annual Christmas Bird Count, in which birdwatchers at various locations around the country see how many different species of birds they can spot. Here are some of the counts reported from sites in Texas during the 1999 event.

228	178	186	162	206	166	163
183	181	206	177	175	167	162
160	160	157	156	153	153	152

a) Create a stem-and-leaf display of these data.
b) Write a brief description of the distribution. Be sure to discuss the overall shape as well as any unusual features.

T 15. Home runs, again. Students were asked to make a histogram of the number of home runs hit by Mark McGwire from 1986 to 2001 (see Exercise 13). One student submitted the following display:

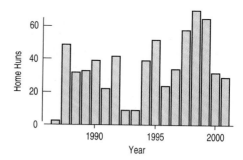

a) Comment on this graph.
b) Create your own histogram of the data.

T 16. Return of the birds. Students were given the assignment to make a histogram of the data reported in Exercise 14, on bird counts. One student submitted the following display:

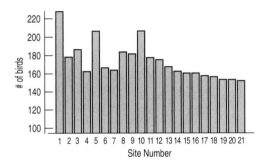

a) Comment on this graph.
b) Create your own histogram of the data.

T 17. Horsepower. Create a stem-and-leaf display for these horsepowers of autos reviewed by *Consumer Reports* one year, and describe the distribution.

155	103	130	80	65
142	125	129	71	69
125	115	138	68	78
150	133	135	90	97
68	105	88	115	110
95	85	109	115	71
97	110	65	90	
75	120	80	70	

T 18. Population growth. Here is a "back-to-back" stem-and-leaf display that shows two data sets at once—one going to the left, one to the right. It compares the percent change in population for two regions of the United States (based on census figures for 1990 and 2000). The fastest growing states were Nevada at 66% and Arizona at 40%. Write a few sentences describing the difference in growth rates for the two regions of the United States. To show the distributions better, this display breaks each stem into two lines, putting leaves 0–4 on one stem and leaves 5–9 on the other.

```
NE/MW States        S/W States
                 6 | 6
                 6 |
                 5 |
                 5 |
                 4 |
                 4 | 0
                 3 |
                 3 | 001
                 2 | 6
                 2 | 001134
                 1 | 578
          2100   1 | 001134444
99998876655      0 | 6999
         4431    0 | 1
```

T 19. Hurricanes. The data below give the number of hurricanes that happened each year from 1944 through 2000 as reported by *Science* magazine.

3, 2, 1, 2, 4, 3, 7, 2, 3, 3, 2, 5, 2, 2, 4, 2, 2, 6, 0, 2, 5, 1, 3, 1, 0, 3, 2, 1, 0, 1, 2, 3, 2, 1, 2, 2, 2, 3, 1, 1, 1, 3, 0, 1, 3, 2, 1, 2, 1, 1, 0, 5, 6, 1, 3, 5, 3

a) Create a dotplot of these data.
b) Describe the distribution.

T 20. Hurricanes, again. A bimodal distribution usually indicates that there are actually two different behaviors present in the data. Investigating those two behaviors separately can produce important insights. Here are the data again, broken into two groups showing the number of hurricanes recorded annually before and after 1970. Create an appropriate visual display and write a few sentences comparing the two distributions.

1944–1969	1970–2000
3, 2, 1, 2, 4, 3, 7, 2, 3, 3, 2,	2, 1, 0, 1, 2, 3, 2, 1, 2, 2,
5, 2, 2, 4, 2, 2, 6, 0, 2, 5, 1,	2, 3, 1, 1, 1, 3, 0, 1, 3, 2,
3, 1, 0, 3	1, 2, 1, 1, 0, 5, 6, 1, 3, 5, 3

T 21. Acid rain. Two researchers measured the pH (a scale on which a value of 7 is neutral and values below 7 are acidic) of water collected from rain and snow over a 6-month period in Allegheny County, Pennsylvania. Describe their data with a graph and a few sentences.

4.57 5.62 4.12 5.29 4.64 4.31 4.30 4.39 4.45
5.67 4.39 4.52 4.26 4.26 4.40 5.78 4.73 4.56
5.08 4.41 4.12 5.51 4.82 4.63 4.29 4.60

T 22. Marijuana. In 1995 the Council of Europe published a report entitled *The European School Survey Project on Alcohol and Other Drugs*. Among other issues, the survey investigated the percentages of 9th graders who had used marijuana. Here are the results for 20 Western European countries. Create an appropriate graph of these data, and describe the distribution.

Austria	10%	Italy	19%
Belgium	19%	Luxembourg	6%
Denmark	17%	Netherlands	31%
England	40%	No. Ireland	23%
Finland	5%	Norway	6%
France	12%	Portugal	7%
Germany	21%	Scotland	53%
Greece	2%	Spain	15%
Iceland	10%	Sweden	6%
Ireland	37%	Switzerland	27%

23. Hospital stays. The U.S. National Center for Health Statistics compiles data on the length of stay by patients in short-term hospitals, and publishes its finding in *Vital and Health Statistics*. Data from a sample of 39 male patients and 35 female patients on length of stay (in days) are displayed in these histograms.

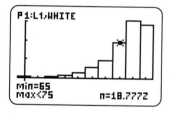

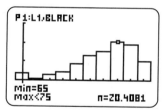

a) What would you suggest be changed about these histograms to make them easier to compare?
b) Describe these distributions by writing a few sentences comparing the duration of hospitalization for men and women.
c) Can you suggest a reason for the peak in women's length of stay?

24. Deaths. A National Vital Statistics Report indicated that nearly 300,000 black Americans died in 1999, compared with just over 2 million white Americans. Here are calculator histograms displaying the distributions of their ages at death.

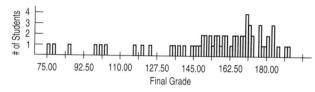

Most of the bars in these histograms display ten-year age groups. For example, the first histogram shows that for white Americans about 19% of the deaths were of people between 65 and 74 years old. The leftmost bars represent the percentage of total deaths that were children aged 0 through 4 years and the rightmost bars people over 85. Write a brief comparison of the distributions.

25. Final grades. A professor (of something other than Statistics!) distributed the following histogram to show the distribution of grades on his 200-point final exam. Comment on the display.

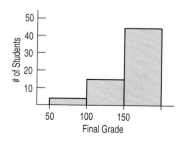

T 26. Cities. Here's a histogram of the cost of living in 25 international cities. Costs are given in U.S. dollars.

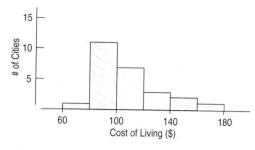

a) Write a few sentences describing this distribution.
b) Does the most expensive city included here (Tokyo) appear to be an outlier? Explain.

27. Final grades revisited. After receiving many complaints about his final grade histogram from students currently taking a Statistics course, the professor from Exercise 25 distributed the following revised histogram.

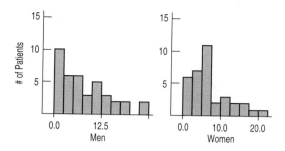

a) Comment on this display.
b) Describe the distribution of grades.

T 28. Cities revisited. Here is a picture of the costs of living in 25 international cities rescaled to bars that are $12 wide, rather than the $20 bins you saw in Exercise 26.

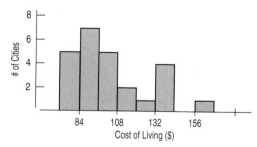

a) Describe what you see in this histogram.
b) Now would you consider Tokyo, the most expensive city, to be an outlier? Defend your opinion.

29. Zip codes. Holes-R-Us, an Internet company that sells piercing jewelry, keeps transaction records on its sales. At a recent sales meeting, one of the staff presented a histogram of the zip codes of the last 500 customers so that they might understand where sales are coming from.

Comment on the usefulness and appropriateness of the display.

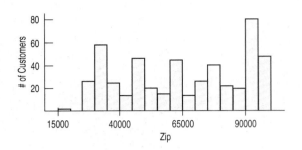

30. CEO data revisited. For each CEO, a code is listed that corresponds to the industry of the CEO's company. Here are a few of the codes and the industries to which they correspond.

Industry	Industry Code
Financial services	1
Food/drink/tobacco	2
Health	3
Insurance	4
Retailing	6
Forest products	9
Aerospace/defense	11
Energy	12
Capital goods	14
Computers/communications	16
Entertainment/information	17
Consumer nondurables	18
Electric utilities	19

A recently hired investment analyst has been assigned to examine the industries and the compensations of the CEOs. To start the analysis, he produces the following histogram of industry codes.

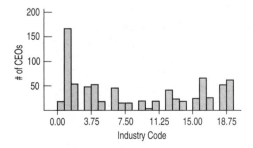

a) What might account for the gaps seen in the histogram?
b) Is the histogram unimodal?
c) What advice might you give the analyst about the appropriateness of this display?

31. Productivity study. The National Center for Productivity releases information on the efficiency of workers. In a recent report, they included the following graph show-

ing a rapid rise in productivity. What questions do you have about this display?

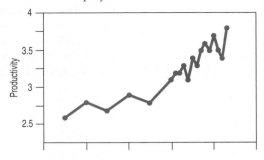

32. Productivity revisited. A second report by the National Center for Productivity analyzed the relationship between productivity and wages. Comment on the graph they used.

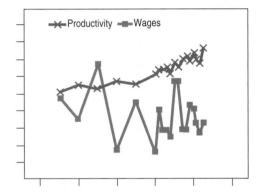

T 33. Law enforcement. Some federal employees have the authority to carry firearms and make arrests. Obviously some danger is associated with these jobs, but how much? The table below summarizes the rates of assault and injury (or death) for these employees for 5 years, 1995–1999.

Agency	Assaults (per 1000)	Killed-Injured (per 1000)
Bureau of Alcohol, Tobacco, and Firearms (BATF)	31.1	2.2
Capitol Police	5.0	3.6
Customs Service	9.7	5.1
Drug Enforcement Agency (DEA)	17.9	1.1
Federal Bureau of Investigation (FBI)	3.9	1.2
Immigration and Naturalization Services (INS)	14.1	2.5
Internal Revenue Service (IRS)	1.7	0.2
U.S. Marshal Service	9.7	3.0
National Park Service	38.7	15.0
Postal Service	5.7	2.9
Secret Service	9.7	3.0

a) Create a visual display of these data.
b) Describe these data (shape, center, spread, unusual features).
c) Which agencies are outliers?

T 34. Cholesterol. A study examining the health risks of smoking measured the cholesterol levels of people who had smoked for at least 25 years and people of similar ages who had smoked for no more than 5 years and then stopped. Create histograms for both groups and write a brief report comparing their cholesterol levels.

Smokers				Ex-smokers		
225	211	209	284	250	134	300
258	216	196	288	249	213	310
250	200	209	280	175	174	328
225	256	243	200	160	188	321
213	246	225	237	213	257	292
232	267	232	216	200	271	227
216	243	200	155	238	163	263
216	271	230	309	192	242	249
183	280	217	305	242	267	243
287	217	246	351	217	267	218
200	280	209		217	183	228

T 35. MPG. A consumer organization compared gas mileage figures for several models of cars made in the United States with autos manufactured in other countries. The data are shown in the table.

U.S. Models	Others
16.9	16.2
15.5	20.3
19.2	31.5
18.5	30.5
30.0	21.5
30.9	31.9
20.6	37.3
20.8	27.5
18.6	27.2
18.1	34.1
17.0	35.1
17.6	29.5
16.5	31.8
18.2	22.0
26.5	17.0
21.9	21.6
27.4	
28.4	
28.8	
26.8	
33.5	
34.2	

a) Create a back-to-back stem-and-leaf display for these data.
b) Write a few sentences comparing the distributions.

T 36. Baseball. American League baseball teams play their games with the designated hitter rule, meaning that pitchers do not bat. The League believes that replacing the pitcher, traditionally a weak hitter, with another player in the batting order produces more runs and generates more interest among fans. Below are the average number of runs scored in American League and National League stadiums for the first half of the 2001 season.

American				National			
11.1	10.8	10.8	10.3	14.0	11.6	10.4	10.3
10.3	10.1	10.0	9.5	10.2	9.5	9.5	9.5
9.4	9.3	9.2	9.2	9.5	9.1	8.8	8.4
	9.0	8.3		8.3	8.2	8.1	7.9

a) Create a back-to-back stem-and-leaf display of these data.
b) Write a few sentences comparing the average number of runs scored per game in the two leagues. (Remember: shape, center, spread, unusual features!)
c) Coors Field, in Denver, stands a mile above sea level, an altitude far greater than that of any other major league ball park. Some believe that the thinner air makes it harder for pitchers to throw curve balls and easier for batters to hit the ball a long way. Do you see any evidence that the 14 runs scored per game there is unusually high? Explain.

T 37. Nuclear power. For a while in the 20th century, many nuclear-powered electrical generating plants were built, but then growing environmental concerns and construction costs led to increasing reliance on other forms of energy. The table shows the dates of completion (in months after January 1967) and costs (in thousands of dollars per megawatt) of 12 nuclear generators.

Time of Completion (months after Jan 1, 1967)	Construction Cost ($1000/mW)
2	35
3	28
10	32
12	60
17	56
19	63
21	62
26	81
30	84
32	79
41	88
47	80

a) Create a stem-and-leaf display of the costs.
b) Describe the distribution.
c) Create a timeplot of the costs.
d) What information about the construction of nuclear plants can you see from the timeplot that is not obvious in the stem-and-leaf display?

T 38. Drunk driving. Accidents involving drunk drivers account for about 40% of all deaths on the nation's highways. The table tracks the number of alcohol-related fatalities for 20 years.

Year	Deaths (thousands)	Year	Deaths (thousands)
1982	25.2	1992	17.9
1983	23.6	1993	17.5
1984	23.8	1994	16.6
1985	22.7	1995	17.2
1986	24.0	1996	17.2
1987	23.6	1997	16.5
1988	23.6	1998	16.0
1989	22.4	1999	16.0
1990	22.0	2000	16.7
1991	19.9	2001	16.7

a) Create a stem-and-leaf display or a histogram of these data.
b) Create a timeplot.
c) Using features apparent in the stem-and-leaf display (or histogram) and the timeplot, write a few sentences about deaths caused by drunk driving.

T 39. Assets. Here is a histogram of the assets (in millions of dollars) of 79 companies chosen from the *Forbes* list of the nation's top corporations.

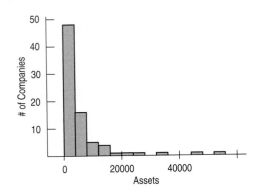

a) What aspect of this distribution makes it difficult to summarize, or to discuss, center and spread?
b) What would you suggest doing with these data if we want to understand them better?

T 40. Music library. Students were asked how many songs they had in their digital music library. Here's a display of the responses:

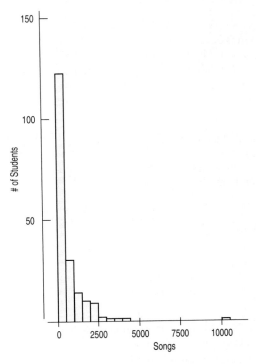

a) What aspect of this distribution makes it difficult to summarize or to discuss center and spread?
b) What would you suggest doing with these data if we want to understand them better?

T 41. Assets again. Here are the same data you saw in Exercise 39 after re-expressions as the square root of assets and the logarithm of assets.

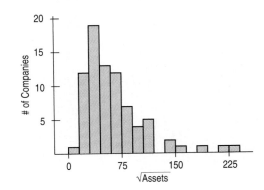

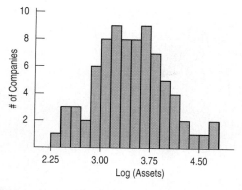

a) Which re-expression do you prefer? Why?
b) In the square root re-expression, what does the value 50 actually indicate about the company's assets?
c) In the logarithm re-expression, what does the value 3 actually indicate about the company's assets?

42. Rainmakers. The table lists the amount of rainfall (in acre-feet) from 26 clouds seeded with silver iodide.

2745	200
1697	198
1656	129
978	119
703	118
489	115
430	92
334	40
302	32
274	31
274	17
255	7
242	4

a) Why is "acre-feet" a good way to measure the amount of precipitation produced by cloud seeding?
b) Plot these data, and describe the distribution.
c) Create a re-expression of these data that produces a more advantageous distribution.
d) Explain what your re-expressed scale means.

just checking

Answers

(Thoughts will vary.)

1. Roughly symmetric, slightly skewed to the right. Center around 3 miles? Few over 10 miles.

2. Bimodal. Center between 1 and 2 hours? Many people watch no football, others watch most of one or more games. Probably only a few values over 5 hours.

3. Strongly skewed to the right, with almost everyone at $0; a few small prizes, with the winner an outlier.

4. Fairly symmetric, somewhat uniform, perhaps slightly skewed to the right. Center in the 40s? Few ages below 25 or above 70.

5. Uniform, symmetric. Center near 5. Roughly equal counts for each digit 0–9.

CHAPTER 5

Describing Distributions Numerically

WHO	191 countries (not WHO!)
WHAT	HALE (healthy life expectancy)
UNIT	Years
WHEN	Data are for babies born in 2001
WHERE	Earth
WHY	Annual report by World Health Organization

The World Health Organization (WHO) collects health data worldwide on every member country of the United Nations. One traditional measure of the overall health of a country has been the life expectancy of its citizens—the number of years that a newborn can expect to live. Starting in 1999, WHO scientists introduced a revised measure to take account of the fact that illness and injuries can affect the quality of life. The "healthy life expectancy" (HALE) adjusts for years of ill health to give a measure of years of healthy life.

Here is a histogram of the HALEs for all 191 United Nations countries:

When distributions are mixed together, the result is often a multimodal and/or skewed distribution. These data are combined from countries of different economic and social conditions, so a skewed distribution is no surprise.

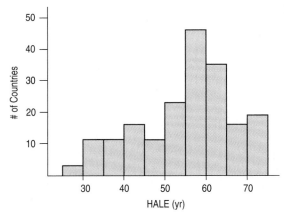

HALEs for the 191 United Nations countries. In 2001, Japan had the top HALE of 73.6 years, while Sierra Leone had the lowest at 26.5 years. **Figure 5.1**

The distribution is possibly multimodal and is clearly skewed to the left.

Finding the Center: The Median

We always use n to indicate the number of values. Some people even say, "How big is the n?" when they mean the number of data values.

Using the Midrange

There actually is *one* everyday use of the midrange. Whenever you make a plot of data, you're likely to center the plot at the midrange, exactly halfway between the two extremes of the data.

What is a typical HALE? Try to put your finger on the histogram at the value you think is typical. (Read the value from the horizontal axis and remember it.) When we think of a typical value, we usually look for the **center** of the distribution. Where do you think the center of this distribution is? For a unimodal, symmetric distribution, it's easy. We'd all agree on the center of symmetry, where we would fold the histogram to match the two sides. But when the distribution is skewed and possibly multimodal, as this one is, it's not immediately clear what we even mean by the center.

You might think of taking the average of the maximum and minimum values as a way of finding the center. This is called the **midrange,** but it's too sensitive to the outlying values to be safe for summarizing the whole distribution. A more reasonable choice of typical value is the value that is literally in the middle, with half the values below it and half above it.

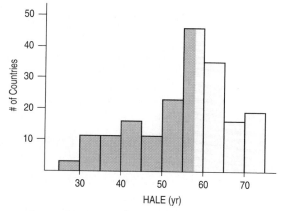

The median splits the histogram into two halves of equal area. Notice that the halves of this histogram have very different shapes.   **Figure 5.2**

Finding the median by hand

Finding the median of a batch of n numbers is easy as long as you remember to order the values first. If n is odd, the median is the middle value. Counting in from the ends, we find this value in the $\frac{n+1}{2}$ position.

When n is even, there are two middle values. So, in this case, the median is the average of the two values in positions $\frac{n}{2}$ and $\frac{n}{2} + 1$.

Here are two examples:

Suppose the batch has the values 14.1, 3.2, 25.3, 2.8, −17.5, 13.9, and 45.8. First we order the values: −17.5, 2.8, 3.2, 13.9, 14.1, 25.3, and 45.8. Since there are 7 values, the median is the $(7 + 1)/2 = 4$th value counting from the top or bottom: 13.9.

Suppose we had the same batch with another value at 35.7. Then the ordered values are −17.5, 2.8, 3.2, 13.9, 14.1, 25.3, 35.7, and 45.8. The median is the average of the 8/2, or 4th, and the $(8/2) + 1$, or 5th, values. So the median is $(13.9 + 14.1)/2 = 14.0$.

Histograms follow the area principle, and each half of the data has about 95 countries, so each colored region has the same area in the display. The middle value that divides the histogram into two equal areas is called the **median.**

The median has the same units as the data. Be sure to include the units whenever you discuss the median.

For the HALEs, there are 191 countries, so the median is found at the $(191 + 1)/2 = 96$th place in the sorted data. This median HALE is 57.7 years.

The median is one way to find the center of the data. But there are many others. We'll look at an even more important measure later in this chapter.

Knowing the median, we could say that a typical healthy life expectancy, worldwide, was about 57.7 years. How much does that really say? How well does the median describe the data? After all, not all countries have HALEs near 57.7 years. Whenever we find the center of data, the next step is always to ask how well it actually summarizes the data.

Spread: Home on the Range

Statistics pays close attention to what we *don't* know as well as what we do know. Understanding how spread out the data are is a first step in understanding what a summary *cannot* tell us about the data. It's the beginning of telling us what we don't know.

If every country had a HALE of 57.7, knowing the median would tell us everything about the distribution of life expectancy worldwide. The more the data vary, however, the less the median alone can tell us. So we need to measure how much the data values vary around the center. In other words, how spread out are they? When we describe a distribution numerically, we always report a measure of its **spread** along with its center.

How should we measure the spread? We could simply look at the extent of the data. How far apart are the two extremes? The **range** of the data is defined as the *difference* between the maximum and minimum values,

$$\text{Range} = max - min.$$

Notice that the range is a *single number, not* an interval of values, as you might think from its use in common speech. The maximum HALE is 73.6 years and the minimum is 26.5 years, so the *range* is $73.6 - 26.5 = 47.1$ years.

The range (like the midrange) has the disadvantage that a single extreme value can make it very large, giving a value that doesn't really represent the data overall. For example, in the CEO compensations from Chapter 4, the range is $202,991,184 - \$28,816 = \$202,962,368$! Most of the compensations were between \$0 and \$5,000,000, so the range doesn't give a very accurate impression of the spread.

Finding quartiles by hand

A simple way to find the quartiles is to split the batch into two halves at the median. (When *n* is odd, include the median in both halves.) The lower quartile is the median of the lower half, and the upper quartile is the median of the upper half.

Here are our two examples again.

The ordered values of the first batch were $-17.5, 2.8, 3.2, 13.9,$ $14.1, 25.3,$ and $45.8,$ with a median of 13.9. Since 7 is odd, we include the median in both halves to get $-17.5, 2.8, 3.2, 13.9$ and $13.9, 14.1, 25.3, 45.8.$

Each half has 4 values, so the median of each is the average of its 2nd and 3rd values. So, the lower quartile is $(2.8 + 3.2)/2 = 3.0$ and the upper quartile is $(14.1 + 25.3)/2 = 19.7.$

The second batch of data had the ordered values $-17.5, 2.8, 3.2,$ $13.9, 14.1, 25.3, 35.7,$ and $45.8.$

Here *n* is even, so the two halves of 4 values are $-17.5, 2.8, 3.2, 13.9$ and $14.1, 25.3, 35.7, 45.8.$

Now the lower quartile is $(2.8 + 3.2)/2 = 3.0$ and the upper quartile is $(25.3 + 35.7)/2 = 30.5.$

Spread: The Interquartile Range

A better way to describe the spread of a variable might be to ignore the extremes and concentrate on the middle of the data. We could, for example, find the range of just the middle half of the data. What do we mean by the middle half? Divide the data in half at the median. Now divide both halves in half again, cutting the data into four quarters. We call these new dividing points **quartiles.** One quarter of the data lies below the **lower quartile,** and one quarter of the data lies above the **upper quartile,** so half the data lies between them. The quartiles border the middle half of the data.

The difference between the quartiles tells us how much territory the middle half of the data covers and is called the **interquartile range.** It's commonly abbreviated IQR (and pronounced "eye-cue-are," not "ikker"):

$$IQR = upper\ quartile - lower\ quartile.$$

For the HALE data, there are 95 values below and 95 values above the median. Including the median with each half of the data gives 96 values in each half. To find the median of each half, we'd average the 48th and 49th values. For the HALEs, the lower quartile is 48.9 years and the upper quartile is 62.65 years. The *difference* between the quartiles gives the IQR:

$$IQR = 62.65 - 48.9\ years = 13.75\ years.$$

Now we know that the middle half of the countries (in terms of HALE) extend for a (interquartile) range of 13.75 years. This seems like a reasonable summary of the spread of the distribution, as we can see from the histogram:

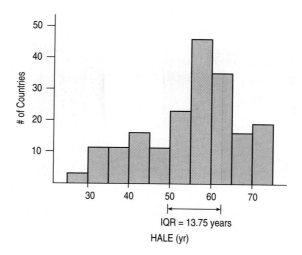

The IQR contains the middle 50% of the values of the distribution. It also gives a visual indication of the spread of the data. Here we see that the IQR is 13.75 years.   **Figure 5.3**

The IQR is often a reasonable summary of the spread of a distribution. For the CEO compensations, the median is $1,304,470 and the quartiles are at $787,304 and $2,518,628, so the IQR is $2,518,628 − $787,304 = $1,731,324. The IQR gives a different impression of how spread out CEOs' salaries are than we might get from the range of $202,962,368.

The lower and upper quartiles are also known as the 25th and 75th **percentiles** of the data, respectively, since the lower quartile falls above 25% of the data and the upper quartile falls above 75% of the data. If we count this way, the median is the 50th percentile. We could, of course, define and calculate any percentile that we want. For example, the 10th percentile would be the number that falls above 10% of the data values.

● **So, what is a quartile anyway?** Finding the quartiles sounds easy, but surprisingly, the quartiles are not well defined. It's not always clear how to find values so that exactly one quarter of the data lies above or below that value. We offer a simple rule in the Finding Quartiles by Hand box: Find the median of each half of the data split by the median. When *n* is odd, we include the median with each of the halves. Some other texts omit the median before finding medians of each half of the data. Both methods are commonly used. If you are willing to do a bit more calculating, there are several other methods that locate a quartile somewhere between adjacent data values. We know of at least six different rules for finding quartiles. Remarkably, each one is in use in some software package or calculator.

Don't worry too much about getting the "exact" value for a quartile. All of the methods agree pretty closely when the data set is large. When the data set is small, different rules will disagree more, but in that case there's little need to summarize the data anyway. Remember, the quartiles are just summary values of the data set. In fact, the main use for quartiles is in calculating the IQR, a measure of how spread out the data are. ●

5-Number Summary

NOTATION ALERT:

We always use Q1 to label the lower (25%) quartile and Q3 to label the upper (75%) quartile. We skip the number 2 because the median would, by this system, naturally be labeled Q2—but we don't usually call it that.

The **5-number summary** of a distribution reports its median, quartiles, and extremes (maximum and minimum). The 5-number summary for the HALE data looks like this.

Max	73.6 years
Q3	62.65
Median	57.7
Q1	48.9
Min	26.5

It's a good idea to report the number of data values and the identity of the cases (the *Who*). Here there are 191 countries.

Rock Concert Deaths: Making Boxplots

WHO	Rock concert goers who died from being crushed
WHAT	Age at death
UNITS	Years
WHERE	Internationally
WHEN	1999–2000
WHY	To increase awareness about concert safety

In their "Rock and Roll Wall of Shame," Crowd Management Strategies lists the names of people who have died at rock concerts from lax safety controls. The Crowd-safe® Database lists the ages, names, causes, and locations of these unfortunate concert-goers. During the period 1999–2000 there were 66 people who died from "crowd crush." How old were these victims? Here's a 5-number summary of their ages:

Max	47 years
Q3	22
Median	19
Q1	17
Min	13

A S **Boxplots.** Dancing boxplots? Not quite, but you can watch the construction in action. (To find this activity, turn the *ActivStats* lesson book back a few pages.)

Whenever we have a 5-number summary of a (quantitative) variable, we can display the information in a **boxplot**. To make a boxplot of the ages of rock concert victims, follow these steps:

1. Draw a single vertical axis spanning the extent of the data.[1] Draw short horizontal lines at the lower and upper quartiles and at the median. Then connect them with vertical lines to form a box. The box can have any width that looks OK.[2]

2. To help us construct the boxplot, we erect "fences" around the main part of the data. We place the upper fence 1.5 IQRs above the upper quartile and the lower fence 1.5 IQRs below the lower quartile. For the rock concert data, we compute

$$Upper\ fence = Q3 + 1.5\ IQRs = 22 + 1.5 \times 5 = 29.5\ years.$$

and

$$Lower\ fence = Q1 - 1.5\ IQRs = 17 - 1.5 \times 5 = 9.5\ years.$$

The fences are just for construction and are not part of the display. We show them here with dotted lines for illustration. You should never include them in your boxplot.

---

[1] The axis could also run horizontally.

[2] Some computer programs draw wider boxes for larger data sets. That can be useful when comparing groups.

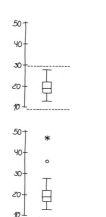

3. We use the fences to grow "whiskers." Draw lines from the ends of the box up and down to *the most extreme data values found within the fences.* If a data value falls outside one of the fences, we do *not* connect it with a whisker.

4. Finally, we add the outliers by displaying any data values beyond the fences with special symbols. (We often use a different symbol for "far outliers"—data values farther than 3 IQRs from the quartiles.)

Now that we've drawn the boxplot, let's summarize what it shows. The center of a boxplot is (remarkably enough) a box that shows the middle half of the data, between the quartiles. The height of the box is equal to the IQR. If the median is roughly centered between the quartiles, then the middle half of the data is roughly symmetric. If it is not centered, the distribution is skewed. The whiskers show skewness as well if they are not roughly the same length. Any outliers are displayed individually, both to keep them out of the way for judging skewness and to encourage you to give them special attention. They may be mistakes, or they may be the most interesting cases in your data.

From the boxplot we see that half of the victims were between 17 and 22 years old. The boxplot makes it look like the distribution of ages is roughly symmetric, with most of the victims between about 13 and 28 years old, but the histogram shows that the distribution is slightly skewed to the right. It is both an advantage and a disadvantage of boxplots that they simplify our view of the distribution like this. Boxplots are particularly good at pointing out outliers. Here, there are two victims who were substantially older. In a careful analysis of these data, we'd want to learn more about them.

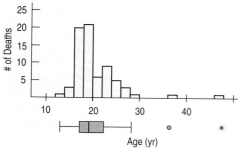

By turning a boxplot and putting it on the same scale, we can compare the boxplot and histogram of the rock concert deaths and see how each represents the distribution. **Figure 5.4**

Boxplots complement histograms by providing more specific information about the center, the quartiles, and outliers. When looking at one variable, it's a good idea to look at the boxplot and histogram together. The main use for boxplots is to compare groups. That's when they really start to shine.

Comparing Groups with Boxplots

Histograms show a lot about the shape of the distribution, but get a little unwieldy when we want to look at more than one group at a time. Boxplots work well for comparing groups because they let the fundamental story show through. When we place them side-by-side, we can easily see which group has the higher median, which has the greater IQR, where the central 50% of the data are located, and which has the greater overall range. And we can get a general idea of symmetry from whether the medians are centered within their boxes and whether the whiskers extend roughly the same distance on either side of the boxes. Equally important, we can see past any outliers in making these comparisons because they've been separated from the rest of the data.

The prominent statistician John W. Tukey, the originator of the boxplot, was asked (by one of the authors) why the outlier nomination rule cut at 1.5 IQRs beyond each quartile. He answered that the reason was that 1 IQR would be too small and 2 IQRs would be too large. That works for us.

Comparing Groups **Step-By-Step**

A student designed an experiment to test the efficiency of various coffee containers by placing hot (180 °F) liquid in each of 4 different container types 8 different times. After 30 minutes she measured the temperature again and recorded the difference in temperature. Because these are temperature *differences*, smaller differences mean that the liquid stayed hot—probably what we want in a coffee mug.

What can we say about the effectiveness of these four mugs? Let's see what story the data tell us.

Think

Plan State what you want to find out.

Variables Identify the *variables* and report the W's.

Be sure to check the appropriate condition.

I want to use data from an experiment to compare the effectiveness of the 4 different mugs in maintaining temperature. I have 8 measurements of *Temperature change* for each of the mugs.

✔ **Quantitative Data Condition:** The temperature changes are quantitative, with units of °F. Boxplots are appropriate displays for comparing the groups. Numerical summaries of each group are appropriate as well.

Show

Mechanics Report the 5-number summaries of the four groups. Including the IQR is a good idea as well.

	Min	Q1	Median	Q3	Max	IQR
CUPPS	6 °F	6	8.25	14.25	18.50	8.25
Nissan	0	1	2	4.50	7	3.50
SIGG	9	11.50	14.25	21.75	24.50	10.25
Starbucks	6	6.50	8.50	14.25	17.50	7.75

Make a picture. Because we want to compare the distributions for four groups, boxplots are an obvious choice.

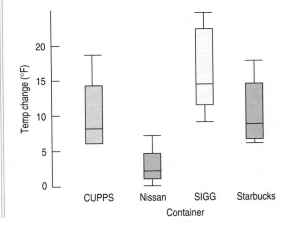

Tell

Conclusion Interpret what the boxplots and summaries say about the ability of these mugs to maintain heat. Compare the shapes, centers, and spreads, and note any outliers.

The individual distributions are all slightly skewed to the high end. The Nissan cup does the best job of keeping liquids hot, with a median loss of only 2 °F, and the SIGG cup does the worst, typically losing 14 °F. The difference is large enough to be important; a coffee drinker would be likely to notice a 14° drop in temperature. And the mugs are clearly different; 75% of the Nissan tests showed less heat loss than any of the other mugs in the study. The IQR of results for the Nissan cup is also the smallest of these test cups, indicating that it is a consistent performer.

Summarizing Symmetric Distributions

In everyday language, sometimes "average" *does* mean what we want it to mean. We don't talk about your grade point mean or a baseball player's batting mean or the Dow Jones Industrial mean. So, we'll continue to say "average" when that seems most natural. When we do, though, you may assume that what we mean is the mean.

Medians do a good job of locating the center of a distribution even when the shape is skewed. But when we have symmetric data, there's another alternative. You probably already know how to average values. In fact, to find the median when *n* is even, we said you should average the two middle values, and you didn't even flinch. In general, to average values, add them up and divide by *n*, the number of values.

Averaging is a common thing to do with data. Once we've averaged some data, what do you think the result is called? The *average*? No, that would be too easy. Informally, we talk about the "average person" or the "average family," but we don't actually add up families and divide by *n*. So, to be more precise we call this summary the mean.

Pulse rates are useful for monitoring medical conditions. Resting heart rates depend on age. For children more than 10 years old and for adults, 60 to 100 beats per minute is considered normal. Here are a histogram and summaries for the pulse rates of 52 adults.

WHO	52 adults
WHAT	Resting heart rates
UNITS	Beats per minute

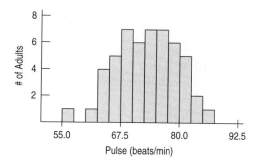

Mean	72.7 beats/min
Min	57
Q1	68
Median	73
Q3	78
Max	86
IQR	10

Pulse rates of 52 adults. **Figure 5.5**

The histogram shows a generally symmetric distribution, and the mean and median agree quite closely. That's what we expect for symmetric data. When the shape of the distribution is symmetric, there's no numerical reason to prefer the median or the mean. As we'll see, it turns out that there's more we can do and say with the mean than with the median. The mean is appropriate, however, only when the shape is symmetric and there are no outliers. (How do you check? Plot the data!)

The Formula for Averaging (Say It in Greek)

You already know how to average values, but this is a good opportunity to introduce some notation that will make it easier to describe calculations later on. Here's the formula:

$$\bar{y} = \frac{Total}{n} = \frac{\sum y}{n}.$$

The *y* with a line over it is pronounced "y-bar." In general, a bar over any symbol or variable name in Statistics denotes finding its mean. The symbol Σ is the Greek letter capital sigma—equivalent to an "S," as in "sum"—and means just

The median splits a histogram so that the *areas* of the bars on either side of the median are equal—regardless of how far they are from the center. The mean balances the histogram, taking into account both the size of the bars and their distance from the center, but as a result, it may not have equal numbers of data values on either side.

How to Say "Square Root"
We usually write a square root with the $\sqrt{}$ symbol. Especially on computers, though, you may see it as a special function, usually called SQRT(y).

You might recall from Algebra that we can also write the square root as the 1/2 power. On the computer, powers are indicated with an up arrow or carat: $y\text{^}0.5$.

That may not look like a square root at first, but it is, and you should recognize it if you happen to see it written like this.

that; add up all the observations. The formula says that to find the **mean,** add up all the numbers and divide by n—but you knew that.

Mean or Median?

Does it make a difference whether we choose a mean or a median? Well, sometimes it does. The median of the HALEs is 57.7 years. The mean is only 55.26 years. Why are they different? The answer lies in the shape of the distribution. The **mean** is the point at which the histogram would balance.

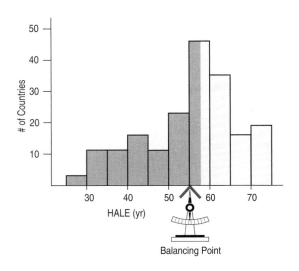

The mean is located at the *balancing point* of the histogram. Since this distribution is skewed to the left, the mean is *lower* than the median. The points at the left have pulled the mean toward them, away from the median.   **Figure 5.6**

Balancing Point

Just like a child who moves away from the center of a see-saw, a bar of the histogram far from the center has more leverage, pulling the mean in its direction. If the skewness is strong or if there are straggling outliers, the mean can be pulled quite far from the median. The skewness on the left pulls the mean somewhat to the left of the median here. For the CEO compensation data from Chapter 4, the skewness is to the right. The median compensation is $1,304,470, but the mean is $2,818,743—more than *twice* the median. It's hard to argue that a value pulled in this way is what we meant by the center of the data. For skewed data, it's better to report the median than the mean as a measure of center.

What About Spread? The Standard Deviation

The IQR is always a reasonable summary of spread, but because it uses only the two quartiles of the data, it ignores much of the information about how individual values vary. A more powerful approach uses the standard deviation, which takes into account how far *each* value is from the mean. Like the mean, the standard deviation is appropriate only for symmetric data.

One way to think about spread is to examine how far each data value is from the mean. This difference is called a *deviation*. We could just average the deviations, but the positive and negative differences always cancel each other out. So the average deviation is always zero—not very helpful.

Finding the standard deviation by hand

To find the standard deviation, you start with the mean, $\bar{y}$. Then you find the *deviations* by taking $\bar{y}$ from each value: $(y - \bar{y})$. Square each deviation: $(y - \bar{y})^2$.

Now you're nearly home. Just add these up and divide by $n - 1$. That gives you the variance, s^2. To find the standard deviation, s, take the square root. Here we go:

Suppose the batch of values is 4, 3, 10, 12, 8, 9, and 3.

The mean is $\bar{y} = 7$. So the deviations are found by subtracting 7 from each value:

Original Values	Deviations	Squared Deviations
4	4 − 7 = −3	$(-3)^2 =$ 9
3	3 − 7 = −4	$(-4)^2 =$ 16
10	10 − 7 = 3	9
12	12 − 7 = 5	25
8	8 − 7 = 1	1
9	9 − 7 = 2	4
3	3 − 7 = −4	16

Add up the squared deviations:
9 + 16 + 9 + 25 + 1 + 4 + 16 = 80.
Now, divide by $n - 1$:   80/6 = 13.33.
Finally, take the square root:
$$s = \sqrt{13.33} = 3.65$$

To keep them from canceling out, we *square* each deviation. Squaring always gives a positive value, so the sum won't be zero. That's great. Squaring also emphasizes larger differences—a feature that turns out to be both good and bad.

When we add up these squared deviations and find their average (almost), we call the result the **variance:**

$$s^2 = \frac{\sum(y - \bar{y})^2}{n - 1}.$$

Why almost? It *would* be a mean if we divided the sum by n. Instead, we divide by $n - 1$. Why? The simplest explanation is "to drive you crazy." But there are good technical reasons, some of which we'll see later.

The variance will play an important role later in this book, but it has a problem as a measure of spread. Whatever the units of the original data are, the variance is in *squared* units. We want measures of spread to have the same units as the data. And we probably don't want to talk about squared dollars, or mpg^2. So, to get back to the original units, we take the square root of s^2. The result, s, is the **standard deviation.**

Putting it all together, the standard deviation of the data is found by the following formula:

$$s = \sqrt{\frac{\sum(y - \bar{y})^2}{n - 1}}.$$

You will almost always rely on a calculator or computer to do the calculating.

Thinking About Variation

A S **Displaying Spread.** What does the standard deviation look like on a histogram? How about the IQR? It's surprisingly simple— find out here.

Statistics is about variation, so spread is an important fundamental concept in Statistics. Measures of spread help us to be precise about what we *don't* know. If many data values are scattered far from the center, the IQR and the standard deviation will be large. If the data values are close to the center, then these measures of spread will be small. If all our data values were exactly the same, we'd have no question about summarizing the center, and all measures of spread would be zero—and we wouldn't need Statistics. You might think this would be a big plus, but it would make for a boring world. Fortunately (at least for Statistics), data do vary.

Measures of spread tell how well other summaries describe the data. That's why we always (always!) report a spread along with any summary of the center.

1. The U.S. Census Bureau reports the median family income in its summary of census data. Why do you suppose they use the median instead of the mean? What might be the disadvantages of reporting the mean?

2. You've just bought a new car that claims to get a highway fuel efficiency of 31 miles per gallon. Of course, your mileage will "vary." If you had to guess, would you expect the IQR of gas mileage attained by all cars like yours to be 30 mpg, 3 mpg, or 0.3 mpg? Why?

A company selling a new MP3 player advertises that the player has a mean lifetime of 5 years. If you were in charge of quality control at the factory, would you prefer that the standard deviation of lifespans of the players you produce be 2 years or 2 months? Why?

Shape, Center, and Spread

A S **Playing with Summaries.** Here's a Statistics game about summaries. Really. Even some experienced statisticians find some of these challenges . . . well, challenging. Your intuition may be better. Give it a try!

How "Accurate" Should We Be? Don't think you should report means and standard deviations to a zillion decimal places; such implied accuracy is really meaningless. Although there is no ironclad rule, statisticians commonly report summary statistics to one or two decimal places more than the original data.

What should you tell about a quantitative variable? Report the shape of its distribution, and include a center and a spread. But which measure of center and which measure of spread? The rules are pretty easy:

- If the shape is skewed, report the median and IQR. You may want to include the mean and standard deviation, but you should point out why the mean and median differ. The fact that the mean and median do not agree is a sign that the distribution may be skewed. A histogram will help you make that point.
- If the shape is symmetric, report the mean and standard deviation and possibly the median and IQR as well. For symmetric data, the IQR is usually a bit larger than the standard deviation. If that's not true for your data set, look again to make sure the distribution isn't skewed and there are no outliers.
- If there are any clear outliers and you are reporting the mean and standard deviation, report them with the outliers present and with the outliers removed. The differences may be revealing. (Of course, the median and IQR are not likely to be affected by the outliers.)

We always pair the median with the IQR and the mean with the standard deviation. It's not useful to report one without the other. Reporting a center without a spread is dangerous. You may think you know more than you do about the distribution. Reporting only the spread leaves us wondering where we are.

Summarizing a Distribution Step-By-Step

One of the authors owned a 1989 Nissan Maxima for 8 years. Being a statistician, he recorded the car's fuel efficiency (in mpg) each time he filled the tank. He wanted to know what fuel efficiency to expect as "ordinary" for his car. (Hey, he's a statistician, what would you expect?[3]) Knowing this, he was able to predict when he'd need to fill the tank again, and to notice if the fuel efficiency suddenly got worse, which could be a sign of trouble. What do the data say?

 Think

Plan State what you want to find out.

Variable Identify the variable and report the W's.

Be sure to check the appropriate condition.

I want to summarize the distribution of Nissan fuel efficiency.

The data are the fuel efficiency values in miles per gallon for the first 100 fill-ups of a 1989 Nissan Maxima between 1989 and 1992.

✔ **Quantitative Data Condition:** The fuel efficiencies are quantitative with units miles per gallon. Histograms and boxplots are appropriate displays for displaying the distribution. Numerical summaries are appropriate as well.

---

[3] He also recorded the time of day, temperature, price of gas, and phase of the moon. (OK, maybe not phase of the moon.)

Show Mechanics Make a histogram and boxplot. Based on the shape, choose appropriate numerical summaries.

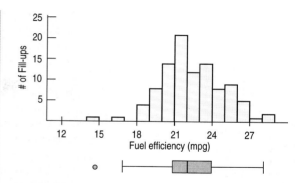

REALITY CHECK A value of 22 mpg seems reasonable for such a car. The spread is reasonable, although the range looks a bit large.

A histogram of the data shows a fairly symmetric distribution with a low outlier.

Count	100
Mean	22.4 mpg
StdDev	2.45
Q1	20.8
Median	22.0
Q3	24.0
IQR	3.2

The mean and median are close, so the outlier doesn't seem to be a problem. I can use the mean and standard deviation.

Tell Conclusion Summarize and interpret your findings in context.

The distribution of mileage is unimodal and symmetric with a mean of 22.4 mpg. There is a low outlier that should be investigated, but it does not influence the mean very much. The standard deviation is 2.4 mpg. The boxplot shows that half of the time, the car had a fuel efficiency between about 21 and 24 mpg.

What Can Go Wrong?

The task of summarizing a quantitative variable is relatively simple, and there is a simple path to follow. However, you need to watch out for certain features of the data that make summarizing them with a number dangerous. Here's some advice:

- ***Don't forget to do a reality check.*** Don't let the computer (or calculator) do your thinking for you. Make sure the calculated summaries make sense. For example, does the mean look like it is in the center of the histogram? Think about the spread: An IQR of 50 mpg would clearly be wrong for gas mileage. And no measure of spread can be negative. The standard deviation can take the value 0, but only in the very unusual case that all the data values equal the same number. If you see the IQR or standard deviation equal to 0, it's probably a sign that something's wrong with the data.

- **Don't forget to sort the values before finding the median or percentiles.** It seems obvious, but when you work by hand, it's easy to forget to sort the data first before counting in to find medians, quartiles, or other percentiles. Don't report that the median of the five values 194, 5, 1, 17, 893 is 1 just because 1 is the middle number.
- **Don't compute numerical summaries of a categorical variable.** The mean zip code or the standard deviation of social security numbers is not meaningful. If the variable is categorical, you should instead report summaries such as percentages of individuals in each category. It is easy to make this mistake when using technology to do the summaries for you. After all, the computer doesn't care what the numbers mean.
- **Watch out for multiple modes.** If the distribution—as seen in a histogram, for example—has multiple modes, consider separating the data into different groups. If you cannot separate the data in a meaningful way, you should not summarize the center and spread of the variable.
- **Be aware of slightly different methods.** Finding the 10th percentile or the lower quartile in a data set sounds easy enough. But it turns out that the definitions are not exactly clear. If you compare different statistics packages or calculators, you may find that they give slightly different answers for the same data. These differences, though, are unlikely to be important in interpreting the data, the quartiles, or the IQR, so don't let them worry you.
- **Beware of outliers.** If the data have outliers but are otherwise unimodal, consider holding the outliers out of the further calculations and reporting them individually. If you can find a simple reason for the outlier (for instance, a data transcription error) you should remove or correct it. If you cannot do either of these, then choose the median and IQR to summarize the center and spread.
- **Make a picture (make a picture, make a picture).** The sensitivity of the mean and standard deviation to outliers is one reason why you should *always* make a picture of the data. Summarizing a variable with its mean and standard deviation when you have not looked at a histogram or dotplot to check for outliers or severe skewness invites disaster. You may find yourself drawing absurd or dangerously wrong conclusions about the data. And, of course, you should demand no less of others. Don't accept a mean and standard deviation blindly without some evidence that the variable they summarize has no outliers or severe skewness.
- **Be careful when comparing groups that have very different spreads.** For example, look at these boxplots. Researchers measured the concentration (nanograms per milliliter) of cotinine in the blood of three groups of people: nonsmokers who have not been exposed to smoke, nonsmokers who have been exposed to smoke (ETS), and smokers. Cotinine is left in the blood when the body metabolizes nicotine, so this measure gives a direct measurement of the effect of passive smoke exposure. The boxplots of the cotinine levels of the three groups tell us that the smokers have higher cotinine levels, but if we want to compare the levels of the passive smokers to those of the nonsmokers, we're in trouble, because on this scale the cotinine levels for both nonsmoking groups are too low to be seen.

WHO	Smokers, nonsmokers, and passive smokers
WHAT	Blood cotinine levels
UNITS	nanograms per milliliter (ng/ml)

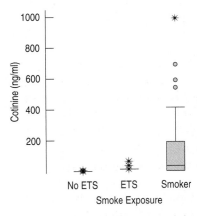

Cotinine levels (nanograms per milliliter) for three groups with different exposures to tobacco smoke. Can you compare the ETS (exposed to smoke) and No ETS groups? **Figure 5.7**

Blood cotinine levels after taking logs.
What a difference a log makes!
Figure 5.8

*Re-expressing to Equalize the Spread of Groups

We can often alleviate the problem of comparing groups that have very different spreads by re-expressing the data values. For measurements like the cotinine data, whose values can't be negative and whose distributions are skewed to the high end, a good first guess at a re-expression is the logarithm.

After taking logs we can compare the groups and see that the nonsmokers exposed to environmental smoke (the ETS group) do show increased levels of (log) cotinine, although not the high levels found in the blood of smokers. Notice that the same re-expression has also improved the symmetry of the cotinine distribution for smokers and pulled in most of the apparent outliers in all of the groups. It is not unusual for a re-expression that improves one aspect of data to improve others as well. We'll talk about other ways to re-express data as the need arises throughout the book.

CONNECTIONS

We discussed the value of summarizing a distribution with shape, center, and spread in Chapter 4. Now we have numerical summaries for center and spread to go with our general observations. The shape of the distribution helps us to decide which numerical summaries are most appropriate.

We compared distributions in Chapter 4 by just looking at histograms. Working with numerical summaries and boxplots adds depth to our comparisons.

What have we learned?

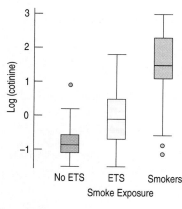

We've learned how to summarize distributions of quantitative variables numerically.
- The 5-number summary displays the two quartiles, median, min, and max for a variable.
- Measures of center for a distribution include the mean and median.
- Measures of spread include the range, IQR, and standard deviation.
- We'll report the median and IQR when the distribution is skewed. If it's symmetric, we'll summarize with the mean and standard deviation (and possibly the median and IQR, as well).

We've also learned to display distributions with boxplots.
- A boxplot reveals some features of a distribution not easily seen in a histogram—the center, the middle 50%, and outliers. Histograms, though, are better at showing the shape.
- Boxplots are very effective for comparing groups graphically. As always, when we compare groups we'll discuss their shapes, centers, spreads, and unusual features.

TERMS

Center	We summarize the center of a distribution with the mean or the median.
Median	The median is the middle value with half of the data above and half below it.
Spread	We summarize the spread of a distribution with the standard deviation, interquartile range, and range.
Range	The difference between the lowest and highest values in a data set. Range = max − min.
Quartile	The lower quartile (Q1) is the value with a quarter of the data below it. The upper quartile (Q3) has a quarter of the data above it. The median and quartiles divide data into four equal parts.
Interquartile range (IQR)	The IQR is the difference between the first and third quartiles. IQR = Q3 − Q1.
Percentile	The ith percentile is the number that falls above i% of the data.
5-number summary	A 5-number summary for a variable consists of: • The minimum and maximum • The quartiles: Q1 and Q3 • The median
Boxplot	A boxplot displays the 5-number summary as a central box with whiskers that extend to the non-outlying data values. Boxplots are particularly effective for comparing groups.
Mean	The mean is found by summing all the data values and dividing by the count.
Variance	The variance is the sum of squared deviations from the mean, divided by the count minus one.
Standard deviation	The standard deviation is the square root of the variance.
Comparing distributions	When comparing the distribution of several groups, consider their • Shape • Center • Spread
Comparing boxplots	When comparing groups with boxplots, • Compare the medians; which group has the higher center? • Compare the IQRs; which group is more spread out? • Judged by the size of the IQRs, are the medians very different? • Check for possible outliers. Identify them if you can.

SKILLS

When you complete this lesson you should:

- Be able to select a suitable measure of center and a suitable measure of spread for a variable based on information about its distribution.
- Know the basic properties of the median: The median divides the data into the half of the data values that are below the median and the half that are above the median.
- Know the basic properties of the mean: The mean is the point at which the histogram balances.
- Know that the standard deviation summarizes how spread out all the data are around the mean.

- Know how to compute the mean and median of a set of data.
- Know how to compute the standard deviation and IQR of a set of data.

- Be able to create a 5-number summary of a variable.

- Be able to construct a boxplot by hand from a 5-number summary.

- Understand that the median and IQR resist the effects of outliers, while the mean and standard deviation do not.

- Understand that in a skewed distribution, the mean is pulled in the direction of the skewness (toward the longer tail) relative to the median.

Tell

- Know how to describe summary measures in a sentence. In particular, know that the common measures of center and spread have the same units as the variable that they summarize, and should be described in those units.

- Be able to describe the distribution of a quantitative variable with a description of the shape of the distribution, a numerical measure of center, and a numerical measure of spread. Be sure to note any unusual features, such as outliers, too.

- Know how to compare the distributions of two or more groups by comparing their shapes, centers, and spreads.

- Be able to compare two or more groups by comparing their boxplots.

- Know how to use the 1.5 IQR rule to identify possible outliers. Interpret outliers found in boxplots made on a computer.

Numerical Summaries on the Computer

Many statistics packages offer a pre-packaged collection of summary measures. The result might look like this:

```
Variable: Weight
N = 234
Mean = 143.3          Median = 139
St. Dev = 11.1         IQR = 14
```

A S Case Study.
Who's safer in a crash—passengers or the driver? Now you have the tools to find out. Investigate with your statistics package.

Alternatively, a package might make a table for several variables and summary measures.

Variable	N	mean	median	stdev	IQR
Weight	234	143.3	139	11.1	14
Height	234	68.3	68.1	4.3	5
Score	234	86	88	9	5

It is usually easy to read the results and identify each computed summary. You should be able to read the summary statistics produced by any computer package.

Packages often provide many more summary statistics than you need. Of course, some of these may not be appropriate when the data are skewed or have outliers. It is your responsibility to check a histogram or stem-and-leaf display and decide which summary statistics to use.

It is common for packages to report summary statistics to many decimal places of "accuracy." Of course, it is rare data that have such accuracy in the original measurements. The ability to calculate to six or seven digits beyond the decimal point doesn't mean that those digits have any meaning. Generally, it's a good idea to round these values, allowing perhaps one more digit of precision than was given in the original data.

Summary statistics are easy to find in most packages.

DATA DESK

Select the variable(s) to summarize. In the **Calc** menu open the **Summaries** submenu. **Options** offer separate tables, a single unified table, and other formats.

EXCEL

Click on an empty cell. Type an equals sign and choose **"Average"** from the popup list of functions that appears to the left of the text editing box. Enter the data range in the box that says **"Number 1."** Click the **OK** button.

To compute the standard deviation of a column of data directly, use the **STDEV** from the popup list of functions in the same way.

Comments

Excel's STDEV function should not be used for data values larger in magnitude than 100,000 or for lists of more than a few thousand values. It is programmed with an unstable formula that can generate rounding errors when these limits are exceeded.

JMP

Choose **Distribution** from the **Analyze** menu. In the **Distribution** dialog, drag the name of the variable that you wish to analyze into the empty window beside the label "Y, Columns." Click

OK. JMP computes standard summary statistics along with displays of the variable.

MINITAB

Choose **Basic Statistics** from the **Stat** menu. From the **Basic Statistics** submenu, choose **Display Descriptive Statistics.**

Assign variables from the variable list box to the Variables box. MINITAB makes a Descriptive Statistics table.

SPSS

Choose **Descriptive Statistics** from the **Analyze** menu and then choose **Explore. . . .** In the **Explore** dialog, assign one or more

variables from the Source List to the Dependent List and click the **OK** button. SPSS will generate several plots and tables.

TI-83/84 PLUS

Choose **1-VarStats** from the **STAT CALC** menu and specify the List where the data are stored. You must scroll down to see the 5-number summary.

To make a boxplot, set up a **STAT PLOT** using the boxplot icon.

Comments

Note that the standard deviation is identified as Sx; don't use σ_x by mistake, because it divides by n instead of $n - 1$.

TI-89

- To compute summary statistics press F4 **(Calc).** Input the name of the list using VAR-LINK. Press ENTER.
  Use the down arrow to scroll through the output.
- To create a boxplot, press F2 **(Plots)** then ENTER. Select a plot to define and press F1. Select either 3: **Box Plot** or 4: **Mod Box Plot** (to identify outliers). Select the mark type of your choice (for outliers). Press ENTER to finish.
  Press F5 to display the graph.

Comments

If the data are stored as a frequency table (say, with data values in list1 and frequencies in list2), use VAR-LINK to select list2 as the frequency variable in 1-Var Stats.

For the plot, change Use Freq and Categories to YES and use VAR-LINK to select list2 as the frequency variable on the plot definition screen.

EXERCISES

1. In the news. Find an article in a newspaper or a magazine that discusses an "average."
a) Does the article discuss the W's for the data?
b) What are the units for the variable?
c) Is the average used the median or the mean? How can you tell?
d) Is the choice of median or mean appropriate for the situation? Explain.

2. In the news II. Find an article in a newspaper or a magazine that discusses a measure of spread.
a) Does the article discuss the W's for the data?
b) What are the units for the variable?
c) Does the article use the range, IQR, or standard deviation?
d) Is the choice of measure of spread appropriate for the situation? Explain.

3. Summaries. Here are costs of 10 electric smoothtop ranges rated very good or excellent by *Consumer Reports* in August 2002.

$850 900 1400 1200 1050 1000 750 1250 1050 565

Find these statistics *by hand* (no calculator!):
a) mean
b) median and quartiles
c) range and IQR

4. More summaries. Here are the annual numbers of deaths from tornadoes in the United States from 1990 through 2000. (Source: NOAA)

   53   39   39   33   69   30   25   67   130   94   40

Find these statistics *by hand* (no calculator!):
a) mean
b) median and quartiles
c) range and IQR

5. Mistake. A clerk entering salary data into a company spreadsheet accidentally put an extra "0" in the boss's salary, listing it as $2,000,000 instead of $200,000. Explain how this error will affect these summary statistics for the company payroll:
a) measures of center: median and mean.
b) measures of spread: range, IQR, and standard deviation.

6. Sick days. During contract negotiations, a company seeks to change the number of sick days employees may take, saying that the annual "average" is 7 days of absence per employee. The union negotiators counter that the "average" employee misses only 3 days of work each year. Explain how both sides might be correct, identifying the measure of center you think each side is using and why the difference might exist.

7. Payroll. A small warehouse employs a supervisor at $1200 a week, an inventory manager at $700 a week, six stock boys at $400 a week, and four drivers at $500 a week.
a) Find the mean and median wage.
b) How many employees earn more than the mean wage?
c) Which measure of center best describes a typical wage at this company, the mean or the median?
d) Which measure of spread would best describe the payroll, the range, the IQR, or the standard deviation? Why?

T 8. Singers. The frequency table shows the heights (in inches) of 130 members of a choir.

Height	Count	Height	Count
60	2	69	5
61	6	70	11
62	9	71	8
63	7	72	9
64	5	73	4
65	20	74	2
66	18	75	4
67	7	76	1
68	12		

a) Find the 5-number summary for these data.
b) Display these data with a boxplot.
c) Find the mean and standard deviation.
d) Display these data with a histogram.
e) Write a few sentences describing the distribution of heights.

9. Standard deviation. For each lettered part, a through c, examine the two given sets of numbers. Without doing any calculations, decide which set has the larger standard deviation and explain why. Then check by finding the standard deviations *by hand*.

	Set 1	Set 2
a)	3, 5, 6, 7, 9	2, 4, 6, 8, 10
b)	10, 14, 15, 16, 20	10, 11, 15, 19, 20
c)	2, 6, 6, 9, 11, 14	82, 86, 86, 89, 91, 94

10. Standard deviation. For each lettered part, a through c, examine the two given sets of numbers. Without doing any calculations, decide which set has the larger standard deviation and explain why. Then check by finding the standard deviations *by hand*.

	Set 1	Set 2
a)	4, 7, 7, 7, 10	4, 6, 7, 8, 10
b)	100, 140, 150, 160, 200	10, 50, 60, 70, 110
c)	10, 16, 18, 20, 22, 28	48, 56, 58, 60, 62, 70

11. States. The stem-and-leaf display shows populations of the 50 states and Washington, DC, in millions of people, according to the 2000 census.

```
3 | 4
2 |
2 | 1
1 | 69
1 | 0122
0 | 5555666667888
0 | 1111111111111122222333333344444
```

State Populations (1|2 means 12 million)

a) What measures of center and spread are most appropriate?
b) Without doing any calculations, which must be larger—the median or the mean? Explain how you know.
c) From the stem-and-leaf display, find the median and the interquartile range.
d) Write a few sentences describing this distribution.

12. Wayne Gretzky. In Chapter 4 (Exercise 12) you examined the number of games played by hockey great Wayne Gretzky during his 20-year career in the NHL. Here is the stem-and-leaf display:

```
8 | 000000122
7 | 8899
7 | 0344
6 |
6 | 4
5 |
5 |
4 | 58
4 |
```

Games Played (4|5 = 45 games)

a) Would you use the median or the mean to describe the center of this distribution? Why?
b) Find the median.
c) Without actually finding the mean, would you expect it to be higher or lower than the median? Explain.

13. Home runs. In 1961 Roger Maris made baseball headlines by hitting 61 home runs, breaking a famous record held by Babe Ruth. Here are Maris's home run totals for his 10 seasons in the American League. Would you consider his record-setting year to be an outlier? Explain.

8, 13, 14, 16, 23, 26, 28, 33, 39, 61

14. Gretzky returns. Look once more at the stem-and-leaf display of hockey games played each season by Wayne Gretzky, seen in Exercise 12.

a) Find the range.
b) Find the interquartile range.
c) Using the Outlier Rule, explain why the two seasons when Gretzky played only 45 and 48 games could be considered outliers.
d) Do you consider the 64-game season an outlier, too? Explain.

15. Population growth. The back-to-back stem-and-leaf display compares the percentage change between the 1990 census and 2000 census in the populations of northeastern and midwestern states with the changes in population of southern and western states. The fastest growing states were Nevada at 66% and Arizona at 40%. Use the data displayed in the stem-and-leaf display to construct comparative boxplots.

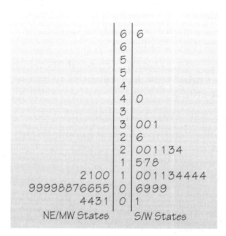

```
              6 | 6
              6 |
              5 |
              5 |
              4 |
              4 | 0
              3 |
              3 | 001
              2 | 6
              2 | 001134
              1 | 578
        2100  1 | 001134444
99998876655  0 | 6999
        4431  0 | 1
   NE/MW States    S/W States
```

16. Camp sites. Shown below and on the next page are the histogram and summary statistics for the number of camp sites at public parks in Vermont.

a) Which statistics would you use to identify the center and spread of this distribution? Why?
b) How many parks would you classify as outliers? Explain.
c) Create a boxplot for these data.
d) Write a few sentences describing the distribution.

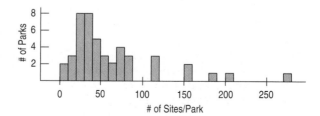

Count	46
Mean	62.8 sites
Median	43.5
StdDev	56.2
Min	0
Max	275
Q1	28
Q3	78

17. Sip size. Researchers in Cornell University's Food Sciences department study how people experience foods. One study considered how much liquid people typically take into their mouths in one "sip" (in milliliters). The researchers also recorded the height (meters) and weight (kilograms) of the participants. Here are histograms of three of the variables from that study:

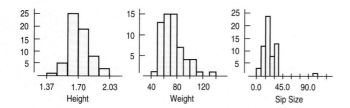

a) For which of the variables depicted in the histograms would you be most satisfied to summarize the center with a mean? Explain.
b) For which of the variables depicted in the histograms would you most strenuously insist on using an IQR rather than a standard deviation to summarize spread? Explain.

T 18. How tall? Students in a large class were asked to estimate the teacher's height in inches. Here's a histogram of their estimates:

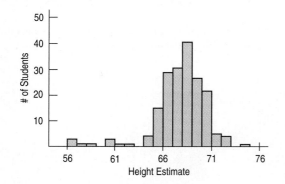

You want a good guess of the professor's true height (or at least a good guess of how tall the students think the

professor is). How would you summarize these data? Why? What might explain guesses as low as 56"?

T 19. Women's basketball. Here are boxplots of the points scored during the first 10 games of the season for both Scyrine and Alexandra.
a) Summarize the similarities and differences in their performance so far.
b) The coach can take only one player to the state championship. Which one should she take? Why?

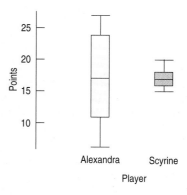

20. Gas prices. Here are boxplots of weekly gas prices at a service station in the Midwest United States (prices in $ per gallon).

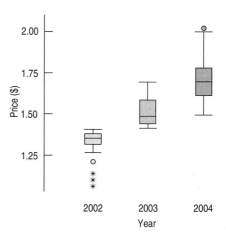

a) Compare the distribution of prices over the three years.
b) In which year were the prices least stable? Explain.

T 21. Marriage age. Do men and women marry at the same age? Here are boxplots of the age at first marriage for a sample of U.S. citizens. Write a brief report discussing what these data show.

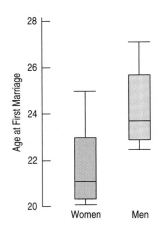

 22. Fuel economy. Describe what these boxplots tell you about the relationship between the number of cylinders a car's engine has and the car's fuel economy (mpg).

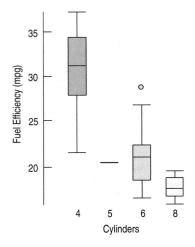

23. Wines. The boxplots display case prices (in dollars) of wines produced by vineyards along three of the Finger Lakes.

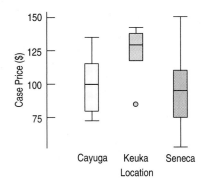

a) Which lake region produces the most expensive wine?
b) Which lake region produces the cheapest wine?
c) In which region are the wines generally more expensive?
d) Write a few sentences describing these wine prices.

24. Ozone. Ozone levels (in parts per million, ppm) were recorded at sites in New Jersey monthly between 1926 and 1971. Here are boxplots of the data for each month (over the 46 years) lined up in order (January = 1).

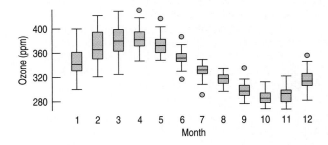

a) In what month was the highest ozone level ever recorded?
b) Which month has the largest IQR?
c) Which month has the smallest range?
d) Write a brief comparison of the ozone levels in January and June.
e) Write a report on the annual patterns you see in the ozone levels.

25. Wild card Summer Olympics. Seventy-one swimmers finished the qualifying first day of the men's 100-m swim in the 2000 Olympics in Sydney. The average time was 52.65 seconds with a standard deviation of 7.66 seconds. The median time was 51.34 seconds and the IQR was 2.58 seconds.
a) Without looking at a graphical display, what shape would you expect the distributions of times to have?
b) What might account for the difference between these two sets of statistics?

Eric Moussambani of Equatorial Guinea was eligible for the 100-m swim in the 2000 Olympics, thanks to a special program.

c) Here is the histogram of the actual times. Write a couple of sentences summarizing what you see.[4]

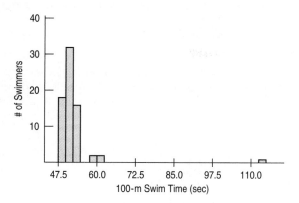

26. Unemployment. In May of 2001, the U.S. Bureau of Labor Statistics (BLS) issued a news release that said, in part:

> *In April, 223 metropolitan areas recorded unemployment rates below the U.S. average of 4.2 percent (not seasonally adjusted), while 99 areas registered higher rates.*

Sketch what the distribution of unemployment rates for the 322 metropolitan areas reported on by BLS probably looks like.

27. Test scores. Three Statistics classes all took the same test. Histograms of the scores for each class are shown below.

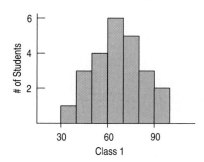

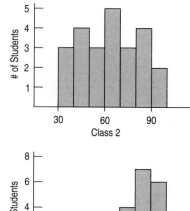

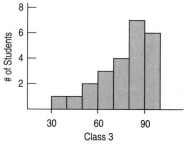

a) Which class had the highest mean score?
b) Which class had the highest median score?
c) For which class are the mean and median most different? Which is higher? Why?
d) Which class had the smallest standard deviation?
e) Which class had the smallest IQR?

28. Test scores. Look again at the histograms of test scores for the three Statistics classes in Exercise 27.
a) Overall, which class do you think performed better on the test? Why?
b) How would you describe the shape of each distribution?
c) Match each class with the corresponding boxplot below.

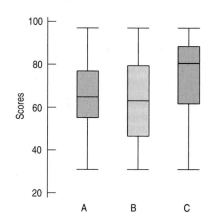

T 29. Still rockin'. On pages 76–77, you read about the 66 deaths attributed to "crowd crush" at rock concerts during the years 1999 and 2000. Here are the histogram and boxplot of the victims' ages that we saw earlier:

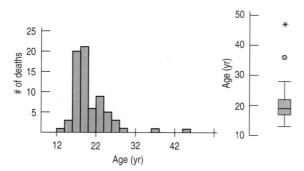

a) What features of the distribution can you see in both the histogram and the boxplot?
b) What features of the distribution can you see in the histogram that you could not see in the boxplot?
c) What summary statistic would you choose to summarize the center of this distribution? Why?
d) What summary statistic would you choose to summarize the spread of this distribution? Why?

T 30. Golf courses. One measure of the difficulty of a golf course is its length: the total distance (in yards) from tee to hole for all 18 holes. Below are the histogram and summary statistics for the lengths of all the golf courses in Vermont.

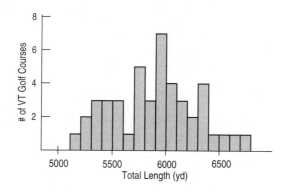

Count	45
Mean	5892.91 yd
StdDev	386.59
Min	5185
Q1	5585.75
Median	5928
Q3	6131
Max	6796

a) What is the range of these lengths?

b) Between what lengths do the central 50% of these courses lie?
c) What summary statistics would you use to describe these data?
d) Write a brief description of these data (shape, center, and spread).

31. Graduation? A survey of major universities asked what percentage of incoming freshmen usually graduate "on time" in 4 years. Use the summary statistics given to answer these questions.

	% on time
Count	48
Mean	68.35
Median	69.90
StdDev	10.20
Min	43.20
Max	87.40
Range	44.20
25th %tile	59.15
75th %tile	74.75

a) Would you describe this distribution as symmetric or skewed? Explain.
b) Are there any outliers? Explain.
c) Create a boxplot of these data.
d) Write a few sentences about the graduation rates.

32. Wineries. Here are summary statistics for the sizes (in acres) of Finger Lakes wineries.

Count	36
Mean	46.50 acres
StdDev	47.76
Median	33.50
IQR	36.50
Min	6
Q1	18.50
Q3	55
Max	250

a) Would you describe this distribution as symmetric or skewed? Explain.
b) Are there any outliers? Explain.
c) Create a boxplot of these data.
d) Write a few sentences about the sizes of the wineries.

33. Caffeine. A student study of the effects of caffeine asked volunteers to take a memory test 2 hours after drinking soda. Some drank caffeine-free cola, some drank regular cola (with caffeine), and others drank a mixture of the two (getting a half-dose of caffeine). Here are the 5-number summaries for each group's scores (number of items recalled correctly) on the memory test:

	n	Min	Q1	Median	Q3	Max
No caffeine	15	16	20	21	24	26
Low caffeine	15	16	18	21	24	27
High caffeine	15	12	17	19	22	24

a) Describe the W's for these data.
b) Name the variables and classify each as categorical or quantitative.
c) Create parallel boxplots to display these results as best you can with this information.
d) Write a few sentences comparing the performances of the three groups.

34. SAT scores. Here are the summary statistics for Verbal SAT scores for a high-school graduating class.

	n	Mean	Median	SD	Min	Max	Q1	Q3
Male	80	590	600	97.2	310	800	515	650
Female	82	602	625	102.0	360	770	530	680

a) Create parallel boxplots comparing the scores of boys and girls as best you can from the information given.
b) Write a brief report on these results. Be sure to discuss shape, center, and spread of the scores.

35. Derby speeds. How fast do horses run? Kentucky Derby winners top 30 miles per hour, as shown in the graph below. In fact, this graph shows the percentage of Derby winners that have run *slower* than a given speed. Note that few have won running less than 33 miles per hour, but about 95% of the winning horses have run less than 37 miles per hour. (A cumulative frequency graph like this is called an "ogive.")

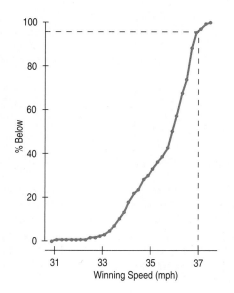

a) Estimate the median winning speed.
b) Estimate the quartiles.
c) Estimate the range and the IQR.
d) Create a boxplot of these speeds.
e) Write a few sentences about the speeds of the Kentucky Derby winners.

36. Cholesterol. The Framingham Heart Study recorded the cholesterol levels of more than 1400 men. Here is an ogive of the distribution of these cholesterol measures. (Recall that an ogive shows the percentage of cases at or below a certain value.) Construct a boxplot for these data and write a few sentences describing the distribution.

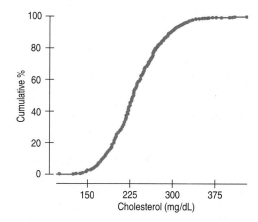

37. Reading scores. A class of fourth graders takes a diagnostic reading test, and the scores are reported by reading grade level. The 5-number summaries for the 14 boys and 11 girls are shown:

Boys: 2.0 3.9 4.3 4.9 6.0
Girls: 2.8 3.8 4.5 5.2 5.9

a) Which group had the highest score?
b) Which group had the greatest range?
c) Which group had the greatest interquartile range?
d) Which group's scores appear to be more skewed? Explain.
e) Which group generally did better on the test? Explain.
f) If the mean reading level for boys was 4.2 and for girls was 4.6, what is the overall mean for the class?

38. Rainmakers? In an experiment to determine whether seeding clouds with silver iodide increases rainfall, 52 clouds were randomly assigned to be seeded or not. The amount of rain they generated was then measured (in acre-feet).

	n	Mean	Median	SD	IQR	Q1	Q3
Unseeded	26	164.59	44.20	278.43	138.60	24.40	163
Seeded	26	441.98	221.60	650.79	337.60	92.40	430

a) Which of the summary statistics are most appropriate for describing these distributions. Why?
b) Do you see any evidence that seeding clouds may be effective? Explain.

T 39. Phone calls. In an advertisement in *USA Today* (July 9, 2001), the company Net2Phone listed its long distance rates to 24 of the 250 countries to which it offers service.

Country	Cost per Minute (cents)	Country	Cost per Minute (cents)
Belgium	7.9	Italy	9.9
Chile	17.0	Japan	7.9
Canada	3.9	Mexico	16.0
Colombia	9.9	Pakistan	49.0
Dominican Republic	15.0	Philippines	21.0
		Puerto Rico	6.9
Finland	9.9	Singapore	11.0
France	7.9	South Korea	9.9
Germany	7.9	Taiwan	9.9
Hong Kong	7.9	United Kingdom	7.9
India	49.0	United States	3.9
Ireland	7.9	Venezuela	22.0
Israel	8.9		

a) Make a display of these rates.
b) Find the mean and the median. Which is a more appropriate measure of center? Why?
c) Find the IQR and the standard deviation. Which is the more appropriate measure of spread? Why?
d) Would you consider any of these to be outliers? Carefully explain how you reached your decision.
e) Write a brief description of these rates. Don't forget to mention shape, center, and spread as well as any unusual features of the distribution.
f) What can you conclude about Net2Phone's rates to the 250 countries the ad says they service?

T 40. Job growth. In 1996 the firm Standard and Poor's DRI predicted that the cities listed below would experience the fastest growing job markets in the United States over the next 3 years and predicted their growth rates, given here.

a) Make a suitable display of the growth rates.
b) Summarize the central growth rate with a median and mean. Why do they differ?
c) Given what you know about the distribution, which of these measures does the better job of summarizing the growth rates? Why?
d) Summarize the spread of the growth rate distribution with a standard deviation and with an IQR.
e) Given what you know about the distribution, which of these measures does the better job of summarizing the growth rates? Why?

City	Growth (%)
Las Vegas, NV-AZ	3.72
Raleigh-Durham-Chapel Hill, NC	2.69
Austin-San Marcos, TX	2.64
Riverside-San Bernardino, CA	2.62
Boise, ID	2.61
Orlando, FL	2.51
Phoenix-Mesa, AZ	2.44
West Palm Beach-Boca Raton, FL	2.37
Sacramento, CA	2.26
Atlanta, GA	2.25
Sarasota-Bradenton, FL	2.22
Portland-Vancouver, OR-WA	2.16
Fort Lauderdale, FL	2.13
Charlotte-Gastonia-Rock Hill, NC-SC	2.07
Tucson, AZ	2.07
Vallejo-Fairfield-Napa, CA	2.02
Omaha, NE-IA	1.93
Salt Lake City-Ogden, UT	1.90
Albuquerque, NM	1.87
Fort Worth-Arlington, TX	1.86

f) Suppose we subtract from each of these growth rates the predicted U.S. average growth rate of 1.20%, so that we could look at how much these growth rates exceed the U.S. rate. How would this change the values of the summary statistics you calculated above? (*Hint:* You need not recompute any of the summary statistics from scratch.)
g) If we were to omit Las Vegas from the data, how would you expect the mean, median, standard deviation, and IQR to change? Explain your expectations for each.
h) Write a brief report about these growth rates.

T 41. Math scores. The National Center for Education Statistics reported 1999 average mathematics achievement scores for eighth graders in 38 nations. Singapore led the group, with an average score of 604, while South Africa had the lowest average of 275. The United States scored 502. The average scores for each nation are given below.

604 587 585 582 579 558 540 534 532 531
530 526 525 520 520 519 511 505 502 496
491 482 479 476 472 469 467 466 448 447
429 428 422 403 392 345 337 275

a) Find the 5-number summary, the IQR, the mean, and the standard deviation of these national averages.
b) Write a brief summary of the performance of eighth graders worldwide. Be sure to comment on the performance of the United States.

T 42. Prisons. A report from the U.S. Department of Justice gave the following percent increases in federal prison populations in 20 northeastern and midwestern states during 1999.

5.9, 1.3, 3.0, 5.9, 4.5, 5.6, 2.1, 6.3, 4.8, 6.9,
5.5, 5.3, 8.0, 4.4, 7.2, 3.2, 4.5, 3.5, 7.2, 6.4

a) Graph these data.
b) Calculate appropriate summary statistics.
c) Write a few sentences about these data. (Remember: shape, center, spread, unusual features.)

T 43. Gasoline usage. The U.S. Department of Transportation collects data on the amount of gasoline sold in each state. The following data show the per capita (gallons used per person) consumption in the year 2000. Using appropriate graphical displays and summary statistics, write a report on the gasoline use by state in the year 2000.

AL	544.71	LA	522.12	OH	574.83
AK	433.08	ME	542.36	OK	457.63
AZ	452.82	MD	452.82	OR	520.42
AR	532.96	MA	438.10	PA	441.44
CA	422.65	MI	502.77	RI	410.31
CO	461.90	MN	528.06	SC	381.86
CT	431.04	MS	559.29	SD	555.06
DE	481.45	MO	563.56	TN	586.58
FL	542.36	MT	548.50	TX	515.17
GA	452.82	NE	508.28	UT	498.66
HI	327.27	NV	446.17	VT	456.27
ID	500.34	NH	542.86	VA	584.03
IL	406.66	NJ	474.28	WA	506.92
IN	518.70	NM	474.28	WV	450.40
IA	534.70	NY	551.18	WI	462.00
KS	511.34	NC	296.66	WY	462.67
KY	510.90	ND	513.30		

T 44. Industrial experiment. Engineers at a computer production plant tested two methods for accuracy in drilling holes into a PC board. They tested how fast they could set the drilling machine by running 10 boards at each of two different speeds. To assess the results, they measured the distance (in inches) from the center of a target on the board to the center of the hole. The data and summary statistics are shown in the table:

	Fast	Slow
	0.000101	0.000098
	0.000102	0.000096
	0.000100	0.000097
	0.000102	0.000095
	0.000101	0.000094
	0.000103	0.000098
	0.000104	0.000096
	0.000102	0.975600
	0.000102	0.000097
	0.000100	0.000096
Mean	0.000102	0.097647
StdDev	0.000001	0.308481

Write a report summarizing the findings of the experiment. Include appropriate visual and verbal displays of the distributions, and make a recommendation to the engineers if they are most interested in the accuracy of the method.

45. Customer database. A philanthropic organization has a database of millions of donors that they contact by mail to raise money for charities. One of the variables in the database, *Title,* contains the title of the person or persons printed on the address label. The most common are Mr., Ms., Miss, and Mrs., but there are also Ambassador and Mrs., Your Imperial Majesty, and Cardinal to name a few others. In all there are over 100 different titles, each with a corresponding numeric code. Here are a few of them:

Code	Title
000	MR.
001	MRS.
002	MR. and MRS.
003	MISS
004	DR.
005	MADAME
006	SERGEANT
009	RABBI
010	PROFESSOR
126	PRINCE
127	PRINCESS
128	CHIEF
129	BARON
130	SHEIK
131	PRINCE AND PRINCESS
132	YOUR IMPERIAL MAJESTY
1035	M. ET MME.
1210	PROF.

An intern who was asked to analyze the organization's fundraising efforts presented these summary statistics for the variable *Title:*

Mean	54.41
StdDev	957.50
Median	1
IQR	2
n	94649

a) What does the mean of 54.41 mean?
b) What are the typical reasons that cause measures of center and spread to be as different as those in this table?
c) Is that why these are so different?

46. Zip codes revisited. Here are some summary statistics to go with the histogram of the zip codes of 500 customers from the Holes-R-Us Internet Jewelry Salon that we saw in Exercise 29 of Chapter 4.

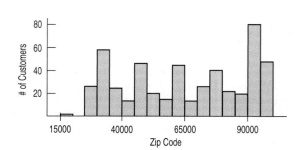

Count	500
Mean	64970.0
StdDev	23523.0
Median	64871
IQR	44183
Q1	46050
Q3	90233

a) Is the mean or median a "better" summary of the center of the zip code distribution? Why?
b) Is the standard deviation or the IQR a better summary of the spread? Why?
c) What can these statistics tell you about the company's sales?

47. Eye and hair color. A survey of 1021 school-age children was conducted by randomly selecting children from several large urban elementary schools. Two of the questions concerned eye and hair color. In the survey, the following codes were used:

Hair Color	Eye Color
1 = Blond	1 = Blue
2 = Brown	2 = Green
3 = Black	3 = Brown
4 = Red	4 = Grey
5 = Other	5 = Other

The Statistics students analyzing the data were asked to study the relationship between eye and hair color. They produced this plot:

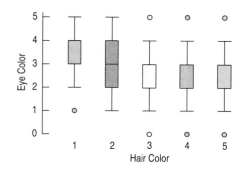

Is their graph appropriate? If so, summarize the findings. If not, explain why not.

48. Stereograms. Stereograms appear to be composed entirely of random dots. However, they contain separate images that a viewer can "fuse" into a three-dimensional (3D) image by staring at the dots while defocusing the eyes. An experiment was performed to determine whether knowledge of the embedded image affected the time required for subjects to fuse the images. One group of subjects (group NV) received no information or just verbal information about the shape of the embedded object. A second group (group VV) received both verbal information and visual information (specifically, a drawing of the object). The experimenters measured how many seconds it took for the subject to report that he or she saw the 3D image.

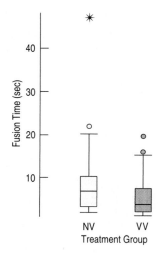

a) What two variables are discussed in this description?
b) For each variable, is it quantitative or categorical? If quantitative, what are the units?
c) Here are boxplots comparing the fusion times for the two treatment groups. Write a few sentences comparing these distributions. What does the experiment show?

49. Stereograms, revisited. Because of the skewness of the distributions of fusion times, we might consider a re-expression. On the next page are the boxplots of the *log* of fusion times. Is it better to analyze the original fusion times or the log fusion times? Explain.

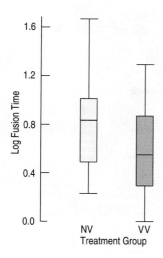

50. Stereograms, yet again. Here are the boxplots of the *reciprocal* of fusion times. Is it better to analyze the original fusion times, the log fusion times, or the reciprocal? Explain.

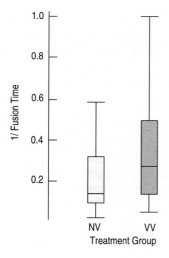

CHAPTER 6

The Standard Deviation as a Ruler and the Normal Model

The women's heptathlon in the Olympics consists of seven track and field events: the 200-m and 800-m runs, 100-m high hurdles, shot put, javelin, high jump, and long jump. Somehow, the performances in all seven events have to be combined into one score. How can performances in such different events be compared? They don't even have the same units; the races are recorded in seconds and the throwing and jumping events in meters. In the 2000 Olympics, the best 800-m time, run by Getrud Bacher of Italy, was 8 seconds faster than the mean. The winning long jump by the Russian Yelena Prokhorova was 60 centimeters longer than the mean. Which performance deserves more points?

You might think at first that this is a silly question, but it's exactly the kind of question that has to be answered to compute heptathlon scores. And comparisons of values with different units show up in many places.

The Standard Deviation as a Ruler

Grading on a Curve

If you score 79% on an exam, what grade should you get? One teaching philosophy looks only at the raw percentage, 79, and bases the grade on that alone. Another looks at your *relative* performance and bases the grade on how you did compared with the rest of the class. Teachers and students still debate which method is better.

The trick in comparing very different-looking values is to use standard deviations. The standard deviation tells us how the whole collection of values varies, so it's a natural ruler for comparing an individual value to the group. Over and over during this course, we will ask questions such as "How far is this value from the mean?" or "How different are these two statistics?" The answer in every case will be to measure the distance or difference in standard deviations.

The concept of the standard deviation as a ruler is not special to this course. You'll find statistical distances measured in standard deviations throughout Statistics, up to the most advanced levels.[1] This approach is one of the basic tools of statistical thinking.

---

[1] Other measures of spread could be used as well, but the standard deviation is the most common measure, and it is almost always used as the ruler.

Bacher's winning 800-m time of 129 seconds was 8 seconds faster than the mean of 137 seconds. How many *standard deviations* better than the mean is that? The standard deviation of all 27 qualifying times was 5 seconds, so her time was $(129 - 137)/5 = -8/5 = -1.6$, or 1.6 *standard deviations* better than the mean. Prokhorova's winning long jump was 60 cm *longer* than the average 6-m jump. The standard deviation was 30 cm, so the winning jump was $(60/30) = 2$ standard deviations better than the mean. The long jump performance was better because it was a greater improvement over its mean than was the winning 800-m time, as measured in *standard deviations*.

Standardizing with z-scores

To understand how an athlete performed in a heptathlon event, we *standardized* her result, finding how many standard deviations from the event mean she performed. To do this, we took her event result and first subtracted the mean of *all* the event results from it. Then, we divided this difference by the standard deviation of the event results. We can write this calculation more simply as:

$$z = \frac{(y - \bar{y})}{s}.$$

We call the resulting values **standardized values,** and denote them with the letter z. Usually we just call them **z-scores.**

Standardized values have no units because z-scores measure the distance of each data value from the mean in standard deviations. A z-score of 2 tells us that a data value is 2 standard deviations above the mean. It doesn't matter whether the original variable was measured in inches, dollars, or seconds. Data values below the mean have negative z-scores, so a z-score of -1.6 means that the data value was 1.6 standard deviations *below* the mean. Of course, regardless of the direction, the farther a data value is from the mean, the more unusual it is, so a z-score of -3 is more surprising than a z-score of 1.2.

By using the standard deviation as a ruler to measure statistical distance from the mean, we can compare values that are measured on different variables, with different scales, with different units, or for different populations. To determine the winner of the heptathlon, the judges must combine performances on seven very different events. Although they actually use predetermined tables, they could combine scores by standardizing each, and then adding the z-scores together to reach a total score. The only trick is that they'll have to switch the sign of the z-score for running events, because unlike throwing and jumping, it's better to have a running time below the mean (have a negative z-score).

For example, suppose we want to *combine* the scores that Bacher and Prokhorova earned in the long jump and the 800-m run. Prokhorova won the long jump and Bacher won the 800. So, are they even after these two events? Shouldn't it matter how well they did in each event, and not just who won? Even though one event is measured in seconds and the other in meters, we can use the z-scores to combine their scores on the two events. In the long jump, Bacher jumped 5.84 m. The mean for all competitors in the long jump was 5.98 m and the standard deviation was 0.32 m, so her z-score for that event is $(5.84 - 5.98)/0.32 = -0.44$. Bacher ran the 800-m in 129.08 seconds. That's 1.59 standard deviations lower (better) than the mean of all the 800-m results, making her z-score for that event -1.59. Because

NOTATION ALERT:

There goes another letter. We always use the letter z to denote values that have been standardized with the mean and standard deviation.

A S　**Changing the Baseline.**
What happens when we shift data? Do measures of center and spread change? Find out by watching it happen.

A S　**Changing the Units.**
Here's a chance to change the center and spread values for a distribution and watch the summaries change (or not, as the case may be). Investigate here.

lower is better for runs, we'll switch signs and give her a score of +1.59 for her 800-m run. What's her total score for the two events? We just combine the 1.59 from running and the −0.44 for jumping to get a total score of 1.59 − 0.44 = 1.15 for the two events.

On the other hand, Prokhorova ran the 800-m in 130.32 seconds—slower than Bacher, but still 1.34 standard deviations faster than the mean 800-m time (a z-score of −1.34), and her winning long jump of 6.59 m was 1.94 standard deviations better than the mean long jump distance. So her total score is 1.34 + 1.94 = 3.28 for the two events, quite a bit better than Bacher for the two events combined. Is this the result we wanted? Yes! Prokhorova ran nearly as fast as Bacher in the 800 (and finished third), but Bacher's jump was near the bottom of the pack. The z-scores take into account how far each result is from the event mean in standard deviation units. And because they are both in standard deviation units, we can combine them. Not coincidentally, Prokhorova went on to win the silver medal, while Bacher finished 14th.

When we standardize data to get a z-score, we do two things. First, we shift the data by subtracting the mean. Then, we rescale the values by dividing by their standard deviation. We often shift and rescale data. What happens to the grade distribution if *everyone* gets a five-point bonus? If we switch from feet to meters, what happens to the distribution of heights of students in your class? Before we can talk more about using z-scores, we need to see exactly how shifting and rescaling work.

A S **Standardizing.** What if we both shift and rescale? The result is so nice that we give it a name. See it work here.

just checking

1 Your Statistics teacher has announced that the lower of your two midterm tests will be dropped. You got a 90 on midterm 1 and an 80 on midterm 2. You're all set to drop the 80 until she announces that she grades "on a curve." She standardized the scores in order to decide which is the lower one. If the mean on the first midterm is 88 with a standard deviation of 4 and the mean on the second was a 75 with a standard deviation of 5,
   a) Which one will be dropped?
   b) Does this seem "fair"?

Shifting Data

WHO	80 male participants of the NHANES survey between the ages of 19 and 24 who measured between 68 and 70 inches tall
WHAT	Their weights
UNIT	Kilograms
WHEN	2001–2002
WHERE	United States
WHY	To study nutrition and health issues and trends
HOW	National survey

Since the 1960s the Centers for Disease Control's National Center for Health Statistics (NCHS) has been collecting health and nutritional information on people of all ages and backgrounds. A recent survey, the National Health and Nutrition Examination Survey (NHANES) 2001–2002, measured a wide variety of variables, including body measurements, cardiovascular fitness, blood chemistry, and demographic information on more than 11,000 individuals.

Included in this group were 80 men between 19 and 24 years old of average height (between 5′8″ and 5′10″ tall). Here are a histogram and boxplot of their weights:

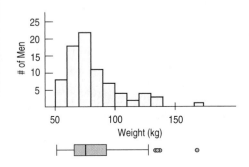

Histogram and boxplot for the men's weights. The shape is skewed to the right with several high outliers. **Figure 6.1**

Their mean weight is 82.36 kg. For this age and height group, the National Institutes of Health recommends a maximum healthy weight of 74 kg, but we can see that some of the men are heavier than the recommended weight. To compare their weights to the recommended maximum, we could subtract 74 kg from each of their weights. What would that do to the center, shape and spread of the histogram? Here's the picture:

Doctors' height and weight charts sometimes give ideal weights for various heights that include 2-inch heels. If the mean height of adult women is 66 inches including 2-inch heels, what is the mean height of women without shoes? Each woman is shorter by 2 inches when barefoot, so the mean is decreased by 2 inches, to 64 inches.

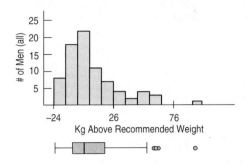

Subtracting 74 kilograms shifts the entire histogram down, but leaves the spread and the shape exactly the same. **Figure 6.2**

On average, they weigh 82.36 kg, so on average they're 8.36 kg overweight. It's not surprising that the mean of the distribution after subtracting 74 kg from each man's weight is 8.36 kg. In fact, when adding (or subtracting) a constant to each value, all measures of position (center, percentiles, min, max) will increase (or decrease) by the same constant.

What about the spread? What does adding or subtracting a constant value do to the spread of the distribution? Look at the two histograms again. Adding or subtracting a constant changes each data value equally, so the entire distribution just shifts. Its shape doesn't change and neither does the spread. None of the measures of spread we've discussed—not the range, not the IQR, not the standard deviation—changes.

> *Adding (or subtracting) a constant to every data value adds (or subtracts) the same constant to measures of position, but leaves measures of spread unchanged.*

Rescaling Data

Not everyone thinks naturally in metric units. Suppose we want to look at the weights in pounds instead. We'd have to **rescale** the data. Because there are about 2.2 pounds in every kilogram, we'd convert the weights by multiplying each value by 2.2. Multiplying or dividing each value by a constant changes the measurement units. Here are histograms of the two distributions, plotted on the same scale, so you can see the effect of multiplying:

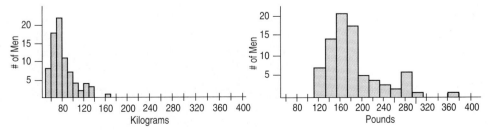

Men's weights in both kilograms and pounds. How do the distributions and numerical summaries change? **Figure 6.3**

What happens to the mean weight? Not too surprisingly, it gets multiplied by 2.2 as well. The mean weight of 82.36 kg is 82.36 × 2.2 = 181.19 lb. All the measures of position get multiplied by 2.2. What happens to the spread? Take a look at the histograms. Clearly the spread in pounds is larger. Consider the range, for example. The min weight of 54.30 kg rescales to 119.46 lb. The max weight of 161.50 kg rescales to 355.30 lb. So, the range, which was 107.20 in kg, also rescales to 107.20 kg × 2.2 lb/kg = 235.84 lb. Of course, it's not really "larger." It's just that the units have changed. We can write:

$$New\ range = max \times 2.2 - min \times 2.2 = (max - min) \times 2.2 = range \times 2.2$$

What about the other measures of spread? The same thing happens to them and for much the same reasons. By the way, the maximum recommended weight of 74 kg is 162.8 lb.

> When we multiply (or divide) all the data values by any constant, all measures of position (such as the mean, median, and percentiles) and measures of spread (such as the range, the IQR, and the standard deviation) are multiplied (or divided) by that same constant.

just checking

2 In 1995 the Educational Testing Service (ETS) adjusted the scores of SAT tests. Before ETS recentered the SAT Verbal test, the mean of all test scores was 450.
   a) How would adding 50 points to each score affect the mean?
   b) The standard deviation was 100 points. What would the standard deviation be after adding 50 points?
   c) Suppose we drew boxplots of test takers' scores a year before and a year after the re-centering. How would the boxplots of the two years differ?

3 A company manufactures wheels for roller blades. The diameter of the wheels has a mean of 3 inches and a standard deviation of 0.1 inches. Because so many of its customers use the metric system, the company decided to report their production statistics in millimeters (1 inch = 25.4 mm). They report that the standard deviation is now 2.54 mm. A corporate executive is worried about this increase in variation. Should they be concerned? Explain.

Back to *z*-scores

Standardizing data into *z*-scores is just shifting them by the mean and rescaling them by the standard deviation. Now we can see how standardizing affects the distribution. When we subtract the mean of the data from every data value, we shift the mean to zero. As we have seen, such a shift doesn't change the standard deviation.

When we *divide* each of these shifted values by *s*, however, the standard deviation should be divided by *s* as well. Since the standard deviation was *s* to start with, the new standard deviation becomes 1.

How, then, does standardizing affect the distribution of a variable? Let's consider the three aspects of a distribution: the shape, center, and spread.

> *z*-scores have mean 0 and standard deviation 1.

- *Standardizing into z-scores does not change the **shape** of the distribution of a variable.*
- *Standardizing into z-scores changes the **center** by making the mean 0.*
- *Standardizing into z-scores changes the **spread** by making the standard deviation 1.*

Working with Standardized Variables Step-By-Step

Many colleges and universities require applicants to submit scores on standardized tests such as the SAT Math and Verbal tests. The college your little sister wants to apply to says that while there is no minimum score required, the middle 50% of their students have combined SAT-I scores between 1020 and 1220. You'd feel confident if you knew her score was in their top 25%, but unfortunately she took the ACT test, an alternative standardized test. How high does her ACT need to be to make it into the top quarter of equivalent SAT scores?

To answer that question you'll have to standardize all the scores, so you'll need to know the mean and standard deviations of scores for some group on both tests. The college doesn't report the mean or standard deviation for their applicants on either test, so we'll use the group of all test takers nationally. For college-bound seniors, the average combined SAT score is about 1000 and the standard deviation is about 200 points. For the same group, the ACT average is 20.8 with a standard deviation of 4.8.

Think

Plan State what you want to find out.

Variables Identify the variables and report the W's.

I want to know what ACT score corresponds to the upper quartile SAT score. I know the mean and standard deviation for both the SAT and ACT scores based on all test takers, but I have no individual data values.

Check the appropriate conditions.

✔ **Quantitative Variable Condition:** Scores for both tests are quantitative but have no meaningful units other than points.

Show

Mechanics Standardize the variables.

The middle 50% of SAT scores at this college fall between 1020 and 1220 points. To be in the top quarter, my sister would have to have a score of at least 1220. That's a z-score of

$$z = \frac{(1220 - 1000)}{200} = 1.10$$

The *y*-value that is *z* standard deviations above the mean is $y = \mu + z\sigma$.

So an SAT score of 1220 is 1.10 standard deviation above the mean of all test takers.

For the ACT, 1.10 standard deviation above the mean is $20.8 + 1.10(4.8) = 26.08$.

Tell

Conclusion Interpret your results in context.

To be in the top quarter of applicants in terms of combined SAT score, she'd need to have an ACT score of at least 26.08.

When Is a z-score BIG?

A z-score gives us an indication of how unusual a value is because it tells us how far it is from the mean. If the data value sits right at the mean, it's not very far at all and its z-score is 0. A z-score of 1 tells us the data value is 1 standard deviation above the mean, while a z-score of −1 tells us that the value is 1 standard deviation below the mean. How far from 0 does a z-score have to be to be interesting or

A S **The Normal Model.**
Learn more about the Normal model and see what data drawn at random from a Normal model might look like.

NOTATION ALERT:

$N(\mu, \sigma)$ always denotes a Normal model. The μ, pronounced "mew," is the Greek letter for "m," and always represents the mean in a model. The σ is the lowercase Greek letter for "s," sigma, and always represents the standard deviation in a model.

σ, not 6
If your σ's look like 6's, here's a trick for writing the symbol σ. Start from the inside top and go clockwise, making sure the top is flat and parallel to the line you're writing on. Try it!

"All models are wrong—but some are useful."
—George Box, famous statistician

unusual? There is no universal standard, but the larger the score is (negative or positive) the more unusual it is. We know that 50% of the data lie between the quartiles. For symmetric data, the standard deviation is usually a bit smaller than the IQR, and it's not uncommon for at least half of the data to have z-scores between −1 and 1. But no matter what the shape of the distribution, a z-score of 3 (plus or minus) or more is rare, and a z-score of 6 or 7 shouts out for attention.

To say more about how big we expect a z-score to be, we need to *model* the data's distribution. A model will let us say much more precisely how often we'd be likely to see z-scores of different sizes. Of course, like all models of the real world, the model will be wrong—wrong in the sense that it can't match reality exactly. But it can still be useful. Like a physical model, it's something we can look at and manipulate in order to learn more about the real world.

Models help our understanding in many ways. Just as a model of an airplane in a wind tunnel can give insights even though it doesn't show every rivet,[2] models of data give us summaries that we can learn from and use even though they don't fit each data value exactly. It's important to remember that they're only models of reality and not reality itself. But without models, what we can learn about the world at large is limited to only what we can say about the data we have at hand.

There is no universal standard for z-scores, but there is a model that shows up over and over in Statistics. You may have heard of "bell-shaped curves." Statisticians call them Normal models. **Normal models** are appropriate for distributions whose shapes are unimodal and roughly symmetric. For these distributions, they provide a measure of how extreme a z-score is. Fortunately, there is a Normal model for every possible combination of mean and standard deviation. We write $N(\mu, \sigma)$ to represent a Normal model with a mean of μ and a standard deviation of σ. Why the Greek? Well, *this* mean and standard deviation are not numerical summaries of data. They are part of the model. They don't come from the data. Rather they are numbers that we choose to help specify the model. Such numbers are called **parameters** of the model.

We don't want to confuse the parameters with summaries of the data such as $\bar{y}$ and s, so we use special symbols. In Statistics, we almost always use Greek letters for parameters. By contrast, summaries of data are called **statistics** and are usually written with Latin letters.

If we model data with a Normal model and standardize them using the corresponding μ and σ, we still call the standardized value a **z-score,** and we write

$$z = \frac{y - \mu}{\sigma}.$$

Usually it's easier to standardize data first (using its mean and standard deviation). Then we need only the model $N(0,1)$. The Normal model with mean 0 and standard deviation 1 is called the **standard Normal model** (or the **standard Normal distribution**).

But be careful. You shouldn't use a Normal model for just any data set. Remember that standardizing won't change the shape of the distribution. If the distribution is not unimodal and symmetric to begin with, standardizing won't make it Normal.

---

[2] In fact, the model is useful *because* it doesn't have every rivet. It is because models offer a simpler view of reality that they are so useful as we try to understand reality.

Is the Standard Normal a standard?

Yes. We call it the "Standard Normal," because it models standardized values. It is also a "standard" because this is the particular Normal model that we almost always use.

When we use the Normal model, we assume that the distribution of the data is, well, Normal. Practically speaking, there's no way to check whether this is true. In fact, it almost certainly is not true. Real data don't behave like mathematical models. Models are idealized; real data are real. The good news, however, is that to use a Normal model, it's sufficient to check the following condition:

Nearly Normal Condition. The shape of the data's distribution is unimodal and symmetric. Check this by making a histogram, or a Normal probability plot (which we'll explain later).

Never use the Normal model without checking whether the condition is satisfied.

The 68-95-99.7 Rule

One in a Million

These magic 68, 95, 99.7 values come from the Normal model. As a model, it can give us corresponding values for any *z*-score. For example, it tells us that fewer than 1 out of a million values have *z*-scores smaller than −5.0 or larger than 5.0. So if someone tells you you're "one in a million" they must really admire your *z*-score.

Normal models give us an idea of how extreme a value is by telling us how likely it is to find one that far from the mean. We'll soon show how to find these numbers precisely—but one simple rule is usually all we need.

It turns out that in a Normal model, about 68% of the values fall within 1 standard deviation of the mean, about 95% of the values fall within 2 standard deviations of the mean, and about 99.7%—almost all—of the values fall within 3 standard deviations of the mean. These facts are summarized in a rule that we call (let's see . . .) the **68-95-99.7 Rule.**[3]

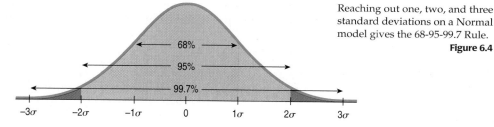

Reaching out one, two, and three standard deviations on a Normal model gives the 68-95-99.7 Rule.

Figure 6.4

just checking

4 As a group, the Dutch are among the tallest people in the world. The average Dutch man is 184 cm tall—just over 6 feet (and the average Dutch woman is 170.8 cm tall—just over 5'7"). If a Normal model is appropriate and the standard deviation for men is about 8 cm, what percentage of all Dutch men will be over 2 meters (6'6") tall?

5 Suppose it takes you 20 minutes, on average, to drive to campus, with a standard deviation of 2 minutes. Suppose a Normal model is appropriate for the distributions of driving times.
   a) How often will you arrive on campus in less than 22 minutes?
   b) How often will it take you more than 24 minutes?
   c) Do you think the distribution of your driving times is unimodal and symmetric?
   d) What does this say about the accuracy of your predictions? Explain.

---

[3] This rule is called the "Empirical Rule" in older texts, because it originally came from observation. The rule was first published by Abraham De Moivre in 1733, 75 years before the Normal model was discovered. Maybe it should be called "De Moivre's Rule," but that wouldn't help us remember the important numbers, 68, 95, and 99.7.

The First Three Rules for Working with Normal Models

Make a picture.

Make a picture.

Make a picture.

Although we're thinking about models, not histograms of data, the three rules don't change. To help you think clearly, a simple hand-drawn sketch is all you need. Even experienced statisticians sketch pictures to help them think about Normal models. You should too.

Of course, when we have data, we'll also need to make a histogram to check the Nearly Normal Condition to be sure we can use the Normal model to model the data's distribution. Other times, we may be told that a Normal model is appropriate based on prior knowledge of the situation or on theoretical considerations.

Working with the 68-95-99.7 Rule Step-By-Step

The SAT-I test comprises both a Verbal and a Math test. Each test has a distribution that is roughly unimodal and symmetric and is designed to have an overall mean of about 500 and a standard deviation of 100 for all test takers. In any one year, the mean and standard deviation may differ from these target values by a small amount, but they are a good overall approximation. Suppose you earned a 600 on one part of your SAT test. Where do you stand among all students who took that SAT? You could calculate your z-score and find out that it's $z = (600 - 500)/100 = 1.0$, but what does that tell you about your percentile? You'll need the Normal model and the 68-95-99.7 Rule to answer that question.

Think	**Plan** State what you want to know.	I want to see how my SAT score compares with all other students. To do that, I'll need to model the distribution.
	Variables Identify the variables and report the W's.	Let y = my SAT score. Scores are quantitative but have no meaningful units other than points.
	Be sure to check the appropriate assumptions and conditions.	✔ **Nearly Normal Condition:** If I had data, I would check the histogram. I have no data but I am told that the SAT scores are roughly unimodal and symmetric.
	Specify the parameters of your model.	I will model SAT score with a N(500,100) model.
Show	**Mechanics** Make a picture of this Normal model. (A simple sketch is all you need.)	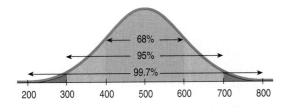
	Locate your score.	My score of 600 is 1 standard deviation above the mean. That corresponds to one of the points of the 68-95-99.7% Rule.

Tell   **Conclusion**   Interpret your result in context.

> About 68% of those who took the test had scores that fell no more than 1 standard deviation from the mean, so 100% − 68% = 32% of all students had scores more than 1 standard deviation away. Only half of those were on the high side, so about 16% (half of 32%) of the test scores were better than my score of 600.

The bounds of SAT scoring at 200 and 800 can also be explained by the 68-95-99.7 Rule. Since 200 and 800 are three standard deviations from 500, it hardly pays to extend the scoring any farther on either side. We'd get more information only on 100 − 99.7 = 0.3% of students.

Finding Normal Percentiles by Hand

An SAT score of 600 is easy to assess, because we can think of it as one standard deviation above the mean. If your score was 680, though, where do you stand among the rest of the people tested? Your z-score is 1.80, so you're somewhere between 1 and 2 standard deviations above the mean. We figured out that no more than 16% of people score better than 600. By the same logic, no more than 2.5% of people score better than 700. Can we be more specific than "between 16% and 2.5%"?

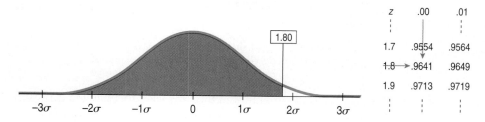

z	.00	.01
⋮	⋮	⋮
1.7	.9554	.9564
1.8 →	.9641	.9649
1.9	.9713	.9719
⋮	⋮	⋮

A table of Normal percentiles (Table Z in Appendix E) lets us find the percentage of individuals in a Standard Normal distribution falling below any specified z-score value.   **Figure 6.5**

When the value doesn't fall exactly 1, 2, or 3 standard deviations from the mean, we can look it up in a table of **Normal percentiles.**[4] Nobody wants to lug around a bunch of tables, so we make a table only for the *standard Normal.* That means we have to convert our data to z-scores before using the table. Your score of 680 has a z-score of (680 − 500)/100 = 1.80. In the piece of the table shown, we find your z-score by looking down the left column for the first two digits, 1.8, and across the top row for the third digit, 0. The table gives the percentile as 0.9641. That means that 96.4% of the z-scores are less than 1.80. Only 3.6% of people, then, scored better than 680 on the SAT.

Finding Normal Percentiles Using Technology

A S **Working with Normal Models.** Well, actually playing with them. This interactive tool lets you do what this chapter's figures can't do, because they don't move when you push on them!

These days, finding percentiles from a Normal probability table is a "desert island" method—something we might do if we desperately needed a Normal per-

---

[4] See Table Z in Appendix E. Many calculators and statistics computer packages do this, too.

centile and were stranded miles from the mainland with only a Normal probability table. (Of course, you might feel just that way during a Statistics exam, so it's a good idea to know how to do it.) Fortunately, most of the time, we can use a calculator or computer. Graphing calculators, such as the TI-84, find Normal percentiles and offer to draw the picture as well. And most statistics programs have functions to find Normal percentiles.

The *ActivStats* Multimedia Assistant provided on the CD-ROM that accompanies this book offers two methods. The "Normal Model Tool" introduced along with the Normal model makes it easy to see how areas under parts of the Normal model correspond to particular cut points. The tool is especially useful for problems in which you want to find the area *between* two z-scores.

The tool also allows you to work in the original units if you wish without converting everything to z-scores. This can be particularly helpful for understanding statements about relative frequencies.

ActivStats also offers a Normal table in which the picture of the Normal model is interactive.[5] Grab the z-score cut point with your mouse and drag it to the value you are interested in. The table will adjust accordingly.

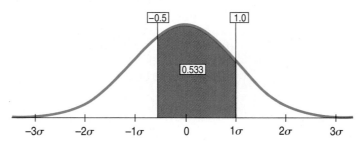

The *ActivStats* Normal model tool makes it easy to see how areas under parts of the Normal model correspond to particular cut points.

Figure 6.6

With a tool like the *ActivStats* Normal model tool, it's easy to answer a question like what percentage of all test takers score between 950 and 1080 on the combined SAT. You can leave the units as they are and just enter the mean and standard deviation on the Normal model tool, sliding the cut points to 950 and 1080. Alternatively, you could convert both values to z-scores and slide the area to match those scores on the Standard Normal model tool. In either case, you would read the percentage right off the picture.

The Normal model is our first model for data. It's the first in a series of modeling situations where we step away from the data at hand to make more general statements about the world. We'll become more practiced in thinking about and learning the details of models as we progress through the book. To give you some practice in thinking about the Normal model, here are several problems that ask you to find percentiles in detail. While you won't need the exact skills used in these problems in the rest of this book, they provide good problem-solving practice and will help you to understand the Normal model better.

---

[5] Now it's time to open the CD that accompanies the book if you haven't done so already. You'll find the instructions for using the tools there. You can also get to the tools directly in the Appendix following the last chapter of the *ActivStats* Lesson Book.

A S Your Pulse z-Score. Is your pulse rate high or low? Find its z-score with the Normal Model Tool.

A S The Normal Table. Table Z just sits there, but this version of the Normal table changes so it always Makes a Picture that fits. A great way to learn to use the table.

A S Normal Models. Normal models have several interesting properties—see them here.

Working with Normal Models I Step-By-Step

What proportion of SAT scores fall between 450 and 600?

Think

Plan State the problem.

I want to know the proportion of SAT scores between 450 and 600.

Variables Name the variable.

Let y = SAT score.

Check the appropriate conditions and specify which Normal model to use.

✔ **Nearly Normal Condition:** We are told that SAT scores are nearly Normal.

I'll model SAT scores with a N(500,100) model, using the mean and standard deviation specified for them.

Show

Mechanics Make a picture of this Normal model. Locate the desired values and shade the region of interest.

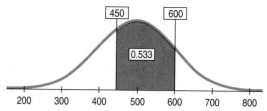

I can see directly from this picture that about 53.3% of scores fall between 450 and 600 under the Normal model.

Locate the desired values. The easiest way is to use a technology that allows you to input the raw data values directly.

Alternatively, you can convert 450 and 600 to z-scores before using technology or a table to find the two areas. Then you need to subtract them to find the area *between* the two values.

Standardizing the two scores, I find

$$z = \frac{(y - \mu)}{\sigma} = \frac{(600 - 500)}{100} = 1.00$$

and

$$z = \frac{(450 - 500)}{100} = -0.50$$

From Table Z, the area (z < 1.0) = 0.8413 and area (z < −0.5) = 0.3085, so the proportion of z-scores between them is 0.8413 − 0.3085 = 0.5328, or 53.28%.

Tell

Conclusion Interpret your result in context.

The Normal model estimates that about 53.3% of SAT scores fall between 450 and 600.

From Percentiles to Scores: *z* in Reverse

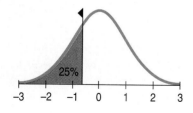

Finding areas from z-scores is the simplest way to work with the Normal model. But sometimes we start with areas and are asked to work backward to find the corresponding z-score or even the original data value.

For instance, what z-score represents the first quartile in a Normal model? In our first set of examples we knew the z-score and used the table or technology to find the percentile. Now we want to find the cut point for the 25th percentile.

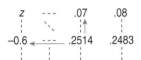

Make a picture like the one at the bottom of page 111, shading the leftmost 25% of the area. Look in Table Z for an area of 0.2500. The exact area is not there, but 0.2514 is pretty close. That shows up in the table with −0.6 in the left margin and .07 in the top margin. The z-score for Q1, then, is approximately z = −0.67.

Computers and calculators will determine the cut point more precisely (and more easily).

Working with Normal Models II Step-By-Step

Suppose a college says it admits only people with SAT scores among the top 10%. How high an SAT score does it take to be eligible?

Think **Plan** State the problem.

How high an SAT score do I need to be in the top 10% of all test takers?

Variables Define the variable.

Check to see if a Normal model is appropriate, and specify which Normal model to use.

Let y = my SAT score.

✔ **Nearly Normal Condition:** I am told that SAT scores are nearly Normal. I'll model them with N(500, 100).

Show **Mechanics** Make a picture of this Normal model. Locate the desired percentile approximately by shading the rightmost 10% of the area.

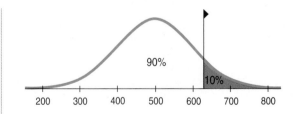

From technology the cutoff appears to be near 625. I'll also use Table Z:

The college takes the top 10%, so their cutoff score is the 90th percentile. Using Table Z, locate 0.90 (or as close to it as you can) in the *interior* of the table and find the corresponding z-score. Here the 1.2 is in the left margin and the .08 is in the margin above the entry. Putting them together gives 1.28.

The cut point is z = 1.28.

Convert the z-score back to the original units.

A z-score of 1.28 is 1.28 standard deviations above the mean. Since the SD is 100, that's 128 SAT points. The cutoff is 128 points above the mean of 500, or 628.

Tell **Conclusion** Interpret your results in the proper context.

Because the school wants SAT scores in the top 10%, the cutoff is 628. (Actually, since SAT scores are reported only in multiples of 10, I'd have to score at least a 630.)

More Working with Normal Models Step-By-Step

Working with Normal percentiles can be a little tricky, depending on how the problem is stated. Here are a few more worked examples of the kind you're likely to see.

A cereal manufacturer has a machine that fills the boxes. Boxes are labeled "16 ounces," so the company wants to have that much cereal in each box, but since no packaging process is perfect, there will be minor variations. If the machine is set at exactly 16 ounces and the Normal model applies (or at least the distribution is roughly symmetric), then about half of the boxes will be underweight, making consumers unhappy and exposing the company to bad publicity and possible lawsuits. To prevent underweight boxes, the manufacturer has to set the mean a little higher than 16.0 ounces.

Based on their experience with the packaging machine, the company believes that the amount of cereal in the boxes fits a Normal model with a standard deviation of 0.2 ounces. The manufacturer decides to set the machine to put an average of 16.3 ounces in each box. Let's use that model to answer a series of questions about these cereal boxes.

Question 1: What fraction of the boxes will be underweight?

Think **Plan** State the problem.

What proportion of boxes weigh less than 16 ounces?

Variable Name the variable.

Let y = weight of cereal in a box.

Check to see if a Normal model is appropriate.

✔ **Nearly Normal Condition:** I have no data, so I cannot make a histogram, but I am told that the company believes the distribution of weights from the machine is Normal.

Specify which Normal model to use.

I'll use a $N(16.3, 0.2)$ model.

Show **Mechanics** Make a picture of this Normal model. Locate the value you're interested in on the picture, label it, and shade the appropriate region.

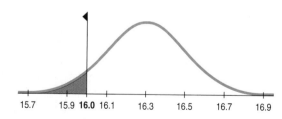

REALITY CHECK Estimate from the picture the percentage of boxes that are underweight. (This will be useful later to check that your answer makes sense.) It looks like a low percentage. Less than 20% for sure.

I want to know what fraction of the boxes will weigh less than 16 ounces.

Convert your cutoff value into a z-score.

$$z = \frac{y - \mu}{\sigma} = \frac{16 - 16.3}{0.2} = -1.50$$

Look up the area in the Normal table, or use technology.

Area $(y < 16)$ = Area $(z < -1.50)$ = 0.0668

Tell **Conclusion** State your conclusion, and check that it's consistent with your earlier guess. It's below 20%—seems okay.

I estimate that approximately 6.7% of the boxes will contain less than 16 ounces of cereal.

Question 2: The company's lawyers say that 6.7% is too high. They insist that no more than 4% of the boxes can be underweight. So the company needs to set the machine to put a little more cereal in each box. What mean setting do they need?

Think   **Plan** State the problem.

What mean weight will reduce the proportion of underweight boxes to 4%?

Variable Name the variable.

Let y = weight of cereal in a box.

Check to see if a Normal model is appropriate.

✔ **Nearly Normal Condition:** I am told that a Normal model applies.

Specify which Normal model to use. This time you are not given a value for the mean!

I don't know μ, the mean amount of cereal. The standard deviation for this machine is 0.2 ounces. The model is $N(\mu, 0.2)$.

REALITY CHECK We found out earlier that setting the machine to $\mu = 16.3$ ounces made 6.7% of the boxes too light. We'll need to raise the mean a bit to reduce this fraction.

No more than 4% of the boxes can be below 16 ounces.

Show   **Mechanics** Make a picture of this Normal model. Center it at μ (since you don't know the mean) and shade the region below 16 ounces.

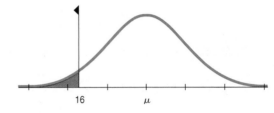

Using the Normal table or technology, find the z-score that cuts off the lowest 4%.

The z-score that has 0.04 area to the left of it is $z = -1.75$.

Use this information to find μ. It's located 1.75 standard deviations to the right of 16. Since σ is 0.2, that's 1.75×0.2, or 0.35 ounces more than 16.

For 16 to be 1.75 standard deviations below the mean, the mean must be

$$16 + 1.75 \times 0.2 = 16.35 \text{ ounces.}$$

Tell   **Conclusion** Interpret your result in context.

The company must set the machine to average 16.35 ounces of cereal per box.

(This makes sense; we knew it would have to be just a bit higher than 16.3.)

Question 3: The company president vetoes that plan, saying the company should give away less free cereal, not more. Her goal is to set the machine no higher than 16.2 ounces and still have only 4% underweight boxes. The only way to accomplish this is to reduce the standard deviation. What standard deviation must the company achieve, and what does that mean about the machine?

Think   **Plan** State the problem.

What standard deviation will allow the mean to be 16.2 ounces and still have only 4% of boxes underweight?

Variable Name the variable.

Let y = weight of cereal in a box.

Check conditions to be sure that a Normal model is appropriate.

✔ **Nearly Normal Condition:** The company believes that the weights are described by a Normal model.

Specify which Normal model to use. This time you don't know σ.

REALITY CHECK We know the new standard deviation must be less than 0.2 ounces.

I know the mean, but not the standard deviation, so my model is $N(16.2, \sigma)$.

Show **Mechanics** Make a picture of this Normal model. Center it at 16.2, and shade the area you're interested in. We want 4% of the area to the left of 16 ounces.

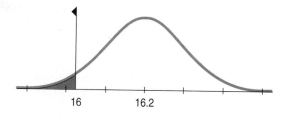

Find the z-score that cuts off the lowest 4%.

I know that the z-score with 4% below it is $z = -1.75$.

Solve for σ. (We need 16 to be 1.75 σ's below 16.2, so 1.75 σ must be 0.2 ounces. You could just start with that equation.)

$$z = \frac{y - \mu}{\sigma}$$

$$-1.75 = \frac{16 - 16.2}{\sigma}$$

$$1.75\,\sigma = 0.2$$

$$\sigma = 0.114$$

Tell **Conclusion** Interpret your result in context.

As we expected, the standard deviation is lower than before—actually, quite a bit lower.

The company must get the machine to box cereal with a standard deviation of only 0.114 ounces. This means the machine must be more consistent (by nearly a factor of 2) in filling the boxes.

Are You Normal? How Can You Tell?

In the examples we've just worked through, we've assumed that the underlying data distribution was roughly unimodal and symmetric, so that using a Normal model makes sense. When you have data, you must *check* to see whether a Normal model is reasonable. How? Make a picture, of course! Drawing a histogram of the data and looking at the shape is one good way to see if a Normal model might work.

There's a more specialized graphical display that can help you to decide whether the Normal model is appropriate: the **Normal probability plot.** If the distribution of the data is roughly Normal, the plot is roughly a diagonal straight line. Deviations from a straight line indicate that the distribution is not Normal. This plot is usually able to show deviations from Normality more clearly than the corresponding histogram, but it's usually easier to understand how a distribution fails to be Normal by looking at its histogram.

Histogram and Normal probability plot for the Nissan gas mileage data we saw in Chapter 5 (p. 83). The vertical axes are the same, so each dot in the probability plot would fall into the bar of the histogram immediately to its left. **Figure 6.7**

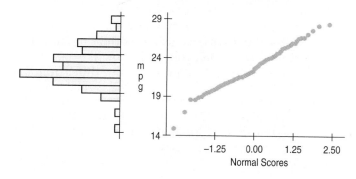

The Nissan car gas mileage data provide an example of data that are nearly Normal. The overall pattern of the Normal probability plot is straight. The two trailing low values correspond to the values in the histogram that trail off the low end. They're not quite in line with the rest of the data set. The Normal probability plot shows us that they're a bit lower than we'd expect of the lowest two values in a Normal model.

Histogram and Normal probability plot for men's weights. Note how a skewed distribution corresponds to a bent probability plot. **Figure 6.8**

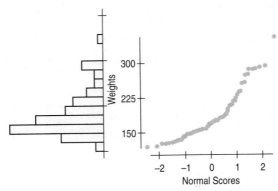

By contrast, the men's weights' Normal probability plot is far from straight. The weights are skewed to the high end, and the plot is curved. We'd conclude from these pictures that approximations using the 68-95-99.7% Rule for these data would not be very accurate.

How Does a Normal Probability Plot Work?

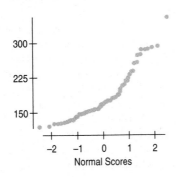

A S **Assessing Normality.**
You can check the Nearly Normal Condition most easily with your computer. This activity guides you through the process using your statistics package.

Why does the Normal probability plot work like that? We have 100 fuel efficiency measures for the Nissan car. The smallest of these has a z-score of −3.16. The Normal model can tell us what value to expect for the smallest z-score in a batch of 100 if a Normal model were appropriate. That turns out to be −2.58. So our first data value is smaller than we would expect from the Normal.

We can continue this and ask a similar question for each value. For example, the 14th smallest fuel efficiency has a z-score of almost exactly −1, and that's just what we should expect (well, −1.1 to be exact). A Normal probability plot makes all of these comparisons and plots them to make it easy to see how they come out.

When the values match up well, the line is straight. If one or two points are surprising from the Normal's point of view, they don't line up. When the entire distribution is skewed or different from the Normal in some other way, the values don't match up very well at all and the plot bends.

It turns out to be tricky to find the values we expect. They're called *Normal scores*, but you can't easily look them up in the tables. That's why probability plots are best made with technology and not by hand.

The best advice on using Normal probability plots is to see whether they are straight. If so, then your data look like data from a Normal model. If not, make a histogram to understand how they differ from the model.

What Can Go Wrong?

- **Don't use a Normal model when the distribution is not unimodal and symmetric.** Normal models are so easy and useful that it is tempting to use them even when they don't describe the data very well. That can lead to wrong conclusions. Don't use a Normal model without first checking the Nearly Normal Condition. Look at a picture of the data to check that it is unimodal and symmetric. A histogram, or a Normal probability plot, can help you tell whether a Normal model is appropriate.

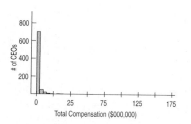

The CEOs (p. 56) had a mean total compensation of $2,818,743.10 with a standard deviation of $8,320,052.70. Using the Normal model rule, we should expect about 68% of the CEOs to have compensations between −$5,501,309.60 and $11,138,795.80. In fact, fewer than 5% of the CEOs' total compensations fall outside this range. What went wrong? The distribution is *skewed*, not symmetric. Using the 68-95-99.7% Rule for data like these will lead to silly results.

- **Don't use the mean and standard deviation when outliers are present.** Both means and standard deviations can be distorted by outliers, and no model based on distorted values will do a good job. It's always a good idea to check for outliers. How? Make a picture.

- **Don't round off too soon.** When we first looked at the athletes' performances in the heptathlon, we said that Bacher ran the 800 in 129 seconds. Later we gave her time as 129.08 seconds. How much precision do we need? There's no firm rule. We didn't want to clutter a simple story with many unnecessary digits. But when we sit down to calculate—especially when we have technology to do the grunt work—we should use all the precision available in the data.

  That can have surprising consequences. For example, to six digits, the mean time for the 800-m run was 136.994074 seconds. We reported that as 137 seconds. The extra digits beyond about the third one don't mean anything (after all, the times were only measured to the nearest 100th of a second), but we carry them during the calculation. For example, with all the precision we can muster, the z-score calculation is really

$$z = \frac{(129.08 - 136.994074)}{4.9782622} = -1.5897262, \text{ not } -1.6$$

  So, what does that mean for calculations? We have some advice . . .

- **Don't round your results in the middle of a calculation.** That can generate errors that grow during the rest of the calculation, especially when multiplying or dividing other calculated quantities. When we calculated the z-scores we didn't use the reported values of 137 for the mean and 5 for the standard deviation, but instead carried the additional digits.

  In some of the Step-by-Step examples in later chapters we'll make the story simple by showing the *rounded* numbers in a calculation and report the final answer as if we had used those rounded numbers. That often makes the story clearer, but if you take the original data and run it through a calculator or other technology you may see a *slightly* different number. So we have some more advice . . .

- **Don't worry about minor differences in results.** Because various calculators and programs may carry different precision in calculations, your answers may differ slightly from those we show in the text and in the Step-by-

Steps, or even from the values given in the answers in the back of the book. Those differences aren't anything to worry about. Our general rule when we *Tell* about calculated values is to report one digit of precision more than we got from the data.

CONNECTIONS

Changing the center and spread of a variable is equivalent to changing its *units*. Indeed, the only part of the data's context changed by standardizing is the *units*. All other aspects of the context do not depend on the choice or modification of measurement units. This fact points out an important distinction between the numbers the data provide for calculation and the meaning of the variables and the relationships among them. Standardizing can make the numbers easier to work with, but it does not alter the meaning.

Another way to look at this is to note that standardizing may change the center and spread values, but it does not affect the *shape* of a distribution. A histogram or boxplot of standardized values looks just the same as the histogram or boxplot of the original values except, perhaps, for the numbers on the axes.

When we summarized *shape, center,* and *spread* for histograms, we compared them to unimodal, symmetric shapes. You couldn't ask for a nicer example than the Normal model. And if the shape *is* like a Normal, we'll use the the mean and standard deviation to standardize the values.

What have we learned?

We've learned that the story data can tell may be easier to understand after shifting or rescaling the data.

- Shifting data by adding or subtracting the same amount from each value affects measures of center and position but not measures of spread.
- Rescaling data by multiplying or dividing every value by a constant changes all the summary statistics—center, position, and spread.

We've learned the power of standardizing data.

- Standardizing uses the standard deviation as a ruler to measure distance from the mean, creating *z*-scores.
- Using these *z*-scores we can compare apples and oranges—values from different distributions or values based on different units.
- And a *z*-score can identify unusual or surprising values among data.

We've learned that a Normal model can sometimes provide a useful way to understand data.

- Again we've seen the importance of *Thinking* about whether a method will work. We can decide whether a Normal model is appropriate by checking the Nearly Normal Condition with a histogram or Normal probability plot.
- Normal models follow the 68-95-99.7 Rule, and we can use technology or tables for a more detailed analysis.

TERMS

Shifting   Adding a constant to each data value adds the same constant to the mean, the median, and the quartiles, but does not change the standard deviation or IQR.

Rescaling	Multiplying each data value by a constant multiplies both the measures of position (mean, median, and quartiles) and the measures of spread (standard deviation and IQR) by that constant.
Standardizing	We standardize to eliminate units. Standardized values can be compared and combined even if the original variables had different units and magnitudes.
Standardized value	A value found by subtracting the mean and dividing by the standard deviation.
Normal model	A useful family of models for unimodal, symmetric distributions.
Parameter	A numerically valued attribute of a model. For example, the values of μ and σ in a $N(\mu,\sigma)$ model are parameters.
Statistic	A value calculated from data to summarize aspects of the data. For example, the mean, $\bar{y}$, and standard deviation, s, are statistics.
z-score	A z-score tells how many standard deviations a value is from the mean; z-scores have a mean of zero and a standard deviation of one.
Standard Normal model	A Normal model, $N(\mu,\sigma)$, with mean $\mu = 0$ and standard deviation $\sigma = 1$.
68-95-99.7 Rule	In a Normal model, about 68% of values fall within 1 standard deviation of the mean, about 95% fall within 2 standard deviations of the mean, and about 99.7% fall within 3 standard deviations of the mean.
Normal percentile	The Normal percentile corresponding to a z-score gives the percentage of values in a standard Normal distribution found at that z-score or below.
Normal probability plot	A display to help assess whether a distribution of data is approximately Normal. If the plot is nearly straight, the data satisfy the Nearly Normal Condition.
Changing center and spread	Changing the center and spread of a variable is equivalent to changing its *units*.

S K I L L S

When you complete this lesson you should:

Think

- Understand how adding (subtracting) a constant or multiplying (dividing) by a constant changes the center and/or spread of a variable.
- Recognize when standardization can be used to compare values.
- Understand that standardizing uses the standard deviation as a ruler.
- Recognize when a Normal model is appropriate.

Show

- Know how to calculate the z-score of an observation.
- Know how to compare values of two different variables using their z-scores.
- Be able to use Normal models and the 68-95-99.7 Rule to estimate the percentage of observations falling within 1, 2, or 3 standard deviations of the mean.
- Know how to find the percentage of observations falling below any value in a Normal model using a Normal table or appropriate technology.
- Know how to check whether a variable satisfies the Nearly Normal Condition by making a Normal probability plot or a histogram.

Tell

- Know what z-scores mean.
- Be able to explain how extraordinary a standardized value may be by using a Normal model.

Normal Plots on the Computer

The best way to tell whether your data can be modeled well by a Normal model is to make a picture or two. We've already talked about making histograms. Normal probability plots are almost never made by hand because the values of the Normal scores are tricky to find. But most statistics software make Normal plots, though various packages call the same plot by different names and array the information differently.

DATA DESK

To make a "Normal Probability Plot" in Data Desk,
- Select the Variable.
- Choose **Normal Prob Plot** from the **Plot** menu.

Comments
Data Desk places the ordered data values on the vertical axis and the Normal scores on the horizontal axis.

EXCEL

Excel offers a "Normal probability plot" as part of the Regression command in the Data Analysis extension, but (as of this writing) it is not a correct Normal probability plot and should not be used.

JMP

To make a "Normal Quantile Plot" in JMP,
- Make a histogram using **Distributions** from the **Analyze** menu.
- Click on the drop-down menu next to the variable name.
- Choose **Normal Quantile Plot** from the drop-down menu.
- JMP opens the plot next to the histogram.

Comments
JMP places the ordered data on the vertical axis and the Normal scores on the horizontal axis. The vertical axis aligns with the histogram's axis, a useful feature.

MINITAB

To make a "Normal Probability Plot" in MINITAB:
- Choose **Probability Plot** from the **Graph** menu.
- Select "Single" for the type of plot. Click **OK.**
- Enter the name of the variable in the "Graph variables" box. Click **OK.**

Comments
MINITAB places the ordered data on the horizontal axis and the Normal scores on the vertical axis.

SPSS

To make a Normal "P-P plot" in SPSS,
- Choose **P-P** from the **Graphs** menu.
- Select the variable to be displayed in the source list.
- Click the arrow button to move the variable into the target list.
- Click the **OK** button.

Comments
SPSS places the ordered data on the horizontal axis and the Normal scores on the vertical axis. You may safely ignore the options in the P-P dialog.

TI-83/84 Plus

To create a "Normal Percentile Plot" on the TI-83,
• Set up a **STAT PLOT** using the last of the Types.
• Specify your datalist, and the axis you choose to represent the data.

Although most people wouldn't open a statistics package just to find a Normal model value they could find in a table, you *would* use a calculator for that function. So . . .

To find what percent of a Normal model lies between two z-scores, choose **normalcdf** from the **DISTRibutions** menu and enter the command **normalcdf(zLeft, zRight).**

To find the z-score that corresponds to a given percentile in a Normal model, choose **invNorm** from the **DISTRibutions** menu and enter the command **invNorm(percentile).**

Comments

We often want to find Normal percentages from a certain z-score to infinity. On the calculator, indicate "infinity" as a very large z-score, say, 99. For example, the percentage of a Normal model over 2 standard deviations above the mean can be evaluated with **normalcdf(2, 99).**

TI-89

• To create a "Normal Prob Plot" press [F2] and select choice 2: **Norm Prob Plot.** Select a plot number and use VAR-LINK to enter the data list. Select X or Y for the data axis. Press [ENTER] to calculate the z-scores.
• Press [F2] and select choice 1: **Plot Setup.** Turn off any undesired plots (either [F3] **(Clear)** or [F4](√)). Press [F5] to display the plot.
• To find what percent of a Normal model lies between two z-scores, press [F5] **(Distr),** then select 4: **Normal Cdf.** Enter the lower and upper z-scores, specify mean 0 and standard deviation 1 and press [ENTER].
• To find the z-score for a given percentile, press [F5] **(Distr),** then arrow down to 2: **Inverse** press the right arrow to see the sub menu and select 1: **Inverse Normal.** Enter the area to the left of the desired point, mean 0 and standard deviation 1 and press [ENTER].

Comments

Normal models strictly go to infinity on either end, which is 1EE99 on the calculator. In practice, any "large" number will work. For example, the percentage of the Normal model over two standard deviations above the mean can use Lower Value 2 and Upper Value 99. To find area more than two standard deviations below the mean, use Lower Value −99, and Upper value −2.

EXERCISES

1. Payroll. Here are the summary statistics for the weekly payroll of a small company: lowest salary = $300, mean salary = $700, median = $500, range = $1200, IQR = $600, first quartile = $350, standard deviation = $400.
   a) Do you think the distribution of salaries is symmetric, skewed to the left, or skewed to the right? Explain why.
   b) Between what two values are the middle 50% of the salaries found?
   c) Suppose business has been good and the company gives every employee a $50 raise. Tell the new value of each of the summary statistics.

   d) Instead, suppose the company gives each employee a 10% raise. Tell the new value of each of the summary statistics.

2. Hams. A specialty foods company sells "gourmet hams" by mail order. The hams vary in size from 4.15 to 7.45 pounds, with a mean weight of 6 pounds and standard deviation of 0.65 pounds. The quartiles and median weights are 5.6, 6.2, and 6.55 pounds.
   a) Find the range and the IQR of the weights.
   b) Do you think the distribution of the weights is symmetric or skewed? If skewed, which way? Why?

c) If these weights were expressed in ounces (1 pound = 16 ounces) what would the mean, standard deviation, quartiles, median, IQR, and range be?

d) When the company ships these hams, the box and packing materials add 30 ounces. What are the mean, standard deviation, quartiles, median, IQR, and range of weights of boxes shipped (in ounces)?

e) One customer made a special order of a 10-pound ham. Which of the summary statistics of part d might *not* change if that data value were added to the distribution?

3. SAT or ACT? Each year thousands of high school students take either the SAT or the ACT, standardized tests used in the college admissions process. Combined SAT scores can go as high as 1600, while the maximum ACT composite score is 36. Since the two exams use very different scales, comparisons of performance are difficult. A convenient rule of thumb is $SAT = 40 \times ACT + 150$; that is, multiply an ACT score by 40 and add 150 points to estimate the equivalent SAT score.

An admissions officer reports the following statistics about the ACT scores of 2355 students who applied to her college. Find the summaries of equivalent SAT scores.

Lowest score = 19   Mean = 27   Standard deviation = 3
Q3 = 30   Median = 28   IQR = 6

4. Cold U? A high school senior uses the Internet to get information on February temperatures in the town where he'll be going to college. He finds a Web site with some statistics, but they are given in degrees Celsius. The conversion formula is $°F = 9/5 °C + 32$. Determine the Fahrenheit equivalents for the summary information below.

Maximum temperature = 11°C   Range = 33°   Mean = 1°
Standard deviation = 7°   Median = 2°   IQR = 16°

5. Temperatures. A town's January high temperatures average 36°F with a standard deviation of 10°, while in July the mean high temperature is 74° and the standard deviation is 8°. In which month is it more unusual to have a day with a high temperature of 55°? Explain.

6. Placement exams. An incoming freshman took her college's placement exams in French and mathematics. In French, she scored 82 and in math, 86. The overall results on the French exam had a mean of 72 and a standard deviation of 8, while the mean math score was 68, with a standard deviation of 12. On which exam did she do better compared with the other freshmen?

7. Final exams. Anna, a language major, took final exams in both French and Spanish and scored 83 on both. Her roommate Megan, also taking both courses, scored 77 on the French exam and 95 on the Spanish exam. Overall, student scores on the French exam had a mean of 81 and a standard deviation of 5, and the Spanish scores had a mean of 74 and a standard deviation of 15.

a) To qualify for language honors, a major must maintain at least an 85 average for all language courses taken. So far, which student qualifies?

b) Which student's overall performance was better?

8. MP3s. Two companies market new batteries targeted at owners of personal music players. DuraTunes claims a mean battery life of 11 hours, while RockReady advertises 12 hours.

a) Explain why you would also like to know the standard deviations of the battery lifespans before deciding which brand to buy.

b) Suppose those standard deviations are 2 hours for DuraTunes and 1.5 hours for RockReady. You are headed for 8 hours at the beach. Which battery is most likely to last all day? Explain.

c) If your beach trip is all weekend, and you probably will have the music on for 16 hours, which battery is most likely to last? Explain.

9. Cattle. The Virginia Cooperative Extension reports that the mean weight of yearling Angus steers is 1152 pounds. Suppose that weights of all such animals can be described by a Normal model with a standard deviation of 84 pounds.

a) How many standard deviations from the mean would a steer weighing 1000 pounds be?

b) Which would be more unusual, a steer weighing 1000 pounds, or one weighing 1250 pounds?

T 10. Car speeds. John Beale of Stanford, CA, recorded the speeds of cars driving past his house, where the speed limit read 20 mph. The mean of 100 readings was 23.84 mph, with a standard deviation of 3.56 mph. (He actually recorded every car for a two-month period. These are 100 representative readings.)

a) How many standard deviations from the mean would a car going under the speed limit be?

b) Which would be more unusual, a car traveling 34 mph or one going 10 mph?

11. More cattle. Recall that the beef cattle described in Exercise 9 had a mean weight of 1152 pounds, with a standard deviation of 84 pounds.

a) Cattle buyers hope that yearling Angus steers will weigh at least 1000 pounds. To see how much over (or under) that goal the cattle are, we could subtract 1000 pounds from all the weights. What would the new mean and standard deviation be?

b) Suppose such cattle sell at auction for 40 cents a pound. Find the mean and standard deviation of the sale prices for all the steers.

12. Car speeds again. For the car speed data of Exercise 10, recall that the mean speed recorded was 23.84 mph, with a standard deviation of 3.56 mph. To see how many cars are speeding, John subtracts 20 mph from all speeds.

a) What is the mean speed now? What is the new standard deviation?
b) His friend in Berlin wants to study the speeds, so John converts all the original miles per hour readings to kilometers per hour by multiplying all speeds by 1.609 (km per mile). What is the mean now? What is the new standard deviation?

13. Cattle, part III. Suppose the auctioneer in Exercise 11 sold a herd of cattle whose minimum weight was 980 pounds, median was 1140 pounds, standard deviation 84 pounds, and IQR 102 pounds. They sold for 40 cents a pound, and the auctioneer took a $20 commission on each animal. Then, for example, a steer weighing 1100 pounds would net the owner 0.40(1100) − 20 = $420. Find the minimum, median, standard deviation, and IQR of the net sale prices.

14. Caught speeding. Suppose police set up radar surveillance on the Stanford street described in Exercise 10. They handed out a large number of tickets to drivers going a mean of 28 mph, with a standard deviation of 2.4 mph, a maximum of 33 mph, and an IQR of 3.2 mph. Local law prescribes fines of $100 plus $10 per mile per hour over the 20 mph speed limit. For example, a driver convicted of going 25 mph would be fined 100 + 10(5) = $150. Find the mean, maximum, standard deviation, and IQR of all the potential fines.

15. Professors. A friend tells you about a recent study dealing with the number of years of teaching experience among current college professors. He remembers the mean but can't recall whether the standard deviation was 6 months, 6 years, or 16 years. Tell him which one it must have been, and why.

16. Rock concerts. A popular band on tour played a series of concerts in large venues. They always drew a large crowd, averaging 21,359 fans. While the band did not announce (and probably never calculated) the standard deviation, which of these values do you think is most likely to be correct: 20, 200, 2000, or 20,000 fans? Explain your choice.

17. Guzzlers? Environmental Protection Agency (EPA) fuel economy estimates for automobile models tested recently predicted a mean of 24.8 mpg and a standard deviation of 6.2 mpg for highway driving. Assume that a Normal model can be applied.
a) Draw the model for auto fuel economy. Clearly label it, showing what the 68-95-99.7 Rule predicts about miles per gallon.
b) In what interval would you expect the central 68% of autos to be found?
c) About what percent of autos should get more than 31 mpg?
d) About what percent of cars should get between 31 and 37 mpg?
e) Describe the gas mileage of the worst 2.5% of all cars.

18. IQ. Some IQ tests are standardized to a Normal model, with a mean of 100 and a standard deviation of 16.
a) Draw the model for these IQ scores. Clearly label it, showing what the 68-95-99.7 Rule predicts about the scores.
b) In what interval would you expect the central 95% of IQ scores to be found?
c) About what percent of people should have IQ scores above 116?
d) About what percent of people should have IQ scores between 68 and 84?
e) About what percent of people should have IQ scores above 132?

19. Small steer. In Exercise 9 we suggested the model $N(1152, 84)$ for weights in pounds of yearling Angus steers. What weight would you consider to be unusually low for such an animal? Explain.

20. High IQ. Exercise 18 proposes modeling IQ scores with $N(100, 16)$. What IQ would you consider to be unusually high? Explain.

T 21. Winter Olympics 2002 downhill. Fifty-three men qualified for the men's alpine downhill race in Salt Lake City. The gold medal winner finished in 1 minute 39.13 seconds. All competitors' times (in seconds) are found in the following list:

99.13	99.35	99.41	99.78	99.96
100.00	100.30	100.31	100.37	100.39
100.58	100.74	100.74	100.76	100.81
100.84	100.85	101.05	101.24	101.25
101.27	101.56	101.66	101.69	101.70
101.76	101.84	101.85	101.86	101.86
101.88	102.15	102.31	102.52	102.54
103.04	103.19	103.20	103.33	103.63
103.73	103.75	104.35	105.25	105.34
105.49	106.36	107.63	107.65	108.37
108.84	109.75	114.42		

a) The mean time was 102.71 seconds, with a standard deviation of 3.01 seconds. If the Normal model is appropriate, what percent of times will be less than 99.7 seconds?
b) What is the actual percent of times less than 99.7 seconds?
c) Why do you think the two percentages don't agree?
d) Create a histogram of these times. What do you see?

22. Rivets. A company that manufactures rivets believes the shear strength (in pounds) is modeled by $N(800, 50)$.
a) Draw and label the Normal model.
b) Would it be safe to use these rivets in a situation requiring a shear strength of 750 pounds? Explain.
c) About what percent of these rivets would you expect to fall below 900 pounds?

d) Rivets are used in a variety of applications with varying shear strength requirements. What is the maximum shear strength for which you would feel comfortable approving this company's rivets? Explain your reasoning.

23. Trees. A forester measured 27 of the trees in a large woods that is up for sale. He found a mean diameter of 10.4 inches and a standard deviation of 4.7 inches. Suppose that these trees provide an accurate description of the whole forest and that a Normal model applies.
a) Draw the Normal model for tree diameters.
b) What size would you expect the central 95% of all trees to be?
c) About what percent of the trees should be less than an inch in diameter?
d) About what percent of the trees should be between 5.7 and 10.4 inches in diameter?
e) About what percent of the trees should be over 15 inches in diameter?

ⓣ 24. Car speeds, the picture. For the car speed data of Exercise 10, here is the histogram, boxplot, and Normal probability plot of the 100 readings. Do you think it is appropriate to apply a Normal model here? Explain.

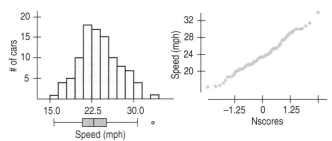

25. Trees, part II. Later on, the forester shows you a histogram of the tree diameters he used in analyzing the woods that was for sale. Do you think he was justified in using a Normal model? Explain, citing some specific concerns.

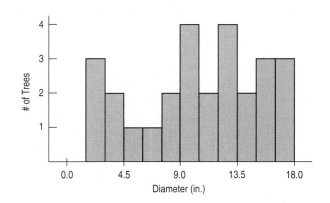

26. Check the model. Recall from Exercise 10 that the mean of the 100 car speeds in Exercise 24 was 23.84 mph, with a standard deviation of 3.56 mph.
a) Using a Normal model, what values should border the middle 95% of all car speeds?
b) Here are some summary statistics.

Percentile		Speed
100%	**Max**	34.060
97.5%		30.976
90.0%		28.978
75.0%	**Q3**	25.785
50.0%	**Median**	23.525
25.0%	**Q1**	21.547
10.0%		19.163
2.5%		16.638
0.0%	**Min**	16.270

From your answer in part a, how well does the model do in predicting those percentiles? Are you surprised? Explain.

ⓣ 27. TV watching. A survey of 200 college students conducted during the week of March 15, 1999, showed the following distribution of the number of hours of TV watched per week:

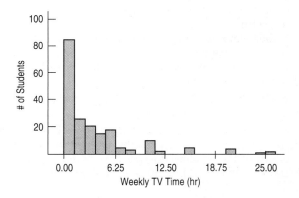

The mean is 3.66 hours, with a standard deviation of 4.93 hours.
a) According to the Normal model, what percent of students will watch fewer than 1 standard deviation below the mean number of hours?
b) For these data, what does that mean? Explain.
c) Explain the problem in using the Normal model for these data.

28. Customer database. A large philanthropic organization keeps records on the people who have contributed to their cause. In addition to keeping records of past giving, the organization buys demographic data on neighborhoods from the U.S. Census Bureau. Eighteen of these variables concern the ethnicity of the neighborhood of the donor. Here is a histogram and summary statistics for the percentage of whites in the neighborhoods of 500 donors:

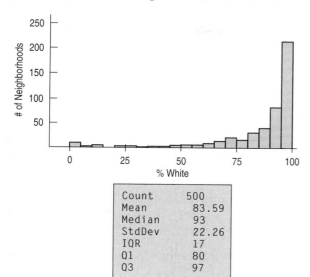

Count	500
Mean	83.59
Median	93
StdDev	22.26
IQR	17
Q1	80
Q3	97

a) Which is a better summary of the percentage of white residents in the neighborhoods, the mean or the median? Explain.
b) Which is a better summary of the spread, the IQR or the standard deviation? Explain.
c) From a Normal model, about what percentage of neighborhoods should have a percent white within one standard deviation of the mean?
d) What percentage of neighborhoods actually have a percent white within one standard deviation of the mean?
e) Explain the discrepancy between parts c and d.

29. Normal models. What percent of a standard Normal model is found in each region? Be sure to draw a picture first.
a) $z > 1.5$
b) $z < 2.25$
c) $-1 < z < 1.15$
d) $|z| > 0.5$

30. Normal models, again. What percent of a standard Normal model is found in each region? Draw a picture first.
a) $z > -2.05$
b) $z < -0.33$
c) $1.2 < z < 1.8$
d) $|z| < 1.28$

31. More Normal models. In a standard Normal model, what value(s) of z cut(s) off the region described? Don't forget to draw a picture.
a) the highest 20%
b) the highest 75%
c) the lowest 3%
d) the middle 90%

32. Yet another Normal model. In a standard Normal model, what value(s) of z cut(s) off the region described? Remember to draw a picture first.
a) the lowest 12%
b) the highest 30%
c) the highest 7%
d) the middle 50%

33. Normal cattle. Using $N(1152,84)$, the Normal model for weights of Angus steers in Exercise 9, what percent of steers weigh
a) over 1250 pounds?
b) under 1200 pounds?
c) between 1000 and 1100 pounds?

34. IQs revisited. Based on the Normal model $N(100,16)$ describing IQ scores, what percent of people's IQs would you expect to be
a) over 80?
b) under 90?
c) between 112 and 132?

35. More cattle. Based on the model $N(1152,84)$, what are the cutoff values for
a) the highest 10% of the weights?
b) the lowest 20% of the weights?
c) the middle 40% of the weights?

36. More IQs. In the Normal model $N(100,16)$ what cutoff value bounds
a) the highest 5% of all IQs?
b) the lowest 30% of the IQs?
c) the middle 80% of the IQs?

37. Cattle, finis. Consider the Angus weights model $N(1152,84)$ one last time.
a) What weight represents the 40th percentile?
b) What weight represents the 99th percentile?
c) What's the IQR of the weights of these Angus steers?

38. IQ, finis. Consider the IQ model $N(100,16)$ one last time.
a) What IQ represents the 15th percentile?
b) What IQ represents the 98th percentile?
c) What's the IQR of the IQs?

39. Parameters. Every Normal model is defined by its parameters, the mean and the standard deviation. For each model described below, find the missing parameter. As always, start by drawing a picture.
a) $\mu = 20$, 45% above 30; $\sigma = ?$
b) $\mu = 88$, 2% below 50; $\sigma = ?$
c) $\sigma = 5$, 80% below 100; $\mu = ?$
d) $\sigma = 15.6$, 10% above 17.2; $\mu = ?$

40. Parameters II. Every Normal model is defined by its parameters, the mean and the standard deviation. For each model described below, find the missing parameter. Don't forget to draw a picture.
a) $\mu = 1250$, 35% below 1200; $\sigma = ?$
b) $\mu = 0.64$, 12% above 0.70; $\sigma = ?$

c) $\sigma = 0.5$, 90% above 10.0; $\mu = ?$
d) $\sigma = 220$, 3% below 202; $\mu = ?$

41. Cholesterol. Assume the cholesterol levels of adult American women can be described by a Normal model with a mean of 188 mg/dL and a standard deviation of 24.
a) Draw and label the Normal model.
b) What percent of adult women do you expect to have cholesterol levels over 200 mg/dL?
c) What percent of adult women do you expect to have cholesterol levels between 150 and 170 mg/dL?
d) Estimate the interquartile range of the cholesterol levels.
e) Above what value are the highest 15% of women's cholesterol levels?

42. Tires. A tire manufacturer believes that the treadlife of its snow tires can be described by a Normal model with a mean of 32,000 miles and standard deviation of 2500 miles.
a) If you buy a set of these tires, would it be reasonable for you to hope they'll last 40,000 miles? Explain.
b) Approximately what fraction of these tires can be expected to last less than 30,000 miles?
c) Approximately what fraction of these tires can be expected to last between 30,000 and 35,000 miles?
d) Estimate the IQR of the treadlifes.
e) In planning a marketing strategy, a local tire dealer wants to offer a refund to any customer whose tires fail to last a certain number of miles. However, the dealer does not want to take too big a risk. If the dealer is willing to give refunds to no more than 1 of every 25 customers, for what mileage can he guarantee these tires to last?

43. Kindergarten. Companies who design furniture for elementary school classrooms produce a variety of sizes for kids of different ages. Suppose the heights of kindergarten children can be described by a Normal model with a mean of 38.2 inches and standard deviation of 1.8 inches.
a) What fraction of kindergarten kids should the company expect to be less than 3 feet tall?
b) In what height interval should the company expect to find the middle 80% of kindergarteners?
c) At least how tall are the biggest 10% of kindergarteners?

44. Body temperatures. Most people think that the "normal" adult body temperature is 98.6°F. That figure, based on a 19th-century study, has recently been challenged. In a 1992 article in the *Journal of the American Medical Association*, researchers reported that a more accurate figure may be 98.2°F. Furthermore, the standard deviation appeared to be around 0.7°F. Assume that a Normal model is appropriate.
a) In what interval would you expect most people's body temperatures to be? Explain.
b) What fraction of people would be expected to have body temperatures above 98.6°F?

c) Below what body temperature are the coolest 20% of all people?

45. Undercover? We learned in the chapter that the average Dutch man is 184 cm tall. The standard deviation of Caucasian adult male heights is about 7 cm. The average Greek 18-year-old in Athens is 167.8 cm tall. How easily could the average Dutch man hide in Athens? (Let's assume he dyes his hair, if necessary.) That is, would his height make him sufficiently extraordinary that he'd stand out easily? Assume heights are nearly Normal.

46. Big mouth! A Cornell University researcher measured the mouth volumes of 31 men and 30 women. She found a mean of 66 cc for men (SD = 17 cc) and a mean of 54 cc for women (SD = 14.5 cc). The man with the largest mouth had a mouth volume of 111.2 cc. The woman with the largest mouth had a mouth volume of 95.8 cc.
a) Which had the more extraordinarily large mouth?
b) If the distribution of mouth volumes is nearly Normal, what percentage of men and of women should have even larger mouths than these?

47. First steps. While only 5% of babies have learned to walk by the age of 10 months, 75% are walking by 13 months of age. If the age at which babies develop the ability to walk can be described by a Normal model, find the parameters (mean and standard deviation).

48. Trout. Wildlife biologists believe that the weights of adult trout can be described by a Normal model. They collect data from fishermen, finding that 22% of the trout caught were thrown back because they were below the 2-pound minimum, and only 6% weighed over 5 pounds. What mean and standard deviation should define the model?

49. Eggs. Hens usually begin laying eggs when they are about 6 months old. Young hens tend to lay smaller eggs, often weighing less than the desired minimum weight of 54 grams.
a) The average weight of the eggs produced by the young hens is 50.9 grams, and only 28% of their eggs exceed the desired minimum weight. If a Normal model is appropriate, what would the standard deviation of the egg weights be?
b) By the time these hens have reached the age of 1 year, the eggs they produce average 67.1 grams, and 98% of them are above the minimum weight. What is the standard deviation for the appropriate Normal model for these older hens?
c) Are egg sizes more consistent for the younger hens or the older ones? Explain.
d) A certain poultry farmer finds that 8% of his eggs are underweight and that 12% weigh over 70 grams. Estimate the mean and standard deviation of his eggs.

50. Tomatoes. Agricultural scientists are working on developing an improved variety of Roma tomatoes. Marketing research indicates that customers are likely to bypass Romas that weigh less than 70 grams. The current variety of Roma plants produces fruit that average 74 grams, but 11% of the tomatoes are too small. It is reasonable to assume that a Normal model applies.

a) What is the standard deviation of the weights of Romas now being grown?

b) Scientists hope to reduce the frequency of undersized tomatoes to no more than 4%. One way to accomplish this is to raise the average size of the fruit. If the standard deviation remains the same, what target mean should they have as a goal?

c) The researchers produce a new variety with a mean weight of 75 grams, which meets the 4% goal. What is the standard deviation of the weights of these new Romas?

d) Based on their standard deviations, compare the tomatoes produced by the two varieties.

just checking

Answers

1. a) On the first test, the mean is 88 and the SD is 4, so $z = (90 - 88)/4 = 0.5$. On the second test, the mean is 75 and the SD is 5, so $z = (80 - 75)/5 = 1.0$. The first midterm has the lower z-score, so it is the one that will be dropped.
b) The second midterm is one standard deviation above the mean, farther away than the first midterm, so it's the better score relative to the class.

2. a) The mean would increase to 500.
b) The standard deviation is still 100 points.
c) The two boxplots would look nearly identical (the shape of the distribution would remain the same), but the later one would be shifted 50 points higher.

3. The standard deviation is now 2.54 millimeters, which is the same as 0.1 inches. Nothing has changed. The standard deviation has "increased" only because we're reporting it in millimeters now, not inches.

4. The mean is 184 centimeters, with a standard deviation of 8 centimeters. 2 meters is 200 centimeters, which is 2 standard deviations above the mean. We expect 5% of the men to be more than 2 standard deviations below or above the mean, so half of those, 2.5%, are likely to be above 2 meters.

5. a) We know that 68% of the time we'll be within 1 standard deviation (2 min) of 20. So 32% of the time we'll arrive in less than 18 or more than 22 minutes. Half of those times (16%) will be greater than 22 minutes, so 84% will be less than 22 minutes.
b) 24 minutes is 2 standard deviations above the mean. Because of the 95% rule, we know 2.5% of the times will be more than 24 minutes.
c) Traffic incidents may occasionally increase the time it takes to get to campus, so the driving times may be skewed to the right, and there may be outliers.
d) If so, the Normal model would not be appropriate and the percentages we predict would not be accurate.

Exploring and Understanding Data

QUICK REVIEW

It's time to put it all together. Real data don't come tagged with instructions for use. So let's step back and look at how the key concepts and skills we've seen work together. This brief list and the review exercises that follow should help you check your understanding of Statistics so far.

▶ We treat data two ways: as categorical and as quantitative.

▶ To describe categorical data:

- Make a picture. Bar graphs work well for comparing counts in categories.
- Summarize the distribution with a table of counts or relative frequencies (percents) in each category.
- Pie charts and segmented bar charts display divisions of a whole.
- Compare distributions with plots side-by-side.
- Look for associations between variables by comparing marginal and conditional distributions.

▶ To describe quantitative data:

- Make a picture. Use histograms, boxplots, stem-and-leaf displays, or dotplots. Stem-and-leaf's are great when working by hand and good for small data sets. Histograms are a good way to see the distribution. Boxplots are best for comparing several distributions.
- Describe distributions in terms of their shape, center, and spread, and note any unusual features such as gaps or outliers.
- The shape of most distributions you'll see will likely be uniform, unimodal, or bimodal. It may be multimodal. If it is unimodal, then it may be symmetric or skewed.
- A 5-number summary makes a good numerical description of a distribution: min, Q1, median, Q3, and max.

- If the distribution is skewed, be sure to include the median and interquartile range (IQR) when you describe its center and spread.
- A distribution that is severely skewed may benefit from re-expressing the data. If it is skewed to the high end, taking logs often works well.
- If the distribution is unimodal and symmetric, describe its center and spread with the mean and standard deviation.
- Use the standard deviation as a ruler to tell how unusual an observed value may be, or to compare or combine measurements made on different scales.
- Shifting a distribution by adding or subtracting a constant affects measures of position but not measures of spread. Rescaling by multiplying or dividing by a constant affects both.
- When a distribution is roughly unimodal and symmetric, we may be able to model it with a Normal model. For Normal models, the 68-95-99.7 Rule is a good rule of thumb.
- If the Normal model fits well (check a histogram or Normal probability plot), then Normal percentile tables or functions found in most statistics technology can provide more detailed values.

Need more help with some of this? It never hurts to reread sections of the chapters! And in the following pages we offer you more opportunities[1] to review these concepts and skills.

The exercises that follow use the concepts and skills you've learned in the first six chapters. To be more realistic and more useful for your review, they don't tell you which of the concepts or methods you need. But neither will the exam.

---

[1] If you doubted that we are teachers, this should convince you. Only a teacher would call additional homework exercises "opportunities."

REVIEW EXERCISES

1. Bananas. Here are the prices (in cents per pound) of bananas reported from 15 markets surveyed by the U.S. Department of Agriculture.

51	52	45
48	53	52
50	49	52
48	43	46
45	42	50

a) Display these data with an appropriate graph.
b) Report appropriate summary statistics.
c) Write a few sentences about this distribution.

2. Prenatal care. Results of a 1996 American Medical Association report about the infant mortality rate for twins carried for the full term of a normal pregnancy are shown below, broken down by the level of prenatal care the mother had received.

Full-Term Pregnancies, Level of Prenatal Care	Infant Mortality Rate Among Twins (deaths per thousand live births)
Intensive	5.4
Adequate	3.9
Inadequate	6.1
Overall	5.1

a) Is the overall rate the average of the other three rates? Should it be? Explain.
b) Do these results indicate that adequate prenatal care is important for pregnant women? Explain.
c) Do these results suggest that a woman pregnant with twins should be wary of seeking too much medical care? Explain.

3. Singers. The boxplots shown display the heights (in inches) of 130 members of a choir.

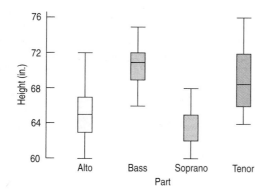

a) It appears that the median height for sopranos is missing, but actually the median and the upper quartile are equal. How could that happen?
b) Write a few sentences describing what you see.

4. Dialysis. In a study of dialysis, researchers found that "of the three patients who were currently on dialysis, 67% had developed blindness and 33% had their toes amputated." What kind of display might be appropriate for these data? Explain.

5. Beanstalks. Beanstalk Clubs are social clubs for very tall people. To join, a man must be over 6'2" tall, and a woman over 5'10". The National Health Survey suggests that heights of adults may be Normally distributed, with mean heights of 69.1" for men and 64.0" for women. The respective standard deviations are 2.8" and 2.5".
a) You are probably not surprised to learn that men are generally taller than women, but what does the greater standard deviation for men's heights indicate?
b) Who are more likely to qualify for Beanstalk membership, men or women? Explain.

6. Bread. Clarksburg Bakery is trying to predict how many loaves to bake. In the last 100 days, they have sold between 95 and 140 loaves per day. Here is a histogram of the number of loaves they sold for the last 100 days.

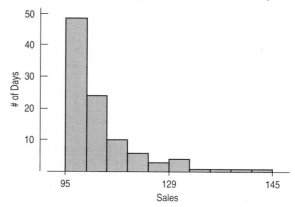

a) Describe the distribution.
b) Which should be larger, the mean number of sales or the median? Explain.
c) Here are the summary statistics for Clarksburg Bakery's bread sales. Use these statistics and the histogram above to create a boxplot. You may approximate the values of any outliers.

Summary of Sales	
Median	100
Min	95
Max	140
25th %tile	97
75th %tile	105.5

d) For these data the mean was 103 loaves sold per day, with a standard deviation of 9 loaves. Do these statistics suggest that Clarksburg Bakery should expect to sell between 94 and 112 loaves on about 68% of the days? Explain.

7. Watsamatta University. Public relations staff at Watsamatta U. collected data on people's opinions of various colleges and universities in their state. They phoned 850 local residents. After identifying themselves, the callers asked the survey participants their ages, whether they had attended college, and whether they had a favorable opinion of the university. The official report to the university's directors claimed that, in general, people had very favorable opinions about Watsamatta U.

a) Identify the W's of these data.
b) Identify the variables, classify each as categorical or quantitative, and specify units if relevant.
c) Are you confident about the report's conclusion? Explain.

8. Acid rain. Based on long-term investigation, researchers have suggested that the acidity (pH) of rainfall in the Shenandoah Mountains can be described by the Normal model $N(4.9, 0.6)$.

a) Draw and carefully label the model.
b) What percent of storms produce rain with pH over 6?
c) What percent of storms produce rainfall with pH under 4?
d) The lower the pH, the more acidic the rain. What is the pH level for the most acidic 20% of all storms?
e) What is the pH level for the least acidic 5% of all storms?
f) What is the IQR for the pH of rainfall?

9. Fraud detection. A credit card bank is investigating the incidence of fraudulent card use. The bank suspects that the type of product bought may provide clues to the fraud. To examine this situation, the bank looks at the Standard Industrial Code (SIC) of the business related to the transaction. This is a code that was used by the U.S. Census Bureau and Statistics Canada to identify the type of business of every registered business in North America.[2] For example, 1011 designates Meat and Meat Products (except Poultry), 1012 is Poultry Products, 1021 is Fish Products, 1031 is Canned and Preserved Fruits and Vegetables, and 1032 is Frozen Fruits and Vegetables.

---

[2] Since 1997 the SIC has been replaced by the NASIC, a code of six letters.

A company intern produces the following histogram of the SIC codes for 1536 transactions:

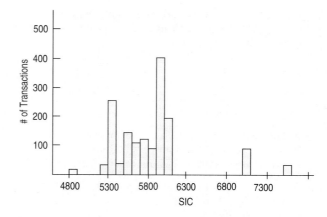

He also reports that the mean SIC is 5823.13 with a standard deviation of 488.17.

a) Comment on any problems you see with the use of the mean and standard deviation as summary statistics.
b) How well do you think the Normal model will work on these data? Explain.

10. Streams. As part of the coursework, a class at an upstate NY college collects data on streams each year. Students record a number of biological, chemical, and physical variables, including the stream name, the substrate of the stream (*limestone, shale,* or *mixed*), the pH, the temperature (°C), and BCI, a measure of biological diversity.

Group	Count	%
Limestone	77	44.8
Mixed	26	15.1
Shale	69	40.1

a) Name each variable, indicating whether it is categorical or quantitative, and giving the units if available.
b) These streams have been classified according to their substrate—the composition of soil and rock over which they flow—as summarized in the table. What kind of graph might be used to display these data?

T 11. Cramming. One Thursday, researchers gave students enrolled in a section of basic Spanish a set of 50 new vocabulary words to memorize. On Friday the students took a vocabulary test. When they returned to class the following Monday they were retested—without advance warning. Both sets of test scores for the 28 students are shown.

Fri	Mon	Fri	Mon
42	36	50	47
44	44	34	34
45	46	38	31
48	38	43	40
44	40	39	41
43	38	46	32
41	37	37	36
35	31	40	31
43	32	41	32
48	37	48	39
43	41	37	31
45	32	36	41
47	44		

a) Create a graphical display to compare the two distributions of scores.
b) Write a few sentences about the scores reported on Friday and Monday.
c) Create a graphical display showing the distribution of the *changes* in student scores.
d) Describe the distribution of changes.

12. Computers and Internet. A U.S. Census Bureau report (August 2000, *Current Population Survey*) found that 51.0% of homes had a personal computer and 41.5% had access to the Internet. A newspaper concluded that 92.5% of homes had either a computer or access to the Internet. Do you agree? Explain.

13. Let's play cards. You pick a card from a deck (see, for example, p. 253) and record its denomination (7, say) and its suit (maybe spades).
a) Is the variable *suit* categorical or quantitative?
b) Name a game you might be playing for which you would consider the variable *denomination* to be categorical. Explain.
c) Name a game you might be playing for which you would consider the variable *denomination* to be quantitative. Explain.

Ⓣ 14. Accidents. In 2001, Progressive Insurance asked customers who had been involved in auto accidents how far they were from home when the accident happened. The data are summarized in the table.

Miles from Home	% of Accidents
Less than 1	23
1 to 5	29
6 to 10	17
11 to 15	8
16 to 20	6
Over 20	17

a) Create an appropriate graph of these data.

b) Do these data indicate that driving near home is particularly dangerous? Explain.

Ⓣ 15. Hard water. In an investigation of environmental causes of disease, data were collected on the annual mortality rate (deaths per 100,000) for males in 61 large towns in England and Wales. In addition, the water hardness was recorded as the calcium concentration (parts per million, ppm) in the drinking water.
a) What are the variables in this study? For each, indicate whether it is quantitative or categorical and what the units are.
b) Here are histograms of calcium concentration and mortality. Describe the distributions of the two variables.

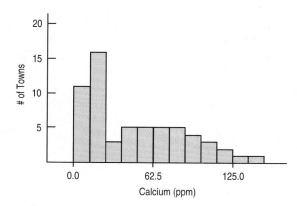

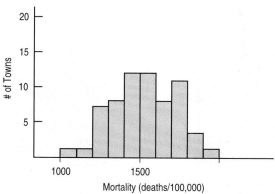

Ⓣ 16. Hard water II. The data set from England and Wales also notes for each town whether it was south or north of Derby. Here are some summary statistics and a comparative boxplot for the two regions.

Summary of Mortality				
Group	Count	Mean	Median	StdDev
North	34	1631.59	1631	138.470
South	27	1388.85	1369	151.114

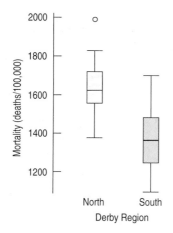

a) What is the overall mean mortality rate for the two regions?

b) Do you see evidence of a difference in mortality rates? Explain.

17. Seasons. Average daily temperatures in January and July for 60 large U.S. cities are graphed in the histograms below.

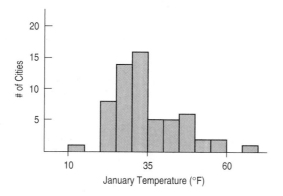

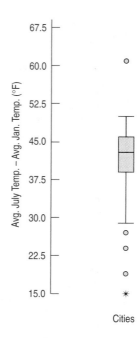

a) What aspect of these histograms makes it difficult to compare the distributions?

b) What differences do you see between the distributions of January and July average temperatures?

c) Differences in temperatures (July–January) for each of the cities are displayed in the boxplot above. Write a few sentences describing what you see.

18. Old Faithful. It is a common belief that Yellowstone's most famous geyser erupts once an hour at very predictable intervals. The histogram below shows the time gaps (in minutes) between 222 successive eruptions. Describe this distribution.

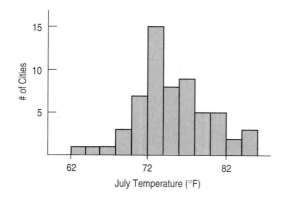

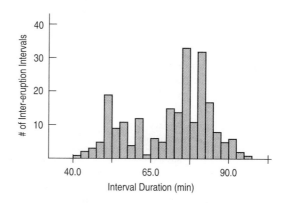

19. Old Faithful? Does the duration of an eruption have an effect on the length of time that elapses before the next eruption?

a) The histogram below shows the duration (in minutes) of those 222 eruptions. Describe this distribution.

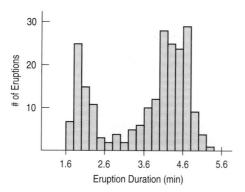

b) Explain why it is not appropriate to find summary statistics for this distribution.
c) Let's classify the eruptions as "long" or "short," depending upon whether or not they last at least 3 minutes. Describe what you see in the comparative boxplots.

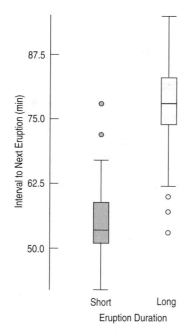

20. Teen drivers. In its *Traffic Safety Facts 2000,* the U.S. Department of Transportation reported that 6.8% of licensed drivers are between the ages of 15 and 20, yet this age group is behind the wheel in 14% of all fatal crashes. Use these statistics to explain the concept of independence.

T 21. Liberty's nose. Is the Statue of Liberty's nose too long? Her nose measures 4'6", but she is a large statue, after all.

Her arm is 42 feet long. That means her arm is 42/4.5 = 9.3 times as long as her nose. Is that a reasonable ratio? Shown in the table are arm and nose lengths of 18 girls in a Statistics class, and the ratio of arm-to-nose length for each.

Arm (cm)	Nose (cm)	Arm/Nose Ratio
73.8	5.0	14.8
74.0	4.5	16.4
69.5	4.5	15.4
62.5	4.7	13.3
68.6	4.4	15.6
64.5	4.8	13.4
68.2	4.8	14.2
63.5	4.4	14.4
63.5	5.4	11.8
67.0	4.6	14.6
67.4	4.4	15.3
70.7	4.3	16.4
69.4	4.1	16.9
71.7	4.5	15.9
69.0	4.4	15.7
69.8	4.5	15.5
71.0	4.8	14.8
71.3	4.7	15.2

a) Make an appropriate plot and describe the distribution of the ratios.
b) Summarize the ratios numerically, choosing appropriate measures of center and spread.
c) Is the ratio of 9.3 for the Statue of Liberty unrealistically low? Explain.

T 22. Winter Olympics 2002 speed skating. The top 25 men's and 25 women's 500-m speed skating times are listed in the table below:

Skater	Country	Time
Casey FitzRandolph	USA	69.23
Hiroyasu Shimizu	Japan	69.26
Kip Carpenter	USA	69.47
Gerard van Velde	Netherlands	69.49
Lee Kyu-Hyuk	South Korea	69.59
Joey Cheek	USA	69.60
Mike Ireland	Canada	69.60
Toyoki Takeda	Japan	69.81
Jan Bos	Netherlands	69.86
Erben Wennemars	Netherlands	69.89
Dmitry Lobkov	Russia	70.10
Kuniomi Haneishi	Japan	70.11
Sergey Klevchenya	Russia	70.28
Manabu Horii	Japan	70.32

(continued)

Janne Hanninen	Finland	70.33
Pawel Abratkiewicz	Poland	70.44
Choi Jae-Bong	South Korea	70.57
Dmitry Dorofeyev	Russia	70.75
Michael Kuenzel	Germany	70.84
Patrick Bouchard	Canada	70.88
Li Yu	China	70.97
Tomasz Swist	Poland	71.27
Davide Carta	Italy	71.39
Eric Brisson	Canada	71.54
Park Jae-Man	South Korea	71.96
Catriona LeMay Doan	Canada	74.75
Monique Garbrecht-Enfeldt	Germany	74.94
Sabine Voelker	Germany	75.19
Andrea Nuyt	Netherlands	75.37
Anzhelika Kotyuga	Belarus	75.39
Tomomi Okazaki	Japan	75.64
Svetlana Zhurova	Russia	75.64
Marianne Timmer	Netherlands	76.17
Yukari Watanabe	Japan	76.2
Svetlana Kaykan	Russia	76.31
Eriko Sannmiya	Japan	76.37
Sayuri Osuga	Japan	76.42
Wang Manli	China	76.62
Chris Witty	USA	76.73
Jenny Wolf	Germany	76.73
Chiara Simionato	Italy	76.92
Marieke Wijsman	Netherlands	77.1
Choi Seung-Yong	South Korea	77.14
Marion Wohlrab	Germany	77.37
Becky Sundstrom	USA	77.6
Susan Auch	Canada	77.6
Eli Ochowicz	USA	77.71
Jin Hua	China	78.26
Yang Chunyuan	China	78.63
Cho Seon-Yeon	South Korea	78.78

a) The mean finishing time was 73.46 seconds, with a standard deviation of 3.33 seconds. If the Normal model is appropriate, what percent of the times should be within 1.67 seconds of 73.46?

b) What percent of the times actually fall within this range?

c) Explain the discrepancy between a and b.

23. Sample. A study in South Africa focusing on the impact of health insurance identified 1590 children at birth and then sought to conduct follow-up health studies 5 years later. Only 416 of the original group participated in the 5-year follow-up study. This made researchers concerned that the follow-up group might not accurately resemble the total group in terms of health insurance. The table below summarizes the two groups by race and by presence of medical insurance when the child was born. Carefully

explain how this study demonstrates Simpson's paradox. (*Birth to Ten Study*, Medical Research Council, South Africa)

		NUMBER (%) INSURED	
		Follow-up	**Not Traced**
Race	**Black**	36 of 404 (8.9%)	91 of 1048 (8.7%)
	White	10 of 12 (83.3%)	104 of 126 (82.5%)
	Overall	46 of 416 (11.1%)	195 of 1174 (16.6%)

24. Sluggers. Roger Maris's 1961 home run record stood until Mark McGwire hit 70 in 1998. Listed below are the home run totals for each season McGwire played. Also listed are Babe Ruth's home run totals.

McGwire: 3*, 49, 32, 33, 39, 22, 42, 9*, 9*, 39, 52, 58, 70, 65, 32*, 27*

Ruth: 54, 59, 35, 41, 46, 25, 47, 60, 54, 46, 49, 46, 41, 34, 22

a) Find the 5-number summary for McGwire's career.

b) Do any of his seasons appear to be outliers? Explain.

c) McGwire played in only 18 games at the end of his first big league season, and missed major portions of some other seasons because of injuries to his back and knees. Those seasons might not be representative of his abilities. They are marked with asterisks in the list above. Omit these values and make parallel boxplots comparing McGwire's career to Babe Ruth's.

d) Write a few sentences comparing the two sluggers.

e) Create a side-by-side stem-and-leaf display comparing the careers of the two players.

f) What aspects of the distributions are apparent in the stem-and-leaf displays that did not clearly show in the boxplots?

25. Be quick! Avoiding an accident when driving can depend on reaction time. That time, measured from the moment the driver first sees the danger until he or she gets his foot on the brake pedal, is thought to follow a Normal model with a mean of 1.5 seconds and a standard deviation of 0.18 seconds.

a) Use the 68-95-99.7 Rule to draw the Normal model.

b) Write a few sentences describing driver reaction times.

c) What percent of drivers have a reaction time less than 1.25 seconds?

d) What percent of drivers have reaction times between 1.6 and 1.8 seconds?

e) What is the interquartile range of reaction times?

f) Describe the reaction times of the slowest 1/3 of all drivers.

26. Music and memory. Is it a good idea to listen to music when studying for a big test? In a study conducted by some Statistics students, 62 people were randomly assigned to listen to rap music, Mozart, or no music while attempting to memorize objects pictured on a page. They were then asked to list all the objects they could remember. Here are the 5-number summaries for each group:

	n	Min	Q1	Median	Q3	Max
Rap	29	5	8	10	12	25
Mozart	20	4	7	10	12	27
None	13	8	9.5	13	17	24

a) Describe the W's for these data: *Who, What, Where, Why, When, How.*
b) Name the variables and classify each as categorical or quantitative.
c) Create parallel boxplots as best you can from these summary statistics to display these results.
d) Write a few sentences comparing the performances of the three groups.

Ⓣ **27. Wine prices.** Here are the case prices for 36 wines produced in the Finger Lakes region of New York State.

123	70	90
80	78	72
52	103	138
112	92	93
118	118	106
95	131	59
151	115	97
100	128	130
66	135	76
143	100	88
110	75	60
115	105	85

a) Plot these data.
b) Find appropriate summary statistics.
c) Write a brief description of these wine prices.
d) What percent of the prices actually lie within one standard deviation of the mean? Comment.

Ⓣ **28. Birth order.** Are first-born children more likely to be interested in science, or perhaps younger siblings in, say, the humanities? A Statistics professor at a large university polled his students to find out what their majors were and what position they held in the family birth order. The results are summarized in the table.

a) What percent of these students are oldest or only children?
b) What percent of Humanities majors are oldest children?
c) What percent of oldest children are Humanities students?
d) What percent of the students are oldest children majoring in the Humanities?

		Birth Order*				
		1	**2**	**3**	**4+**	**Total**
Major	Math/Science	34	14	6	3	57
	Agriculture	52	27	5	9	93
	Humanities	15	17	8	3	43
	Other	12	11	1	6	30
	Total	**113**	**69**	**20**	**21**	**223**

* 1 = oldest or only child

29. Engines. One measure of the size of an automobile engine is its "displacement," the total volume (in liters or cubic inches) of its cylinders. Summary statistics for several models of new cars are shown. These displacements were measured in cubic inches.

Summary of Displacement	
Count	38
Mean	177.29
Median	148.5
StdDev	88.88
Range	275
25th %tile	105
75th %tile	231

a) How many cars were measured?
b) Why might the mean be so much larger than the median?
c) Describe the center and spread of this distribution with appropriate statistics.
d) Your neighbor is bragging about the 227-cubic-inch engine he bought in his new car. Is that engine unusually large? Explain.
e) Are there any engines in this data set that you would consider to be outliers? Explain.
f) Is it reasonable to expect that about 68% of car engines measure between 88 and 266 cubic inches? (That's 177.289 ± 88.8767.) Explain.
g) We can convert all the data from cubic inches to cubic centimeters (cc) by multiplying by 16.4. For example, a 200-cubic-inch engine has a displacement of 3280 cc. How would such a conversion affect each of the summary statistics?

Ⓣ **30. Birth order revisited.** Consider again the data on birth order and college majors in Exercise 28.
a) What is the marginal distribution of majors?
b) What is the conditional distribution of majors for the oldest children?
c) What is the conditional distribution of majors for the children born second?
d) Do you think that college major appears to be independent of birth order? Explain.

31. Herbal medicine. Researchers for the Herbal Medicine Council collected information on people's experiences with a new herbal remedy for colds. They went to a store selling natural health products. There they asked 100 customers whether they had taken the cold remedy and, if so, to rate its effectiveness (on a scale from 1 to 10) in curing their symptoms. The Council concluded that this product was highly effective in treating the common cold.

a) Identify the W's of these data.

b) Identify the variables, classify each as categorical or quantitative, and specify units if relevant.

c) Are you confident about the Council's conclusion? Explain.

32. Pay. According to the *1999 National Occupational Employment and Wage Estimates for Management Occupations*, the mean hourly wage for Chief Executives was $48.67 and the median hourly wage was $52.08. By contrast, for General and Operations Managers, the mean hourly wage was $31.69 and the median was $27.23. Are these wage distributions likely to be symmetric, skewed left, or skewed right? Explain. (Bureau of Labor Statistics)

T 33. Age and party. The Gallup Poll conducted a representative telephone survey during the first quarter of 1999. Among their reported results was the following table concerning the preferred political party affiliation of respondents and their ages.

		Party		
		Rep.	**Dem.**	**Ind.**	**Total**
	18–29	241	351	409	**1001**
Age	**30–49**	299	330	370	**999**
	50–64	282	341	375	**998**
	65+	279	382	343	**1004**
	Total	**1101**	**1404**	**1497**	**4002**

a) What percent of people surveyed were Republicans?

b) Do you think this might be a reasonable estimate of the percentage of all voters who are Republicans? Explain.

c) What percent of people surveyed were under 30 or over 65?

d) What percent of people were Independents under the age of 30?

e) What percent of Independents were under 30?

f) What percent of people under 30 were Independents?

34. Engines, again. Horsepower is another measure commonly used to describe auto engines. Here are the summary statistics and histogram displaying horsepowers of the same group of 38 cars.

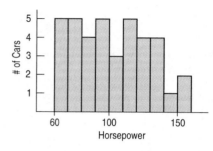

Summary of Horsepower	
Count	38
Mean	101.7
Median	100
StdDev	26.4
Range	90
25th %tile	78
75th %tile	125

a) Describe the shape, center, and spread of this distribution.

b) What is the interquartile range?

c) Are any of these engines outliers in terms of horsepower? Explain.

d) Do you think the 68-95-99.7 Rule applies to the horsepower of auto engines? Explain.

e) From the histogram, make a rough estimate of the percentage of these engines whose horsepower is within one standard deviation of the mean.

f) A fuel additive boasts in its advertising that it can "add 10 horsepower to any car." Assuming that is true, what would happen to each of these summary statistics if this additive were used in all the cars?

T 35. Age and party II. Consider again the Gallup Poll results on age and political party in Exercise 33.

a) What is the marginal distribution of party affiliation?

b) Create segmented bar graphs displaying the conditional distribution of party affiliation for each age group.

c) Summarize these poll results in a few sentences that might appear in a newspaper article about party affiliation in the United States.

d) Do you think party affiliation is independent of the voter's age? Explain.

T 36. Public opinion. For many years Martha Stewart was a popular expert in home decorating, an arbiter of good taste, and a very successful businesswoman. In June 2002 she first came under attack amidst rumors of insider stock trading. At that time a series of Gallup polls each contacted over 1000 people to ask about their overall opinion of her. Those polled could answer favorable, unfavorable, or that they did not know who she was. Results are summarized in the following table.

Poll Results

Opinion		Oct. 1999	June 2002	July 2002
	Favorable	49%	46%	30%
	Unfavorable	16%	27%	39%
	Don't know	32%	23%	27%

a) Each poll should total 100%. Can you think of a reason why these do not?
b) How could the number of people who do not know who Martha Stewart is *increase* from June to July? Did people forget her? Or perhaps the poll is flawed? Explain how these results could be valid.
c) Display these results in a bar graph.
d) Display these results with pie charts.
e) Display these results with a timeplot.
f) Which display do you think best depicts these data? Why?
g) Write a few sentences describing Martha Stewart's public image at that time.

T 37. Bike safety. The Massachusetts Governor's Highway Safety Bureau's report on bicycle injuries for the years 1991–2000 included the counts shown in the table.

Year	Bicycle Injuries Reported
1991	1763
1992	1522
1993	1452
1994	1370
1995	1380
1996	1343
1997	1312
1998	1275
1999	1030
2000	1118

a) What are the W's for these data?
b) Display the data in a stem-and-leaf display.
c) Display the data in a timeplot.
d) What is apparent in the stem-and-leaf display that is hard to see in the timeplot?
e) What is apparent in the timeplot that is hard to see in the stem-and-leaf display?
f) Write a few sentences about bicycle injuries in Massachusetts.

T 38. Profits. Here is a stem-and-leaf display showing profits as a percent of sales for 29 of the *Forbes* 500 largest U.S.

corporations. The stems are split; each stem represents a span of 5%, from a loss of 9% to a profit of 25%.

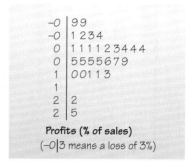

```
-0 | 99
-0 | 1 2 3 4
 0 | 1 1 1 1 2 3 4 4 4
 0 | 5 5 5 5 6 7 9
 1 | 0 0 1 1 3
 1 |
 2 | 2
 2 | 5
```

Profits (% of sales)
(−0|3 means a loss of 3%)

a) Find the 5-number summary.
b) Draw a boxplot for these data.
c) Find the mean and standard deviation.
d) Describe the distribution of profits for these corporations.

39. Some assembly required. A company that markets build-it-yourself furniture sells a computer desk that is advertised with the claim "less than an hour to assemble." However, through postpurchase surveys the company has learned that only 25% of their customers succeeded in building the desk in under an hour; 5% said it took them over 2 hours. The company assumes that consumer assembly time follows a Normal model.
a) Find the mean and standard deviation of the assembly time model.
b) One way the company could solve this problem would be to change the advertising claim. What assembly time should the company quote in order that 60% of customers succeed in finishing the desk by then?
c) Wishing to maintain the "less than an hour" claim, the company hopes that revising the instructions and labeling the parts more clearly can improve the 1-hour success rate to 60%. If the standard deviation stays the same, what new lower mean time does the company need to achieve?
d) Months later, another postpurchase survey shows that new instructions and part labeling did lower the mean assembly time, but only to 50 minutes. Nonetheless, the company did achieve the 60%-in-an-hour goal, too. How was that possible?

40. Crime and punishment. Because of the development of statistics and the methods of analysis you are learning about in this book, the 20th century has been called "The First Measured Century" in a series of documentaries on PBS. On the following page are two of their graphs.

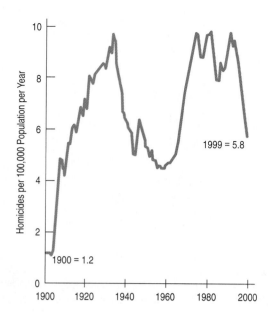

a) Write a few sentences describing the homicide rate in the United States during the 20th century.
b) Write a few sentences describing the use of capital punishment in the United States.
c) Could these two graphs be used to support the argument that capital punishment serves as a deterrent? If yes, expain how the graphs demonstrate that support. If no, explain why not.

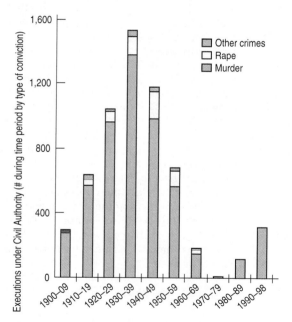

PART II

Exploring Relationships Between Variables

Scatterplots, Association, and Correlation

WHO	U.S. voters
WHAT	Percentage saying they would vote for a woman for president
UNITS	Percent
WHEN	1937–1999
WHERE	United States
HOW	Gallup Poll

Regularly, since 1937, the Gallup Poll has asked likely U.S. voters whether they would vote for a qualified woman for president if their preferred political party nominated one. Are people more likely to say yes to this question now than they were 70 years ago? If so, has the increase been consistent, or do you think there might have been periods when the "yes's" didn't increase at all, or even decreased? Here we show the percentage saying they would vote for a woman, plotted against the year in which the survey took place:

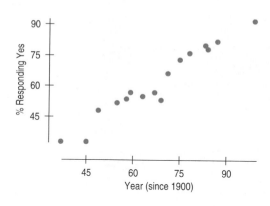

A scatterplot of percentage saying they would vote for a woman plotted against the year of the survey. Has the increase in willingness to vote for a woman been constant over the entire time period? What features in the trend do you see? **Figure 7.1**

Clearly, attitudes have changed. The plot shows fairly steady growth since 1937, reaching a level of 90% of voters saying "yes" by the year 1999. We can also see that there was a period of no growth in the 1960s and early 1970s.

This timeplot is an example of a more general kind of display called a **scatterplot.** Scatterplots may be the most common and most effective display for data. By just looking at them, you can see patterns, trends, relationships, and even the occasional extraordinary value sitting apart from the others. As the great philosopher Yogi Berra[1] once said, "You can observe a lot by watching."[2] Scatterplots are the best way to start observing the relationship between two *quantitative* variables.

A S **Heights of Husbands and Wives.** Husbands are usually taller than their wives. Or are they? Here's a chance to find out for yourself.

---

[1] Hall of Fame catcher and manager of the New York Mets and Yankees.
[2] But then he also said "I really didn't say everything I said." So we can't really be sure.

Relationships between variables are often at the heart of what we'd like to learn from data:

- Are grades actually higher now than they used to be?
- Do people tend to reach puberty at a younger age than in previous generations?
- Does applying magnets to parts of the body relieve pain? If so, are stronger magnets more effective?
- Do students learn better with the use of computer technology?

Questions such as these relate two quantitative variables and ask whether there is an **association** between them. Scatterplots are the ideal way to *picture* such associations.

Looking at Scatterplots

WHO	70 U.S. cities
WHAT	Cost per person ($ per year) and peak period freeway speed (mph)
WHEN	2000
WHY	Annual report from the Texas Transportation Institute

The Texas Transportation Institute studies the mobility provided by the nation's transportation system. It issues an annual report on traffic congestion and its costs. Here's a scatterplot of the annual cost per person of traffic delays (in dollars) in 70 cities in the United States against the peak period freeway speed (mph).

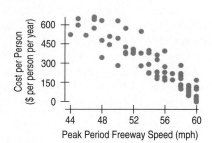

Cost per person ($ per year) of traffic delays vs. peak period freeway speed (mph) for 70 U.S. cities.   **Figure 7.2**

A S **Making and Understanding Scatterplots.** The best way to make scatterplots is with a computer. See how here.

Everyone looks at scatterplots. But, if asked, many people would find it hard to say what to look for in a scatterplot. What do *you* see? Try to describe the scatterplot congestion cost against freeway speed.

You might say that the **direction** of the association is important. As the peak freeway speed goes up, the cost of congestion goes down. A pattern that runs from the upper left to the lower right is said to be **negative.** A pattern running the other way is called **positive.**

Look for **Direction:** What's my sign—positive, negative, or neither?

The second thing to look for in a scatterplot is its **form.** If there is a straight line relationship, it will appear as a cloud or swarm of points stretched out in a generally consistent, straight form. For example, the scatterplot of traffic congestion has such an underlying **linear** form, although some points stray away from it.

Scatterplots can reveal many different kinds of patterns. Often they will not be straight, but straight line patterns are both the most common and the most useful for statistics.

Look for **Form**: straight, curved, something exotic, or no pattern?

If the relationship isn't straight, but curves gently, while still increasing or decreasing steadily, we can often find ways to make it more nearly straight. But if it curves sharply—up and then down, for example, —there is much less we can say about it with the methods of this book.

The third feature to look for in a scatterplot is how strong the relationship is. At

Look for **Strength**: how much scatter?

one extreme, do the points appear tightly clustered in a single stream (whether straight, curved, or bending all over the place)? Or, at the other extreme, does the swarm of points seem to form a vague cloud through which we can barely discern

any trend or pattern? The traffic congestion plot shows moderate scatter around a generally straight form. That indicates that there's a moderately strong linear relationship between cost and speed.

Look for **Unusual Features**: Are there outliers or subgroups?

Finally, always look for the unexpected. Often the most interesting thing to see in a scatterplot is something you never thought to look for. One example of such a surprise is an **outlier** standing away from the overall pattern of the scatterplot. Such a point is almost always interesting and always deserves special attention. Clusters or subgroups that stand away from the rest of the plot or show a trend in a different direction than the rest of the plot should raise questions about why they are different. They may be a clue that you should split the data into subgroups instead of looking at it all together.

Scatterplot Details

Descartes was a philosopher, famous for his statement *cogito ergo sum:* I think, therefore I am.

Scatterplots were among the first modern mathematical displays. The idea of using two axes at right angles to define a field on which to display values can be traced back to René Descartes (1596–1650), and the playing field he defined in this way is formally called a *Cartesian plane,* in his honor.

The two axes Descartes specified characterize the scatterplot. The axis that runs up and down is, by convention, called the *y*-axis, and the one that runs from side to side is called the *x*-axis. You can count on these names. If someone refers to the *y*-axis, you may be sure they mean the vertical, up-and-down axis, and similarly with the *x*-axis.[3]

---

[3] The axes are also called the "ordinate" and the "abscissa"—but we can never remember which is which because statisticians don't use these terms. In Statistics (and in all statistics computer programs) the axes are generically called "*x*" and "*y*," but are usually labeled with the names of the corresponding variables.

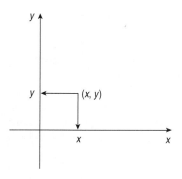

To make a scatterplot of two quantitative variables, assign one to the *y*-axis and the other to the *x*-axis. Be sure to label the axes clearly, and indicate the scales of the axes with numbers. Scatterplots display *quantitative* variables. Each variable has units, which should appear with the display to define what it's showing.

Each point is placed on a scatterplot at a position that corresponds to values of these two variables. Its horizontal location is specified by its *x*-value and its vertical location is specified by its *y*-value variable. Together, these are known as *coordinates* and written (x, y).

Scatterplots made by computer programs (such as the two we've seen in this chapter) often do not—and usually should not—show the *origin*, the point at $x = 0$, $y = 0$ where the axes meet. If both variables have values near or on both sides of zero, then the origin will be part of the display. If the values are far from zero, though, there's no reason to include the origin. In fact, it's far better to focus on the part of the Cartesian plane that contains the data. In our example about freeways, none of the speeds was anywhere near 0 mph, so the computer drew the scatterplot (Figure 7.2) with axes that don't quite meet.

Roles for Variables

Which variable should go on the *x*-axis and which on the *y*-axis? What we want to know about the relationship can tell us how to make the plot. We often have questions such as:

- Do heavier smokers develop lung cancer at younger ages?
- Is birth order an important factor in predicting future income?
- Can we estimate a person's % body fat more simply by just measuring girth or wrist size?

In each of these examples, the two variables play different roles. One plays the role of the **explanatory** or **predictor variable,** while the other takes on the role of the **response variable.** When the roles are clear, we place the explanatory variable on the *x*-axis and the response variable on the *y*-axis. When you make a scatterplot, you can assume that those who view it will think this way, so choose which variables to assign to which axes carefully.

The roles that we choose for variables are more about how we *think* about them than about the variables themselves. Just placing a variable on the *x*-axis doesn't necessarily mean that it explains or predicts *anything*. And the variable on the *y*-axis may not respond to it in any way. We plotted cost per person against peak freeway speed, thinking that the slower you go, the more it costs in delays. But maybe *spending* $500 per person in freeway improvement would increase speed. If we were examining that option, we might choose to plot cost per person as the explanatory variable and speed as the response.

A S **Scatterplot Check.**
Can you identify a scatterplot's direction, form, and strength? Test your skills here.

Older textbooks and disciplines other than Statistics sometimes refer to the *x*- and *y*-variables as the *independent* and *dependent* variables, respectively. The idea was that the *y*-variable depended on the *x*-variable and the *x*-variable acted independently to make *y* respond. These names, however, conflict with other uses of the same terms in Statistics. We'll use the terms "explanatory" and "response" when we're thinking about a relationship in those terms, but we'll often just say *x-variable* and *y-variable*.

Correlation

WHO	Students
WHAT	Height (inches), weight (pounds)
WHERE	Ithaca, NY
WHY	Data for class
HOW	Survey

Data collected from students in Statistics classes included their heights (in inches) and weights (in pounds; see Figure 7.3). It's no great surprise to discover that there is a positive association between the two. As you might suspect, taller students tend to weigh more. (If we had reversed the roles and chosen height as the explanatory variable, we might say that heavier students tend to be taller.) And the form of the scatterplot is fairly straight as well, although there seems to be a high outlier, as the plot shows.

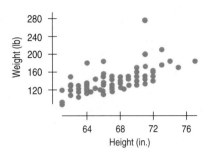

Weight vs. height of Statistics students. **Figure 7.3**

There is clearly a positive association, but how strong is it? If you had to put a number (say, between 0 and 1) on the strength, what would it be? Whatever measure we use shouldn't depend on our choice of units for the variables. After all, if we had measured heights and weights in centimeters and kilograms instead, it wouldn't change the direction, form, or strength, so it shouldn't change the number, as Figure 7.4 shows.

A S **Correlation.** Here's a good example of how correlation works to summarize the strength of a linear relationship and disregard scaling.

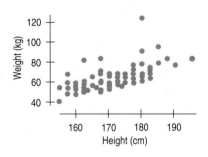

Plotting weight vs. height in different units doesn't change the shape of the pattern. **Figure 7.4**

Since the units don't matter, why not just remove them altogether? If we standardize both variables, we'll turn the coordinates of each point into a pair of z-scores. We could label the standardized variables z_x and z_y and write the coordinates of a point as (z_x, z_y). Remember that to standardize values we subtract the mean of each variable and then divide by its standard deviation, so

$$(z_x, z_y) = \left(\frac{x - \bar{x}}{s_x}, \frac{y - \bar{y}}{s_y}\right).$$

Because standardizing makes the means of both variables 0, the center of the new scatterplot is at the origin. The scales on both axes are now standard deviation units. (Look at Figure 7.5.)

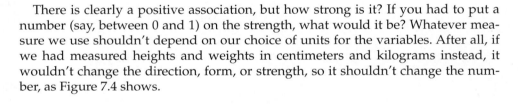

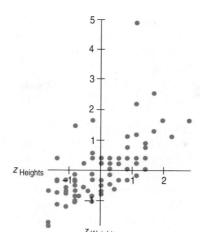

A scatterplot of standardized heights and weights.   **Figure 7.5**

Is this the only difference between these plots? Well, no. The underlying linear pattern seems steeper in the standardized plot. That's because we made the scales of the axes the same. Now the length of one standard deviation is the same vertically and horizontally. When we worked in the original units, we were free to make the plot as tall and thin

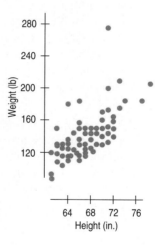

or as squat and wide

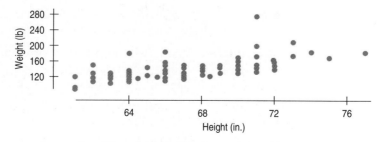

as we wanted to. Equal scaling gives a neutral way of drawing the scatterplot and a fairer impression of the strength of the association.[4]

---

[4] When we draw a scatterplot, what often looks best is to make the length of the *x*-axis slightly larger than the length of the *y*-axis. This is an aesthetic choice, probably related to the Golden Ratio of the Greeks.

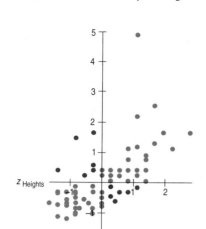

Points are colored by how they affect the association: green for positive, red for negative, and blue for neutral.

A S   Correlation. What does a correlation of 0.8 look like? How about 0.3? Watch the scatterplot change as you play with the correlation.

Which points in the scatterplot of the z-scores give the impression of a positive association? In a positive association, y tends to increase as x increases. So, the points in the upper right and lower left (colored green) strengthen that impression. For these points z_x and z_y have the same sign. If we multiplied them together, each point would have a positive product. Points far from the origin (which make the association look more positive) have an even bigger product.

The red points in the upper left and lower right quadrants tend to weaken the positive association (or support a negative association). For these points, z_x and z_y have opposite signs. So the product $z_x z_y$ for these points is negative. Points far from the origin (which make the association look more negative) have an even more negative product.

Points with z-scores of zero on either variable don't vote either way, because $z_x z_y = 0$. We've colored them blue.

We can turn these products into a measure of the strength of the association. We just add up the $z_x z_y$ products for every point in the scatterplot:

$$\sum z_x z_y.$$

This summarizes the direction *and* strength of the association for all the points. If most of the points are in the green quadrants, the sum will tend to be positive. If most are in the red quadrants, it will tend to be negative.

But the *size* of this sum gets bigger the more data we have. To adjust for this we divide the sum by $n - 1$.[5] The ratio is the famous **correlation coefficient:**

$$r = \frac{\sum z_x z_y}{n - 1}.$$

For the students' heights and weights, the correlation is 0.644. There are a number of alternative formulas for the correlation coefficient, using x and y in their original units. You may find them written elsewhere.[6] They can be more convenient for computing correlation by hand. But the form given above is best for understanding what it means.

Correlation Conditions

Correlation measures the strength of the *linear* association between two *quantitative* variables. Before you use correlation, you must check several *conditions:*

• **Quantitative Variables Condition:** Correlation applies only to quantitative variables. Don't apply correlation to categorical data masquerading as quantitative. Check that you know the variables' units and what they measure.

---

[5] Yes, the same $n - 1$ we saw for the standard deviation. And we offer the same promise to explain it later.

[6] Like here, for example: $r = \dfrac{\sum(x - \bar{x})(y - \bar{y})}{\sqrt{\sum(x - \bar{x})^2 \sum(y - \bar{y})^2}} = \dfrac{\sum(x - \bar{x})(y - \bar{y})}{(n - 1) s_x s_y}.$

A S **Correlation and Linearity.** How much does straightness matter? See for yourself as you bend the scatterplot.

A S **Case Study: Mortality and Education.** Is the mortality rate lower in cities with higher median education levels? Find out with scatterplots and correlation.

- **Straight Enough Condition:** Sure, you can *calculate* a correlation coefficient for any pair of variables. But correlation measures the strength only of the *linear* association, and will be misleading if the relationship is not linear. What is "straight enough"? How non-straight would the scatterplot have to be to fail the condition? This is another vague concept. It's a judgment call that you just have to think about. Do you think that the underlying relationship is curved? If so, then summarizing its strength with a correlation would be misleading.

- **Outlier Condition:** Outliers can distort the correlation dramatically. An outlier can make an otherwise small correlation look big or hide a large correlation. It can even give an otherwise positive association a negative correlation coefficient (and vice versa). When you see an outlier, it's often a good idea to report the correlations with and without the point.

Each of these conditions is easy to check with a scatterplot. Many correlations are reported without supporting data or plots. You should still think about the conditions. And you should be cautious in interpreting (or accepting others' interpretations of) the correlation when you can't check the conditions for yourself.

Finding the correlation coefficient by hand

To find the correlation coefficient by hand, we'll use the second formula in footnote 6 (p. 146). This will save us the work of having to standardize each individual data value first. Start with the summary statistics for both variables: $\bar{x}$, $\bar{y}$, s_x, and s_y. Then find the deviations as we did for the standard deviation, but now in both x and y: $(x - \bar{x})$ and $(y - \bar{y})$. For each data pair, multiply these deviations together: $(x - \bar{x})(y - \bar{y})$. Add the products up for all data pairs. Finally, divide the sum by the product of $(n - 1) \times s_x \times s_y$ to get the correlation coefficient.

Here we go:
Suppose the data pairs are:

x	6	10	14	19	21
y	5	3	7	8	12

Then $\bar{x} = 14$, $\bar{y} = 7$, $s_x = 6.20$, and $s_y = 3.39$

Deviations in x	Deviations in y	Product
$6 - 14 = -8$	$5 - 7 = -2$	$-8 \times -2 = 16$
$10 - 14 = -4$	$3 - 7 = -4$	16
$14 - 14 = 0$	$7 - 7 = 0$	0
$19 - 14 = 5$	$8 - 7 = 1$	5
$21 - 14 = 7$	$12 - 7 = 5$	35

Add up the products: $16 + 16 + 0 + 5 + 35 = 72$

Finally, we divide by $(n - 1) \times s_x \times s_y = (5 - 1) \times 6.20 \times 3.39 = 84.07$

The ratio is the correlation coefficient:

$$r = 72/84.07 = 0.856$$

Your Statistics teacher tells you that the correlation between the scores (points out of 50) on Exam 1 and Exam 2 was 0.75.

1. Before answering any questions about the correlation, what would you like to see? Why?

2. If she adds 10 points to each Exam 1 score, how will this change the correlation?

3. If she standardizes both scores, how will this affect the correlation?

4. In general, if someone does poorly on Exam 1, are they likely to do poorly or well on Exam 2? Explain.

5. If someone does poorly on Exam 1, will they definitely do poorly on Exam 2 as well?

Looking at Association Step-By-Step

When your blood pressure is measured, it is reported as two values, systolic blood pressure and diastolic blood pressure. How are these variables related to each other? Do they tend to both be high or low? Let's examine their relationship with a scatterplot.

 Think

Plan State what you are trying to investigate.

Variables Identify two quantitative variables whose relationship we wish to examine. Report the W's and be sure both variables are recorded for the same individuals.

Plot Make the scatterplot. Use a computer program or graphing calculator if you can.

I'll examine the relationship between two measures of blood pressure.

The variables are systolic and diastolic blood pressure (SBP and DBP) recorded (in millimeters of mercury, or mm Hg) for each of 1406 participants in a famous health study in Framingham, MA.

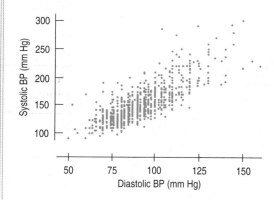

Check the conditions.

✓ **Quantitative Variables Condition:** Both SBP and DBP are quantitative and measured in mm Hg.

✓ **Straight Enough Condition:** The scatterplot looks straight.

✓ **Outlier Condition:** There are a few straggling points, but none far enough from the body of the data to be called outliers.

REALITY CHECK ⟩ Looks like a strong positive linear association. We shouldn't be surprised if the correlation coefficient is positive and fairly large.

I have two quantitative variables that satisfy the conditions, so correlation is a suitable measure of association.

Show   **Mechanics** We usually calculate correlations with technology. Here we have 1406 cases, so we'd never try it by hand.

The correlation coefficient is $r = 0.792$.

Tell   **Conclusion** Describe the direction, form, and strength you see in the plot, along with any unusual points or features. Be sure to state your interpretations in the proper context.

The scatterplot shows a positive direction, with higher SBP going with higher DBP. The plot is generally straight with a moderate amount of scatter. The correlation of 0.792 is consistent with what I saw in the scatterplot. A few cases stand out with unusually high SBP compared with their DBP. It seems far less common for the DBP to be high by itself.

Correlation Properties

A S **Construct Scatterplots with a Given Correlation.**
Take the correlation challenge. Try to make a scatterplot that has a given correlation: How close can you get?

Here's a useful list of facts about the correlation coefficient:

- The sign of a correlation coefficient gives the direction of the association.
- Correlation is always between -1 and $+1$. Correlation *can* be exactly equal to -1.0 or $+1.0$, but these values are unusual in real data because they mean that all the data points fall *exactly* on a single straight line.
- Correlation treats x and y symmetrically. The correlation of x with y is the same as the correlation of y with x.
- Correlation has no units. This fact can be especially appropriate when the data's units are somewhat vague to begin with (IQ score, personality index, socialization, and so on). Correlation is sometimes given as a percentage, but we discourage that because it suggests a percentage of *something*—and correlation, lacking units, has no "something" of which to be a percentage.
- Correlation is not affected by changes in the center or scale of either variable. Changing the units or baseline of either variable has no effect on the correlation coefficient. Correlation depends only on the z-scores, and they are unaffected by changes in center or scale.
- Correlation measures the strength of the *linear* association between the two variables. Variables can be strongly associated but still have a small correlation if the association isn't linear.
- Correlation is sensitive to outliers. A single outlying value can make a small correlation large or make a large one small.

Height and Weight, Again
We could have measured the students' weights in stones—a stone is a measure in the now outdated UK system of measures equal to 14 pounds. And we could have measured heights in hands—hands are still commonly used to measure the heights of horses. A hand is 4 inches. But no matter what *units* we use to measure the two variables, the *correlation* stays the same.

● **How strong is strong?** Some writers like to characterize correlations as "weak," "moderate," or "strong," but there's no agreement on what those terms mean. What's more, the same numerical correlation might be strong in one context and weak in another. You might be thrilled to discover a correlation of 0.7 between the new summary of the economy you've come up with and stock market prices, but finding "only" a correlation of 0.7 between two tests intended to measure the same skill might be viewed as a design failure. Using general terms like "weak," "moderate," or "strong" to describe a linear association adds a value judgment to the numerical summary that correlation provides. This can be useful, but if you tell the correlation and show a scatterplot, others can judge for themselves. ●

Correlation Tables

It is common in some fields to compute the correlations between each pair of variables in a collection of variables and arrange these correlations in a table. The rows and columns of the table name the variables, and the cells hold the correlations.

Correlation tables are compact and give a lot of summary information at a glance. They can be an efficient way to start to look at a large data set, but a dangerous one. By presenting all of these correlations without any checks for linearity and outliers, the correlation table risks showing truly small correlations that have been inflated by outliers, truly large correlations that are hidden by outliers, and correlations of any size that may be meaningless because the underlying form is not linear.

	Assets	Sales	Market Value	Profits	Cash Flow	Employees
Assets	1.000					
Sales	0.746	1.000				
Market Value	0.682	0.879	1.000			
Profits	0.602	0.814	0.968	1.000		
Cash Flow	0.641	0.855	0.970	0.989	1.000	
Employee	0.594	0.924	0.818	0.762	0.787	1.000

A correlation table of data reported by *Fortune* magazine for large companies. From this table, can you be sure that the variables are linearly associated and free from outliers?   **Table 7.1**

The diagonal cells of a correlation table always show correlations of exactly 1 (Can you see why?). Correlation tables are commonly offered by statistics packages on computers. These same packages often offer simple ways to make all the scatterplots you need to look at.[7]

*Straightening Scatterplots

Straight line relationships are the ones that we can measure with correlation. When a scatterplot shows a bent form that consistently increases or decreases, we can often straighten the form of the plot by re-expressing one or both variables.

Some camera lenses have an adjustable aperture, the hole that lets the light in. The size of the aperture is expressed in a mysterious number called the f/stop. Each increase of one f/stop number corresponds to a halving of the light that is allowed to come through. The f/stops of one digital camera are

f/stop:	2.8	4	5.6	8	11	16	22	32

---

[7] A table of scatterplots arranged just like a correlation table is sometimes called a *scatterplot matrix*, sometimes abbreviated to *SPLOM*. You might see these terms in a statistics package.

When you increase the f/stop one notch, you cut down the light, so you have to increase the time the shutter is open. We could experiment to find the best shutter speed for each f/stop value. A table of recommended shutter speeds and f/stops for a camera lists the relationship like this:

f/stop:	2.8	4	5.6	8	11	16	22	32
Shutter speed:	1/1000	1/500	1/250	1/125	1/60	1/30	1/15	1/8

The correlation of these f/stops and shutter speeds is 0.979. That sounds pretty high. You might assume that there must be a strong linear relationship. But when we check the scatterplot (we *always* check the scatterplot) it shows that something is not quite right:

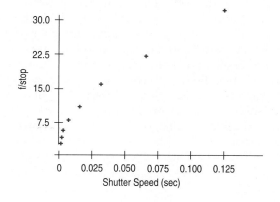

A scatterplot of *f/stop* vs. *Shutter Speed* shows a bent relationship.   **Figure 7.6**

We can see that the f/stop is not *linearly* related to the shutter speed. Can we find a transformation of f/stop that straightens out the line? What if we look at the *square* of the f/stop against the shutter speed?

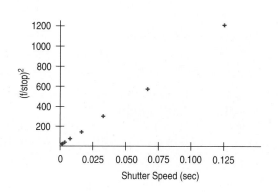

Re-expressing *f/stop* by squaring straightens the plot.   **Figure 7.7**

The second plot looks much more nearly straight. In fact, the correlation is now 0.998, but the increase in correlation is not important. (The original value of 0.979 should please almost anyone who sought a large correlation.) What is important is that the *form* of the plot is now straight, so the correlation is now an appropriate measure of association.[8]

We can often find transformations that straighten a scatterplot's form. Here, we found the square. Chapter 10 discusses simple ways to find a good re-expression.

What Can Go Wrong?

Did you know that there's a strong correlation between playing an instrument and drinking coffee? No? One reason might be that the statement doesn't make sense. Correlation is valid only for *quantitative* variables.

- ***Don't say "correlation" when you mean "association."*** How often have you heard the word "correlation"? Chances are pretty good that when you've heard the term, it's been misused. When people want to sound scientific, they often say "correlation" when talking about the relationship between two variables. It's one of the most widely misused Statistics terms, and given how often statistics are misused, that's saying a lot. One of the problems is that many people use the specific term *correlation* when they really mean the more general term *association*. Association is a deliberately vague term describing the relationship between two variables.

   Correlation is a precise term describing the strength and direction of the linear relationship between quantitative variables.

- ***Don't correlate categorical variables.*** People who misuse "correlation" to mean "association" often fail to notice whether the variables they discuss are quantitative. Be sure to check the Quantitative Variables Condition.

- ***Be sure the association is linear.*** Not all associations between quantitative variables are linear. Correlation can miss even a strong nonlinear association. A student project evaluating the quality of brownies baked at different temperatures reports a correlation of −0.05 between judges' scores and baking temperature. That seems to say there is no relationship until we look at the scatterplot:

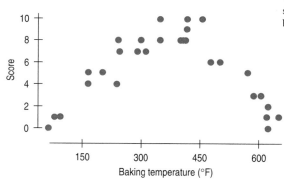

The relationship between brownie taste score and baking temperature is strong, but not at all linear.   **Figure 7.8**

There is a strong association, but the relationship is not linear. Don't forget to check the Straight Enough Condition.

- ***Beware of outliers.*** You can't interpret a correlation coefficient safely without a background check for outliers. Here's a silly example:

---

[8] Sometimes we can do a "reality check" on our choice of re-expression. In this case, a bit of research reveals that f/stops are related to the diameter of the open shutter. Since the amount of light that enters is determined by the *area* of the open shutter, which is related to the diameter by squaring, the square re-expression seems reasonable. Not all re-expressions have such nice explanations, but it's a good idea to think about them.

The relationship between IQ and shoe size among comedians shows a surprisingly strong positive correlation of 0.50. To check assumptions, we look at the scatterplot.

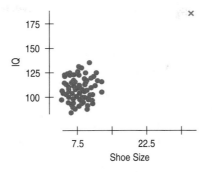

A scatterplot of *IQ* vs. *Shoe Size*. From this "study," what is the relationship between the two? The correlation is 0.50. Who *does* that point in the upper right-hand corner belong to?

Figure 7.9

The outlier is Bozo the Clown, known for his large shoes, and widely acknowledged to be a comic "genius." Without Bozo the correlation is near zero.

Even a single outlier can dominate the correlation value. That's why you need to check the Outlier Condition.

• ***Don't confuse correlation with causation.*** Once we have a strong correlation, it's tempting to try to explain it by imagining that the predictor variable has *caused* the response to change. Humans are like that; we tend to see causes and effects in everything.

Sometimes we can play with this tendency. Below is a scatterplot that shows the human population (*y*) of Oldenburg, Germany, in the beginning of the 1930s plotted against the number of storks nesting in the town (*x*).

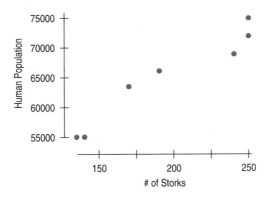

A scatterplot of the number of storks in Oldenburg, Germany, plotted against the population of the town for 7 years in the 1930s. The association is clear. How about the causation? (*Ornithologishe Monatsberichte*, 44, no. 2) **Figure 7.10**

Anyone who has seen the beginning of the movie *Dumbo* remembers Mrs. Jumbo anxiously awaiting the arrival of the stork to bring her new baby. Even though you know it's silly, you can't help but think for a minute that this plot shows that storks are the culprits. The two variables are obviously related to each other (the correlation is 0.97!), but that doesn't prove that storks bring babies.

It turns out that storks nest on house chimneys. More people means more houses, more nesting sites, and so more storks. The causation is actually in the *opposite* direction, but you can't tell from the scatterplot or correlation.

You need additional information—not just the data—to determine the real mechanism.

● **Does cancer cause smoking?** Even if the correlation of two variables is due to a causal relationship, the correlation itself cannot tell us what causes what.

Sir Ronald Aylmer Fisher (1890–1962) was one of the greatest statisticians of the 20th century. Fisher testified in court (paid by the tobacco companies) that a causal relationship might underlie the correlation of smoking and cancer:

"Is it possible, then, that lung cancer . . . is one of the causes of smoking cigarettes? I don't think it can be excluded . . . the pre-cancerous condition is one involving a certain amount of slight chronic inflammation . . .

A slight cause of irritation . . . is commonly accompanied by pulling out a cigarette, and getting a little compensation for life's minor ills in that way. And . . . is not unlikely to be associated with smoking more frequently."

Ironically, the proof that smoking indeed is the cause of many cancers came from experiments conducted following the principles of experiment design and analysis that Fisher himself developed—and that we'll see in Chapters 13 and 28. ●

Scatterplots and correlation coefficients *never* prove causation. This is, for example, partly why it took so long for the U.S. Surgeon General to get warning labels on cigarettes. Although there was plenty of evidence that increased smoking was *associated* with increased levels of lung cancer, it took years to provide evidence that smoking actually *causes* lung cancer. (The tobacco companies used this to great advantage.)

- ● *Watch out for lurking variables.* A scatterplot of the damage (in dollars) caused to a house by fire would show a strong correlation with the number of firefighters at the scene. Surely the damage doesn't cause firefighters. And firefighters do seem to cause damage, spraying water all around and chopping holes. Does that mean we shouldn't call the fire department? Of course not. There is an underlying variable that leads to both more damage and more firefighters—the size of the blaze.

A hidden variable that stands behind a relationship and determines it by simultaneously affecting the other two variables is called a **lurking variable.** You can often debunk claims made about data by finding the lurking variable behind the scenes.

CONNECTIONS

Scatterplots are the basic tool for examining the relationship between two quantitative variables. We start with a picture when we want to understand the distribution of a single variable, and we always make a scatterplot to begin to understand the relationship between two quantitative variables.

We used z-scores as a way to measure the statistical distance of data values from their means. Now we've seen the z-scores of x and y working together to build the correlation coefficient. Correlation is a summary statistic like the mean and standard deviation—only it summarizes the strength of a linear relationship. And we interpret it as we did z-scores, using the standard deviations as our rulers in both x and y.

What have we learned?

In recent chapters we learned how to listen to the story told by data from a single variable. Now we've turned our attention to the more complicated (and more interesting) story we can discover in the association between two quantitative variables.

We've learned to begin our investigation by looking at a scatterplot. We're interested in the *direction* of the association, the *form* it takes, and its *strength*.

We've learned that, although not every relationship is linear, when the scatterplot is straight enough, the *correlation coefficient* is a useful numerical summary.
- The sign of the correlation tells us the direction of the association.
- The magnitude of the correlation tells us of the *strength* of a linear association. Strong associations have correlations near +1 or −1, and very weak associations near 0.
- Correlation has no units, so shifting or scaling the data, standardizing, or even swapping the variables has no effect on the numerical value.

Once again we've learned that doing Statistics right means we have to *Think* about whether our choice of methods is appropriate.
- Before finding or talking about a correlation, we'll always check the Straight Enough Condition.
- And, as always, we'll watch out for outliers!

Finally, we've learned not to make the mistake of assuming that a high correlation or strong association is evidence of a cause-and-effect relationship. Beware of lurking variables!

A S **Correlation, Center, and Scale.** Do you have any lingering doubts that shifting and rescaling the data won't change the correlation? This activity is for you.

TERMS

Scatterplots
A scatterplot shows the relationship between two quantitative variables measured on the same cases.

Association
- **Direction:** A positive direction or association means that, in general, as one variable increases, so does the other. When increases in one variable generally correspond to decreases in the other, the association is negative.
- **Form:** The form we care about most is straight, but you should certainly describe other patterns you see in scatterplots.
- **Strength:** A scatterplot is said to show a strong association if there is little scatter around the underlying relationship.

Explanatory variable, Response variable, *x*-variable, *y*-variable
In a scatterplot, you must choose a role for each variable. Assign to the *y*-axis the response variable that you hope to predict or explain. Assign to the *x*-axis the explanatory or predictor variable that accounts for, explains, predicts, or is otherwise responsible for the *y*-variable.

Correlation
Correlation is a numerical measure of the direction and strength of a linear association.

Outlier
A point that does not fit the overall pattern seen in the scatterplot.

Lurking variable
A variable other than *x* and *y* that simultaneously affects both variables, accounting for the correlation between the two.

SKILLS

Think

When you complete this lesson you should:

- Recognize when interest in the pattern of possible relationship between two quantitative variables suggests making a scatterplot.

- Know how to identify the roles of the variables and to place the response variable on the *y*-axis and the explanatory variable on the *x*-axis.

- Know the conditions for correlation and how to check them.

- Know that correlations are between −1 and +1, and that each extreme indicates a perfect linear association.

- Understand how the magnitude of the correlation reflects the strength of a linear association as viewed in a scatterplot.

- Know that the correlation has no units.

- Know that the correlation coefficient is not changed by changing the center or scale of either variable.

- Understand that causation cannot be demonstrated by a scatterplot or correlation.

Show

- Know how to make a scatterplot by hand (for a small set of data) or with technology.

- Know how to compute the correlation of two variables.

- Know how to read a correlation table produced by a statistics program.

Tell

- Be able to describe the direction, form, and strength of a scatterplot.

- Be prepared to identify and describe points that deviate from the overall pattern.

- Be able to use correlation as part of the description of a scatterplot.

- Be alert to misinterpretations of correlation.

- Understand that finding a correlation between two variables does not indicate a causal relationship between them. Beware the dangers of suggesting causal relationships when describing correlations.

Scatterplots and Correlation on the Computer

Statistics packages generally make it easy to look at a scatterplot to check whether the correlation is appropriate. Some packages make this easier than others.

Many packages allow you to modify or enhance a scatterplot, altering the axis labels, the axis numbering, the plot symbols, or the colors used. Some options, such as color and symbol choice, can be used to display additional information on the scatterplot.

DATA DESK

To make a scatterplot of two variables, select one variable as Y and the other as X and choose **Scatterplot** from the **Plot** menu. Then find the correlation by choosing **Correlation** from the scatterplot's HyperView menu.

Alternatively, select the two variables and choose **Pearson Product-Moment** from the **Correlations** submenu of the **Calc** menu.

Comments

We prefer that you look at the scatterplot first and then find the correlation. But if you've found the correlation first, click on the correlation value to drop down a menu that offers to make the scatterplot.

EXCEL

To make a Scatterplot with the Excel Chart Wizard:

- Click on the **Chart Wizard** Button in the menu bar. Excel opens the Chart Wizard's Chart Type Dialog window.
- Make sure the **Standard Types** tab is selected, and select **XY (Scatter)** from the choices offered.
- Specify the **scatterplot without lines** from the choices offered in the Chart sub-type selections. The **Next** button takes you to the Chart Source Data dialog.
- If it is not already frontmost, click on the **Data Range** tab, and enter the data range in the space provided.
- By convention, we always represent variables in columns. The Chart Wizard refers to variables as Series. Be sure the **Column** option is selected.
- Excel places the leftmost column of those you select on the *x*-axis of the scatterplot. If the column you wish to see on the *x*-axis is not the leftmost column in your spreadsheet, click on the **Series** tab and edit the specification of the individual axis series.
- Click the **Next** button. The Chart Options dialog appears.
- Select the **Titles** tab. Here you specify the title of the chart and names of the variables displayed on each axis.
- Type the chart title in the **Chart title:** edit box.
- Type the *x*-axis variable name in the **Value (X) Axis:** edit box. Note that you must name the columns correctly here. Naming another variable will not alter the plot, only mislabel it.
- Type the *y*-axis variable name in the **Value (Y) Axis:** edit box.

- Click the **Next** button to open the chart location dialog.
- Select the **As new sheet:** option button.
- Click the **Finish** button.

Often, the resulting scatterplot will not be useful. By default, Excel includes the origin in the plot even when the data are far from zero. You can adjust the axis scales.

To change the scale of a plot axis in Excel:

- Double-click on the axis. The **Format Axis Dialog** appears.
- If the **scale tab** is not the frontmost, select it.
- Enter new minimum or new maximum values in the spaces provided. You can drag the dialog box over the scatterplot as a straightedge to help you read the maximum and minimum values on the axes.
- Click the **OK** button to view the rescaled scatterplot.
- Follow the same steps for the *x*-axis scale.

Compute a correlation in Excel with the **CORREL** function from the drop-down menu of functions. If CORREL is not on the menu, choose **More Functions** and find it among the statistical functions in the browser.

In the dialog that pops up, enter the range of cells holding one of the variables in the space provided.

Enter the range of cells for the other variable in the space provided.

JMP

To make a scatterplot and compute correlation, choose **Fit Y by X** from the **Analyze** menu.

In the Fit Y by X dialog, drag the Y variable into the "Y, Response" box, and drag the X variable into the "X, Factor" box. Click the **OK** button.

Once JMP has made the scatterplot, click on the red triangle next to the plot title to reveal a menu of options. Select **Density Ellipse** and select .95. JMP draws an ellipse around the data and reveals the **Correlation** tab. Click the blue triangle next to Correlation to reveal a table containing the correlation coefficient.

MINITAB

To make a scatterplot, choose **Scatterplot** from the **Graph** menu. Choose "Simple" for the type of graph. Click **OK.** Enter variable names for the Y-variable and X-variable into the table. Click **OK.**

To compute a correlation coefficient, choose **Basic Statistics** from the **Stat** menu. From the Basic Statistics submenu, choose **Correlation.** Specify the names of at least two quantitative variables in the "Variables" box. Click **OK** to compute the correlation table.

SPSS

To make a scatterplot, choose **Interactive** from the **Graphs** menu. From the Interactive Graphs submenu, choose **Scatterplot.** In the Create Scatterplot dialog, drag variable names from the source list into the targets on the right. Each target describes a specific part or aspect of the plot. For example, drag a variable name to the *y*-axis target to specify the variable to display on the *y*-axis.

Similarly, the *x*-axis target gets the name of the variable to display on the *x*-axis.

To compute a correlation coefficient, choose **Correlate** from the **Analyze** menu. From the Correlate submenu, choose **Bivariate.** In the Bivariate Correlations dialog, use the arrow button to move variables between the source and target lists.

Make sure the **Pearson** option is selected in the Correlation Coefficients field.

TI-83/84 Plus

To create a scatterplot, set up the **STAT PLOT** by choosing the **scatterplot** icon (the first option). Specify the lists where the data are stored as Xlist and Ylist. Set the graphing **WINDOW** to the appropriate scale and **GRAPH** (or take the easy way out and just ZoomStat!).

To find the correlation, go to **STAT CALC** menu and select **8: LinReg(a + bx).** Then specify the lists where the data are stored. The final command you will enter should look like **LinReg(a + bx) L1, L2.**

Comments

Notice that if you **TRACE** the scatterplot, the calculator will tell you the X and Y value at each point.

If the calculator does not tell you the correlation after you enter a LinReg command, try this: Hit **2nd CATALOG.** You now see a list of everything the calculator knows how to do. Scroll down until you find **DiagnosticOn.** Hit **ENTER** twice. (It should say Done.) Now and forevermore (or until you change batteries) you can find a correlation using your calculator.

TI-89

To create a scatterplot, press ⎡F2⎤ **(Plots).** Select choice 1: **Plot Setup.** Select a plot to define and press ⎡F1⎤. Select **Plot Type 1: Scatter.** Select a mark type. Specify the lists where the data are stored as Xlist and Ylist using VAR-LINK. Press ⎡ENTER⎤ to finish. Press ⎡F5⎤ to display the plot.

To find the correlation, press ⎡F4⎤ **(CALC),** then arrow to **3: Regressions** press the right arrow and select **1:LinReg(a + bx).** Then specify the lists where the data are stored. You can also select a *y*-function to store the equation of the line.

Comments

Notice that if you **TRACE** (press ⎡F3⎤) the scatterplot the calculator will tell you the *x*- and *y*-value at each point.

EXERCISES

1. Association. Suppose you were to collect data for each pair of variables. You want to make a scatterplot. Which variable would you use as the explanatory variable, and which as the response variable? Why? What would you expect to see in the scatterplot? Discuss the likely direction, form, and strength.
   a) Apples: weight in grams, weight in ounces
   b) Apples: circumference (inches), weight (ounces)
   c) College freshmen: shoe size, grade point average
   d) Gasoline: number of miles you drove since filling up, gallons remaining in your tank

2. Association. Suppose you were to collect data for each pair of variables. You want to make a scatterplot. Which variable would you use as the explanatory variable, and which as the response variable? Why? What would you expect to see in the scatterplot? Discuss the likely direction, form, and strength.
   a) T-shirts at a store: price each, number sold
   b) Skin diving: depth, water pressure
   c) Skin diving: depth, visibility
   d) All elementary-school students: weight, score on a reading test

3. Association. Suppose you were to collect data for each pair of variables. You want to make a scatterplot. Which variable would you use as the explanatory variable, and which as the response variable? Why? What would you expect to see in the scatterplot? Discuss the likely direction, form, and strength.
   a) When climbing a mountain: altitude, temperature
   b) For each week: Ice cream cone sales, air conditioner sales
   c) People: age, grip strength
   d) Drivers: blood alcohol level, reaction time

4. Association. Suppose you were to collect data for each pair of variables. You want to make a scatterplot. Which variable would you use as the explanatory variable, and which as the response variable? Why? What would you expect to see in the scatterplot? Discuss the likely direction, form, and strength.
   a) Long-distance calls: time (minutes), cost
   b) Lightning strikes: distance from lightning, time delay of the thunder
   c) A streetlight: its apparent brightness, your distance from it
   d) Cars: weight of car, age of owner

5. Scatterplots. Which of the scatterplots below show
   a) little or no association?
   b) a negative association?
   c) a linear association?
   d) a moderately strong association?
   e) a very strong association?

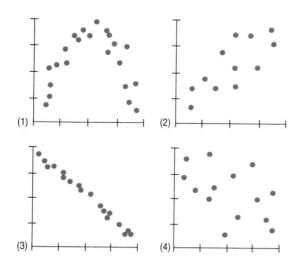

6. Scatterplots. Which of the scatterplots below show
   a) little or no association?
   b) a negative association?
   c) a linear association?
   d) a moderately strong association?
   e) a very strong association?

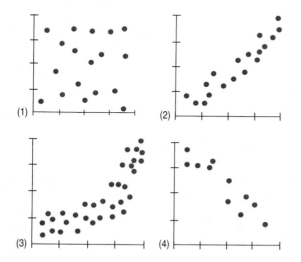

T 7. Performance IQ scores vs. brain size. A study examined brain size (measured as pixels counted in a digitized magnetic resonance image [MRI] of a cross-section of the brain) and IQ (4 Performance scales of the Weschler IQ test) for college students. The scatterplot shows the Performance IQ scores vs. the brain size. Comment on the association between brain size and IQ as seen in this scatterplot.

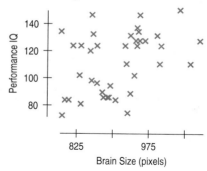

T 8. Kentucky Derby. The fastest horse in Kentucky Derby history was Secretariat in 1973. The scatterplot shows speed (in miles per hour) of the winning horses each year. What do you see? In most sporting events, performances have improved and continue to improve, so surely we anticipate a positive direction. But what of the form? Has the performance increased at the same rate throughout the last 125 years?

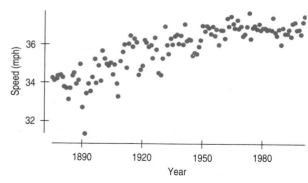

9. Firing pottery. A ceramics factory can fire eight large batches of pottery a day. Sometimes in the process a few of the pieces break. In order to understand the problem better, the factory records the number of broken pieces in each batch for 3 days and then creates the scatterplot shown.

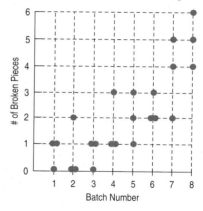

a) Make a histogram showing the distribution of the number of broken pieces in the 24 batches of pottery examined.
b) Describe the distribution as shown in the histogram. What feature of the problem is more apparent in the histogram than in the scatterplot?
c) What aspect of the company's problem is more apparent in the scatterplot?

10. Coffee sales. Owners of a new coffee shop tracked sales for the first 20 days, and displayed the data in a scatterplot (by day):

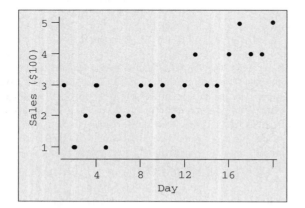

a) Make a histogram of the daily sales since the shop has been in business.
b) State one fact that is obvious from the scatterplot, but not from the histogram.
c) State one fact that is obvious from the histogram, but not from the scatterplot.

11. Matching. Here are several scatterplots. The calculated correlations are −0.923, −0.487, 0.006, and 0.777. Which is which?

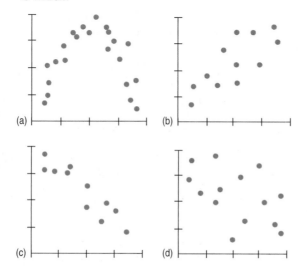

12. Matching. Here are several scatterplots. The calculated correlations are −0.977, −0.021, 0.736, and 0.951. Which is which?

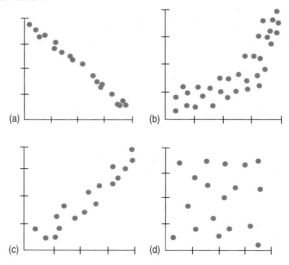

T 13. Roller coasters. Roller coasters get all their speed by dropping down a steep initial incline, so it makes sense that the height of that drop might be related to the speed of the coaster. Here's a scatterplot of top *speed* and largest *drop* for 75 roller coasters around the world.
a) Does the scatterplot indicate that it is appropriate to calculate the correlation? Explain.
b) In fact, the correlation of *speed* and *drop* is 0.91. Describe the association.

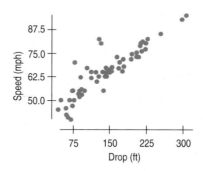

T 14. Antidepressants. A study compared the effectiveness of several antidepressants by examining the experiments in which they had passed the FDA requirements. Each of those experiments compared the active drug to a placebo, an inert pill given to some of the subjects. In each experiment some patients treated with the placebo had improved, a phenomenon called the *placebo effect*. Patients' depression levels were evaluated on the Hamilton Depression Rating Scale, where larger numbers indicate greater improvement. The scatterplot below compares mean improvement levels for the antidepressants and placebos for several experiments.

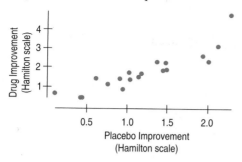

a) Is it appropriate to calculate the correlation? Explain.
b) The correlation is 0.898. Explain what we have learned about the results of these experiments.

15. Lunchtime. Does how long children remain at the lunch table help predict how much they eat? The table gives data on 20 toddlers observed over several months at a nursery school. "Time" is the average number of minutes a child spent at the table when lunch was served. "Calories" is the average number of calories the child consumed during lunch, calculated from careful observation of what the child ate each day.

Calories	Time
472	21.4
498	30.8
465	37.7
456	33.5
423	32.8
437	39.5
508	22.8
431	34.1
479	33.9
454	43.8
450	42.4
410	43.1
504	29.2
437	31.3
489	28.6
436	32.9
480	30.6
439	35.1
444	33.0
408	43.7

a) Find the correlation for these data.
b) Suppose we were to record time at the table in hours rather than in minutes. How would the correlation change? Why?

c) Write a sentence or two explaining what this correlation means for these data. Remember to write about food consumption by toddlers rather than about correlation coefficients.
d) One analyst concluded, "It is clear from this correlation that toddlers who spend more time at the table eat less. Evidently something about being at the table causes them to lose their appetites." Explain why this explanation is not an appropriate conclusion from what we know about the data.

16. Vehicle weights. The Minnesota Department of Transportation hoped that they could measure the weights of big trucks without actually stopping the vehicles by using a newly developed "weight-in-motion" scale. To see if the new device was accurate, they conducted a calibration test. They weighed several trucks when stopped (static weight), assuming that this weight was correct. Then they weighed the trucks again while they were moving to see how well the new scale could estimate the actual weight. Their data are given in the table.

WEIGHT OF A TRUCK (THOUSANDS OF POUNDS)	
Weight-in-Motion	**Static Weight**
26.0	27.9
29.9	29.1
39.5	38.0
25.1	27.0
31.6	30.3
36.2	34.5
25.1	27.8
31.0	29.6
35.6	33.1
40.2	35.5

a) Make a scatterplot for these data.
b) Describe the direction, form, and strength of the plot.
c) Write a few sentences telling what the plot says about the data. (*Note:* The sentences should be about weighing trucks, not about scatterplots.)
d) Find the correlation.
e) If the trucks were weighed in kilograms, how would this change the correlation? (1 kilogram = 2.2 pounds)
f) Do any points deviate from the overall pattern? What does the plot say about a possible recalibration of the weight-in-motion scale?

17. Fuel economy. Here are advertised horsepower ratings and expected gas mileage for several 2001 vehicles.

Vehicle	Horsepower	Gas Mileage
Audi A4	170 hp	22 mpg
Buick LeSabre	205	20
Chevy Blazer	190	15
Chevy Prizm	125	31
Ford Excursion	310	10
GMC Yukon	285	13
Honda Civic	127	29
Hyundai Elantra	140	25
Lexus 300	215	21
Lincoln LS	210	23
Mazda MPV	170	18
Olds Alero	140	23
Toyota Camry	194	21
VW Beetle	115	29

a) Make a scatterplot for these data.
b) Describe the direction, form, and strength of the plot.
c) Find the correlation between horsepower and miles per gallon.
d) Write a few sentences telling what the plot says about fuel economy.

T 18. Drug abuse. A survey was conducted in the United States and 10 countries of Western Europe to determine the percentage of teenagers who had used marijuana and other drugs. The results are summarized in the table.

	Percent Who Have Used	
Country	**Marijuana**	**Other Drugs**
Czech Rep.	22	4
Denmark	17	3
England	40	21
Finland	5	1
Ireland	37	16
Italy	19	8
No. Ireland	23	14
Norway	6	3
Portugal	7	3
Scotland	53	31
USA	34	24

a) Create a scatterplot.
b) What is the correlation between the percent of teens who have used marijuana and the percent who have used other drugs?
c) Write a brief description of the association.
d) Do these results confirm that marijuana is a "gateway drug," that is, that marijuana use leads to the use of other drugs? Explain.

T 19. Burgers. Fast food is often considered unhealthy because much of it is high in both fat and sodium. But are the two related? Here are the fat and sodium contents of several brands of burgers. Analyze the association between fat content and sodium.

Fat (g)	19	31	34	35	39	39	43
Sodium (mg)	920	1500	1310	860	1180	940	1260

T 20. Burgers. In the previous exercise you analyzed the association between the amounts of fat and sodium in fast food hamburgers. What about fat and calories? Here are data for the same burgers.

Fat (g)	19	31	34	35	39	39	43
Calories	410	580	590	570	640	680	660

T 21. Attendance. American League baseball games are played under the designated hitter rule, meaning that weak-hitting pitchers do not come to bat. Baseball owners believe that the designated hitter rule means more runs scored, which in turn means higher attendance. Is there evidence that more fans attend games if the teams score more runs? Data collected midway through the 2001 season indicate a correlation of 0.74 between runs scored and the number of people at the game.

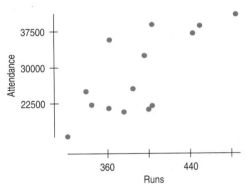

a) Does the scatterplot indicate that it's appropriate to calculate a correlation? Explain.
b) Describe the association between attendance and runs scored.
c) Does this prove that the owners are right, that more fans will come to games if the teams score more runs?

T 22. Second inning. Perhaps fans are just more interested in teams that win. Are the teams that win necessarily those that score the most runs?

	CORRELATION		
	Wins	**Runs**	**Attend**
Wins	1.000		
Runs	0.680	1.000	
Attend	0.557	0.740	1.000

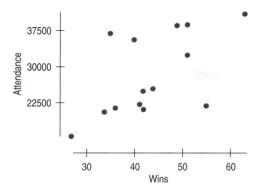

a) Do winning teams generally enjoy greater attendance at their home games? Describe the association.

b) Is attendance more strongly associated with winning or scoring runs? Explain.

c) How strongly is scoring more runs associated with winning more games?

23. Politics. A candidate for office claims that "there is a correlation between television watching and crime." Criticize this statement in statistical terms.

24. Association. A researcher investigating the association between two variables collected some data and was surprised when he calculated the correlation. He had expected to find a fairly strong association, yet the correlation was near 0. Discouraged, he didn't bother making a scatterplot. Explain to him how the scatterplot could still reveal the strong association he anticipated.

25. Height and reading. A researcher studies children in elementary school and finds a strong positive linear association between height and reading scores.

a) Does this mean that taller children are generally better readers?

b) What might explain the strong correlation?

26. Cellular telephones and life expectancy. A survey of the world's nations in 2004 shows a strong positive correlation between percentage of the country using cell phones and life expectancy in years at birth.

a) Does this mean that cell phones are good for your health?

b) What might explain the strong correlation?

27. Hard water. In a study of streams in the Adirondack Mountains, the following relationship was found between the pH of the water and the water's hardness (measured in grains):

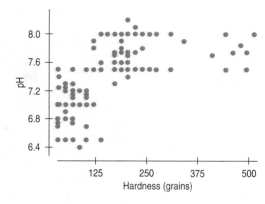

Is it appropriate to summarize the strength of association with a correlation? Explain.

28. Traffic headaches. A study of traffic delays in 68 U.S. cities found the following relationship between total delays (in total hours lost) and mean highway speed:

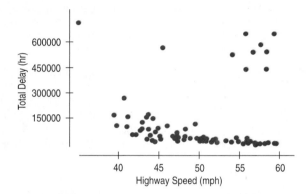

Is it appropriate to summarize the strength of association with a correlation? Explain.

29. Correlation errors. Your Economics instructor assigns your class to investigate factors associated with the gross domestic product (*GDP*) of nations. Each student examines a different factor (such as *life expectancy, literacy rate,* etc.) for a few countries and reports to the class. Apparently some of your classmates do not understand Statistics very well because you know several of their conclusions are incorrect. Explain the mistakes in their statements below.

a) "My correlation of −0.772 shows that there is almost no association between *GDP* and *infant mortality rate.*"

b) "There was a correlation of 0.44 between *GDP* and *continent.*"

30. More correlation errors. Students in the Economics class discussed in Exercise 29 also wrote these conclusions. Explain the mistakes they made.

a) "There was a very strong correlation of 1.22 between *life expectancy* and *GDP*."

b) "The correlation between *literacy rate* and *GDP* was 0.83. This shows that countries wanting to increase their standard of living should invest heavily in education."

31. Baldness and heart disease. Medical researchers followed 1435 middle-aged men for a period of 5 years, measuring the amount of *baldness* present (none = 1, little = 2, some = 3, much = 4, extreme = 5) and presence of *heart disease* (No = 0, Yes = 1). They found a correlation of 0.089 between the two variables. Comment on their conclusion that this shows that baldness is not a possible cause of heart disease.

32. Sample survey. A polling organization is checking its database to see if the two data sources they used sampled the same zip codes. The variable *datasource* = 1 if the data source is MetroMedia, 2 if the data source is DataQwest, and 3 if it's RollingPoll. The organization finds that the correlation between five digit zip code and *datasource* is −0.0229. It concludes that the correlation is low enough to state that there is no dependency between *zip code* and *source of data*. Comment.

33. Thrills. People who responded to a July 2004 Discovery Channel poll named the 10 best roller coasters in the United States. The table below shows the length of the initial drop (in feet) and the duration of the ride (in seconds). What do these data indicate about the height of a roller coaster and the length of the ride you can expect?

Roller Coaster	State	Drop (ft)	Duration (sec)
Incredible Hulk	FL	105	135
Millennium Force	OH	300	105
Goliath	CA	255	180
Nitro	NJ	215	240
Magnum XL-2000	OH	195	120
The Beast	OH	141	65
Son of Beast	OH	214	140
Thunderbolt	PA	95	90
Ghost Rider	CA	108	160
Raven	IN	86	90

T 34. Oil production. The following table shows the oil production of the United States from 1949 to 2000 (in thousands of barrels per year).

Year	Oil	Year	Oil
1949	1,841,940	1975	3,056,779
1950	1,973,574	1976	2,976,180
1951	2,247,711	1977	3,009,265
1952	2,289,836	1978	3,178,216
1953	2,357,082	1979	3,121,310
1954	2,314,988	1980	3,146,365
1955	2,484,428	1981	3,128,624
1956	2,617,283	1982	3,156,715
1957	2,616,901	1983	3,170,999
1958	2,448,987	1984	3,249,696
1959	2,574,590	1985	3,274,553
1960	2,574,933	1986	3,168,252
1961	2,621,758	1987	3,047,378
1962	2,676,189	1988	2,979,123
1963	2,752,723	1989	2,778,773
1964	2,786,822	1990	2,684,687
1965	2,848,514	1991	2,707,039
1966	3,027,763	1992	2,624,632
1967	3,215,742	1993	2,499,033
1968	3,329,042	1994	2,431,476
1969	3,371,751	1995	2,394,268
1970	3,517,450	1996	2,366,017
1971	3,453,914	1997	2,354,831
1972	3,455,368	1998	2,281,919
1973	3,360,903	1999	2,146,732
1974	3,202,585	2000	2,135,062

a) Find the correlation between *year* and *production*.

b) A reporter concludes that a low correlation between *year* and *production* shows that oil production has remained steady over the 50-year period. Do you agree with this interpretation? Explain.

T 35. Planets. Is there any pattern to the locations of the planets in our solar system? The table shows the average distance of each of the nine planets from the sun.

Planet	Position Number	Distance from Sun (million miles)
Mercury	1	36
Venus	2	67
Earth	3	93
Mars	4	142
Jupiter	5	484
Saturn	6	887
Uranus	7	1784
Neptune	8	2796
Pluto	9	3666

a) Make a scatterplot and describe the association. (Remember: direction, form, and strength!)

b) Why would you not want to talk about the correlation between planet *position* and *distance* from the sun?

c) Make a scatterplot showing the logarithm of *distance* vs. *position*. What is better about this scatterplot?

T 36. Internet journals. The rapid growth of Internet publishing is seen in a number of electronic academic journals made available during the last decade.

Year	Number of Journals
1991	27
1992	36
1993	45
1994	181
1995	306
1996	1093
1997	2459

a) Make a scatterplot and describe the trend.

b) Re-express the data in order to make the association more nearly linear.

just checking

Answers

1. We know the scores are quantitative. We should check to see if the *Straight Enough Condition* and the *Outlier Condition* are satisfied by looking at a scatterplot of the two scores.

2. It won't change.

3. It won't change.

4. They are more likely to do poorly. The positive correlation means that low scores on Exam 1 are associated with low scores on Exam 2 (and similarly for high scores).

5. No. The general association is positive, but individual performances may vary.

CHAPTER 8

Linear Regression

<table>
<tr><td>WHO</td><td>Items on the Burger King menu</td></tr>
<tr><td>WHAT</td><td>Protein content and total fat content</td></tr>
<tr><td>UNITS</td><td>grams of protein grams of fat</td></tr>
<tr><td>HOW</td><td>Supplied by BK on request or at their Web site</td></tr>
</table>

The Whopper™ has been Burger King's signature sandwich since 1957. One Double Whopper with cheese provides 53 grams of protein—all the protein you need in a day. It also supplies 1020 calories and 65 grams of fat. The Daily Value (based on a 2000-calorie diet) for fat is 65 grams. So after a double Whopper you'll want the rest of your calories that day to be fat-free.[1]

There are other items on the Burger King menu. How are their fat and protein related? The scatterplot of the *fat* (in grams) versus the *protein* (in grams) for foods sold at Burger King shows a positive, moderately strong, linear relationship.

A S **Manatees and Motorboats.** Are motorboats killing more manatees in Florida? Here's the story, on video. We'll use the data throughout this lesson.

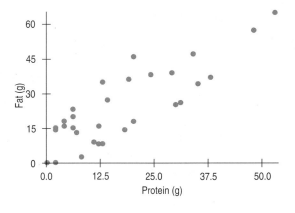

Total *fat* versus *protein* for 30 items on the Burger King menu. The Double Whopper is in the upper right corner. It's extreme, but is it out of line? **Figure 8.1**

---
[1] Sorry about the fries.

Linear Equations. For a quick review of linear equations, view this activity and play with the interactive tool.

If you want to get 25 grams of protein in your lunch, how much fat should you expect to consume at Burger King? The correlation between *fat* and *protein* is 0.83. That tells you that there's a strong linear association. But *strength* of the relationship is only part of the picture. The correlation says, "There seems to be a linear association between these two variables," but it doesn't tell us *what the line is*.

The Linear Model

> "Statisticians, like artists, have the bad habit of falling in love with their models."
>
> —George Box, famous statistician

Of course, we *can* say more. We can **model** the relationship with a line and give its equation. The equation will let us predict the fat content for any Burger King food, given its amount of protein. Clearly no line can go through all the points. Like all models of the real world, the line will be wrong—wrong in the sense that it can't match reality *exactly*. But it can still be useful. Like a physical model, it's something we can look at and manipulate in order to learn more about the real world.

We met our first model, the Normal model, in Chapter 6. We saw there that we can specify a Normal model with its mean (μ) and standard deviation (σ). These are the parameters of the Normal model.

For the Burger King foods, we might choose a linear model to describe the relationship between *protein* and *fat*. The **linear model** is just an equation of a straight line through the data. The points in the scatterplot don't all line up, but a straight line can summarize the general pattern with only a few parameters. This model can help us understand how the variables are associated.

 Residuals. Residuals are the basis for fitting lines to scatterplots. See how they work.

Residuals

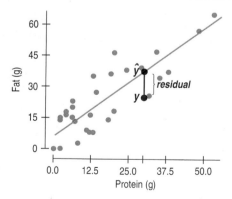

A *negative* residual means the predicted value's too big—an **overestimate**. And a *positive* residual shows the model makes an **underestimate**. These may seem backwards until you think about them.

Of course, the model won't be perfect. No matter what line we draw, it won't go through many of the points. The best line might not even hit any of the points. Then how can it be the "best" line? We want to find the line that somehow comes *closer* to all the points than any other line. Some of the points will be above the line and some below. For example, the line might suggest that a BK Broiler chicken sandwich with 30 grams of protein should have 36 grams of fat when, in fact, it actually has only 25 grams of fat. We call the estimate made from a model the **predicted value,** and write it as $\hat{y}$ (called y-hat) to distinguish it from the true value y (called, well, y). The difference between the observed value and its associated predicted value is called the **residual.** The residual value tells us how far off the model's prediction is at that point. The BK Broiler chicken residual would be $y - \hat{y} = 25 - 36 = -11$ g of fat.

To find the residuals, we always subtract the predicted value from the observed one. The negative residual tells us that the actual fat content of the BK Broiler chicken is about 11 grams *less* than the model predicts for a typical Burger King menu item with 30 grams of protein.

Our question now is how to find the right line to use.

"Best Fit" Means Least Squares

A S **The Least Squares Criterion.** Does your sense of "best fit" look like the least squares line? Drag your own line around a scatterplot, then check it.

When we draw a line through a scatterplot, some residuals are positive and some negative. We can't assess how well the line fits by adding up all the residuals—the positive and negative ones would just cancel each other out. We faced the same issue when we calculated a standard deviation to measure spread. And we deal with it the same way here: by squaring the residuals. Squaring makes them all positive. Now we can add them up. Squaring also emphasizes the large residuals. After all, points near the line are consistent with the model, but we're more concerned about points far from the line. When we add all the squared residuals together, that sum indicates how well the line we drew fits the data—the smaller the sum, the better the fit. A different line will produce a different sum, maybe bigger, maybe smaller.

The **line of best fit** is the line for which the sum of the squared residuals is smallest.

Our line has the special property that the variation of the data from the model, as seen in the residuals, is the smallest it can be for any straight line model for these data. No other line has this property. Speaking mathematically, we say that this line minimizes the sum of the squared residuals. You might think that finding this "least squares line" would be pretty hard. Surprisingly, it's not, although it was an exciting mathematical discovery when Legendre published it in 1805 (see margin note).

Who's on First

In 1805, Legendre was the first to publish the "least squares" solution to the problem of fitting a line to data when the points don't all fall exactly on the line. The main challenge was how to distribute the errors "fairly." After considerable thought, he decided to minimize the sum of the squares of what we now call the residuals. When Legendre published his paper, though, Gauss claimed he had been using the method since 1795. Gauss later referred to the "least squares" solution as "*our* method" (principium *nostrum*), which certainly didn't help his relationship with Legendre.

The Least Squares Line

You may remember from Algebra that a straight line can be written as:

$$y = mx + b.$$

If you plot all the (x, y) pairs that satisfy this equation, they'll all fall exactly on a straight line. We'll use this form for our linear model, but, of course, with real data the points won't all fall on the line. If we write our model as:

$$\hat{y} = b_0 + b_1 x,$$

then the *predictions* from our model follow a straight line.[2] If the model is a good one, the data values will scatter closely around it.

For the Burger King menu items, the best fit line is

$$\widehat{fat} = 6.8 + 0.97 \, protein.$$

NOTATION ALERT:

"Putting a hat on it" is standard Statistics notation to indicate that something has been predicted by a model. Whenever you see a hat over a variable name or symbol, you can assume it is the predicted version of that variable or symbol (and look around for the model).

In a linear model, we use b_1 for the slope and b_0 for the y-intercept.

---

[2] Besides using $\hat{y}$ to represent the *predicted* y-value now, we also changed from $mx + b$ to $b_0 + b_1 x$. We did this for a reason—not just to be difficult. Eventually we'll want to add more x's to the model to make it more realistic and we don't want to use up the entire alphabet. What would we use after m? The next letter is n and that one's already taken. o? See our point? Sometimes subscripts are the best approach.

A S **Interpreting Equations.**
This animated discussion shows how to use and interpret linear equations.

What does this mean? The **slope**, 0.97, says that an additional gram of protein is associated with an additional 0.97 grams of fat, on average. Less formally, we might say that Burger King sandwiches pack about 0.97 grams of fat per gram of protein. Slopes are always expressed in y-units per x-unit. They tell how the y-variable changes (in its units) for a one-unit change in the x-variable. When you see a phrase like "students per teacher" or "kilobytes per second" think slope.

How about the **intercept**, 6.8? Algebraically, that's the value the line takes when x is zero. Here, our model predicts that even a BK item with no protein would have, on average, about 6.8 grams of fat. Is that reasonable? Well, the apple pie, with 2 grams of protein, has 14 grams of fat, so it's not impossible. But often 0 is not a plausible value for x (the year 0, a baby born weighing 0 grams, . . .). Then the intercept serves only as a starting value for our predictions and we don't interpret it as a meaningful predicted value.

How do we find the slope and intercept of the least squares line? The formulas are simple. The model is built from the summary statistics we've used over and over: the correlation (to tell us the strength of the linear association), the standard deviations (to give us the units), and the means (to tell us where to put the line). The slope of the line is just

Slope
$$b_1 = r\frac{s_y}{s_x}$$

$$b_1 = r\frac{s_y}{s_x}.$$

We've already seen that the correlation tells us the *sign* as well as the strength of the relationship, so it's no surprise to see that the slope inherits this sign as well. If the correlation is positive, the scatterplot runs from lower left to upper right and the slope of the line is positive as well.

Changing the units of the variables doesn't change the *correlation*, but for the *slope*, units do matter. We may know that age and height in children are positively correlated, but the *value* of the slope depends on the units. If children grow an average of 3 inches per year, that's the same as 0.21 millimeters per day. For the slope, it matters whether you express age in days or years and whether you measure height in inches or millimeters. How you choose to express x and y—what units you use—affects the slope directly. Why? We know changing units doesn't change the correlation, but does change the standard deviations. The slope introduces the units into the equation by multiplying the correlation by the ratio of s_y to s_x. The units of the **slope** are always the units of y per unit of x.

Units of *y* per unit of *x*
Get into the habit of identifying the units by writing down "*y*-units per *x*-unit," with the unit names put in place. You'll find it'll really help you to Tell about the line in context.

What about the intercept? If you had to predict the y-value for a data point whose x-value was average, what would you say? The best fit line always predicts $\bar{y}$ for points whose x-value is $\bar{x}$. Putting that into our equation, we find:

$$\bar{y} = b_0 + b_1\bar{x}.$$

Or, rearranging the terms,

$$b_0 = \bar{y} - b_1\bar{x}.$$

Knowing the slope and the fact that the line goes through the point $(\bar{x}, \bar{y})$ tells us how to find the intercept.

Intercept
$$b_0 = \bar{y} - b_1\bar{x}$$

Let's try out the slope and intercept formulas on the "BK data" and see if we can find the line. We checked the conditions when we calculated the correlation. We'll need the summary statistics (located conveniently in the margin). So, the slope is

Protein	Fat
$\bar{x} = 17.2$ g	$\bar{y} = 23.5$ g
$s_x = 14.0$ g	$s_y = 16.4$ g
$r = 0.83$	

$$b_1 = r\frac{s_y}{s_x} = 0.83 \times \frac{16.4 \text{ g fat}}{14.0 \text{ g protein}} = 0.97 \text{ grams of fat } per \text{ gram of protein.}$$

The intercept is

$$b_0 = \bar{y} - b_1\bar{x} = 23.5 \text{ g fat} - \frac{0.97 \text{ g fat}}{\text{g protein}} \times 17.2 \text{ g protein} = 6.8 \text{ g fat.}$$

Putting these results back into the equation gives:

$$\widehat{fat} = 6.8 \text{ g fat} + \frac{0.97 \text{ g fat}}{\text{g protein}} protein$$

or more simply

$$\widehat{fat} = 6.8 + 0.97 \, protein.$$

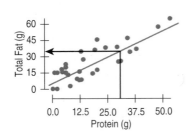

With the estimated linear model it's easy to predict fat content for any menu item we want. For example, for the BK Broiler chicken sandwich with 30 grams of *protein*, we can plug in 30 grams for the amount of *protein* and see that the *predicted fat* content is 6.8 + 0.97(30) = 35.9 grams of fat. Because the BK Broiler chicken sandwich actually has 25 grams of fat, its residual is

Burger King menu items in their natural units with the regression line. **Figure 8.2**

$$fat - \widehat{fat} = 25 - 35.9 = -10.9 \text{ g.}$$

Least squares lines are commonly called regression lines. In a few pages we'll see that this name is an accident of history. For now, you just need to know that "regression" almost always means "the linear model fit by least squares."

Clearly, regression and correlation are closely related. We'll need to check the same conditions for regressions as we did for correlation: the **Quantitative Variables Condition**, the **Straight Enough Condition**, and the **Outlier Condition**.

Calculating a Regression Equation Step-By-Step

In Chapter 7 we examined a scatterplot of annual cost per person due to traffic delays, against peak period freeway speed, and saw that slower speeds were linearly associated with higher costs. That's all very well, but it doesn't answer the all-important (when it comes to costs) question "how much?" Given a city's peak period freeway *speed*, what would we predict the annual *cost* per person to be? That's the kind of question that might be raised when a city considers budget items for highway improvement or for public transportation. The answer lies in the regression model. So let's find it.

 Think

Plan State the problem.

Variables Identify the variables and report the W's.

I want to estimate costs per person associated with traffic delays.

✔ **Quantitative Variables Condition:** I know two quantitative variables, the annual cost per person ($) and the peak period freeway speed (mph) for 70 U.S. cities in 2000. These data come from the 2002 Urban Mobility Report issued by the Texas Transportation Institute.

Just as we did for correlation, check the conditions for a regression by making a picture. Never fit a regression without looking at the scatterplot first.

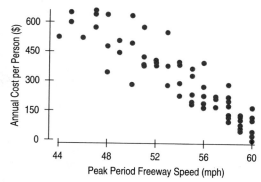

✔ The **Straight Enough Condition** is satisfied, as the scatterplot shows.

✔ **Outlier Condition:** No outliers are evident in the scatterplot.

It's okay to model this relationship with a regression line.

Show

Mechanics Find the equation of the regression line. Summary statistics give the building blocks of the calculation.

(We generally report summary statistics to one more digit of accuracy than the data. We do the same for intercepts and predicted values, but for slopes we usually report an additional digit. Remember, though, not to round off until you finish computing an answer.)

Annual cost per person:
Mean = $298.96
SD = $180.830

Peak period freeway speed:
Mean = 54.34 mph
SD = 4.494 mph

Correlation
$r = -0.90$

Find the slope, b_1.

$$b_1 = \frac{rs_y}{s_x} = \frac{(-0.90)(180.83)}{4.494} = -36.21$$

Find the intercept, b_0.

$$b_0 = \bar{y} - b_1\bar{x}$$
$$= 298.96 - (-36.21)(54.34)$$
$$= \$2266.61$$

Write the equation of the model using meaningful variable names.

The model is

$$\widehat{cost} = 2266.61 - 36.21\ speed.$$

Tell

Conclusion Interpret what you have found in the proper context.

The model suggests traffic delays cost each urban area resident about $36 per year for each mile per hour the freeways are slowed at peak periods.

A S **Find a Regression Equation.** Now that we've done it by hand, try it with technology. That's the way we'll usually find regressions.

Correlation and the Line

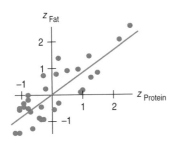

The Burger King scatterplot in z-scores. **Figure 8.3**

How can we get from the correlation coefficient to the equation of the linear model? It turns out that it's not a very big step.

We learned a lot about how correlation worked in Chapter 7 by looking at a scatterplot of the standardized variables. Let's standardize the protein and fat content of the BK items to see what we can learn about the linear model. Here's a scatterplot of z_y (standardized *fat*) vs. z_x (standardized *protein*) along with the least squares line. What's the slope of this line?

We know that $b_1 = r\frac{s_y}{s_x}$, but here we are working with standardized variables whose standard deviations are both 1. So for the regression of z_y on z_x, $b_1 = r$.

What about the intercept? Look at the plot. We know that $b_0 = \bar{y} - b_1\bar{x}$. But standardized variables have zero means, so we just get $b_0 = 0$; the line goes through the origin.

Wow! This line has an equation that's about as simple as we could possibly hope for:

$$\hat{z}_y = rz_x.$$

Great. It's simple, but what does it tell us? It says that in moving one standard deviation from the mean in x, we can expect to move about r standard deviations away from the mean in y. Now that we're thinking about regression lines, the correlation is more than just a vague measure of strength of association. It's a great way to think about what regression tells us.

Let's be more specific. For the sandwiches, the correlation is 0.83. If we standardize both protein and fat, we can write:

$$\hat{z}_{fat} = 0.83\, z_{protein}.$$

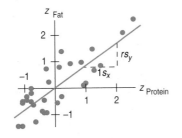

Standardized fat vs. standardized protein with the regression line. Each one standard deviation change in *protein* results in a predicted change of r standard deviations in *fat*. **Figure 8.4**

But we don't have to standardize the two variables to get the benefit of this regression. It tells us that for every standard deviation above (or below) the mean a sandwich is in protein, we'll predict that its fat content is 0.83 standard deviations above (or below) the mean fat content.

A double hamburger has 31 grams of protein, about 1 SD above the mean.

How much fat should you expect it to have? Putting 1.0 in for z_x in the model gives a $\hat{z}_y$ value of 0.83. If you trust the model, you'd expect the fat content to be about 0.83 fat SDs above the mean fat level. Moving one standard deviation away from the mean in x moves us r standard deviations away from the mean in y.

If $r = 0$, there's no linear relationship. The line is horizontal, and no matter how many standard deviations you move in x, the predicted value for y doesn't change. On the other hand, if $r = 1.0$ or -1.0, there's a perfect linear association. In that case, moving any number of standard deviations in x moves exactly the same number of standard deviations in y. In general, moving any number of standard deviations in x moves r times that number of standard deviations in y.

just checking

A scatterplot of house prices (in thousands of dollars) vs. house size (in thousands of square feet) shows a relationship that is straight, with only moderate scatter and no outliers. The correlation between house price and house size is 0.85 and the equation of the regression model is:

$$\widehat{Price} = 9.564 + 122.74\ size.$$

1 What does the slope of 122.74 mean?

2 What are the units?

3 How much can a homeowner expect the value of his house to increase if he builds on an additional 2000 sq ft?

4 If a house is 1 SD above the mean in size (making it about 2170 sq ft), how many SDs above the mean would you predict its price to be?

How Big Can Predicted Values Get?

Sir Francis Galton was the first to speak of "regression," although others had fit lines to data by the same method.

The First Regression

Sir Francis Galton related the heights of sons to the heights of their fathers with a regression line. The slope of his line was less than 1. That is, sons of tall fathers were tall, but not as much above the average height as their fathers had been above their mean. Sons of short fathers were short, but generally not as far from their mean as their fathers. Galton interpreted the slope correctly as indicating a "regression" toward the mean height—and "regression" stuck as a description of the method he had used to find the line.

Suppose you were told that a new male student was about to join the class, and you were asked to guess his height in inches. What would be your guess? A good guess would be the mean height of male students. Now suppose you are also told that this student has a grade point average (*GPA*) of 3.9—about 2 SDs above the mean *GPA*. Would that change your guess? Probably not. There's no correlation between *GPA* and *height*, so knowing the *GPA* value doesn't tell you anything and doesn't move your guess. (And the formula tells us that as well, since it says that we should move 0×2 SDs from the mean.)

On the other hand, if you were told that, measured in centimeters, the student's height was 2 SDs above the mean, you'd know his height in inches. There's a perfect correlation between *height in inches* and *height in centimeters*, so you know he's 2 SDs above mean height in inches as well. (The formula would tell us to move 1.0×2 SDs from the mean.)

What if you're told that the student is 2 SDs above the mean in *shoe size*? Would you still guess that he's of average *height*? You might guess that he's taller than average, since there's a positive correlation between *height* and *shoe size*. But would you guess that he's 2 SDs above the mean? When there was no correlation, we didn't move away from the mean at all. With a perfect correlation, we moved our guess the full 2 SDs. Any correlation between these extremes should lead us to move somewhere between 0 and 2 SDs above the mean. (To be exact, the formula tells us to move $2 \times r$ standard deviations away from the mean.)

Notice that we can't ever move more than 2 SDs away, since r can't be bigger than 1.0. So, each predicted y tends to be closer to its mean (in standard deviations) than its corresponding x was. This property of the linear model is called **regression to the mean,** and that's where we got the term **regression line.**

● **A tale of two regressions** Regression slopes may not behave exactly the way you'd expect at first. If we have a regression model,

$$\hat{y} = b_0 + b_1 x,$$

it might seem natural to think that by solving the equation for *x* we'd get a model for predicting *x* from *y*. But that doesn't work. Such a model needs to evaluate an $\hat{x}$ based on a value of *y*. We don't have *y* in our model, only $\hat{y}$, and that makes all the difference. The model doesn't fit the data values perfectly, and the least squares criterion focuses on the vertical errors the model makes in using $\hat{y}$ to model y—not on horizontal errors related to x.

A quick look at the equations reveals why. Simply solving our equation for x would give a new line whose slope is the reciprocal of ours. To model y in terms of x, our slope is $b_1 = r\frac{s_y}{s_x}$. To model x in terms of y, we'd need to use the slope $b_1 = r\frac{s_x}{s_y}$. Notice that it's *not* the reciprocal of ours. Sure, if the correlation, r, were 1.0 or −1.0 the slopes *would* be reciprocals, but that would happen only if we had a perfect fit. Real data don't follow perfect straight lines, so in the real world y and $\hat{y}$ aren't the same, r is a fraction, and the slopes of the two models are not simple reciprocals of one another.

If the standard deviations were equal—for example, if we standardize both variables—the two slopes would be *the same.* Far from being reciprocals, both would be equal to the correlation. But we already knew that the correlation of x with y is the same as the correlation of y with x.

Moral of the story: *Think.* (Where have you heard *that* before?) Decide which variable you want to use to predict values for the other. Then find the model that does that. If, later, you want to make predictions in the other direction, you'll need to start over and create the other model from scratch. ●

Residuals Revisited

The linear model we are using assumes that the relationship between the two variables is a perfect straight line. The residuals are the part of the data that *hasn't* been modeled. We can write

$$Data = Model + Residual$$

or, equivalently,

$$Residual = Data - Model.$$

Or, in symbols,

$$e = y - \hat{y}.$$

When we want to know how well the model fits, we can ask instead what the model missed. To see that, we look at the residuals.

Residuals help us to see whether the model makes sense. When a regression model is appropriate, it should model the underlying relationship. Nothing interesting should be left behind. So after we fit a regression model, we usually plot the residuals in the hope of finding . . . nothing.

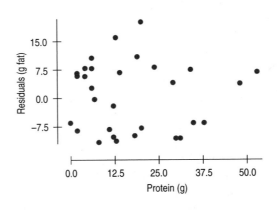

The residuals for the BK menu regression look appropriately boring. **Figure 8.5**

We find the standard deviation of the residuals in almost the way you'd expect:

$$s_e = \sqrt{\frac{\sum e^2}{n-2}}$$

We don't need to subtract the mean because $\bar{e} = 0$. Why $n - 2$ rather than $n - 1$? We used $n - 1$ for s when we estimated the mean. Now we're estimating both a slope and an intercept. Looks like a pattern—and it is. We subtract one more for each parameter we estimate.

A scatterplot of the residuals versus the x-values should be the most boring scatterplot you've ever seen. It shouldn't have any interesting features, like a direction or shape. It should stretch horizontally, with about the same amount of scatter throughout. It should show no bends, and it should have no outliers. If you see any of these features, find out what the regression model missed.

Most computer statistics packages plot the residuals against the predicted values $\hat{y}$, rather than against x. When the slope is negative, the two versions are mirror images. When the slope is positive, they're virtually identical except for the axis labels. Since all we care about is the patterns (or, better, lack of patterns) in the plot, it really doesn't matter which way we plot the residuals.

If the residuals show no interesting pattern when we plot them against x, we can look at how big they are. After all, we're trying to make them as small as possible. Since their mean is always zero, though, it's only sensible to look at how much they vary. The standard deviation of the residuals, s_e, gives us a measure of how much the points spread around the regression line.

For the Burger King foods, the standard deviation of the residuals is 9.2 grams of fat. That looks about right in the scatterplot of residuals. The residual for the BK Broiler chicken was −11 grams, just over one standard deviation.

R^2—The Variation Accounted For

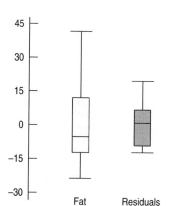

Compare the variability of total fat with the residuals from the regression. The means have been subtracted to make it easier to compare spreads. The variation left in the residuals is unaccounted for by the model, but it's less than the variation in the original data.   **Figure 8.6**

The variation in the residuals is the key to assessing how well the model fits. Let's compare the variation of the response variable with the variation of the residuals. The total *fat* has a standard deviation of 16.4 grams. The standard deviation of the residuals is 9.2 grams. If the correlation were 1.0 and the model predicted the *fat* values perfectly, the residuals would all be zero and have no variation. We couldn't possibly do any better than that.

On the other hand, if the correlation were zero, the model would simply predict 23.5 grams of *fat* (the mean) for all menu items. The residuals from that prediction would just be the observed *fat* values minus their mean. These residuals would have the same variability as the original data because, as we know, just subtracting the mean doesn't change the spread.

How well does the BK regression model do? Look at the boxplots. The variation in the residuals is smaller than in the data, but certainly bigger than zero. That's nice to know, but how much of the variation is still left in the residuals? If you had to put a number between 0% and 100% on the fraction of the variation left in the residuals, what would you say?

All regression models fall somewhere between the two extremes of zero correlation and perfect correlation. We'd like to gauge where our model falls. Can we use the correlation to do that? Well, a regression model with correlation −0.5 is doing as well as one with correlation +0.5. They just have different directions. But if we *square* the correlation coefficient, we'll get a value between 0 and 1, and the direction won't matter. The squared correlation, r^2, gives the fraction of the data's variation accounted for by the model, and $1 - r^2$ is the fraction of the original variation left in the residuals. For the Burger King model, $r^2 = 0.83^2 = 0.69$, and $1 - r^2$ is 0.31, so 31% of the variability in total *fat* has been left in the residuals. How close was that to your guess?

All regression analyses include this statistic, although by tradition, it is written with a capital letter, R^2, and pronounced "R-squared." An R^2 of 0 means that none of the variance in the data is in the model; all of it is still in the residuals. It would be hard to imagine using that model for anything.

Because R^2 is a fraction of a whole, it is often given as a percentage.[3] For the Burger King data, R^2 is 69%.

When interpreting a regression model, you need to *Tell* what R^2 means. According to our linear model, 69% of the variability in the fat content of Burger King sandwiches is accounted for by variation in the protein content.

> ● **How can we see that R^2 is really the fraction of variance accounted for by the model?** It's a simple calculation. The variance of the fat content of the Burger King foods is $16.4^2 = 268.42$. If we treat the residuals as data, the variance of the residuals is 83.195.[4] As a fraction, that's $83.195/268.42 = 0.31$, or 31%. That's the fraction of the variance that is *not* accounted for by the model. The fraction that is accounted for is $100\% - 31\% = 69\%$, just the value we got for R^2. ●

Back to our regression of house price (in thousands of $) on house size (in thousands of square feet):

$$\widehat{Price} = 9.564 + 122.74\ size.$$

The R^2 value is reported as 71.4%.

just checking

⑤ What does the R^2 value mean about the relationship of price and size?

⑥ Is the correlation of price and size positive or negative? How do you know?

⑦ If we measured the size in thousands of square meters instead of thousands of square feet, would the R^2 value change? How about the slope?

Is a correlation of 0.80 twice as strong as a correlation of 0.40? Not if you think in terms of R^2. A correlation of 0.80 means an R^2 of $0.80^2 = 64\%$. A correlation of 0.40 means an R^2 of $0.40^2 = 16\%$ —only a quarter as much of the variability accounted for. A correlation of 0.80 gives an R^2 four times as strong as a correlation of 0.40 and accounts for four times as much of the variability.

How Big Should R^2 Be?

Some Extreme Tales
One major company developed a method to differentiate between proteins. To do so, they had to distinguish between regressions with R^2 of 99.99% and 99.98%. For this application, 99.98% was not high enough.

The president of a financial services company reports that although his regressions give R^2 below 2%, they are highly successful because those used by his competition are even lower.

R^2 is always between 0% and 100%. But what's a "good" R^2 value? The answer depends on the kind of data you are analyzing and on what you want to do with it. Just as with correlation, there is no value for R^2 that automatically determines that the regression is "good." Data from scientific experiments often have R^2 in the 80% to 90% range and even higher. Data from observational studies and surveys, though, often show relatively weak associations because it's so difficult to measure reliable responses. An R^2 of 50% to 30% or even lower might be taken as evidence of a useful regression. The standard deviation of the residuals can give us more information about the usefulness of the regression by telling us how much scatter there is around the line.

As we've seen, an R^2 of 100% is a perfect fit, with no scatter around the line. The s_e would be zero. All of the variance is accounted for by the model and none is left in the residuals at all. This sounds great, but it's too good to be true for real data.[5]

Along with the slope and intercept for a regression, you should always report R^2 so that readers can judge for themselves how successful the regression is at fitting the data. Statistics is about variation, and R^2 measures the success of the regression model in terms of the fraction of the variation of y accounted for by the

---

[3] By contrast, we usually give correlation coefficients as decimal values between -1.0 and 1.0.

[4] This isn't quite the same as squaring the s_e that we discussed on the previous page, but it's very close. We'll deal with the distinction in Chapter 27.

[5] If you see an R^2 of 100%, it's a good idea to figure out what happened. You may have discovered a new law of Physics, but it's much more likely that you accidentally regressed two variables that measure the same thing.

regression. R^2 is the first part of a regression that many people look at because, along with the scatterplot, it tells whether the regression model is even worth thinking about.

Assumptions and Conditions

The linear regression model is perhaps the most widely used model in all of Statistics. It has everything we could want in a model: two easily estimated parameters, a meaningful measure of how well the model fits the data, and the ability to predict new values. It even provides a self-check in plots of the residuals, to help us avoid silly mistakes.

Like all models, though, linear models apply only when certain assumptions are true, so we'd better think about whether they're reasonable. It makes no sense to make a scatterplot of categorical variables, and even less to perform a regression on them. Always check the **Quantitative Variables Condition** to be sure a regression is appropriate.

The linear model assumes that the relationship between the variables is linear. If you try to model a curved relationship with a straight line, you'll usually get exactly what you deserve. But a scatterplot will let you check that the assumption is reasonable.

> **Make a Picture**
>
> Check the scatterplot. The shape must be linear or we can't use regression at all. And watch out for outliers.

The **Straight Enough Condition** is satisfied if the scatterplot looks reasonably straight. It's a good idea to check linearity again *after* computing the regression, when we can examine the residuals. You should also check for outliers, which could change the regression. If the data seem to clump or cluster in the scatterplot, that could be a sign of trouble worth looking into.

If the scatterplot is not straight enough, stop here. You can't use a linear model for *any* two variables, even if they are related. They must have a *linear* association or the model won't mean a thing. Some nonlinear relationships can be saved by re-expressing the data to make the scatterplot more linear.

And you must watch out for outliers. Check the **Outlier Condition** to make sure no point should be singled out for special attention. Outlying values may have large residuals, and squaring makes their influence that much greater. Outlying points can dramatically change a regression model. Outliers can even change the sign of the slope, misleading us about the underlying relationship between the variables. We'll see examples in the next chapter.

Regression Step-By-Step

Even if you hit the fast food joints for lunch, you should have a good breakfast. Researchers recorded facts about 77 breakfast cereals, including the *calories* and *sugar* content (in grams) of a serving. Let's build a linear model to understand how calories are related to sugar content.

Think

Plan State the problem.

I am interested in the relationship between sugar content and calories in cereals.

Variables Name the variables and report the W's.

✔ **Quantitative Variables Condition:** I have two quantitative variables, calories and sugar content, measured on 77 breakfast cereals.

Check the conditions for a regression by making a picture. Never fit a regression without looking at the scatterplot first.

The units of measurement are calories and grams of sugar, respectively.

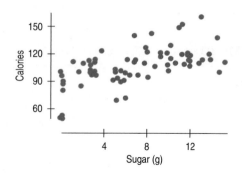

✔ **Outlier Condition:** There are no obvious outliers or groups.

✔ The **Straight Enough Condition** is satisfied; I will fit a regression model to these data.

Show

Mechanics If there are no clear violations of the condition, fit a straight line model of the form $\hat{y} = b_0 + b_1 x$ to the data. Summary statistics give the building blocks of the calculation.

Calories
$$\bar{y} = 107.0 \text{ calories}$$
$$s_y = 19.5 \text{ calories}$$

Sugars
$$\bar{x} = 7.0 \text{ grams}$$
$$s_x = 4.4 \text{ grams}$$

Correlation
$$r = 0.564$$

Find the slope.

$$b_1 = \frac{r s_y}{s_x} = \frac{0.564(19.5)}{4.4}$$

$$= 2.50 \text{ calories per gram of sugar.}$$

Find the intercept.

$$b_0 = \bar{y} - b_1\bar{x} = 107 - 2.50(7) = 89.5 \text{ calories.}$$

So the least squares line is

$$\hat{y} = 89.5 + 2.50\,x,$$

Write the equation using meaningful variable names.

or $\widehat{calories} = 89.5 + 2.50 \text{ sugar.}$

State the value of R^2.

Squaring the correlation gives

$$R^2 = 0.564^2 = 0.318 \text{ or } 31.8\%.$$

Tell

Conclusion Describe what the model says in words and numbers. Be sure to use the names of the variables and their units.

The scatterplot shows a positive, linear relationship and no outliers. The least squares regression line fit through these data has the equation

$\widehat{calories} = 89.5 + 2.50 \text{ sugar.}$

The key to interpreting a regression model is to start with the phrase "b_1 *y*-units per *x*-unit," substituting the estimated value of the slope for b_1 and the names of the respective units. The intercept is then a starting or base value.

The slope says that cereals gain about 2.50 calories per gram of sugar.

The intercept predicts that sugar-free cereals would average about 89.5 calories.

R^2 gives the fraction of the variability of y accounted for by the linear regression model.

The R^2 says that 31.8% of the variability in *calories* is accounted for by variation in *sugar* content.

Find the residuals standard deviation, s_e, and compare it to the original s_y.

$s_e = 16.2$ calories. That's smaller than the original SD of 19.5, but still fairly large.

Think AGAIN

Check Again Even though we looked at the scatterplot *before* fitting a regression model, a plot of the residuals is essential to any regression analysis because it is the best check for additional patterns and interesting quirks in the data.

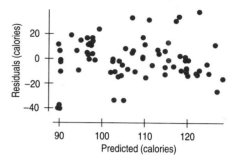

The residuals show a horizontal direction, a shapeless form, and roughly equal scatter for all predicted values. The linear model appears to be appropriate.

Reality Check: Is the Regression Reasonable?

Statistics don't come out of nowhere. They are based on data. The results of a statistical analysis should reinforce your common sense, not fly in its face. If the results are surprising, then either you've learned something new about the world or your analysis is wrong.

Whenever you perform a regression, think about the coefficients and ask whether they make sense. Is a slope of 2.5 calories per gram of sugar reasonable? That's hard to say right off. We know from the summary statistics that a typical cereal has about 100 calories and 7 grams of sugar. A gram of sugar contributes some calories (actually, 4, but you don't need to know that), so calories should go up with increasing sugar. The direction of the slope seems right.

To see if the *size* of the slope is reasonable, a useful trick is to consider its order of magnitude. We'll start by asking if decreasing the slope by a factor of 10 seems reasonable. Is 0.25 calories per gram of sugar enough? The 7 grams of sugar found in the average cereal would contribute less than 2 calories. That seems too small.

Now let's try inflating the slope by a factor of 10. Is 25 calories per gram reasonable? Then the average cereal would have 175 calories from sugar alone. The average cereal has only 100 calories per serving, though, so that slope seems too big.

We have tried inflating the slope by a factor of 10 and deflating it by 10 and found both to be unreasonable. So, like Goldilocks, we're left with the value in the middle that's just right. And an increase of 2.5 calories per gram of sugar is certainly *plausible*.

The small effort of asking yourself whether the regression equation is plausible is repaid whenever you catch errors or avoid saying something silly or absurd

> **Adjective, Noun, or Verb**
> You may see the term *"regression"* used in different ways. There are many ways to fit a line to data, but the term "regression line" or "regression" without any other qualifiers always means least squares. People also use regression as a verb when they speak of *regressing* a *y*-variable on an *x*-variable to mean fitting a linear model.

about the data. It's too easy to take something that comes out of a computer at face value and assume that it makes sense.

Always be skeptical and ask yourself if the answer is reasonable.

What Can Go Wrong?

There are many ways in which data that appear at first to be good candidates for regression analysis may be unsuitable. And there are ways that people use regression that can lead them astray. Here's an overview of the most common problems. We'll discuss these at length in the next chapter.

- **Don't fit a straight line to a nonlinear relationship.** Linear regression is suited only to relationships that are, well, *linear*. Fortunately, we can often improve the linearity easily by using re-expression. We'll come back to this topic in Chapter 10.
- **Beware of extraordinary points.** Data values can be extraordinary in a regression in two ways. They can have *y*-values that stand off from the linear pattern suggested by the bulk of the data, or extreme *x*-values. Both kinds of extraordinary points require attention.
- **Don't extrapolate beyond the data.** A linear model will often do a reasonable job of summarizing a relationship in the narrow range of observed *x*-values. Once we have a working model for the relationship, it's tempting to use it. But beware of predicting *y*-values for *x*-values that lie outside the range of the original data. The model may no longer hold there, so such **extrapolations** too far from the data are dangerous.
- **Don't infer that x causes y just because there is a good linear model for their relationship.** When two variables are strongly correlated, it is often tempting to assume a causal relationship between them. Putting a regression line on a scatterplot tempts us even further, but it doesn't make the assumption of causation any more valid.
- **Don't choose a model based on R^2 alone.** Although R^2 measures the *strength* of the linear association, a high R^2 does not demonstrate the *appropriateness* of the regression. A single outlier, or data that separate into two groups rather than a single cloud of points, can make the R^2 seem quite large when, in fact, the linear regression model is simply inappropriate. Conversely, a low R^2 value may be due to a single outlier as well. It may be that most of the data fall roughly along a straight line, with the exception of a single point. Always look at the scatterplot.

> The R^2 does **not** mean that *protein* accounts for 69% of the *fat* in a BK food item. It is the *variation* in fat content that is accounted for by the linear model.

CONNECTIONS

We've talked about the importance of models before, but have seen only the Normal model as an example. The linear model is one of the most important models in Statistics. Chapter 7 talked about the assignment of variables to the *y*- and *x*-axes. That didn't matter to correlation, but it does matter to regression because *y* is predicted by *x* in the regression model.

The connection of R^2 to correlation is obvious, although it may not be immediately clear that just by squaring the correlation we can learn the fraction of the variability of *y* accounted for by a regression on *x*. We'll return to this in subsequent chapters.

We made a big fuss about knowing the units of your quantitative variables. We didn't need units for correlation, but without the units we can't define the slope of a regression. A regression makes no sense if you don't know the *Who*, the *What*, and the *Units* of both your variables.

We've summed squared deviations before when we computed the standard deviation and variance. That's not coincidental. They are closely connected to regression.

When we first talked about models, we noted that deviations away from a model were often interesting. Now we have a formal definition of these deviations as residuals.

What have we learned?

We've learned that when the relationship between quantitative variables is fairly straight, a linear model can help summarize that relationship and give us insights about it.

- The regression (best fit) line doesn't pass through all the points, but it is the best compromise in the sense that the sum of squares of the residuals is the smallest possible.

We've learned several ways the correlation, r, tells us about the regression:

- The slope of the line is based on the correlation, adjusted for the units of x and y. We've learned to interpret that slope in context.
- For each SD in x that we are away from the x mean, we expect to be r SDs in y away from the y mean.
- Because r is always between -1 and $+1$, each predicted y is fewer SDs away from its mean than the corresponding x was, a phenomenon called regression to the mean.
- The square of the correlation coefficient, R^2, gives us the fraction of the variation of the response accounted for by the regression model. The remaining $1 - R^2$ of the variation is left in the residuals.

The residuals also reveal how well the model works.

- If a plot of residuals against predicted values shows a pattern, we should re-examine the data to see why.
- The standard deviation of the residuals, s_e, quantifies the amount of scatter around the line.

TERMS

Model   An equation or formula that simplifies and represents reality.

Linear model   A linear model is an equation of the form

$$\hat{y} = b_0 + b_1 x.$$

To interpret a linear model we need to know the variables (along with their W's) and their units.

Residuals   Residuals are the differences between data values and the corresponding values predicted by the regression model—or, more generally, values predicted by any model.

$$\text{Residual} = \text{observed value} - \text{predicted value} = y - \hat{y}$$

Predicted value   The value of $\hat{y}$ found for each x-value in the data. A predicted value is found by substituting the x-value in the regression equation. The predicted values are the values on the fitted line; the points $(x, \hat{y})$ all lie exactly on the fitted line.

Slope   The slope gives a value in "y-units per x-unit." Changes of one unit in x are associated with changes of b_1 units in predicted values of y.

Regression to the mean   Because the correlation is always less than 1.0 in magnitude, each predicted $\hat{y}$ tends to be fewer standard deviations from its mean than its corresponding x was from its mean. This is called regression to the mean.

Regression line **Line of best fit**	The particular linear equation

$$\hat{y} = b_0 + b_1 x$$

that satisfies the least squares criterion is called the least squares regression line. Casually, we often just call it the regression line, or the line of best fit.

Intercept	The intercept, b_0, gives a starting value in *y*-units. It's the $\hat{y}$-value when *x* is 0.
Least squares	The least squares criterion specifies the unique line that minimizes the variance of the residuals or, equivalently, the sum of the squared residuals.
R^2	• R^2 is the square of the correlation between *y* and *x*. • R^2 gives the fraction of the variability of *y* accounted for by the least squares linear regression on *x*. • R^2 is an overall measure of how successful the regression is in linearly relating *y* to *x*.

S K I L L S

When you complete this lesson you should:

Think

- Be able to identify response (*y*) and explanatory (*x*) variables in context.
- Understand how a linear equation summarizes the relationship between two variables.
- Recognize when a regression should be used to summarize a linear relationship between two quantitative variables.
- Be able to judge whether the slope of a regression makes sense.
- Know how to examine your data for violations of the Straight Enough Condition that would make it inappropriate to compute a regression.
- Understand that the least squares slope is easily affected by extreme values.
- Know that residuals are the differences between the data values and the corresponding values predicted by the line and that the *least squares criterion* finds the line that minimizes the sum of the squared residuals.
- Know how to use a plot of residuals against predicted values to check the Straight Enough Condition or look for outliers.

Show

- Know how to find a regression equation from the summary statistics for each variable and the correlation between the variables.
- Know how to find a regression equation using your statistics software and how to find the slope and intercept values in the regression output table.
- Know how to use regression to predict a value of *y* for a given *x*.
- Know how to compute the residual for each data value and how to display them.

Tell

- Be able to write a sentence explaining what a linear equation says about the relationship between *y* and *x*, basing it on the fact that the slope is given in *y-units per x-unit*.
- Understand how the correlation coefficient and the regression slope are related. Know how R^2 describes how much of the variation in *y* is accounted for by its linear relationship with *x*.
- Be able to describe a prediction made from a regression equation, relating the predicted value to the specified *x*-value.

Regression on the Computer

All statistics packages make a table of results for a regression. These tables may differ slightly from one package to another, but all are essentially the same—and all include much more than we need to know for now. Every computer regression table includes a section that looks something like this:

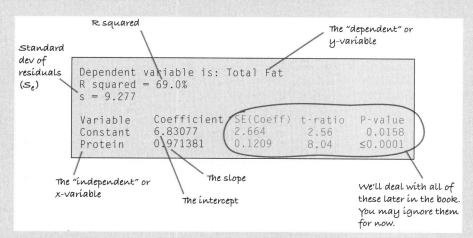

R squared

The "dependent" or y-variable

Standard dev of residuals (S_e)

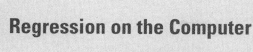 **Finding Least Squares Lines.** We almost always use technology to find regressions. Practice now—just in time for the exercises.

```
Dependent variable is: Total Fat
R squared = 69.0%
s = 9.277

Variable    Coefficient   SE(Coeff)   t-ratio   P-value
Constant    6.83077       2.664       2.56      0.0158
Protein     0.971381      0.1209      8.04      ≤0.0001
```

The "independent" or x-variable

The slope

The intercept

We'll deal with all of these later in the book. You may ignore them for now.

The slope and intercept coefficient are given in a table such as this one. Usually the slope is labeled with the name of the *x*-variable, and the intercept is labeled "Intercept" or "Constant." So the regression equation shown here is

$$\widehat{Fat} = 6.83077 + 0.97138 \; Protein.$$

It is not unusual for statistics packages to give many more digits of the estimated slope and intercept than could possibly be estimated from the data. (The original data were reported to the nearest gram.) Ordinarily, you should round the reported numbers to one digit more than the precision of the data, and the slope to two. We will learn about the other numbers in the regression table later in the book. For now, all you need to be able to do is find the coefficients and the R^2 value.

DATA DESK

Select the *y*-variable and the *x*-variable. In the **Plot** menu choose **Scatterplot**. From the scatterplot HyperView menu choose **Add Regression Line** to display the line. From the HyperView menu choose **Regression** to compute the regression.

Comments

Alternatively, find the regression first with the **Regression** command in the **Calc** menu. Click on the *x*-variable's name to open a menu that offers the scatterplot.

EXCEL

Make a scatterplot of the data. With the scatterplot frontmost, select **Add Trendline . . .** from the **Chart** menu. Click the **Options** tab and select **Display Equation on Chart.** Click **OK.**

Comments

The computer section for Chapter 7 shows how to make a scatterplot. We don't repeat those steps here.

JMP

Choose **Fit Y by X** from the **Analyze** menu. Specify the *y*-variable in the Select Columns box and click the **"Y, Response"** button. Specify the *x*-variable and click the **"X, Factor"** button. Click **OK** to make a scatterplot. In the scatterplot window, click on the red triangle beside the heading labeled "Bivariate Fit . . ." and choose **"Fit Line."** JMP draws the least squares regression line on the scatterplot and displays the results of the regression in tables below the plot.

MINITAB

Choose **Regression** from the **Stat** menu. From the Regression submenu, choose **Fitted Line Plot.** In the Fitted Line Plot dialog, click in the **Response Y** box, and assign the *y*-variable from the Variable list. Click in the **Predictor X** box, and assign the *x*-variable from the Variable list. Make sure that the Type of Regression Model is set to Linear. Click the **OK** button.

SPSS

Choose **Interactive** from the **Graphs** menu. From the Interactive Graphs submenu, choose **Scatterplot.** In the Create Scatterplot dialog, drag the *y*-variable into the **y-axis target,** and the *x*-variable into the **x-axis target.** Click on the **Fit** tab. Choose **Regression** from the **Method** popup menu. Click the **OK** button.

TI-83/84 Plus

To find the equation of the regression line (add the line to a scatterplot) choose **LinReg(a+bx),** tell it the list names, and then add a comma to specify a function name (from **VARS Y-Vars 1:Function**). The final command looks like

$$\text{LinReg (a+bx) L1, L2, Y1.}$$

To make a residuals plot, set up a **STATPLOT** as a scatterplot. Specify your explanatory datalist as Xlist. For Ylist import the name **RESID** from the LIST NAMES menu. ZoomStat will now create the residuals plot.

Comments

Each time you execute a **LinReg** command, the calculator automatically computes the residuals and stores them in a data list named RESID. If you want to see them, go to **STAT EDIT.** Space through the names of the lists until you find a blank. Import RESID from the LIST NAMES menu. Now every time you have the calculator compute a regression analysis, it will show you the residuals.

TI-89

To find the equation of the regression line (and add the line to a scatterplot) choose **LinReg (a+bx)** from the **Calc Regressions** menu, tell it the list names, and a function to store the equation. To make a residuals plot, define a **PLOT** as a scatterplot.

Specify your explanatory datalist as Xlist. For Ylist find the list name **resid** from VAR-LINK by arrowing to the **STATVARS** portion, then press [2] **(r)** and locate the list. Press [ENTER] to finish the plot definition and [F5] to display the plot.

Comments

Each time you execute a **LinReg** command, the calculator automatically computes the residuals and stores them in a data list named RESID. If you don't want to see this (or any other calculator-generated list) any more, press [F1] (Tools) and select choice 3: Setup Editor. Leaving the box for lists to display blank will reset the calculator to show only lists 1 through 6.

EXERCISES

1. Regression equations. Fill in the missing information in the table below.

	$\overline{x}$	s_x	$\overline{y}$	s_y	r	$\hat{y} = b_0 + b_1 x$
a)	10	2	20	3	0.5	
b)	2	0.06	7.2	1.2	-0.4	
c)	12	6			-0.8	$\hat{y} = 200 - 4x$
d)	2.5	1.2		100		$\hat{y} = -100 + 50x$

2. More regression equations. Fill in the missing information in the table below.

	$\overline{x}$	s_x	$\overline{y}$	s_y	r	$\hat{y} = b_0 + b_1 x$
a)	30	4	18	6	-0.2	
b)	100	18	60	10	0.9	
c)		0.8	50	15		$\hat{y} = -10 + 15x$
d)			18	4	-0.6	$\hat{y} = 30 - 2x$

3. Residuals. Tell what each of the residual plots below indicates about the appropriateness of the linear model that was fit to the data.

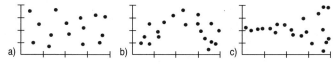

4. Residuals. Tell what each of the residual plots below indicates about the appropriateness of the linear model that was fit to the data.

5. Least squares. Consider the four points (10, 10), (20, 50), (40, 20), and (50, 80). The least squares line is $\hat{y} = 7.0 + 1.1x$. Explain what "least squares" means using these data as a specific example.

6. Least squares. Consider the four points (200, 1950), (400, 1650), (600, 1800), and (800, 1600). The least squares line is $\hat{y} = 1975 - 0.45x$. Explain what "least squares" means using these data as a specific example.

7. Real estate. A random sample of records of sales of homes from Feb. 15 to Apr. 30, 1993, from the files maintained by the Albuquerque Board of Realtors gives the *price* and *size* (in square feet) of 117 homes. A regression to predict *price* (in thousands of dollars) from *size* has an R-squared of 71.4%. The residuals plot indicated that a linear model is appropriate.
a) What are the variables and units in this regression?
b) What units does the slope have?
c) Do you think the slope is positive or negative? Explain.

8. Roller coaster. People who responded to a July 2004 Discovery Channel poll named the 10 best roller coasters in the United States. A table in the last chapter's exercises shows the length of the initial drop (in feet) and the duration of the ride (in seconds). A regression to predict *duration* from *drop* has $R^2 = 12.4\%$.
a) What are the variables and units in this regression?
b) What units does the slope have?
c) Do you think the slope is positive or negative? Explain.

9. Real estate again. The regression of *price* on *size* of homes in Albuquerque had $R^2 = 71.4\%$, as described in Exercise 7. Write a sentence (in context, of course) summarizing what the R^2 says about this regression.

10. Coasters again. Exercise 8 examined the association between the *duration* of a roller coaster ride and the height of its initial *drop*, reporting that $R^2 = 12.4\%$. Write a sentence (in context, of course) summarizing what the R^2 says about this regression.

11. Real estate redux. The regression of *price* on *size* of homes in Albuquerque had $R^2 = 71.4\%$, as described in Exercise 7.
a) What is the correlation between *size* and *price*?
b) What would you predict about the *price* of a home one standard deviation above average in *size*?
c) What would you predict about the *price* of a home two standard deviations below average in *size*?

12. Another ride. The regression of *duration* of a roller coaster ride on the height of its initial *drop*, described in Exercise 8, had $R^2 = 12.4\%$.
a) What is the correlation between *drop* and *duration*?
b) What would you predict about the *duration* of the ride on a coaster whose initial *drop* was 1 standard deviation below the mean *drop*?
c) What would you predict about the *duration* of the ride on a coaster whose initial *drop* was 3 standard deviations above the mean *drop*?

13. More real estate. Consider the Albuquerque home sales from Exercise 7 again. The regression analysis gives

the model $\widehat{price}$ = 47.82 + 0.061 *size*.
a) Explain what the slope of the line says about housing prices and house size.
b) What price would you predict for a 3000-square-foot house in this market?
c) A real estate agent shows a potential buyer a 1200-square-foot home, saying that the asking price is $6000 less than what one would expect to pay for a house of this size. What is the asking price, and what is the $6000 called?

14. Last ride. Consider the roller coasters described in Exercise 8 again. The regression analysis gives the model $\widehat{duration}$ = 91.033 + 0.242 *drop*.
a) Explain what the slope of the line says about how long a roller coaster ride may last and the height of the coaster.
b) A new roller coaster advertises an initial drop of 200 feet. How long would you predict the rides last?
c) Another coaster with a 150-foot initial drop advertises a 2-minute ride. Is this longer or shorter than you'd expect? By how much? What's that called?

15. Cigarettes. Is the nicotine content of a cigarette related to the "tars"? A collection of data (in milligrams) on 29 cigarettes produced the scatterplot, residuals plot, and partial regression analysis shown:

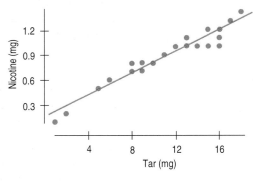

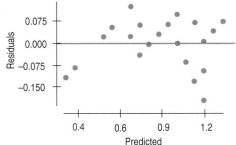

Dependent variable is: nicotine
R squared = 92.4%

Variable	Coefficient
Constant	0.154030
Tar	0.065052

a) Do you think a linear model is appropriate here? Explain.
b) Explain the meaning of R^2 in this context.

16. Baseball. In the last chapter you looked at the relationship between the number of *games won* by American League baseball teams and the *average attendance* at their home games for the first half of the 2001 season. Here are the scatterplot, the residuals plot, and part of the regression analysis.

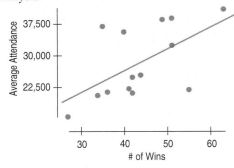

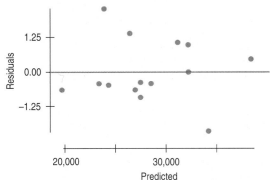

Dependent variable is: Attendance
R^2 = 33.3%

Variable	Coefficient
Intercept	5773.27
Wins	517.609

a) Do you think a linear model is appropriate here? Explain.
b) Interpret the meaning of R^2 in this context.

17. Another cigarette. Consider again the regression of *nicotine* content on *tar* (both in milligrams) for the cigarettes examined in Exercise 15.
a) What is the correlation between *tar* and *nicotine*?
b) What would you predict about the average *nicotine* content of cigarettes that are 2 standard deviations below average in *tar* content?
c) If a cigarette is 1 standard deviation above average in *nicotine* content, what do you suspect is true about its *tar* content?

18. Second inning. Consider again the regression of *average attendance* on *wins* for the baseball teams examined in Exercise 16.

a) What is the correlation between *wins* and *average attendance?*

b) What would you predict about the *average attendance* for a team that is 2 standard deviations above average in games won?

c) If a team is 1 standard deviation below average in attendance, what would you predict about the number of games the team has won?

19. Last cigarette. Take another look at the regression analysis of tar and nicotine content of the cigarettes in Exercise 15.

a) Write the equation of the regression line.

b) Estimate the *nicotine* content of cigarettes with 4 milligrams of *tar.*

c) Interpret the meaning of the slope of the regression line in this context.

d) What does the *y*-intercept mean?

e) If a new brand of cigarette contains 7 milligrams of tar and a nicotine level whose residual is –0.5 mg, what is the nicotine content?

20. Last inning. Refer again to the regression analysis for average attendance and games won by American League baseball teams, seen in Exercise 16.

a) Write the equation of the regression line.

b) Estimate the *average attendance* for a team with 50 *wins.*

c) Interpret the meaning of the slope of the regression line in this context.

d) In general, what would a negative residual mean in this context?

e) The St. Louis Cardinals are not included in these data because they are a National League team. At the time these data were collected, the Cardinals had won 43 games and averaged 38,988 fans at their home games. Calculate the residual for this team, and explain what it means.

21. What slope? If you create a regression model for predicting the *weight* of a car (in pounds) from its *length* (in feet), is the slope most likely to be 3, 30, 300, or 3000? Explain.

22. What slope? If you create a regression model for estimating the *height* of a pine tree (in feet) based upon the *circumference* of its trunk (in inches), is the slope most likely to be 0.1, 1, 10, or 100? Explain.

23. Misinterpretations. A Biology student who created a regression model to use a bird's *height* when perched for predicting its *wingspan* made these two statements. Assuming the calculations were done correctly, explain what is wrong with each interpretation.

a) My R^2 of 93% shows that this linear model is appropriate.

b) A bird 10 inches tall will have a wingspan of 17 inches.

24. More misinterpretations. A Sociology student investigated the association between a country's *literacy rate* and *life expectancy,* then drew the conclusions listed below. Explain why each statement is incorrect. (Assume that all the calculations were done properly.)

a) The *literacy rate* determines 64% of the *life expectancy* for a country.

b) The slope of the line shows that an increase of 5% in *literacy rate* will produce a 2-year improvement in *life expectancy.*

25. ESP. People who claim to "have ESP" participate in a screening test in which they have to guess which of several images someone is thinking of. You and a friend both took the test. You scored 2 standard deviations above the mean, and your friend scored 1 standard deviation below the mean. The researchers offer everyone the opportunity to take a retest.

a) Should you choose to take this retest? Explain.

b) Now explain to your friend what his decision should be and why.

26. SI jinx. Players in any sport who are having great seasons, turning in performances that are much better than anyone might have anticipated, often are pictured on the cover of *Sports Illustrated.* Frequently, their performances then falter somewhat, leading some athletes to believe in a "*Sports Illustrated* jinx." Similarly, it is common for phenomenal rookies to have less stellar second seasons—the so-called "sophomore slump." While fans, athletes, and analysts have proposed many theories about what leads to such declines, a statistician might offer a simpler (statistical) explanation. Explain.

Ⓣ 27. SAT scores. The SAT is a test often used as part of an application to college. SAT scores are between 200 and 800, but have no units. Tests are given in both Math and Verbal areas. Doing the SAT-Math problems also involves the ability to read and understand the questions, but can a person's verbal score be used to predict the math score? Verbal and math SAT scores of a high school graduating class are displayed in the scatterplot, with the regression line added.

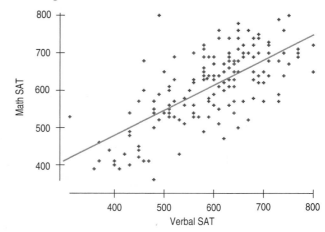

a) Describe the relationship.
b) Are there any students whose scores do not seem to fit the overall pattern?
c) For these data, $r = 0.685$. Interpret this statistic.
d) These verbal scores averaged 596.3, with a standard deviation of 99.5, and the math scores averaged 612.2, with a standard deviation of 96.1. Write the equation of the regression line.
e) Interpret the slope of this line.
f) Predict the math score of a student with a verbal score of 500.
g) Every year some student scores a perfect 1600. Based on this model, what would that student's residual be for her math score?

28. Success in college. Colleges use SAT scores in the admissions process because they believe these scores provide some insight into how a high school student will perform at the college level. Suppose the entering freshmen at a certain college have mean combined SAT scores of 1222, with a standard deviation of 83. In the first semester these students attained a mean GPA of 2.66, with a standard deviation of 0.56. A scatterplot showed the association to be reasonably linear, and the correlation between SAT score and GPA was 0.47.
a) Write the equation of the regression line.
b) Explain what the y-intercept of the regression line indicates.
c) Interpret the slope of the regression line.
d) Predict the GPA of a freshman who scored a combined 1400.
e) Based upon these statistics, how effective do you think SAT scores would be in predicting academic success during the first semester of the freshman year at this college? Explain.
f) As a student, would you rather have a positive or a negative residual in this context? Explain.

29. SAT, take 2. Suppose we wanted to use SAT math scores to estimate verbal scores based on the information in Exercise 27.
a) What is the correlation?
b) Write the equation of the line of regression predicting verbal scores from math scores.
c) In general, what would a positive residual mean in this context?
d) A person tells you her math score was 500. Predict her verbal score.
e) Using that predicted verbal score and the equation you created in Exercise 27, predict her math score.
f) Why doesn't the result in part e come out to 500?

30. Success, part 2. Based on the statistics for college freshmen given in Exercise 28, what SAT score might be expected among freshmen who attained a first-semester GPA of 3.0?

T 31. Used cars. Classified ads in the *Ithaca Journal* offered several used Toyota Corollas for sale. Listed below are the ages of the cars and the advertised prices.

Age (yr)	Price Advertised ($)
1	12995, 10950
2	10495
3	10995, 10995
4	6995, 7990
5	8700, 6995
6	5990, 4995
9	3200, 2250, 3995
11	2900, 2995
13	1750

a) Make a scatterplot for these data.
b) Describe the association between age and price of a used Corolla.
c) Do you think a linear model is appropriate?
d) Computer software says that $R^2 = 0.894$. What is the correlation between age and price?
e) Explain the meaning of R^2 in this context.
f) Why doesn't this model explain 100% of the variability in the price of a used Corolla?

32. Drug abuse. In the exercises of the last chapter you examined results of a survey conducted in the United States and 10 countries of Western Europe to determine the percentage of teenagers who had used marijuana and other drugs. Below is the scatterplot. Summary statistics showed that the mean percent that had used marijuana was 23.9%, with a standard deviation of 15.6%. An average of 11.6% of teens had used other drugs, with a standard deviation of 10.2%.

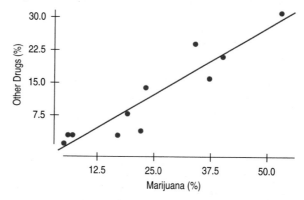

a) Do you think a linear model is appropriate? Explain.
b) For this regression, R^2 is 87.3%. Interpret this statistic in this context.
c) Write the equation you would use to estimate the percentage of teens who use other drugs from the percentage who have used marijuana.
d) Explain in context what the slope of this line means.

e) Do these results confirm that marijuana is a "gateway drug," that is, that marijuana use leads to the use of other drugs?

T 33. More used cars. Use the advertised prices for Toyota Corollas given in Exercise 31 to create a linear model for the relationship between a car's *age* and its *price*.
a) Find the equation of the regression line.
b) Explain the meaning of the slope of the line.
c) Explain the meaning of the *y*-intercept of the line.
d) If you want to sell a 7-year-old Corolla, what price seems appropriate?
e) You have a chance to buy one of two cars. They are about the same age and appear to be in equally good condition. Would you rather buy the one with a positive residual or a negative residual? Explain.
f) You see a "For Sale" sign on a 10-year-old Corolla stating the asking price as $1500. What is the residual?
g) Would this regression model be useful in establishing a fair price for a 20-year-old car? Explain.

34. Veggie burgers. Recently Burger King introduced a meat-free burger. The nutrition label is shown here.

Nutrition Facts

Calories	330
Fat	10g*
Sodium	760g
Sugars	5g
Protein	14g
Carbohydrates	43g
Dietary Fiber	4g
Cholesterol	0

* (2 grams of saturated fat)

RECOMMENDED DAILY VALUES *
(based on a 2,000-calorie/day diet)

Iron	20%
Vitamin A	10%
Vitamin C	10%
Calcium	6%

a) Use the regression model created in this chapter, $\widehat{fat} = 6.8 + 0.97\ protein$, to predict the fat content of this burger from its protein content.
b) What is its residual? How would you explain the residual?
c) Write a brief report about the *fat* and *protein* content of this new menu item. Be sure to talk about the variables by name and in the correct units.

35. Burgers. In the last chapter you examined the association between the amounts of *fat* and *calories* in fast-food hamburgers. Here are the data:

Fat (g)	19	31	34	35	39	39	43
Calories	410	580	590	570	640	680	660

a) Create a scatterplot of *calories* vs. *fat content*.
b) Interpret the value of R^2 in this context.
c) Write the equation of the line of regression.
d) Use the residuals plot to explain whether your linear model is appropriate.
e) Explain the meaning of the *y*-intercept of the line.
f) Explain the meaning of the slope of the line.
g) A new burger containing 28 grams of fat is introduced. According to this model, its residual for calories is +33. How many calories does the burger have?

36. Chicken. Chicken sandwiches are often advertised as a healthier alternative to beef because many are lower in fat. Tests on 11 brands of fast food chicken sandwiches produced the following summary statistics and scatterplot from a graphing calculator:

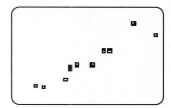

	Fat (g)	Calories
Mean	20.6	472.7
St. Dev.	9.8	144.2
Correlation		0.947

a) Do you think a linear model is appropriate in this situation?
b) Describe the strength of this association.
c) Write the equation of the regression line.
d) Explain the meaning of the slope.
e) Explain the meaning of the *y*-intercept.
f) What does it mean if a certain sandwich has a negative residual?
g) If a chicken sandwich and a burger each advertised 35 grams of fat, which would you expect to have more calories? (See Exercise 35.)
h) McDonald's Filet-O-Fish sandwich has 26 grams of fat and 470 calories. Does the fat/calorie relationship in this sandwich appear to be very different from that found in chicken sandwiches or in burgers (see Exercise 35)? Explain.

37. A second helping of burgers. In Exercise 35 you created a model that can estimate the number of *calories* in a burger when the *fat* content is known.
a) Explain why you cannot use that model to estimate the *fat* content of a burger with 600 *calories*.
b) Using an appropriate model, estimate the *fat* content of a burger with 600 *calories*.

T 38. Cost of living. The *Worldwide Cost of Living Survey City Rankings* determine the cost of living in the 25 most expensive cities in the world. These rankings scale New

York City as 100, and express the cost of living in other cities as a percentage of the New York cost. For example, the table indicates that in Tokyo the cost of living was 65% higher than New York in 2000, but dropped to only 34% higher than NY in 2001.

City	2001	2000	City	2001	2000
Tokyo	134.0	164.9	Shenzhen	90.8	100.1
Moscow	132.4	136.1	Ho Chi Minh City	90.4	92.7
Hong Kong	130.0	141.5	Singapore	86.4	94.7
Beijing	124.4	138.3	White Plains, NY	85.5	84.0
Osaka	116.7	143.6	Tel Aviv	85.0	88.7
Shanghai	114.3	128.0	San Francisco	84.4	83.7
St. Petersburg	106.5	109.7	Chicago	84.3	82.7
New York	100.0	100.0	Kiev	83.8	78.2
Guangzhou	97.4	107.9	Beirut	83.8	86.2
Seoul	95.3	111.1	Buenos Aires	83.6	91.0
Hanoi	94.3	94.9	Los Angeles	83.4	83.1
Taipei	92.9	102.8	Miami	83.0	83.8
London	92.9	106.9			

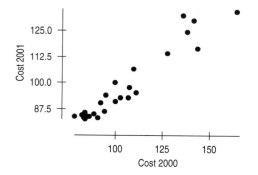

a) Here is a scatterplot. Describe the association between costs of living in 2000 and 2001.
b) The correlation is 0.957. Find and interpret the value of R^2.
c) The regression equation predicting the 2001 cost of living from the 2000 figure is $\widehat{cost01} = 25.41 + 0.69 \times cost00$. Use this equation to find the residual for Moscow.
d) Explain what the residual means.

39. Cell phones. Most people prefer digital cell phone service to analog because it is clearer and more private. Nonetheless, many calls are switched to analog channels because digital lines are busy, or because the calls are between people with different carriers. These analog calls require more power, so they are harder on phone batteries. How much battery time do you lose when making analog calls?

Consumer Reports tested seven popular brands of cell phones, comparing digital and analog talk times. The

scatterplot appears to show a linear relationship with $r = 0.94$ and regression equation $\widehat{An} = -5.1 + 0.44\,dig$.
a) If you have a cell phone claiming to provide 200 minutes of digital calls between battery chargings, how many minutes of analog calling would you expect to get?
b) How much faith do you place in that prediction? Explain.
c) Explain what the slope of the line means in this context.
d) If a certain brand of battery has a positive residual, what does that mean in this context?

T 40. Candy. The table shows the increase in Halloween candy sales over a 7-year period as reported by the National Confectioners Association. Using these data, predict the amount of sales for 2002. Discuss the appropriateness of your model and your faith in the prediction. (Enter *Year* as 95, 96, . . . , 101.)

Year	Halloween Candy Sales (millions of dollars)
1995	1.474
1996	1.660
1997	1.708
1998	1.787
1999	1.896
2000	1.985
2001	2.035

41. El Niño. Concern over the weather associated with El Niño has increased interest in the possibility that the climate on earth is getting warmer. The most common theory relates an increase in atmospheric levels of carbon dioxide (CO_2), a greenhouse gas, to increases in temperature. Here is a scatterplot showing the mean annual CO_2 concentration in the atmosphere measured in parts per million (ppm) at the top of Mauna Loa in Hawaii, and the mean annual air temperature over both land and sea across the globe, in degrees Celsius (C).

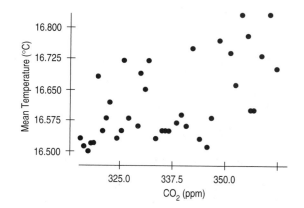

A regression predicting *temperature* from CO_2 *level* produces the following output table (in part):

```
Dependent variable is: Temperature
R-squared = 33.4%
```

Variable	Coefficient
Intercept	15.3066
CO_2	0.004

a) What is the correlation between CO_2 *level* and *temperature?*
b) Explain the meaning of *R*-squared in this context.
c) Give the regression equation.
d) What is the meaning of the slope in this equation?
e) What is the meaning of the *y*-intercept of this equation?
f) Here is a scatterplot of the residuals vs. CO_2 *level*. Does this plot show evidence of the violation of any assumptions behind the regression? If so, which ones?

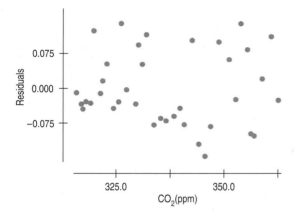

g) CO_2 *levels* will probably reach 364 ppm this year. What *mean temperature* does the regression predict from that information?

42. Birth rates. The table shows the number of live births per 1000 women aged 15–44 years in the United States, starting in 1965. (*National Vital Statistics Report*, April 2001)

Year	1965	1970	1975	1980	1985	1990	1995	1999
Rate	19.4	18.4	14.8	15.9	15.6	16.4	14.8	14.5

a) Make a scatterplot and describe the general trend in *birth rates*. (Enter *Year* as 65, 70, 75, etc.)
b) Find the equation of the regression line.
c) Check to see if the line is an appropriate model. Explain.
d) Interpret the slope of the line.
e) The table gives rates only at 5-year intervals. Estimate what the rate was in 1978.

f) In 1978 the birth rate was actually 15.0. How close did your model come?
g) Predict what the *birth rate* will be in 2005. Comment on your faith in this prediction.
h) Predict the *birth rate* for 2020. Comment on your faith in this prediction.

T 43. Body fat. It is difficult to accurately determine a person's body fat percentage without immersing him or her in water. Researchers hoping to find ways to make a good estimate immersed 20 male subjects, then measured their waists and recorded their weights.

Waist (in.)	Weight (lb)	Body Fat (%)
32	175	6
36	181	21
38	200	15
33	159	6
39	196	22
40	192	31
41	205	32
35	173	21
38	187	25
38	188	30
33	188	10
40	240	20
36	175	22
32	168	9
44	246	38
33	160	10
41	215	27
34	159	12
34	146	10
44	219	28

a) Create a model to predict *%body fat* from *weight*.
b) Do you think a linear model is appropriate? Explain.
c) Interpret the slope of your model.
d) Is your model likely to make reliable estimates? Explain.
e) What is the residual for a person who weighs 190 pounds and has 21% body fat?

T 44. Body fat, again. Would a model that uses the person's *waist* size be able to predict the *%body fat* more accurately than one that uses *weight*? Using the data in Exercise 43, create and analyze that model.

T 45. Heptathlon. We discussed the women's 2000 Olympic heptathlon in Chapter 6. Here are the results from the high jump, 800-meter run, and long jump for the 26 women who successfully completed all three events:

Name		Country	High Jump (m)	800-m (sec)	Long Jump (m)
Y	Azizi	(Alg)	1.60	141.82	5.88
G	Bacher	(Ita)	1.75	129.08	5.84
S	Biswas	(Ind)	1.63	142.17	5.64
L	Blonska	(Ukr)	1.72	133.52	5.57
S	Braun	(Ger)	1.81	139.14	6.22
K	Ertl	(Ger)	1.78	136.25	6.22
PG	Ganapathy	(Ind)	1.69	140.86	5.96
M	Garcia	(Cub)	1.66	139.64	5.92
T	Hautala	(Fin)	1.78	134.90	6.12
J	Jamieson	(Aus)	1.81	136.57	6.09
S	Kabanova	(Uzb)	1.72	140.11	5.22
S	Kazanina	(Kaz)	1.75	130.45	5.84
D	Koritskaia	(Rus)	1.72	129.77	5.56
D	Lewis	(Gbr)	1.75	136.83	6.48
M	Mark	(Tri)	1.66	152.36	5.90
D	Nathan	(USA)	1.78	136.67	6.06
I	Naumenko	(Kaz)	1.84	138.49	5.88
L	Necheporuk	(Ukr)	1.72	139.94	5.89
Y	Prokhorova	(Rus)	1.81	130.32	6.59
S	Rajamaki	(Fin)	1.66	138.47	6.36
N	Roshchupkina	(Rus)	1.84	132.24	5.47
N	Sazanovich	(Blr)	1.84	136.41	6.50
A	Skujyte	(Lit)	1.78	140.25	5.97
N	Teppe	(Fra)	1.72	138.56	5.94
VS	Tigau	(Rom)	1.72	139.65	6.01
U	Wlodarczyk	(Pol)	1.78	132.15	6.31

Let's examine the association among these events. Perform a regression to predict high jump performance from the 800-meter results.

a) What is the regression equation? What does the slope mean?

b) What is the R^2 value?

c) Do good high jumpers tend to be fast runners? (Be careful—low times are good for running events and high distances are good for jumps.)

d) What does the residuals plot reveal about the model?

e) Do you think this is a useful model? Would you use it to predict high jump performance? (Compare the residual standard deviation to the standard deviation of the high jumps.)

T 46. Heptathlon again. We saw the data for the women's 2000 Olympic heptathlon in Exercise 45. Are the two jumping events associated? Perform a regression of the long jump results on the high jump results.

a) What is the regression equation? What does the slope mean?

b) What is the R^2 value?

c) Do good high jumpers tend to be good long jumpers?

d) What does the residuals plot reveal about the model?

e) Do you think this is a useful model? Would you use it to predict long jump performance? (Compare the residual standard deviation to the standard deviation of the long jumps.)

T 47. Hard water. In an investigation of environmental causes of disease, data were collected on the annual mortality rate (deaths per 100,000) for males in 61 large towns in England and Wales. In addition, the water hardness was recorded as the calcium concentration (parts per million, ppm) in the drinking water. The following display shows the relationship between *mortality* and *calcium* concentration for these towns:

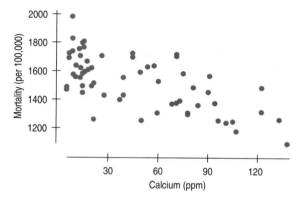

a) Describe what you see in this scatterplot, in context.

b) Here is the regression analysis of *mortality* and *calcium* concentration. What is the regression equation?

Dependent variable is: Mortality
R-squared = 43%
s = 143.0

Variable	Coefficient	SE(Coeff)	t-ratio	P-value
Intercept	1676	29.30	57.2	<0.0001
Calcium	−3.23	0.48	−6.66	<0.0001

c) Interpret the slope and y-intercept of the line, in context.

d) The largest residual, with a value of −348.6, is for the town of Exeter. Explain what this value means.

e) The hardness of Derby's municipal water is about 100 ppm of calcium. Use this equation to predict the mortality rate in Derby.

f) Explain the meaning of R-squared in this situation.

48. Gators. Wildlife researchers monitor many wildlife populations by taking aerial photographs. Can they estimate the weights of alligators accurately from the air? Here is a regression analysis of the *weight* of alligators (in pounds) and their *length* (in inches) based on data collected from captured alligators.

Dependent variable is: Weight
R-squared = 83.6%
s = 54.01

Variable	Coefficient	SE(Coeff)	t-ratio	P-value
Intercept	−393	47.53	−8.27	<0.0001
Length	5.9	0.5448	10.8	<0.0001

a) Did they choose the correct variable to use as the dependent variable and the predictor? Explain.
b) What is the correlation between an alligator's length and weight?
c) Write the regression equation.
d) Interpret the slope of the equation in this context.
e) Do you think this equation will allow the scientists to make accurate predictions about alligators? What part of the regression analysis indicates this? What additional concerns do you have?

T 49. Internet. The rapid growth of Internet publishing is seen in the number of electronic academic journals made available during the last decade. Explain why you cannot use the methods of this chapter to model this growth. (Enter *Year* as 91, 92, etc.)

Year	Number of Journals
1991	27
1992	36
1993	45
1994	181
1995	306
1996	1093
1997	2459

T 50. Oil production. In the exercises for Chapter 7, you looked at the oil production of the United States from 1949 to 2000. We provide the data again.
a) Fit a least squares regression line to oil production by year. (If you type the data in yourself, you should enter *Year* as 49, 50, . . . , 100.)
b) Using this regression line, predict U.S. oil production in the year 2001.
c) Does the prediction in part b look reasonable? Comment.
d) Do you think the regression line is an appropriate model? Comment.

Year	Oil	Year	Oil
1949	1,841,940	1975	3,056,779
1950	1,973,574	1976	2,976,180
1951	2,247,711	1977	3,009,265
1952	2,289,836	1978	3,178,216
1953	2,357,082	1979	3,121,310
1954	2,314,988	1980	3,146,365
1955	2,484,428	1981	3,128,624
1956	2,617,283	1982	3,156,715
1957	2,616,901	1983	3,170,999
1958	2,448,987	1984	3,249,696
1959	2,574,590	1985	3,274,553
1960	2,574,933	1986	3,168,252
1961	2,621,758	1987	3,047,378
1962	2,676,189	1988	2,979,123
1963	2,752,723	1989	2,778,773
1964	2,786,822	1990	2,684,687
1965	2,848,514	1991	2,707,039
1966	3,027,763	1992	2,624,632
1967	3,215,742	1993	2,499,033
1968	3,329,042	1994	2,431,476
1969	3,371,751	1995	2,394,268
1970	3,517,450	1996	2,366,017
1971	3,453,914	1997	2,354,831
1972	3,455,368	1998	2,281,919
1973	3,360,903	1999	2,146,732
1974	3,202,585	2000	2,135,062

just checking

Answers

1. An increase of home size of 1000 square feet is associated with an increase of $122,740, on average, in price.

2. Units are thousands of dollars per thousands of square feet.

3. About $245,480, on average

4. About 0.85 SDs above the mean price

5. Differences in the size of houses account for about 71.4% of the variation in the house prices.

6. It's positive. The correlation and the slope have the same sign.

7. The correlation won't change, so neither will R^2. The slope will change because the units have changed.

Regression Wisdom

Regression may be the most widely used statistics method. It is used every day throughout the world to predict customer loyalty, numbers of admissions at hospitals, sales of automobiles, and many other things. Because regression is so widely used, it's also widely abused and misinterpreted. This chapter presents examples of regressions in which things are not quite as simple as they may have seemed at first, and shows how you can still use regression to discover what the data have to say.

Sifting Residuals for Groups

No regression analysis is complete without a display of the residuals to check that the linear model is reasonable. Because the residuals are what is "left over" after the model describes the relationship, they often reveal subtleties that were not clear from a plot of the original data. Sometimes these are additional details that help confirm or refine our understanding. Sometimes they reveal violations of the regression conditions that require our attention.

In the Step-by-Step analysis of the cereal data from Chapter 8, we examined a scatterplot of the residuals. Our first impression was that it had no particular structure—a conclusion that supported using the regression model. But let's look again.

It's often a good idea to look at a histogram of the residuals. How would you describe the shape of this one? It looks like there might be small modes on both sides of the central body of the data. A group of cereals seems to stand out as having large negative residuals, with fewer calories than we might have predicted. The calories in these cereals were overestimated by the model. Whenever we suspect multiple modes, we ask whether they are somehow different.

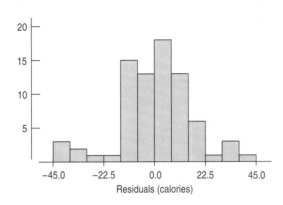

A histogram of the regression residuals shows small modes both above and below the central large mode. These may be worth a second look. **Figure 9.1**

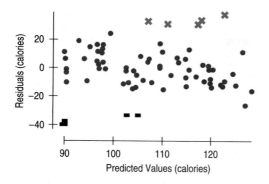

A scatterplot of the residuals vs. predicted values for the cereal regression. The green "x" points are cereals whose calorie content is higher than the linear model predicts. The red "–" points show cereals with fewer calories than the model predicts. Is there something special about these cereals? **Figure 9.2**

Let's look more carefully at the residual scatterplot. Here's the same residual plot from Chapter 8, with the points in those modes marked. Now we can see that those two groups stand away from the central pattern in the scatterplot. If we examine the data set, we find that the high-residual cereals are Just Right Fruit & Nut; Muesli Raisins, Dates & Almonds; Peaches & Pecans; Mueslix Crispy Blend; and Nutri-Grain Almond Raisin. Do these cereals seem to have something in common? These high-calorie cereals all present themselves as "healthy." This might be surprising, but in fact, "healthy" cereals often contain more fat and therefore more calories than we might expect from looking at their sugar content alone.

The low-residual cereals are Puffed Rice, Puffed Wheat, three bran cereals, and Golden Crisps. These cereals have fewer calories than we would expect based on their sugar content. We might not have grouped these cereals together before. What they have in common is a low calorie count *relative to their sugar content*—even though their sugar contents are quite different.

These observations may not lead us to question the overall linear model, but they do help us understand that other factors may be part of the story. An examination of residuals often leads us to discover groups of observations that are different from the rest.

When we discover that there is more than one group in a regression, we may decide to analyze the groups separately, using a different model for each group. Or we can stick with the original model and simply note that there are groups that are a little different. Either way, the model will be wrong, but useful, and so it will improve our understanding of the data.

Subsets

> Here's an important unstated condition for fitting models: **All the data must come from the same group.**

The residuals plot suggests that there may be different "types" of breakfast cereals. Cereal manufacturers aim cereals at different segments of the market. Supermarkets and cereal manufacturers try to attract different customers by placing different types on certain shelves. Cereals for kids tend to be on the "kid's shelf," at their eye level. Toddlers wouldn't be likely to grab a box from this shelf and beg, "Mom, can we please get this All-Bran with Extra Fiber?"

Calories and *sugars* colored according to the shelf on which the cereal was found in a supermarket, with regression lines fit for each shelf individually. Do these data appear homogeneous? That is, do the cereals seem to all be from the same population of cereals? Or are there different kinds of cereals that we might want to consider separately? **Figure 9.3**

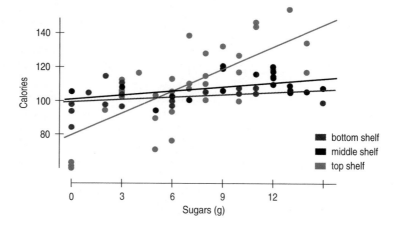

Should we take this extra information into account in our analysis? Figure 9.3 shows a scatterplot of *calories* and *sugar*, colored according to the shelf on which the cereals were found and with a separate regression line fit for each. Now we can see that the top shelf is different. We might want to report two regressions, one for the top shelf and one for the bottom two shelves.[1]

Getting the "Bends"

We can't *know* if the **Linearity Assumption** is true, but we can see if it's *plausible* by checking the **Straight Enough Condition.**

The fundamental assumption for working with a linear model is that the relationship you are modeling is, in fact, linear. That sounds obvious, but when you fit a regression, you can't take it for granted. Often it's hard to tell from the scatterplot you looked at before you fit the regression model. Sometimes you can't see a bend in the relationship until you plot the residuals.

Fuel efficiency (mpg) vs. *weight* (thousands of pounds) shows a strong, apparently linear, negative trend. **Figure 9.4**

Here are the *weight* (in thousands of pounds) and *fuel efficiency* (measured in miles per gallon) of 38 cars. We know, from common sense and from physics, that heavier cars generally need more fuel to move. The scatterplot looks fairly linear with a moderately strong negative association ($R^2 = 81.6\%$). The linear regression equation

$$\widehat{mpg} = 48.7 - 8.4\ weight$$

says that *fuel efficiency* drops by 8.4 mpg per 1000 pounds of *weight*, starting from a value of 48.7 mpg (presumably for a weightless car).

The scatterplot of the residuals against *weight* holds a surprise. This plot should have no pattern. Instead, there's a bend, starting high on the left, dropping down in the middle of the display, and then rising again at the right. Graphs of residuals often reveal patterns such as this that were hard to see in the original scatterplot.

Look back at the original scatterplot. The scatter of points isn't really straight. There's a slight bend to the plot, but the bend is much easier to see in the residuals. Even though it means rechecking the Straight Enough Condition *after* we find the regression, it's always a good idea to plot the residuals.

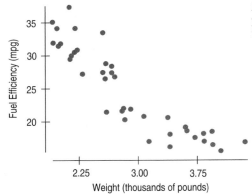

Plotting residuals against *weight* reveals a bend. It was also in the original scatterplot, but here it's easier to see. **Figure 9.5**

---

[1] More complex models can take into account both sugar content and shelf information. This kind of *multiple regression* model (Chapter 29 on the CD) is a natural extension of the model we're using here.

Extrapolation: Reaching Beyond the Data

Predicting Manatee Kills.
Manatees are often killed by collisions with powerboats, and the number killed has risen over time. Can we use regression to predict manatee kills?

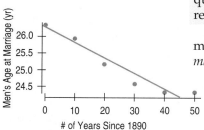

The average age at which men married fell at the rate of about a year every 25 years from 1890 to 1940.

When the Data Are Years . . .

. . . we usually don't enter them as 4-digit numbers. Here we used 0 for 1890, 10 for 1900, and so on. Or we may simply enter two digits, using 88 for 1988, for instance. Rescaling years like this often makes calculations easier and equations simpler. We recommend you do it, too. But be careful; if 1988 is 88, then 2004 is 104 (not 4), right?

Linear models give a predicted value for each case in the data. Put a new x-value into the equation and it gives a predicted value, $\hat{y}$, to go with it. But when the new x-value lies far from the data we used to build the regression, how trustworthy is the prediction?

The simple answer is that the farther the new x-value is from $\bar{x}$, the less trust we should place in the predicted value. Once we venture into new x territory, such a prediction is called an **extrapolation**. Extrapolations are dubious because they require the additional—and very questionable—assumption that nothing about the relationship between x and y changes even at extreme values of x and beyond.

Extrapolations can get us into deep trouble. The U.S. Census Bureau reports the median age at first marriage for men and women. Here's a regression of *age at first marriage* for men against *year* at every census from 1890 to 1940:

R-squared = 92.6%
s = 0.2417

Variable	Coefficient
Intercept	25.7
Year	−0.04

The regression equation is

$$\widehat{age} = 25.7 - 0.04\ year.$$

The slope of −0.04 years of age per year of the century shows that first marriage age for men was falling at a rate of about 4 years per century. That is, men were getting married for the first time at younger and younger ages. In these data, the intercept can be interpreted as the predicted mean age of men at first marriage in 1890 (counted as year 0). The high R^2 suggests a strong linear pattern. Can we extrapolate?

When *year* counts from 0 in 1890, the year 2000 is "110." Substituting 110 for *year*, we find that the model predicts a first marriage age of $25.7 - 0.04 \times 110 = 21.3$ years old. In fact, though, by the year 2000, the median age at first marriage for men was almost 27 years. What's gone wrong?

Here's a scatterplot of the *mean age* at first marriage for men for all the data from 1890 to 1998:

Mean age at first marriage (years of age) for men in the United States vs. *year*. The regression line is fit only to the first 50 years of the data (shown in blue), which looked nicely linear. But the linear pattern could not have continued, and in fact it changed in direction, steepness, and strength.   **Figure 9.6**

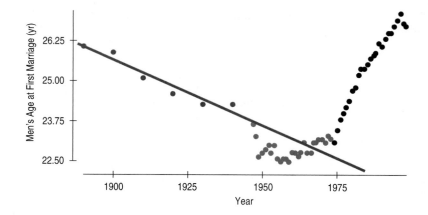

Now we can see why the extrapolation failed. Although the trend in age at first marriage was linear for parts of the century, it did not follow the same linear pattern over the entire century.

Predicting the Future

"Prediction is difficult, especially about the future."
—*Niels Bohr, Danish physicist*

Extrapolation is always dangerous. But when, as in the marriage age predictions, the *x*-variable in a linear model is *time*, extrapolation becomes an attempt to peer into the future. People have always wanted to see into the future. We can foresee that they will always want to see into the future. In the past, seers, oracles, and wizards were called on to predict the future. Today mediums, fortune-tellers, and Tarot card readers still find many customers. The clever ones usually phrase their predictions ambiguously in such a way that they can claim to be right no matter how things actually turn out.

Those with a more scientific outlook may spurn such methods—and then turn around and use a linear model to predict the future. That's just using a digital crystal ball. Linear models are based on the *x*-values of the data at hand and cannot be trusted beyond that span. Some phenomena do exhibit a kind of "inertia" that allows us to guess that current systematic behavior will continue outside this range. When *x* is time, though, you should be especially wary. Such regularity can't be counted on in phenomena such as stock prices, sales figures, hurricane tracks, or public opinion.

Extrapolating from current trends is a mistake made not only by regression beginners or the naïve. Professional forecasters are prone to the same mistakes, and sometimes the errors are striking. In the mid 1970s, in the midst of an energy crisis, oil prices surged and long lines at gas stations were common. In 1970, oil cost about $3 a barrel. A few years later it had surged to $15. In 1975, a survey of 15 top econometric forecasting models (built by groups that included Nobel prize–winning economists) found predictions for 1985 oil prices that ranged from $50 to $200 a barrel (in 1975 dollars). How close were these forecasts? Well, oil prices were actually *lower* in 1985 than in 1975, after accounting for inflation. No one predicted that. No one. The forecasts had all been made *assuming* oil prices would continue to rise at the same rate or even faster. By the year 2000, oil prices averaged about $7 a barrel in 1975 dollars ($28 a barrel in 2000 dollars).

Of course, knowing that extrapolation is dangerous doesn't stop people. The temptation to see into the future is hard to resist. So our more realistic advice is this:

If you must extrapolate into the future, at least don't believe that the prediction will come true.

Outliers, Leverage, and Influence

The outcome of the year 2000 U.S. presidential election was determined in Florida amid much controversy. The main race was between George W. Bush and Al Gore, but two minor candidates played a significant role. To the political right of the main party candidates was Pat Buchanan, while to the political left was Ralph Nader. Generally, Nader earned more votes than Buchanan throughout the state. We would expect counties with larger vote totals to give more votes to each candidate. Here's a regression relating *Buchanan's* vote totals by county in the state of Florida to *Nader's*:

Dependent variable is: Buchanan vote
R-squared = 42.8%

Variable	Coefficient
Intercept	50.3
Nader vote	0.14

The regression model,

$$\widehat{Buchanan} = 50.3 + 0.14 \, Nader,$$

says that in each county, Buchanan received about 0.14 times (or 14% of) the vote Nader received, starting from a base of 50.3 votes.

This seems like a reasonable regression with an R^2 of almost 43%. But we've violated all three Laws of Data Analysis by going straight to the regression table without making a picture.

Here's a scatterplot that shows the vote for Buchanan in each county of Florida plotted against the vote for Nader. The outlying point is Palm Beach County.

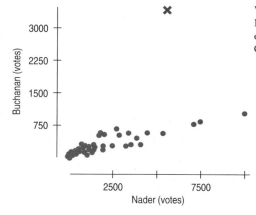

Votes received by Buchanan against votes for Nader in all Florida counties in the presidential election of 2000. The red "x" point is Palm Beach County, home of the "butterfly ballot."   **Figure 9.7**

The so-called "butterfly ballot," used only in Palm Beach County, was a source of controversy. It has been claimed that the format of this ballot confused voters so that some who intended to vote for the Democrat, Al Gore, punched the wrong hole next to his name and, as a result, voted for Buchanan.

The scatterplot shows a strong, positive, linear association, and one striking point. With Palm Beach removed from the regression, the R^2 jumps from 42.8% to 82.1% and the slope of the line changes to 0.1, suggesting that Buchanan received only about 10% of the vote that Nader received. With more than 82% of the variability of the Buchanan vote accounted for, the model when Palm Beach is omitted certainly fits better. Palm Beach County now stands out, not as a Buchanan stronghold, but rather as a clear violation of the model that begs for explanation.

The red line shows the effect that one unusual point can have on a regression.

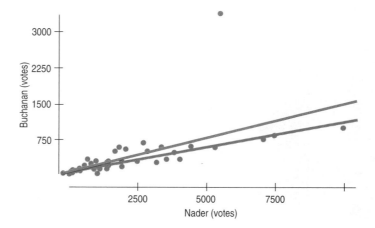

One of the great values of models is that, by establishing an idealized behavior, they help us to see when and how data values are unusual. In regression, we'll focus on two ways in which a point can stand out. First, a data value can have a large residual, as Palm Beach County does in our example. We know that the linear model doesn't fit points with large residuals very well. Because they seem to be different from the other cases, points whose residuals are large always deserve special attention.

A data point can also be unusual if its x-value is far from the mean of the x-values. Such a point is said to have high **leverage.** The physical image of a lever is

A S **Leverage.** You may be surprised to see how sensitive a regression line is as you drag around a single influential point.

exactly right. We know the line must pass through $(\bar{x}, \bar{y})$, so you can picture that point as the fulcrum of the lever. Just as sitting farther from the hinge on a see-saw gives you more leverage to pull it your way, points with values far from $\bar{x}$ pull more strongly on the regression line.

A point with high leverage has the potential to change the regression line. But it doesn't always use that potential. If the point lines up with the pattern of the other points, then including it doesn't change our estimate of the line. By sitting so far from $\bar{x}$, though, it may strengthen the relationship, inflating the correlation and R^2. How can we tell if a high-leverage point actually changes the model? We fit the linear model twice, both with and without the point in question. We say that a point is **influential** if omitting it from the analysis gives a very different model.[2]

Influence depends on both leverage and residual; a case with high leverage whose y-value sits right on the line fit to the rest of the data is not influential. Removing that case won't change the slope, even if it does affect R^2. A case with modest leverage but a very large residual (such as Palm Beach County) can be influential. Of course, if a point has enough leverage, it can pull the line right to it. Then it's highly influential but its residual is small. The only way to be sure is to fit both regressions.

Unusual points in a regression often tell us more about the data and the model than any other points. Whenever you have—or suspect that you have—influential points, you should fit the linear model to the other points alone and then compare the two regression models to understand how they differ. On one hand, a model dominated by a single point is unlikely to be useful for understanding the rest of the cases. On the other hand, the best way to understand unusual points is against the background of the model established by the other data values. (That insight's at least 400 years old. See the sidebar.) Don't give in to the temptation to simply delete points that don't fit the line. You can take points out and discuss what the model looks like with and without them, but arbitrarily deleting points can give a false sense of how well the model fits the data and artificially inflate the R^2.

In 2000 George W. Bush won Florida (and thus the presidency) by only a few hundred votes, so Palm Beach County's residual is big enough to be meaningful. It's the rare unusual point that determines a presidency, but all are worth examining and trying to understand.

A point with so much influence that it pulls the regression line close to it can make its residual deceptively small. Influential points like that can have a shocking effect on the regression. Here's a plot of *IQ* against *shoe size* again from the fanciful study of intelligence and foot size in comedians we saw in Chapter 7. The linear regression output shows:

"For whoever knows the ways of Nature will more easily notice her deviations; and, on the other hand, whoever knows her deviations will more accurately describe her ways."

—Francis Bacon (1561–1626)

Dependent variable is: IQ
R-squared = 24.8%

Variable	Coefficient
Intercept	93.3265
Shoe size	2.08318

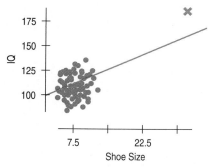

Bozo's extraordinarily large shoes give his data point high leverage in the regression. Wherever Bozo's IQ falls, the regression line will follow. **Figure 9.8**

Although this is a silly example, it illustrates an important and common potential problem. Almost all of the variance accounted for ($R^2 = 24.8\%$) is

---

[2] Some textbooks use the term *influential point* for any observation that influences the slope, intercept, or R^2. We'll reserve the term for points that influence the slope.

due to *one* point, namely, Bozo. Without Bozo, there is little correlation between shoe size and IQ. Look what happens to the regression when we take him out:

Dependent variable is: IQ
R-squared = 0.7%

Variable	Coefficient
Intercept	105.458
Shoe size	0.460194

The R^2 value is now 0.7%—a very weak linear relationship (as one might expect!). One single point exhibits a great influence on the regression analysis.

What would have happened if Bozo hadn't shown his comic genius on IQ tests? Suppose his measured *IQ* had been only 50. The slope of the line drops from 0.96 points/shoe size to −0.69 points/shoe size. No matter where Bozo's *IQ* is, the line tends to follow it because his *shoe size*, being so far from the mean shoe size, makes this a high-leverage point.

You can often see high-leverage points best in the original scatterplot of the data. But it isn't always clear what to do with them. Sometimes these are data values that you had to work hard to get, and they may say more about the relationship between *y* and *x* than any of the other data values. At other times, high-leverage points are values that really don't belong with the rest of the data. Such points should probably be omitted, and a linear model found without them for comparison. The situation may lie between these two extremes. A data value may be far from the other *x*-values, and therefore a point with high leverage, but it may not be clear whether it belongs. When in doubt, it's usually best to perform and compare the two regressions.

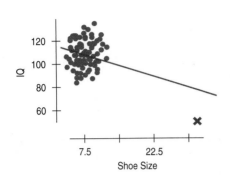

If Bozo's IQ were low, the regression slope would change from positive to negative. A single influential point can change a regression model drastically. **Figure 9.9**

● **Warning:** Influential points can hide in plots of residuals. Points with high leverage pull the line close to them, so they often have small residuals. You'll see influential points more easily in scatterplots of the original data or by finding a regression model with and without the points. ●

Each of these scatterplots shows an unusual point. For each, tell whether the point is a high-leverage point, would have a large residual, or is influential.

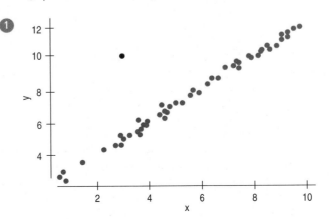

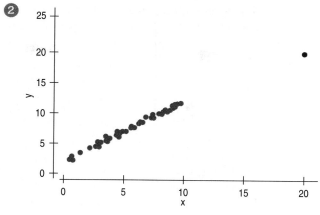

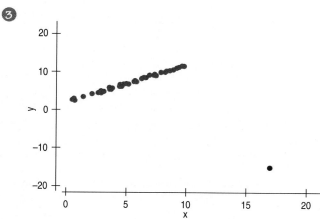

Lurking Variables and Causation

In Chapter 7, we tried to make it clear that no matter how strong the correlation is between two variables, there's no simple way to show that one variable causes the other. Putting a regression line through a cloud of points just increases the temptation to think and to say that the *x*-variable *causes* the *y*-variable. Just to make sure, let's repeat the point again. No matter how strong the association, no matter how large the R^2 value, no matter how straight the line, there is no way to conclude from a regression alone that one variable *causes* the other. There's always the possibility that some third variable is driving both of the variables you have observed. With observational data, as opposed to data from a designed experiment, there is no way to be sure that a **lurking variable** is not the cause of any apparent association.

Here's an example. The scatterplot shows the *life expectancy* (average between men and women, in years) for each of 41 countries of the world, plotted against the square root of the number of *doctors* per person in the country. (We've taken the square root to make the relationship linear. It's not a big deal. Without this re-expression, the plot is too curved to fit a straight line. The square root is not the point. The point is that *life expectancy* goes up with availability of *doctors*. You can learn more about re-expression in regression in more detail in the next chapter.)

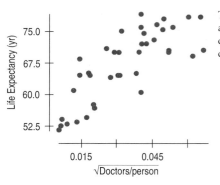

The relationship between *life expectancy* (years) and availability of *doctors* (measured as √doctors per person) for countries of the world shows a strong, positive, linear association. **Figure 9.10**

The strong positive association ($r = 0.79$; $R^2 = 62.4\%$) seems to confirm our expectation that more *doctors* per person improves healthcare, leading to longer lifetimes and a larger *life expectancy*. The strength of the association would *seem* to argue that we should send more doctors to developing countries to increase life expectancy.

That conclusion is about the consequences of a change. If we increase the number of doctors, should we predict that the life expectancy will increase? This is a causal explanation, arguing that adding doctors *causes* greater life expectancy. But these are observed data. Could there be another explanation of the association?

Let's consider another variable. Here's a very similar-looking scatterplot, again with *life expectancy* as the *y*-variable, but this time the *x*-variable is the square root of the number of *televisions* per person in each country. The positive association shown in this scatterplot is even *stronger* than the association in the previous plot ($r = 0.85$; $R^2 = 72.3\%$). We can fit the linear model, and quite possibly use the number of TVs as a way to predict life expectancy. Should we conclude, however, that increasing the number of TVs actually extends lifetimes? If so, we should send TVs instead of doctors to developing countries. Not only is the correlation with life expectancy higher, but TVs are much cheaper than doctors.

What's wrong with this reasoning? Maybe we were a bit hasty earlier when we concluded that doctors *cause* longer lives. Maybe there's a lurking variable here. Countries with higher standards of living have both longer life expectancies *and* more doctors. Could higher living standards *cause* changes in the other variables? If so, then improving living standards might be expected to prolong lives, increase the number of doctors, and increase the number of TVs.

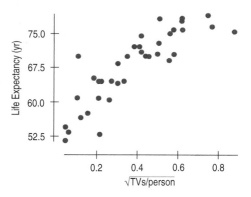

To increase *life expectancy*, don't send doctors, send TVs; they're cheaper and more fun. Or maybe that's not the right interpretation of this scatterplot of *life expectancy* against availability of *TVs* (as √TVs per person). **Figure 9.11**

From this example, you can see how easy it is to fall into the trap of mistakenly inferring causality from a regression. For all we know, doctors (or TVs!) *do* increase life expectancy. But we can't tell that from data like these no matter how much we'd like to. Resist the temptation to conclude that *x* causes *y* from a regression, no matter how obvious that conclusion seems to you.

Working with Summary Values

Scatterplots of statistics summarized over groups tend to show less variability than we would see if we measured the same variable on individuals. This is because the summary statistics themselves vary less than the data on the individuals do—a fact we will make more specific in coming chapters.

In Chapter 7 we looked at the heights and weights of individual students. There we saw a correlation of 0.644, so R^2 is 41.5%.

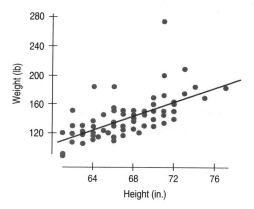

Weight (lb) against *height* (in.) for a sample of men. There's a strong, positive, linear association.

Figure 9.12

Suppose, instead of data on individuals, we had been given only the mean weight for each height value. The scatterplot of mean weight by height would show less scatter. And the R^2 would increase to 80.1%.

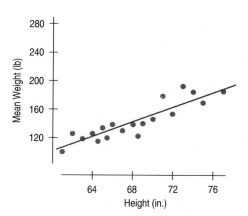

Mean weight (lb) shows a stronger linear association with *height* than do the weights of individuals. Means vary less than individual values. **Figure 9.13**

Scatterplots of summary statistics show less scatter than the baseline data on individuals, and can give a false impression of how well a line summarizes the data. There's no simple correction for this phenomenon. Once we're given summary data, there's no simple way to get the original values back.

In the life expectancy and TVs example, we have no good measure of exposure to doctors or to TV on an individual basis. But if we did, we should expect the scatterplot to show more variability and the corresponding R^2 to be smaller. The bottom line is that you should be a bit suspicious of conclusions based on regressions of summary data. They may look better than they really are.

What Can Go Wrong?

This entire chapter has held warnings about things that can go wrong in a regression analysis. So let's just recap. When you make a linear model:

- *Make sure the relationship is straight.* Check the Straight Enough Condition. Always examine the residuals for evidence that the Linearity Assumption has failed. It's often easier to see deviations from a straight line in the residuals plot than in the scatterplot of the original data. Pay special attention to the most extreme residuals, because they may have something to add to the story told by the linear model.

- *Be on guard for different groups in your regression.* Check for evidence that the data consist of separate subsets. If you find subsets that behave differently, consider fitting a different linear model to each subset.

- *Beware of extrapolating.* Beware of extrapolation beyond the x-values that were used to fit the model. Although it's common to use linear models to extrapolate, the practice is dangerous.

- *Beware especially of extrapolating into the future!* Be especially cautious about extrapolating into the future with linear models. A linear model assumes that changes over time will continue forever at the same rate we've observed in the past. Predicting the future is particularly tempting and particularly dangerous.

- *Look for unusual points.* Unusual points always deserve attention, and may well reveal more about your data than the rest of the points combined. Always look for them and try to understand why they stand apart. A scatterplot of the data is a good way to see high leverage and influential points. A scatterplot of the residuals against the predicted values is a good tool for finding points with large residuals.

- *Beware of high-leverage points, and especially of those that are influential.* Influential points can alter the regression model a great deal. The resulting model may say more about one or two points than about the overall relationship.

- *Consider comparing two regressions.* To see the impact of outliers on a regression, it's often wise to run two regressions, one with and one without the extraordinary points, and then to discuss the differences.

- *Treat unusual points honestly.* If you remove enough carefully selected points, you can always get a regression with a high R^2, eventually. But it won't give you much understanding. Some data are not simple enough for a linear model to fit very well. When that happens, report the failure and stop.

- *Beware of lurking variables.* Think about lurking variables before interpreting a linear model. It's particularly tempting to explain a strong regression by thinking that the x-variable *causes* the y-variable. A linear model fitted to observational data (coming in Chapter 13) can never demonstrate such causation, in part because it cannot eliminate the chance that a lurking variable has caused the variation in both x and y.

- *Watch out when dealing with data that are summaries.* Be cautious in working with data values that are themselves summaries, such as means or medians. Such statistics are less variable than the data on which they are based, so they tend to inflate the impression of the strength of a relationship.

CONNECTIONS

We are always alert to things that can go wrong if we use statistics without thinking carefully. Regression opens new vistas of potential problems. But each one relates to issues we've thought about before.

It is always important that our data be from a single homogeneous group and not made up of disparate groups. We looked for multiple modes in single variables. Now we check scatterplots for evidence of subgroups in our data. As with modes, it's often best to split the data and analyze the groups separately.

Our concern with unusual points and their potential influence also harks back to our earlier concern with outliers in histograms and boxplots—and for many of the same reasons. As we've seen here, regression offers such points new scope for mischief.

The risks of interpreting linear models as causal or predictive arose in Chapters 7 and 8. And they're important enough to mention again in later chapters.

What have we learned?

We've learned that there are many ways in which a data set may be unsuitable for a regression analysis.

- Watch out for more than one group hiding in your regression analysis. If you find subsets of the data that behave differently, consider fitting a different regression model to each subset.
- The **Straight Enough Condition** says that the relationship should be reasonably straight to fit a regression. Somewhat paradoxically, sometimes it's easier to see that the relationship is not straight *after* fitting the regression by examining the residuals.
- The **Outlier Condition** actually means two things: Points with large residuals or high leverage (especially both) can influence the regression model significantly. It's a good idea to perform the regression analysis with and without such points to see their impact.

And we've learned that even a good regression doesn't mean we should believe the model completely:

- Extrapolation far from the mean can lead to silly and useless predictions.
- Even an R^2 near 100% doesn't indicate that x caused y (or the other way around). Watch out for lurking variables that may affect both x and y.
- Watch out for regressions based on *summaries* of the data sets. These regressions tend to look stronger than the regression on the original data.

TERMS

Subset   One unstated condition for finding a linear model is that the data be homogeneous. If, instead, the data consist of two or more groups that have been thrown together, it is usually best to fit different linear models to each group than to try to fit a single model to all of the data. Displays of the residuals can often help you find subsets in the data.

Extrapolation   Although linear models provide an easy way to predict values of y for a given value of x, it is unsafe to predict for values of x far from the ones used to find the linear model equation. Such extrapolation may pretend to see into the future, but the predictions should not be trusted.

Outlier	Any data point that stands away from the others can be called an outlier. In regression, outliers can be extraordinary in two ways, by having a large residual or by having high leverage.
Leverage	Data points whose *x*-values are far from the mean of *x* are said to exert leverage on a linear model. High-leverage points pull the line close to them, and so they can have a large effect on the line, sometimes completely determining the slope and intercept. With high enough leverage, their residuals can appear to be deceptively small.
Influential point	If omitting a point from the data results in a very different regression model, then that point is called an influential point.
Lurking variable	A variable that is not explicitly part of a model but affects the way the variables in the model appear to be related is called a lurking variable. Because we can never be certain that observational data are not hiding a lurking variable that influences both *x* and *y*, it is never safe to conclude that a linear model demonstrates a causal relationship, no matter how strong the linear association.

SKILLS

When you complete this lesson you should:

Think

- Understand that we cannot fit linear models or use linear regression if the underlying relationship between the variables is not itself linear.
- Understand that data used to find a model must be homogeneous. Look for subgroups in data before you find a regression, and analyze each separately.
- Know the danger of extrapolating beyond the range of the *x*-values used to find the linear model, and especially when the extrapolation tries to predict into the future.
- Understand that points can be unusual by having a large residual or by having high leverage.
- Understand that an influential point can change the slope and intercept of the regression line.
- Look for lurking variables whenever considering the association between two variables. Understand that a strong association does not mean that the variables are causally related.

Show

- Know how to display residuals from a linear model by making a scatterplot of residuals against predicted values or against the *x*-variable, and know what patterns to look for in the picture.
- Know how to look for high-leverage and influential points by examining a scatterplot of the data, and how to look for points with large residuals by examining a scatterplot of the residuals against the predicted values or against the *x*-variable. Understand how fitting a regression line with and without influential points can add to understanding of the regression model.
- Know how to look for high-leverage points by examining the distribution of the *x*-values or by recognizing them in a scatterplot of the data, and understand how they can affect a linear model.

Tell

- Include diagnostic information such as plots of residuals and leverages as part of your report of a regression.
- Report any high-leverage points.
- Report any outliers. Consider reporting analyses with and without outliers included to assess their influence on the regression.
- Include appropriate cautions about extrapolation when reporting predictions from a linear model.
- Discuss possible lurking variables.

Regression Diagnosis on the Computer

Most statistics technology offers simple ways to check whether your data satisfy the conditions for regression. We have already seen that these programs can make a simple scatterplot. They can also help us check the assumptions; we do that best by plotting residuals.

DATA DESK

Click on the **HyperView** menu on the **Regression** output table. A menu drops down to offer scatterplots of residuals against predicted values, Normal probability plots of residuals, or just the ability to save the residuals and predicted values.

Click on the name of a predictor in the regression table to be offered a scatterplot of the residuals against that predictor.

Comments

If you change any of the variables in the regression analysis, Data Desk will offer to update the plots of residuals.

EXCEL

The Data Analysis add-in for Excel includes a Regression command. The dialog box it shows offers to make plots of residuals.

Comments

Do not use the Normal probability plot offered in the regression dialog. It is not what it claims to be and is wrong.

JMP

From the **Analyze** menu choose **Fit Y by X.** Select **Fit Line.** Under Linear Fit, Select **Plot Residuals.** You can also choose to **Save Residuals.** Subsequently, from the **Distribution** menu choose **Normal quantile plot** or **histogram** for the residuals.

MINITAB

From the **Stat** menu choose **Regression.** From the **Regression** submenu, select **Regression** again. In the Regression dialog, enter the response variable name in the "Response" box and the predictor variable name in the "Predictor" box. To specify saved results, in the Regression dialog, click **Storage.** Check "Residuals" and "Fits." Click **OK.** To specify displays, in the Regression dialog, click **Graphs.** Under "Residual Plots," select "Individual plots," and check "Residuals versus fits." Click **OK.** Now back in the Regression dialog, Click **OK.** Minitab computes the regression and requested saved values and graphs.

SPSS

From the **Analyze** menu choose **Regression.** From the Regression submenus choose **Linear.** After assigning variables to their roles in the regression, click the **"Plots . . ."** button.

In the Plots dialog you can specify a Normal probability plot of residuals and scatterplots of various versions of standardized residuals and predicted values.

Comments

A plot of ***ZRESID** against ***PRED** will look most like the residual plots we've discussed. SPSS standardizes the residuals by dividing by their standard deviation. (There's no need to subtract their mean; it must be zero.) The standardization doesn't affect the scatterplot.

TI-83/84 Plus

To make a residuals plot, set up a **STATPLOT** as a scatterplot. Specify your explanatory datalist as **Xlist.** For **Ylist** import the name **RESID** from the **LIST NAMES** menu. **ZoomStat** will now create the residuals plot.

Comments

Each time you execute a **LinReg** command, the calculator automatically computes the residuals and stores them in a data list named **RESID.** If you want to see them, go to **STAT EDIT.** Space through the names of the lists until you find a blank. **Import RESID** from the **LIST NAMES** menu. Now every time you have the calculator compute a regression analysis, it will show you the residuals.

EXERCISES

① 1. Marriage age. We've looked at the ages of men at first marriage. How about women? Is there evidence that the age at which women get married has changed over the past 100 years? The scatterplot shows the trend in age at first marriage for American women.

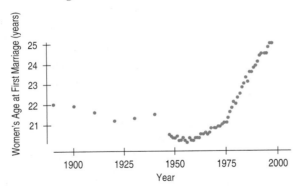

a) Do you think there is a clear pattern? Describe the trend.
b) Is the association strong?
c) Is the correlation high? Explain.
d) Do you think a linear model is appropriate for these data? Explain.

① 2. Ages of couples. The graph shows the ages of both men and women at first marriage.

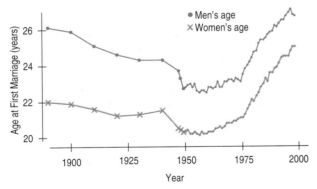

Clearly, the pattern for men is very similar to the pattern for women. But are the two lines getting closer together?

Here's a timeplot showing the *difference* in average age (men's age − women's age) at first marriage, the regression analysis, and the associated residuals plot.

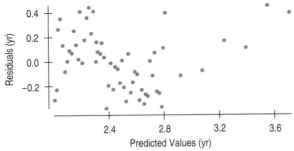

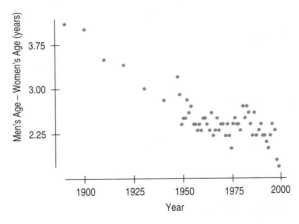

Dependent variable is: Men − Women
R-squared = 71.6%

Variable	Coefficient	SE(Coeff)
Constant	33.4830	2.606
Year	−0.015756	0.0013

a) What is the correlation between *age difference* and *year?*

b) Interpret the slope of this line.
c) Predict the average age difference in 2010.
d) Describe reasons why you might not place much faith in that prediction.

T **3. Marriage age revisited.** Suppose you wanted to predict the trend in marriage age for American women into the early part of this century.
 a) How could you use the data graphed in Exercise 1 to get a good prediction? Marriage ages in selected years starting in 1900 are listed below. Use all or part of these data to create an appropriate model for predicting the average age at which women will first marry in 2005.

1900–1950 (10-yr intervals): 21.9, 21.6, 21.2, 21.3, 21.5, 20.3
1955–1995 (5-yr intervals): 20.2, 20.2, 20.6, 20.8, 21.1, 22.0, 23.3, 23.9, 24.5

 b) How much faith do you place in this prediction? Explain.
 c) Do you think your model would produce an accurate prediction about your grandchildren, say, 50 years from now? Explain.

T **4. Ages of couples, again.** Is the trend of decreasing difference in age at first marriage seen in Exercise 2 stronger recently? Here are the scatterplot and residual plot for the data from 1975 through 1998, along with a regression for just those years.

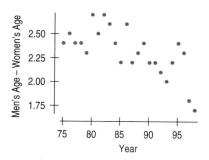

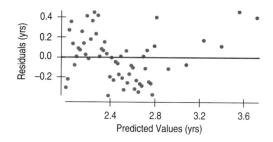

Dependent variable is: Men − Women
R-squared = 46.3%

Variable	Coefficient
Intercept	49.9021
Year	−0.023957

a) Why is R^2 higher for the first model (in Exercise 2)?
b) Is this linear model appropriate for the post-1975 data? Explain.
c) What does the slope say about marriage ages since 1975?
d) Explain why it's not reasonable to interpret the *y*-intercept.

5. Good model? In justifying his choice of a model, a student wrote, "I know this is the correct model because $R^2 = 99.4\%$."
a) Is this reasoning correct? Explain.
b) Does this model allow the student to make accurate predictions? Explain.

6. Bad model? A student who has created a linear model is disappointed to find that her R^2 value is a very low 13%.
a) Does this mean that a linear model is not appropriate? Explain.
b) Does this model allow the student to make accurate predictions? Explain.

7. Reading. To measure progress in reading ability, students at an elementary school take a reading comprehension test every year. Scores are measured in "grade level" units; that is, a score of 4.2 means that a student is reading at slightly above the expected level for a fourth grader. The school principal prepares a report to parents that includes a graph showing the mean reading score for each grade. In his comments he points out that the strong positive trend demonstrates the success of the school's reading program.

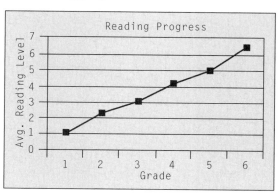

a) Does this graph indicate that students are making satisfactory progress in reading? Explain.
b) What would you estimate the correlation between *grade* and *average reading level* to be?
c) If, instead of this plot showing average reading levels, the principal had produced a scatterplot of the reading levels for all the individual students, would you expect the correlation to be the same, higher, or lower? Explain.
d) Although the principal did not do a regression analysis, someone as statistically astute as you might do

that. (But don't bother.) What value of the slope of that line would you view as demonstrating acceptable progress in reading comprehension? Explain.

8. Grades. A college admissions officer, defending the college's use of SAT scores in the admissions process, produced the graph below. It shows the mean GPAs for last year's freshmen, grouped by SAT scores. How strong is the evidence that *SAT score* is a good predictor of *GPA*? What concerns you about the graph, the statistical methodology, or the conclusions reached?

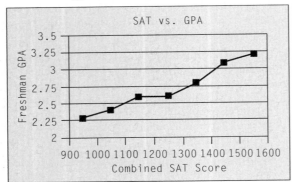

9. Heating. After keeping track of his heating expenses for several winters, a homeowner believes he can estimate the monthly *cost* from the average daily Fahrenheit temperature using the model $\widehat{cost} = 133 - 2.13 \, temp$. The residuals plot for his data is shown.

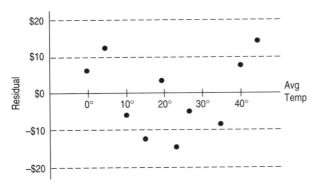

a) Interpret the slope of the line in this context.
b) Interpret the *y*-intercept of the line in this context.
c) During months when the temperature stays around freezing, would you expect cost predictions based on this model to be accurate, too low, or too high? Explain.
d) What heating cost does the model predict for a month that averages 10°?
e) During one of the months on which the model was based, the temperature did average 10°. What were the actual heating costs for that month?
f) Do you think the homeowner should use this model? Explain.

g) Would this model be more successful if created using degrees Celsius? Explain.

10. Speed. How does the speed at which you drive impact your fuel economy? To find out, researchers drove a compact car for 200 miles at speeds ranging from 35 to 75 miles per hour. From their data, they created the model $\widehat{mpg} = 32 - 0.1 \, mph$ and created this residual plot:

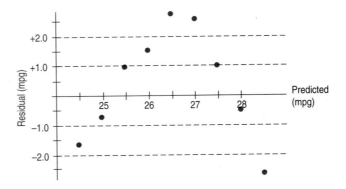

a) Interpret the slope of this line in context.
b) Explain why it's silly to attach any meaning to the *y*-intercept.
c) When this model predicts high gas mileage, what can you say about those predictions?
d) What gas mileage does the model predict when the car is driven at 50 mph?
e) What was the actual gas mileage when the car was driven at 45 mph?
f) Do you think there appears to be a strong association between speed and fuel economy? Explain.
g) Do you think this is the appropriate model for that association? Explain.

11. Unusual points. Each of the scatterplots below shows a cluster of points and one "stray" point. For each, answer these questions:
1) In what way is the point unusual? Does it have high leverage, a large residual, or both?
2) Do you think that point is an influential point?
3) If that point were removed from the data, would the correlation become stronger or weaker? Explain.
4) If that point were removed from the data, would the slope of the regression line increase or decrease? Explain.

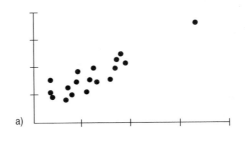

a)

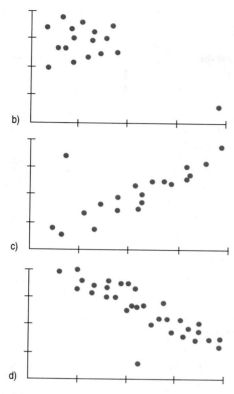

b)

c)

d)

12. More unusual points. Each of the scatterplots below shows a cluster of points and one "stray" point. For each, answer these questions:
1) In what way is the point unusual? Does it have high leverage, a large residual, or both?
2) Do you think that point is an influential point?
3) If that point were removed from the data, would the correlation become stronger or weaker? Explain.
4) If that point were removed from the data, would the slope of the regression line increase or decrease? Explain.

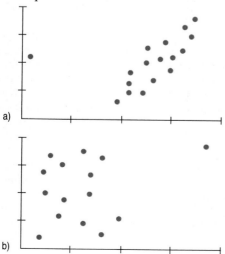

a)

b)

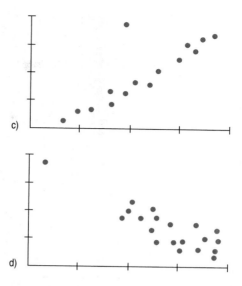

c)

d)

13. The extra point. The scatterplot shows five blue data points at the left. Not surprisingly, the correlation for these points is $r = 0$. Suppose *one* additional data point is added at one of the five positions suggested below in green. Match each point (a–e) with the correct new correlation from the list given.
1) −0.90
2) −0.40
3) 0.00
4) 0.05
5) 0.75

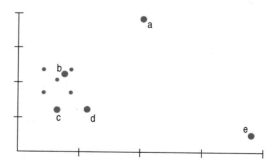

14. The extra point revisited. The original five points in Exercise 13 produce a regression line with slope 0. Match each of the green points (a–e) with the slope of the line after that one point is added.
1) −0.45
2) −0.30
3) 0.00
4) 0.05
5) 0.85

T 15. Gestation. For women, pregnancy lasts about 9 months. In other species of animals, the length of time from conception to birth varies. Is there any evidence that gestation period is related to the animal's lifespan? The first scatterplot shows *gestation period* (in days) vs. *lifespan* (in years) for 18 species of mammals. The highlighted point at the far right represents humans.

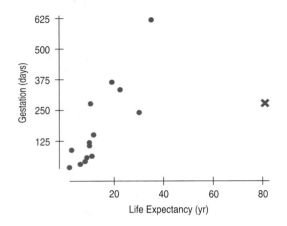

a) For these data $r = 0.54$, not a very strong relationship. Do you think the association would be stronger or weaker if humans were removed? Explain.

b) Is there reasonable justification for removing humans from the data set? Explain.

c) Here are the scatterplot and regression analysis for the 17 nonhuman species. Comment on the strength of the association.

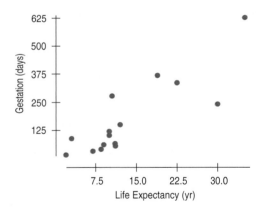

Dependent variable is: Gestation
R-squared = 72.2%

Variable	Coefficient
Constant	−39.5172
LifExp	15.4980

d) Interpret the slope of the line.

e) Some species of monkeys have a life expectancy of about 20 years. Estimate the expected gestation period of one of these monkeys.

16. Elephants and hippos. We removed humans from the scatterplot in Exercise 15 because our species was an outlier in life expectancy. The resulting scatterplot shows two points that now may be of concern. The point in the upper right corner of this scatterplot is for elephants, and the other point at the far right is for hippos.

a) By removing one of these points, we could make the association appear to be stronger. Which point? Explain.

b) Would the slope of the line increase or decrease?

c) Should we just keep removing animals to increase the strength of the model? Explain.

d) If we remove elephants from the scatterplot, the slope of the regression line becomes 11.6 days per year. Do you think elephants were an influential point? Explain.

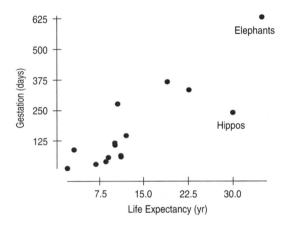

17. What's the cause? Suppose a researcher studying health issues measures blood pressure and the percentage of body fat for several adult males, finding a strong positive association. Describe three different possible cause-and-effect relationships that might be present.

18. What's the effect? A researcher studying violent behavior in elementary-school children asks the children's parents how much time each child spends playing computer games and has their teachers rate each child on the level of aggressiveness they display while playing with other children. Suppose that the researcher finds a moderately strong positive correlation. Describe three different possible cause-and-effect explanations for this relationship.

T 19. Law enforcement. Federal employees with the authority to carry firearms and make arrests are sometimes assaulted, injured, or killed. The table on the next page summarizes the rates of *assault* and *injury* (or death) for these employees for 5 years, 1995–1999. Can the assault rate be used to predict injuries or deaths?

Agency	Assaults (per 1000)	Killed or Injured (per 1000)
Bureau of Alcohol, Tobacco, and Firearms (BATF)	31.1	2.2
Capitol Police	5.0	3.6
Customs Service	9.7	5.1
Drug Enforcement Agency (DEA)	17.9	1.1
Federal Bureau of Investigation (FBI)	3.9	1.2
Immigration and Naturalization Service (INS)	14.1	2.5
Internal Revenue Service (IRS)	1.7	0.2
U.S. Marshal Service	9.7	3.0
National Park Service	38.7	15.0
Postal Service	5.7	2.9
Secret Service	9.7	3.0

T 20. Swim the lake. People swam across Lake Ontario 37 times between 1974 and 2004. We might be interested in whether they are getting any faster or slower. Here are the regression of the crossing times (minutes) against the year of the crossing and the residuals plot:

Dependent variable is: Time
R squared = 7.0%

Variable	Coefficient
Constant	−29161.9
Year	15.3323

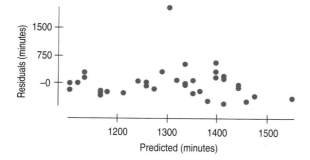

a) What does the R^2 mean for this regression?
b) Are the swimmers getting faster or slower? Explain.
c) The outlier seen in the residuals plot is a crossing by Vicki Keith in 1987 in which she swam a round trip, North to South and then back again. Clearly, this swim doesn't belong with the others. Do you think that removing it would change the model a lot? Explain.
d) Here is the new regression after the unusual point is removed:

Dependent variable is: Time
R squared = 15.9%

Variable	Coefficient
Intercept	−28399.9
Year	14.9198

Now would you be willing to say that the swimmers were getting faster or slower?

T 21. Illegitimate births. The National Center for Health Statistics reported the data below, showing the percentage of all births that are to unmarried women for selected years between 1980 and 1998. Create a model that describes this trend. Justify decisions you make about how to best use these data.

Year	1980	1985	1990	1991	1992	1993	1994	1995	1996	1997	1998
%	18.4	22.0	28.0	29.5	30.1	31.0	32.6	32.2	32.4	32.4	32.8

T 22. Smoking. The Centers for Disease Control and Prevention track cigarette smoking in the United States. How has the percentage of people who smoke changed since the danger became clear during the last half of the 20th century? The table below shows percentages of smokers among men 18–24 years of age, as estimated by surveys. Create a model to describe the changes in smoking rate. Justify decisions you make in using these data to construct your model.

Year	Percent	Year	Percent
1965	54.1	1992	28.0
1974	42.1	1993	28.8
1979	35.0	1994	29.8
1983	32.9	1995	27.8
1985	28.0	1997	31.7
1990	26.6	1998	31.3

T 23. Inflation. The Consumer Price Index (CPI) tracks the prices of consumer goods in the United States, as shown in the table. It indicates, for example, that the average item costing $17.90 in 1928 cost $168.80 in the year 2000.

Year	CPI	Year	CPI
1916	10.4	1960	29.3
1920	16.5	1964	30.9
1924	17.3	1968	34.1
1928	17.9	1972	41.1
1932	12.9	1976	55.6
1936	14.2	1980	77.8
1940	13.9	1984	101.9
1944	17.4	1988	115.7
1948	23.7	1992	138.1
1952	26.5	1996	154.4
1956	26.8	2000	168.8

a) Make a scatterplot showing the trend in consumer prices. Describe what you see.

b) Be an economic forecaster: Project increases in the cost of living over the next decade. Justify decisions you make in creating your model.

T 24. Life expectancy. Data from the World Bank for 26 Western Hemisphere countries can be used to examine the association between female *life expectancy* and the average *number of children* women give birth to.

Country	Births/ Woman	Life Exp.	Country	Births/ Woman	Life Exp.
Argentina	2.5	77	Guatemala	4.7	68
Bahamas	2.2	77	Honduras	4.0	72
Barbados	1.8	78	Jamaica	2.5	77
Belize	3.5	74	Mexico	2.8	75
Bolivia	4.0	64	Nicaragua	3.6	71
Brazil	2.2	71	Panama	2.5	76
Canada	1.5	82	Paraguay	4.0	72
Chile	2.2	79	Peru	3.1	71
Colombia	2.7	74	Puerto Rico	3.1	71
Costa Rica	25.0	79	United States	2.1	80
Dom. Rep.	2.8	73	Uruguay	2.3	78
Ecuador	3.1	71	Venezuela	2.9	76
El Salvador	3.2	72	Virgin Islands	2.4	79

a) Create a scatterplot relating these two variables and describe the association.

b) Are there any countries that do not seem to fit the overall pattern?

c) Find the correlation, and interpret the value of R^2.

d) Find the equation of the regression line.

e) Is the line an appropriate model? Describe what you see in the residuals plot.

f) Interpret the slope and the *y*-intercept of the line.

g) If government leaders wanted to increase life expectancy, in their country, should they encourage women to have fewer children? Explain.

T 25. Inflation again. In Exercise 23 we looked at the average CPI of all U.S. cities. In the complete data set, we have an annual CPI index from 1970 to 2004 for New York and Chicago.

a) Make a scatterplot of Chicago's CPI against the CPI in New York, and describe what you see.

b) Perform a regression of Chicago's CPI against New York's. Interpret the regression equation and the R^2 value.

c) Plot the residuals against *Year*. Is there a pattern? What was different between the late 70s and the 80s in the relationship between the two cities?

T 26. Tour de France. We met the Tour de France data set in Chapter 2 (in Just Checking). One hundred years ago, the fastest rider finished the course at an average speed of about 25.3 kph (around 15.8 mph). In 2003, Lance Armstrong averaged 40.94 kph (25.59 mph) for the fastest average winning speed in history.

a) Make a scatterplot of *avg speed* against *year*. Describe the relationship of *avg speed* by *year*, being careful to point out any unusual features in the plot.

b) Find the regression equation of *avg speed* by *year*.

c) Are the conditions for regression met? Comment.

T 27. Inflation, finis. Let's take one more look at the trend in Consumer Price Index shown in the table in Exercise 23. Sometimes re-expressing the data can reveal interesting information concealed there. Create a new variable containing the logarithm of the CPI. Plot *log(CPI)* against *year*. What historical events had a clear impact on the cost of living? (The next chapter explores more advantages that may arise by re-expressing data.)

T 28. Second stage. Look once more at the data from the Tour de France. In Exercise 26 we looked at the whole history of the race, but now let's consider just the post–World War II era.

a) Find the regression of *avg speed* by *year* only for years 1947 to the present. Are the conditions for regression met?

b) Interpret the slope.

c) In 1979 Bernard Hinault averaged 39.8 kph, while in 2001 Lance Armstrong averaged 40.02 kph. Which was the more remarkable performance and why?

---

just checking

✔ **Answers**

1. Not high leverage, not influential, large residual

2. High leverage, not influential, small residual

3. High leverage, influential, not large residual

10

Re-expressing Data:
It's Easier Than You Think

A S **Re-expressing Data.**
Should you re-express data?
Actually, you already do. See
some examples here.

We have seen several cases in which a simple re-expression of the data, such as taking a logarithm or looking at the reciprocal ($1/y$), makes the data much easier to understand. Many of our tools for displaying and summarizing data work only when the data meet certain conditions. Linear models raise the stakes even further. We cannot use a linear model effectively unless the underlying relationship between the two variables is linear. Often re-expression can save the day, straightening bent relationships so that we can fit and use a simple linear model. It turns out that finding a useful re-expression is probably easier than you think. This chapter shows how to find re-expressions in a simple, systematic way.

Sometimes when we use re-expressions, we hear people say, "Wait a minute. If you're allowed to do anything you want to the data, you can make them show anything you want. The original data weren't measured in square roots! You're cheating." The fact is, it's not cheating, and re-expression is something that you do everyday. The point of this chapter is to show you how to find re-expressions easily and to show you that not only is it incredibly useful, but it's easy and natural as well. The material in this chapter is not needed explicitly for any subsequent chapter. But it makes all of them more useful by allowing more data to satisfy the conditions of each method we will discuss.

Straightening Relationships

The relationship between *fuel efficiency* measured in miles per gallon and *weight* measured in pounds for late model cars looks fairly linear at first. It isn't surprising that lighter cars get better gas mileage. With an R^2 of 81.6%, the regression line might seem like a good fit. A look at the residuals plot, though, shows a problem.

217

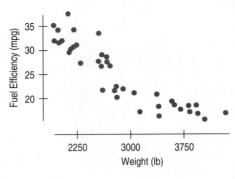

Fuel efficiency (city mpg) vs. *weight* for 38 cars as reported by *Consumer Reports.* The plot shows a negative direction, roughly linear shape, and strong relationship. **Figure 10.1**

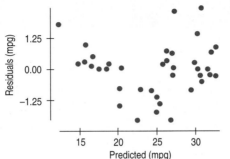

The residuals from a regression of *fuel efficiency* on *weight* reveal a bent shape when plotted against the predicted values. Looking back at the original scatterplot, you may be able to see the bend. **Figure 10.2**

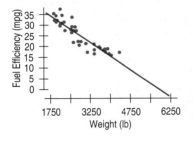

Extrapolating the regression line gives an absurd answer for vehicles that weigh as little as 6000 pounds. **Figure 10.3**

The shape is clearly bent. Looking back at the first scatterplot, you can probably see the slight bending. Think about the regression line through the points. How heavy would a car have to be to have a predicted gas mileage of 0? It looks like the *fuel efficiency* would go negative at about 6000 pounds. A Hummer H2 weighs about 6400 pounds. The H2 is hardly known for fuel efficiency, but it does get more than the *minus* 2 mpg this regression would predict. Extrapolation is always dangerous, but it's more dangerous the more the model is wrong, because wrong models tend to do even worse the farther you get from the middle of the data.

The bend in the relationship between *fuel efficiency* and *weight* is the kind of failure to satisfy the conditions for an analysis that we have repaired before by re-expressing the data. That method works here too. Instead of looking at miles per gallon, we can take the reciprocal and work with gallons per hundred miles.[1]

● **"Gallons per hundred miles—what an absurd way to measure fuel efficiency! Who would ever do it that way?"** Not all re-expressions are easy to understand, but in this case, the answer is "Everyone except U.S. drivers." Most of the world measures fuel efficiency in liters per 100 kilometers (L/100 km). This is the same reciprocal form (fuel amount per distance driven) and differs from gal/100 mi only by a constant multiple of about 2.38.

It has been suggested that most of the world says, "I've got to go 100 km, how much gas do I need?" But Americans say, "I've got 10 gallons in the tank. How far can I drive?"

In much the same way, re-expressions "think" about the data differently but don't change what they mean. ●

---

[1] The reciprocal is the part of this re-expression that makes the relationship more linear. Multiplying by 100 to get gallons per 100 miles simply makes the numbers easier to think about. You might have a good idea of how many gallons your car needs to drive 100 miles, but probably a much poorer sense of how much gas you need to go just 1 mile.

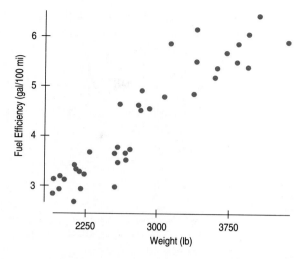

The reciprocal (1/y) is measured in gallons per mile. *Gallons per 100 miles* gives more meaningful numbers. The reciprocal is more nearly linear against *weight* than the original variable, but the re-expression changes the direction of the relationship.
Figure 10.4

The direction of the association is positive now, since we're measuring gas consumption and heavier cars consume more gas per mile. The relationship is much straighter, as we can see from a scatterplot of the regression residuals.

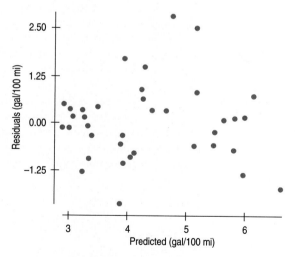

The residuals from the regression of *gallons per 100 miles* on *weight* show less of a pattern than before. **Figure 10.5**

This is more the kind of boring scatterplot (no direction, no particular shape, no outliers, no bends) that we hope to see in a scatterplot of residuals, so we have reason to think that the Straight Enough Condition is now satisfied.

What does the reciprocal model say about the Hummer? The regression line fit to the scatterplot of *gallons per 100 miles* vs. *weight* predicts somewhere near 9.7

for a car weighing 6400 pounds. What does this mean? It means the car is predicted to use 9.7 gallons for every 100 miles, or in other words

$$\frac{100 \ miles}{9.7 \ gallons} = 10.3 \ mpg.$$

That's a much more reasonable prediction, and very close to the true value of 11.0 miles per gallon.

Everybody Does It

> Scientific laws often include simple re-expressions. For example, in Psychology, Fechner's Law states that sensation increases as the logarithm of stimulus intensity, $S = k \log R$.

> Scan through any Physics book. You're unlikely to see equations that don't have powers, reciprocals, or logs.

> **How Fast Can a Human Being Run?**
> The world record for 200 meters is 19.32 seconds (Michael Johnson, 1996). That's 23.4 mph (37.3 kph)! Most people can't even *bicycle* faster than about 19 or 20 mph. Of course, sprinters can't keep up that speed for very long. If they could, they could run a 2:34 mile—or a marathon in an hour and 7 minutes!

You may not think about it, but you use re-expressions in everyday life. How fast can you go on a bicycle? If you measure your speed, you probably do it in distance per time (miles per hour or kilometers per hour). In 2000, during a 25-mile-long time trial in the Tour de France, Lance Armstrong *averaged* over 33.5 mph (53.9 kph). You probably realize that's a tough act to follow. It's fast. You can tell that at a glance because you have no trouble thinking in terms of distance covered per time.

OK, then, if you averaged 12.5 mph (20.1 kph) for a mile *run*, would that be fast? Would it be fast for a 100-m dash? Even if you run the mile often, you probably have to stop and calculate. Although we measure speed of bicycles in distance per time, we don't usually measure running speed that way. Instead, we use the *reciprocal*, time per distance (minutes per mile, seconds per 100 meters, etc.). Running a mile in under 5 minutes (12 mph) is fast. A mile at 16 mph would be a world record (that's a 3 minute 45 second mile).

The point is that there is no single *natural* way to measure speed. In some cases we use distance traveled per time, and in other cases we use the reciprocal. It's just because we're used to thinking that way in each case, not because one way is correct. It's important to realize that the way these quantities are measured is not sacred. It's usually just convenience or custom. When we re-express a quantity to make it satisfy certain conditions, we may either leave it in those new units when we explain the analysis to others, or convert it back to the original units.

Other examples of common variables that re-express a crude measurement include the Richter scale of earthquake strength, the decibel scale for sound intensity, the f/stop scale for camera aperture openings, and the gauges of shotguns.

Careful re-expression is a common practice that can simplify patterns and relationships.

Goals of Re-expression

We re-express data for several reasons. Each of these goals helps make the data more suitable for analysis by our methods. We'll illustrate each goal with an example from data about large companies.

Goal 1

Make the distribution of a variable (as seen in its histogram, for example) more symmetric. It's easier to summarize the center of a symmetric distribution, and for nearly symmetric distributions, we can use the mean and standard deviation. If the distribution is unimodal, then the resulting distribution may be closer to the Normal model, allowing us to use the 68-95-99.7 Rule.

Here are the *assets* of the companies we first saw in Chapter 4 (in $100,000):

WHO	77 large companies
WHAT	Assets, sales, and market sector
UNITS	$100,000
HOW	Public records
WHEN	1986
WHY	By *Forbes* magazine in reporting on the Forbes 500 for that year

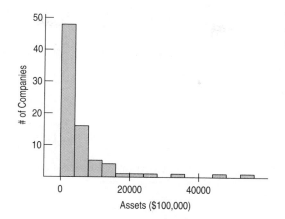

The distribution of the *assets* of large companies is skewed to the right. Data on wealth often look like this.   **Figure 10.6**

The skewed distribution is made much more symmetric by taking logs.

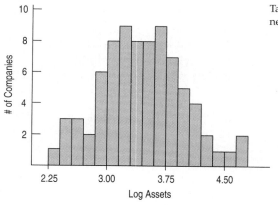

Taking logs makes the distribution more nearly symmetric.   **Figure 10.7**

A S **Re-expression in Action.**
Slide the re-expression power and watch the histogram change. There's no better way to see re-expression at work on distribution shape.

Goal 2

Make the spread of several groups (as seen in side-by-side boxplots) more alike, even if their centers differ. Groups that share a common spread are easier to compare. We'll see methods later in the book that can be applied only to groups with a common standard deviation. We saw an example of re-expression for comparing groups with boxplots in Chapter 5.

Here are the *assets* of these companies by *market sector:*

Assets of large companies by *market sector.* It's hard to compare centers or spreads, and there seem to be a number of high outliers.   **Figure 10.8**

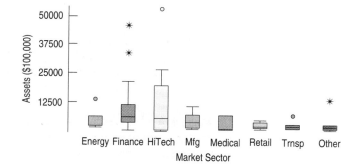

Taking logs makes the individual boxplots more symmetric and gives them spreads that are more nearly equal.

After re-expressing by logs, it's much easier to compare across market sectors. The boxplots are more nearly symmetric, most have similar spreads, and the companies that seemed to be outliers before are no longer extraordinary. Two new outliers have appeared in the finance sector. They are the only companies in that sector that are not banks. Perhaps they don't belong there.
Figure 10.9

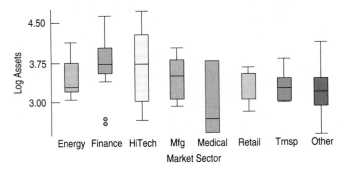

This makes it easier to compare assets across market sectors. It can also reveal problems in the data. Some companies that looked like outliers on the high end turned out to be more typical. But two companies in the Finance sector now stick out. They are not banks. Unlike the rest of the companies in that sector, they may have been placed in the wrong sector, but we couldn't see that in the original data.

Goal 3

Make the form of a scatterplot more nearly linear. Linear scatterplots are easier to describe. We saw an example of scatterplot straightening in Chapter 7. The greater value of re-expression to straighten a relationship is that we can fit a linear model once the relationship is straight.

Here are *assets* plotted against the logarithm of *sales*:

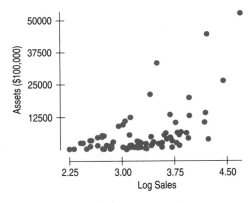

Assets vs. log *sales* shows a positive association (bigger sales goes with bigger assets) but a bent shape. Note also that the points go from tightly bunched at the left to widely scattered at the right—a "fan" shape. **Figure 10.10**

The plot's shape is bent. Taking logs makes things much more linear.

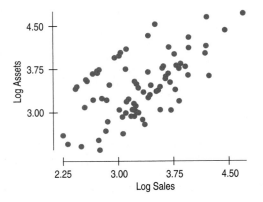

Log *assets* vs. log *sales* shows a clean, positive, linear association. And the variability at each value of x is about the same. **Figure 10.11**

Goal 4

Make the scatter in a scatterplot spread out evenly rather than following a fan shape. Having an even scatter is a condition of many methods of Statistics, as we'll see in later chapters. This goal is closely related to Goal 2, but it often comes along with Goal 3. Indeed, a glance back at the scatterplot (Figure 10.10) shows that the plot for *assets* is much more spread out on the right than on the left, while the plot for log *assets* has roughly the same variation in log *assets* for any *x*-value.

The Ladder of Powers

 All About Powers.
Here's the animated version of the Ladder of Powers.

We've seen that taking logs or reciprocals can really simplify an analysis. How do we know which re-expression to use? We could use trial and error to choose a re-expression, but there's an easier way. We can choose our re-expressions from a family of simple re-expressions that move data toward our goals in a consistent way. This family includes the most common ways to re-express data. More important, the members of the family line up in order, so that the farther you move away from the original data (the "1" position), the greater the effect on the data. This fact lets you search systematically for a re-expression that works, stepping a bit farther from "1" or taking a step back toward "1" as you see the results.

Where to start? It turns out that certain kinds of data are more likely to be helped by particular re-expressions. Knowing that gives you a good place to start your search for a re-expression.

We call this collection of re-expressions the **Ladder of Powers**. Each power is specified as a single value, as follows:

Power	Name	Comment
2	The square of the data values, y^2.	Try this for unimodal distributions that are skewed to the left.
1	The raw data—no change at all. This is "home base." The farther you step from here up or down the ladder, the greater the effect.	Data that can take on both positive and negative values with no bounds are less likely to benefit from re-expression.
1/2	The square root of the data values, $\sqrt{y}$.	Counts often benefit from a square root re-expression. For counted data, start here.
"0"	Although mathematicians define the "0-th" power differently,[2] for us the place is held by the logarithm. You may feel uneasy about logarithms. Don't worry, the computer or calculator does the work.[3]	Measurements that cannot be negative, and especially values that grow by percentage increases such as salaries or populations, often benefit from a log re-expression. When in doubt, start here. If your data have zeros, try adding a small constant to all values before finding the logs.
−1/2	The (negative) reciprocal square root $-1/\sqrt{y}$.	An uncommon re-expression, but sometimes useful. Changing the sign to take the *negative* of the reciprocal square root preserves the direction of relationships, which can be a bit simpler.
−1	The (negative) reciprocal, $-1/y$.	Ratios of two quantities (miles per hour, for example) often benefit from a reciprocal. (You have about a 50-50 chance that the original ratio was taken in the "wrong" order for simple statistical analysis and would benefit from re-expression.) Often, the reciprocal will have simple units (hours per mile). Change the sign if you want to preserve the direction of relationships. If your data have zeros, try adding a small constant to all values before finding the reciprocal.

---

[2] You may remember that for any non-zero number, *y*, $y^0 = 1$. This is not a very exciting transformation for data; every data value would be the same. We use the logarithm in its place.

[3] Your calculator or software package probably gives you a choice between "base 10" logarithms and "natural (base *e*)" logarithms. Don't worry about that. It doesn't matter at all which you use; they have exactly the same effect on the data. If you want to choose, then base 10 logarithms can be a bit easier to interpret.

The Ladder of Powers orders the *effects* that the re-expressions have on data. If you try, say, taking the square roots of all the values in a variable and it helps, but not enough, then moving farther down the ladder to the logarithm or reciprocal root will have a similar effect on your data, but even stronger. If you go too far, you can always back up. But don't forget—when you take a negative power, the *direction* of the relationship will change. That's OK. You can always change the sign of the response if you want to keep the same direction.

A generation ago, before desktop computers and powerful calculators were available, finding a suitable re-expression was a major undertaking, often given as an end-of-semester project. Now, it's usually no harder than the push of a button, making it easy to search for a suitable re-expression.

Re-expressing to Straighten a Scatterplot Step-By-Step

We know why we want to re-express variables: Straightening a scatterplot lets us fit a linear model. Now let's work through a simple example to see how it's done.

Standard (monofilament) fishing line comes in a range of strengths, usually expressed as "test pounds." Five-pound test line, for example, can be expected to withstand a pull of up to five pounds without breaking. The convention in selling fishing line is that the price of a spool doesn't vary with strength. Instead, the length of line on the spool varies. Because higher test pound line is thicker, though, spools of fishing line hold about the same amount of material. Some spools hold line that is thinner and longer, some fatter and shorter. Let's look at the *length* and *strength* of spools of monofilament line manufactured by the same company and sold for the same price at one store. How are the *length* on the spool and the *strength* related? And what re-expression will straighten the relationship?

Think

Plan State the problem.

Variables Identify the variables and report the W's.

I want to fit a linear model for the length and strength of monofilament fishing line.

I have the *length* and "pound test" *strength* of monofilament fishing line sold by a single vendor at a particular store. Each case is a different strength of line, but all spools of line sell for the same price.

Let *length* = length (in yards) of fishing line on the spool

strength = the test strength (in pounds).

Plot Check that even if there is a curve, the overall pattern does not reach a minimum or maximum, and then turn around and go back. An up-and-down curve can't be fixed by re-expression.

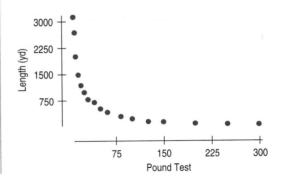

The plot shows a negative direction and an association that has little scatter but is not straight.

Show Mechanics

Try a re-expression.

The lesson of the Ladder of Powers is that if we're moving in the right direction but have not had sufficient effect, we should go farther along the ladder. This example shows improvement, but is still not straight.

(Because *length* is an amount of something and cannot be negative, we probably should have started with logs. This plot is here in part to illustrate how the Ladder of Powers works.)

Here's a plot of the square root of *length* against *strength*:

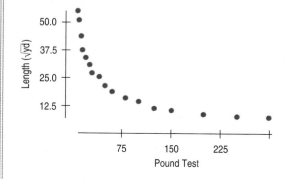

The plot is less bent, but still not straight.

Stepping from the 1/2 power to the "0" power, we try the logarithm of *length* against *strength*.

The scatterplot of the logarithm of *length* against *strength* is still less bent:

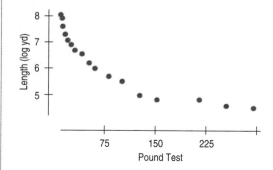

This is much better, but still not straight, so I'll take another step to the "−1" power, or reciprocal.

The straightness is improving, so we know we're moving in the right direction. But since the plot of the logarithms is not yet straight, we know we haven't gone far enough. To keep the direction consistent, change the sign and re-express to −1/length.

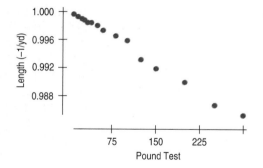

Maybe now I moved too far along the ladder.

We may have to choose between two adjacent re-expressions. For most data analyses, it really doesn't matter which we choose.

A half-step back is the −1/2 power: the reciprocal square root.

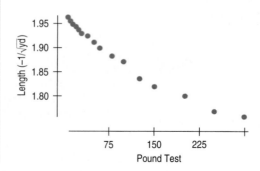

Tell

Conclusion Specify your choice of re-expression. If there's some natural interpretation (as for *gallons per 100 miles*), give that.

It's hard to choose between the last two alternatives. Either of the last two choices is good enough. I'll choose the −1/2 power.

Now that the re-expressed data satisfy the Straight Enough Condition, we can fit a linear model by least squares. We find

$$\frac{-1}{\sqrt{length}} = -0.023 - 0.000373 \; strength.$$

We can use this model to predict the length of a spool of, say, 35-pound test line:

$$\frac{-1}{\sqrt{length}} = -0.023 - 0.000373 \times 35 = -0.036$$

We could leave the result in these units ($-1/\sqrt{yards}$). Sometimes the new units may be as meaningful as the original, but here we want to transform the predicted value back into yards. Fortunately, each of the re-expressions in the Ladder of Powers can be reversed.

To reverse the process, we first take the reciprocal: $\sqrt{length} = 1/0.036 = 27.778$. Then squaring gets us back to the original units:

$$length = 27.778^2 = 771.6 \; yards.$$

This may be the most painful part of the re-expression. Getting back to the original units can sometimes be a little work. Nevertheless, it's worth the effort to always consider re-expression. Re-expressions can often vastly improve a statistical analysis or model.

just checking

❶ You want to model the relationship between the number of birds counted at a nesting site and the temperature (in degrees Celsius). The scatterplot of counts vs. temperature shows an upwardly curving pattern with more birds spotted at higher temperatures. What transformation (if any) of the bird counts might you start with?

❷ You want to model the relationship between prices for various items in Paris and Hong Kong. The scatterplot of Hong Kong prices vs. Parisian prices shows a generally straight pattern with a small amount of scatter. What transformation (if any) of the Hong Kong prices might you start with?

③ You want to model the population growth of the United States over the past 200 years. The scatterplot shows a strongly upwardly curved pattern. What transformation (if any) of the population might you start with?

Plan B: Attack of the Logarithms

We've suggested using the Ladder of Powers as one approach to finding a re-expression that straightens a curved scatterplot. That's often successful. Sometimes, though, the curvature is more stubborn, and we're not satisfied with the residuals plots. What then?

When none of the data values is zero or negative, logarithms can be a helpful ally in the search for a useful model. Try taking the logs of **both** the x- and y-variable. Then re-express the data using some combination of x or $\log(x)$ vs. y or $\log(y)$. You may find that one of these works pretty well.

Model Name	x-axis	y-axis	Comment
Exponential	x	$\log(y)$	This model is the "0" power in the ladder approach, useful for values that grow by percentage increases.
Logarithmic	$\log(x)$	y	A wide range of x-values, or a scatterplot descending rapidly at the left but leveling off toward the right, may benefit from trying this model.
Power	$\log(x)$	$\log(y)$	The Goldilocks model: When one of the ladder's powers is too big and the next is too small, this one may be just right.

When we tried to model the relationship between the length of fishing line and its strength, we were torn between the "−1" power and the "−1/2" power. The first showed slight upward curvature, and the second downward. Maybe there's a better power between those values.

The scatterplot shows what happens when we graph the logarithm of *length* against the logarithm of *strength*. Technology reveals that the equation of our log-log model is

$$\log(\widehat{length}) = 3.65 - 0.90 \log(strength).$$

It's interesting that the slope of this line (−0.90) is a power we didn't try. After all, the ladder can't have every imaginable rung.

A warning, though! Don't expect to be able to straighten every curved scatterplot you find. It may be that there just isn't a very effective re-expression to be had. You'll certainly encounter situations when nothing seems to work the way you wish it would. Don't set your sights too high—you won't find a perfect model. Keep in mind: We seek a *useful* model, not perfection (or even "the best").

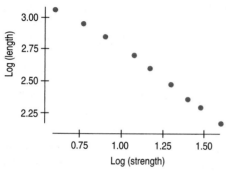

Plotting log (*length*) against log (*strength*) gives a straighter shape.   **Figure 10.12**

Multiple Benefits

We often choose a re-expression for one reason and then discover that it has helped other aspects of an analysis. For example, we might re-express a variable to make its histogram more nearly symmetric. It wouldn't be surprising to find that the same re-expression also straightens scatterplots or makes spreads more nearly equal. This phenomenon is one reason we encourage that data re-expression be used more often. Sometimes there's an obvious "best" or "right" re-expression for a variable. For example, it makes sense that things that tend to grow by a constant percentage, so that larger values grow faster (populations, bacteria counts, wealth), will grow exponentially. Logarithms straighten out the exponential trend and pull in the long right tail in the histogram.

We saw just this phenomenon in the companies example earlier. Each of our goals was met by re-expressing *assets* with logarithms. That single re-expression improved all four of the goals at the same time. That turns out not to be all that unusual.

Measurement errors are often larger when measuring larger quantities than when measuring smaller ones. (The error in your height may be only a centimeter or two, but the error in the height of a tree could be ten times that much.) Again, logarithms are likely to help. Measurements of rates (time to complete a task) are often plagued by infinities for those who never finish. The reciprocal of "minutes per task" is "tasks per minute"—a speed measure. And the unfinished tasks that once took "infinite" time now simply rate a zero speed. Here once more the re-expression seems natural.

In other cases, the only evidence we have to favor re-expression is that it seems to work well and that it leads to simpler models. Often we can find a re-expression for a variable that simplifies our analysis in several ways at once, making the distribution symmetric, making the relationship linear in terms of other variables of interest, or stabilizing its variance. If so, re-expressing the variable certainly simplifies our efforts to analyze and understand it.

> **Occam's Razor**
>
> If you think that simpler explanations and simpler models are more likely to give a true picture of the way things work, then you should look for opportunities to re-express your data and simplify your analyses.
>
> The general principle that simpler explanations are likely to be the better ones is known as Occam's Razor after the English philosopher and theologian William of Occam (1284–1347).

Why Not Just Use a Curve?

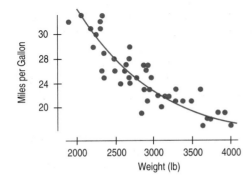

When a clearly curved pattern shows up in the scatterplot, why not just fit a curve to the data? We saw earlier that the association between the *weight* of a car and its *fuel economy* was not a straight line. Instead of trying to find a way to straighten the plot, why not find a curve that seems to describe the pattern well?

We can find "curves of best fit" using essentially the same approach that led us to linear models. You won't be surprised, though, to learn that the mathematics and the calculations are considerably more difficult for curved models. Many calculators and computer packages do have the ability to fit curves to data, but this approach has many drawbacks.

Straight lines are easy to understand. We know how to think about the slope and the *y*-intercept, for example. We often want some of the other benefits mentioned earlier, such as making the spread around the model more nearly the same everywhere. In later chapters you will learn more advanced statistical methods for analyzing linear associations.

We give all of that up when we fit a model that is not linear. For many reasons, then, it is usually better to re-express the data to straighten the plot.

What Can Go Wrong?

- **Don't expect your model to be perfect.** In Chapter 6 we quoted statistician George Box: "All models are wrong, but some are useful." Be aware that the real world is a messy place, and data can be uncooperative. Don't expect to find one elusive re-expression that magically irons out every kink in your scatterplot and produces perfect residuals. You aren't looking for the Right Model, because that mythical creature doesn't exist. Find a useful model and use it wisely.

- **Don't choose a model based on R^2 alone.** You've tried re-expressing your data to straighten a curved relationship, and found a model with a high R^2. Beware: That doesn't mean the pattern is straight now. Here's a plot of a relationship with an R^2 of 98.3%.

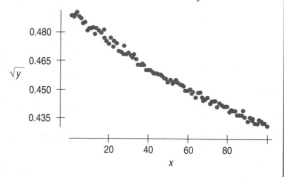

  The R^2 is about as high as we could ask for, but if you look closely, you'll see that there's a consistent bend. Plotting the residuals from the least squares line makes the bend much easier to see.

  Remember the basic rule of data analysis: *make a picture.* Before you fit a line, always look at the pattern in the scatterplot. After you fit the line, check for linearity again by plotting the residuals.

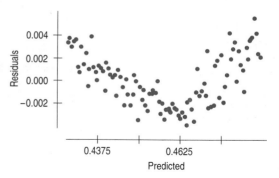

- **Beware of multiple modes.** Re-expression can often make a skewed unimodal histogram more nearly symmetric, but it cannot pull separate modes together. A suitable re-expression may, however, make the separation of the modes clearer, simplifying their interpretation and making it easier to separate than to analyze individually.

- **Watch out for scatterplots that turn around.** Re-expression can straighten many bent relationships, but not those that go up and then down or down and then up. You should refuse to analyze such data with methods that require a linear form.

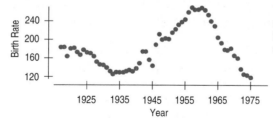

The shape of the scatterplot of *birth rates* (births per 100,000 women) in the United States shows an oscillation that cannot be straightened by re-expressing the data.

Figure 10.13

- ***Watch out for negative data values.*** It's impossible to re-express negative values by any power that is not a whole number on the Ladder of Powers or to re-express values that are zero for negative powers. Most statistics programs will just mark the result of trying to re-express such values "missing" if they can't be re-expressed. But that might mean that when you try a re-expression, you inadvertently lose a bunch of data values. The effect of that loss may be surprising and may substantially change your analysis. Because you are likely to be working with a computer package or calculator, take special care that you do not lose otherwise good data values when you choose a re-expression.

  One possible cure if the negative values are not very far below zero is to add a small constant (1/2 and 1/6 are both used) to all the data values to bring the minimum up above zero.

- ***Watch for data far from 1.*** Data values that are all very far from 1 may not be much affected by re-expression unless the range is very large. Re-expressing numbers between 1 and 100 will have a much greater effect than re-expressing numbers between 100,001 and 100,100. When all your data values are large (for example, working with years), consider subtracting a constant to bring them back near 1. (For example, consider "years since 1950" as an alternative variable for re-expression. Unless your data start at 1950—then avoid creating a zero by using "years since 1949.")

- ***Don't stray too far from the ladder.*** Finally, it's wise not to stray too far from the powers that we suggest in the Ladder of Powers. Taking the y-values to an extremely high power may artificially inflate R^2, but it won't give a useful or meaningful model, so it doesn't really simplify anything. It's better to stick to powers between 2 and -2. Even in that range, you should prefer the simpler powers in the ladder to those in the cracks. A square root is easier to understand than the 0.413 power. That simplicity may compensate for a slightly less straight relationship.

CONNECTIONS

We have seen several ways to model or summarize data. Each requires that the data have a particular simple structure. We seek symmetry for summaries of center and spread and to use a Normal model. We seek equal variation across groups when we compare groups with boxplots or want to compare their centers. We seek linear shape in a scatterplot so that we can use correlation to summarize the scatter and regression to fit a linear model.

Data do often satisfy the requirements to use statistics methods. But often they do not. Our choice is to stop with just displays, to use much more complex methods, or to re-express the data so that we can use the simpler methods we have developed.

In this fundamental sense, this chapter connects to everything we have done thus far and to all of the methods we will introduce throughout the rest of the book. Re-expression greatly extends the reach and applicability of all of these methods.

What have we learned?

We've learned that when the conditions for regression are not met, a simple re-expression of the data may help. There are several reasons to consider a re-expression:

* To make the distribution of a variable more symmetric (as we saw in Chapter 4)
* To make the spread across different groups more similar
* To make the form of a scatterplot straighter
* To make the scatter around the line in a scatterplot more consistent

We've learned that when seeking a useful re-expression, taking logs is often a good, simple starting point. To search further, the Ladder of Powers or the log-log approach can help us find a good re-expression.

We've come to understand that our models won't be perfect, but that re-expression can lead us to a useful model.

TERMS

Re-express data   We re-express data by taking the logarithm, the square root, the reciprocal, or some other mathematical operation on all values in the data set.

Ladder of Powers   The Ladder of Powers places in order the effects that many re-expressions have on the data.

SKILLS

When you complete this lesson you should:

Think
* Recognize when a well-chosen re-expression may help you improve and simplify your analysis.
* Understand the value of re-expressing data to improve symmetry, to make the scatter around a line more constant, or to make a scatterplot more linear.
* Recognize when the pattern of the data indicates that no re-expression can improve the structure of the data.

Show
* Know how to re-express data with powers and how to find an effective re-expression for your data using your statistics software or calculator.
* Be able to reverse any of the common re-expressions to put a predicted value or residual back into the original units.

Tell
* Be able to describe a summary or display of a re-expressed variable making clear how it was re-expressed and giving its re-expressed units.
* Be able to describe a regression model fit to re-expressed data in terms of the re-expressed variables.

Re-expression on the Computer

Computers and calculators make it easy to re-express data. Most statistics packages offer a way to re-express and compute with variables. Some packages permit you to specify the power of a re-expression with a slider or other moveable control, possibly while watching the consequences of the re-expression on a plot or analysis. This, of course, is a very effective way to find a good re-expression.

DATA DESK

To re-express a variable in Data Desk, select the variable and Choose the function to re-express it from the **Manip > Transform** menu. Square root, log, reciprocal, and reciprocal root are immediately available. For others, make a derived variable and type the function. Data Desk makes a new derived variable that holds the re-expressed values. Any value changed in the original variable will immediately be re-expressed in the derived variable.

Comments

Or choose **Manip > Transform > Dynamic > Box-Cox** to generate a continuously changeable variable and a slider that specifies the power. Set plots to **Automatic Update** in their HyperView menus and watch them change dynamically as you drag the slider.

EXCEL

To re-express a variable in Excel, use Excel's built-in functions as you would for any calculation. Changing a value in the original column will change the re-expressed value.

JMP

To re-express a variable in JMP, double-click to the right of the last column of data to create a new column. Name the new column and select it. Choose **Formula** from the **Cols** menu. In the Formula dialog, choose the transformation and variable that you wish to assign to the new column. Click the **OK** button. JMP places the re-expressed data in the new column.

Comments

The log and square root re-expressions are found in the **Transcendental** menu of functions in the formula dialog.

MINITAB

To re-express a variable in MINITAB, choose **Calculator** from the **Calc** menu. In the Calculator dialog, specify a name for the new re-expressed variable. Use the **Functions List,** the calculator buttons, and the **Variables list** box to build the expression. Click **OK**.

SPSS

To re-express a variable in SPSS, Choose **Compute** from the **Transform** menu. Enter a name in the Target Variable field. Use the calculator and Function List to build the expression. Move a variable to be re-expressed from the source list to the Numeric Expression field. Click the **OK** button.

TI-83/84 Plus

To re-express data stored in a list, perform the re-expression on the whole list and store it in another list. For example, to use the logarithms of the data in L1, enter the command **log(L1) STO L2.**

TI-89

To re-express data stored in a list, perform the re-expression on the whole list and store it in another list. For example, to use the common (base 10) logarithms of the data in list1, on the home screen, enter the command **log(list1)** STO► **list2.**

Comments

• To find the log command press CATALOG then 4 (L) arrow to log and press ENTER.
• Natural logs are **LN** (press 2nd X).
• For square roots, press 2nd x

Player	Year	Salary (million $)
Nolan Ryan	1980	1
George Foster	1982	2.04
Kirby Puckett	1990	3
Jose Canseco	1990	4.7
Roger Clemens	1991	5.3
Ken Griffey, Jr.	1996	8.5
Albert Belle	1997	11
Pedro Martinez	1998	12.5
Mike Piazza	1999	12.5
Mo Vaughn	1999	13.3
Kevin Brown	1999	15
Carlos Delgado	2001	17
Alex Rodriguez	2001	25.2

a) Re-express the data to straighten the scatterplot.
b) Create an appropriate model for the trend in salaries.
c) Predict a superstar salary for 2005.

T 12. Planet distances and years. The table below shows the average *distance* of each of the nine planets from the sun, and the *length* of the year (in earth years).

	Position Number	Distance from Sun (million miles)	Length of Year (earth years)
Mercury	1	36	0.24
Venus	2	67	0.61
Earth	3	93	1.00
Mars	4	142	1.88
Jupiter	5	484	11.86
Saturn	6	887	29.46
Uranus	7	1784	84.07
Neptune	8	2796	164.82
Pluto	9	3666	247.68

a) Plot the *length* of the year against the *distance* from the sun. Describe the shape of your plot.
b) Re-express one or both variables to straighten the plot. Use the re-expressed data to create a model describing the length of a planet's year based on its distance from the sun.
c) Comment on how well your model fits the data.

T 13. Planet distances and order. Let's look again at the pattern in the locations of the planets in our solar system seen in the table in Exercise 12.
a) Use re-expressed data to create a model for the *distance* from the sun based on the planet's *position*.
b) There is some debate among astronomers as to whether Pluto is truly a planet or actually a large member of the Kuiper Belt of comets and other icy bodies. Does your model suggest that Pluto may not belong in the planet group? Explain.

T 14. Planets, part 3. The asteroid belt between Mars and Jupiter may be the remnants of a failed planet. If so, then Jupiter is really in position 6, Saturn is in 7, and so on. Repeat Exercise 13, using this revised method of numbering the positions. Which method seems to work better?

T 15. Quaoar: Planets, part 4. In June 2002, Caltech astronomers Chad Trujillo and Mike Brown discovered a new large body in orbit about the sun, a billion miles beyond Jupiter. Named Quaoar (pronounced "kwa-whar") after a Native American god, it is about one-tenth the diameter of earth. Quaoar orbits the sun once every 288 years at a distance of about 4 billion miles. Astronomers have classified it as a member of the Kuiper Belt, rather than as a planet.

There are reasons for suspecting that Pluto is unlike other planets. For example, its orbit is tilted relative to the plane in which other planetary orbits are found. And its orbit is very eccentric, passing both inside the orbit of Neptune and outside the orbit of Quaoar. Omit Pluto from your count of planets, and consider Quaoar as a candidate for the next planet beyond Neptune.
a) Based on its *position*, how does Quaoar's *distance* from the sun (re-expressed to logs) compare with the prediction made by your model from Exercise 13?
b) Refit the model using Quaoar's *distance* and *position* in the model instead of Pluto's. Now how well does your model to predict (re-expressed) distance from position number fit?

T 16. Models and laws: Planets part 5. The model you found in Exercise 12 is a relationship noted in the 17th century by Kepler as his Third Law of Planetary Motion. It was subsequently explained as a consequence of Newton's Law of Gravitation. The models for Exercises 13–15 relate to what is sometimes called the Titius-Bode "law," a pattern noticed in the 18th century but lacking any scientific explanation.

Compare how well the re-expressed data are described by their respective linear models. What aspect of the model of Exercise 12 suggests that we have found a physical law? In the future, we may learn enough about a planetary system around another star to tell whether the Titius-Bode pattern applies there. If you discovered that another planetary system followed the same pattern, how would it change your opinion about whether this is a real natural "law"? What would you think if the next system we find does not follow this pattern?

T 17. Logs (not logarithms). The value of a log is based on the number of "board feet" of lumber the log may contain. (A board foot is the equivalent of a piece of wood 1 inch thick, 12 inches wide, and 1 foot long. For example, a 2" × 4" piece that is 12 feet long contains 8 board feet.) To estimate the amount of lumber in a log, buyers measure the diameter inside the bark at the smaller end. Then they look in a table based on the Doyle Log Scale. The table below shows the estimates for logs 16 feet long.

Diameter of Log	8"	12"	16"	20"	24"	28"
Board Feet	16	64	144	256	400	576

a) What model does this scale use?
b) How much lumber would you estimate that a log 10 inches in diameter contains?
c) What does this model suggest about logs 36 inches in diameter?

T 18. Weight lifting. Listed below are the gold medal–winning men's weightlifting performances at the 2000 Olympics.

Weight Class (kg)	Winner (country)	Weight Lifted (kg)
56	Mutli (Turkey)	305
62	Pechalov (Croatia)	325
69	Boevski (Bulgaria)	357.5
77	Xugang (China)	367.5
85	Dimas (Greece)	390
94	Kakiasvilas (Greece)	405
105	Tavakoli (Iran)	425

a) Create a linear model for the *amount* lifted in each *weight class*.
b) Check the residuals plot. Is your linear model appropriate?
c) Create a better model.
d) Explain why you think your model is better.
e) Based on your model, which of the medalists turned in the most surprising performance? Explain.

T 19. Life expectancy. The data below list the *life expectancy* for white males in the United States every decade during the last century (1 = 1900 to 1910, 2 = 1911 to 1920, etc.). Create a model to predict future increases in life expectancy. (National Vital Statistics Report)

Decade	1	2	3	4	5	6	7	8	9	10
Life exp.	48.6	54.4	59.7	62.1	66.5	67.4	68.0	70.7	72.7	74.9

T 20. Lifting more weight. In Exercise 18 you examined the winning weight-lifting performances for the 2000 Olympics. One of the competitors turned in a performance that appears not to fit the model you created.
a) Consider that competitor to be an outlier. Eliminate that data point and recreate your model.
b) Using this revised model, how much would you have expected the outlier competitor to lift?
c) Explain the meaning of the residual from your new model for that competitor.

T 21. Slower is cheaper? Researchers studying how a car's *gas mileage* varies with its *speed* drove a compact car 200 miles at various speeds on a test track. Their data are shown in the table.

Speed (mph)	35	40	45	50	55	60	65	70	75
Miles per gal	25.9	27.7	28.5	29.5	29.2	27.4	26.4	24.2	22.8

Create a linear model for this relationship and report any concerns you may have about the model.

T 22. Orange production. The table below shows that as the number of oranges on a tree increases, the fruit tend to get smaller. Create a model for this relationship, and express any concerns you may have.

Number of Oranges/Tree	Average Weight/Fruit (lb)
50	0.60
100	0.58
150	0.56
200	0.55
250	0.53
300	0.52
350	0.50
400	0.49
450	0.48
500	0.46
600	0.44
700	0.42
800	0.40
900	0.38

T 23. Years to live. Insurance companies and other organizations use actuarial tables to estimate the remaining lifespan of their customers. Below are the estimated additional years of life for black males in the United States, according to a 1999 National Vital Statistics Report.

Age	10	20	30	40	50	60	70	80	90	100
Years Left	59.2	49.6	40.7	31.9	24.0	17.2	11.6	7.2	4.4	2.8

a) Create an appropriate model.
b) Predict the lifespan of an 18-year-old black man.

T 24. Oil production (again). Here are the data on U.S. oil production first seen in Exercise 34 of Chapter 7.

How successfully might a model based on these data predict the future of U.S. oil production? Explain.

Year	Oil	Year	Oil
1949	1,841,940	1975	3,056,779
1950	1,973,574	1976	2,976,180
1951	2,247,711	1977	3,009,265
1952	2,289,836	1978	3,178,216
1953	2,357,082	1979	3,121,310
1954	2,314,988	1980	3,146,365
1955	2,484,428	1981	3,128,624
1956	2,617,283	1982	3,156,715
1957	2,616,901	1983	3,170,999
1958	2,448,987	1984	3,249,696
1959	2,574,590	1985	3,274,553
1960	2,574,933	1986	3,168,252
1961	2,621,758	1987	3,047,378
1962	2,676,189	1988	2,979,123
1963	2,752,723	1989	2,778,773
1964	2,786,822	1990	2,684,687
1965	2,848,514	1991	2,707,039
1966	3,027,763	1992	2,624,632
1967	3,215,742	1993	2,499,033
1968	3,329,042	1994	2,431,476
1969	3,371,751	1995	2,394,268
1970	3,517,450	1996	2366,017
1971	3,453,914	1997	2,354,831
1972	3,455,368	1998	2,281,919
1973	3,360,903	1999	2,146,732
1974	3,202,585	2000	2,135,062

T 26. Tree growth. A 1996 study examined the growth of grapefruit trees in Texas, determining the average trunk *diameter* (in inches) for trees of varying *ages*.

Age (yr)	2	4	6	8	10	12	14	16	18	20
Diameter (in.)	2.1	3.9	5.2	6.2	6.9	7.6	8.3	9.1	10.0	11.4

a) Fit a linear model to these data. What concerns do you have about the model?
b) If data had been given for individual trees instead of averages, would you expect the fit to be stronger, less strong, or about the same? Explain.

just checking

Answers

1. Counts are often best transformed by using the square root.

2. None. The relationship is already straight.

3. Even though, technically, the population values are counts, you should probably try a stronger transformation like log(population).

T 25. Internet. It's often difficult to find the ideal model for situations in which the data are strongly curved. The table below shows the rapid growth of the number of academic journals published on the Internet during the last decade.

Year	Number of Journals
1991	27
1992	36
1993	45
1994	181
1995	306
1996	1093
1997	2459

a) Try to create a good model to describe this growth.
b) Use your model to estimate the number of electronic journals in the year 2000.
c) Comment on your faith in this estimate.

QUICK REVIEW

You have now survived your second major unit of Statistics. Here's a brief summary of the key concepts and skills:

▶ We treat data two ways: as categorical and as quantitative.

▶ To explore relationships in categorical data, check out Chapter 3.

▶ To explore relationships in quantitative data:

- Make a picture. Use a scatterplot. Put the explanatory variable on the *x*-axis and the response variable on the *y*-axis.
- Describe the association between two quantitative variables in terms of direction, form, and strength.
- The amount of scatter determines the strength of the association.
- If as one variable increases so does the other, the association is positive. If one increases as the other decreases, it's negative.
- If the form of the association is linear, calculate a correlation to measure its strength numerically, and do a regression analysis to model it.
- Correlations closer to −1 or +1 indicate stronger linear associations. Correlations near 0 indicate weak linear relationships, but other forms of association may still be present.
- The line of best fit is also called the least squares regression line because it minimizes the sum of the squared residuals.
- The regression line predicts values of the response variable from values of the explanatory variable.

- A residual is the difference between the true value of the response variable and the value predicted by the regression model.
- The slope of the line is a rate of change, best described in "*y*-units" per "*x*-unit."
- R^2 gives the percentage of the variation in the response variable that is accounted for by the model.
- The standard deviation of the residuals measures the amount of scatter around the line.
- Outliers and influential points can distort any of our models.
- If you see a pattern (a curve) in the residuals plot, your chosen model is not appropriate; use a different model. You may, for example, straighten the relationship by re-expressing one of the variables.
- To straighten bent relationships, re-express the data using logarithms or a power (squares, square roots, reciprocals, etc.).
- Always remember that an association is not necessarily an indication that one of the variables causes the other.

Need more help with some of this? Try rereading some sections of Chapters 7 through 10. And go on to the next page for more opportunities to review these concepts and skills.

"One must learn by doing the thing; though you think you know it, you have no certainty until you try."
—Sophocles (495–406 B.C.E.)

REVIEW EXERCISES

1. College. Every year *US News and World Report* publishes a special issue on many U.S. colleges and universities. The scatterplots below have *student/faculty ratio* (number of students per faculty member) for the colleges and universities on the *y*-axes plotted against 4 other variables. The correct correlations for these scatterplots appear in this list. Match them.

$$-0.98 \quad -0.71 \quad -0.51 \quad 0.09 \quad 0.23 \quad 0.69$$

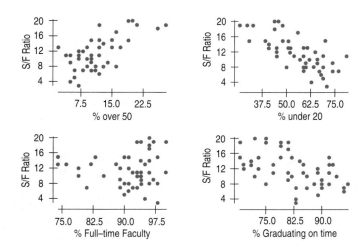

2. Togetherness. Are good grades in high school associated with family togetherness? A simple random sample of 142 high-school students was asked how many meals per week their families ate together. Their responses produced a mean of 3.78 meals per week, with a standard deviation of 2.2. Researchers then matched these responses against the students' grade point averages (GPA). The scatterplot appeared to be reasonably linear, so they created a line of regression. No apparent pattern emerged in the residuals plot. The equation of the line was $\widehat{gpa} = 2.73 + 0.11$ *meals*.
a) Interpret the *y*-intercept in this context.
b) Interpret the slope in this context.
c) What was the mean GPA for the students in this study?
d) If a student in this study had a negative residual, what did that mean?
e) Upon hearing of this study, a counselor recommended that parents who want to improve the grades their children get should get the family to eat together more often. Do you agree with this interpretation? Explain.

3. Vineyards. Shown below are the scatterplot and regression analysis for *case prices* of 36 wines from vineyards in

the Finger Lakes region of New York State and the *ages* of the vineyards.

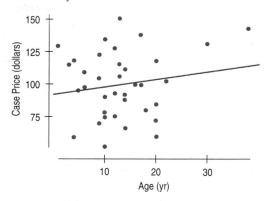

Dependent variable is: Case Price
R-squared = 2.7%

Variable	Coefficient
Constant	92.7650
Age	0.567284

a) Does it appear that vineyards in business longer get higher prices for their wines? Explain.
b) What does this analysis tell us about vineyards in the rest of the world?
c) Write the regression equation.
d) Explain why that equation is essentially useless.

T 4. Vineyards again. Instead of *age,* perhaps the *size* of the vineyard (in acres) is associated with the value of the wines. Look at the scatterplot:

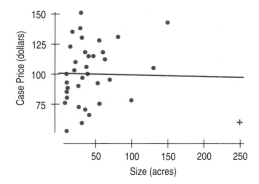

a) Do you see any evidence of an association?
b) What concern do you have about this scatterplot?
c) If the red "+" data point is removed, would you expect the correlation to become stronger or weaker? Explain.
d) If the red "+" data point is removed, would you expect the slope of the line to go up or down? Explain.

T **5. More twins?** As the table shows, the number of twins born in the United States has been increasing.

Year	Twin Births	Year	Twin Births
1981	70,049	1990	93,865
1982	71,631	1991	94,779
1983	72,287	1992	95,372
1984	72,949	1993	96,445
1985	77,102	1994	97,064
1986	79,485	1995	96,736
1987	81,778	1996	100,750
1988	85,315	1997	104,137
1989	90,118		

a) Find the equation of the regression line for predicting the number of twin births.
b) Explain in this context what the slope of this line means.
c) Predict the number of twin births in the United States for the year 2002. Comment on your faith in that prediction.
d) Comment on the residuals plot.

6. Dow Jones. When the Dow Jones stock index first reached 10,000, the *New York Times* reported the dates on which the Dow first crossed each of the "thousand" marks, starting with reaching 1000 in 1972. A regression of the Dow prices on year looks (in part) like this:

Dependent variable is: Dow
R-squared = 65.8%

Variable	Coefficient
Intercept	−603335
Year	305.471

a) What is the correlation between the *Dow index* and the *year*?
b) Write the regression equation.
c) Explain in this context what the equation says.
d) Here's a scatterplot of the residuals. Which assumption(s) of the regression analysis appear to be violated?

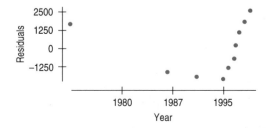

7. Acid rain. Biologists studying the effects of acid rain on wildlife collected data from 163 streams in the Adiron-

dack Mountains. They recorded the *pH* (acidity) of the water and the *BCI*, a measure of biological diversity, and they calculated $R^2 = 27\%$. Here's a scatterplot of *BCI* against *pH*.

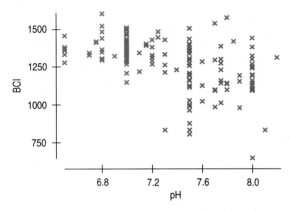

a) What is the correlation between *pH* and *BCI*?
b) Describe the association between these two variables.
c) If a stream has average *pH*, what would you predict about the *BCI*?
d) In a stream where the *pH* is 3 standard deviations above average, what would you predict about the *BCI*?

T **8. Manatees.** Marine biologists warn that the growing number of powerboats registered in Florida threatens the existence of manatees. The February 12, 2002, *New York Times* reported these data:

Year	Manatees Killed	Powerboat Registrations (in 1000s)
1982	13	447.0
1983	21	459.6
1984	24	481.0
1985	16	497.9
1986	24	512.6
1987	20	512.3
1988	15	526.5
1989	34	558.6
1990	33	585.3
1992	33	613.5
1993	39	645.5
1994	43	675
1995	50	711
1996	47	719
1997	53	716
1998	38	716
1999	35	716
2000	49	735
2001	81	860

a) In this context, which do you think is the explanatory variable?
b) Make a scatterplot of these data and describe the association you see.
c) Find the correlation between *boat registrations* and *manatee deaths.*
d) Interpret the value of R^2.
e) Does your analysis prove that powerboats are killing manatees?

9. A manatee model. Continue your analysis of the manatee situation from the previous exercise.
a) Create a linear model of the association between *manatee deaths* and *powerboat registrations.*
b) Interpret the slope of your model.
c) Interpret the *y*-intercept of your model.
d) How accurately did your model predict the high number of manatee deaths in 2001?
e) Which is better for the manatees, positive residuals or negative residuals? Explain.
f) What does your model suggest about the future for the manatee?

10. Grades. A Statistics instructor created a linear regression equation to predict students' final exam scores from their midterm exam scores. The regression equation was $\widehat{fin} = 10 + 0.9\, mid$.
a) If Susan scored a 70 on the midterm, what did the instructor predict for her score on the final?
b) Susan got an 80 on the final. How big is her residual?
c) Suppose that the standard deviation of the final was 12 points and the standard deviation of the midterm was 10 points. What is the correlation between the two tests?
d) How many points would someone need to score on the midterm to have a predicted final score of 100?
e) Suppose someone scored 100 on the final. Explain why you can't estimate this student's midterm score from the information given.
f) One of the students in the class scored 100 on the midterm but got overconfident, slacked off, and scored only 15 on the final exam. What is the residual for this student?
g) No other student in the class "achieved" such a dramatic turnaround. If the instructor decides not to include this student's scores when constructing a new regression model, will the R^2 value of the regression increase, decrease, or remain the same? Explain briefly.
h) Will the slope of the new line increase or decrease?

11. Traffic. Highway planners investigated the relationship between *traffic density* (number of automobiles per mile) and the *average speed* of the traffic on a moderately large city thoroughfare. The data were collected at the same location at 10 different times over a span of 3 months. They found a mean *traffic density* of 68.6 cars per mile (cpm) with standard deviation of 27.07 cpm. Overall, the

cars' *average speed* was 26.38 mph, with standard deviation of 9.68 mph. These researchers found the regression line for these data to be $\widehat{speed} = 50.55 - 0.352\, density$.
a) What is the value of the correlation coefficient between *speed* and *density?*
b) What percent of the variation in *average speed* is explained by *traffic density?*
c) Predict the *average speed* of traffic on the thoroughfare when the *traffic density* is 50 cpm.
d) What is the value of the residual for a *traffic density* of 56 cpm with an observed *speed* of 32.5 mph?
e) The data set initially included the point *density* = 125 cpm, *speed* = 55 mph. This point was considered an outlier and was not included in the analysis. Will the slope increase, decrease, or remain the same if we redo the analysis and include this point?
f) Will the correlation become stronger, weaker, or remain the same if we redo the analysis and include this point (125, 55)?
g) A European member of the research team measured the *speed* of the cars in kilometers per hour (1 km ≈ 0.62 miles) and the *traffic density* in cars per kilometer. Find the value of his calculated correlation between speed and density.

12. Cramming. One Thursday, researchers gave students enrolled in a section of basic Spanish a set of 50 new vocabulary words to memorize. On Friday the students took a vocabulary test. When they returned to class the following Monday, they were retested—without advance warning. Both sets of test scores for the 25 students are shown.

Fri.	Mon.	Fri.	Mon.	Fri.	Mon.
42	36	48	37	39	41
44	44	43	41	46	32
45	46	45	32	37	36
48	38	47	44	40	31
44	40	50	47	41	32
43	38	34	34	48	39
41	37	38	31	37	31
35	31	43	40	36	41
43	32				

a) What is the correlation between *Friday* and *Monday* scores?
b) What does a scatterplot show about the association between the scores?
c) What does it mean for a student to have a positive residual?
d) What would you predict about a student whose *Friday* score was one standard deviation below average?
e) Write the equation of the regression line.
f) Predict the *Monday* score of a student who earned a 40 on Friday.

13. Correlations. What factor most explains differences in *fuel efficiency* among cars? Here's a correlation matrix exploring that relationship for the car's *weight, horsepower, engine size* (displacement), and *number of cylinders.*

	MPG	Weight	Horse-power	Displace-ment	Cylinders
MPG	1.000				
Weight	−0.903	1.000			
Horsepower	−0.871	0.917	1.000		
Displacement	−0.786	0.951	0.872	1.000	
Cylinders	−0.806	0.917	0.864	0.940	1.000

a) Which factor seems most strongly associated with *fuel efficiency* (miles per gallon)?
b) What does the negative correlation indicate?
c) Explain the meaning of R^2 for that relationship.

14. Autos revisited. Look again at the correlation table for cars in the last exercise.
a) Which two variables in the table exhibit the strongest association?
b) Is that strong association necessarily cause-and-effect? Offer at least two explanations why that association might be so strong.
c) Engine displacements for U.S.-made cars are often measured in cubic inches. For many foreign cars the units are either cubic centimeters or liters. How would changing from cubic inches to liters affect the calculated correlations involving *displacement*?
d) What would you predict about the *fuel efficiency* of a car whose engine *displacement* is one standard deviation above the mean?

15. Cars, one more time! Can we predict the *horsepower* of the engine manufacturers will put in a car by knowing the weight of the car? Here are the regression analysis and residuals plot.

Dependent variable is: Horsepower
R-squared = 84.1%

Variable	Coefficient
Intercept	3.49834
Weight	34.3144

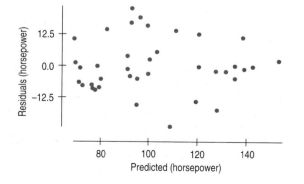

a) Write the equation of the regression line.
b) Do you think the car's *weight* is measured in pounds, or thousands of pounds? Explain.
c) Do you think this linear model is appropriate? Explain.
d) The highest point in the residuals plot, representing a residual of 22.5 horsepower, is for a Chevy weighing 2595 pounds. How many horsepower does this car have?

16. Colorblind. Although some women are colorblind, this condition is primarily found in men. Why is it wrong to say there's a strong correlation between *gender* and *colorblindness*?

17. Old Faithful. There is evidence that eruptions of Old Faithful can best be predicted by knowing the duration of the previous eruption.
a) Describe what you see in the scatterplot of *intervals* between eruptions vs. *duration* of the previous eruption.

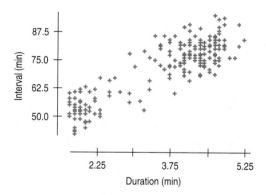

b) Write the equation of the line of best fit. Here's the regression analysis:

Dependent variable is: Interval
R-squared = 77.0%

Variable	Coefficient
Intercept	33.9668
Duration	10.3582

c) Carefully explain what the slope of the line means in this context.
d) How accurate do you expect predictions based on this model to be? Cite evidence from the regression analysis.
e) If you just witnessed an eruption that lasted 4 minutes, how long do you predict you'll have to wait to see the next eruption?
f) So you waited, and the next eruption came in 79 minutes. Use this as an example to define a residual.

18. Which croc? The ranges inhabited by the Indian gharial crocodile and the Australian saltwater crocodile overlap in Bangladesh. Suppose a very large crocodile skeleton is found there, and we wish to determine the species of the animal. Wildlife scientists have measured the lengths of the heads and the complete bodies of several crocs (in centimeters) of each species, creating the regression analyses below:

Indian Crocodile

Dependent variable is: IBody
R-squared = 97.2%

Variable	Coefficient
Intercept	−69.3693
IHead	7.40004

Australian Crocodile

Dependent variable is: ABody
R-squared = 98.0%

Variable	Coefficient
Intercept	−20.2245
A Head	7.71726

a) Do the associations between the sizes of the heads and bodies of the two species appear to be strong? Explain.

b) In what ways are the two relationships similar? Explain.

c) What is different about the two models? What does that mean?

d) The crocodile skeleton found had a head length of 62 cm and a body length of 380 cm. Which species do you think it was? Explain why.

T 19. How old is that tree? One can determine how old a tree is by counting its rings, but that requires cutting the tree down. Can we estimate the age simply from its diameter? A forester measured 27 trees of the same species that had been cut down, and counted the rings to determine the ages of the trees.

Diameter (in.)	Age (yr)	Diameter (in.)	Age (yr)
1.8	4	10.3	23
1.8	5	14.3	25
2.2	8	13.2	28
4.4	8	9.9	29
6.6	8	13.2	30
4.4	10	15.4	30
7.7	10	17.6	33
10.8	12	14.3	34
7.7	13	15.4	35
5.5	14	11.0	38
9.9	16	15.4	38
10.1	18	16.5	40
12.1	20	16.5	42
12.8	22		

a) Find the correlation between *diameter* and *age*. Does this suggest that a linear model may be appropriate? Explain.

b) Create a scatterplot and describe the association.

c) Create the linear model.

d) Check the residuals. Explain why a linear model is probably not appropriate.

e) If you used this model, would it generally overestimate or underestimate the ages of very large trees? Explain.

T 20. Improving trees. In the last exercise you saw that the linear model had some deficiencies. Let's create a better model.

a) Perhaps the cross-sectional area of a tree would be a better predictor of its age. Since area is measured in square units, try re-expressing the data by squaring the diameters. Does the scatterplot look better?

b) Create a model that predicts *age* from the square of the *diameter*.

c) Check the residuals plot for this new model. Is this model more appropriate? Why?

d) Estimate the age of a tree 18 inches in diameter.

21. New homes. A real estate agent collects data to develop a model that will use the *size* of a new home (in square feet) to predict its *sale price* (in thousands of dollars). Which of these is most likely to be the slope of the regression line: 0.008, 0.08, 0.8, or 8? Explain.

T 22. Smoking and pregnancy. The organization Kids Count monitors issues related to children. The table shows a 50-city average of the percent of expectant mothers who smoked cigarettes during their pregnancies.

Year	% Smoking While Pregnant	Year	% Smoking While Pregnant
1990	17.7	1995	12.7
1991	17.0	1996	11.9
1992	16.0	1997	11.2
1993	14.9	1998	10.8
1994	13.9	1999	10.4

a) Create a scatterplot and describe the trend you see.

b) Find the correlation.

c) How is the value of the correlation affected by the fact that the data are averages rather than percentages for each of the 50 cities?

d) Write a linear model and interpret the slope in this context.

T 23. No smoking? The downward trend in smoking you saw in the last exercise is good news for the health of babies, but will it ever stop?

a) Explain why you can't use the linear model you created in Exercise 22 to see when smoking during pregnancy will cease altogether.

b) Create a model that could estimate the year in which the level of smoking would be 0%.

c) Comment on the reliability of such a prediction.

24. Tips. It's commonly believed that people use tips to reward good service. A researcher for the hospitality industry examined tips and ratings of service quality from 2645 dining parties at 21 different restaurants. The correlation between ratings of service and tip percentages was 0.11. (M. Lynn and M. McCall, "Gratitude and Gratuity." *Journal of Socio-Economics* 29: 203–214.)
a) Describe the relationship between *quality of service* and *tip size*.
b) Find and interpret the value of R^2 in this context.

25. Move south? Data from 50 large U.S. cities show the mean *January temperature* and the *latitude*. Describe what you see in the scatterplot.

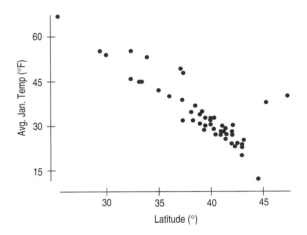

26. Correlations. A study of living conditions in 55 large U.S. cities found the mean *January temperature* (degrees Fahrenheit), *altitude* (feet above sea level), and *latitude* (degrees north of the equator). Here's the correlation matrix.

	Jan. Temp	Latitude	Altitude
Jan. Temp	1.000		
Latitude	−0.848	1.000	
Altitude	−0.369	0.184	1.000

a) Which seems to be more useful in predicting *January temperature*—altitude or latitude? Explain.
b) If the temperature were measured in degrees Celsius, what would be the correlation between temperature and latitude?
c) If the temperature were measured in degrees Celsius and the altitude in meters, what would be the correlation? Explain.
d) What would you predict about the January temperatures in a city whose altitude is two standard deviations higher than the average altitude?

27. Winter in the city. Summary statistics for the data relating the latitude and average January temperature for 55 large U.S. cities are given below.

Variable	Mean	StdDev
Latitude	39.02	5.42
JanTemp	26.44	13.49

Correlation = −0.848

a) What percent of the variation in January *temperatures* can be explained by variation in *latitude*?
b) What is indicated by the fact that the correlation is negative?
c) Write the equation of the line of regression for predicting January *temperature* from *latitude*.
d) Explain what the slope of the line means.
e) Do you think the *y*-intercept of the line is meaningful? Explain.
f) The latitude of Denver is 40° N. Predict the mean January temperature there.
g) If the residual for a city is positive, what does that mean?

28. Depression. The September 1998 issue of the *American Psychologist* published an article by Kraut et al. that reported on an experiment examining "the social and psychological impact of the Internet on 169 people in 73 households during their first 1 to 2 years online." In the experiment, 73 households were offered free Internet access for one or two years in return for allowing their time and activity online to be tracked. The members of the households who participated in the study were also given a battery of tests at the beginning and again at the end of the study. The conclusion of the study made news headlines: Those who spent more time online tended to be more depressed at the end of the experiment. Although the paper reports a more complex model, the basic result can be summarized in the following regression of *Depression* (at the end of the study, in "depression scale units") vs. *Internet use* (in mean hours per week.)

Dependent variable is: Depression
R-squared = 4.6%
s = 0.4563

Variable	Coefficient
Intercept	0.5655
Internet use	0.0199

The news reports about this study clearly concluded that using the Internet causes depression. Discuss whether such a conclusion can be drawn from this regression. If so, discuss the supporting evidence. If not, say why not.

29. Jumps. How are Olympic performances in various events related? The plot shows winning long jump and high jump distances, in inches, for the 20th century Olympic Games.

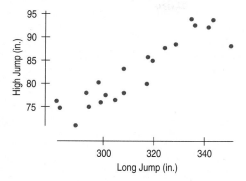

a) Describe the association.
b) Do long jump performances somehow influence the high-jumpers? How do you account for the relationship you see?
c) The correlation for the given scatterplot is 0.92, but at the Olympics these jumps are actually measured in meters rather than inches. Does that make the actual correlation higher or lower?
d) What would you predict about the long jump in a year when the high-jumper jumped one standard deviation better than the average high jump?

30. Modeling jumps. Here are the summary statistics for the Olympic long jumps and high jumps displayed in the scatterplot above.

Event	Mean	StdDev
Long Jump	314.10	20.71
High Jump	83.04	7.26

Correlation = 0.917

a) Write the equation of the line of regression for estimating *high jump* from *long jump*.
b) Interpret the slope of the line.
c) In a year when the long jump is 340 inches, what high jump would you predict?
d) Why can't you use this line to estimate the long jump for a year when you know the high jump was 85 inches?
e) Write the equation of the line you need to make that prediction.

31. French. Consider the association between a student's score on a French vocabulary test and the weight of the student. What direction and strength of correlation would you expect in each of the following situations? Explain.
a) The students are all in third grade.
b) The students are in third through twelfth grades in the same school district.

c) The students are in tenth grade in France.
d) The students are in third through twelfth grades in France.

32. Twins. Twins are often born after a pregnancy that lasts less than 9 months. The graph from the *Journal of the American Medical Association (JAMA)* shows the rate of pre-term twin births in the United States over the past 20 years. In this study, *JAMA* categorized mothers by the level of prenatal medical care they received: inadequate, adequate, or intensive.
a) Describe the overall trend in preterm twin births.
b) Describe any differences you see in this trend, depending on the level of prenatal medical care the mother received.
c) Should expectant mothers be advised to cut back on the level of medical care they seek in the hope of avoiding preterm births? Explain.

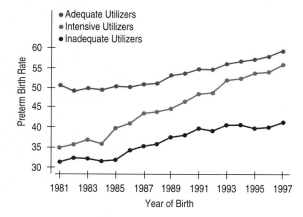

Preterm birth rate per 100 live twin births among U.S. twins by intensive, adequate, and less than adequate prenatal care utilization, 1981–1997. (JAMA 284[2000]: 335–341)

T 33. Lunchtime. Create and interpret a model for the toddlers' lunchtime data presented in Chapter 7. The table and graph show the number of minutes the kids stayed at the table and the number of calories they consumed.

Calories	Time	Calories	Time
472	21.4	450	42.4
498	30.8	410	43.1
465	37.7	504	29.2
456	33.5	437	31.3
423	32.8	489	28.6
437	39.5	436	32.9
508	22.8	480	30.6
431	34.1	439	35.1
479	33.9	444	33.0
454	43.8	408	43.7

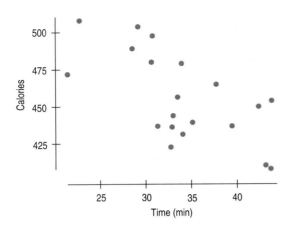

34. Gasoline. Since clean air regulations have dictated the use of unleaded gasoline, the supply of leaded gas in New York state has diminished. The table below was given on the August 2001 New York State Math B exam, a statewide achievement test for high-school students.

Year	1984	1988	1992	1996	2000
Gallons (1000's)	150	124	104	76	50

a) Create a linear model to predict the number of gallons that will be available in 2005.
b) The exam then asked students to estimate the year when leaded gasoline will first become unavailable, expecting them to use the model from part a to answer the question. Explain why that method is incorrect.
c) Create a model that *would* be appropriate for that task, and make the estimate.
d) The "wrong" answer from the other model is fairly accurate in this case. Why?

Ⓣ 35. Tobacco and alcohol. Are people who use tobacco products more likely to consume alcohol? Here are data on household spending (in pounds) taken by the British Government on 11 regions in Great Britain. Do tobacco and alcohol spending appear to be related? What questions do you have about these data? What conclusions can you draw?

Region	Alcohol	Tobacco
North	6.47	4.03
Yorkshire	6.13	3.76
Northeast	6.19	3.77
East Midlands	4.89	3.34
West Midlands	5.63	3.47
East Anglia	4.52	2.92
Southeast	5.89	3.20
Southwest	4.79	2.71
Wales	5.27	3.53
Scotland	6.08	4.51
Northern Ireland	4.02	4.56

Ⓣ 36. Football weights. The Sears Cup was established in 1993 to honor institutions that maintain a broad-based athletic program, achieving success in many sports, both men's and women's. Since its Division III inception in 1995, the cup has been won by Williams College in every year except one. Their football team has a 85.3% winning record under their current coach. Why does the football team win so much? Is it because they're heavier than their opponents? The table shows the average team weights for selected years from 1973 to 1993.

Year	Weight (lb)	Year	Weight (lb)
1973	185.5	1983	192.0
1975	182.4	1987	196.9
1977	182.1	1989	202.9
1979	191.1	1991	206.0
1981	189.4	1993	198.7

a) Fit a straight line to the relationship between *weight* and *year*.
b) Does a straight line seem reasonable?
c) Predict the average weight of the team for the year 2003. Does this seem reasonable?
d) What about the prediction for the year 2103? Explain.
e) What about the prediction for the year 3003? Explain.

37. Models. Find the predicted value of y using each model for $x = 10$.
a) $\hat{y} = 2 + 0.8 \ln x$
b) $\log \hat{y} = 5 - 0.23x$
c) $\dfrac{1}{\sqrt{\hat{y}}} = 17.1 - 1.66x$

Ⓣ 38. Williams vs. Texas. Here are the average weights of the football team for the University of Texas for various years in the 20th century.

Year	Weight (lb)
1905	164
1919	163
1932	181
1945	192
1955	195
1965	199

a) Fit a straight line to the relationship of *weight* by *year* for Texas football players.
b) According to these models, in what year will the predicted weight of the Williams College team from Exercise 36 first be more than the University of Texas team?
c) Do you believe this? Explain.

39. Vehicle weights. The Minnesota Department of Transportation hoped that they could measure the weights of big trucks without actually stopping the vehicles by using a newly developed "weigh-in-motion" scale. After installation of the scale, a study was conducted to find out whether the scale's readings correspond to the true weights of the trucks being monitored. In Exercise 14 of Chapter 7, you examined the scatterplot for the data they collected, finding the association to be approximately linear with $R^2 = 93\%$. Their regression equation is $\widehat{Wt} = 10.85 + 0.64\,scale$, where both the scale reading and the predicted weight of the truck are measured in thousands of pounds.

a) Estimate the weight of a truck if this scale read 31,200 pounds.

b) If that truck actually weighed 32,120 pounds, what was the residual?

c) If the scale reads 35,590 pounds, and the truck has a residual of −2440 pounds, how much does it actually weigh?

d) In general, do you expect estimates made using this equation to be reasonably accurate? Explain.

e) If the police plan to use this scale to issue tickets to trucks that appear to be overloaded, will negative or positive residuals be a greater problem? Explain.

40. Profit. How are a company's profits related to its sales. Let's examine data from 71 large U.S. corporations. All amounts are in millions of dollars.

a) Here are histograms of *profits* and *sales* and histograms of the logarithms of *profits* and *sales*. Why are the re-expressed data better for regression?

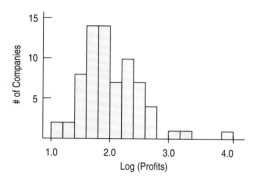

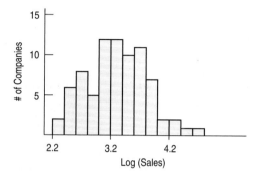

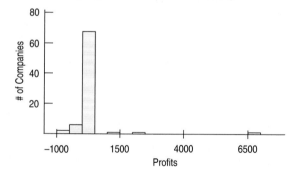

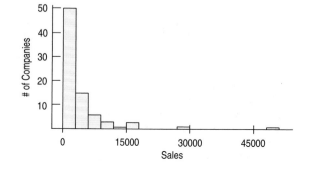

b) Here are the scatterplot and residuals plot for the regression of logarithm of *profits* vs. log of *sales*. Do you think this model is appropriate? Explain.

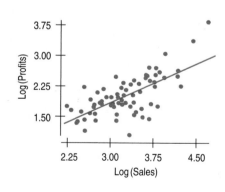

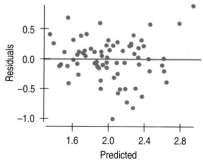

c) Here's the regression analysis. Write the equation.

Dependent variable is: Log Profit
R-squared = 48.1%

Variable	Coefficient
Intercept	−0.106259
LogSales	0.647798

d) Use your equation to estimate profits earned by a company with sales of 2.5 billion dollars. (That's 2500 million.)

T 41. Down the drain. Most water tanks have a drain plug so that the tank may be emptied when it's to be moved or repaired. How long it takes a certain size tank to drain depends on the size of the plug, as shown in the table. Create a model.

Diameter of Plug (in.)	Drain Time (min.)
$\frac{3}{8}$	140
$\frac{1}{2}$	80
$\frac{3}{4}$	35
1	20
$1\frac{1}{4}$	13
$1\frac{1}{2}$	10
2	5

T 42. Chips. A start-up company has developed an improved electronic chip for use in laboratory equipment. The company needs to project the manufacturing cost, so it develops a spreadsheet model that takes into account the purchase of production equipment, overhead, raw materials, depreciation, maintenance, and other business costs. The spreadsheet estimates the cost of producing 10,000 to 200,000 chips per year, as seen in the table. Develop a regression model to predict *costs* based on the *level* of production.

Chips Produced (1000s)	Cost per Chip ($)
10	146.10
20	105.80
30	85.75
40	77.02
50	66.10
60	63.92
70	58.80
80	50.91
90	47.22
100	44.31
120	42.88
140	39.05
160	37.47
180	35.09
200	34.04

43. Least squares. Consider the four points (12, 800), (24, 680), (36, 740), and (48, 660). The line of best fit is $\hat{y} = 810 - 3x$. Use these data to explain what "least squares" means.

PART III

Gathering Data

Understanding Randomness

W e all know what it means for something to be random. Or do we? Many children's games rely on chance outcomes. Rolling dice, spinning spinners, and shuffling cards all select at random. Adult games use randomness as well, from card games to lotteries to Bingo. What's the most important aspect of the randomness in these games? It must be fair.

What is it about random selection that makes it seem fair? It's really two things. First, nobody can guess the outcome before it happens. Second, when we want things to be fair, usually some underlying set of outcomes will be equally likely (although in many games some combinations of outcomes are more likely than others).

Randomness is not always what we might think of as "at random." Random outcomes have a lot of structure, especially when viewed in the long run. You can't predict how a fair coin will land on any single toss, but you're pretty confident that if you flipped it thousands of times you'd see about 50% heads. As we will see, randomness is an essential tool of Statistics. Statisticians don't think of randomness as the annoying tendency of things to be unpredictable or haphazard. Statisticians use randomness as a tool. In fact, without deliberately applying randomness we couldn't do most of Statistics and this book would stop right about here.[1]

But truly random values are surprisingly hard to get. Just to see how fair humans are at selecting, pick a number at random from the top of the next page. Go ahead. Turn the page, look at the numbers quickly, and pick a number at random.

Ready?

Go.

---

[1] Don't get your hopes up.

It's Not Easy Being Random

"The generation of random numbers is too important to be left to chance."
—Robert R. Coveyou,
Oak Ridge National Laboratory

A S   **Random Behavior.**
ActivStats' Random Experiment Tool lets you experiment with truly random outcomes. We'll use it a lot in the coming chapters.

Did you pick 3? If so, you've got company. Almost 75% of all people pick the number 3. About 20% pick either 2 or 4. If you picked 1, well, consider yourself a little different. Only about 5% choose 1. Psychologists have proposed reasons for this phenomenon, but for us, it simply serves as a lesson that we've got to find a better way to choose things at random.

So how should we generate **random numbers?** It's surprisingly difficult to get random values even when they're equally likely. Computers have become a popular way to generate random numbers. Even though they often do much better than humans, computers can't generate truly random numbers, either. Computers follow programs. Start a computer from the same place, and it will always follow exactly the same path. So numbers generated by a computer program are not truly random. Technically, "random" numbers generated this way are *pseudorandom* numbers. Pseudorandom values are generated in a fixed sequence and because computers can represent only a finite number of distinct values, the sequence of pseudorandom numbers must eventually repeat itself. Fortunately, pseudorandom values are good enough for most purposes because they are virtually indistinguishable from truly random numbers.

A S **Truly Random Values on the Internet.** This activity takes you to an Internet site that generates all the truly random numbers you could want.

There *are* ways to generate random numbers so that they are both equally likely and truly random. In the past, entire books of carefully generated random numbers were published. The books never made the best-seller lists and probably didn't make for great reading, but they were quite valuable to those who needed truly random values.[2] Today, we have a choice. We can use these books or find genuinely random digits from several Internet sites. The sites use methods like timing of the decay of a radioactive element or even the random changes of lava lamps to generate truly random digits. (See this book's Web site for URLs.) In either case, a string of random digits might look like this:

2217726304387410092537086270581997622725849795907032825001108963
3217535822643800292254644943760642389043766557204107354186024508
8906427308645681412198226653885873285580169902784311038042067664
8740522639824530519902027044464984322000946238678577902639002954
8887003319933147508331265192321413908608674496383528968974910533
6944182713168919406022181281304751019321546303870481407676636740
6070204916508913632855351361361043794293428486909462881431793360
7706356513310563210508993624272872250535395513645991015328128202

You probably have more interesting things to download than a few million random digits, but we'll discuss ways to use such random digits to apply randomness to real situations soon. The best ways we know to generate data that give a fair and accurate picture of the world rely on randomness, and the ways in which we draw conclusions from those data depend on the randomness, too.

An ordinary deck of playing cards, like the ones used in bridge and many other card games, consists of 52 cards. There are numbered cards (2 through 10), and face cards (Jack, Queen, King, Ace) whose value depends on the game you are playing. Each card is also marked by one of four suits (clubs, diamonds, hearts, or spades), whose significance is also game-specific.

● **Aren't you done shuffling yet?** Even something as common as card shuffling may not be as random as you might think. If you shuffle cards by the usual method in which you split the deck in half and try to let cards fall roughly alternately from each half, you're doing a "riffle shuffle."

How many times should you shuffle cards to make the deck random? A surprising fact was discovered by statisticians Persi Diaconis, Ronald Graham, and W. M. Kantor. It takes seven riffle shuffles. Fewer than seven leaves order in the deck, but after that, more shuffling does little good. Most people, though, don't shuffle that many times.

When computers were first used to generate hands in bridge tournaments, some professional bridge players complained that the computer was making too many "weird" hands—hands with 10 cards of one suit, for example. Suddenly these hands were appearing more often than players were used to when cards were shuffled by hand. The players assumed that the computer was doing something wrong. But it turns out that it's humans who hadn't been shuffling enough to make the decks really random and have those "weird" hands appear as often as they should. ●

Practical Randomness

Suppose a cereal manufacturer puts pictures of famous athletes on cards in boxes of cereal in the hope of boosting sales. The manufacturer announces that 20% of the boxes contain a picture of Tiger Woods, 30% a picture of Lance Armstrong,

---

[2] You'll find a table of random digits of this kind in the back of this book.

and the rest a picture of Serena Williams. You want all three pictures. How many boxes of cereal do you expect to have to buy in order to get the complete set?

How can we answer questions like this? Well, one way is to go buy hundreds of boxes of cereal to see what might happen. But let's not. Instead, we'll consider using a random model. Why random? When we pick a box of cereal off the shelf, we don't know what picture is inside. We'll assume that the pictures are randomly placed in the boxes and then the boxes are distributed randomly to stores around the country.

Why a model? We'll use a model to imitate cards found in random boxes of cereal. To do that, we use the random digits 0, 1, 2, . . . , 9, assuming that they are equally likely to occur and that we don't know which one will come up next. We identify 20% of the digits to represent a Tiger Woods picture, 30% to represent Lance, and the other 50% Serena. With this model we can pretend we've bought sequences of cereal boxes and simulate the outcomes—even though these outcomes (which card we get) are not equally likely—to see what happens.

A Simulation

We want to use this simulation to give us some insight into how many boxes of cereal we might have to open until we get all three cards. We'll pretend to buy cereal, and the random numbers will tell us which card to pretend we got. Like any model, the simulation will be imperfect, but we hope the results we get will help us understand the situation. A **simulation** consists of a sequence of random outcomes that model a situation. In this sequence the most basic event is called a **component** of the simulation. Each component has a set of possible **outcomes,** one of which will occur at random. In our cereal example, the component is the selection of a particular box of cereal and the outcome is the type of card in the box. The sequence of events we want to investigate is called a **trial.** Trials usually involve several components; here we pretend to buy several boxes of cereal. After the trial, we record what happened—our **response variable.** By repeating this process many times—a run of the simulation—we can get an idea of what might happen if we really did try buying boxes of cereal in search of Tiger's picture. Now let's look at these steps for making a simulation in more detail.

1. **Identify the component to be repeated.** In this case, our component is the selection of a box of cereal.

2. **Explain how you will model the outcome.** The digits from 0 to 9 are equally likely to occur. Because 20% of the boxes contain Tiger's picture, we'll use 2 of the 10 digits to represent that outcome. Three of the 10 digits can model the 30% of boxes with Lance Armstrong cards, and the remaining 5 digits can represent the 50% of boxes with Serena. One possible assignment of the digits is

$$0, 1 = \text{Woods} \qquad 2, 3, 4 = \text{Armstrong} \qquad 5, 6, 7, 8, 9 = \text{Williams}$$

3. **Explain how you will simulate the trial.** A trial is the sequence of events that we are pretending will take place. In this case we want to pretend to open cereal boxes until we have one of each picture. We do this by looking at each random number and indicating what outcome it represents. We continue until we've encountered all three pictures.

For example, the random number sequences 29240 would mean you get Lance's picture (2) in the first box you open, Serena's picture (9) in the next box, two more Armstrong pictures (2, 4) in the next two boxes, and then the Tiger Woods picture (0) you need to complete your collection. Since we've gotten all three pictures, we've finished one trial of the simulation.

4. **State clearly what the response variable is.** What are we interested in? We want to know how many boxes it takes to get all three pictures. This is the response variable. In the sample trial here, the response value is 5 boxes.

5. **Run several trials.** A simulation is cheaper than really buying cereal, and the more trials you perform, the better. For example, consider the third line of random digits shown earlier:

89064 2730 8645681 41219 822665388587328580 16990278431103804200067...

Let's create a chart to keep track of what happened.

Trial number	Outcomes	y = Number of boxes
1	89064 = Serena, Serena, **Tiger**, Serena, Lance	5
2	2730 = Lance, Serena, Lance, **Tiger**	4
3	8645681 = Serena, Serena, Lance, ..., **Tiger**	7
4	41219 = Lance, **Tiger**, Lance, Tiger, Serena	5
5	822665388587328580 = Serena, Lance, ..., **Tiger**	18

6. **Analyze the response variable.** We wanted to know how many boxes we might expect to buy to get all three cards. To answer the question, we need to analyze the response variable in a number of ways. We know how to do this. In our first 5 trials we needed 5, 4, 7, 5, and then 18 for an average of 7.8 boxes.

 Bigger Samples Are Better. Learn a new trick: The random simulation tool can generate lots of outcomes with a single click, so you can see more of the long run with less effort.

7. **State your conclusion (in the context of the problem, as always).** Based on our simulation, we estimate that customers who want the complete set of sports star pictures will buy an average of 7.8 boxes of cereal.

If you fear that this may not be an accurate estimate because we ran only five trials, you are absolutely correct. The more trials the better, and five is woefully inadequate. Twenty trials is probably a reasonable minimum if you are doing this by hand. Even better, use a computer and run a few hundred trials.

just checking

The baseball World Series consists of up to seven games. The first two are played at one team's home ballpark, the next three at the other team's park, and the final two (if needed) are played back at the first park. Records over the past century show that there is a home field advantage; the home team has about a 55% chance of winning. Is it an advantage to play the first two games on your home field? Or would you rather have the three middle games at home with the opportunity to win it all? In other words, do the two teams have an equal chance to win the four games needed to win the Series?

Let's set up the simulation:

1. What is the component to be repeated?

2. How will you model the outcome?

3. How will you simulate the trial?

4. What is the response variable?

5. How will you analyze the response variable?

Simulation Step-By-Step

Fifty-seven students participated in a lottery for a particularly desirable dorm room—a triple with a fireplace and private bath in the tower. Twenty of the participants were members of the same varsity team. When all three winners were members of the team, the other students cried foul. Use a simulation to determine whether an all-team outcome could reasonably be expected to happen if everyone had a fair shot at the room.

Think

Plan State the problem. Identify the important parts of your simulation.

I'll use a simulation to investigate whether it's unlikely that three varsity athletes would get the great room in the dorm if the lottery were fair.

Components Identify the components.

A component is the selection of a student.

Outcomes State how you will model the random occurrence of an outcome. You can't just use the digits from 0 to 9 because the outcomes you are simulating are not multiples of 10%.

I'll look at two-digit random numbers.

Let 00–19 represent the 20 varsity applicants.

Let 20–56 represent the other 37 applicants.

There are 20 and 37 students in the two groups. This time you must use *pairs* of random digits (and ignore some of them) to represent the 57 students.

Skip 57–99. If I get a number in this range, I'll throw it away and go back for another two-digit random number.

Trial Define a trial. Be sure that you don't use a repeated number this time.

Each trial consists of identifying pairs of digits as V (varsity) or N (nonvarsity) until 3 people are chosen, ignoring out-of-range or repeated numbers (X)—I can't put the same person in the room twice.

Response Variable Define your response variable.

The response variable is whether or not all selected students are on the varsity team.

Show

Mechanics

Run several trials. Carefully record the random numbers, indicating what each represents. Indicate the value of the response variable for each trial.

Trial Number	Outcomes	All Varsity?
1	74 02 94 39 02 77 55 X V X N X X N	No
2	18 63 33 25 V X N N	No
3	05 45 88 91 56 V N X X N	No
4	39 09 07 N V V	No
5	65 39 45 95 43 X N N X N	No
6	98 95 11 68 77 12 17 X X V X X V V	Yes
7	26 19 89 93 77 27 N V X X X N	No

8	23 52 37				No
	N N N				
9	16 50 83 44				No
	V N X N				
10	74 17 46 85 09				No
	X V N X V				

Estimate

Summarize the results across all trials.

"All varsity" occurred once, or 10% of the time.

Tell

Conclusion

Describe what the simulation shows and interpret your results in the context of the real world.

In my simulation of "fair" room draws, the three people chosen were all varsity team members only 10% of the time. While this result *could* happen by chance, it is not particularly likely. I'm suspicious, but I'd need many more trials and a smaller frequency of the all-varsity outcome before I would make an accusation of unfairness.

What Can Go Wrong?

- ***Don't overstate your case.*** Let's face it: In some sense a simulation is *always* wrong. After all, it's not the real thing. We didn't buy any cereal, or run a room draw. So beware of confusing what really *happens* with what a simulation suggests *might* happen. Never forget that future results will not match your simulated results exactly.
- ***Model the outcome chances accurately.*** A common mistake in constructing a simulation is to adopt a strategy that may appear to produce the right kind of results, but that does not accurately model the situation. For example, in our room draw we could have gotten 0, 1, 2, or 3 team members. Why not just see how often these digits occur in random digits from 0 to 9, ignoring the digits 4 and up?

  3 2 1 7 9 0 0 5 9 7 3 7 9 2 5 2 4 1 3 8

  3 2 1 x x 0 0 x x x 3 x x 2 x 2 x 13 x

  This "simulation" makes it seem fairly likely that three team members would be chosen. There's a big problem with this approach, though. The digits 0, 1, 2, and 3 occur with equal frequency among random digits, making each outcome appear to happen 25% of the time. In fact, the selection of 0, 1, 2, or all 3 team members are not all equally likely outcomes. In our correct simulation, we estimated that all 3 would be chosen only about 10% of the time. If your simulation overlooks important aspects of the real situation, your model will not be accurate.
- ***Run enough trials.*** Simulation is cheap and fairly easy to do. Don't try to draw conclusions based on 5 or 10 trials (even though we did for illustration purposes here). We'll make precise how many trials to use in later chapters. For now, err on the side of large numbers of trials.

A S **Estimating Summaries from Random Outcomes.** This activity throws you a curve. See how well you can estimate something you can't know (just by generating random outcomes).

CONNECTIONS

Simulations often generate many outcomes of a response variable, and we are often interested in the distribution of these responses. The tools we use to display and summarize the distribution of any real variable are appropriate for displaying and summarizing randomly generated responses as well.

Make histograms, boxplots, and Normal probability plots of the response variables from simulations, and summarize them with measures of center and spread. Be especially careful to report the variation of your response variable.

Don't forget to think about your analyses. Simulations can hide subtle errors. A careful analysis of the responses can save you from erroneous conclusions based on a faulty simulation.

You may be less likely to find an outlier in simulated responses, but if you find one, you should certainly determine how it happened.

What have we learned?

We've learned to harness the power of randomness. We've learned that a simulation model can help us investigate a question for which many outcomes are possible, we can't (or don't want to) collect data, and a mathematical answer is hard to calculate. We've learned how to base our simulation on random values generated by a computer, generated by a randomizing device such as a die or spinner, or found on the Internet. Like all models, simulations can provide us with useful insights about the real world.

TERMS

Random
An event is random if we know what outcomes could happen, but not which particular values will happen.

Random numbers
Random numbers are hard to generate. Nevertheless, several Internet sites offer an unlimited supply of equally likely random values.

Simulation
A simulation models random events by using random numbers to specify event outcomes with relative frequencies that correspond to the true real-world relative frequencies we are trying to model.

Simulation component
The most basic situation in a simulation in which something happens at random.

Outcome
An individual result of a component of a simulation is its outcome.

Trial
The sequence of several components representing events that we are pretending will take place.

Response variable
Values of the response variable record the results of each trial with respect to what we were interested in.

SKILLS

When you complete this lesson you should:

 Think
- Be able to recognize random outcomes in a real-world situation.
- Be able to recognize when a simulation might usefully model random behavior in the real world.

Show
- Know how to perform a simulation either by generating random numbers on a computer or calculator, or by using some other source of random values such as dice, a spinner, or a table of random numbers.

 Tell
- Be able to describe a simulation so that others could repeat it.
- Be able to discuss the results of a simulation study and draw conclusions about the question being investigated.

 A S **Creating Random Variables.** Play with your statistics package as you learn to generate random outcomes.

Simulation on the Computer

Simulations are best done with the help of technology simply because more trials makes a better simulation, and computers are fast. There are special computer programs designed for simulation, but most statistics packages can generate random numbers and support a simulation.

All technology-generated random numbers are *pseudorandom*. The random numbers available on the Internet may technically be better, but the differences won't matter for any simulation of modest size. Pseudorandom numbers generate the next random value from the previous one by a specified algorithm. But they have to start somewhere. This starting point is called the "seed." Most programs let you set the seed. There's usually little reason to do this, but if you wish to, go ahead. If you reset the seed to the same value, the programs will generate the same sequence of "random" numbers.

DATA DESK

Generate random numbers in Data Desk with the **Generate Random Numbers . . .** command in the **Manip** menu. A dialog guides you in specifying the number of variables to fill, the number of cases, and details about the values. For most simulations, generate random uniform values.

Comments

Bernoulli Trials generate random values that are 0 or 1, with a specified chance of a 1.
Binomial Experiments automatically generate a specified number of Bernoulli trials and count the number of 1's.

EXCEL

The **RAND** function generates a random value between 0 and 1. You can multiply to scale it up to any range you like and use the INT function to turn the result into an integer.

Comments

Published tests of Excel's random number generation have declared it to be inadequate. However, for simple simulations, it should be OK. Don't trust it for important large simulations.

JMP

In a new column, in the **Cols** menu choose **Column Info . . .** In the dialog, click the **New Property** button, and choose **Formula** from the drop-down menu.

Click the **Edit Formula** button and in the **Functions(grouped)** window, click on **Random. Random Integer (10),** for example, will generate a random integer between 1 and 10.

MINITAB

In the **Calc** menu choose **Random Data . . .**
In the Random Data submenu choose **Uniform . . .**

A dialog guides you in specifying details of range and number of columns to generate.

SPSS

The **RV.UNIFORM(min, max)** function returns a random value that is equally likely between the min and max limits.

TI-83/84 Plus

To generate random numbers use **5:RandInt** from the **Math PRB** menu. This command will produce any number of random integers in a specified range.

Comments

Some examples:

RandInt(0,1) randomly chooses a 0 or a 1. This is an effective simulation of a coin toss.

RandInt(1,6,2) randomly returns two integers between 1 and 6. This is a good way to simulate rolling two dice. To run several trials, just hit ENTER repeatedly.

RandInt(0,57,3) produces three random integers between 0 and 57—a nice way to simulate the chapter's dorm room lottery.

TI-89

To generate random numbers, move the cursor to highlight the name of a blank list. Use **5:RandInt** from the F4 (Calc) Probability menu. This command will produce any number of random integers in the specified range.

Comments

Some examples:

RandInt(0,10) randomly chooses a 0 or a 1. This is an effective simulation of 10 coin tosses.

RandInt(1,6,2) randomly returns two integers between 1 and 6. This is a good way to simulate rolling two dice.

RandInt(0,56,3) produces three random integers between 0 and 57, a nice way to simulate the chapter's dorm room lottery.

EXERCISES

1. Coin toss. Is a coin flip random? Why or why not, in your opinion?

2. Casino. A casino claims that its electronic "video roulette" machine is truly random. What should that claim mean?

3. The lottery. Many states run lotteries, giving away millions of dollars if you match a certain set of winning numbers. How are those numbers determined? Do you think this method guarantees randomness? Explain.

4. Games. Many kinds of games people play rely on randomness. Cite three different methods commonly used in the attempt to achieve this randomness, and discuss the effectiveness of each.

5. Bad simulations. Explain why each of the following simulations fails to model the real situation properly.
   a) Use a random integer from 0 through 9 to represent the number of heads that appear when 9 coins are tossed.

   b) A basketball player takes a foul shot. Look at a random digit, using an odd digit to represent a good shot and an even digit to represent a miss.
   c) Use five random digits from 1 through 13 to represent the denominations of the cards in a poker hand.

6. More bad simulations. Explain why each of the following simulations fails to model the real situation properly.
   a) Use random numbers 2 through 12 to represent the sum of the faces when two dice are rolled.
   b) Use a random integer from 0 through 5 to represent the number of boys in a family of 5 children.
   c) Simulate a baseball player's time at bat by letting 0 = an out, 1 = a single, 2 = a double, 3 = a triple, and 4 = a home run.

7. Wrong conclusion. A Statistics student properly simulated the length of checkout lines in a grocery store and then reported, "The average length of the line will be 3.2 people." What's wrong with this conclusion?

8. Another wrong conclusion. After simulating the spread of a disease, a researcher wrote, "24% of the people contracted the disease." What should the correct conclusion be?

9. Election. You're pretty sure that your candidate for class president has about 55% of the votes in the entire school. But you're worried that only 100 students will show up to vote. How often will the underdog (the one with 45% support) win? To find out you set up a simulation.
   a) Describe how you will simulate a component and its outcomes.
   b) Describe how you will simulate a trial.
   c) Describe the response variable.

10. Two pair, or three of a kind? When drawing five cards randomly from a deck, which is more likely, two pairs or three of a kind? A pair is exactly two of the same denomination. Three of a kind is exactly 3 of the same denomination. (Don't count three 8's as a pair—that's 3 of a kind. And don't count 4 of the same kind as two pair—that's 4 of a kind, a very special hand.) How could you simulate 5-card hands? Be careful; once you've picked the 8 of spades for a hand, you can't get it again until the next hand.
   a) Describe how you will simulate a component and its outcomes.
   b) Describe how you will simulate a trial.
   c) Describe the response variable.

11. Cereal. In the chapter's example, 20% of the cereal boxes contained a picture of Tiger Woods, 30% Lance Armstrong, and the rest Serena Williams. Suppose you buy five boxes of cereal. Estimate the probability that you end up with a complete set of the pictures. Your simulation should have at least 20 runs.

12. Cereal, again. Suppose you really want the Tiger Woods picture. How many boxes of cereal do you need to buy to be pretty sure of getting at least one? Your simulation should use at least 10 runs.

13. Multiple choice. You take a quiz with 6 multiple choice questions. After you studied, you estimated that you would have about an 80% chance of getting any individual question right. What are your chances of getting them all right? Your simulation should use at least 20 runs.

14. Lucky guessing? A friend of yours who took that same multiple choice quiz got all 6 questions right, but now claims to have guessed blindly on every question. If each question offered 4 possible answers, do you believe her? Explain, basing your argument on a simulation involving at least 10 runs.

15. Beat the lottery. Many states run lotteries to raise money. A Web site advertises that it knows "how to increase YOUR chances of Winning the Lottery." They offer several systems and criticize others as foolish. One

system is called *Lucky Numbers*. People who play the *Lucky Numbers* system just pick a "lucky" number to play, but maybe some numbers are luckier than others. Let's use a simulation to see how well this system works.

To make the situation manageable, simulate a simple lottery in which a single digit from 0 to 9 is selected as the winning number. Pick a single value to bet, such as 1, and keep playing it over and over. You'll want to run at least 100 trials. (If you can program the simulations on a computer or programmable calculator, run several hundred. Or generalize the questions to a lottery that chooses two- or three-digit numbers—for which you'll need thousands of trials.)
   a) What proportion of the time do you expect to win?
   b) Would you expect better results if you picked a "luckier" number, such as 7? (Try it, if you don't know.) Explain.

16. Random is as random does. The "beat the lottery" Web site discussed in Exercise 15 suggests that because lottery numbers are random, it is better to select your bet randomly. For the same simple lottery in Exercise 15 (random values from 0 to 9), generate each bet by choosing a separate random value between 0 and 9. Play many games. What proportion of the time do you win?

17. It evens out in the end. The "beat the lottery" Web site of Exercise 15 notes that in the long run we expect each value to turn up about the same number of times. That leads to their recommended betting strategy. First, watch the lottery for a while, recording all the winners. Then bet the value that has turned up the least, because we expect it will need to turn up more often to even things out. If there is more than one "rarest" value, just take the lowest one (since it doesn't matter). Simulating the simplified lottery described in Exercise 15, play many games with this system. What proportion of the time do you win?

18. Play the winner? Another strategy for beating the lottery is the reverse of the system described in Exercise 17. Simulate the simplified lottery described in Exercise 15. Each time, bet the number that just turned up. The Web site suggests that this method should do worse. Does it? Play many games and see.

19. Driving test. You are about to take the road test for your driver's license. You hear that only 34% of candidates pass the test the first time, but the percentage rises to 72% on subsequent retests. Estimate the average number of tests drivers take in order to get a license. Your simulation should use at least 20 runs.

20. Still learning? As in Exercise 19, assume that your chance of passing the driver's test is 34% the first time and 72% for subsequent retests. Estimate the percentage of those tested who still do not have a driver's license after two attempts.

21. **Basketball strategy.** Late in a basketball game, the team that is behind often fouls someone in an attempt to get the ball back. Usually the opposing player will get to shoot foul shots "one and one," meaning he gets a shot, and then a second shot only if he makes the first one. Suppose the opposing player has made 72% of his foul shots this season. Estimate the number of points he will score in a one-and-one situation.

22. **Blood donors.** A person with type O-positive blood can receive blood only from other type O donors. About 44% of the U.S. population has type O blood. At a blood drive, how many potential donors do you expect to examine in order to get three units of type O blood?

23. **Free groceries.** To attract shoppers, a supermarket runs a weekly contest that involves "scratch-off" cards. With each purchase, customers get a card with a black spot obscuring a message. When the spot is scratched away, most of the cards simply say, "Sorry—please try again." But during the week, 100 customers will get cards that make them eligible for a drawing for free groceries. Ten of the cards say they may be worth $200, 10 others say $100, 20 may be worth $50, and the rest could be worth $20. To register those cards, customers write their names on them and put them in a barrel at the front of the store. At the end of the week the store manager draws cards at random, awarding the lucky customers free groceries in the amount specified on their card. The drawings continue until the store has given away more than $500 of free groceries. Estimate the average number of winners each week.

24. **Find the ace.** A new electronics store holds a contest to attract shoppers. Once an hour someone in the store is chosen at random to play the Music Game. Here's how it works: An ace and four other cards are shuffled and placed face down on a table. The customer gets to turn cards over one at a time, looking for the ace. The person wins $100 worth of free CDs or DVDs if the ace is the first card, $50 if it is the second card, and $20, $10, or $5 if it is the third, fourth, or fifth card chosen. What is the average dollar amount of music the store will give away?

25. **The family.** Many couples want to have both a boy and a girl. If they decide to continue to have children until they have one child of each gender, what would the average family size be? Assume that boys and girls are equally likely.

26. **A bigger family.** Suppose a couple will continue having children until they have at least two children of each gender (two boys *and* two girls). How many children might they expect to have?

27. **Dice game.** You are playing a children's game in which the number of spaces you get to move is determined by the rolling of a die. You must land exactly on the final space in order to win. If you are 10 spaces away, how many turns might it take you to win?

28. **Parcheesi.** You are three spaces from a win in Parcheesi. On each turn you will roll two dice. To win, you must roll a total of 3, or roll a 3 on one of the dice. How many turns might you expect this to take?

29. **The hot hand.** A basketball player with a 65% shooting percentage has just made 6 shots in a row. The announcer says this player "is hot tonight! She's in the zone!" Assume the player takes about 20 shots per game. Is it unusual for her to make 6 or more shots in a row during a game?

30. **The World Series.** The World Series ends when a team wins 4 games. Suppose that sports analysts consider one team a bit stronger, with a 55% chance to win any individual game. Estimate the likelihood that the underdog wins the series.

31. **Teammates.** Four couples at a dinner party play a board game after the meal. They decide to play as teams of two and to select the teams randomly. All eight people write their names on slips of paper. The slips are thoroughly mixed, then drawn two at a time. How likely is it that every person will be teamed with someone other than the person he or she came to the party with?

32. **Second team.** Suppose the couples in Exercise 31 choose the teams by having one member of each couple write their names on the cards and the other people each pick a card at random. How likely is it that every person will be teamed with someone other than the person he or she came with?

33. **Job discrimination?** A company with a large sales staff announces openings for three positions as regional managers. Twenty-two of the current salespersons apply, 12 men and 10 women. After the interviews, when the company announces the newly appointed managers, all three positions go to women. The men complain of job discrimination. Do they have a case? Simulate a random selection of three people from the applicant pool and make a decision about the likelihood that a fair process would result in hiring all women.

34. **Cell phones.** A proud legislator claims that your state's new law against talking on a cell phone while driving has reduced cell phone use to less than 12% of all drivers. While waiting for your bus the next morning, you notice that 4 of the 10 people who drive by are using their cell phones. Does this cast doubt on the legislator's figure of 12%? Use a simulation to estimate the likelihood of seeing at least 4 of 10 randomly selected drivers talking on their cell phones if the actual rate of usage is 12%. Explain your conclusion clearly.

35. Freshmen. A certain college estimates that SAT scores of students who apply for admission can be described by a Normal model with a mean of 1050 and a standard deviation of 120. Admissions officers search the pile of envelopes, opening them at random to look for three applicants with SAT scores over 1200. How many envelopes do you think they will need to open?

36. Tires. A tire manufacturer believes that the tread life of its snow tires can be described by a Normal model with mean of 32,000 miles and a standard deviation of 2500 miles. You buy four of these tires, hoping to drive them at least 30,000 miles. Estimate the chances that all four last at least that long.

Just Checking
Answers

1. The component is one World Series game.

2. From a list of two-digit random values, assign the values 00 to 54 to "home team wins" and the remaining values to "visiting team wins".

3. Record wins for teams "A" and "B," letting team A begin play in its home ballpark. Remember to swap the chances of winning after the first two games and again after the next three games. Continue until one team wins four games.

4. Record which team wins four games.

5. Determine the fraction of times each team wins.

CHAPTER 12

Sample Surveys

We have learned ways to display, describe, and summarize data. Up to now, though, our conclusions have been limited to the particular batch of data we are examining. That's OK as far as it goes, but it doesn't go very far. We usually aren't satisfied with conclusions based only on *this* group of customers, or the people who answered the survey on *this* particular day. To make business decisions, to do science, to choose wise investments, or to understand what voters think they'll do in the next election, we need to stretch beyond the data at hand to the world at large.

To make that stretch, we need three ideas. You'll find the first one natural. The second may be more surprising. The third is one of the strange but true facts that often confuse those who don't know Statistics.

Idea 1: Examine a Part of the Whole

The first idea is to draw a sample. We'd like to know about an entire **population** of individuals, but examining all of them is usually impractical, if not impossible. So we settle for examining a smaller group of individuals—a **sample**—selected from the population.

You do this every day. For example, suppose you wonder how the vegetable soup you're cooking for dinner tonight is going to go over with your friends. To decide whether it meets your standards, you only need to try a small amount. You might taste just a spoonful or two. You certainly don't have to consume the whole pot. You trust that the taste will *represent* the flavor of the entire pot. The idea behind your tasting is that a small sample, if selected properly, can represent the entire population.

> **The W's and Sampling**
> The population we are interested in is usually determined by the *Why* of our study. The sample we draw will be the *Who*. *When* and *How* we draw the sample may depend on what is practical.

It's hard to go a day without hearing about the latest opinion poll. These polls are examples of **sample surveys,** designed to ask questions of a small group of people in the hope of learning something about the entire population. Most likely, you've never been selected to be part of one of these national opinion polls. That's true of most people. So how can the pollsters claim that a sample is representative of the entire population? The answer is that professional pollsters work quite hard to ensure that the "taste"—the sample that they take—represents the population. If not, the sample can give misleading information about the population.

Bias

In 1936, a young pollster named George Gallup used a subsample of only 3000 of the 2.4 million responses that the *Literary Digest* received to reproduce the wrong prediction of Landon's victory over Roosevelt. He then used an entirely different sample of 50,000 and predicted that Roosevelt would get 56% of the vote to Landon's 44%. His sample was apparently much more *representative* of the actual voting populace. The Gallup Organization went on to become one of the leading polling companies.

Selecting a sample to represent the population fairly is more difficult than it sounds. The most common failure of a poll or survey is in using a sample that fails to represent the population in some important way. The sample may overlook subgroups that are harder to find (such as the homeless or those who use only cell phones), or overrepresent others (such as Internet users who like to respond to online surveys). Samples that don't represent every individual in the population fairly are said to be **biased.** Bias is the bane of sampling—the one thing above all to avoid. Conclusions based on biased samples are inherently flawed. There is usually no way to fix a biased sample and no way to salvage useful information from it.

What are the basic techniques for making sure that a sample is representative? Sometimes the best way to see how to do something is to study a really dismal failure. Here's a famous one. By the beginning of the 20th century, it was common for newspapers to ask readers to return "straw" ballots on a variety of topics. (Today's Internet surveys are the same idea, gone electronic.) The earliest known example of such a straw vote in the United States dates back to 1824.

The success of these regional polls in the early 1900s inspired national magazines to try their luck. Although the *Farm Journal* was probably the first, the *Literary Digest* was at the top of the heap. During the period 1916 to 1936, it regularly surveyed public opinion, and forecast election results correctly. During the 1936 presidential campaign between Alf Landon and Franklin Delano Roosevelt, the *Literary Digest* mailed more than 10 million ballots. The magazine got back an astonishing 2.4 million. (Polls were still a relatively novel idea, and many people thought it was important to send back their opinions.) The results from the millions of responses were clear. Alf Landon would be the next president by a landslide: 57% to 43%. You remember President Landon, don't you? In fact, Landon carried only two states. Roosevelt won, 62% to 37%, and, perhaps coincidentally, the *Digest* went bankrupt soon afterward.

What went wrong? The problem was that the *Digest* sample was not representative. The pollsters made some mistakes that are now considered classics. First, let's look at how they got the list of 10 million names to start with. Where would you go to get such a list? You might think of using phone numbers as a way to select people—and that's just what the *Digest* did. But in 1936, at the height of the Great Depression, telephones were real luxuries. Any list of phone owners would include far more rich than poor people. In fact, it wasn't until 1986 that enough families in the United States had telephones so that phoning became a reliable way of surveying people.[1] The other lists available to the *Digest* were even less representative—drivers' registrations and memberships in organizations such as country clubs.

---

[1] Even today, phone numbers must be computer-generated to make sure that the phone owners are representative. Using only phone book listings would miss people with unlisted numbers, cell phones, or who have recently moved. Leaving these groups out may make the sample unrepresentative.

A S The *Literary Digest* Poll **and the Election of 1936.** Watch the story of one of the most famous polling failures in history. (Turn ahead in the Lesson Book to find this one.)

The main campaign issue in 1936 was the economy. Roosevelt's core supporters, who tended to be less well off, were not well represented in the *Digest*'s sample, so the results of a survey based on that sample did not reflect the opinions of the overall population. It did not matter how well the sample was measured, nor how many people responded.

How can we avoid the *Digest*'s errors? To avoid bias and make the sample as representative as possible, you might be tempted to handpick the individuals included in the sample with care and precision. The best strategy is to do something quite different. We should select individuals for the sample *at random.* The value of deliberately introducing randomness is one of the great insights of Statistics.

Idea 2: Randomize

Think back to the soup sample. Suppose you add some salt to the pot. If you sample it from the top before stirring, what will happen? With the salt sitting on top, you'll get the misleading idea that the whole pot is salty. If you sample from the bottom, you'll get an equally misleading idea that the whole pot is bland. By stirring, you *randomize* the amount of salt throughout the pot, making each taste more typical in terms of the amount of salt in the whole pot.

Randomization can protect you against factors that you know are in the data. It can also help protect against factors that you aren't even aware of. Suppose, while you weren't looking, a friend added a handful of peas to the soup. They are down at the bottom of the pot, mixing with the other vegetables. If you don't randomize the soup by stirring, your test spoonful from the top won't have any peas. By stirring in the salt, you *also* randomize the peas throughout the pot, making your sample taste more typical of the overall pot *even though you didn't know the peas were there.* So randomizing protects us by giving us a representative sample even over effects we were unaware of.

How would we "stir" people in our survey? We'd try to select them at random. **Randomizing** protects us from the influences of *all* the features of our population, even ones that we may not have thought about. It does that by making sure that *on average* the sample looks like the rest of the population.

● **Why not match the sample to the population?** Rather than randomizing, we could try to design our sample so that the people we choose are typical in terms of every characteristic we can think of. In the 1936 vote, rich and poor voted differently as in no previous election. So, we'd like the income levels of those we sample to match the population. How about age? Do young and old vote alike? Political affiliation? Marital status? Having children? Living in the suburbs? We can't possibly think of all the things that might be important. Even if we could, we wouldn't be able to match our sample to the population for all these characteristics. ●

A S **Sampling from Some Real Populations.** Draw random samples for yourself to see how closely they resemble each other and the population.

Not only does randomizing protect us from bias, it actually makes it possible for us to draw inferences about the population when we see only a sample. Such inferences are among the most powerful things we can do with Statistics, and we'll spend much of the rest of the book discussing them. Keep in mind, though, that it's all made possible because we deliberately choose things randomly.

Here's an example from a company's database of 3.5 million customers. We've taken two samples of size 8000 at random from the population. Here's how the means and proportions match up on seven variables.

Age (yr)	White (%)	Female (%)	# of Children	Income Bracket (1–7)	Wealth Bracket (1–9)	Homeowner? (% Yes)
61.4	85.12	56.2	1.54	3.91	5.29	71.36
61.2	84.44	56.4	1.51	3.88	5.33	72.30

Notice how well randomizing has stirred the population. We didn't look at these variables when we drew the samples, but randomizing has automatically matched them pretty closely. We can reasonably assume that since the two samples don't differ too much from each other, they don't differ much from the rest of the population either.

Idea 3: It's the Sample Size

A friend who knows that you are taking Statistics asks your advice on her study. What can you possibly say that will be helpful? Just say: "If you could just get a larger sample size, it would probably improve your study." Even though a larger sample might not be worth the cost, it will almost always make the results more precise.

How large a random sample do we need for the sample to be reasonably representative of the population? Obviously, if your sample is too small, it can't give much information. You might think that we need a large percentage, or *fraction*, of the population. That's what most people think, but it turns out that all that matters is the *number* of individuals in the sample. The size of the population doesn't matter at all.[2] A random sample of 100 students in a college represents the student body just about as well as a random sample of 100 voters represents the entire electorate of the United States. This is the *third* idea and probably the most surprising one in designing surveys.

How can it be that only the number in the sample, and not how big the population is, matters? Well, let's return one last time to that pot of soup. If you're cooking for a banquet rather than just for a few people, your pot will be bigger, but do you need a bigger spoon to decide how the soup tastes? Of course not. The same size spoonful is probably enough to make a decision about the entire pot, no matter how large the pot. The *fraction* of the population that you've sampled doesn't matter. It's the *sample size* itself that's important.

This somewhat surprising idea is of key importance to the design of any sample survey, because it determines the balance between how well the survey can measure the population and how much the survey costs.

How big a sample do you need? That depends on what you're estimating. If you're just interested in the broth, then you can just taste a sip from a spoon. But to get an idea of what's really in the soup, you need a large enough taste to be a *representative* sample from the pot, including a selection of the vegetables. For a survey that tries to find the proportion of the population that fall into a category, you'll usually need a large enough sample to see several respondents in each category—usually at least several hundred respondents—to say anything precise enough to be useful.[3]

---

[2] Well, that's not exactly true. If the population is small enough and the sample is more than 10% of the whole population, it *can* matter. It doesn't matter whenever, as usual, our sample is a very small fraction of the population.

[3] Chapter 19 gives the details behind this statement and shows how to decide on a sample size for a survey.

● **What do the pollsters do?** How do professional polling agencies do their work? The most common polling method today is to contact respondents by telephone. Computers generate random telephone numbers for telephone exchanges known to include residential customers; this way pollsters will even contact some people with unlisted phone numbers. The person who answers the phone will be invited to respond to the survey—if that person qualifies. (For example, only adults are usually surveyed and the respondent usually must live at the residence phoned.) If the person answering doesn't qualify, the caller will ask for an appropriate alternative. When they conduct the interview, the pollsters often list possible responses (such as candidates' names) in randomized orders to avoid biases that might favor the first name on the list.

Do these methods work? The Pew Research Center for the People and the Press, reporting on one survey, says that

> *Across five days of interviewing, surveys today are able to make some kind of contact with the vast majority of households (76%), and there is no decline in this contact rate over the past seven years. But because of busy schedules, skepticism and outright refusals, interviews were completed in just 38% of households that were reached using standard polling procedures.*

A S **Does the Population Size Matter?** Here's the narrated version of this important idea about sampling.

Nevertheless, studies indicate that those actually sampled can give a good snapshot of larger populations from which the surveyed households were drawn. ●

Does a Census Make Sense?

Why bother determining the right sample size? Wouldn't it be better to just include everyone and "sample" the entire population? Such a special sample is called a **census.** Although a census would appear to provide the best possible information about the population, there are a number of reasons why it might not.

First, it can be difficult to complete a census. There always seem to be some individuals who are hard to locate or hard to measure. The cost of locating the last few cases may far exceed your budget. It can also be just plain impractical to take a census. If you were a taste tester for the Hostess Company, you probably wouldn't want to census *all* the Twinkies on the production line. Aside from the fact that you couldn't possibly eat that many Twinkies, it would defeat the purpose of your job. You wouldn't have any left to sell. And a wine tasting wouldn't be very practical if we needed to drink the whole bottle to get an idea of its quality. So, it's often not practical to attempt a census.

A S **Frito-Lay Sampling for Quality.** How does a potato chip manufacturer make sure to cook only the best potatoes? See sampling at work in this video.

Second, populations rarely stand still. In populations of people, babies are born and folks die or leave the country. In opinion surveys, events may cause a shift in opinion during the survey. Even if you could take a census, the population changes while you work, so it's never possible to get a perfect measure. A sample surveyed in just a few days may give more accurate information.

Third, taking a census can be more complex than sampling. Often a census requires a team effort or the cooperation of the population members. Because it tries to count everyone, the U.S. Census records too many college students. Many are included by their families and are then also counted a second time in a report filed by their schools. Other errors of this sort can be found throughout the Census.

● **The undercount** It's particularly difficult to compile a complete census of a population as large, complex, and spread out as the U.S. population. The U.S. Census is known to miss some residents. On occasion the undercount has been striking. For example, there have been blocks in inner cities in which the number of residents recorded by the Census was smaller than the number of electric meters for which bills were being paid. What makes the problem particularly important is that some groups have a higher probability of being missed than others—the homeless, the poor, the indigent. The Census Bureau proposed the use of random sampling to estimate the number of residents missed by the ordinary census. Unfortunately, the resulting debate has become more political than statistical. ●

Populations and Parameters

Any quantity that we calculate from data could be called a "statistic." But in practice, we usually use a statistic to estimate a population parameter.

A S **Statistics and Parameters.** Explore the difference between statistics and parameters.

Remember: Population model parameters are not just unknown—usually they are *unknowable*. We have to settle for sample statistics.

A study found that teens were less likely to "buckle up." The National Center for Chronic Disease Prevention and Health Promotion reports that 21.7% of U.S. teens never or rarely wear seatbelts. What does this statement mean? We're sure they didn't take a census. Even if they had, they must have missed some teens. So what *does* the 21.7% mean? We can't possibly know what percentage of teenagers wear seatbelts. Reality is just too complex. But, as we've seen many times before, we can simplify the question by building a model.

Models use mathematics to represent reality. Parameters are the key numbers in those models. All kinds of models have parameters, so sometimes a parameter used in a model for a population is called (redundantly) a **population parameter.**

But let's not forget about the data. We use the data to try to estimate the population parameters. As we know, any summary found from the data is a **statistic.** Those statistics that estimate population parameters are particularly interesting. Sometimes—and especially when we match statistics with the parameters they estimate—you'll see the (also redundant) term **sample statistic.**[4]

We've already met two parameters in Chapter 6: the mean, μ, and the standard deviation, σ. We'll try to keep denoting population model parameters with Greek letters and the corresponding statistics with Latin letters. Usually, but not always, the letter used for the statistic and the parameter correspond in a natural way. So the standard deviation of the data is s, and the corresponding parameter is σ (Greek for s). In Chapter 7, we used r to denote the sample correlation. The corresponding correlation in a model for the population would be called ρ (rho). In Chapter 8, b_1 represented the slope of a linear regression estimated from the data. But when we think about a (linear) *model* for the population, we denote the slope parameter β_1 (beta).

Get the pattern? Good. Now it breaks down. We denote the mean of a population model with μ (because μ is the Greek letter for m). It might make sense to denote the sample mean with m, but longstanding convention is to put a bar over anything when we average it, so we write $\bar{y}$. What about proportions? Suppose we want to talk about the proportion of teens who don't wear seatbelts. If we use p to denote the proportion from the data, what is the corresponding model parameter? By all rights it should be π. Some books do this, but statements like $\pi = 0.25$ might be confusing because π has been equal to $3.1415926\ldots$ for so long, and it's worked so *well*. So, once again we violate the rule. We'll use p for the population model parameter and $\hat{p}$ for the proportion from the data. Here's a table summarizing the notation:

---

[4] Where else besides a sample *could* a statistic come from?

Name	Statistic	Parameter
Mean	$\bar{y}$	μ (mu, pronounced "meeoo," not "moo")
Standard deviation	s	σ (sigma)
Correlation	r	ρ (rho)
Regression coefficient	b	β (beta, pronounced "baytah"[5])
Proportion	$\hat{p}$	p (pronounced "pee"[6])

1 Various claims are often made for surveys. Why are each of the following claims *not* correct?

   a) It is always better to take a census than to draw a sample.

   b) Stopping students on their way out of the cafeteria is a good way to sample if we want to know about the quality of the food there.

   c) We drew a sample of 100 from the 3000 students in a school. To get the same level of precision for a town of 30,000 residents, we'll need a sample of 1000.

   d) A poll taken at our favorite Web site (*www.statsisfun.org*) garnered 12,357 responses. The majority said they enjoy doing statistics homework. With a sample size that large, we can be pretty sure that most Statistics students feel this way, too.

   e) The true percentage of all Statistics students who enjoy the homework is called a "population statistic."

Simple Random Samples

We draw samples because we can't work with the entire population. We need to be sure that the statistics we compute from the sample reflect the corresponding parameters accurately. A sample that does this is said to be **representative.**

How would you select a representative sample? Most people would say that every individual in the population should have an equal chance to be selected, and certainly that seems fair. But it's not sufficient. There are many ways to give everyone an equal chance that still wouldn't give a representative sample. Consider, for example, a school that has equal numbers of males and females. We could sample like this. Flip a coin. If it comes up heads, select 100 female students at random. If it comes up tails, select 100 males at random. Everyone has an equal chance of selection, but every sample is of only a single sex—hardly representative.

We need to do better. Suppose we insist that every possible *sample* of the size we plan to draw has an equal chance to be selected. This ensures that situations like the one just described are not likely to occur and still guarantees that each person has an equal chance of being selected. What's different is that with this method each *combination* of people has an equal chance of being selected as well. A sample drawn in this way is called a **Simple Random Sample,** usually abbreviated **SRS.** An SRS is the standard against which we measure other sampling methods, and the sampling method on which the theory of working with sampled data is based.

---

[5] If you're American. If you're British or Canadian, it's "beetah."

[6] Just in case you weren't sure.

To select a sample at random, we first need to define where the sample will come from. The **sampling frame** is a list of individuals from which the sample is drawn. For example, to draw a random sample of students at a college, we might obtain a list of all registered full-time students and sample from that list. In defining the sampling frame, we must deal with the details of defining the population. Are part-time students included? How about those who are attending school elsewhere and transferring credits back to the college?

Once we have a sampling frame, the easiest way to choose an SRS is with random numbers. We already talked about some good ways to get random numbers in Chapter 11. We can assign a random number to each individual in the sampling frame. As in a simulation, we then select only those whose random numbers satisfy some rule. Let's look at a couple of examples:

- We want to select 5 students from the 80 enrolled in an Introductory Statistics class. We start by numbering the students from 00 to 79. Now we get a sequence of random digits from a table, technology, or the Internet, finding 05166 29305 77482. Taking those random numbers two digits at a time gives us 05, 16, 62, 93, 05, 77, and 48. We ignore 93 because no one had a number that high. And we don't want to pick the same person twice, so we also skip the repeated number 05. Our simple random sample consists of students with the numbers 05, 16, 62, 77, and 48.

- Often the sampling frame is so large that it would be too tedious to number everyone consecutively. If our intended sample size is approximately 10% of the sampling frame, we assign each individual a single random digit 0 to 9. Then we select only those with a specific random digit, say, 5.

Samples drawn at random generally differ one from another. Each draw of random numbers selects *different* people for our sample. These differences lead to different values for the variables we measure. We call these sample-to-sample differences **sampling variability.** Surprisingly, we'll see sampling variability not as a problem, but as an opportunity. If different samples from a population vary little from each other, then most likely the underlying population harbors little variation. If the samples show much sampling variability, the underlying population probably varies a lot. In the coming chapters, we'll spend much time and attention working with sampling variability and using it to better understand what we are trying to measure.

Stratified Sampling

Simple random sampling is not the only fair way to sample. More complicated designs may save time or money or help avoid sampling problems. All statistical sampling designs have in common the idea that chance, rather than human choice, is used to select the sample.

Designs that are used to sample from large populations—especially populations residing across large areas—are often more complicated than simple random samples. Sometimes the population is first sliced into homogeneous groups, called **strata**, before the sample is selected. Then simple random sampling is used within each stratum before the results are combined. This common sampling design is called **stratified random sampling.**

Why would we want to complicate things? Here's an example. Suppose we want to survey how students feel about funding for the football team at a large university. The campus is 60% men and 40% women. We suspect that men and women have different views on the funding and we want to protect ourselves from getting a really bad sample, one that has too many men or women, by chance. If we select 100 people for the survey at random we might get 35 men and 65 women, which could bias our result. To avoid this, we can decide to force a representative gender balance by selecting 60 men at random and 40 women at random. This would guarantee that the proportions of men and women within the sample match the proportions in the population. You can imagine the importance of stratifying by race, income, age, and other characteristics depending on the questions in the survey.

In addition to reducing bias, stratifying can reduce the variability of our results. When we restrict by strata, additional samples are more like one another, so statistics calculated for the sampled values will vary less from one sample to another. This reduced variability is an important added benefit of stratifying.

Cluster and Multistage Sampling

Sometimes dividing the sample into homogeneous strata isn't practical. And even simple random sampling may be difficult. For example, suppose we wanted to assess the reading level of this textbook based on the length of the sentences. Simple random sampling could be awkward; we'd have to number each sentence, and then find, for example, the 576th sentence or the 2482nd sentence, and so on. Doesn't sound like much fun, does it?

We could make our task much easier by picking a few *pages* at random and then counting the lengths of the sentences on those pages. That's easier than picking individual sentences, and works if we believe that the pages are all reasonably similar to one another in terms of reading level. Splitting the population into similar parts or **clusters** can make sampling more practical. Then we could simply select one or a few clusters at random and perform a census within each of them. This sampling design is called **cluster sampling.** If each cluster fairly represents the full population, cluster sampling will give us an unbiased sample.

What's the difference between cluster sampling and stratified sampling? We stratify to ensure that our sample represents different groups in the population, and sample randomly within each stratum. Strata are homogeneous, but differ from one another. By contrast, clusters are more or less alike, each heterogeneous and resembling the overall population. We select clusters to make sampling more practical or affordable.

Sometimes we use a variety of sampling methods together. In trying to assess the reading level of this book, we might worry that it starts out easy and then gets harder as the concepts become more difficult. If so, we'd want to avoid samples that selected heavily from early or from late chapters. To guarantee a fair mix of chapters, we could randomly choose one chapter from each of the seven parts of the book. Then we would randomly select a few pages from each of those chapters. If altogether that made too many sentences, we might select a few sentences at random from each of the chosen pages. So, what is our sampling strategy? First we stratify by the part of the book, and randomly choose a chapter to represent each stratum. Within each selected chapter, we choose pages as clusters. Finally, we consider an SRS of sentences within each cluster. Sampling schemes that combine several methods are called **multistage samples.** Most

Strata, or Clusters?
We may split a population into strata or clusters. What's the difference? We create strata by dividing the population into groups of similar individuals so that each stratum is different from the others. (For example, we often stratify by age, race, or sex.) By contrast, we create clusters that all look pretty much alike, each representing the wide variety of individuals seen in the population.

surveys conducted by professional polling organizations use some combination of stratified and cluster sampling as well as simple random samples.

Systematic Samples

Sometimes we draw a sample by selecting individuals systematically. For example, you might survey every 10th person on an alphabetical list of students. To make it random, you still must start the systematic selection from a randomly selected individual. When there is no reason to believe that the order of the list could be associated in any way with the responses sought, **systematic sampling** can give a representative sample. Systematic sampling can be much less expensive than true random sampling. When you use a systematic sample, you should justify the assumption that the systematic method is not associated with any of the measured variables.

Think about the reading level sampling example again. Suppose we have chosen a chapter of the book at random, then three pages at random from that chapter, and now we want to select a sample of 10 sentences from the 73 sentences found on those pages. Instead of numbering each sentence so we can pick a simple random sample, it would be easier to sample systematically. A quick calculation shows $73/10 = 7.3$, so we can get our sample by just picking every seventh sentence on the page. But where should you start? At random, of course. We've accounted for $10 \times 7 = 70$ of the sentences, so we'll throw the extra 3 into the starting group and choose a sentence at random from the first 10. Then we pick every seventh sentence after that and record its length.

 We need to survey a random sample of the 300 passengers on a flight from San Francisco to Tokyo. Name each sampling method described below.

a) Pick every 10th passenger as people board the plane.
b) From the boarding list, randomly choose 5 people flying first class and 25 of the other passengers.
c) Randomly generate 30 seat numbers and survey the passengers who sit there.
d) Randomly select a seat position (right window, right center, right aisle, etc.) and survey all the passengers sitting in those seats.

Sampling Step-By-Step

The assignment says, "Conduct your own sample survey to find out how many hours per week students at your school spend watching TV during the school year." Let's see how we might do this step by step.

 Think

Plan State what you want to know.

Population and Parameter

Identify the W's of the study. The *Why* determines the population and the associated sampling frame. The *What* identifies the parameter of interest and the variables measured.

I wanted to design a study to find out how many hours of TV students at my school watch.

The population studied was students at our school. I obtained a list of all students currently enrolled and used it as the sampling frame. The parameter of interest was the number of TV hours watched per week during the

The *Who* is the sample we actually draw. The *How*, *When*, and *Where* are given by the sampling plan.

Often thinking about the *Why* will help us see whether the sampling frame and plan are adequate to learn about the population.

Sampling Plan

Specify the sampling method and the sample size, *n*. Specify how the sample was actually drawn. What is the sampling frame?

The description should, if possible, be complete enough to allow someone to replicate the procedure, drawing another sample from the same population in the same manner. A good description of the procedure is essential, even if it could never practically be repeated.

Terms like "simple random sample," "cluster sample," and "stratified sample" often appear in this part of the discussion. Be sure to describe how the randomization was performed.

Show

Sampling Practice

Specify *When*, *Where*, and *How* the sampling was performed. Specify any other details of your survey, such as how respondents were contacted, what incentives were offered to encourage them to respond, how nonrespondents were treated, and so on.

Tell

Summary and Conclusion

This report should include a discussion of all the elements. In addition, it's good practice to discuss any special circumstances. Professional polling organizations report the *When* of their samples, but will also note, for example, any important news that might have changed respondents' opinions during the sampling process. In this survey, perhaps, a major news story or sporting event might change students' TV viewing behavior.

The question you ask also matters. It's better to be specific ("How many hours did you watch TV last week?") than to ask a general question ("How many hours of TV do you usually watch in a week?").

The report should provide and interpret the statistics from the sample and state the conclusions that you reached about the population.

---

school year, which I attempted to measure by asking students how much TV they watched during the previous week.

I decided against stratifying by class or sex because I didn't think TV watching would differ much between males and females or across classes. I selected a simple random sample of 200 students from the list. I obtained a list of random digits, matched it to an alphabetically arranged list of students, and selected all students who were assigned a "4." This method generated a sample of 212 students from the population of 2133 students.

The survey was taken over the period Oct. 15 to Oct. 25. Surveys were sent to selected students by e-mail with the request that they respond by e-mail as well. Students who could not be reached by e-mail were handed the survey in person.

During the period Oct. 15 to Oct. 25, 212 students were randomly selected, using a simple random sample from a list of all students currently enrolled. The survey they received asked the following question: "How many hours did you spend watching television last week?"

Of the 212 students surveyed, 110 responded. It's possible that the nonrespondents differ in the number of TV hours watched from those who responded, but I was unable to follow up on them due to limited time and funds. The 110 respondents reported an average 3.62 hours of TV watching per week. The median was only 2 hours per week. A histogram of the data shows that the data are highly right-skewed, indicating that the median might be a more appropriate summary of the typical TV watching of the students. Most of the students (90%) watch between 0 and 10 hours per week, while 30% reported watching less than 1 hour per week.

A few watch much more. About 3% reported watching more than 20 hours per week.

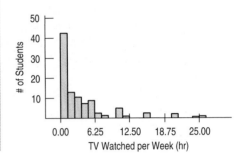

Who's Who?

> The population is determined by the *Why* of the study. Unfortunately, the sample is just those we can reach to obtain responses—the *Who* of the study. This difference could undermine even a well-designed study.

The *Who* of a survey can refer to different groups, and the resulting ambiguity can tell you a lot about the success of a study. To start, you should think about the population of interest. Often, you'll find that this is not really a well-defined group (who, exactly, is a student?). Even if the population is clear, it may not be a practical group to study. For example, election polls want to sample from all those who will vote in the next election—a population that is particularly tricky to identify before election day.

Second, you must specify the sampling frame. (Do you have a list of students to sample from? How about a list of registered voters?) Usually, the sampling frame is not the group you *really* want to know about. (All those registered to vote are not equally likely to show up.) The sampling frame limits what your survey can find out.

Then there's your target sample. These are the individuals for whom you *intend* to measure responses. You're not likely to get responses from all of them. ("I know it's dinner time, but I'm sure you wouldn't mind answering a few questions. It'll only take 20 minutes or so. Oh, you're busy?") Nonresponse is a problem in many surveys.

Finally, there is your sample—the actual respondents. These are the individuals about whom you *do* get data and can draw conclusions. Unfortunately, they might not be representative of the sampling frame or the population.

At each step, the group we can study may be constrained further. The *Who* keeps changing, and each constraint can introduce biases. A careful study should address the question of how well each group matches the population of interest. One of the main benefits of simple random sampling is that it never loses its sense of who's *Who*. The *Who* in an SRS is the population of interest from which we've drawn a representative sample. That's not always true for other kinds of samples.

What Can Go Wrong?—or, How to Sample Badly

Bad sample designs yield worthless data. Many of the most convenient forms of sampling can be seriously biased. And there is no way to correct for the bias from a bad sample. So it's wise to pay attention to sample design—and to beware of reports based on poor samples.

Sample Badly with Volunteers

One of the most common dangerous sampling methods is a voluntary response sample. In a **voluntary response sample,** a large group of individuals is invited to respond, and all who do respond are counted. This method is used by call-in shows, 900 numbers, Internet polls, and letters written to members of Congress. Voluntary response samples are almost always biased, and so conclusions drawn from them are almost always wrong.

It's often hard to define the sampling frame of a voluntary response study. Practically, the frames are groups such as Internet users who frequent a particular Web site or those who happen to be watching a particular TV show at the moment. But those sampling frames don't correspond to interesting populations.

Even within the sampling frame, voluntary response samples are often biased toward those with strong opinions or those who are strongly motivated. People with very negative opinions tend to respond more often than those with equally strong positive opinions. The sample is not representative, even though every individual in the population may have been offered the chance to respond. The resulting **voluntary response bias** invalidates the survey.

How often do people write to their congressional representative when they're pretty happy about how things are going? People respond when they feel strongly. Which survey would *you* be more likely to respond to—one that asked, "Should the minimum age to drive a car be raised to 25?" or one that asked, "Should shoe sizes be adjusted so women's and men's sizes correspond?"

A S **Sources of Sampling Bias.** Here's a narrated exploration of sampling bias.

● **If you had it to do over again, would you have children?** Ann Landers, the advice columnist, asked parents this question. The overwhelming majority—70% of the more than 10,000 people who wrote in—said no, kids weren't worth it. A more carefully designed survey later showed that about 90% of parents actually are happy with their decision to have children. What accounts for the striking difference in these two results? What kind of parents do you think are most likely to respond to the original question? ●

Sample Badly, but Conveniently

Another sampling method that doesn't work is convenience sampling. As the name suggests, in **convenience sampling** we simply include the individuals who are convenient. Unfortunately, this group may not be representative of the population. A recent survey of 437 potential home buyers in Orange County, California, found, among other things that

All but 2 percent of the buyers have at least one computer at home, and 62 percent have two or more. Of those with a computer, 99 percent are connected to the Internet (Jennifer Hieger, "Portrait of Homebuyer Household: 2 Kids and a PC," Orange County Register, 27 July 2001).

Later in the article, we learn that the survey was conducted via the Internet! That was a convenient way to collect data and surely easier than drawing a

Do you use the Internet?
Click here ○ for yes
Click here ○ for no

Internet convenience surveys are worthless. As voluntary response surveys, they have no well-defined sampling frame (all those who use the Internet and visit their site?) and thus report no useful information. Do not believe them.

simple random sample, but perhaps home builders shouldn't conclude from this study that *every* family has a computer and an Internet connection.

Many surveys conducted at shopping malls suffer from the same problem. People in shopping malls are not necessarily representative of the population of interest. Mall shoppers tend to be more affluent and include a larger percentage of teenagers and retirees than the population at large. To make matters worse, survey interviewers tend to select individuals who look "safe," or easy to interview.

You may think that convenience sampling is only a problem for students or other beginning samplers. In fact, convenience sampling is a widespread problem in the business world. When a company wants to find out what people think about its products or services, whom does it survey? The easiest people to sample are its own customers. After all, the company has a list of its own customers with addresses, phone numbers, and other information—or at least a list of those customers who bothered to send in their registration cards. The company can easily stratify by age, sex, or other information it has, but no matter how it selects a sample from its customers, the sample will suffer from convenience sample bias. The company will find out only how its own customers feel about its services and, unless it takes the extra cost and effort to go outside its own customer base, will never learn how those who *don't* buy its product feel about it.

Sample from a Bad Sampling Frame

A simple SRS from an incomplete sampling frame introduces bias because the individuals included may differ from the ones not in the frame. People in prison, homeless people, students, and long-term travelers are all likely to be missed. In telephone surveys, selecting numbers at random from the phone book misses those with unlisted numbers (who might be wealthier), so pollsters often generate phone numbers randomly from *all* possible residential numbers. The growing number of people who have only cell phones are missing from the sampling frame.

Undercoverage

Many survey designs suffer from **undercoverage,** in which some portion of the population is not sampled at all or has a smaller representation in the sample than it has in the population. Undercoverage can arise for a number of reasons, but it's always a potential source of bias.

Telephone survey takers used to be able to find someone at home in the middle of the day. But now that most people work outside the home, telephone surveys are more likely to be conducted when you are likely to be home, interrupting your dinner. Of course, if you eat out often you may be less likely to be surveyed, a possible source of undercoverage.

You may have already won a free copy of a leading Statistics textbook!!

What Else Can Go Wrong? • ***Watch out for nonrespondents.*** A common and serious potential source of bias for most surveys is **nonresponse bias.** No survey succeeds in getting responses from everyone. The problem is that those who don't respond may differ from those who do. And they may differ on just the variables we care about. The lack of response will bias the results. Rather than sending out a large number of surveys for which the response rate will be low, it is often better to design a smaller randomized survey for which you have the resources to ensure a high response rate. One of the problems with nonresponse bias is that it's usually impossible to tell what the nonrespondents might have said.

⚫ **Do people like filling out questionnaires?** A business school student wanted to study how people feel about filling out questionnaires. To do this, he designed—you guessed it—a questionnaire and gave it to a sample of all the other students at his business school. He was surprised and happy to learn that nearly all the questionnaires he got back indicated a positive attitude toward answering surveys. From this he concluded that people in general must not mind filling out questionnaires. ⚫

- *Don't bore respondents with surveys that go on and on and on and on ...* Surveys that are too long are more likely to be refused, reducing the response rate and biasing *all* the results. When designing a survey, always keep in mind the purpose (the *Why*) of the survey and the population to be sampled (and about whom you hope to draw conclusions). For each question you consider including in the survey, ask yourself, "What would I do if I knew the answer to this question?" If you don't have a use for the answer, then don't ask the question.
- *Work hard to avoid influencing responses.* **Response bias** refers to anything in the survey design that influences the responses. Response bias is not the opposite of nonresponse bias. (We don't make these terms up; we just try to explain them.) Response biases include the tendency of respondents to tailor their responses to try to please the interviewer, the ways in which the wording of the questions can change responses, and the natural unwillingness of respondents to reveal personal facts or admit to illegal or unapproved behavior.

  People responding to the survey questions often want to please the interviewer either consciously or unconsciously. The sex, race, attire, or behavior of the interviewer can influence the answers by subtle (or not so subtle) indications that certain answers are more desirable than others.

A S **Biased Question Wording.** Watch a hapless interviewer make every mistake in the book.

THE WIZARD OF ID   parker and hart

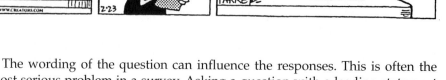

The wording of the question can influence the responses. This is often the most serious problem in a survey. Asking a question with a leading statement

is a good way to bias the response. Many surveys, especially those conducted by special interest groups, present one side of an issue before the question itself. For example, asking a question like

> *Given that the threat of nuclear war is higher now than it has ever been in human history, and the fact that a nuclear war poses a threat to the very existence of the human race, would you favor an all-out nuclear test ban?*

will probably result in a higher percentage of people in favor of the ban than the simpler

> *Are you in favor of or opposed to a nuclear test ban?*

A Short Survey
Given the fact that those who understand Statistics are smarter and better looking than those who don't, don't you think it is important to take a course in Statistics?

● Remember the *Literary Digest* Survey? It turns out that they were wrong on *two* counts. First, their list of 10 million people was not representative. There was a selection bias in their sampling frame. There was also a nonresponse bias. We know this because the *Digest* also surveyed a *systematic* sample in Chicago, sending the same question used in the larger survey to every 3rd registered voter. They *still* got a result in favor of Landon, even though Chicago voted overwhelmingly for Roosevelt in the election. This suggests that the Roosevelt supporters were less likely to respond to the *Digest* survey. There's a modern version of this problem. It's been suggested that those who screen their calls with caller ID or an answering machine, and so might not talk to a pollster, may differ in wealth or political views from those who just answer the phone. And college students may be under-represented in telephone polls because, as a group, they are more likely than the general public to rely only on cell phones. ●

How to Think About Biases

A S **Can a Large Sample Protect Against Bias?** Explore questions about how we can learn about the population from large or repeated samples.

- *Look for biases in any survey you encounter.* If you design one of your own, ask someone else to help look for biases that may not be obvious to you. And do this *before* you collect your data. There's no way to recover from a biased sample or a survey that asks biased questions. Sorry, it just can't be done.
  A bigger sample size for a biased study just gives you a bigger useless study. A really big sample gives you a really big useless study. (Think of the 2.4 million *Literary Digest* responses.)
- *Spend your time and resources reducing biases.* No other use of resources is as worthwhile as reducing the biases.

A researcher distributed a survey to an organization before some economizing changes were made. She asked how people felt about a proposed cutback in secretarial and administrative support on a seven-point scale from Very Happy to Very Unhappy.

But virtually all respondents were very unhappy about the cutbacks, so the results weren't particularly useful. If she had pretested the question, she might have chosen a scale that ran from Unhappy to Outraged.

- *If you possibly can, pretest your survey.* Administer the survey in the exact form that you intend to use it to a small sample drawn from the population you intend to sample. Look for misunderstandings, misinterpretation, confusion, or other possible biases. Then redesign your survey instrument.
- *Always report your sampling methods in detail.* Others may be able to detect biases where you did not expect to find them.

CONNECTIONS

With this chapter we take our first formal steps to relate our sample data to a larger population. Some of these ideas have been lurking in the background as we sought patterns and summaries for data. Even when we only worked with the data at hand, we often thought about implications for a larger population of individuals.

Notice the ongoing central importance of models. We've seen models in several ways in previous chapters. Here we recognize the value of a model for a population. The parameters of such a model are values we will often want to estimate using statistics such as those we've been calculating. The connections to summary statistics for center, spread, correlation, and slope are obvious.

We now have a specific application for random numbers. The idea of applying randomness deliberately showed up in Chapter 11 for simulation. Now we need randomization to get good quality data from the real world.

What have we learned?

We've learned that a representative sample can offer us important insights about populations. It's the size of the sample—and not its fraction of the larger population—that determines the precision of the statistics it yields.

We've learned several ways to draw samples, all based on the power of randomness to make them representative of the population of interest:

- A Simple Random Sample (SRS) is our standard. Every possible group of *n* individuals has an equal chance of being our sample. That's what makes it *simple*.
- Stratified samples can reduce sampling variability by identifying homogeneous subgroups and then randomly sampling within each.
- Cluster samples randomly select among heterogeneous subgroups that each resemble the population at large, making our sampling tasks more manageable.
- Systematic samples can work in some situations and are often the least expensive method of sampling. But we still want to start them randomly.
- Multistage samples combine several random sampling methods.

We've learned that bias can destroy our ability to gain insights from our sample:

- Nonresponse bias can arise when sampled individuals will not or cannot respond.
- Response bias arises when respondents' answers might be affected by external influences, such as question wording or interviewer behavior.

We've learned that bias can also arise from poor sampling methods:

- Voluntary response samples are almost always biased and should be avoided and distrusted.
- Convenience samples are likely to be flawed for similar reasons.
- Even with a reasonable design, sample frames may not be representative. Undercoverage occurs when individuals from a subgroup of the population are selected less often than they should be.

Finally, we've learned to look for biases in any survey we find and to be sure to report our methods whenever we perform a survey so that others can evaluate the fairness and accuracy of our results.

TERMS

Population	The entire group of individuals or instances about whom we hope to learn.
Sample	A (representative) subset of a population, examined in hope of learning about the population.
Sample survey	A study that asks questions of a sample drawn from some population in the hope of learning something about the entire population. Polls taken to assess voter preferences are common sample surveys.
Bias	Any systematic failure of a sample to represent its population is bias. It is almost impossible to recover from a biased sample, so efforts to avoid bias are well spent. Common errors include
	• relying on voluntary response.
	• undercoverage of the population.
	• nonresponse bias.
	• response bias.
Randomization	The best defense against bias is randomization, in which each individual is given a fair, random chance of selection.
Matching	Any attempt to force a sample to resemble specified attributes of the population is a form of matching. Matching may help make better samples, but it is no substitute for randomizing.
Sample size	The number of individuals in a sample. The sample size determines how well the sample represents the population, not the fraction of the population sampled.
Census	A sample that consists of the entire population is called a census.
Population parameter	A numerically valued attribute of a model for a population. We rarely expect to know the true value of a population parameter, but we do hope to estimate it from sampled data. For example, the mean income of all employed people in the country is a population parameter.
Statistic, sample statistic	Statistics are values calculated for sampled data. Those that correspond to, and thus estimate, a population parameter, are of particular interest. For example, the mean income of all employed people in a representative sample can provide a good estimate of the corresponding population parameter. The term "sample statistic" is sometimes used, usually to parallel the corresponding term, "population parameter."
Representative	A sample is said to be representative if the statistics computed from it accurately reflect the corresponding population parameters.
Simple random sample (SRS)	A simple random sample of sample size n is one in which each set of n elements in the population has an equal chance of selection.
Sampling frame	A list of individuals from whom the sample is drawn is called the sampling frame. Individuals who may be in the population of interest but who are not in the sampling frame cannot be included in any sample.
Sampling variability	The natural tendency of randomly drawn samples to differ, one from another. Sometimes, unfortunately, called *sampling error,* sampling variability is no error at all, but just the natural result of random sampling.
Stratified random sample	A sampling design in which the population is divided into several subpopulations, or **strata,** and random samples are then drawn from each stratum. If the strata are homogeneous but are different from each other, a stratified sample may yield more consistent results.

Cluster sample	A sampling design in which entire groups, or **clusters,** are chosen at random. Cluster sampling is usually selected as a matter of convenience, practicality, or cost. Each cluster should be heterogeneous (and representative of the population), so all the clusters should be similar to each other.
Multistage sample	Sampling schemes that combine several sampling methods are called multistage samples. For example, a national polling service may stratify the country by geographical regions, select a random sample of cities from each region, and then interview a cluster of residents in each city.
Systematic sample	A sample drawn by selecting individuals systematically from a sampling frame. When there is no relationship between the order of the sampling frame and the variables of interest, a systematic sample can be representative.
Voluntary response bias	Bias introduced to a sample when individuals can choose on their own whether to participate in the sample. Samples based on voluntary response are always invalid and cannot be recovered, no matter how large the sample size.
Convenience sample	A convenience sample consists of the individuals who are conveniently available. Convenience samples often fail to be representative because every individual in the population is not equally convenient to sample.
Undercoverage	A sampling scheme that biases the sample in a way that gives a part of the population less representation than it has in the population, suffers from undercoverage.
Nonresponse bias	Bias introduced to a sample when a large fraction of those sampled fails to respond. Those who do respond are likely to not represent the entire sample. Voluntary response bias is a form of nonresponse bias, but nonresponse may occur for other reasons. For example, those who are at work during the day won't respond to a telephone survey conducted only during working hours.
Response bias	Anything in a survey design that influences responses falls under the heading of response bias. One typical response bias arises from the wording of questions, which may suggest a favored response. Voters, for example, are more likely to express support of "the president" than support of the particular person holding that office at the moment.

S K I L L S

When you complete this lesson you should:

Think

- Know the basic concepts and terminology of sampling (see the preceding list).
- Recognize population parameters in descriptions of populations and samples.
- Understand the value of randomization as a defense against bias.
- Understand the value of sampling to estimate population parameters from statistics calculated on representative samples drawn from the population.
- Understand that the size of the sample (not the fraction of the population) determines the precision of estimates.

Show

- Know how to draw a simple random sample from a master list of a population, using a computer or a table of random numbers.

Tell

- Know what to report about a sample as part of your account of a statistical analysis.
- Report possible sources of bias in samples. Recognize voluntary response and nonresponse as sources of bias in a sample survey.

Sampling on the Computer

Computer-generated pseudorandom numbers are usually quite good enough for drawing random samples. But there is little reason not to use the truly random values available on the Internet.

Here's a convenient way to draw an SRS of a specified size using a computer-based sampling frame. The sampling frame can be a list of names or of identification numbers arrayed, for example, as a column in a spreadsheet, statistics program, or database:

1. Generate random numbers of enough digits so that each exceeds the size of the sampling frame list by several digits. This makes duplication unlikely.
2. Assign the random numbers arbitrarily to individuals in the sampling frame list. For example, put them in an adjacent column.
3. Sort the list of random numbers, *carrying* along the sampling frame list.
4. Now the first *n* values in the sorted sampling frame column are an SRS of *n* values from the entire sampling frame.

EXERCISES

1–10. What did they do? For the following reports about statistical studies, identify the following items (if possible). If you can't tell, then say so—this often happens when we read about a survey.

   a) The population
   b) The population parameter of interest
   c) The sampling frame
   d) The sample
   e) The sampling method, including whether or not randomization was employed
   f) Any potential sources of bias you can detect and any problems you see in generalizing to the population of interest

1. A business magazine mailed a questionnaire to the human resource directors of all of the Fortune 500 companies, and received responses from 23% of them. Those responding reported that they did not find that such surveys intruded significantly on their workday.

2. A question posted on the Lycos Web site on 18 June 2000 asked visitors to the site to say whether they thought that marijuana should be legally available for medicinal purposes.

3. Consumers Union asked all subscribers whether they had used alternative medical treatments and, if so, whether they had benefited from them. For almost all of the treatments, approximately 20% of those responding reported cures or substantial improvement in their condition.

4. The Gallup Poll interviewed 1423 randomly selected American citizens September 10–14, 1999, and reported that when "asked which type of content bothers them most on TV, 44% of Americans identify 'violence,' 23% choose 'lewd and profane language,' while 22% say 'sexual situations.'"

5. Researchers waited outside a bar they had randomly selected from a list of such establishments. They stopped every 10th person who came out of the bar and asked whether he or she thought drinking and driving was a serious problem.

6. Hoping to learn what issues may resonate with voters in the coming election, the campaign director for a mayoral candidate selects one block from each of the city's election districts. Staff members go there and interview all the residents they can find.

7. The Environmental Protection Agency took soil samples at 16 locations near a former industrial waste dump and checked each for evidence of toxic chemicals. They found no elevated levels of any harmful substances.

8. State police set up a roadblock to check cars for up-to-date registration, insurance, and safety inspections. They usually find problems with about 10% of the cars they stop.

9. A company packaging snack foods maintains quality control by randomly selecting 10 cases from each day's production and weighing the bags. Then they open one bag from each case and inspect the contents.

10. Dairy inspectors visit farms unannounced and take samples of the milk to test for contamination. If the milk is found to contain dirt, antibiotics, or other foreign matter, the milk will be destroyed and the farm re-inspected until purity is restored.

11. **Parent opinion, part 1.** In a large city school system with 20 elementary schools, the school board is considering the adoption of a new policy that would require elementary students to pass a test in order to be promoted to the next grade. The PTA wants to find out whether parents agree with this plan. Listed below are some of the ideas proposed for gathering data. For each, indicate what kind of sampling strategy is involved and what (if any) biases might result.
    a) Put a big ad in the newspaper asking people to log their opinions on the PTA Web site.
    b) Randomly select one of the elementary schools and contact every parent by phone.
    c) Send a survey home with every student, and ask parents to fill it out and return it the next day.
    d) Randomly select 20 parents from each elementary school. Send them a survey, and follow up with a phone call if they do not return the survey within a week.

12. **Parent opinion, part 2.** Let's revisit the school system described in Exercise 11. Four new sampling strategies have been proposed to help the PTA determine whether parents favor requiring elementary students to pass a test in order to be promoted to the next grade. For each, indicate what kind of sampling strategy is involved and what (if any) biases might result.
    a) Run a poll on the local TV news, asking people to dial one of two phone numbers to indicate whether they favor or oppose the plan.
    b) Hold a PTA meeting at each of the 20 elementary schools and tally the opinions expressed by those who attend the meetings.
    c) Randomly select one class at each elementary school and contact each of those parents.
    d) Go through the district's enrollment records, selecting every 40th parent. PTA volunteers will go to those homes to interview the people chosen.

13. **Churches.** For your political science class, you'd like to take a survey from a sample of all the Catholic Church members in your city. A list of churches shows 17 Catholic churches within the city limits. Rather than try to obtain a list of all members of all these churches, you decide to pick 3 churches at random. For those churches, you'll ask to get a list of all current members and contact 100 members at random.
    a) What kind of design have you used?
    b) What could go wrong with the design that you have proposed?

14. **Fish.** The U.S. Fish and Wildlife Service plans to study the kinds of fish being taken out of Saginaw Bay. To do that, they decide to randomly select 5 fishing boats at the end of a randomly chosen fishing day and to count the numbers and types of all the fish on those boats.
    a) What kind of design have they used?
    b) What could go wrong with the design that they have proposed?

15. **Roller coasters.** An amusement park has opened a new roller coaster. It is so popular that people are waiting for up to 3 hours for a 2-minute ride. Concerned about how patrons (who paid a large amount to enter the park and ride on the rides) feel about this, they survey every 10th person on the line for the roller coaster, starting from a randomly selected individual.
    a) What kind of sample is this?
    b) Is it likely to be representative?
    c) What is the sampling frame?

16. **Playground.** Some people have been complaining that the children's playground at a municipal park is too small and is in need of repair. Managers of the park decide to survey city residents to see if they believe the playground should be rebuilt. They hand out questionnaires to parents who bring children to the park. Describe possible biases in this sample.

17. **Wording the survey.** Two members of the PTA committee in Exercises 11 and 12 have proposed different questions to ask in seeking parents' opinions.

    Question 1: Should elementary school–age children have to pass high stakes tests in order to remain with their classmates?

    Question 2: Should schools and students be held accountable for meeting yearly learning goals by testing students before they advance to the next grade?

    a) Do you think responses to these two questions might differ? How? What kind of bias is this?
    b) Propose a question with more neutral wording that might better assess parental opinion.

18. **Banning ephedra.** An online poll at a popular Web site asked:

    A nationwide ban of the diet supplement ephedra went into effect recently. The herbal stimulant has been linked to 155 deaths and many more heart attacks and strokes. Ephedra manufacturer NVE Pharmaceuticals, claiming that the FDA lacked proof that ephedra is dangerous if used as directed, was denied a temporary restraining order on the ban yesterday by a federal judge. Do you think that ephedra should continue to be banned nationwide?

    65% of 17,303 respondents said "yes." Comment on each of the following statements about this poll:
    a) With a sample size that large, we can be pretty certain we know the true proportion of Americans who think ephedra should be banned.
    b) The wording of the question is clearly very biased.

c) The sampling frame is all Internet users.

d) This is a voluntary response survey, so the results can't be reliably generalized to any population of interest.

19. Another ride. The survey of patrons waiting in line for the roller coaster in Exercise 15 asks whether they think it is worthwhile to wait a long time for the ride and whether they'd like the amusement park to install still more roller coasters. What biases might cause a problem for this survey?

20. Playground, act two. The survey described in Exercise 16 asked:

Many people believe this playground is too small and in need of repair. Do you think the playground should be repaired and expanded even if that means raising the entrance fee to the park?

Describe two ways this question may lead to response bias.

21. Survey questions. Examine each of the following questions for possible bias. If you think the question is biased, indicate how and propose a better question.

a) Should companies that pollute the environment be compelled to pay the costs of cleanup?

b) Given that 18-year-olds are old enough to vote and to serve in the military, is it fair to set the drinking age at 21?

22. More survey questions. Examine each of the following questions for possible bias. If you think the question is biased, indicate how and propose a better question.

a) Do you think high-school students should be required to wear uniforms?

b) Given humanity's great tradition of exploration, do you favor continued funding for space flights?

23. Phone surveys. Anytime we conduct a survey we must take care to avoid undercoverage. Suppose we plan to select 500 names from the city phone book, call their homes between noon and 4 p.m., and interview whoever answers, anticipating contacts with at least 200 people.

a) Why is it difficult to use a simple random sample here?

b) Describe a more convenient, but still random, sampling strategy.

c) What kinds of households are likely to be included in the eventual sample of opinion? Who will be excluded?

d) Suppose, instead, that we continue calling each number, perhaps in the morning or evening, until an adult is contacted and interviewed. How does this improve the sampling design?

e) Random digit dialing machines can generate the phone calls for us. How would this improve our design? Is anyone still excluded?

24. Cell phone survey. What about drawing a random sample only from cell phone exchanges? Discuss the advantages and disadvantages of such a sampling method as compared with surveying randomly generated telephone numbers from non–cell phone exchanges. Do you think these advantages and disadvantages have changed over time? How do you expect they'll change in the future?

25. Arm length. How long is your arm compared with your hand size? Put your right thumb at your left shoulder bone, stretch your hand open wide, and extend your hand down your arm. Put your thumb at the place where your little finger is and extend down the arm again. Repeat this a third time. Now your little finger will probably have reached the back of your left hand. If the fourth hand width goes past the end of your middle finger, turn your hand sideways and count finger widths to get there.

a) How many hand and finger widths is your arm?

b) Suppose you repeat your measurement 10 times, and average your results. What parameter would this average estimate? What is the population?

c) Suppose you now collect arm lengths measured in this way from 9 friends and average these 10 measurements. What is the population now? What parameter would this average estimate?

d) Do you think these 10 arm lengths are likely to be representative of the population of arm lengths in your community? In the country? Why or why not?

26. Fuel economy. Occasionally when I fill my car with gas, I figure out how many miles per gallon my car got. I wrote down those results after 6 fill-ups in the past few months. Overall, it appears my car gets 28.8 miles per gallon.

a) What statistic have I calculated?

b) What is the parameter I'm trying to estimate?

c) How might my results be biased?

d) When the Environmental Protection Agency (EPA) checks a car like mine to predict its fuel economy, what parameter is it trying to estimate?

27. Accounting. Between quarterly audits, a company likes to check on its accounting procedures to address any problems before they become serious. The accounting staff processes payments on about 120 orders each day. The next day, the supervisor rechecks 10 of the transactions to be sure they were processed properly.

a) Propose a sampling strategy for the supervisor.

b) How would you modify that strategy if the company makes both wholesale and retail sales, requiring different bookkeeping procedures?

28. Happy workers? A manufacturing company employs 14 project managers, 48 foremen, and 377 laborers. In an effort to keep informed about any possible sources of employee discontent, management wants to conduct job satisfaction interviews with a sample of employees every month.

a) Do you see any danger of bias in the company's plan? Explain.

b) Propose a sampling strategy that uses a simple random sample.

c) Why do you think a simple random sample might not provide the representative opinion the company seeks?

d) Propose a better sampling strategy.

e) Listed below are the last names of the project managers. Use random numbers to select two people to be interviewed. Be sure to explain your method carefully.

Barrett	Bowman	Chen
DeLara	DeRoos	Grigorov
Maceli	Mulvaney	Pagliarulo
Rosica	Smithson	Tadros
Williams	Yamamoto	

29. Quality control. Sammy's Salsa, a small local company, produces 20 cases of salsa a day. Each case contains 12 jars and is imprinted with a code indicating the date and batch number. To help maintain consistency, at the end of each day Sammy selects three bottles of salsa, weighs the contents, and tastes the product. Help Sammy select the sample jars. Today's cases are coded 07N61 through 07N80.

a) Carefully explain your sampling strategy.

b) Show how to use random numbers to pick the three jars for testing.

c) Did you use a simple random sample? Explain.

30. A fish story. Concerned about reports of discolored scales on fish caught downstream from a newly sited chemical plant, scientists set up a field station in a shoreline public park. For one week they asked fishermen there to bring any fish they caught to the field station for a brief inspection. At the end of the week, the scientists said that 18% of the 234 fish that were submitted for inspection displayed the discoloration. From this information, can the researchers estimate what proportion of fish in the river have discolored scales? Explain.

31. Sampling methods. Consider each of these situations. Do you think the proposed sampling method is appropriate? Explain.

a) We want to know what percentage of local doctors accept Medicaid patients. We call the offices of 50 doctors randomly selected from local Yellow Page listings.

b) We want to know what percentage of local businesses anticipate hiring additional employees in the upcoming month. We randomly select a page in the Yellow Pages, and call every business listed there.

32. More sampling methods. Consider each of these situations. Do you think the proposed sampling method is appropriate? Explain.

a) We want to know if there is neighborhood support to turn a vacant lot into a playground. We spend a Saturday afternoon going door-to-door in the neighborhood, asking people to sign a petition.

b) We want to know if students at our college are satisfied with the selection of food available on campus. We go to the largest cafeteria and interview every 10th person in line.

just checking

Answers

1. a) It can be hard to reach all members of a population, and it can take so long that circumstances change, affecting the responses. A well-designed sample is often a better choice.

b) This sample is probably biased—students who didn't like the food at the cafeteria might not choose to eat there.

c) No, only the sample size matters, not the fraction of the overall population.

d) Students who frequent this Web site might be more enthusiastic about Statistics than the overall population of Statistics students. A large sample cannot compensate for bias.

e) It's the population "parameter." "Statistics" describe samples.

2. a) systematic

b) stratified

c) simple

d) cluster

Experiments and Observational Studies

W ho gets good grades? And, more importantly, why? Is there something schools and parents could do to help weaker students improve their grades? Some people think they have an answer: music! No, not your portable MP3 player, but an instrument. In a 1981 study conducted at Mission Viejo High School, in California, researchers compared the scholastic performance of music students with that of non-music students. Guess what? The music students had a much higher overall grade point average than the non-music students, 3.59 to 2.91. Not only that, a whopping 16% of the music students had all A's compared with only 5% of the non-music students.

As a result of this study and others, many parent groups and educators pressed for expanded music programs in the nation's schools. They argued that the work ethic, discipline, and feeling of accomplishment fostered by learning to play an instrument also enhance a person's ability to succeed in school. They thought that involving more students in music would raise academic performance. What do you think? Does this study provide solid evidence? Or are there other possible explanations for the difference in grades? Is there any way to really prove such a conjecture?

Observational Studies

For rare illnesses, it's not practical to draw a large enough sample to see many ill respondents, so the only option remaining is to develop retrospective data. For example, researchers can interview those who have become ill. The likely causes of both legionnaires' disease and HIV were initially identified from such retrospective studies of the small populations who were initially infected. But to confirm the causes, researchers needed laboratory-based experiments.

This research tried to show an association between music education and grades. But it wasn't a survey. Nor did it assign students to get music education. Instead, it simply observed students "in the wild," recording the choices they made and the outcome. Such studies are called **observational studies.** In observational studies, researchers don't *assign* choices; they simply observe them. In addition, this was a **retrospective study,** because researchers first identified subjects who studied music and then collected data on their past grades.

What's wrong with concluding that music education causes good grades? One high school during one academic year may not be representative of the whole United States. That's true, but the real problem is that the claim that music study *caused* higher grades depends on there being *no other differences* between the groups that could account for the differences in grades, and studying music was not the *only* difference between the two groups of students.

We can think of lots of lurking variables that might cause the groups to perform differently. Students who study music may have better work habits to start with, and this makes them successful in both music and course work. Music students may have more parental support (someone had to pay for all those lessons), and that support may have enhanced their academic performance, too. Maybe they came from wealthier homes and had other advantages. Or it could be that smarter kids just like to play musical instruments.

Observational studies are valuable for discovering trends and possible relationships. They are used widely in public health and marketing. Observational studies that try to discover variables related to rare outcomes, such as specific diseases, are often retrospective. They first identify people with the disease and then look into their history and heritage in search of things that may be related to their condition. But retrospective studies have a restricted view of the world because they are usually restricted to a small part of the entire population. And because retrospective records are based on historical data, they can have errors. (Do you recall *exactly* what you ate even yesterday? How about last Wednesday?)

A somewhat better approach is to observe individuals over time, recording the variables of interest and ultimately seeing how things turn out. For example, we might start by selecting young students who have not begun music lessons. We could then track their academic performance over several years, comparing those who later choose to study music with those who do not. Identifying subjects in advance and collecting data as events unfold would make this a **prospective study.**

Although an observational study may identify important variables related to the outcome we are interested in, there is no guarantee that we have found the right or the most important related variables. Students who choose to study an instrument might still differ from the others in some important way that we failed to observe. It may be this difference—whether we know what it is or not—rather than music itself that leads to better grades. It's just not possible for observational studies, whether prospective or retrospective, to demonstrate a causal relationship.

Randomized, Comparative Experiments

"He that leaves nothing to chance will do few things ill, but he will do very few things."
—Lord Halifax
(1633–1695)

Is it *ever* possible to prove a cause-and-effect relationship? Well, yes it is, but we would have to take a different approach. We could take a group of third graders, randomly assign half to take music lessons, and forbid the other half to do so. Then we could compare their grades several years later. This kind of study design is called an **experiment.**

An experiment requires a **random assignment** of subjects to treatments. Only an experiment can justify a claim like "music lessons cause higher grades." Questions such as "Does taking vitamin C reduce the chance of getting a cold?" and "Does working with computers improve performance in Statistics class?" and "Is

An Experiment:

Manipulates the factor levels to create treatments.

Randomly assigns subjects to these treatment levels.

Compares the responses of the subject groups across treatment levels.

Experimental design was advanced in the 19th century by work in psychophysics by Gustav Fechner (1801–1887), the founder of experimental psychology. Fechner designed ingenious experiments that exhibited many of the features of modern designed experiments. Fechner was careful to control for the effects of factors that might affect his results. For example, he cautioned readers in his 1860 book *Elemente der Psychophysik* to group experiment trials together to minimize the possible effects of time of day and fatigue.

No drug can be sold in the United States without first showing, in a suitably designed experiment approved by the FDA (Food and Drug Administration), that it's safe and effective. The small print on the booklet that comes with many prescription drugs usually describes the outcomes of that experiment.

this drug a safe and effective treatment for that disease?" require a designed experiment to establish cause and effect.

Experiments study the relationship between two or more variables. An experimenter must identify at least one explanatory variable, called a **factor,** to manipulate and at least one **response variable** to measure. What distinguishes an experiment from other types of investigation is that the experimenter actively and deliberately manipulates the factors to control the details of the possible treatments, and assigns the subjects to those treatments *at random.* The experimenter then observes the response variable and *compares* responses for different groups of subjects who have been treated differently. For example, we might design an experiment to see whether the amount of sleep and exercise you get affects your performance.

The individuals on whom or which we experiment are known by a variety of terms. Humans who are experimented on are commonly called **subjects** or **participants.** Other individuals (rats, days, petri dishes of bacteria) are commonly referred to by the more generic term **experimental unit.** When we recruit subjects for our sleep deprivation experiment by advertising in Statistics class, we'll probably have better luck if we invite them to be participants than if we advertise that we need experimental units.

The specific values that the experimenter chooses for a factor are called the **levels** of the factor. We might assign our participants to sleep for 4, 6, or 8 hours. Often there are several factors at a variety of levels. (Our subjects will also be assigned to a treadmill for 0 or 30 minutes.) The combination of specific levels from all the factors that an experimental unit receives is known as its **treatment.** (Our subjects could have any one of six different treatments—three sleep levels, each at two exercise levels.)

How should we assign our participants to these treatments? Some students prefer 4 hours of sleep, while others need 8. Some exercise regularly; others are couch potatoes. Should we let the students choose the treatments they'd prefer? No. That would not be a good idea. To have any hope of drawing a fair conclusion, we must assign our participants to their treatments *at random.*

It may be obvious to you that we shouldn't let the students choose the treatment they'd prefer, but the need for random assignment is a lesson that was once hard for some to accept. For example, physicians might naturally prefer to assign patients to the therapy that they think best rather than have a random element such as a coin flip determine the treatment. But we've known for more than a century that for the results of an experiment to be valid, there is no way to avoid deliberate randomization.

 ◉ **The Women's Health Initiative** is a major 15-year research program funded by the National Institutes of Health to address the most common causes of death, disability, and poor quality of life in older women. It consists of both an observational study with more than 93,000 participants and several randomized comparative experiments. The goals of this study include:

- giving reliable estimates of the extent to which known risk factors predict heart disease, cancers, and fractures;
- identifying "new" risk factors for these and other diseases in women;
- comparing risk factors, presence of disease at the start of the study, and new occurrences of disease during the study across all study components; and

A S **An Industrial Experiment.** Manufacturers often use designed experiments to help them perfect new products. Here's a video about one such experiment.

• creating a future resource to identify biological indicators of disease, especially substances and factors found in blood.

That is, the study seeks to identify possible risk factors and assess how serious they might be. It seeks to build up data that might be checked retrospectively as the women in the study continue to be followed. There would be no way to find out these things with an experiment because the task includes identifying new risk factors. If we don't know those risk factors, we could never control them as factors in an experiment.

By contrast, one of the clinical trials (randomized experiments) that received much press attention randomly assigned postmenopausal women to take either hormone replacement therapy or an inactive pill. The results published in 2002 and 2004 concluded that hormone replacement with estrogen carried increased risks of stroke. ●

The Four Principles of Experimental Design

The deep insight that experiments should use random assignment is quite an old one. It is due to the American philosopher and scientist C. S. Peirce in his experiments with J. Jastrow, published in 1885.

A S **Rules of Experimental Design.** Watch an animated discussion of the three rules of design.

1. **Control.** We control sources of variation other than the factors we are testing by making conditions as similar as possible for all treatment groups. For human subjects, we try to treat them alike. However, there is always a question of degree and practicality. Controlling extraneous sources of variation reduces the variability of the responses, making it easier to detect differences among the treatment groups.

   Making generalizations from the experiment to other levels of the controlled factor can be risky. For example, suppose we test two laundry detergents and carefully control the water temperature at 180°F. This would reduce the variation in our results due to water temperature, but what could we say about the detergents' performance in cold water? Not much. It would be hard to justify extrapolating the results to other temperatures.

   Although we control both experimental factors and other sources of variation, we think of them very differently. We control a factor by assigning subjects to different factor levels because we want to see how the response will change at those different levels. We control other sources of variation to *prevent* them from changing and affecting the response variable.

2. **Randomize.** As in sample surveys, **randomization** allows us to equalize the effects of unknown or uncontrollable sources of variation. It does not eliminate the effects of these sources, but it spreads them out across the treatment levels so that we can see past them. If experimental units are not assigned to treatments at random, you will not be able to use the powerful methods of Statistics to draw conclusions from your study. Assigning subjects to treatments at random reduces bias due to uncontrolled sources of variation. Randomization protects us even from effects we didn't know about. There's an adage that says "control what you can, and randomize the rest." This may be good advice, but we should be sure when we control that we don't want to generalize to other conditions.

3. **Replicate.** Two kinds of replication show up in comparative experiments. First, we should repeat the experiment, applying the treatments to a number of subjects. Only with such replication can we estimate the variability of responses. If we have not assessed the variation, the experiment is not complete. The outcome of an experiment on a single subject is an anecdote, not data.

A S **Perform an Experiment.**
How well can you read pie charts
and bar charts? Find out as you
serve as the subject in your own
experiment.

A second kind of replication shows up when the experimental units are not a representative sample from the population of interest. We may believe that what is true of the students in Psych 101 who volunteered for the sleep experiment is true of all humans, but we'll feel more confident if our results for the experiment are *replicated* in another part of the country, with people of different ages, and at different times of the year. **Replication** of an entire experiment with the controlled sources of variation at different levels is an essential step in science.

● **Replicating cold fusion** "The origins of Cold Fusion have been loudly and widely documented in the press and popular literature. Pons and Fleischmann, fearing they were about to be scooped by a competitor named Steven Jones from nearby Brigham Young University, and with the encouragement of their own administration, held a press conference on March 23, 1989, at the University of Utah, to announce what seemed to be the scientific discovery of the century. Nuclear fusion, producing usable amounts of heat, could be induced to take place on a table-top by electrolyzing heavy water, using electrodes made of palladium and platinum, two precious metals. If so, the world's energy problems were at an end, to say nothing of the fiscal difficulties of the University of Utah. What followed was a kind of feeding frenzy, science by press conference and e-mail, confirmations and disconfirmations, claims and retractions, ugly charges and obfuscation, science gone berserk. For all practical purposes, it ended a mere 5 weeks after it began, on May 1st, 1989, at a dramatic session of The American Physical Society, in Baltimore. Although there were numerous presentations at this session, only two really counted. Steven Koonin and Nathan Lewis, speaking for himself and Charles Barnes, all three from Caltech, executed between them a perfect slam-dunk that cast Cold Fusion right out of the arena of mainstream science. . . . seasoned experimentalists like Lewis and Barnes refused to believe what they couldn't reproduce in their own laboratories."—David Goodstein, "Whatever Happened to Cold Fusion?" ●

4. **Block.** The ability of randomizing to equalize variation across treatment groups works best in the long run. For example, if we're allocating players to two 6-player soccer teams from a pool of 12 children, we might do so at random to equalize the talent. But what if there were two 12-year-olds and ten 6-year-olds in the group? Randomizing may place both 12-year-olds on the same team. In the long run, if we did this over and over, it would all equalize. But wouldn't it be better to assign one 12-year-old to each group (at random) and five 6-year-olds to each team (at random)? By doing this we would improve fairness in the short run. This approach makes the division more fair by recognizing the variation in *age* and allocating the players at random

within each age level. When we do this, we call the variable *age* a **blocking variable.** The levels of *age* are called blocks.

Sometimes, attributes of the experimental units that we are not studying and that we can't control may nevertheless affect the outcomes of an experiment. If we group similar individuals together and then randomize within each of these **blocks,** we can remove much of the variability due to the difference among the blocks. Blocking is an important compromise between randomization and control. However, unlike the first three principles, blocking is not *required* in an experimental design.

Diagrams

An experiment is carried out over time with specific actions occurring in a specified order. A diagram of the procedure can help in thinking about experiments.[1]

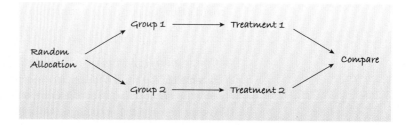

The diagram emphasizes the random allocation of subjects to treatment groups, the separate treatments applied to these groups, and the ultimate comparison of results. It's best to specify the responses that will be compared. A good way to start comparing results for the treatment groups is with boxplots.

Designing an Experiment   Step-By-Step

A **completely randomized experiment** is the ideal simple design, just as a *simple random sample* is the ideal simple sample—and for many of the same reasons.

An ad for OptiGro plant fertilizer claims that with this product you will grow "juicier, tastier" tomatoes. You'd like to test this claim, and wonder whether you might be able to get by with half the specified dose. How can you set up an experiment to check out the claim?

Of course, you'll have to get some tomatoes, try growing some plants with the product and some without, and see what happens. But you'll need a clearer plan than that. How should you design your experiment?

Let's work through the design, step by step. We'll design the simplest kind of experiment, a **completely randomized experiment in one factor.** Since this is a *design* for an experiment, most of the steps are part of the *Think* stage. The statements in the right column are the kinds of things you would need to say in *proposing* an experiment. You'd need to include them in the "methods" section of a report once the experiment is run.

---

[1] Diagrams of this sort were introduced by David Moore in his textbooks and are widely used.

Think

Plan State what you want to know.

I want to know whether tomato plants grown with OptiGro yield juicier, tastier tomatoes than plants raised in otherwise similar circumstances but without the fertilizer.

Response Specify the response variable.

I'll evaluate the juiciness and taste of the tomatoes by asking a panel of judges to rate them on a scale from 1 to 7 in juiciness and in taste.

Treatments Specify the factor levels and the treatments.

The factor is fertilizer, specifically OptiGro fertilizer. I'll grow tomatoes at three different factor levels: some with no fertilizer, some with half the specified amount of OptiGro, and some with the full dose of OptiGro. These are the three treatments.

Experimental Units Specify the experimental units.

I'll obtain 24 tomato plants of the same variety from a local garden store.

Experimental Design Observe the principles of design:

Control any sources of variability you know of and can control.

I'll locate the farm plots near each other so that the plants get similar amounts of sun and rain and experience similar temperatures. I will weed the plots equally and otherwise treat the plants alike.

Randomly assign experimental units to treatments, to equalize the effects of unknown or uncontrollable sources of variation.

I'll randomly divide the plants into three groups. I will use random numbers from a table to determine the assignment.

Specify how the random numbers needed for randomization will be obtained.

Replicate results by placing more than one plant in each treatment group.

There are 8 plants in each treatment group.

Make a Picture A diagram of your design can help you think about it clearly.

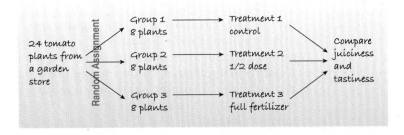

Specify any other experiment details. You must give enough details so that another experimenter could exactly replicate your experiment. It's generally better to include details that might seem irrelevant than to leave out matters that could turn out to make a difference.

> I will grow the plants until the tomatoes are mature, as judged by reaching a standard color.
>
> I'll harvest the tomatoes when ripe and store them for evaluation.

Specify how to **measure the response.**

> I'll set up a numerical scale of juiciness and one of tastiness for the taste testers. Several people will taste slices of tomato and rate them.

Show   Once you collect the data, you'll need to display them and compare the results for the three treatment groups.

> I will display the results with side-by-side box-plots to compare the three treatment groups.
>
> I will compare the means of the groups.

Tell   To answer the initial question, we ask whether the differences we observe in the means of the three groups are meaningful.

Because this is a randomized experiment, we can attribute significant differences to the treatments. To do this properly, we'll need methods from what is called "statistical inference," the subject of the rest of this book.

> If the differences in taste and juiciness among the groups are greater than I would expect by knowing the usual variation among tomatoes, I may be able to conclude that these differences can be attributed to treatment with the fertilizer.

Does the Difference Make a Difference?

A S   **Graph the Data.** Do you think there's a significant difference in your perception of pie charts and bar charts? Explore the data from your plot perception experiment.

We've said that if the differences among the treatment groups are big enough, we'll attribute the differences to the treatments. How will we know whether the differences are big enough?

Would we expect the group means to be identical? Not really. Even if the treatment made no difference whatever, there would still be some variation. We assigned the plants to treatments at random. But a different random assignment would have led to different results. Even a repeat of the *same* treatment on a different randomly assigned set of plants would lead to a different mean. The real question is whether the differences we observed are about as big as we'd get just from the randomization alone, or whether they're bigger than that. If we decide that they're bigger, we'll attribute the differences to the treatments. In that case we say the differences are **statistically significant.**

How will we decide if something is different enough to be considered statistically significant? We'll find a precise answer to that question in a later chapter, but to get some intuition, think about deciding whether a coin is fair. If we flip a fair coin 100 times, we expect, *on average,* to get 50 heads. Suppose we get 54 heads out of 100. That doesn't seem very surprising. It's well within the bounds of ordinary random fluctuations. What if we'd seen 94 heads? That's clearly outside the bounds. We'd be pretty sure that the coin flips were not random. But what about 74 heads? Is that far enough from 50% to arouse our suspicions? That's the sort of question we need to ask of our experiment results.

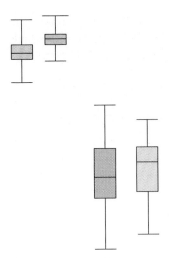

The boxplots in both pairs have centers the same distance apart, but when the spreads are large, the observed difference may be just from random fluctuation.

Figure 13.1

In Statistics terminology, 94 heads would be a statistically significant difference from 50 and 54 heads would not. Whether 74 is *statistically significant* or not would depend on the chance of getting 74 heads in 100 flips of a fair coin and on our tolerance for believing that rare events can happen to us.

Back at the tomato stand, we ask whether the differences we see among the treatment groups are the kind of differences we'd expect from randomization. A good way to get a feeling for that is to look at how much our results vary among plants that get the same treatment. Boxplots of our results by treatment group can give us a general idea.

For example, Figure 13.1 shows two pairs of boxplots whose centers differ by exactly the same amount. In the upper set, that difference appears to be larger than we'd expect just by chance. Why? Because the variation is quite small *within* treatment groups, so the larger difference *between* the groups is unlikely to be just from the randomization. In the bottom pair, that same difference between the centers looks less impressive. There the variation *within* each group swamps the difference *between* the two means. We'd say the difference is statistically significant in the upper pair and not statistically significant in the lower pair.

Later we'll see statistical tests that quantify this intuition. For now, the important point is that a difference is statistically significant if we don't believe that it's likely to have occurred only by chance.

1. At one time a method called "gastric freezing" was used to treat people with peptic ulcers. An inflatable bladder was inserted down the esophagus and into the stomach, and then a cold liquid was pumped into the bladder. Now you can find the following notice on the Internet site of a major insurance company:

   [Our company] does not cover gastric freezing (intragastric hypothermia) for chronic peptic ulcer disease. . . .

   Gastric freezing for chronic peptic ulcer disease is a non-surgical treatment which was popular about 20 years ago but now is seldom performed. It has been abandoned due to a high complication rate, only temporary improvement experienced by patients, and a lack of effectiveness when tested by double-blind, controlled clinical trials.

   What did that "controlled clinical trial" (experiment) probably look like? (Don't worry about "double-blind"; we'll get to that soon.)

   a) What was the factor in this experiment?
   b) What was the response variable?
   c) What were the treatments?
   d) How did researchers decide which subjects received which treatment?
   e) Were the results statistically significant?

Experiments and Samples

Both experiments and sample surveys use randomization to get unbiased data. But they do so in different ways and for different purposes. Sample surveys try to estimate population parameters, so the sample needs to be as representative of the population as possible. By contrast, experiments try to assess the effects of treatments. Experimental units are not always drawn randomly from a population. For example, a medical experiment may deal only with patients who have the disease under study. The randomization is in the assignment of their therapy. This is appropriate because our focus is on differences in the effects of the treatments.

Unless the experimental units are chosen from the population at random, you should be cautious about generalizing experiment results to larger populations until the experiment has been repeated under different circumstances. Results become more persuasive if they remain the same in completely different settings, such as in a different season, in a different country, or for a different species, to name a few.

Even without choosing experimental units from a population at random, experiments can draw stronger conclusions than surveys. By looking only at the differences across treatment groups, experiments cancel out many sources of bias. For example, the entire pool of subjects may be biased and not representative of the population. (College students may need more sleep on average than the general population.) When we assign subjects randomly to treatment groups, all the groups are still biased, but *in the same way.* When we consider the differences in their responses, these biases cancel out, allowing us to see the *differences* due to treatment effects more clearly.

Experiments are rarely performed on random samples from a population. Don't describe the subjects in an experiment as a random sample unless they really are. More likely, the randomization was in assigning subjects to treatments.

Control Treatments

A S **Control Groups in Experiments.** Is a control group really necessary? Here's a story about a medical study.

Suppose you wanted to test a $300 piece of software designed to shorten download times. You could just try it on several files and record the download times, but you probably want to *compare* the speed with what would happen *without* the software installed. Such a baseline measurement is called a **control** treatment, and the experimental units to whom it is applied is called a **control group.**

This is a use of the word "control" in an entirely different context. Previously, we controlled extraneous sources of variation by keeping them constant. Here, we use a control treatment as another *level* of the factor in order to compare the treatment results to a situation in which "nothing happens." That's what we did in the tomato experiment when we used no fertilizer on the 8 tomatoes in Group 1.

Blinding

Humans are notoriously susceptible to errors in judgment.[2] All of us. When we know what treatment was assigned, it's difficult not to let that knowledge influence our assessment of the response, even when we try to be careful.

Suppose you were trying to advise your school on which brand of cola to stock in the school's vending machines. You set up an experiment to see which of the three competing brands students prefer (or whether they can tell the difference at all). But people have brand loyalties. You probably prefer one brand already. So if you knew which brand you were tasting, it might influence your

---

2 For example, here we are in Chapter 13 and you're still reading the footnotes.

rating. To avoid this bias, it would be better to disguise the brands as much as possible. This strategy is called **blinding** the participants to the treatment.[3]

But it isn't just the subjects who should be blind. Experimenters themselves often subconsciously behave in ways that favor what they believe. Even technicians may treat plants or test animals differently if, for example, they expect them to die. An animal that starts doing a little better than others by showing an increased appetite may get fed a bit more than the experimental protocol specifies.

People are so good at picking up subtle cues about treatments that the best (in fact, the *only*) defense against such biases in experiments on human subjects is to keep *anyone* who could affect the outcome or the measurement of the response from knowing which subjects have been assigned to which treatments. So, not only should your cola-tasting subjects be blinded, but also *you*, as the experimenter, shouldn't know which drink is which, either—at least until you're ready to analyze the results.

There are two main classes of individuals who can affect the outcome of the experiment:

- those who could influence the results (the subjects, treatment administrators, or technicians)
- those who evaluate the results (judges, treating physicians, etc.)

When all the individuals in either one of these classes is blinded, an experiment is said to be **single-blind.** When everyone in *both* classes is blinded, we call the experiment **double-blind.** Even if several individuals in one class are blinded—for example, both the patients and the technicians who administer the treatment—the study would still be just single-blind. If only some of the individuals in a class are blind—for example, if subjects are not told of their treatment, but the administering technician is not blind—there is a substantial risk that subjects can discern their treatment from subtle cues in the technician's behavior or that the technician might inadvertently treat subjects differently. Such experiments cannot be considered truly blind.

In our tomato experiment, we certainly don't want the people judging the taste to know which tomatoes got the fertilizer. That makes the experiment single-blind. We might also not want the people caring for the tomatoes to know which ones were being fertilized, in case they might treat them differently in other ways, too. We can accomplish this double-blinding by having some fake fertilizer for them to put on the other plants. Read on.

A S Blinded Experiments.
This narrated account of blinding isn't a placebo!

Placebos

Often simply applying *any* treatment can induce an improvement. Every parent knows the medicinal value of a kiss to make a toddler's scrape or bump stop hurting. Some of the improvement seen with a treatment—even an effective treatment—can be due simply to the act of treating. To separate these two effects, we can use a control treatment that mimics the treatment itself.

---

[3] C. S. Peirce, in the same 1885 work in which he introduced randomization, also recommended blinding.

The placebo effect is stronger when placebo treatments are administered with authority or by a figure who appears to be an authority. "Doctors" in white coats generate a stronger effect than salespeople in polyester suits. But the placebo effect is not reduced much, even when subjects know that the effect exists. People often suspect that they've gotten the placebo if nothing at all happens. So, recently, drug manufacturers have gone so far in making placebos realistic that they sometimes even give them the same side effects as the drug being tested! Such "active placebos" are usually more effective. But when those side effects include loss of appetite or hair, the practice may raise ethical questions.

A "fake" treatment that looks just like the treatments being tested is called a **placebo.** Placebos are the best way to blind subjects from knowing whether they are receiving the treatment or not. One common version of a placebo in drug testing is a "sugar pill." Especially when psychological attitude can affect the results, control group subjects treated with a placebo may show an improvement.

The fact is that subjects treated with a placebo sometimes improve. It's not unusual for 20% or more of subjects given a placebo to report reduction in pain, improved movement, or greater alertness, or even to demonstrate improved health or performance. This **placebo effect** highlights both the importance of effective blinding and the importance of comparing treatments with a control. Placebo controls are so effective that you should use them as an essential tool for blinding whenever possible.

The best experiments are usually
- randomized.
- comparative.
- double-blind.
- placebo-controlled.

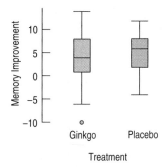

● **Does ginkgo biloba improve memory?**   Researchers investigated the purported memory-enhancing effect of ginkgo biloba tree extract (P. R. Solomon, F. Adams, A. Silver, J. Zimmer, R. De Veaux, "Ginkgo for Memory Enhancement. A Randomized Controlled Trial." *JAMA* 288 [2002]: 835–840). In a randomized, comparative, double-blind, placebo-controlled study, they administered treatments to 230 elderly community members. One group received Ginkoba™ according to the manufacturer's instructions. The other received a similar-looking placebo. Thirteen different tests of memory were administered before and after treatment. The placebo group showed greater improvement on 7 of the tests; the treatment group on the other 6. None showed any significant differences. Here are boxplots of one measure. ●

Blocking

We wanted to use 18 tomato plants of the same variety for our experiment, but suppose the garden store had only 12 plants left. So we drove down to the nursery and bought 6 more plants of that variety. We worry that the tomato plants from the two stores are different somehow, and, in fact, they don't really look the same.

How can we design the experiment so that the differences between the stores don't mess up our attempts to see differences among fertilizer levels? We can't measure the effect of a store the same way as the fertilizer because we can't assign it as we would a factor in the experiment. You can't tell a tomato what store to come from.

Because stores may vary in the care they give plants or in the sources of their seeds, the plants from one store are likely to be more like each other than they are like the plants from the other store. When groups of experimental units are similar, it's often a good idea to gather them together into **blocks.** By blocking we isolate the variability attributable to the differences between the blocks, so that we can see the differences caused by the treatments more clearly. Here, we would define the plants from each store to be a block. The randomization is introduced when we randomly assign treatments within each block.

In a completely randomized design, each of the 18 plants would have an equal chance to land in each of the three treatment groups. But we realize that the store may have an effect. To isolate the store effect, we block on store by assigning the plants from each store to treatments at random. So we now have six treatment groups, three for each block. Within each block, we'll randomly assign the same number of plants to each of the three treatments. The experiment is still fair because each treatment is still applied (at random) to the same number of plants and to the same proportion from each store: 4 from store A and 2 from store B. Because the randomization occurs only within the blocks (plants from one store cannot be assigned to treatment groups for the other), we call it a **randomized block design.**

In effect, we conduct two parallel experiments, one for tomatoes from each store, and then combine the results. The picture tells the story.

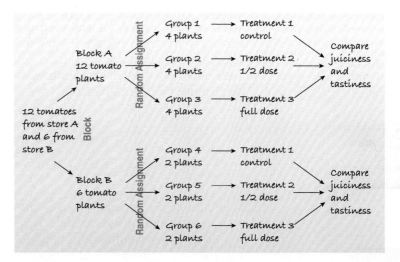

In a retrospective or prospective study, subjects are sometimes paired because they are similar in ways *not* under study. **Matching** subjects in this way can reduce variation in much the same way as blocking. For example, a retrospective study of music education and grades might match each student who studies an instrument with someone of the same sex who is similar in family income but didn't study an instrument. When we compare grades of music students with those of non-music students, the matching would reduce the variation due to income and gender differences.

Blocking is the same idea for experiments as stratifying is for sampling. Both methods group together subjects that are similar and randomize within those groups as a way to remove unwanted variation. (But be careful to keep the terms straight. Don't say that we "stratify" an experiment or "block" a sample.) We use blocks to reduce variability so we can see the effects of the factors; we're not usually interested in studying the effects of the blocks themselves.

2 Recall the experiment about gastric freezing, an old method for treating peptic ulcers that you read about in the first Just Checking. Doctors would insert an inflatable bladder down the patient's esophagus and into the stomach and then pump in a cold liquid. A major insurance company now states that it doesn't cover this treatment because "double-blind, controlled clinical trials" failed to demonstrate that gastric freezing was effective.

a) What does it mean that the experiment was double-blind?
b) Why would you recommend a placebo control?

c) Suppose that researchers suspected that the effectiveness of the gastric freezing treatment might depend on whether a patient had recently developed the peptic ulcer or had been suffering from the condition for a long time. How might the researchers have designed the experiment?

*Adding More Factors

There are two kinds of gardeners. Some water frequently, making sure that the plants are never dry. Others let Mother Nature take her course and leave the watering to her. The makers of OptiGro want to ensure that their product will work under a wide variety of watering conditions. Maybe we should include the amount of watering as part of our experiment. Can we study a second factor at the same time and still learn as much about fertilizer?

We now have two factors (fertilizer at three levels and irrigation at two levels). We combine them in all possible ways to yield six treatments:

	No Fertilizer	Half Fertilizer	Full Fertilizer
No Added Water	1	2	3
Daily Watering	4	5	6

If we allocate the original 12 plants, the experiment now assigns 2 plants to each of these six treatments at random. This experiment is a **completely randomized two-factor experiment** because any plant could end up assigned at random to any of the six treatments (and we have two factors).

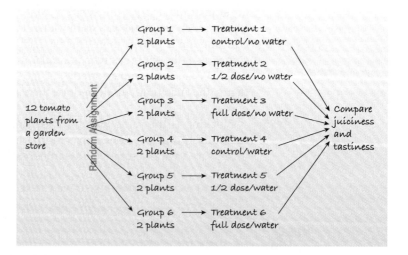

Think Like a Statistician
With two factors we can account for more of the variation. That lets us see the underlying patterns more clearly.

It's often important to include several factors in the same experiment in order to see what happens when the factor levels are applied in different *combinations*. A common misconception is that applying several factors at once makes it difficult to separate the effects of the individual factors. You may hear people say that experiments should always be run "one factor at a time." In fact, just the opposite is true. Experiments with more than one factor are both more efficient and provide more information than one-at-a-time experiments. There are many ways to design efficient multifactor experiments, and you can take a whole course on the design and analysis of such experiments.

Confounding

Professor Stephen Ceci of Cornell University performed an experiment to investigate the effect of a teacher's classroom style on student evaluations. He taught a class in developmental psychology during two successive terms to a total of 472 students in two very similar classes. He kept everything about his teaching identical (same text, same syllabus, same office hours, etc.) and modified only his style in class. During the fall term he maintained a subdued demeanor. During the spring term, he used expansive gestures and lectured with more enthusiasm, varying his vocal pitch and using more hand gestures. He administered a standard student evaluation form at the end of each term.

The students in the fall term class rated him only an average teacher. Those in the spring term class rated him an excellent teacher, praising his knowledge and accessibility, and even the quality of the textbook. On the question "How much did you learn in the course?" the average response changed from 2.93 to 4.05 on a 5-point scale.[4]

How much of the difference he observed was due to his difference in manner, and how much might have been due to the season of the year? Fall term in Ithaca, NY (home of Cornell University), starts out colorful and pleasantly warm but ends cold and bleak. Spring term starts out bitter and snowy and ends with blooming flowers and singing birds. Might students' overall happiness have been affected by the season and reflected in their evaluations?

Unfortunately, there's no way to tell. Nothing in the data enables us to tease apart these two effects, because all the students who experienced the subdued manner did so during the fall term and all who experienced the expansive manner did so during the spring. When the levels of one factor are associated with the levels of another factor, we say that these two factors are **confounded.**

In some experiments, such as this one, it's just not possible to avoid some confounding. Professor Ceci could have randomly assigned students to one of two classes during the same term, but then we might question whether mornings or afternoons were better, or whether he really delivered the same class the second time (after practicing on the first class). Or he could have had another professor deliver the second class, but that would have raised more serious issues about differences in the two professors and concern over more serious confounding.

● ***A two-factor example*** Confounding can also arise from a badly designed multifactor experiment. Here's a classic. A credit card bank wanted to test the sensitivity of the market to two factors: the annual fee charged for a card and the annual percentage rate charged. Not wanting to scrimp on sample size, the bank selected 100,000 people at random from a mailing list. It sent out 50,000 offers with a low rate and no fee, and 50,000 offers with a higher rate and a $50 annual fee. Guess what happened? That's right—people preferred the low-rate, no-fee card. No surprise. In fact, they signed up for that card at over twice the rate as the other offer. And because of the large sample size, the bank was able

---

[4] But the two classes performed almost identically well on the final exam.

to estimate the difference precisely. But the question the bank really wanted to answer was "How much of the change was due to the rate, and how much was due to the fee?" Unfortunately, there's simply no way to separate out the two effects. If the bank had sent out all four possible different treatments—low rate with no fee; low rate with $50 fee; high rate with no fee, and high rate with $50 fee—each to 25,000 people, it could have learned about both factors and could have also seen what happens when the two factors occur in combination. ●

Lurking or Confounding?

Confounding may remind you of the problem of lurking variables we discussed back in Chapters 7 (p. 154) and 9 (p. 203). Confounding variables and lurking variables are alike in that they interfere with our ability to interpret our analyses simply. Each can mislead us, but there are important differences in both how and where the confusion may arise.

A lurking variable creates an association between two other variables that tempts us to think that one may cause the other. This can happen in a regression analysis or an observational study when a lurking variable influences both the explanatory and response variables. Recall that countries with more TV sets per capita tend to have longer life expectancies. We shouldn't conclude it's the TVs "causing" longer life. We suspect instead that a generally higher standard of living may mean that people can afford more TVs and get better health care, too. Our data revealed an association between TVs and life expectancy, but economic conditions were a likely lurking variable. We also saw a positive association between the number of doctors per capita and life expectancy. Does that mean the presence of more doctors might lead to longer life? That seems more plausible than the "TV effect," but it may well be due to the same lurking variable. A lurking variable, then, is usually thought of as a prior cause of both y and x that makes it appear that x may be causing y.

Confounding can arise in experiments when some other variable associated with a factor has an effect on the response variable. However, in a designed experiment, the experimenter, *assigns* treatments (at random) to subjects rather than just observing them. A confounding variable can't be thought of as causing that assignment. Professor Ceci's choice of teaching styles was not caused by the weather, but because he used one style in the fall and the other in spring, he was unable to tell how much of his students' reactions were attributable to his teaching and how much to the weather. A confounding variable, then, is associated in a noncausal way with a factor and affects the response. Because of the confounding, we find that we can't tell whether any effect we see was caused by our factor or by the confounding variable—or even by both working together.

Both confounding and lurking variables are outside influences that make it harder to understand the relationship we are modeling. However, the nature of the causation is different in the two situations. In regression and observational studies, we can only observe associations between variables. Although we can't demonstrate a causal relationship, we often imagine whether x could cause y. We can be misled by a lurking variable that causes changes in both. In a designed experiment, we often hope to show that the factor causes a response. Here we can be misled by a confounding variable that's associated with the factor and causes the differences we observe in the response.

What Can Go Wrong?

- ***Don't give up just because you can't run an experiment.*** Sometimes we can't run an experiment because we can't identify or control the factors. Sometimes it would simply be unethical to run the experiment. (Consider randomly assigning each of two identical twins to be placed in an enriched, stimulating environment or a barren, boring one to try to assess the effects of environment on intelligence. Or consider randomly assigning students to take—and be graded in—a Statistics course deliberately taught to be boring and difficult or one that had an unlimited budget to use multimedia and real-world examples and field trips to make the subject more interesting.) If we can't perform an experiment, often an observational study is a good choice.

- ***Beware of confounding.*** Use randomization whenever possible to ensure that the factors not in your experiment are not confounded with your treatment levels. Be alert to confounding that cannot be avoided, and report it along with your results.

- ***Bad things can happen even to good experiments.*** Protect yourself by recording additional information. An experiment in which the air conditioning failed for 2 weeks, affecting the results, was saved by recording the temperature (although that was not originally one of the factors) and estimating the effect the higher temperature had on the response.[5]

  It's generally good practice to collect as much information as possible about your experimental units and the circumstances of the experiment. For example, in the tomato experiment, it would be wise to record details of the weather (temperature, rainfall, sunlight) that might affect the plants and any facts available about their growing situation. (Is one side of the field in shade sooner than the other as the day proceeds? Is one area lower and a bit wetter?) Sometimes we can use this extra information during the analysis to reduce biases.

- ***Don't spend your entire budget on the first run.*** Just as it's a good idea to pretest a survey, it's always wise to try a small pilot experiment before running the full-scale experiment. You may learn, for example, how to choose factor levels more effectively, about effects you forgot to control, and about unanticipated confoundings.

CONNECTIONS

The fundamental role of randomization in experiments clearly points back to our discussions of randomization, to our experiments with simulations, and to our use of randomization in sampling. The similarities and differences between experiments and samples are important to keep in mind and can make each concept clearer.

If you think that blocking in an experiment resembles stratifying in a sample, you're quite right. Both are ways of removing variation we can identify to help us see past the variation in the data.

Experiments compare groups of subjects that have been treated differently. Graphics such as box-plots that help us compare groups are closely related to these ideas. Think about what we look for in

---

[5] R. D. DeVeaux and M. Szelewski, "Optimizing Automatic Splitless Injection Parameters for Gas Chromatographic Environmental Analysis." *Journal of Chromatographic Science* 27, no. 9 [1989]: 513–518.

a boxplot to tell whether two groups look really different, and you'll be thinking about the same issues as experiment designers.

Generally, we're going to consider how different the mean responses are for different treatment groups. And we're going to judge whether those differences are large by using standard deviations as rulers. (That's why we needed to replicate results for each treatment; we need to be able to estimate those standard deviations.) The discussion in Chapter 6 introduced this fundamental statistical thought, and it's going to keep coming back over and over again. Statistics is about variation.

We'll see a number of ways to analyze results from experiments in subsequent chapters.

What have we learned?

We've learned to recognize sample surveys, observational studies, and randomized comparative experiments. We know that these methods collect data in different ways and lead us to different conclusions.

We've learned to identify retrospective and prospective observational studies and understand the advantages and disadvantages of each.

We've learned that only well-designed experiments can allow us to reach cause-and-effect conclusions. We manipulate levels of treatments to see if the factor we have identified produces changes in our response variable.

We've learned the principles of experimental design:
- We want to be sure that variation in the response variable can be attributed to our factor, so we identify as many other sources of variability as possible.
- We control the sources of variability we can, and consider blocking to reduce variability from sources we recognize but cannot control.
- Because there are many possible sources of variability that we cannot identify, we try to equalize those by randomly assigning experimental units to treatments.
- We replicate the experiment on as many subjects as possible.

We've learned the value of having a control group and of using blinding and placebo controls.

Finally, we've learned to recognize the problems posed by confounding variables in experiments and lurking variables in observational studies.

TERMS

Observational study   A study based on data in which no manipulation of factors has been employed.

Retrospective study   An observational study in which subjects are selected and then their previous conditions or behaviors are determined. Because retrospective studies are not based on random samples, they usually focus on estimating differences between groups or associations between variables.

Prospective study   An observational study in which subjects are followed to observe future outcomes. Because no treatments are deliberately applied, a prospective study is not an experiment. Nevertheless, prospective studies typically focus on estimating differences among groups that might appear as the groups are followed during the course of the study.

Experiment An experiment *manipulates* factor levels to create treatments, *randomly assigns* subjects to these treatment levels, and then *compares* the responses of the subject groups across treatment levels.

Random assignment To be valid, an experiment must assign experimental units to treatment groups at random. This is called random assignment.

Factor A variable whose levels are controlled by the experimenter. Experiments attempt to discover the effects that differences in factor levels may have on the responses of the experimental units.

Response A variable whose values are compared across different treatments. In a randomized experiment, large response differences can be attributed to the effect of differences in treatment level.

Experimental units Individuals on whom an experiment is performed. Usually called **subjects** or **participants** when they are human.

Level The specific values that the experimenter chooses for a factor are called the levels of the factor.

Treatment The process, intervention, or other controlled circumstance applied to randomly assigned experimental units. Treatments are the different levels of a single factor or are made up of combinations of levels of two or more factors.

Principles of experimental design
- **Control** aspects of the experiment that we know may have an effect on the response, but that are not the factors being studied.
- **Randomize** subjects to treatments to even out effects that we cannot control.
- **Replicate** over as many subjects as possible. Results for a single subject are just anecdotes. If, as often happens, the subjects of the experiment are not a representative sample from the population of interest, replicate the entire study with a different group of subjects, preferably from a different part of the population.
- **Block** to reduce the effects of identifiable attributes of the subjects that cannot be controlled.

Statistically significant When an observed difference is too large for us to believe that it is likely to have occurred naturally, we consider the difference to be statistically significant. Subsequent chapters will show specific calculations and give rules, but the principle remains the same.

Control group The experimental units assigned to a baseline treatment level, typically either the default treatment, which is well understood, or a null, placebo treatment. Their responses provide a basis for comparison.

Blinding Any individual associated with an experiment who is not aware of how subjects have been allocated to treatment groups is said to be blind.

Single-blind Double-blind There are two main classes of individuals who can affect the outcome of an experiment:
- those who could *influence the results* (the subjects, treatment administrators, or technicians).
- those who *evaluate the results* (judges, treating physicians, etc.).

When every individual in *either* of these classes is blinded, an experiment is said to be single-blind. When everyone in *both* classes is blinded, we call the experiment double-blind.

Placebo A treatment known to have no effect, administered so that all groups experience the same conditions. Many subjects respond to such a treatment (a response known as a placebo effect). Only by comparing with a placebo can we be sure that the observed effect of a treatment is not due simply to the placebo effect.

Placebo effect	The tendency of many human subjects (often 20% or more of experiment subjects) to show a response even when administered a placebo.
Block	When groups of experimental units are similar, it is often a good idea to gather them together into blocks. By blocking we isolate the variability attributable to the differences between the blocks so that we can see the differences caused by the treatments more clearly.
Matching	In a retrospective or prospective study, subjects who are similar in ways not under study may be matched and then compared with each other on the variables of interest. Matching, like blocking, reduces unwanted variation.
Designs	In a **randomized block design,** the randomization occurs only within blocks.
	In a **completely randomized design,** all experimental units have an equal chance of receiving any treatment.
Confounding	When the levels of one factor are associated with the levels of another factor so their effects cannot be separated, we say that these two factors are confounded.

SKILLS

When you complete this lesson you should:

Think

- Recognize when an observational study would be appropriate.
- Be able to identify observational studies as retrospective or prospective, and understand the strengths and weaknesses of each method.
- Know the four basic principles of sound experiment design: control, randomize, replicate, and block, and be able to explain each.
- Be able to recognize the factors, the treatments, and the response variable in a description of a designed experiment.
- Understand the essential importance of randomization in assigning treatments to experimental units.
- Understand the importance of replication to move from anecdotes to general conclusions.
- Understand the value of blocking so that variability due to differences in attributes of the subjects can be removed.
- Understand the importance of a control group and the need for a placebo treatment in some studies.
- Understand the importance of blinding and double-blinding in studies on human subjects, and be able to identify blinding and the need for blinding in experiments.
- Understand the value of a placebo in experiments with human participants.

Show

- Be able to design a completely randomized experiment to test the effect of a single factor.
- Be able to design an experiment in which blocking is used to reduce variation.
- Know how to use graphical displays to compare responses for different treatment groups. Understand that you should *never* proceed with any other analysis of a designed experiment without first looking at boxplots or other graphical displays.

Tell

- Know how to report the results of an observational study. Identify the subjects, how the data were gathered, and any potential biases or flaws you may be aware of. Identify the factors known and those that might have been revealed by the study.

- Know how to compare the responses in different treatment groups to assess whether the differences are larger than could be reasonably expected from ordinary sampling variability.

- Know how to report the results of an experiment. Tell who the subjects are and how their assignment to treatments was determined. Report on how the response variable was measured and in what measurement units.

- Understand that your description of an experiment should be sufficient for another researcher to replicate the study with the same methods.

- Be able to report on the statistical significance of the result in terms of whether the observed group-to-group differences are larger than could be expected from ordinary sampling variation.

Experiments on the Computer

Most experiments are analyzed with a statistics package. You should almost always display the results of a comparative experiment with side-by-side boxplots. You may also want to display the means and standard deviations of the treatment groups in a table.

The analyses offered by statistics packages for comparative randomized experiments fall under the general heading of Analysis of Variance, usually abbreviated ANOVA. These analyses are beyond the scope of this chapter. You'll find a discussion of ANOVA in Chapter 28 on the CD-ROM that comes with this book.

EXERCISES

1–20. What's the design? Read each brief report of statistical research, and identify:
   a) whether it was an observational study or an experiment.

If it was an investigative study, identify (if possible)
   b) whether it was retrospective or prospective.
   c) the subjects studied, and how they were selected.
   d) the parameter of interest.
   e) the nature and scope of the conclusion the study can reach.

If it was an experiment, identify (if possible)
   b) the subjects studied.
   c) the factor(s) in the experiment, and the number of levels for each.
   d) the number of treatments.
   e) the response variable measured.
   f) the design (completely randomized, blocked, or matched).
   g) whether it was blind (or double-blind).
   h) the nature and scope of the conclusion the experiment can reach.

1. Over a 4-month period, among 30 people with bipolar disorder, patients who were given a high dose (10 g/day) of omega-3 fats from fish oil improved more than those given a placebo. (*Archives of General Psychiatry* 56 [1999]: 407)

2. The leg muscles of men aged 60 to 75 were 50% to 80% stronger after they participated in a 16-week, high-intensity resistance-training program twice a week. (*Journal of Gerontology* 55A [2000]: B336)

3. In a test of roughly 200 men and women, those with moderately high blood pressure (averaging 164/89 mm Hg) did worse on tests of memory and reaction time than those with normal blood pressure. (*Hypertension* 36 [2000]: 1079)

4. Among a group of disabled women aged 65 and older who were tracked for several years, those who had a vitamin B_{12} deficiency were twice as likely to suffer severe depression as those who did not. (*American Journal of Psychology* 157 [2000]: 715)

5. An examination of the medical records of more than 360,000 Swedish men showed that those who were overweight or who had high blood pressure had a higher risk of kidney cancer. (*New England Journal of Medicine* 3434 [2000]: 1305)

6. To research the effects of "dietary patterns" on blood pressure in 459 subjects, subjects were randomly assigned to three groups and had their meals prepared by dieticians. Those who were fed a diet low in fat and cholesterol and high in fruits, vegetables, and low-fat dairy foods (known as the DASH diet) lowered their systolic blood pressure by an average of 6.7 points when compared with subjects fed a control diet.

7. After menopause many women take supplemental estrogen. There is some concern that if these women also drink alcohol, their estrogen levels will rise too high. Twenty-four volunteers, 12 who were receiving supplemental estrogen and 12 who were not, were randomly divided into two groups. One group drank an alcoholic beverage, the other a nonalcoholic beverage. An hour later everyone's estrogen level was checked. Only those on supplemental estrogen who drank alcohol showed a marked increase.

8. Is diet or exercise effective in combating insomnia? Some believe that cutting out desserts can help alleviate the problem, while others recommend exercise. Forty volunteers suffering from insomnia agreed to participate in a month-long test. Half were randomly assigned to a special no-desserts diet; the others continued desserts as usual. Half of the people in each of these groups were randomly assigned to an exercise program, while the others did not exercise. Those who ate no desserts and engaged in exercise showed the most improvement.

9. Some gardeners prefer to use nonchemical methods to control insect pests in their gardens. Researchers have designed two kinds of traps, and want to know which design will be more effective. They randomly choose 10 locations in a large garden and place one of each kind of trap at each location. After a week they count the number of bugs in each trap.

10. Researchers have linked an increase in the incidence of breast cancer in Italy to dioxin released by an industrial accident in 1976. The study identified 981 women who lived near the site of the industrial explosion and were under age 40 at the time. Fifteen of the women had developed breast cancer at an unusually young average age of 45. Medical records showed that these women had heightened concentrations of dioxin in their blood, and that each 10-fold increase in dioxin level was associated with a doubling of the risk of breast cancer. (*Science News*, Aug. 3, 2002)

11. In 2002 the journal *Science* reported that a study of women in Finland indicated that having sons shortened the lifespans of mothers by about 34 weeks per son, but that daughters helped to lengthen the mothers' lives. The data came from church records from the period 1640 to 1870.

12. In 2001 a report in the *Journal of the American Cancer Institute* indicated that women who work nights have a 60% greater risk of developing breast cancer. Researchers based these findings on the work histories of 763 women with breast cancer and 741 women without the disease.

13. The May 4, 2000, issue of *Science News* reported that, contrary to popular belief, depressed individuals cry no more often in response to sad situations than nondepressed people. Researchers studied 23 men and 48 women with major depression, and 9 men and 24 women with no depression. They showed the subjects a sad film about a boy whose father has died, noting whether or not the subjects cried. Women cried more often than men, but there were no significant differences between the depressed and nondepressed groups.

14. Scientists at a major pharmaceutical firm investigated the effectiveness of an herbal compound to treat the common cold. They exposed each subject to a cold virus, then gave him or her either the herbal compound or a sugar solution known to have no effect on colds. Several days later they assessed the patient's condition, using a cold severity scale ranging 0 to 5. They found no evidence of benefits associated with the compound.

15. Scientists examined the glycogen content of rats' brains at the rats' normal bedtimes and after they had been kept awake for an extra 6, 12, or 24 hours. The scientists found that glycogen was 38% lower among rats that had been sleep-deprived for 12 hours or more, and that the levels recovered during subsequent sleep. These researchers speculate that we may need to sleep in order to restore the brain's energy fuel. (*Science News*, July 20, 2002)

16. Some people who race greyhounds give the dogs large doses of vitamin C in the belief that the dogs will run faster. Investigators at the University of Florida tried three different diets in random order on each of five racing greyhounds. They were surprised to find that when the dogs ate high amounts of vitamin C they ran more slowly. (*Science News*, July 20, 2002)

17. Some people claim they can get relief from migraine headache pain by drinking a large glass of ice water. Researchers plan to enlist several people who suffer from migraines in a test. When a participant experiences a migraine headache, he or she will take a pill that may be a standard pain reliever or a placebo. Half of each group will also drink ice water. Participants will then report the level of pain relief they experience.

18. Weight is an issue for both humans and their pets. A dog food company wants to compare a new lower-calorie

food with their standard dog food to see if it's effective in helping inactive dogs maintain a healthy weight. They have found several dog owners willing to participate in the trial. The dogs have been classified as small, medium, or large breeds, and the company will supply some owners of each size of dog with one of the two foods. The owners have agreed not to feed their dogs anything else for a period of 6 months, after which the dogs' weights will be checked.

19. Athletes who had suffered hamstring injuries were randomly assigned to one of two exercise programs. Those who engaged in static stretching returned to sports activity in a mean of 37.4 days (SD = 27.6 days). Those assigned to a program of agility and trunk stabilization exercises returned to sports in a mean of 22.2 days (SD = 8.3 days). (*Journal of Orthopaedic & Sports Physical Therapy* 34 [March 2004]: 3)

20. Pew Research compared respondents to an ordinary 5-day telephone survey with respondents to a 4-month-long rigorous survey designed to generate the highest possible response rate. They were especially interested in identifying any variables for which those who responded to the ordinary survey were different from those who could be reached only by the rigorous survey.

21. **Tomatoes.** Describe a strategy to randomly split the 24 tomatoes into the three groups for the chapter's completely randomized single factor test of OptiGro fertilizer.

22. **Tomatoes II.** The chapter also described a completely randomized two-factor experiment testing OptiGro fertilizer in conjunction with two different routines for watering the plants. Describe a strategy to randomly assign the 24 tomato plants to the six treatments.

23. **Shoes.** A running shoe manufacturer wants to test the speed of its new sprinting shoe on the 100-meter dash times. The company sponsors 5 athletes who are running the 100-meter dash in the 2004 Summer Olympic games. To test the shoe, they have all 5 runners run the 100-meter dash with a competitor's shoe and then again with their new shoe. They use the difference in times as the response variable.
a) Suggest some improvements to the design.
b) Why might the shoe manufacturer not be able to generalize the results they find to all runners?

24. **Swimsuits.** A swimsuit manufacturer wants to test the speed of its newly designed suit. They design an experiment by having 6 randomly selected Olympic swimmers swim as fast as they can with their old swim suit first and then swim the same event again with the new, expensive swim suit. They'll use the difference in times as the response variable. Criticize the experiment and point out some of the problems with generalizing the results.

25. **Hamstrings.** Exercise 19 discussed an experiment to see if the time it took athletes with hamstring injuries to be able to return to sports was different depending on which of two exercise programs they engaged in.
a) Explain why it was important to assign the athletes to the two different treatments randomly.
b) There was no control group of athletes who did not participate in a special exercise program. Explain the advantage of including such a group in this experiment.
c) How might blinding have been used in this experiment?
d) One group returned to sports activity in a mean of 37.4 days (SD = 27.6 days), and the other in a mean of 22.2 days (SD = 8.3 days). Do you think this difference is statistically significant? Explain.

26. **Diet and blood pressure.** Exercise 6 reports on an experiment that showed that subjects fed the DASH diet were able to lower their blood pressure by an average of 6.7 points when compared to a group fed a "control diet." All meals were prepared by dieticians.
a) Why were the subjects randomly assigned to the diets instead of letting people pick what they wanted to eat?
b) Why were the meals prepared by dieticians?
c) Why did the researchers need the control group? If the DASH diet group's blood pressure was lower at the end of the experiment than at the beginning, wouldn't that prove the effectiveness of that diet?
d) What additional information would you want to know in order to decide whether an average reduction in blood pressure of 6.7 points was statistically significant?

27. **Mozart.** Will listening to a Mozart piano sonata make you smarter? In a 1995 study, Rauscher, Shaw, and Ky reported that when students were given a spatial reasoning section of a standard IQ test, those who listened to Mozart for 10 minutes improved their scores more than those who simply sat quietly.

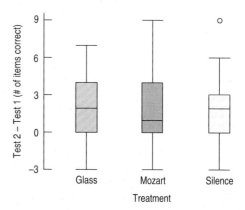

a) These researchers said the differences were statistically significant. Explain what that means in this context.

b) Steele, Bass, and Crook tried to replicate the original study. The subjects were 125 college students who participated in the experiment for course credit. Subjects first took the test. Then they were assigned to one of three groups: listening to a Mozart piano sonata, listening to music by Philip Glass, and sitting for 10 minutes in silence. Three days after the treatments, they were retested. Draw a diagram displaying the design of this experiment.

c) The boxplots show the differences in score before and after treatment for the three groups. Did the Mozart group show improvement?

d) Do you think the results prove that listening to Mozart is beneficial? Explain.

28. **More Mozart.** An advertisement selling specially designed music CDs of Mozart's music specifically because they will "strengthen your mind, heal your body, and unlock your creative spirit" claims (we *swear!*) that "In Japan, a brewery reports that their best sake is made when Mozart is played near the yeast." Suppose, just for the sake (as it were) of discussion, you wished to design an experiment to test whether this is true. Assume you have the full cooperation of the sake brewery. Specify how you would design the experiment. Indicate factors and response and how they would be measured, controlled, or randomized.

29. **Frumpies.** The makers of Frumpies, "the breakfast of rug rats," want to improve their marketing, so they consult you:

a) They first want to know what fraction of children, ages 10 to 13, like their celery-flavored cereal. What kind of study should they perform?

b) They are thinking of introducing a new flavor, maple-marshmallow Frumpies, and want to know whether children will prefer the new flavor to the old one. Design a completely randomized experiment to investigate this question.

c) They suspect that children who regularly watch the Saturday morning cartoon show starring Frump, the flying teenage warrior rabbit who eats Frumpies in every episode, may respond differently to the new flavor. How would you take that into account in your design?

30. **Full moon.** It's a common belief that people behave strangely when there's a full moon, and that as a result police and emergency rooms are busier than usual. Design a way you could find out if there is any merit to this belief. Will you use an observational study or an experiment? Why?

31. **Wine.** A 2001 Danish study published in the *Archives of Internal Medicine* casts significant doubt on suggestions that adults who drink wine have higher levels of "good" cholesterol and fewer heart attacks. These researchers followed a group of individuals born at a Copenhagen hospital between 1959 and 1961 for 40 years. Their study found that in this group the adults who drank wine were richer and better educated than those who did not.

a) What kind of study was this?

b) It is generally true that people with high levels of education and high socioeconomic status are healthier than others. How does this call into question the supposed health benefits of wine?

c) Can studies such as these prove causation (that wine helps prevent heart attacks, that drinking wine makes one richer, that being rich helps prevent heart attacks, etc.)? Explain.

32. **Swimming.** Recently, a group of adults who swim regularly for exercise were evaluated for depression. It turned out that these swimmers were less likely to be depressed than the general population. The researchers said the difference was statistically significant.

a) What does "statistically significant" mean in this context?

b) Is this an experiment or an observational study? Explain.

c) News reports claimed this study proved that swimming can prevent depression. Explain why this conclusion is not justified by the study. Include an example of a possible confounding variable.

d) But perhaps it is true. We wonder if exercise can ward off depression, and whether anaerobic exercise (like weight training) is as effective as aerobic exercise (like swimming). We find 120 volunteers not currently engaged in a regular program of exercise. Design an appropriate experiment.

33. **Dowsing.** A water dowser claims to be able to sense the presence of water using a forked stick. Suppose we wish to set up an experiment to test his ability. We get 20 identical containers, fill some with water, and ask the dowser to tell which ones are full and which empty.

a) How will we randomize this procedure?

b) The dowser correctly identifies the contents of 12 out of 20 containers. Do you think this level of success is statistically significant? Explain.

c) How many correct identifications (out of 20) would the dowser have to make to convince you that the forked stick trick works? Explain.

34. **Healing.** A medical researcher suspects that giving postsurgical patients large doses of vitamin E will speed their recovery times by helping their incisions heal more quickly. Design an experiment to test this conjecture. Be sure to identify the factors, levels, treatments, response variable, and the role of randomization.

35. **Reading.** Some schools teach reading using phonics (the sounds made by letters) and others using whole language (word recognition). Suppose a school district wants to

know which method works better. Suggest a design for an appropriate experiment.

36. Gas mileage. Do cars get better gas mileage with premium instead of regular unleaded gasoline? While it might be possible to test some engines in a laboratory setting, we'd rather use real cars and real drivers in real day-to-day driving, so we get 20 volunteers. Design the experiment.

37. Weekend deaths. A study published in the *New England Journal of Medicine* (Aug. 2001) suggests that it's dangerous to enter a hospital on a weekend. During a 10-year period, researchers tracked over 4 million emergency admissions to hospitals in Ontario, Canada. Their findings revealed that patients admitted on weekends had a much higher risk of death than those who went to the emergency room on weekdays.
a) The researchers said the difference in death rates was "statistically significant." Explain in this context what that means.
b) What kind of study was this? Explain.
c) If you think you're quite ill on a Saturday, should you wait until Monday to seek medical help? Explain.
d) Suggest some possible explanations for this troubling finding.

38. Shingles. A research doctor has discovered a new ointment that she believes will be more effective than the current medication in the treatment of shingles (a painful skin rash). Eight patients have volunteered to participate in the initial trials of this ointment. You are the statistician hired as a consultant to help design a completely randomized experiment.
a) Describe how you will conduct this experiment.
b) Suppose the eight patients' last names start with the letters A to H. Using the random numbers listed below, show which patients you will assign to each treatment. Explain your randomization procedure clearly.

41098   18329   78458   31685   55259

c) Can you make this experiment double-blind? If so, explain how.
d) The initial experiment revealed that males and females may respond differently to the ointment. Further testing of the drug's effectiveness is now planned, and many patients have volunteered. What changes in your first design, if any, would you make for this second stage of testing?

39. Beetles. Hoping to learn how to control crop damage by a certain species of beetle, a researcher plans to test two different pesticides in small plots of corn. A few days after application of the chemicals, he'll check the number of beetle larvae found on each plant. The researcher wants to know if either pesticide works, and whether

there is a significant difference in effectiveness between them. Design an appropriate experiment.

40. SAT Prep. Can special study courses actually help raise SAT scores? One organization says that the 30 students they tutored achieved an average gain of 60 points when they retook the test.
a) Explain why this does not necessarily prove that the special course caused the scores to go up.
b) Propose a design for an experiment that could test the effectiveness of the tutorial course.
c) Suppose you suspect that the tutorial course might be more helpful for students whose initial scores were particularly low. How would this affect your proposed design?

41. Safety switch. An industrial machine requires an emergency shutoff switch that must be designed so that it can be easily operated with either hand. Design an experiment to find out whether workers will be able to deactivate the machine as quickly with their left hands as with their right hands. Be sure to explain the role of randomization in your design.

42. Washing clothes. A consumer group wants to test the effectiveness of a new "organic" laundry detergent and make recommendations to customers about how to best use the product. They intentionally get grass stains on 30 white T-shirts in order to see how well the detergent will clean them. They want to try the detergent in cold water and in hot water on both the "regular" and "delicates" wash cycles. Design an appropriate experiment, indicating the number of factors, levels, and treatments. Explain the role of randomization in your experiment.

43. Skydiving, anyone? A humor piece published in the *British Medical Journal* notes ("Parachute use to prevent death and major trauma related to gravitational challenge: systematic review of randomized control trials," Gordon, Smith, and Pell, *BMJ*, 2003:327) that we can't tell for sure whether parachutes are safe and effective because there has never been a properly randomized, double-blind, placebo-controlled study of parachute effectiveness in sky diving. (Yes, this is the sort of thing statisticians find funny. . . .) Suppose you were designing such a study:
a) What is the factor in this experiment?
b) What experimental units would you propose?[6]
c) Explain what would serve as a placebo for this study.
d) What would the treatments be?
e) What would be the response variable for such a study?
f) What sources of variability would you control?
g) How would you randomize this "experiment"?
h) How would you make the experiment double-blind?

---

[6] Don't include your Statistics instructor!

just checking

Answers

1. a) The factor was type of treatment for peptic ulcer.
   b) The response variable could be a measure of relief from gastric ulcer pain or an evaluation by a physician of the state of the disease.
   c) Treatments would be gastric freezing and some alternative control treatment.
   d) Treatments should be assigned randomly.
   e) No. The Web site reports "lack of effectiveness," indicating that no large differences in patient healing were noted.

2. a) Neither the patient who received the treatment nor the doctor who evaluated them to see if they had improved knew what treatment they had received.
   b) The placebo is needed to accomplish blinding. Best alternatives would be using body-temperature liquid rather than the freezing liquid.
   c) The researchers should block the subjects by the length of time they had had the ulcer, then randomly assign subjects in each block to the freezing and placebo groups.

REVIEW OF PART **III** **Gathering Data**

QUICK REVIEW

Before you can make a boxplot, calculate a mean, describe a distribution, or fit a line, you must have meaningful data to work with. Getting good data is essential to any investigation. No amount of clever analysis can make up for badly collected data. Here's a brief summary of the key concepts and skills:

▶ The way you gather data depends both on what you want to discover and on what is practical.

▶ To get some insight into what might happen in a real situation, model it with a **simulation** using random numbers.

▶ To answer questions about a target population, collect information from a sample with a **survey** or poll.

- Choose the sample randomly. Random sampling designs include simple, stratified, systematic, cluster, and multistage.
- A simple random sample draws without restriction from the entire target population.
- When there are subgroups within the population that may respond differently, use a stratified sample.
- Avoid bias, a systematic distortion of the results. Sample designs that allow undercoverage or response bias and designs such as voluntary response or convenience samples don't faithfully represent the population.
- Samples will naturally vary one from another.

This sample-to-sample variation is called sampling error. Each sample only approximates the target population.

▶ **Observational studies** collect information from a sample drawn from a target population.

- Retrospective studies examine existing data. Prospective studies identify subjects in advance, then follow them to collect data as the data are created, perhaps over many years.
- Observational studies can spot associations between variables, but cannot establish cause and effect. It's impossible to eliminate the possibility of lurking or confounding variables.

▶ To see how different treatments influence a response variable, design an **experiment.**

- Assign subjects to treatments randomly. If you don't assign treatments randomly, your experiment is not likely to yield valid results.
- Control known sources of variation as much as possible. Reduce variation that cannot be controlled by using blocking, if possible.
- Replicate the experiment, assigning several subjects to each treatment level.
- If possible, replicate the entire experiment with an entirely different collection of subjects.
- A well-designed experiment can provide evidence that changes in the factors cause changes in the response variable.

Now for more opportunities to review these concepts and skills . . .

REVIEW EXERCISES

1–18. What design? Analyze the design of each research example reported. Is it a sample, a study, or an experiment? If a sample, what are the population, the parameter of interest, and the sampling procedure? If a study, was it retrospective or prospective? If an experiment, describe the factors, treatments, randomization, response variable, and any blocking, matching, or blinding that may be present. In each, what kind of conclusions can be reached?

1. Researchers identified 242 children in the Cleveland area who had been born prematurely (at about 29 weeks). They examined these children at age 8 and again at age 20, comparing them to another group of 233 children not born prematurely. According to their report, published

in the *New England Journal of Medicine*, the "preemies" engaged in significantly less risky behavior than the others. Differences between the groups showed up in the use of alcohol and marijuana, conviction of crimes, and teenage pregnancy.

2. The journal *Circulation* reported that among 1900 people who had heart attacks, those who drank an average of 19 cups of tea a week were 44% more likely than nondrinkers to survive at least 3 years after the attack.

3. Researchers at the Purina Pet Institute studied Labrador retrievers for evidence of a relationship between diet and longevity. At 8 weeks of age, 2 puppies of the same gender and weight were randomly assigned to one of two groups—a total of 48 dogs in all. One group was allowed to eat all they wanted, while the other group was fed a low-calorie diet (about 75% as much as the others). The median lifespan of dogs fed the restricted diet was 22 months longer than that of other dogs. (*Science News* 161, no. 19)

4. Radon is a radioactive gas found in some homes that poses a health risk to residents. To assess the level of contamination in their area, a county health department wants to test a few homes. If the risk seems high, they will publicize the results to emphasize the need for home testing. Officials plan to use the local property tax list to randomly choose 25 homes from various areas of the county.

5. Almost 90,000 women participated in a 16-year study of the role of the vitamin folate in preventing colon cancer. Some of the women had family histories of colon cancer in close relatives. In this at-risk group the incidence of colon cancer was cut in half among those who maintained a high folate intake. No such difference was observed in those with no family-based risk. (*Science News,* Feb. 9, 2002)

6. In a study appearing in the journal *Science,* a research team reports that plants in southern England are flowering earlier in the spring. Records of the first flowering dates for 385 species over a period of 47 years indicate that flowering has advanced an average of 15 days per decade, an indication of climate warming, according to the authors.

7. Fireworks manufacturers face a dilemma. They must be sure that the rockets work properly, but test firing a rocket essentially destroys it. On the other hand, not testing the product leaves open the danger that they sell a bunch of duds, leading to unhappy customers and loss of future sales. The solution, of course, is to test a few of the rockets produced each day, assuming that if those tested work properly the others are ready for sale.

8. Can makeup damage fetal development? Many cosmetics contain a class of chemicals called phthalates. Studies that exposed some laboratory animals to these chemicals found a heightened incidence of damage to male reproductive systems. Since traces of phthalates are found in the urine of women who use beauty products, there is growing concern that they may present a risk to male fetuses. (*Science News,* July 20, 2002)

9. Can long-term exposure to strong electromagnetic fields cause cancer? Researchers in Italy tracked down 13 years of medical records for people living near Vatican Radio's powerful broadcast antennas. A disproportionate share of the leukemia cases occurred among men and children who lived within 6 kilometers of the antennas. (*Science News,* July 20, 2002)

10. Some doctors have expressed concern that men who have vasectomies seemed more likely to develop prostate cancer. Medical researchers used a national cancer registry to identify 923 men who had had prostate cancer and 1224 men of similar ages who had not. Roughly one quarter of the men in each group had undergone a vasectomy, many more than 25 years before the study. The study's authors concluded that there is strong evidence that having the operation presents no long-term risk for developing prostate cancer. (*Science News,* July 20, 2002)

11. Researchers investigating appetite control as a means of losing weight found that female rats ate less and lost weight after injections of the hormone leptin, while male rats responded better to insulin. (*Science News,* July 20, 2002)

12. An artisan wants to create pottery that has the appearance of age. He prepares several samples of clay with four different glazes and test fires them in a kiln at three different temperature settings.

13. Tests of gene therapy on laboratory rats have raised hopes of stopping the degeneration of tissue that characterizes chronic heart failure. Researchers at the University of California, San Diego, used hamsters with cardiac disease, randomly assigning 30 to receive the gene therapy and leaving the other 28 untreated. Five weeks after treatment the gene therapy group's heart muscles stabilized, while those of the untreated hamsters continued to weaken. (*Science News,* July 27, 2002)

14. Researchers at the University of Bristol (England) investigated reasons why different species of birds begin to sing at different times in the morning. They captured and examined birds of 57 species at seven different sites. They measured the diameter of the birds' eyes and also recorded the time of day at which each species began to sing. These researchers reported a strong relationship between eye diameter and time of singing, saying that birds with bigger eyes tended to sing earlier. (*Science News,* 161, no. 16 [2002])

15. An orange juice processing plant will accept a shipment of fruit only after several hundred oranges selected from various locations within the truck are carefully inspected. If too many of those checked show signs of unsuitability for juice (bruised, rotten, unripe, etc.) the whole truckload is rejected.

16. A soft drink manufacturer must be sure the bottle caps on the soda are fully sealed and will not come off easily. Inspectors pull a few bottles off the production line at regular intervals and test the caps. If they detect any problems, they will stop the bottling process to adjust or repair the machine that caps the bottles.

17. Physically fit people are less likely to die of cancer. A report in the May 2002 issue of *Medicine and Science in Sports and Exercise* followed 25,892 men aged 30 to 87 for 10 years. The most physically fit men had a 55% lower risk of death from cancer than the least fit group.

18. Does the use of computer software in Introductory Statistics classes lead to better understanding of the concepts? A professor teaching two sections of Statistics decides to investigate. She teaches both sections using the same lectures and assignments, but gives one class statistics software to help them with their homework. The classes take the same final exam, and graders do not know which students used computers during the semester. The professor is also concerned that students who have had calculus may perform differently from those who have not, so she plans to compare software vs. no-software scores separately for these two groups of students.

19. **Point spread.** When taking bets on sporting events, bookmakers often include a "point spread" that awards the weaker team extra points. In theory this makes the outcome of the bet a toss-up. Suppose a gambler places a $10 bet and picks the winners of five games. If he's right about fewer than three of the games, he loses. If he gets three, four, or all five correct, he's paid $10, $20, or $50, respectively. Estimate the amount such a bettor might expect to lose over many weeks of gambling.

20. **The lottery.** Many people spend a lot of money trying to win huge jackpots in state lotteries. Let's play a simplified version using only the numbers from 1 to 20. You bet on three numbers. The state picks five winning numbers. If your three are all among the winners, you are rich!
   a) Simulate repeated plays. How long did it take you to win?
   b) In real lotteries there are many more choices (often 54), and you must match all five winning numbers. Explain how these changes affect your chances of hitting the jackpot.

21. **Everyday randomness.** Aside from casinos, lotteries, and games, there are other situations you encounter in which something is described as "random" in some way. Give three different examples. Describe how randomness is (or is not) achieved in each.

22. **Cell phone risks.** Researchers at the Washington University School of Medicine randomly placed 480 rats into one of three chambers containing radio antennas. One group was exposed to digital cell phone radio waves, the second to analog cell phone waves, and the third group to no radio waves. Two years later the rats were examined for signs of brain tumors. In June 2002 the scientists said that differences among the three groups were not statistically significant.
   a) Is this a study or an experiment? Explain.
   b) Explain in this context what "statistically significant" means.

   c) Comment on the fact that this research was supported by funding from Motorola, a manufacturer of cell phones.

23. **Tips.** In restaurants, servers rely on tips as a major source of income. Does serving candy after the meal produce larger tips? To find out, two waiters determined randomly whether or not to give candy to 92 dining parties. They recorded the sizes of the tips, and reported that guests getting candy tipped an average of 17.8% of the bill, compared with an average tip of only 15.1% from those who got no candy. ("Sweetening the Till: The Use of Candy to Increase Restaurant Tipping." *Journal of Applied Social Psychology* 32, no. 2 [2002]: 300–309)
   a) Was this an experiment or an observational study? Explain.
   b) Is it reasonable to conclude that the candy caused guests to tip more? Explain.
   c) The researchers said the difference was statistically significant. Explain in this context what that means.

24. **Tips, take 2.** In another experiment to see if getting candy after a meal would induce customers to leave a bigger tip, a waitress randomly decided what to do with 80 dining parties. Some parties received no candy, some just one piece, and some two pieces. Others initially got just one piece of candy, and then the waitress suggested that they take another piece. She recorded the tips received, finding that, in general, the more candy the higher the tip, but the highest tips (23%) came from the parties who got one piece and then were offered more. ("Sweetening the Till: The Use of Candy to Increase Restaurant Tipping." *Journal of Applied Social Psychology* 32, no. 2 [2002]: 300–309)
   a) Diagram this experiment.
   b) How many factors are there? How many levels?
   c) How many treatments are there?
   d) What is the response variable?
   e) Did this experiment involve blinding? Double blinding?
   f) In what way might the waitress, perhaps unintentionally, have biased the results?

25. **Cloning.** In September 1998, *USA Weekend* magazine asked, "Should humans be cloned?" Readers were invited to register a "Yes" or "No" answer by calling one of two different 900 numbers. Based on 38,023 responses, the magazine reported that "9 out of 10 readers oppose cloning."
   a) Explain why you think the conclusion is not justified. Describe the types of bias that may be present.
   b) Reword the question in a way that you think might create a more positive response.

26. **Laundry.** An experiment to test a new laundry detergent, SparkleKleen, is being conducted by a consumer advocate group. They would like to compare its performance with that of a laboratory standard detergent they have used in previous experiments. They can stain 16 swatches of cloth with 2 tsp of a common staining compound and then use a well-calibrated optical scanner to detect the amount of the stain that is left after washing with detergent. To save time in the experiment, several suggestions have been made. Comment on the possible merits and drawbacks of each one.

a) Since data for the laboratory standard detergent are already available from previous experiments, for this experiment wash all 16 swatches with SparkleKleen, and compare the results with the previous data.

b) Use both detergents with eight separate runs each, but to save time, use only a 10-second wash time with very hot water.

c) To ease bookkeeping, run successively all of the standard detergent washes on eight swatches, then run all of the SparkleKleen washes on the other eight swatches.

d) Rather than run the experiment, use data from the company that produced SparkleKleen, and compare them with past data from the standard detergent.

27. When to Stop? You play a game that involves rolling a die. You can roll as many times as you want, and your score is the total for all the rolls. But . . . if you roll a 6 your score is 0 and your turn is over. What might be a good strategy for a game like this?

a) One of your opponents decides to roll 4 times, then stop (hoping not to get the dreaded 6 before then). Use a simulation to estimate his average score.

b) Another opponent decides to roll until she gets at least 12 points, then stop. Use a simulation to estimate her average score.

c) Propose another strategy that you would use to play this game. Using your strategy, simulate several turns. Do you think you would beat the two opponents?

28. Rivets. A company that manufactures rivets believes the shear strength of the rivets they manufacture follows a Normal model with a mean breaking strength of 950 pounds and a standard deviation of 40 pounds.

a) What percentage of rivets selected at random will break when tested under a 900-pound load?

b) You're trying to improve the rivets and want to examine some that fail. Use a simulation to estimate how many rivets you might need to test in order to find three that fail at 900 pounds (or below).

29. Homecoming. A college Statistics class conducted a survey concerning community attitudes about the college's large homecoming celebration. That survey drew its sample in the following manner: Telephone numbers were generated at random by selecting one of the local telephone exchanges (first three digits) at random and then generating a random four-digit number to follow the exchange. If a person answered the phone and the call was to a residence, then that person was taken to be the subject for interview. (Undergraduate students and those under voting age were excluded, as was anyone who could not speak English.) Calls were placed until a sample of 200 eligible respondents had been reached.

a) Did every telephone number that could possibly occur in that community have an equal chance of being generated?

b) Did this method of generating telephone numbers result in a simple random sample (SRS) of local residences? Explain.

c) Did this method generate a SRS of local voters? Explain.

d) Did this method generate an unbiased sample of households? Explain.

30. Youthful appearance. *Readers' Digest* reported results of several surveys that asked graduate students to examine photographs of men and women and try to guess their ages. Researchers compared these guesses with the number of times the people in the pictures reported having sexual intercourse. It turned out that those who had been more sexually active were judged as looking younger, and that the difference was described as "statistically significant." Psychologist David Weeks, who compiled the research, speculated that lovemaking boosts hormones that "reduce fatty tissue and increase lean muscle, giving a more youthful appearance."

a) What does "statistically significant" mean in this context?

b) Explain in statistical terms why you might be skeptical about Dr. Weeks' conclusion. Propose an alternative explanation for these results.

31. Smoking and Alzheimer's. Medical studies indicate that smokers are less likely to develop Alzheimer's disease than people who never smoked.

a) Does this prove that smoking may offer some protection against Alzheimer's? Explain.

b) Offer an alternative explanation for this association.

c) How would you conduct a study to investigate this issue?

32. Antacids. A researcher wants to compare the performance of three types of antacid in volunteers suffering from acid reflux disease. Because men and women may react differently to this medication, the subjects are split into two groups, by gender. Subjects in each group are randomly assigned to take one of the antacids or to take a sugar pill made to look the same. The subjects will rate their level of discomfort 30 minutes after eating.

a) What kind of design is this?

b) The experiment uses volunteers rather than a random sample of all people suffering from acid reflux disease. Does this make the results invalid? Explain.

c) How may the use of the placebo confound this experiment? Explain.

33. Sex and violence. Does the content of a television program impact viewers' memory of the products advertised in commercials? Design an experiment to compare the ability of viewers to recall brand names of items featured in commercials during programs with violent content, sexual content, or neutral content.

34. Pubs. In England, a Leeds University researcher said that the local watering hole's welcoming atmosphere helps men get rid of the stresses of modern life and is vital for their psychological well-being. Author of the report, Dr. Colin Gill, said rather than complain, women should encourage men to "pop out for a swift half." "Pub-time allows men to bond with friends and colleagues," he said. "Men need break-out time as much as women and are mentally healthier for it." Gill added that men might feel unfulfilled or empty if they had not been to the pub for a week. The report, commissioned by alcohol-free beer brand Kaliber, surveyed 900 men on their reasons for going to the pub. More than 40% said they went for the conversation, with relaxation and a friendly atmosphere being the other most common reasons. Only 1 in 10 listed alcohol as the overriding reason.

Let's examine this news story from a statistical perspective.
a) What are the W's: *Who, What, When, Where, Why?*
b) What population does the researcher think the study applies to?
c) What is the most important thing about the selection process that the article does *not* tell us?
d) How do *you* think the 900 respondents were selected? (Name a method of drawing a sample that is likely to have been used.)
e) Do you think the report that only 10% of respondents listed alcohol as an important reason for going to the pub might be a biased result? Why?

35. Age and party. The Gallup Poll conducted a representative telephone survey during the first quarter of 1999. Among its reported results was the following table concerning the preferred political party affiliation of respondents and their ages.

	Party			
	Republican	**Democratic**	**Independent**	**Total**
18–29	241	351	409	1001
30–49	299	330	370	999
50–64	282	341	375	998
65+	279	382	343	1004
Total	1101	1404	1497	**4002**

(Age labels down the left side)

a) What sampling strategy do you think the pollsters used? Explain.

b) What percentage of the people surveyed were Democrats?
c) Do you think this is a good estimate of the percentage of voters in the United States who are registered Democrats? Why or why not?
d) In creating this sample design, what question do you think the pollsters were trying to answer?

36. Bias? Political analyst Michael Barone has written that "conservatives are more likely than others to refuse to respond to polls, particularly those polls taken by media outlets that conservatives consider biased" (*The Weekly Standard*, March 10, 1997). The Pew Research Foundation tested this assertion by asking the same questions in a national survey run by standard methods and in a more rigorous survey that was a true SRS with careful follow-up to encourage participation. The response rate in the "standard survey" was 42%. The response rate in the "rigorous survey" was 71%.

a) What kind of bias does Barone claim may exist in polls?
b) What is the population for these surveys?
c) On the question of political position, the Pew researchers report the following table:

	Standard Survey	Rigorous Survey
Conservative	37%	35%
Moderate	40%	41%
Liberal	19%	20%

What makes you think these results are incomplete?
d) The Pew researchers report that differences between opinions expressed on the two surveys were not statistically significant. Explain what "statistically significant" means in this context.

37. Save the grapes. Vineyard owners have problems with birds that like to eat the ripening grapes. Grapes damaged by birds cannot be used for winemaking (or much of anything else). Some vineyards use scarecrows to try to keep birds away. Others use netting that covers the plants. Owners really would like to know if either method works and, if so, which one is better. One owner has offered to let you use his vineyard this year for an experiment. Propose a design. Carefully indicate how you would set up the experiment, specifying the factor(s) and response variable.

38. Bats. It's generally believed that baseball players can hit the ball farther with aluminum bats than with the traditional wooden ones. Is that true? And, if so, how much farther? Players on your local high school baseball team have agreed to help you find out. Design an appropriate experiment.

39. Knees. Research reported in the spring of 2002 cast doubt on the effectiveness of arthroscopic knee surgery for patients with arthritis. Patients suffering from arthritis pain who volunteered to participate in the study were randomly divided into groups. One group received arthroscopic knee surgery. The other group underwent "placebo surgery" during which incisions were made in their knees, but no surgery was actually performed. Follow-up evaluations over a period of 2 years found that differences in the amount of pain relief experienced by the two groups were not statistically significant.

a) Why did the researchers feel it was necessary to have some of the patients undergo "placebo surgery"?

b) Because patients had to consent to participate in this experiment, the subjects were essentially self-selected—a kind of voluntary response group. Explain why that does not invalidate the findings of the experiment.

c) What does "statistically significant" mean in this context?

40. NBA draft lottery. Professional basketball teams hold a "draft" each year in which they get to pick the best available college and high-school players. In an effort to promote competition, teams with the worst records get to pick first, theoretically allowing them to add better players. To combat the fear that teams with no chance to make the playoffs might try to get better draft picks by intentionally losing late-season games, the NBA's Board of Governors adopted a weighted lottery system in 1990. Under this system the 11 teams that did not make the playoffs were eligible for the lottery. The NBA prepared 66 cards, each naming one of the teams. The team with the worst win-loss record was named on 11 of the cards, the second-worst team on 10 cards, and so on, with the team having the best record among the nonplayoff clubs getting only one chance at having the first pick. The cards were mixed, then drawn randomly to determine the order in which the teams could draft players. (Since 1995, 13 teams have been involved in the lottery, using a complicated system with 14 numbered Ping-Pong balls drawn in groups of four.) Suppose there are two exceptional players available in this year's draft and your favorite team had the third-worst record. Use a simulation to find out how likely it is that your team gets to pick first or second. Describe your simulation carefully.

41. Security. There are 20 first-class passengers and 120 coach passengers scheduled on a flight. In addition to the usual security screening, 10% of the passengers will be subjected to a more complete search.

a) Describe a sampling strategy to randomly select those to be searched.

b) Here is the first-class passenger list and a set of random digits. Select two passengers to be searched, carefully demonstrating your process.

65436   71127   04879   41516   20451   02227   94769   23593

Bergman	Cox	Fontana	Perl
Bowman	DeLara	Forester	Rabkin
Burkhauser	Delli-Bovi	Frongillo	Roufaiel
Castillo	Dugan	Furnas	Swafford
Clancy	Febo	LePage	Testut

c) Explain how you would use a random number table to select the coach passengers to be searched.

42. Profiling? Among the 20 first-class passengers on the flight described in Exercise 41 there were four businessmen from the Middle East. Two of them were the two passengers selected to be searched. They complained of profiling, but the airline claims that the selection was random. What do you think? Support your conclusion with a simulation.

43. Par 4. In theory, a golfer playing a par-4 hole tees off, hitting the ball in the fairway, then hits an approach shot onto the green. The first putt (usually long) probably won't go in, but the second putt (usually much shorter) should. Sounds simple enough, but how many strokes might it really take? Use a simulation to estimate a pretty good golfer's score based on these assumptions:

• The tee shot hits the fairway 70% of the time.
• A first approach shot lands on the green 80% of the time from the fairway, but only 40% of the time otherwise.
• Subsequent approach shots land on the green 90% of the time.
• The first putt goes in 20% of the time, and subsequent putts go in 90% of the time.

44. The back nine. Use simulations to estimate more golf scores, similar to the procedure in Exercise 43.

a) On a par 3, the golfer hopes the tee shot lands on the green. Assume that the tee shot behaves like the first approach shot described in Exercise 43.

b) On a par 5, the second shot will reach the green 10% of the time and hit the fairway 60% of the time. If it does not hit the green, the golfer must play an approach shot as described in Exercise 43.

c) Create a list of assumptions that describe your golfing ability and then simulate your score on a few holes. Explain your simulation clearly.

Randomness and Probability

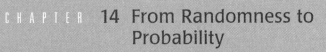

14

From Randomness to Probability

What's the difference between randomness and chaos? At first glance they might seem to be the same. Neither of their outcomes can be anticipated with certainty, but random phenomena have another important feature. In the long run, they settle down in a way that's actually consistent and predictable. Chaotic processes don't do this. It's this property of randomness that enables us to do Statistics. We'll show what we mean in this chapter and the next as we tame randomness and turn it into something that we can use.

Dealing with Random Phenomena

Every day you drive through the intersection at College and Main. Even though it may seem that the light is always red when you get there, you know this can't really be true. In fact, if you try really hard, you can recall just sailing through the green light once in a while.

What's random here? The light itself is governed by a timer. Its pattern isn't haphazard. In fact, the light may even be red at precisely the same times each day. It's the pattern of *your driving* that is random. No, we're certainly not insinuating that you can't keep the car on the road. At the precision level of the 30 seconds or so that the light spends being red or green, the time you arrive at the light *is random*. Even if you try to leave your house at exactly the same time every day, whether the light is red or green as *you* reach the intersection is a **random phenomenon.**

Is the color of the light completely unpredictable? When you stop to think about it (maybe while waiting for the green light), it's clear that we do expect some kind of *regularity* in your long-run experience. Some *fraction* of the time the light will be red as you get to the intersection. How can we figure out what that fraction is?

You might record what happens at the intersection each day and graph the *accumulated percentage* of red lights like this:

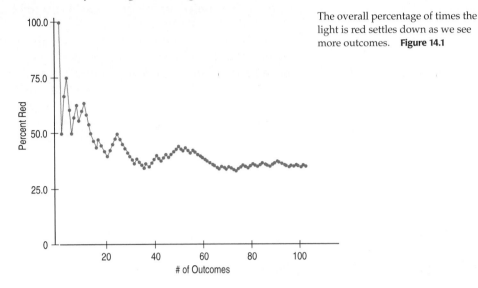

The overall percentage of times the light is red settles down as we see more outcomes.   **Figure 14.1**

Day	Light	% Red
1	Red	100
2	Green	50
3	Red	66.7
4	Red	75
5	Green	60
6	Green	50
⋮	⋮	⋮

The first day you recorded the light it was red. Then on the next five days, it was in turn green, then red again, then red, green, and green. If we plot the percentage red against the day, the graph would start at 100% because the first time the light was red (1 out of 1, for 100%). Then the next day it was green, so the accumulated percentage dropped to 50% (1 out of 2). The third day it was red again (2 out of 3, or 67% red), then red (3 out of 4, or 75%), then green twice in a row (3 out of 5, for 60% red, and then 3 out of 6, for 50%), and so on. As you collect a new data value for each day, each new datum becomes a smaller and smaller *fraction* of the accumulated experience, so, in the long run, the graph settles down. As it settles down, we can see that, in fact, the light is red only about 35% of the time.

Probability

A S **What Is Probability?**
The best way to get a feeling for probabilities is to experiment with them. We'll use this random outcomes tool many more times, so this is a good chance to try it out.

A **phenomenon** consists of **trials.** Each **trial** has an **outcome. Outcomes** combine to make **events.**

If you were asked for the *probability* that you would be stopped at the traffic light at College and Main, you might say that it was 35%. You would base that on knowing (from your data collection) that, in the long run, the fraction of the time the light is red is 35%. We could also say that the long-run *relative frequency* of red lights settles down to about 35%.

That's exactly what we mean by probability. The **probability** of an event is its long-run relative frequency.

Although we may not be able to predict a *particular* individual outcome, such as how a flipped coin will land, we're reasonably sure that a fair coin will fall heads about half the time *in the long run.* Of course, in the real world, we usually can't see the "true" proportions hiding behind the random outcomes. This forces us to reason backwards from the outcomes we observe to estimate what the probabilities of those outcomes must have been, as we did with the traffic light.

When talking about long-run behavior, it helps to define our terms. For any random phenomenon, each attempt, or **trial,** generates an **outcome.** Something happens on each trial, and we call whatever happens the outcome. These outcomes are *individual* possibilities, such as the number we see on top when we roll a die. Often, instead of individual possibilities, we want to talk about *combinations* of outcomes such as "The number on the die is less than 4 (that is, it is 1, 2, or 3)." We call such a combination of outcomes an **event.**

In order to think about what happens with a series of trials, it really simplifies things if the individual trials are **independent.** Roughly speaking, this means that the outcome of one trial doesn't influence or change the outcome of another. (In Chapter 3, we called two *variables* independent for similar reasons.) For example, if when you drove to work there was a friend right behind you in another car, you couldn't just add his observations to yours. Most of the time, his experience will just duplicate yours. Except for those few times when the light turns red just after you enter the intersection, the light will be the same for him as for you. So his observations would not be independent of yours. Because they're almost repeats of your observations, they wouldn't supply much more information. In order for us to make statements about the long-run behavior of random phenomena, the trials have to be independent.

The Law of Large Numbers

"For even the most stupid of men . . . is convinced that the more observations have been made, the less danger there is of wandering from one's goal."
—Jacob Bernoulli, 1713, discoverer of the LLN

Don't let yourself think that there's a Law of Averages that promises short-term compensation for recent deviations from expected behavior. A belief in such a "Law" can lead to money lost in gambling and to poor business decisions.

Do random phenomena always behave so well? If we're going to use long-run relative frequency as our definition of probability, we should first be sure that it will, in fact, get closer to a particular value. Could it be that the probability just bounces back and forth between two values forever, never settling on just one number? Fortunately, a principle called the **Law of Large Numbers** (LLN) gives us the guarantee that we need. It says that the long-run *relative frequency* of repeated independent events settle down to the *true* probability as the number of trials increases.

Although the LLN wasn't proven until the 18th century, everyone expects the kind of long-run regularity that the Law describes from everyday experience. If the light at College and Main is set by the traffic engineers to be red 35% of the time for the drivers on College Avenue, then no one is surprised to find out that in the long run, when you arrive at the intersection from College Avenue, *you'll* encounter a red light 35% of the time.

Yet the Law is often misunderstood because the idea of the *long run* is hard to grasp. Many gamblers believe, for example, that a number that has not come up on the roulette wheel or in a lottery for a long time, is "due" to occur—and they give the Law of Large Numbers as their reason. (Well, they usually call it the "Law of Averages" or some similar term, but the LLN is what they really mean.) But the LLN doesn't apply to short-run behavior. The problem is that probabilities even out *only in the long run.* And (according to the LLN) the long run must be *infinitely* long to give them enough time to even out.

In fact, the so-called Law of Averages doesn't exist at all. The common (mis)understanding of this "Law" is that random phenomena are supposed to *compensate* somehow for whatever happened in the past. So if recent results have fallen to one side of what is expected, somehow results on the other side are "due" because the results must "average out" to the right value. Is a .300 hitter in baseball who has struck out the last 6 times *due* for a hit his next time up? If you've just flipped a fair coin 5 heads in a row, is the next flip more likely to be tails because the coin *owes us* a tail?

The Law of Averages in Everyday Life

"Dear Abby: My husband and I just had our eighth child. Another girl, and I am really one disappointed woman. I suppose I should thank God she was healthy, but, Abby, this one was supposed to have been a boy. Even the doctor told me that the law of averages was in our favor 100 to one." (Abigail Van Buren, 1974. Quoted in Karl Smith, *The Nature of Mathematics.* 6th ed. Pacific Grove, CA: Brooks/Cole, 1991, p. 589)

A S **Multiple Discrete Outcomes.** The world isn't all heads or tails. Experiment with an event with 4 random alternative outcomes.

No. This is not the way random phenomena work. The coin can't *remember* what happened and make things come out right. In fact, if you flipped a fair coin several thousand times, you would find lots of long streaks of all heads. And if we looked at *all* streaks of 5 (or 6 or 10 . . .) heads, we'd see that, on the average, the next flip is *just as likely to be heads or tails* even after a streak. The Law of Large Numbers promises that given a very large number of trials (a *long run*), the distribution of subsequent results will *eventually* overwhelm any recent drift away from what is expected. The long run is a long time.

If a fair coin has landed heads 6 times in a row, is the next flip more likely to be a tail? Let's do a thought experiment. Suppose I flip a quarter 6 times and it comes up heads all 6 times. Now I spend it and you happen to get it in change from the candy machine. When you start flipping the coin, do you expect a run of tails? Does the coin "owe" some tails? Of course not. Each flip is a new flip. The coin doesn't "remember" what it gave me and so it can't "owe" you any particular outcomes. Of course, if you continue to flip the coin 100 times, you'd expect to get about 50 heads and 50 tails. So even if you knew of the earlier run of 6 heads, you'd never notice them; the totals would be about 56 heads and 50 tails—nothing unusual about that. The lesson of the LLN is that random processes don't need to compensate in the *short* run to get back to the right long-run probabilities. If the probabilities don't change and the events are independent, the probability of the next trial is *always* the same, no matter what has happened up to then. There is *no* Law of Averages for short runs.

● Keno and the Law of Averages Of course, sometimes an apparent drift from what we expect means that the probabilities are, in fact, *not* what we thought. If you get 10 heads in a row, maybe the coin has heads on both sides!

Keno is a simple casino game in which numbers from 1 to 80 are chosen. The numbers, as in most lottery games, are supposed to be equally likely. Payoffs are made depending on how many of those numbers you match on your card. A group of graduate students from a Statistics department decided to take a field trip to Reno. They (*very* discreetly) wrote down the outcomes of the games for a couple of days, then drove back to test whether the numbers were, in fact, equally likely. It turned out that some numbers were *more likely* to come up than others. Rather than bet on the Law of Averages and put their money on the numbers that were "due," the students put their faith in the LLN—and all their (and their friend's) money on the numbers that had come up before. After they pocketed more than $50,000, they were escorted off the premises and invited never to show their faces in that casino again. ●

① One common proposal for beating the lottery is to note which numbers have come up lately, eliminate those from consideration, and bet on numbers that have not come up for a long time. Proponents of this method argue that in the long run, every number should be selected equally often, so those that haven't come up are due. Explain why this is faulty reasoning.

Probability

Now that we know, thanks to the Law of Large Numbers, that relative frequencies settle down in the long run, we can officially give the name **probability** to that value. If the relative frequency of red lights settles down to 35%, we say that the *probability* of a red light is 0.35, and write

$$P(Red) = 0.35$$

We can't record more "red lights" than the number of times we hit the intersection (or fewer than none) so our probability must be a value between 0 and 1:

$$0 \le P \le 1.$$

A probability of zero indicates impossibility. A probability of one indicates certainty. Remember that we said "in the long run." We may have to wait infinitely long to be sure that an event is impossible or certain.

Equally Likely Outcomes

Probability was first studied by a group of French mathematicians interested in games of chance.[1] To make things simple, they started by looking at games in which all the possible outcomes were *equally likely*. It's easy to think of situations where this is true, especially in gambling. It's equally likely to get any one of six outcomes from the roll of a fair die. Any of the 52 cards is equally likely to be picked from a well-shuffled deck. Each slot of a roulette wheel is equally likely (or at least it *should* be).

When outcomes are equally likely, the probability of their occurrence is easy to compute—it's just 1 divided by the number of possible outcomes. So, the probability of rolling a 3 with a fair die is 1/6. The probability of picking the ace of spades from the top of a well-shuffled deck is 1/52.

But don't get trapped into thinking that random events are always equally likely. The chance of winning a lottery—especially lotteries with very large payoffs—is small. Regardless, people continue to buy tickets. In an attempt to understand why, an interviewer asked someone who had just purchased a lottery ticket, "What do you think your chances are of winning the lottery?" The reply was, "Oh, about 50–50." The shocked interviewer asked, "How do you get that?" to which the response was, "Well, the way I figure it, either I win or I don't!"

The moral of this story is that outcomes are *not* always equally likely.

Personal Probability

What's the probability that your grade in this Statistics course will be an A? You may be able to come up with a number that seems reasonable. Of course, no matter how confident or depressed you feel about your chances for success, your probability should be between 0 and 1. How did you come up with this probability? From our discussion of probability, we've said that probability represents the relative frequency, or the fraction of times that the event occurs in the long run. Is

---

[1] OK, gambling.

that what you meant? We doubt that you plan on taking this course over and over and over, so that can't be it.

We use the language of probability in everyday speech to express a degree of uncertainty *without* basing it on long-run relative frequencies. Your personal assessment of your chances of getting an A expresses your uncertainty about the outcome. That uncertainty may be based on how comfortable you're feeling in the course, or on your midterm grade, but it can't be based on long-run behavior. We call this kind of probability a subjective or **personal probability.**

Although personal probabilities may be based on experience, they're not based either on long-run relative frequencies or on equally likely events. So they don't display the kind of consistency that we'll need probabilities to have. For that reason, we'll stick to formally defined probabilities. You should be alert to the difference.

The line between personal probability and relative frequency probability can be a fuzzy one. When a weather forecaster predicts a 40% probability of rain, is this a personal probability or a relative frequency probability? The claim may be that 40% of the time, when the map looks like this, it has rained (over some period of time). Or the forecaster may be stating a personal opinion that is based on years of experience and reflects a sense of what has happened in the past in similar situations. When you hear a probability stated, it's good to try to ascertain what kind of probability is intended.

Formal Probability

For some people, the phrase "50/50" means something vague like "I don't know" or "whatever." But when we discuss probabilities, it takes on the precise meaning of *equally likely.* Speaking vaguely about probabilities will get us into trouble, so whenever we talk about probabilities, we'll need to be precise. And to do that, we'll need to develop some formal rules about how probability works.

The simplest way to deal with probabilities is to base the calculations on individual outcomes. But any interesting analysis is likely to combine individual outcomes into events. In a database containing customer information on zip codes, each individual zip code is an outcome. Often we're interested in questions like the probability that the customer is from Illinois, or from the Midwest, rather than in a particular zip code. We could always compute the probability of each individual outcome separately and add them up, but this might get tedious. Instead, we'll want to develop rules for the probability of events—rules like the following:

1. If the probability is 0, the event *never* occurs, and likewise if it has probability 1, it *always* occurs. Even if you think an event is very unlikely, its probability can't be negative, and even if you're sure it will happen, its probability can't be greater than 1. So we require that

> **A probability is a number between 0 and 1.**
>
> **For any event A, $0 \leq P(A) \leq 1$.**

NOTATION ALERT:

We often represent events with capital letters (**A**, **B**, etc.), So $P(\mathbf{A})$ means "the probability of event **A**."

2. If a random phenomenon has only one possible outcome, it's not very interesting (or very random). So we need to distribute the probabilities among all the outcomes a trial can have. How can we do that so that it makes sense? For example, consider what you're doing as you read this book. The possible outcomes might be:

A: You read to the end of this chapter before stopping.

B: You finish this section but stop reading before the end of the chapter.
C: You bail out before the end of this section.

When we assign probabilities to these outcomes, the first thing to be sure of is that we distribute all of the available probability. Something always occurs, so the probability of *something* happening is 1.

We put all the possible outcomes into a big event called the sample space.

Making this more formal gives the **"Something Has to Happen Rule"**:

The probability of the set of all possible outcomes of a trial must be 1.

$P(\mathbf{S}) = 1$. (**S** represents the set of all possible outcomes.[2])

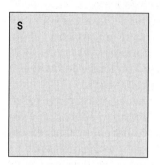

The sample space **S**.

3. Suppose the probability that you get to class on time is 0.8. What's the probability that you *don't* get to class on time? Yes, it's 0.2. The set of outcomes that are *not* in the event **A** is called the **complement** of **A**, and is denoted **A**^{C}. This leads to the **Complement Rule**:

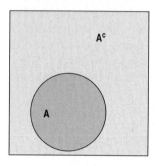

The set **A** and its complement **A**^{C}.

**The probability of an event occurring
is 1 minus the probability that it doesn't occur.**

$$P(\mathbf{A}) = 1 - P(\mathbf{A}^C)$$

4. Suppose the probability that a randomly selected student is a sophomore (**A**) is 0.20, and the probability that he or she is a junior (**B**) is 0.30. What is the probability that the student is *either* a sophomore *or* a junior, written $P(\mathbf{A}\ or\ \mathbf{B})$? If you guessed 0.50 you've deduced the **Addition Rule,** which says that you can add the probabilities of events that are disjoint.[3] To see whether two events are disjoint, we take them apart into their component outcomes and check whether they have any outcomes in common. **Disjoint** (or **mutually exclusive**) events have no outcomes in common. The **Addition Rule** states:

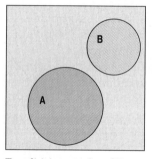

Two disjoint sets, **A** and **B**.

**For two disjoint events A and B, the probability that one *or* the other
occurs is the sum of the probabilities of the two events.**

$P(\mathbf{A}\ or\ \mathbf{B}) = P(\mathbf{A}) + P(\mathbf{B})$, **provided that A and B are disjoint.**

A S **Addition Rule for
Disjoint Events.** Experiment with disjoint events to explore the Addition Rule.

We can always add the probabilities of two events that each consist of only a single outcome. Because they have no outcomes to share, they must be disjoint. This gives us an easy way to check whether the probabilities we've assigned to the possible outcomes are **legitimate**.

---

[2] Most likely, you remember sets from a previous math course. But even if not, don't panic. A set is just a mathematical way of saying a collection.

[3] You may see $P(\mathbf{A}\ or\ \mathbf{B})$ written as $P(\mathbf{A} \cup \mathbf{B})$. The symbol $\cup$ means "union," representing the outcomes in event **A** *or* event **B** (or both). The symbol $\cap$ means "intersection," representing outcomes that are in both event **A** *and* event **B**. You may see $P(\mathbf{A}\ and\ \mathbf{B})$ sometimes written as $P(\mathbf{A} \cap \mathbf{B})$.

By the Something Has to Happen Rule, the total of the probabilities of all possible outcomes must be exactly one. No more, no less. And because individual outcomes are disjoint, we can add their probabilities to check that the sum is exactly 1. For example, if we were told the probabilities of selecting at random a freshman, sophomore, junior, or senior from all the undergraduates at a school were 0.25, 0.23, 0.22, and 0.20, respectively, we would know that something was wrong. These "probabilities" sum to only 0.90, so this is not a legitimate probability assignment. Either a value is *wrong* or we just missed some possible outcomes, like "pre-freshman" or "postgraduate" categories that soak up the remaining 0.10. Similarly, a claim that the probabilities were 0.26, 0.27, 0.29, and 0.30 would be wrong because these "probabilities" sum to more than 1.

But be careful. The Addition Rule doesn't work for events that aren't disjoint. If the probability of owning an MP3 player is 0.50 and the probability of owning a computer is 0.90, the probability of owning either an MP3 player or a computer may be pretty high, but it is *not* 1.40! Why can't you add probabilities like this? Because these events are not disjoint. You *can* own both. In the next chapter, we'll see how to add probabilities for events like these, but we'll need another rule.

"*Baseball is 90% mental. The other half is physical.*"
—*Yogi Berra*

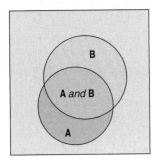

Two sets **A** and **B** that are not disjoint. The event (**A** and **B**) is their intersection.

5. The color of a traffic light as you reach an intersection is a random outcome. Suppose the light spends 35% of its time red in your direction and the other 65% either green or yellow. What's the chance of finding it red two days in a row? That's the same as asking the probability of finding it red today *and* finding it red tomorrow. For independent events, the answer is very simple. And the color of the light today *is* independent of the color yesterday. The **Multiplication Rule** says that for independent events, to find the probability that both events occur, we just multiply the probabilities together. Formally:

For two independent events A and B, the probability that both A *and* B occur is the product of the probabilities of the two events.

$$P(A \text{ } and \text{ } B) = P(A) \times P(B), \text{ provided that A and B are independent.}$$

This rule can be extended to more than two independent events. What's the chance of finding the light red every day this week? We can multiply the probability of it happening each day, which is

$$0.35 \times 0.35 \times 0.35 \times 0.35 \times 0.35 = 0.00525$$

or about 5 times in a thousand. Of course, to calculate this probability we have used the assumption that the five events are independent.

Many Statistics methods require an **Independence Assumption**, but *assuming* independence doesn't make it true. Always *Think* about whether that assumption is reasonable before using the Multiplication Rule.

Putting the Rules to Work

In informal English, you may see "some" used to mean "at least one." "What's the probability that some of the eggs in that carton are broken?" means at least one.

 Probabilities of Compound Events. The Random Tool also lets you experiment with compound random events to see if they are independent.

In most situations where we want to find a probability, we'll use the rules in combination.

For example, what's the chance that we'll hit a red light *at least once* during the week? Hitting at least one red light means that we hit either 1, 2, 3, 4, or 5 red lights during the week. So, we could calculate the probability of hitting at least one red light by first calculating the probability of hitting exactly 1 red light during the week, then exactly 2, then 3, 4, and 5. Then we'd just need to add them up. Aarrgghh! There must be an easier way to do this, and there is.

It can be easier to work with the *complement* of the event we're really interested in. Once we have the complement's probability, we know (from Rule 3) that the probability of the event we're interested in is 1 minus the probability of the complement. What's the complement of getting at least one red light? It's getting *no* red lights, five days in a row. We know the probability of hitting a red light is 0.35 each day, so by the Complement Rule, the probability of *not* hitting a red light is $1 - 0.35$, or 0.65 each day. The probability of making it through five consecutive days without hitting a red light is $(0.65)^5$. This is the probability of the complement of what we wanted. Now, taking the complement of this compound event, we find the probability that it *doesn't* happen (*not* 0 red lights) is $1 - (0.65)^5 = 1 - 0.116 = 0.884$.

just checking

2. Opinion polling organizations contact their respondents by telephone. Random telephone numbers are generated, and interviewers try to contact those households. In the 1990s this method could reach about 69% of U.S. households. According the Pew Research Center for the People and the Press, by 2003 the contact rate had risen to 76%. Each household, of course, is independent of the others.

a) What is the probability that the interviewer will successfully contact the next household on the list?

b) What is the probability that an interviewer successfully contacts both of the next *two* households on her list?

c) What is the probability that the interviewer's first successful contact will be the third household on the list?

d) What is the probability the interviewer will make at least one successful contact among the next 5 households on the list?

Probability Step-By-Step

The five rules we've seen can be used in a number of different combinations to answer a surprising number of questions. Let's try one to see how we might go about it.

In 2001, Masterfoods, the manufacturers of M&M's® milk chocolate candies, decided to add another color to the standard color lineup of brown, yellow, red, orange, blue, and green. To decide which color, they surveyed kids in nearly every country of the world and asked them to vote among purple, pink, and teal. The global winner was purple! In the United States, 42% of those who voted said purple, 37% said teal, and only 19% said pink. But in Japan the percentages were 38% pink, 36% teal, and only 16% purple. Let's use Japan's percentages to ask some questions.

1. What's the probability that a Japanese M&M's survey respondent selected at random preferred either pink or teal?
2. If we pick two respondents at random, what's the probability that they *both* selected purple?
3. If we pick three respondents at random, what's the probability that *at least one* preferred purple?

Think

The probability of an event is its long-term relative frequency. It can be determined in several ways: by looking at many replications of an event, by deducing it from equally likely events, or by using some other information. Here, we are told the relative frequencies of the three responses.

Make sure the probabilities are legitimate. Here, they're not. Either there was a mistake or the other voters must have chosen a color other than the three given. A check of other countries shows a similar deficit, so probably we're seeing those who had no preference or who wrote in another color.

The M&M's Web site reports the proportions of Japanese votes by color. These give the probability of selecting a voter who preferred each of the colors:

$P(\text{pink}) = 0.38$
$P(\text{teal}) = 0.36$
$P(\text{purple}) = 0.16$

Each is between 0 and 1, but these don't add up to 1. The remaining 10% of the voters must have not expressed a preference or written in another color. I'll put them together into "no preference" and add: $P(\text{no preference}) = 0.10$.

With this addition, I have a legitimate assignment of probabilities.

Question 1. What's the probability that a Japanese M&M's survey respondent selected at random preferred either pink or teal?

Think

Plan Decide which rules to use and check the conditions they require.

The events "Pink" and "Teal" are individual outcomes (A respondent can't choose both colors.), so they are disjoint. I can apply the Addition Rule.

Show

Mechanics Show your work.

$P(\text{pink or teal}) = P(\text{pink}) + P(\text{teal})$
$= 0.38 + 0.36 = 0.74$

Tell

Conclusion Interpret your results in the proper context.

The probability that the respondent said pink or teal is 0.74.

Question 2. If we pick two respondents at random, what's the probability that they both said purple?

Think

Plan The word "both" suggests we want $P(\mathbf{A}\ and\ \mathbf{B})$, which calls for the Multiplication Rule. Check the required condition.

✔ **Independence:** It's unlikely that the choice made by one respondent affected the choice of the other, so the events seem to be independent. I can use the Multiplication Rule.

Show

Mechanics Show your work.
For both respondents to pick purple, each one has to pick purple.

$P(\text{both purple})$

$= P(\text{first respondent picks purple and second respondent picks purple})$
$= P(\text{first respondent picks purple}) \times P(\text{second respondent picks purple})$
$= 0.16 \times 0.16 = 0.0256$

Tell

Conclusion Interpret your results in the proper context.

> The probability that both respondents pick purple is 0.0256.

Question 3. If we pick three respondents at random, what's the probability that at least one preferred purple?

Think

Plan The phrase "at least . . ." often flags a question best answered by looking at the complement, and that's the best approach here. The complement of "at least one preferred purple" is "none of them preferred purple."

Check the conditions.

> $P(\text{at least one picked purple})$
>
> $$= P(\{\text{none picked purple}\}^C)$$
> $$= 1 - P(\text{none picked purple}).$$
>
> $P(\text{none picked purple}) = P(\text{not purple and not purple and not purple}).$
>
> ✔ **Independence:** These are independent events because they are choices by three random respondents. I can use the Multiplication Rule:

Show

Mechanics We calculate $P(\text{none purple})$ by using the Multiplication Rule.

> $P(\text{none picked purple}) = P(\text{first not purple}) \times P(\text{second not purple}) \times P(\text{third not purple})$
> $$= [P(\text{not purple})]^3.$$
>
> $P(\text{not purple}) = 1 - P(\text{purple})$
> $$= 1 - 0.16 = 0.84.$$
>
> So $P(\text{none picked purple}) = (0.84)^3 = 0.5927.$

Then we can use the Complement Rule to get the probability we want.

> $P(\text{at least 1 picked purple})$
> $$= 1 - P(\text{none picked purple})$$
> $$= 1 - 0.5927 = 0.4073.$$

Tell

Conclusion Interpret your results in the proper context.

> There's about a 40.7% chance that at least one of the respondents picked purple.

What Can Go Wrong?

- ***Beware of probabilities that don't add up to 1.*** To be a legitimate probability distribution, the sum of the probabilities for all possible outcomes must total 1. If the sum is less than 1, you may need to add another category ("other") and assign the remaining probability to that outcome. If the sum is more than 1, check that the outcomes are disjoint. If they're not, then you can't assign probabilities by just counting relative frequencies.
- ***Don't add probabilities of events if they're not disjoint.*** Events must be disjoint to use the Addition Rule. The probability of being under 80 *or* a female is not the probability of being under 80 *plus* the probability of being female. That sum may be more than 1.
- ***Don't multiply probabilities of events if they're not independent.*** The probability of selecting a student at random who is over 6'10" tall *and* on the basketball team is *not* the probability the student is 6'10" tall *times* the probability he's on the basketball team. Knowing that the student is over 6'10" changes the probability of his being on the basketball team. You can't multiply these probabilities. The multiplication of probabilities of events that are not independent is one of the most common errors people make in dealing with probabilities.

- ***Don't confuse disjoint and independent.*** Disjoint events *can't* be independent. If **A** = {you get an A in this class} and **B** = {you get a B in this class}, **A** and **B** are disjoint. Are they independent? If you find out that **A** is true, does that change the probability of **B?** You bet it does! So they can't be independent. We'll return to this issue in the next chapter.

CONNECTIONS

We saw in the previous three chapters that randomness plays a critical role in gathering data. That fact alone makes it important that we understand how random events behave. The rules and concepts of probability give us a language to talk and think about random phenomena. From here on, randomness will be fundamental to how we think about data, and probabilities will show up in every chapter.

We began thinking about independence back in Chapter 3 when we looked at contingency tables and asked whether the distribution of one variable was the same for each category of another. Then in Chapter 12, we saw that independence was fundamental to drawing a Simple Random Sample. For computing compound probabilities, we again ask about independence. And we'll continue to think about independence throughout the rest of the book.

Our interest in probability extends back to the start of the book. We've talked about "relative frequencies" often. But—let's be honest—that's just a casual term for probability. For example, you can now rephrase the 68-95-99.7 Rule to talk about the *probability* that a random value selected from a Normal model will fall within 1, 2, or 3 standard deviations of the mean.

Why not just say "probability" from the start? Well, we didn't need any of the formal rules of this chapter (or the next one), so there was no point to weighing down the discussion with those rules. And "relative frequency" is the right intuitive way to think about probability in this course, so you've been thinking right all along.

Keep it up.

What have we learned?

We've learned that probability is based on long-run relative frequencies. We've thought about the Law of Large Numbers and noted that it speaks only of long-run behavior. Because the long run is a very long time, we need to be careful not to misinterpret the Law of Large Numbers. Even when we've observed a string of heads, we shouldn't expect extra tails in subsequent coin flips.

Also, we've learned some basic rules for combining probabilities of outcomes to find probabilities of more complex events. These include

- the Something Has to Happen Rule,
- the Complement Rule,
- the Addition Rule for disjoint events, and
- the Multiplication Rule for independent events.

TERMS

Random phenomenon
A phenomenon is random if we know what outcomes could happen, but not which particular values will happen.

Probability
The probability of an event is a number between 0 and 1 that reports the likelihood of the event's occurrence. A probability can be derived from equally likely outcomes, from the long-run relative frequency of the event's occurrence, or from known proportions. We write $P(\mathbf{A})$ for the probability of the event **A.**

Trial
A single attempt or realization of a random phenomenon.

Outcome
The outcome of a trial is the value measured, observed, or reported for an individual instance of that trial.
Outcomes are considered to be either
- discrete if they have distinct values such as heads or tails (even if those values are labeled with numerals), or
- continuous if they take on numeric values in some range of possible values.

Event
A collection of outcomes. Usually, we identify events so that we can attach probabilities to them. We denote events with bold capital letters such as **A, B,** or **C.**

Independence (informally)
Two events are *independent* if knowing whether one event occurs does not alter the probability that the other event occurs.

Law of Large Numbers
The Law of Large Numbers states that the long-run *relative frequency* of repeated independent events settles down to the *true* probability as the number of trials increases.

"Something Has to Happen Rule"
The sum of the probabilities of all possible outcomes of a trial must be 1.

Complement Rule
The probability of an event occurring is 1 minus the probability that it doesn't occur.

$$P(\mathbf{A}) = 1 - P(\mathbf{A}^{\mathbf{C}})$$

Disjoint (Mutually exclusive)
Two events are disjoint if they share no outcomes in common. If **A** and **B** are disjoint, then knowing that **A** occurs tells us that **B** cannot occur. Disjoint events are also called "mutually exclusive."

Addition Rule
If **A** and **B** are disjoint events, then the probability of **A** *or* **B** is

$$P(\mathbf{A} \ or \ \mathbf{B}) = P(\mathbf{A}) + P(\mathbf{B}).$$

Legitimate probability assignment
An assignment of probabilities to outcomes is legitimate if
- each probability is between 0 and 1 (inclusive).
- the sum of the probabilities is 1.

Multiplication Rule
If **A** and **B** are independent events, then the probability of **A** *and* **B** is

$$P(\mathbf{A} \ and \ \mathbf{B}) = P(\mathbf{A}) \times P(\mathbf{B}).$$

SKILLS

When you complete this lesson you should:

Think
- Understand that random phenomena are unpredictable in the short term but show long-run regularity.
- Be able to recognize random outcomes in a real-world situation.
- Know that the relative frequency of an outcome of a random phenomenon settles down as we gather more random outcomes. Be able to state the Law of Large Numbers.

- Know the basic definitions and rules of probability.

- Recognize when events are disjoint and when events are independent. Understand the difference, and that disjoint events cannot be independent.

Show

- Be able to use the facts about probability to determine whether an assignment of probabilities is legitimate. Each probability must be a number between 0 and 1 and the sum of the probabilities assigned to all possible outcomes must be 1.

- Know how and when to apply the Addition Rule. Know that events must be disjoint for the Addition Rule to apply.

- Know how and when to apply the Multiplication Rule. Know that events must be independent for the Multiplication Rule to apply. Be able to use the Multiplication Rule to find probabilities for combinations of independent events.

- Know how to use the Complement Rule to make calculating probabilities simpler. Recognize that probabilities of "at least . . ." are likely to be simplified in this way.

Tell

- Be able to use statements about probability in describing a random phenomenon. You will need this skill soon for making statements about statistical inference.

- Know and be able to use correctly the terms "sample space," "disjoint events," and "independent events."

EXERCISES

1. **Roulette.** A casino claims that its roulette wheel is truly random. What should that claim mean?

2. **Rain.** The weather reporter on TV makes predictions such as a 25% chance of rain. What do you think is the meaning of such a phrase?

3. **Winter.** Comment on the following quotation:

   "What I think is our best determination is it will be a colder than normal winter," said Pamela Naber Knox, a Wisconsin state climatologist. "I'm basing that on a couple of different things. First, in looking at the past few winters, there has been a lack of really cold weather. Even though we are not supposed to use the law of averages, we are due." (Associated Press, fall 1992, quoted by Schaeffer et al.)

4. **Snow.** After an unusually dry autumn, a radio announcer is heard to say, "Watch out! We'll pay for these sunny days later on this winter." Explain what he's trying to say, and comment on the validity of his reasoning.

5. **Cold streak.** A batter who had failed to get a hit in seven consecutive times at bat then hits a game-winning home run. When talking to reporters afterward, he says he was very confident that last time at bat because he knew he was "due for a hit." Comment on his reasoning.

6. **Crash.** Commercial airplanes have an excellent safety record. Nevertheless, there are crashes occasionally, with the loss of many lives. In the weeks following a crash, airlines often report a drop in the number of passengers, probably because people are afraid to risk flying.
   a) A travel agent suggests that, since the law of averages makes it highly unlikely to have two plane crashes within a few weeks of each other, flying soon after a crash is the safest time. What do you think?
   b) If the airline industry proudly announces that it has set a new record for the longest period of safe flights, would you be reluctant to fly? Are the airlines due to have a crash?

7. **Fire insurance.** Insurance companies collect annual payments from homeowners in exchange for paying to rebuild houses that burn down.
   a) Why should you be reluctant to accept a $300 payment from your neighbor to replace his house should it burn down during the coming year?
   b) Why can the insurance company make that offer?

8. **Jackpot.** On January 20, 2000, the International Gaming Technology company issued a press release:

   (LAS VEGAS, Nev.)—Cynthia Jay was smiling ear to ear as she walked into the news conference at The Desert Inn Resort in Las Vegas today, and well she should. Last night, the 37-year-old cocktail waitress won the world's largest slot jackpot—$34,959,458—on a Megabucks machine. She said she

had played $27 in the machine when the jackpot hit. Nevada Megabucks has produced 49 major winners in its 14-year history. The top jackpot builds from a base amount of $7 million and can be won with a 3-coin ($3) bet.

a) How can the Desert Inn afford to give away millions of dollars on a $3 bet?

b) Why did the company issue a press release? Wouldn't most businesses want to keep such a huge loss quiet?

9. Spinner. The plastic arrow on a spinner for a child's game stops rotating to point at a color that will determine what happens next. Which of the following probability assignments are possible?

	Probabilities of . . .			
	Red	Yellow	Green	Blue
a)	0.25	0.25	0.25	0.25
b)	0.10	0.20	0.30	0.40
c)	0.20	0.30	0.40	0.50
d)	0	0	1.00	0
e)	0.10	0.20	1.20	−1.50

10. Scratch off. Many stores run "secret sales": Shoppers receive cards that determine how large a discount they get, but the percentage is revealed by scratching off that black stuff (what *is* that?) only after the purchase has been totaled at the cash register. The store is required to reveal (in the fine print) the distribution of discounts available. Which of these probability assignments are plausible?

	Probabilities of . . .			
	10% off	20% off	30% off	50% off
a)	0.20	0.20	0.20	0.20
b)	0.50	0.30	0.20	0.10
c)	0.80	0.10	0.05	0.05
d)	0.75	0.25	0.25	−0.25
e)	1.00	0	0	0

11. Car repairs. A consumer organization estimates that over a 1-year period 17% of cars will need to be repaired once, 7% will need repairs twice, and 4% will require three or more repairs. What is the probability that a car chosen at random will need

a) no repairs?

b) no more than one repair?

c) some repairs?

12. Stats projects. In a large Introductory Statistics lecture hall, the professor reports that 55% of the students enrolled have never taken a Calculus course, 32% have taken only one semester of Calculus, and the rest have taken two or more semesters of Calculus. The professor randomly assigns students to groups of three to work on

a project for the course. What is the probability that the first groupmate you meet has studied

a) two or more semesters of Calculus?

b) some Calculus?

c) no more than one semester of Calculus?

13. More repairs. Consider again the auto repair rates described in Exercise 11. If you own two cars, what is the probability that

a) neither will need repair?

b) both will need repair?

c) at least one car will need repair?

14. Another project. You are assigned to be part of a group of three students from the Intro Stats class described in Exercise 12. What is the probability that, of your other two groupmates,

a) neither has studied Calculus?

b) both have studied at least one semester of Calculus?

c) at least one has had more than one semester of Calculus?

15. Repairs, again. You used the Multiplication Rule to calculate repair probabilities for your cars in Exercise 13.

a) What must be true about your cars in order to make that approach valid?

b) Do you think this assumption is reasonable? Explain.

16. Final project. You used the Multiplication Rule to calculate probabilities about the Calculus background of your Statistics groupmates in Exercise 14.

a) What must be true about the groups in order to make that approach valid?

b) Do you think this assumption is reasonable? Explain.

17. Energy. A Gallup poll in March 2001 asked 1005 U.S. adults how the United States should deal with the current energy situation: by more production, more conservation, or both? Here are the results:

Response	Number
More production	332
More conservation	563
Both	80
No opinion	30
Total	**1005**

If we select a person at random from this sample of 1005 adults,

a) what is the probability that the person responded "More production"?

b) what is the probability that the person responded "Both" or had no opinion?

18. All about Bill. A Gallup Poll in June 2004 asked 1005 U.S. adults how likely they were to read Bill Clinton's autobiography *My Life*. Here's how they responded:

Response	Number
Will definitely read it	90
Will probably read it	211
Will probably not read it	322
Will definitely not read it	382
Total	**1005**

If we select a person at random from this sample of 1005 adults,

a) what is the probability that the person responded "Will definitely not read it"?

b) what is the probability that the person will probably or definitely read it?

19. More energy. Exercise 17 shows the results of a Gallup Poll about energy. Suppose we select three people at random from this sample.

a) What is the probability that all three responded "More conservation"?

b) What is the probability that none responded "Both"?

c) What assumption did you make in computing these probabilities?

d) Explain why you think that assumption is reasonable.

20. More about Bill. Consider the results of the poll about President Clinton's book, summarized in Exercise 18. Let's call someone who responded that they would definitely or probably read it a "likely reader" and the other two categories, "unlikely reader." If we select two people at random from this sample,

a) what is the probability that both are likely readers?

b) what is the probability that neither is a likely reader?

c) what is the probability that one is a likely reader and one isn't?

d) What assumption did you make in computing these probabilities?

e) Explain why you think that assumption is reasonable.

21. Polling. As mentioned in the chapter, opinion-polling organizations contact their respondents by sampling random telephone numbers. Although interviewers now can reach about 76% of U.S. households, the percentage of those contacted who agree to cooperate with the survey has fallen from 58% in 1997 to only 38% in 2003 (Pew Research Center for the People and the Press). Each household, of course, is independent of the others.

a) What is the probability that the next household on the list will be contacted, but will refuse to cooperate?

b) What is the probability (in 2003) of failing to contact a household or of contacting the household but not getting them to agree to the interview?

c) Show another way to calculate the probability in part b.

22. Polling, part II. According to Pew Research, the contact rate (probability of contacting a selected household)

in 1997 was 69% and in 2003 was 76%. However, the cooperation rate (probability of someone at the contacted household agreeing to be interviewed) was 58% in 1997 and dropped to 38% in 2003.

a) What is the probability (in 2003) of obtaining an interview with the next household on the sample list? (To obtain an interview, an interviewer must both contact the household and then get agreement for the interview.)

b) Was it more likely to obtain an interview from a randomly selected household in 1997 or in 2003?

23. M&M's. The Masterfoods company says that before the introduction of purple, yellow candies made up 20% of their plain M&M's, red another 20%, and orange, blue, and green each made up 10%. The rest were brown.

a) If you pick an M&M at random, what is the probability that

1. it is brown?

2. it is yellow or orange?

3. it is not green?

4. it is striped?

b) If you pick three M&M's in a row, what is the probability that

1. they are all brown?

2. the third one is the first one that's red?

3. none are yellow?

4. at least one is green?

24. Blood. The American Red Cross says that about 45% of the U.S. population has Type O blood, 40% Type A, 11% Type B, and the rest Type AB.

a) Someone volunteers to give blood. What is the probability that this donor

1. has Type AB blood?

2. has Type A or Type B?

3. is not Type O?

b) Among four potential donors, what is the probability that

1. all are Type O?

2. no one is Type AB?

3. they are not all Type A?

4. at least one person is Type B?

25. Disjoint or independent? In Exercise 23 you calculated probabilities of getting various M&M's. Some of your answers depended on the assumption that the outcomes described were *disjoint*; that is, they could not both happen at the same time. Other answers depended on the assumption that the events were *independent*; that is, the occurrence of one of them doesn't affect the probability of the other. Do you understand the difference between disjoint and independent?

a) If you draw one M&M, are the events of getting a red one and getting an orange one disjoint or independent or neither?

b) If you draw two M&M's one after the other, are the events of getting a red on the first and a red on the second disjoint or independent or neither?

c) Can disjoint events ever be independent? Explain.

26. Disjoint or independent? In Exercise 24 you calculated probabilities involving various blood types. Some of your answers depended on the assumption that the outcomes described were *disjoint*; that is, they could not both happen at the same time. Other answers depended on the assumption that the events were *independent*; that is, the occurrence of one of them doesn't affect the probability of the other. Do you understand the difference between disjoint and independent?

a) If you examine one person, are the events that the person is Type A and that the person is Type B disjoint or independent or neither?

b) If you examine two people, are the events that the first is Type A and the second Type B disjoint or independent or neither?

c) Can disjoint events ever be independent? Explain.

27. Dice. You roll a fair die three times. What is the probability that

a) you roll all 6's?

b) you roll all odd numbers?

c) none of your rolls gets a number divisible by 3?

d) you roll at least one 5?

e) the numbers you roll are not all 5's?

28. Slot machine. A slot machine has three wheels that spin independently. Each has 10 equally likely symbols: 4 bars, 3 lemons, 2 cherries, and a bell. If you play, what is the probability

a) you get 3 lemons?

b) you get no fruit symbols?

c) you get 3 bells (the jackpot)?

d) you get no bells?

e) you get at least one bar (an automatic loser)?

29. Champion bowler. A certain bowler can bowl a strike 70% of the time. What is the probability that she

a) goes three consecutive frames without a strike?

b) makes her first strike in the third frame?

c) has at least one strike in the first three frames?

d) bowls a perfect game (12 consecutive strikes)?

30. The train. To get to work, a commuter must cross train tracks. The time the train arrives varies slightly from day to day, but the commuter estimates he'll get stopped on about 15% of work days. During a certain 5-day work week, what is the probability that he

a) gets stopped on Monday and again on Tuesday?

b) gets stopped for the first time on Thursday?

c) gets stopped every day?

d) gets stopped at least once during the week?

31. Voters. Suppose that in your city 37% of the voters are registered as Democrats, 29% as Republicans, and 11%

as members of other parties (Liberal, Right to Life, Green, etc.). Voters not aligned with any official party are termed "Independent." You are conducting a poll by calling registered voters at random. In your first three calls, what is the probability you talk to

a) all Republicans?

b) no Democrats?

c) at least one Independent?

32. Religion. Census reports for a city indicate that 62% of residents classify themselves as Christian, 12% as Jewish, and 16% as members of other religions (Muslims, Buddhists, etc.). The remaining residents classify themselves as nonreligious. A polling organization seeking information about public opinions wants to be sure to talk with people holding a variety of religious views, and makes random phone calls. Among the first four people they call, what is the probability they reach

a) all Christians?

b) no Jews?

c) at least one person who is nonreligious?

33. Tires. You bought a new set of four tires from a manufacturer who just announced a recall because 2% of those tires are defective. What is the probability that at least one of yours is defective?

34. Pepsi. For a sales promotion, the manufacturer places winning symbols under the caps of 10% of all Pepsi bottles. You buy a six-pack. What is the probability that you win something?

35. 9/11? On September 11, 2002, the first anniversary of the terrorist attack on the World Trade Center, the New York State Lottery's daily number came up 9-1-1. An interesting coincidence or a cosmic sign?

a) What is the probability that the winning three numbers match the date on any given day?

b) What is the probability that a whole year passes without this happening?

c) What is the probability that the date and winning lottery number match at least once during any year?

d) If every one of the 50 states has a three-digit lottery, what is the probability that at least one of them will come up 9-1-1 on September 11?

36. Red cards. You shuffle a deck of cards, and then start turning them over one at a time. The first one is red. So is the second. And the third. In fact, you are surprised to get 10 red cards in a row. You start thinking, "The next one is due to be black!"

a) Are you correct in thinking that there's a higher probability that the next card will be black than red? Explain.

b) Is this an example of the Law of Large Numbers? Explain.

---

just checking

Answers

1. The LLN works only in the long run, not in the short run. The random methods for selecting lottery numbers have no memory of previous picks, so there is no change in the probability that a certain number will come up.

2. a) 0.76
   b) $0.76(0.76) = 0.5776$
   c) $(1 - 0.76)^2 (0.76) = 0.043776$
   d) $1 - (1 - 0.76)^5 = 0.9992$

15

Probability Rules!

et's face it. Probabilities of simple events are just not that interesting. When we combine events and think about conditional outcomes, though, things get juicier. We saw back in Chapter 3 that the chance of surviving the *Titanic* disaster changed depending on the ticket class of the passenger. To navigate the stormy seas of probability, we'll need some rules to steer by.

Remember, for any random phenomenon, each **trial** generates an **outcome.** An **event** is *any* set or collection of outcomes. The collection of *all possible* outcomes is called the **sample space,** and denoted S.[1]

If you flip a coin, what is the sample space? All you can get is heads or tails (we'll ignore the possibility that it lands on the edge). So the sample space is just the set {H, T}, H for heads and T for tails. It's no big deal. To make it seem more important, some books use the Greek letter Ω instead of S. But whatever the symbol, the sample space is just the collection of *all the possible outcomes*.

Events

Pull a bill from your wallet or pocket without looking at it. An outcome of this trial is the bill you select. The sample space is all the bills in circulation: $S =$ {$1 bill, $2 bill, $5 bill, $10 bill, $20 bill, $50 bill, $100 bill}. These are *all* the possible outcomes. (In spite of what you may have seen in bank robbery movies, there are no $500 or $1000 bills.)

---

[1] Mathematicians like to use the term "space" as a fancy name for a set. Sort of like referring to that closet they gave you for a dorm room as your "living space." But remember that it's really just a bunch of outcomes.

We can combine possible outcomes of such a trial into events. For example, the event **A** = {$1, $5, $10} represents selecting a $1, $5, or $10 bill. The event **B** = {a bill that does not have a president on it} is the collection of outcomes (Don't look! Can you name them?): {$10 (Hamilton), $100 (Franklin)}. The event **C** = {enough money to pay for a $12 meal with one bill} is the set of outcomes {$20, $50, $100}.

Notice that outcomes don't have to be equally likely. You'd no doubt be more surprised (and pleased) to pull out a $100 bill than a $1 bill—it's not very likely, though. You probably carry many more $1 than $100 bills, but without information about the probability of each outcome, we can't calculate the probability of such events.

When outcomes *are* equally likely, probabilities for events are easy to find just by counting. When the *k* possible outcomes are equally likely, each has a probability of $1/k$. For example, consider the final digit of the serial number of that bill you extracted at random. Final digits are equally likely to be any digit from 0 to 9.

What's the probability of randomly selecting a bill whose serial number ends in an odd digit? We could simulate this event by simply assigning the digits to the random numbers 0 through 9. Because half of the digits from 0 to 9 are odd, you wouldn't be surprised to learn that the probability of finding a bill whose last digit is odd is 5/10. We can make this more formal by saying that for any event, **A,** that is made up of *equally likely* outcomes,

$$P(\mathbf{A}) = \frac{count\ of\ outcomes\ in\ \mathbf{A}}{count\ of\ all\ possible\ outcomes}.$$

Be careful! This rule won't work unless the outcomes are *equally likely*. The probability of the event **C** (getting a bill worth more than $12) is *not* 3/7. There are 7 possible outcomes, and 3 of them exceed $12, but they are *not equally* likely. (Remember the probability that your lottery ticket will win rather than lose still isn't 1/2.)

The First Three Rules for Working with Probability Rules

Make a picture.
Make a picture.
Make a picture.

We're dealing with probabilities now, not data, but the three rules don't change. The most common kind of picture to make is called a Venn diagram. We'll use Venn diagrams throughout the rest of this chapter. Even experienced statisticians make Venn diagrams to help them think about probabilities of compound and overlapping events. You should, too.

John Venn (1834–1923) created the Venn diagram. His book on probability, *The Logic of Chance*, was "strikingly original and considerably influenced the development of the theory of Statistics," according to John Maynard Keynes, one of the founders of Economics.

The General Addition Rule

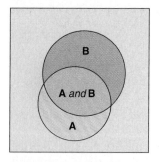

Events **A** and **B** and their intersection.

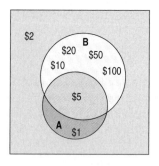

Denominations of bills that are odd (**A**) or that have a building on the reverse side (**B**). The two sets both include the $5 bill and both exclude the $2 bill.

Let's think about a randomly drawn bill again. There are images of famous buildings in the center of the backs of all but two bills in circulation. The $1 bill has the word ONE in the center, and the $2 bill shows the signing of the Declaration of Independence.

What's the probability of randomly selecting **A** = {a bill with an odd-numbered value} *or* **B** = {a bill with a building on the reverse}? We know **A** = {$1, $5} and **B** = {$5, $10, $20, $50, $100}. But $P(\textbf{A} \text{ or } \textbf{B})$ is not simply the sum $P(\textbf{A}) + P(\textbf{B})$, because the events **A** and **B** are not disjoint. The $5 bill is in both sets. So what can we do? We'll need a new probability rule.

As the diagrams show, we can't use the **Addition Rule** and add the two probabilities because the events are not **disjoint;** they overlap. There's an outcome (the $5 bill) in the *intersection* of **A** and **B**. The Venn diagram represents the sample space. Notice that the $2 bill has neither a building nor an odd denomination, so it sits outside both circles.

The $5 bill plays a crucial role here because it is both odd *and* has a building on the reverse. It's in both **A** and **B**, which places it in the *intersection* of the two circles. The reason we can't simply add the probabilities of **A** and **B** is that we'd count the $5 bill twice.

If we did add the two probabilities, we could compensate by *subtracting* out the probability of that $5 bill. So,

P(odd number value *or* building)

$\quad = P(\text{odd number value}) + P(\text{building}) - P(\text{odd number value } and \text{ building})$
$\quad = P(\$1, \$5) + P(\$5, \$10, \$20, \$50, \$100) - P(\$5).$

This method works in general. We add the probabilities of two events and then subtract out the probability of their intersection. This approach gives us the **General Addition Rule,** which does not require disjoint events:

$$P(\textbf{A} \text{ or } \textbf{B}) = P(\textbf{A}) + P(\textbf{B}) - P(\textbf{A} \text{ and } \textbf{B}).$$

● **Would you like dessert or coffee?** Natural language can be ambiguous. In this question, is the answer one of the two alternatives, or simply "yes"? Must you decide between them, or may you have both? That kind of ambiguity can confuse our probabilities.

Suppose we had been asked the different question: What is the probability that the bill we draw has *either* an odd value *or* a building but *not both*? Which bills are we talking about now? The set we're interested in would be {$1, $10, $20, $50, $100}. We don't include the $5 bill in the set because it has both characteristics.

Why isn't this the same answer as before? The problem is that when we say the word "or," we usually mean *either* one *or* both. We don't usually mean the *exclusive* version of "or" as in, "Would you like the steak *or* the vegetarian entrée?" Ordinarily when we ask for the probability that **A** *or* **B** occurs, we mean **A** or **B** or both. And we know *that* probability is $P(\textbf{A}) + P(\textbf{B}) - P(\textbf{A} \text{ and } \textbf{B})$. The General Addition Rule subtracts the probability of the outcomes in **A** *and* **B** because we've counted those outcomes *twice*. But they're still there.

If we really mean **A** or **B,** but NOT both, we have to get rid of the outcomes in {**A** *and* **B**}. So $P(\textbf{A} \text{ or } \textbf{B}, \text{ but } not \text{ both}) = P(\textbf{A} \text{ or } \textbf{B}) - P(\textbf{A} \text{ and } \textbf{B}) = P(\textbf{A}) + P(\textbf{B}) - 2 \times P(\textbf{A} \text{ and } \textbf{B})$. Now we've subtracted $P(\textbf{A} \text{ and } \textbf{B})$ twice—once because we don't want to double-count these events and a second time because we really didn't want to count them at all. At this point, it would be a good idea to make a picture to be sure you see how this works! ●

① Back in Chapter 1 we suggested that you sample some pages of this book at random to see whether they held a graph or other data display. We actually did just that. We drew a representative sample and found the following:

> *48% of pages had some kind of data display,*
> *27% of pages had an equation, and*
> *7% of pages had both a data display and an equation.*

a) Display these results in a Venn diagram.
b) What is the probability that a randomly selected sample page had neither a data display nor an equation.
c) What is the probability that a randomly selected sample page had a data display but no equation?

Using the General Addition Rule Step-By-Step

Police report that 78% of drivers stopped on suspicion of drunk driving are given a breath test, 36% a blood test, and 22% both tests. What is the probability that a randomly selected DWI suspect is given

1. a test?
2. a blood test or a breath test, but not both?
3. neither test?

Think **Plan** Define the events we're interested in. There are no conditions to check; the General Addition Rule works for any events!

Plot Make a picture, and use the given probabilities to find the probability for each region.

The blue region represents **A** but not **B**. The green intersection region represents **A** *and* **B**. Note that since $P(\mathbf{A}) = 0.78$ and $P(\mathbf{A}$ *and* $\mathbf{B})$ = 0.22, the probability of **A** but not **B** must be $0.78 - 0.22 = 0.56$.

The yellow region is **B** but not **A**.

The gray region outside both circles represents the outcome neither **A** nor **B**. All the probabilities must total 1, so you can determine the probability of that region by subtraction.

Then, figure out what you want to know. The probabilities can come from the diagram or a formula. Sometimes translating the words to equations is the trickiest step.

Let **A** = {suspect is given a breath test}.
Let **B** = {suspect is given a blood test}.

I know that

$$P(\mathbf{A}) = 0.78$$
$$P(\mathbf{B}) = 0.36$$
$$P(\mathbf{A} \text{ and } \mathbf{B}) = 0.22$$

So
$$P(\mathbf{A} \text{ and } \mathbf{B}^C) = 0.78 - 0.22 = 0.56$$
$$P(\mathbf{B} \text{ and } \mathbf{A}^C) = 0.36 - 0.22 = 0.14$$
$$P(\mathbf{A}^C \text{ and } \mathbf{B}^C) = 1 - (0.56 + 0.22 + 0.14)$$
$$= 0.08$$

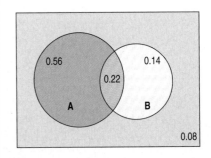

Question 1. What is the probability that the suspect is given a test?

Show | **Mechanics** The probability the suspect is given a test is $P(A \text{ or } B)$. We can use the General Addition Rule, or we can add the probabilities seen in the diagram.

$$P(A \text{ or } B) = P(A) + P(B) - P(A \text{ and } B)$$
$$= 0.78 + 0.36 - 0.22$$
$$= 0.92$$

OR

$$P(A \text{ or } B) = 0.56 + 0.22 + 0.14 = 0.92$$

Tell | **Conclusion** Don't forget to interpret your result in context.

92% of all suspects are given a test.

Question 2. What is the probability that the suspect gets either a blood test or a breath test but NOT both?

Show | **Mechanics** We can use the rule, or just add the appropriate probabilities seen in the Venn diagram.

$$P(A \text{ or } B \text{ but NOT both}) = P(A \text{ or } B) - P(A \text{ and } B)$$
$$= 0.92 - 0.22 = 0.70$$

OR

$$P(A \text{ or } B \text{ but NOT both}) = P(A \text{ and } B^C) + P(B \text{ and } A^C)$$
$$= 0.56 + 0.14 = 0.70$$

Tell | **Conclusion** Interpret your result in context.

70% of the suspects get exactly one of the tests.

Question 3. What is the probability that the suspect gets neither test?

Show | **Mechanics** Getting neither test is the complement of getting one or the other. Use the Complement Rule or just notice that "neither test" is represented by the region outside both circles.

$$P(\text{neither test}) = 1 - P(\text{either test})$$
$$= 1 - P(A \text{ or } B)$$
$$= 1 - 0.92 = 0.08$$

OR

$$P(A^C \text{ and } B^C) = 0.08$$

Tell | **Conclusion** Interpret your result in context.

Only 8% of the suspects get no test.

It Depends . . .

Two psychologists surveyed 478 children in grades 4, 5, and 6 in elementary schools in Michigan. They stratified their sample, drawing roughly 1/3 from rural, 1/3 from suburban, and 1/3 from urban schools. Among other questions, they asked the students whether their primary goal was to get good grades, to be popular, or to be good at sports. One question of interest was whether boys and girls at this age had similar goals.

Here's a *contingency table* giving counts of the students by their goals and sex.

		Goals			
		Grades	Popular	Sports	Total
Sex	Boy	117	50	60	227
	Girl	130	91	30	251
	Total	247	141	90	478

The distribution of goals for boys and girls. **Table 15.1**

A S **Birthweights and Smoking.** Does smoking increase the chance of having a baby with low birthweight? Check the conditional probabilities.

We looked at contingency tables and graphed *conditional distributions* back in Chapter 3. The graphs show the *relative frequencies* with which boys and girls named the three goals. It's only a short step from these relative frequencies to probabilities.

Let's focus on this study and make the sample space just the set of these 478 students. If we select a student at random from this study, the probability we select a girl is just the corresponding relative frequency (since we're equally likely to select any of the 478 students). There are 251 girls in the data out of a total of 478, giving a probability of

$$P(\text{girl}) = 251/478 = 0.525$$

The same method works for more complicated events like intersections. For example, what's the probability of selecting a girl whose goal is to be popular? Well, 91 girls named popularity as their goal, so the probability is

$$P(\text{girl } and \text{ popular}) = 91/478 = 0.190$$

The probability of selecting a student whose goal is to excel at sports is

$$P(\text{sports}) = 90/478 = 0.188$$

We can now build on what we did in Chapter 3. Now that we've defined the sample space as these 478 students, we can recognize the relative frequencies as probabilities.

What if we are given the information that the selected student is a girl? Would that change the probability that the selected student's goal is sports? You bet it would! The pie charts show that girls are much less likely to say their goal is to excel at sports than are boys. When we restrict our focus to girls, we look only at the girls' row of the table, which gives the conditional distribution of goals given "girl." Of the 251 girls, only 30 of them said their goal was to excel at sports.

We write the probability that a selected student wants to excel at sports *given that we have selected a girl* as

$$P(\text{sports}|\text{girl}) = 30/251 = 0.120$$

For boys, we look at the conditional distribution of goals given "boy" shown in the top row of the table. There, of the 227 boys, 60 said their goal was to excel at sports. So, $P(\text{sports}|\text{boy}) = 60/227 = 0.264$, more than twice the girls' probability.

In general, when we want the probability of an event from a *conditional* distribution, we write $P(\mathbf{B}|\mathbf{A})$ and pronounce it "the probability of **B** *given* **A**." A probability that takes into account a given *condition* such as this is called a **conditional probability.**

Let's look at what we did. We worked with the counts, but we could work with the probabilities just as well. There were 30 students who both were girls and had sports as their goal, and there are 251 girls. So we found the probability to be 30/251. To find the probability of the event **B** *given* the event **A**, we restrict our attention to the outcomes in **A**. We then find in what fraction of *those* outcomes **B** also occurred. Formally, we write:

$$P(\mathbf{B}|\mathbf{A}) = \frac{P(\mathbf{A} \text{ and } \mathbf{B})}{P(\mathbf{A})}.$$

Thinking this through, we can see that it's just what we've been doing, but with probabilities rather than with counts. Look back at the girls for whom sports was the goal. How did we calculate $P(\text{sports}|\text{girl})$?

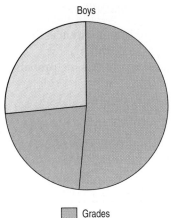

Girls

Boys

☐ Grades
☐ Popular
☐ Sports

The distribution of goals for boys and girls. **Figure 15.1**

NOTATION ALERT:

$P(\mathbf{B}|\mathbf{A})$ is the conditional probability of **B** *given* **A**.

The rule says to use probabilities. It says to find $P(\mathbf{A}$ *and* $\mathbf{B})/P(\mathbf{A})$. The result is the same whether we use counts or probabilities because the total number in the sample cancels out:

$$\frac{P(\text{sports } and \text{ girl})}{P(\text{girl})} = \frac{30/478}{251/478} = \frac{30}{251}.$$

A S **Conditional Probability.**
Simulation is great for seeing conditional probabilities at work. Here's a hands-on way to get a feel for them.

To use the formula for conditional probability, we're supposed to insist on one restriction. The formula doesn't work if $P(\mathbf{A})$ is 0. That would mean that we were "given" the fact that **A** was true even though the probability of **A** is 0, which would be a contradiction. The formula works only when the event that's given has probability greater than 0.

Let's take our rule out for a spin. What's the probability that we have selected a girl *given* that the selected student's goal is popularity? Applying the rule, we get

$$P(\text{girl}|\text{popular}) = \frac{P(\text{girl } and \text{ popular})}{P(\text{popular})}$$

$$= \frac{91/478}{141/478} = \frac{91}{141}.$$

The General Multiplication Rule

Remember the **Multiplication Rule** for the probability of **A** *and* **B**? It said

$$P(\mathbf{A} \ and \ \mathbf{B}) = P(\mathbf{A}) \times P(\mathbf{B}) \text{ when } \mathbf{A} \text{ and } \mathbf{B} \text{ are independent.}$$

Now we can write a more general rule that doesn't require **independence.** In fact, we've *already* written it down. We just need to rearrange the equation a bit.

The equation in the definition for conditional probability contains the probability of **A** *and* **B**. Rearranging the equation gives

A S **The General Multiplication Rule.** The best way to understand the General Multiplication Rule is with an experiment. See the rule at work here.

$$P(\mathbf{A} \ and \ \mathbf{B}) = P(\mathbf{A}) \times P(\mathbf{B}|\mathbf{A}).$$

This is a **General Multiplication Rule** for compound events that does not require the events to be independent. Better than that, it even makes sense. The probability that two events, **A** and **B**, *both* occur is the probability that event **A** occurs multiplied by the probability that event **B** *also* occurs—that is, by the probability that event **B** occurs *given* that event **A** occurs.

Of course, there's nothing special about which set we call **A** and which one we call **B**. We should be able to state this the other way around. And indeed we can. It is equally true that

$$P(\mathbf{A} \ and \ \mathbf{B}) = P(\mathbf{B}) \times P(\mathbf{A}|\mathbf{B}).$$

Independence

Let's return to the question of just what it means for events to be independent. We've said informally that what we mean by independence is that the outcome of

A S **Independence.** Are *Smoking* and *Low Birthweight* independent? The General Multiplication Rule helps us find out.

one event does not influence the probability of the other. With our new notation for conditional probabilities, we can write a formal definition: Events **A** and **B** are **independent** whenever

$$P(\mathbf{B}|\mathbf{A}) = P(\mathbf{B}).$$

Now we can see that the Multiplication Rule for independent events we saw in Chapter 14 is just a special case of the General Multiplication Rule. The general rule says

$$P(\mathbf{A} \ and \ \mathbf{B}) = P(\mathbf{A}) \times P(\mathbf{B}|\mathbf{A})$$

whether the events are independent or not. But when events **A** and **B** are independent, we can write $P(\mathbf{B})$ for $P(\mathbf{B}|\mathbf{A})$ and we get back our simple rule:

$$P(\mathbf{A} \ and \ \mathbf{B}) = P(\mathbf{A}) \times P(\mathbf{B}).$$

Sometimes people use this statement as the definition of independent events, but we find the other definition more intuitive. Either way, the idea is that the probabilities of independent events don't change when you find out that one of them has occurred.

Is the probability of having good grades as a goal independent of the sex of the responding student? Looks like it might be. We need to check whether

$$P(\text{grades}|\text{girl}) = P(\text{grades})$$

$$\frac{130}{251} = 0.52 \stackrel{?}{=} \frac{247}{478} = 0.52$$

To two decimal place accuracy, it looks like we can consider choosing good grades as a goal to be independent of sex.

On the other hand, $P(\text{sports})$ is 90/478, or about 18.8%, but $P(\text{sports}|\text{boy})$ is $60/227 = 26.4\%$. Because these probabilities aren't equal, we can be pretty sure that choosing success in sports as a goal is not independent of the student's sex.

Independent ≠ Disjoint

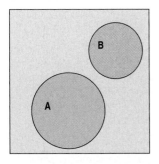

Disjoint events **A** and **B**. These events are mutually exclusive.

Are disjoint events independent? Both concepts seem to have similar ideas of separation and distinctness about them, but in fact disjoint events *cannot* be independent. Let's see why. Consider the two disjoint events {you get an A in this course} and {you get a B in this course}. They're disjoint because they have no outcomes in common. Suppose you learn that you *did* get an A in the course. Now what is the probability that you got a B? You can't get both grades, so it must be 0.

Think about what that means. Knowing that the first event (getting an A) occurred changed your probability for the second event (down to 0). So these events aren't independent.

Mutually exclusive events can't be independent. They have no outcomes in common, so knowing that one occurred means the other didn't. A common error is to treat disjoint events as if they were independent, and apply the Multiplication Rule for independent events. Don't make that mistake.

Depending on Independence

A S **Is There a Hot Hand in Basketball?** Even the experts can be fooled about independent events. Most coaches and fans believe that basketball players sometimes get "hot" and make more of their shots. What do the conditional probabilities say?

A S **Hot Hand Simulation.** Here's a challenge for basketball fans. Can you tell the difference between real and simulated sequences of shot hits and misses?

It's much easier to think about independent events than to deal with conditional probabilities. It seems that most people's natural intuition for probabilities breaks down when it comes to conditional probabilities. Someone may estimate the probability of a compound event by multiplying the probabilities of its component events together without asking seriously whether those probabilities are independent.

For example, experts have assured us that the probability of a major commercial nuclear plant failure is so small that we should not expect such a failure to occur even in a span of hundreds of years. After only a few decades of commercial nuclear power, however, the world has seen two failures (Chernobyl and Three Mile Island). How could the estimates have been so wrong?

One simple part of the failure calculation is to test a particular valve and determine that valves such as this one fail only once in, say, 100 years of normal use. For a coolant failure to occur, several valves must fail. So we need the compound probability, *P*(valve 1 fails *and* valve 2 fails *and* . . .). A simple risk assessment might multiply the small probability of one valve failure together as many times as needed.

But if the valves all came from the same manufacturer, a flaw in one might be found in the others. And maybe when the first fails, it puts additional pressure on the next one in line. In either case, the events aren't independent and so we can't simply multiply the probabilities together.

Whenever you see probabilities multiplied together, stop and ask whether you think they are really independent.

Tables and Conditional Probability

One of the easiest ways to think about conditional probabilities is with contingency tables. We did that earlier in the chapter when we began our discussion. But sometimes we're given probabilities without a table. You can often construct a simple table to correspond to the probabilities.

For example, in the drunk driving data, we are told that 78% of suspect drivers get a breath test, 36% a blood test, and 22% both. That's enough information. Translating percentages to probabilities, what we know looks like this:

		Breath Test		
		Yes	No	Total
Blood Test	Yes	0.22		0.36
	No			
	Total	0.78		1.00

Notice that the 0.78 and 0.36 are *marginal* probabilities and so they go into the *margins.* The 0.22 is the probability of getting both tests—a breath test *and* a blood test, so that's a *joint* probability. Those belong in the interior of the table.

Because the cells of the table show disjoint events, the probabilities always add to the marginal totals going across rows or down columns. So, filling in the rest of the table is quick:

		Breath Test		
		Yes	No	Total
Blood Test	**Yes**	0.22	0.14	**0.36**
	No	0.56	0.08	**0.64**
	Total	**0.78**	**0.22**	**1.00**

Compare this with the Venn Diagram on page 341. Notice which entries in the table match up with the sets in this diagram. Whether a Venn diagram or a table is better to use will depend on what you are given and the questions you're being asked. Try both.

Are the Events Disjoint? Independent? Step-By-Step

Let's take another look at the drunk driving situation. Police report that 78% of drivers are given a breath test, 36% a blood test, and 22% both tests.

1. Are giving a DWI suspect a blood test and a breath test mutually exclusive?
2. Are giving the two tests independent?

Think **Plan** Define the events we're interested in.

Let **A** = {suspect is given a breath test}.
Let **B** = {suspect is given a blood test}.

State the given probabilities.

I know that
$$P(A) = 0.78$$
$$P(B) = 0.36$$
$$P(A \text{ and } B) = 0.22$$

Question 1. Are giving a DWI suspect a blood test and a breath test mutually exclusive?

Show **Mechanics** Disjoint events cannot *both* happen at the same time, so check to see if $P(A \text{ and } B) = 0$.

$P(A \text{ and } B) = 0.22$. Since some suspects are given both tests, $P(A \text{ and } B) \neq 0$. The events are *not* mutually exclusive.

Tell **Conclusion** State your conclusion in context.

22% of all suspects get both tests, so a breath test and a blood test are *not* disjoint events.

Question 2. Are the two tests independent?

Make a table (make a table, make…)

		Breath Test		
		Yes	No	Total
Blood Test	**Yes**	0.22	0.14	**0.36**
	No	0.56	0.08	**0.64**
	Total	**0.78**	**0.22**	**1.00**

Show **Mechanics** Does getting a breath test change the probability of getting a blood test? That is, does $P(B|A) = P(B)$?

Because the two probabilities are *not* the same, the events are not independent.

$$P(B|A) = \frac{P(A \text{ and } B)}{P(A)} = \frac{0.22}{0.78} \approx 0.28$$

$$P(B) = 0.36$$

$$P(B|A) \neq P(B)$$

Tell **Conclusion** Interpret your results in context.

Overall, 36% of the drivers get blood tests, but only 28% of those who get a breath test do. Since suspects who get a breath test are less likely to have a blood test, the two events are not independent.

2 Remember our sample of pages in this book from this chapter's first Just Checking…? We found that

48% of pages had a data display,

27% of pages had an equation, and

7% of pages had both a data display and an equation.

a) Make a contingency table for the variables *display* and *equation*.

b) What is the probability that a randomly selected sample page with an equation also had a data display?

c) Are having an equation and having a data display disjoint events?

d) Are having an equation and having a data display independent events?

Drawing Without Replacement

Room draw is a process for assigning dormitory rooms to students who live on campus. Sometimes, when students have equal priority, they are randomly assigned to the currently available dorm rooms. When it's time for you and your friend to draw, there are 12 rooms left. Three are in Gold Hall, a very desirable dorm with spacious wood-paneled rooms. Four are in Silver Hall, centrally located, but not quite as desirable. And five are in Wood Hall, a new dorm with cramped rooms, located half a mile from the center of campus on the edge of the woods.

You get to draw first, and then your friend will draw. Naturally, you would both like to score rooms in Gold. What are your chances? In particular, what's the chance that you *both* can get rooms in Gold?

When you go first, the chance that *you* will draw one of the Gold rooms is 3/12. Suppose you do. Now, with you clutching your prized room assignment, what chance does your friend have? At this point there are only 11 rooms left and just 2 left in Gold, so your friend's chance is now 2/11.

Using our notation, we write

$$P(\text{friend draws Gold} \mid \text{you draw Gold}) = 2/11.$$

The reason the denominator changes is that we draw these rooms *without replacement*. That is, once one is drawn, it doesn't go back into the pool.

We often sample without replacement. When we draw from a very large population, the change in the denominator is too small to worry about. But when there's a small population to draw from, as in this case, we need to take note and adjust the probabilities.

What are the chances that *both* of you will luck out? Well, now we've calculated the two probabilities we need for the General Multiplication Rule, so we can write:

$$P(\text{you} = \text{Gold } and \text{ friend} = \text{Gold})$$
$$= P(\text{you} = \text{Gold}) \times P(\text{friend} = \text{Gold} \,|\, \text{you} = \text{Gold})$$
$$= 3/12 \times 2/11 = 1/22 = 0.045$$

In this instance, it doesn't matter who went first, or even if the rooms were drawn simultaneously. Even if the room draw was accomplished by shuffling cards containing the names of the dormitories and then dealing them out to 12 applicants (rather than by each student drawing a room in turn) we can still *think* of the calculation as having taken place in two steps:

$$\xrightarrow{3/12} \text{Gold} \xrightarrow{2/11} \text{Gold} \,|\, \text{Gold}$$

That is, one of you has a probability of 3/12 of drawing a Gold room and the other then has a probability of 2/11 of also drawing a Gold room. It doesn't matter whose draw we think of first. The probability changes for the second person nonetheless. The diagram shows this ordering of our thoughts.

Diagramming conditional probabilities leads to a more general way of helping us think with pictures—one that works for calculating conditional probabilities even when they involve different variables.

Tree Diagrams

For men, binge drinking is defined as having five or more drinks in a row, and for women as having four or more drinks in a row. (The difference is because of the average difference in weight.) According to a study by the Harvard School of Public Health (H. Wechsler, G. W. Dowdall, A. Davenport, and W. DeJong, "Binge Drinking on Campus: Results of a National Study"), 44% of college students engage in binge drinking, 37% drink moderately, and 19% abstain entirely. Another study, published in the *American Journal of Health Behavior,* finds that among binge drinkers aged 21 to 34, 17% have been involved in an alcohol-related automobile accident, while among nonbingers of the same age, only 9% have been involved in such accidents.

What's the probability that a randomly selected college student will be a binge drinker who has had an alcohol-related car accident?

To start, we see that the probability of selecting a binge drinker is about 44%. To find the probability of selecting someone who is both a binge drinker and a driver with an alcohol-related accident, we would need to pull out the General Multiplication Rule and multiply the probability of one of the events by the conditional probability of the other given the first.

Or we *could* make a picture. Which would you prefer?

We thought so.

The kind of picture that helps us think through this kind of reasoning is called a **tree diagram,** because it shows sequences of events, like those we had in room draw, as paths that look like branches of a tree. It is a good idea to make a tree diagram almost any time you plan to use the General Multiplication Rule. The number of different paths we can take can get large, so we usually draw the tree starting from the left and growing vine-like across the page, although sometimes you'll see them drawn from the bottom up or top down.

The first branch of our tree separates students according to their drinking habits. We label each branch of the tree with a possible outcome and its corresponding probability.

"Why," said the Dodo, "the best way to explain it is to do it."

—Lewis Carroll

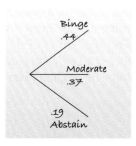

We can diagram the three outcomes of drinking and indicate their respective probabilities with a simple tree diagram.
Figure 15.2

Notice that we cover all possible outcomes with the branches. The probabilities add up to one. But we're also interested in car accidents. The probability of having an alcohol-related accident *depends* on one's drinking behavior. Because the probabilities are *conditional,* we draw the alternatives separately on each branch of the tree:

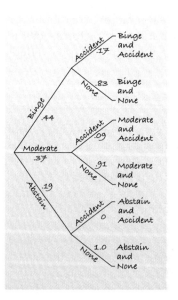

Extending the tree diagram, we can show both drinking and accident outcomes. The accident probabilities are conditional on the drinking outcomes, and they change depending on which branch we follow. Because we are concerned only with *alcohol-related* accidents, the conditional probability $P(\text{accident} \mid \text{abstinence})$ must be 0. **Figure 15.3**

On each of the second set of branches, we write the possible outcomes associated with having an alcohol-related car accident (having an accident or not) and the associated probability. These probabilities are different because they are *conditional* depending on the student's drinking behavior. (It shouldn't be too surprising that those who binge drink have a higher probability of alcohol-related accidents.) The probabilities add up to one, because given the outcome on the first branch, these outcomes cover all the possibilities. Looking back at the General Multiplication Rule, we can see how the tree depicts the calculation. To find the probability that a randomly selected student will be a binge drinker who has had an alcohol-related car accident, we follow the top branches. The probability of selecting a binger is 0.44. The conditional probability of an accident *given* binge drinking is 0.17. The General Multiplication Rule tells us that to find the *joint* probability of being a binge drinker and having an accident, we multiply these two probabilities together:

$$P(\text{binge } and \text{ accident}) = P(\text{binge}) \times P(\text{accident}|\text{binge})$$

$$= 0.44 \times 0.17 = 0.075$$

And we can do the same for each combination of outcomes:

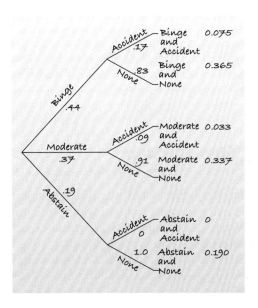

We can find the probabilities of compound events by multiplying the probabilities along the branch of the tree that leads to the event, just the way the General Multiplication Rule specifies.

The probability of abstaining and having an alcohol-related accident is, of course, zero.

Figure 15.4

All the outcomes at the far right are disjoint because at each branch of the tree we chose between disjoint alternatives. And they are *all* the possibilities, so the probabilities on the far right must add up to one.

Because the final outcomes are disjoint, we can add up their probabilities to get probabilities for compound events. For example, what's the probability that a selected student has had an alcohol-related car accident? We simply find *all* the outcomes on the far right in which an accident has happened. There are three and we can add their probabilities: 0.075 + 0.033 + 0 = 0.108—almost an 11% chance.

Reversing the Conditioning

If we know a student has had an alcohol-related accident, what's the probability that the student is a binge drinker? That's an interesting question, but we can't just read it from the tree. The tree gives us $P(\text{accident}|\text{binge})$, but we want

$P(\text{binge} | \text{accident})$—conditioning in the other direction. The two probabilities are definitely *not* the same. We have reversed the conditioning.

We may not have the conditional probability we want, but we do know everything we need to know to find it. To find a conditional probability, we need the probability that both events happen divided by the probability that the given event occurs. We have already found the probability of an alcohol-related accident: $0.075 + 0.033 + 0 = 0.108.$

The joint probability that a student is both a binge drinker and someone who's had an alcohol-related accident is found at the top branch: 0.075. We've restricted the *Who* of the problem to the students with alcohol-related accidents, so we divide the two to find the conditional probability:

$$P(\text{binge} | \text{accident}) = \frac{P(\text{binge } and \text{ accident})}{P(\text{accident})}$$

$$= \frac{0.075}{0.108} = 0.694$$

The chance that a student who has an alcohol-related car accident is a binge drinker is more than 69%! As we said, reversing the conditioning is rarely intuitive, but tree diagrams help us keep track of the calculation when there aren't too many alternatives to consider.

Reversing the Conditioning   Step-By-Step

When the authors were in college, there were only three requirements for graduation that were the same for all students: You had to be able to tread water for 2 minutes, you had to learn a foreign language, and you had to be free of tuberculosis. For the last requirement, all freshmen had to take a TB screening test that consisted of a nurse jabbing what looked like a corn cob holder into your forearm. You were then expected to report back in 48 hours to have it checked. If you were healthy and TB-free, your arm was supposed to look as though you'd never had the test.

Sometime during the 48 hours, one of us had a reaction. When he finally saw the nurse, his arm was about 50% bigger than normal and a very unhealthy red. Did he have TB? The nurse had said that the test was about 99% effective, so it seemed that the chances must be pretty high that he had TB. How high do you think the chances were? Go ahead and guess. Guess low.

We'll call **TB** the event of actually having TB and **+** the event of testing positive. To start a tree, we need to know $P(\mathbf{TB})$, the probability of having TB. This isn't given, so we looked it up. Even today TB is a fairly uncommon disease, with an incidence of about 5 cases per 10,000 in the United States, so $P(\mathbf{TB}) = 0.0005$. We also need to know the conditional probabilities $P(\mathbf{+}|\mathbf{TB})$ and $P(\mathbf{+}|\mathbf{TB}^C)$. Diagnostic tests can make two kinds of errors. They can give a positive result for a healthy person (a *false positive*) or a negative result for a sick person (a *false negative*). Being 99% accurate usually means a false-positive rate of 1%. That is, someone who doesn't have the disease has a 1% chance of testing positive anyway. We can write $P(\mathbf{+}|\mathbf{TB}^C) = 0.01$.

Since a false negative is more serious (because a sick person might not get treatment), tests are usually constructed to have a lower false-negative rate. We don't know exactly, but let's assume a 0.1% false-negative rate. So only 0.1% of sick people test negative. We can write $P(\mathbf{-}|\mathbf{TB}) = 0.001$.

Think   **Plan** Define the events we're interested in and their probabilities.

Let **TB** = {having TB} and **TB**^{C} = {no TB}
+ = {testing positive} and
− = {testing negative}

I know that $P(+|\mathbf{TB}^C) = 0.01$ and $P(-|\mathbf{TB}) = 0.001$. I also know that $P(\mathbf{TB}) = 0.0005$.

Figure out what you want to know in terms of the events. Use the notation of conditional probability to write the event whose probability you want to find.

I'm interested in the probability that the author had TB given that he tested positive: $P(\mathbf{TB}|+)$.

Show   **Plot** Draw the tree diagram. When probabilities are very small like these are, be careful to keep all the significant digits.

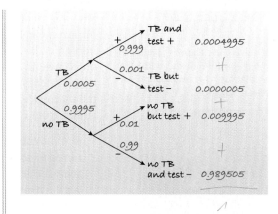

To finish the tree we need $P(\mathbf{TB}^C)$, $P(-|\mathbf{TB}^C)$ and $P(-|\mathbf{TB})$. We can find each of these from the Complement Rule:

$$P(\mathbf{TB}^C) = 1 - P(\mathbf{TB}) = 0.9995$$

$$P(-|\mathbf{TB}^C) = 1 - P(+|\mathbf{TB}^C)$$
$$= 1 - 0.01 = 0.99 \text{ and}$$

$$P(+|\mathbf{TB}) = 1 - P(-|\mathbf{TB})$$
$$= 1 - 0.001 = 0.999$$

Mechanics Multiply along the branches to find the probabilities of the four possible outcomes. It's a good idea to check your work by seeing if they total 1.

(Check: $0.0004995 + 0.0000005 + 0.0099950 + 0.9895050 = 1$)

Add up the probabilities corresponding to the condition of interest—in this case, testing positive. We can add the probabilities from the tree twigs that correspond to testing positive because the tree shows disjoint events.

$$P(+) = P(\mathbf{TB} \text{ and } +) + P(\mathbf{TB}^C \text{ and } +)$$
$$P(+) = 0.0004995 + 0.009995$$
$$= 0.010495$$

Divide the probability of both events occuring (here, having TB given a positive test) by the probability (or frequency) of satisfying the condition (testing positive).

$$P(\mathbf{TB}|+) = \frac{P(\mathbf{TB} \text{ and } +)}{P(+)}$$
$$= \frac{0.0004995}{0.010495}$$
$$= 0.047$$

Tell   **Conclusion** Interpret your result in context.

The chance of having TB even after you test positive is less than 5%.

When we reverse the order of conditioning, we change the *Who* we are concerned with. With events of low probability, the result can be surprising. That's the reason patients who test positive for HIV, for example, are always told to seek medical counseling. They may have only a small chance of actually being infected. That's why global drug or disease testing can have unexpected consequences if people interpret *testing* positive as *being* positive.

Bayes's Rule

The Reverend Thomas Bayes is credited posthumously with the rule that is the foundation of Bayesian Statistics.

Nicholas Saunderson (1682–1739) was a blind English mathematician who invented a tactile board to help other blind people do mathematics. And he may have been the true originator of "Bayes's Rule."

When we have $P(\mathbf{A}|\mathbf{B})$ but want the *reverse* probability $P(\mathbf{B}|\mathbf{A})$, we need to find $P(\mathbf{A} \textit{ and } \mathbf{B})$ and $P(\mathbf{A})$. A tree is often a convenient way of finding these probabilities. It can work even when we have more than two possible events, as we saw in the binge drinking example. Instead of using the tree, we *could* write the calculation algebraically, showing exactly how we found the quantities that we needed: $P(\mathbf{A} \textit{ and } \mathbf{B})$ and $P(\mathbf{A})$. The result is a formula known as Bayes's Rule, after the Reverend Thomas Bayes (1702?–1761), who was credited with the rule after his death, when he could no longer defend himself. Bayes's Rule is quite important in Statistics and is the foundation of an approach to Statistical analysis known as Bayesian Statistics. Although the simple rule deals with two alternative outcomes, the rule can be extended to the situation in which there are more than two branches to the first split of the tree. The principle remains the same (although the math gets more difficult). Bayes's Rule is just a formula for reversing the probability from the conditional probability that you're originally given.[2]

● **Who discovered Bayes's Rule?** Stigler's "Law of Eponymy" states that discoveries named for someone (eponyms) are usually named for the wrong person. Steven Stigler, who admits he didn't originate the law, is an expert on the history of Statistics, and he suspected that the law might apply to Bayes's Rule. He looked at the possibility that another candidate—one Nicholas Saunderson—was the real discoverer, not the Reverend Bayes. He assembled historical evidence and compared probabilities that the historical events would have happened *given* that Bayes was the discoverer of the rule, with the corresponding probabilities *given* that Saunderson was the discoverer. Of course, what he really wanted to know were the probabilities that Bayes or Saunderson was the discoverer *given* the historical events. How did he *reverse* the conditional probabilities? He used Bayes's Rule and concluded that, actually, it's more likely that Saunderson is the real originator of the rule.

But that doesn't change our tradition of naming the rule for Bayes and calling the branch of Statistics arising from this approach Bayesian Statistics. The Bayesians would never stand for being called Saundersonians, anyway. ●

What Can Go Wrong?

- **Don't use a simple probability rule where a general rule is appropriate.** Don't assume independence without reason to believe it. Don't assume that outcomes are disjoint without checking that they are. Remember that the general rules always apply even when outcomes are in fact independent or disjoint.
- **Don't find probabilities for samples drawn without replacement as if they had been drawn with replacement.** Remember to adjust the denominator of your probabilities. This warning applies only when we draw from small populations or draw a large fraction of a finite population. When the population is very large relative to the sample size, the adjustments make very little difference, and we ignore them.
- **Don't reverse conditioning naively.** As we have seen, the probability of **A** *given* **B** may not, and, in general does not, resemble the probability of **B** *given* **A.** The true probability may be counterintuitive.

---

[2] Bayes's Rule for two events says that $P(\mathbf{B}|\mathbf{A}) = \dfrac{P(\mathbf{A}|\mathbf{B})P(\mathbf{B})}{P(\mathbf{A}|\mathbf{B})P(\mathbf{B}) + P(\mathbf{A}|\mathbf{B^c})P(\mathbf{B^c})}$. Try it with the TB testing probabilities. (It's easier to just draw the tree, isn't it?)

- **Don't confuse "disjoint" with "independent."** Disjoint events *cannot* happen at the same time. When one happens, you know the other did not, so $P(\mathbf{B}|\mathbf{A}) = 0$. Independent events *must* be able to happen at the same time. When one happens, you know it has no effect on the other, so $P(\mathbf{B}|\mathbf{A}) = P(\mathbf{B})$.

CONNECTIONS

This chapter shows the unintuitive side of probability. If you've been thinking, "My mind doesn't work this way," you're probably right. Humans don't seem to find conditional and compound probabilities natural and often have trouble with them. Even statisticians make mistakes with conditional probability.

Our central connection is to the guiding principle that Statistics is about understanding the world. The events discussed in this chapter are close to the kinds of real-world situations in which understanding probabilities matters. The methods and concepts of this chapter are the tools you need to understand the part of the real world that deals with the outcomes of complex, uncertain events.

What have we learned?

The last chapter's basic rules of probability are important, but they work only in special cases—when events are disjoint or independent. Now we've learned the more versatile General Addition Rule and General Multiplication Rule. We've also learned about conditional probabilities, and seen that reversing the conditioning can give surprising results.

We've learned the value of Venn diagrams, tables, and tree diagrams to help organize our thinking about probabilities.

Most important, we've learned to think clearly about independence. We've seen how to use conditional probability to determine whether two events are independent and to work with events that are not independent. A sound understanding of independence will be important throughout the rest of this book.

TERMS

Sample space The collection of all possible outcome values. The sample space has a probability of 1.

Addition Rule If **A** and **B** are disjoint events, then the probability of **A** or **B** is

$$P(\mathbf{A} \text{ or } \mathbf{B}) = P(\mathbf{A}) + P(\mathbf{B}).$$

Disjoint events Two events are *disjoint* (or *mutually exclusive*) if they have no outcomes in common.

General Addition Rule For any two events, **A** and **B,** the probability of **A** or **B** is

$$P(\mathbf{A} \text{ or } \mathbf{B}) = P(\mathbf{A}) + P(\mathbf{B}) - P(\mathbf{A} \text{ and } \mathbf{B}).$$

Conditional probability $P(\mathbf{B}|\mathbf{A}) = \dfrac{P(\mathbf{A} \text{ and } \mathbf{B})}{P(\mathbf{A})}$

$P(\mathbf{B}|\mathbf{A})$ is read "the probability of **B** *given* **A**."

Independence (used casually) Two events are *independent* if knowing whether one event occurs does not alter the probability that the other event occurs.

Multiplication Rule	If **A** and **B** are independent events, then the probability of **A** and **B** is
	$$P(\textbf{A } and \textbf{ B}) = P(\textbf{A}) \times P(\textbf{B}).$$
General Multiplication Rule	For any two events, **A** and **B,** the probability of **A** and **B** is
	$$P(\textbf{A } and \textbf{ B}) = P(\textbf{A}) \times P(\textbf{B}\,|\,\textbf{A}).$$
Independence (used formally)	Events **A** and **B** *are* independent when $P(\textbf{B}\,|\,\textbf{A}) = P(\textbf{B})$.
Tree diagram	A display of conditional events or probabilities that is helpful in thinking through conditioning.

SKILLS

When you complete this lesson you should:

Think

- Understand that the probability of an event is the proportion of times it occurs in many repetitions of a random phenomenon.
- Understand the concept of conditional probability as redefining the *Who* of concern, according to the information about the event that is *given.*
- Understand the concept of independence.

Show

- Know how and when to apply the General Addition Rule.
- Know how to find probabilities for compound events as fractions of counts of occurrences in a two-way table.
- Know how and when to apply the General Multiplication Rule.
- Know how to make and use a tree diagram to understand conditional probabilities and reverse conditioning.

Tell

- Be able to make a clear statement about a conditional probability that makes clear how the condition affects the probability.
- Avoid making statements that assume independence of events when there is no clear evidence that they are in fact independent.

EXERCISES

1. Sample spaces. For each of the following, list the sample space and tell whether you think the outcomes are equally likely.
a) Toss 2 coins; record the order of heads and tails.
b) A family has 3 children; record the number of boys.
c) Flip a coin until you get a head or 3 consecutive tails.
d) Roll two dice; record the larger number.

2. Sample spaces. For each of the following, list the sample space and tell whether you think the outcomes are equally likely.
a) Roll two dice; record the sum of the numbers.
b) A family has 3 children; record the genders in order of birth.
c) Toss four coins; record the number of tails.
d) Toss a coin 10 times; record the longest run of heads.

3. Homes. Real estate ads suggest that 64% of homes for sale have garages, 21% have swimming pools, and 17% have both features. What is the probability that a home for sale has
a) a pool or a garage?
b) neither a pool nor a garage?
c) a pool but no garage?

4. Travel. Suppose the probability that a U.S. resident has traveled to Canada is 0.18, to Mexico is 0.09, and to both countries is 0.04. What's the probability that an American chosen at random has
a) traveled to Canada but not Mexico?
b) traveled to either Canada or Mexico?
c) not traveled to either country?

5. Amenities. A check of dorm rooms on a large college campus revealed that 38% had refrigerators, 52% had TVs, and 21% had both a TV and a refrigerator. What's the probability that a randomly selected dorm room has
 a) a TV but no refrigerator?
 b) a TV or a refrigerator, but not both?
 c) neither a TV nor a refrigerator?

6. Workers. Employment data at a large company reveal that 72% of the workers are married, that 44% are college graduates, and that half of the college grads are married. What's the probability that a randomly chosen worker
 a) is neither married nor a college graduate?
 b) is married but not a college graduate?
 c) is married or a college graduate?

7. First lady. A Gallup survey of June 2004 asked 1005 U.S. adults who they think better fits their idea of what a first lady should be, Laura Bush or Hillary Rodham Clinton. Suppose the data break down as follows:

		Age Group			
		18–29	30–49	50–64	Over 65	Total
Response	Clinton	135	158	79	65	437
	Bush	77	237	112	92	518
	Equally/Neither/ No opinion	5	21	14	10	50
	Total	217	416	205	167	1005

If we select a person at random from this sample:
 a) What is the probability that the person thought Laura Bush best fits their first lady ideal?
 b) What is the probability that the person is younger than 50 years old?
 c) What is the probability that the person is younger than 50 *and* thinks Hillary Clinton best fits their ideal?
 d) What is the probability that the person is younger than 50 *or* thinks Hillary Clinton best fits their ideal?

8. Birth order. A survey of students in a large Introductory Statistics class asked about their birth order (1 = oldest or only child) and which college of the university they were enrolled in. Here are the data:

		Birth Order		
		1 or only	2 or more	Total
College	Arts & Sciences	34	23	57
	Agriculture	52	41	93
	Human Ecology	15	28	43
	Other	12	18	30
	Total	113	110	223

Suppose we select a student at random from this class.
 a) What is the probability we select a Human Ecology student?
 b) What is the probability that we select a first-born student?
 c) What is the probability that the person is first-born *and* a Human Ecology student?
 d) What is the probability that the person is first-born *or* a Human Ecology student?

9. Cards. You draw a card at random from a standard deck of 52 cards. Find each of the following conditional probabilities:
 a) The card is a heart, given that it is red.
 b) The card is red, given that it is a heart.
 c) The card is an ace, given that it is red.
 d) The card is a queen, given that it is a face card.

10. Pets. In its monthly report, the local animal shelter states that it currently has 24 dogs and 18 cats available for adoption. Eight of the dogs and 6 of the cats are male. Find each of the following conditional probabilities if an animal is selected at random:
 a) The pet is male, given that it is a cat.
 b) The pet is a cat, given that it is female.
 c) The pet is female, given that it is a dog.

11. Health. The probabilities that an adult American man has high blood pressure and/or high cholesterol are shown in the table.

		Blood Pressure	
		High	OK
Cholesterol	High	0.11	0.21
	OK	0.16	0.52

 a) What's the probability that a man has both conditions?
 b) What's the probability that he has high blood pressure?
 c) What's the probability that a man with high blood pressure has high cholesterol?
 d) What's the probability that a man has high blood pressure if it's known that he has high cholesterol?

12. Death penalty. The table shows the political affiliation of American voters and their positions on the death penalty.

		Death Penalty	
		Favor	Oppose
Party	Republican	0.26	0.04
	Democrat	0.12	0.24
	Other	0.24	0.10

a) What's the probability that a randomly chosen voter favors the death penalty?
b) What's the probability that a Republican favors the death penalty?
c) What's the probability that a voter who favors the death penalty is a Democrat?
d) A candidate thinks she has a good chance of gaining the votes of anyone who is a Republican or in favor of the death penalty. What portion of the voters is that?

13. First lady, take 2. Look again at the data from the Gallup survey on first ladies in Exercise 7.
a) If we select a respondent at random, what's the probability we choose a person between 18 and 29 who picked Clinton?
b) Among the 18- to 29-year-olds, what is the probability that a person responded "Clinton"?
c) What's the probability that a person who chose Clinton was between 18 and 29?
d) If the person responded "Bush," what is the probability that they are over 65?
e) What's the probability that a person over 65 preferred Bush?

14. Birth order, take 2. Look again at the data about birth order of Intro Stats students and their choices of colleges shown in Exercise 8.
a) If we select a student at random, what's the probability the person is an Arts and Sciences student who is a second child (or more)?
b) Among the Arts and Sciences students, what's the probability a student was a second child (or more)?
c) Among second children (or more), what's the probability the student is enrolled in Arts and Sciences?
d) What's the probability that a first or only child is enrolled in the Agriculture College?
e) What is the probability that an Agriculture student is a first or only child?

15. Sick kids. Seventy percent of kids who visit a doctor have a fever, and 30% of kids with a fever have sore throats. What's the probability that a kid who goes to the doctor has a fever and a sore throat?

16. Sick cars. Twenty percent of cars that are inspected have faulty pollution control systems. The cost of repairing a pollution control system exceeds $100 about 40% of the time. When a driver takes her car in for inspection, what's the probability that she will end up paying more than $100 to repair the pollution control system?

17. Cards. You are dealt a hand of three cards, one at a time. Find the probability of each of the following.
a) The first heart you get is the third card dealt.
b) Your cards are all red (that is, all diamonds or hearts).
c) You get no spades.
d) You have at least one ace.

18. Another hand. You pick three cards at random from a deck. Find the probability of each event described below.

a) You get no aces.
b) You get all hearts.
c) The third card is your first red card.
d) You have at least one diamond.

19. Batteries. A junk box in your room contains a dozen old batteries, five of which are totally dead. You start picking batteries one at a time and testing them. Find the probability of each outcome.
a) The first two you choose are both good.
b) At least one of the first three works.
c) The first four you pick all work.
d) You have to pick 5 batteries in order to find one that works.

20. Shirts. The soccer team's shirts have arrived in a big box, and people just start grabbing them, looking for the right size. The box contains 4 medium, 10 large, and 6 extra-large shirts. You want a medium for you and one for your sister. Find the probability of each event described.
a) The first two you grab are the wrong sizes.
b) The first medium shirt you find is the third one you check.
c) The first four shirts you pick are all extra-large.
d) At least one of the first four shirts you check is a medium.

21. Eligibility. A university requires its biology majors to take a course called BioResearch. The prerequisite for this course is that students must have taken either a Statistics course or a computer course. By the time they are juniors, 52% of the Biology majors have taken Statistics, 23% have had a computer course, and 7% have done both.
a) What percent of the junior Biology majors are ineligible for BioResearch?
b) What's the probability that a junior Biology major who has taken Statistics has also taken a computer course?
c) Are taking these two courses disjoint events? Explain.
d) Are taking these two courses independent events? Explain.

22. Benefits. Fifty-six percent of all American workers have a workplace retirement plan, 68% have health insurance, and 49% have both benefits. We select a worker at random.
a) What's the probability he has neither employer-sponsored health insurance nor a retirement plan?
b) What's the probability he has health insurance if he has a retirement plan?
c) Are having health insurance and a retirement plan independent events? Explain.
d) Are having these two benefits mutually exclusive? Explain.

23. For sale. In the real estate ads described in Exercise 3, 64% of homes for sale have garages, 21% have swimming pools, and 17% have both features.
a) If a home for sale has a garage, what's the probability that it has a pool, too?

b) Are having a garage and a pool independent events? Explain.

c) Are having a garage and a pool mutually exclusive? Explain.

24. **On the road again.** According to Exercise 4, the probability that a U.S. resident has traveled to Canada is 0.18, to Mexico is 0.09, and to both countries is 0.04.
   a) What's the probability that someone who has traveled to Mexico has visited Canada, too?
   b) Are travel to Mexico and Canada disjoint events? Explain.
   c) Are travel to Mexico and Canada independent events? Explain.

25. **Cards.** If you draw a card at random from a well-shuffled deck, is getting an ace independent of the suit? Explain.

26. **Pets again.** The local animal shelter in Exercise 10 reported that it currently has 24 dogs and 18 cats available for adoption; 8 of the dogs and 6 of the cats are male. Are the species and gender of the animals independent? Explain.

27. **First lady, final visit.** In Exercises 7 and 13 we looked at results of a Gallup Poll that asked people whether they thought Laura Bush or Hillary Clinton better fits their idea of a first lady.
   a) Are being under 30 and being over 65 disjoint? Explain.
   b) Are being under 30 and being over 65 independent? Explain.
   c) Are answering "Clinton" and being over 65 disjoint? Explain.
   d) Are answering "Clinton" and being over 65 independent? Explain.

28. **Birth order, finis.** In Exercises 8 and 14 we looked at the birth orders and college choices of some Intro Stats students.
   a) Are enrolling in Agriculture and Human Ecology disjoint? Explain.
   b) Are enrolling in Agriculture and Human Ecology independent? Explain.
   c) Are being first-born and enrolling in Human Ecology disjoint? Explain.
   d) Are being first-born and enrolling in Human Ecology independent? Explain.

29. **Men's health, again.** Given the table of probabilities from Exercise 11, are high blood pressure and high cholesterol independent? Explain.

		Blood Pressure	
		High	**OK**
Cholesterol	**High**	0.11	0.21
	OK	0.16	0.52

30. **Politics.** Given the table of probabilities from Exercise 12, are party affiliation and position on the death penalty independent? Explain.

		Death Penalty	
		Favor	**Oppose**
Party	**Republican**	0.26	0.04
	Democrat	0.12	0.24
	Other	0.24	0.10

31. **Phone service.** According to estimates from the federal government's 2003 National Health Interview Survey, based on face-to-face interviews in 16,677 households, approximately 58.2% of U.S. adults have both a land line in their residence and a cell phone, 2.8% have only cell phone service but no land line, and 1.6% have no telephone service at all.
   a) Polling agencies won't phone cell phone numbers because customers object to paying for such calls. What proportion of U.S. households can be reached by a land line call?
   b) Are having a cell phone and having a land line independent? Explain.

32. **Snoring.** After surveying 995 adults, 81.5% of whom were over 30, the National Sleep Foundation reported that 36.8% of all the adults snored. 32% of the respondents were snorers over the age of 30.
   a) What percent of the respondents were under 30 and did not snore?
   b) Is snoring independent of age? Explain.

33. **Montana.** A 1992 poll conducted by the University of Montana classified respondents by gender and political party, as shown in the table. Is party affiliation independent of sex? Explain.

	Democrat	Republican	Independent
Male	36	45	24
Female	48	33	16

34. **Cars.** A random survey of autos parked in student and staff lots at a large university classified the brands by country of origin, as seen in the table. Is country of origin independent of type of driver?

		Driver	
		Student	**Staff**
Origin	**American**	107	105
	European	33	12
	Asian	55	47

35. Luggage. Leah is flying from Boston to Denver with a connection in Chicago. The probability her first flight leaves on time is 0.15. If the flight is on time, the probability that her luggage will make the connecting flight in Chicago is 0.95, but if the first flight is delayed, the probability that the luggage will make it is only 0.65.

a) Are the first flight leaving on time and the luggage making the connection independent events? Explain.

b) What is the probability that her luggage arrives in Denver with her?

36. Graduation. A private college report contains these statistics:

70% of incoming freshmen attended public schools.

75% of public school students who enroll as freshmen eventually graduate.

90% of other freshmen eventually graduate.

a) Is there any evidence that a freshman's chances to graduate may depend upon what kind of high school the student attended? Explain.

b) What percent of freshmen eventually graduate?

37. Late luggage. Remember Leah (Exercise 35)? Suppose you pick her up at the Denver airport, and her luggage is not there. What is the probability that Leah's first flight was delayed?

38. Graduation, part II. What percent of students who graduate from the college in Exercise 36 attended a public high school?

39. Absenteeism. A company's records indicate that on any given day about 1% of their day shift employees and 2% of the night shift employees will miss work. Sixty percent of the employees work the day shift.

a) Is absenteeism independent of shift worked? Explain.

b) What percent of employees are absent on any given day?

40. Lungs and smoke. Suppose that 23% of adults smoke cigarettes. It's known that 57% of smokers and 13% of nonsmokers develop a certain lung condition by age 60.

a) Explain how these statistics indicate that lung condition and smoking are not independent.

b) What's the probability that a randomly selected 60-year-old has this lung condition?

41. Absenteeism, part II. At the company described in Exercise 39, what percent of the absent employees are on the night shift?

42. Lungs and smoke, again. Based on the statistics in Exercise 40, what's the probability that someone with the lung condition was a smoker?

43. Drunks. Police often set up sobriety checkpoints—roadblocks where drivers are asked a few brief questions to allow the officer to judge whether or not the person may have been drinking. If the officer does not suspect a problem, drivers are released to go on their way. Other-

wise, drivers are detained for a Breathalyzer test that will determine whether or not they are arrested. The police say that based on the brief initial stop, trained officers can make the right decision 80% of the time. Suppose the police operate a sobriety checkpoint after 9 p.m. on a Saturday night, a time when national traffic safety experts suspect that about 12% of drivers have been drinking.

a) You are stopped at the checkpoint and, of course, have not been drinking. What's the probability that you are detained for further testing?

b) What's the probability that any given driver will be detained?

c) What's the probability that a driver who is detained has actually been drinking?

d) What's the probability that a driver who was released had actually been drinking?

44. Polygraphs. Lie detectors are controversial instruments, barred from use as evidence in many courts. Nonetheless, many employers use lie detector screening as part of their hiring process in the hope that they can avoid hiring people who might be dishonest. There has been some research, but no agreement, about the reliability of polygraph tests. Based on this research, suppose that a polygraph can detect 65% of lies, but incorrectly identifies 15% of true statements as lies.

A certain company believes that 95% of its job applicants are trustworthy. The company gives everyone a polygraph test, asking, "Have you ever stolen anything from your place of work?" Naturally, all the applicants answer "No," but the polygraph identifies some of those answers as lies, making the person ineligible for a job. What's the probability that a job applicant rejected under suspicion of dishonesty was actually trustworthy?

45. Dishwashers. Dan's Diner employs three dishwashers. Al washes 40% of the dishes and breaks only 1% of those he handles. Betty and Chuck each wash 30% of the dishes, and Betty breaks only 1% of hers, but Chuck breaks 3% of the dishes he washes. (He, of course, will need a new job soon. . . .) You go to Dan's for supper one night and hear a dish break at the sink. What's the probability that Chuck is on the job?

46. Parts. A company manufacturing electronic components for home entertainment systems buys electrical connectors from three suppliers. The company prefers to use supplier A because only 1% of those connectors prove to be defective, but supplier A can deliver only 70% of the connectors needed. The company also must purchase connectors from two other suppliers, 20% from supplier B and the rest from supplier C. The rates of defective connectors from B and C are 2% and 4%, respectively. You buy one of these components, and when you try to use it you find that the connector is defective. What's the probability that your component came from supplier A?

Just Checking

Answers

1. a)

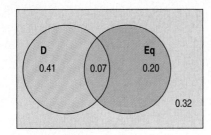

b) 0.32

c) 0.41

2. a)

		Equation		
		Yes	**No**	**Total**
Display	**Yes**	0.07	0.41	**0.48**
	No	0.20	0.32	**0.52**
	Total	**0.27**	**0.73**	**1.00**

b) $P(\mathbf{D}|\mathbf{Eq}) = P(\mathbf{D}\ and\ \mathbf{Eq})/P(\mathbf{Eq}) = 0.07/0.27 = 0.259$

c) No, pages can (and 7% do) have both.

d) To be independent, we'd need $P(\mathbf{D}|\mathbf{Eq}) = P(\mathbf{D})$.
$P(\mathbf{D}|\mathbf{Eq}) = 0.259$, but $P(\mathbf{D}) = 0.48$. Overall, 48% of pages have data displays, but only about 26% of pages with equations do. They do not appear to be independent.

*CHAPTER 16

Random Variables

Insurance companies make bets. They bet that you're going to live a long life. You bet that you're going to die sooner. Both you and the insurance company want the company to stay in business, so it's important to find a "fair price" for your bet. Of course, the right price for *you* depends on many factors, and nobody can predict exactly how long you'll live. But when the company averages over enough customers, it can make reasonably accurate estimates of the amount it can expect to collect on a policy before it has to pay its benefit.

Here's a simple example. An insurance company offers a "death and disability" policy that pays $10,000 when you die or $5000 if you are permanently disabled. It charges a premium of only $50 a year for this benefit. Is the company likely to make a profit selling such a plan? To answer this question, the company needs to know the *probability* that its clients will die or be disabled in any year. From actuarial information like this, the company can calculate the expected value of this policy.

Expected Value: Center

We'll want to build a probability model in order to answer the questions about the insurance company's risk. First we need to define a few terms. The amount the company pays out on an individual policy is called a **random variable** because its value is based on the outcome of a random event. We use a capital letter, like X, to denote a random variable. We'll denote a particular value that it can have by the corresponding lower-case letter, in this case x. For the insurance company, x can be $10,000 (if you die that year), $5000 (if you are disabled), or $0 (if neither occurs). Because we can list all the outcomes, we might formally call this random variable a **discrete** random variable. Otherwise, we'd call it a **continuous** random variable.

A S **Random Variables.** Learn more about random variables from this animated tour.

The collection of all the possible values and the probabilities that they occur is called the **probability model** for the random variable.

Suppose, for example, that the death rate in any year is 1 out of every 1000 people, and that another 2 out of 1000 suffer some kind of disability. Then we can display the probability model for this insurance policy in a table like this:

Policyholder Outcome	Payout x	Probability $P(X = x)$
Death	10,000	$\dfrac{1}{1000}$
Disability	5000	$\dfrac{2}{1000}$
Neither	0	$\dfrac{997}{1000}$

To see what the insurance company can expect, imagine that it insures exactly 1000 people. Further imagine that, in perfect accordance with the probabilities, 1 of the policyholders dies, 2 are disabled, and the remaining 997 survive the year unscathed. The company would pay $10,000 to one client and $5000 to each of 2 clients. That's a total of $20,000, or an average of 20000/1000 = $20 per policy. Since it is charging people $50 for the policy, the company expects to make a profit of $30 per customer. Not bad!

We can't predict what *will* happen during any given year, but we can say what we *expect* to happen. To do this, we (or, rather, the insurance company) need the probability model. The expected value of a policy is a parameter of this model. In fact, it's the mean. We'll signify this with the notation μ (for population mean) or $E(X)$ for expected value. This isn't an average of some data values, so we won't estimate it. Instead, we assume that the probabilities are known and simply calculate the expected value from them.

How did we come up with $20 as the expected value of a policy payout? Here's the calculation. As we've seen, it often simplifies probability calculations to think about some (convenient) number of outcomes. For example, we could imagine that we have exactly 1000 clients. Of those, exactly 1 died and 2 were disabled, corresponding to what the probabilities would say.

$$\mu = E(X) = \frac{10,000(1) + 5000(2) + 0(997)}{1000}$$

So our total payout comes to $20,000, or $20 per policy.

Instead of writing the expected value as one big fraction, we can rewrite it as separate terms each divided by 1000.

$$\mu = E(X)$$
$$= \$10,000\left(\frac{1}{1000}\right) + \$5000\left(\frac{2}{1000}\right) + \$0\left(\frac{997}{1000}\right)$$
$$= \$20$$

How convenient! See the probabilities? For each policy, there's a 1/1000 chance that we'll have to pay $10,000 for a death and a 2/1000 chance that we'll have to

NOTATION ALERT:

The expected value (or mean) of a random variable is written $E(X)$ or μ.

pay $5000 for a disability. Of course, there's a 997/1000 chance that we won't have to pay anything.

Take a good look at the expression now. It's easy to calculate the **expected value** of a (discrete) random variable—just multiply each possible value by the probability that it occurs, and find the sum:

$$\mu = E(X) = \sum x \cdot P(X = x).$$

Be sure that every possible outcome is included in the sum. And verify that you have a valid probability model to start with—the probabilities should each be between 0 and 1 and should sum to one.

just checking

1. One of the authors took his minivan in for repair recently because the air conditioner was cutting out intermittently. The mechanic identified the problem as dirt in a control unit. He said that in about 75% of such cases, drawing down and then recharging the coolant a couple of times cleans up the problem—and costs only $60. If that fails, then the control unit must be replaced at an additional cost of $100 for parts and $40 for labor.

   a) Define the random variable and construct the probability model.
   b) What is the expected value of the cost of this repair?
   c) What does that mean in this context?

   Oh—in case you were wondering—the $60 fix worked!

First Center, Now Spread . . .

Of course, this expected value (or mean) is not what actually happens to any *particular* policyholder. No individual policy actually costs the company $20. We are dealing with random events, so some policyholders receive big payouts, others nothing. Because the insurance company must anticipate this variability, it needs to know the *standard deviation* of the random variable.

For data, we calculated the **standard deviation** by first computing the deviation from the mean and squaring it. We do that with (discrete) random variables as well. First, we find the deviation of each payout from the mean (expected value):

Policyholder Outcome	Payout x	Probability $P(X = x)$	Deviation $(x - \mu)$
Death	10,000	$\frac{1}{1000}$	$(10,000 - 20) = 9980$
Disability	5000	$\frac{2}{1000}$	$(5000 - 20) = 4980$
Neither	0	$\frac{997}{1000}$	$(0 - 20) = -20$

Next, we square each deviation. The **variance** is the expected value of those squared deviations, so we multiply each by the appropriate probability and sum those products. That gives us the variance of X. Here's what it looks like:

$$Var(X) = 9980^2\left(\frac{1}{1000}\right) + 4980^2\left(\frac{2}{1000}\right) + (-20)^2\left(\frac{997}{1000}\right) = 149{,}600.$$

Finally, we take the square root to get the standard deviation:

$$SD(X) = \sqrt{149,600} \approx \$386.78.$$

The insurance company can expect an average payout of $20 per policy, with a standard deviation of $386.78.

Think about that. The company charges $50 for each policy and expects to pay out $20 per policy. Sounds like an easy way to make $30. In fact, most of the time (probability 997/1000) the company pockets the entire $50. But would you consider selling your roommate such a policy? The problem is that occasionally the company loses big. With probability 1/1000, it will pay out $10,000, and with probability 2/1000, it will pay out $5000. That may be more risk than you're willing to take on. The standard deviation of $386.78 gives an indication that it's no sure thing. That's a pretty big spread (and risk) for an average profit of $30.

Here are the formulas for what we just did. Because these are parameters of our probability model, the variance and standard deviation can also be written as σ^2 and σ. You should recognize both kinds of notation.

$$\sigma^2 = Var\ (X) = \sum (x - \mu)^2 \cdot P(X = x)$$
$$\sigma = SD(X) = \sqrt{Var(X)}$$

Expected Values and Standard Deviations for Discrete Random Variables Step-By-Step

As the head of inventory for Knowway computer company, you were thrilled that you had managed to ship 2 computers to your biggest client the day the order arrived. You are horrified, though, to find out that someone had restocked refurbished computers in with the new computers in your storeroom. The shipped computers were selected randomly from the 15 computers in stock, but 4 of those were actually refurbished.

If your client gets 2 new computers, things are fine. If the client gets one refurbished computer, it will be sent back at your expense—$100—and you can replace it. However, if both computers are refurbished, the client will cancel the order this month and you'll lose $1000. What's the expected value and the standard deviation of your loss?

Think

Plan State the problem.

I want to find the company's expected loss for shipping refurbished computers, and the standard deviation.

Variable Define the random variable.

Let X = amount of loss.

Plot Make a picture. This is another job for tree diagrams.

If you prefer calculation to drawing, find $P(NN)$ and $P(RR)$, then use the Complement Rule to find $P(NR\ or\ RN)$.

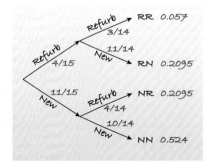

Outcome	x	P(X = x)
Two refurbs	1000	P(**RR**) = 0.057
One refurb	100	P(**NR** or **RN**) = 0.2095 + 0.2095 = 0.419
New/new	0	P(**NN**) = 0.524

Model List the possible values of the random variable, and determine the probability model.

Show

Mechanics Find the expected value.

$$E(X) = 0(0.524) + 100(0.419) + 1000(0.057)$$
$$= \$98.90$$

Find the variance.

$$Var(X) = (0 - 98.90)^2 \times 0.524$$
$$+ (100 - 98.90)^2 \times 0.419$$
$$+ (1000 - 98.90)^2 \times 0.057$$
$$= 51,408.79$$

Find the standard deviation.

$$SD(X) = \sqrt{51,408.79} = \$226.735$$

Tell

Conclusion Interpret your results in context.

I expect this mistake to cost the firm $98.90, with a standard deviation of $226.74. The large standard deviation reflects the fact that there's a pretty large range of possible losses.

REALITY CHECK > Both numbers seem reasonable. The expected value of $98.90 is between the extremes of $0 and $1000, and there's great variability in the outcome values.

More About Means and Variances

Our insurance company expected to pay out an average of $20 per policy, with a standard deviation of about $387. If we take the $50 premium into account, we see the company makes a profit of 50 − 20 = $30 per policy. Suppose the company lowers the premium by $5 to $45. It's pretty clear that the expected profit also drops an average of $5 per policy, to 45 − 20 = $25.

What about the standard deviation? We know that adding or subtracting a constant from data shifts the mean but doesn't change the variance or standard deviation. The same is true of random variables.[1]

$$E(X \pm c) = E(X) \pm c \qquad Var(X \pm c) = Var(X)$$

What if the company decides to double all the payouts—that is, pay $20,000 for deaths and $10,000 for disability? This would double the average payout per policy and also increase the variability in payouts. We have seen that multiplying or dividing all data values by a constant changes both the mean and the standard deviation by the same factor. Variance, being the square of standard deviation, changes by the square of the constant. The same is true of random variables. In general, multiplying each value of a random variable by a constant multiplies the mean by that constant and the variance by the *square* of the constant.

$$E(aX) = aE(X) \qquad Var(aX) = a^2 Var(X)$$

---

[1] The rules in this section are true for both discrete *and* continuous random variables.

This insurance company sells policies to more than just one person. How can we figure means and variances for a collection of customers? For example, how can the company find the total expected value (and standard deviation) of policies taken over all policyholders? Consider a simple case: just two customers, Mr. Ecks and Ms. Wye. With an expected payout of $20 on each policy, we might expect a total of $20 + $20 = $40 to be paid out on the two policies. Nothing surprising there. The expected value of the sum is the sum of the expected values.

The variability is another matter. Is the risk of insuring two people the same as the risk of insuring one person for twice as much? We wouldn't expect both clients to die or become disabled in the same year. Because we've spread the risk, the standard deviation should be smaller. Indeed, this is the fundamental principle behind insurance. By spreading the risk among many policies, a company can keep the standard deviation quite small and predict costs more accurately.

But how much smaller is the standard deviation of the sum? It turns out that, if the random variables are independent, there is a simple Addition Rule for variances: *The variance of the sum of two independent random variables is the sum of their variances.*

For Mr. Ecks and Ms. Wye, the insurance company can expect their outcomes to be independent, so (using X for Mr. Ecks's payout and Y for Ms. Wye's)

$$Var(X + Y) = Var(X) + Var(Y)$$
$$= 149{,}600 + 149{,}600$$
$$= 299{,}200.$$

If they had insured only Mr. Ecks for twice as much, there would only be one outcome rather than two *independent* outcomes, so the variance would have been

$$Var(2X) = 2^2Var(X) = 4 \times 149{,}600 = 598{,}400, \text{ or}$$

twice as big as with two independent policies.

Of course, variances are in squared units. The company would prefer to know standard deviations, which are in dollars. The standard deviation of the payout for two independent policies is $\sqrt{299{,}200} = \$546.99$. But the standard deviation of the payout for a single policy of twice the size is $\sqrt{598{,}400} = \$773.56$, or about 40% more.

If the company has two customers, then, it will have an expected annual total payout of $40 with a standard deviation of about $547.

In general,

- *The mean of the sum of two random variables is the sum of the means.*
- *The mean of the difference of two random variables is the difference of the means.*
- *If the random variables are independent, the variance of their sum or difference is always the sum of the variances.*

$$E(X \pm Y) = E(X) \pm E(Y) \qquad Var(X \pm Y) = Var(X) + Var(Y)$$

Pythagorean Theorem of Statistics

We often use the standard deviation to measure variability, but when we add independent random variables, we use their variances. Think of the Pythagorean Theorem. In a right triangle (only), the *square* of the length of the hypotenuse is the sum of the *squares* of the lengths of the other two sides:

$c^2 = a^2 + b^2$.

For independent random variables (only), the *square* of the standard deviation of their sum is the sum of the *squares* of their standard deviations:

$SD^2(X + Y) = SD^2(X) + SD^2(Y)$.

It's simpler to write this with *variances:*

$Var(X + Y) = Var(X) + Var(Y)$.

Variances add—just ask Pythagoras!

Wait a minute! Is that third part correct? Do we always *add* variances? Yes. Think about the two insurance policies. Suppose we want to know the mean and standard deviation of the *difference* in payouts to the two clients. Since each policy has an expected payout of $20, the expected difference is 20 − 20 = $0. If we also subtract variances, we get $0, too, and that surely doesn't make sense. Note that if the outcomes for the two clients are independent, the difference in payouts could range from $10,000 − $0 = $10,000 to $0 − $10,000 = −$10,000, a spread of $20,000. The variability in differences increases as much as the variability in sums. If the company has two customers, the difference in payouts has a mean of $0 and a standard deviation of about $547 (again).

● **For random variables, does $X + X + X = 3X$?** Maybe. But as we've just seen, insuring one person for $30,000 is not the same risk as insuring three people for $10,000 each. When each instance represents a different outcome for the same random variable, though, it's easy to fall into the trap of writing all of them with the same symbol. Don't make this common mistake. Make sure you write each instance as a *different* random variable. Just because each random variable describes a similar situation doesn't mean that each random outcome will be the same.

These are *random* variables, not the variables you saw in Algebra. Being random, they take on different values each time they're evaluated. So what you really mean is $X_1 + X_2 + X_3$. Written this way, it's clear that the sum shouldn't necessarily equal 3 times *anything*. ●

just checking

2 Suppose the time it takes a customer to get and pay for seats at the ticket window of a baseball park is a random variable with a mean of 100 seconds and a standard deviation of 50 seconds. When you get there, you find only two people in line in front of you.

a) How long do you expect to wait for your turn to get tickets?
b) What's the standard deviation of your wait time?
c) What assumption did you make about the two customers in finding the standard deviation?

Hitting the Road Step-By-Step

You're planning to spend next year wandering through the mountains of Kyrgyzstan. You plan to sell your used Isuzu Trooper so you can purchase an off-road Honda motor scooter when you get there. Used Isuzus of the year and mileage of yours are selling for a mean of $6940 with a standard deviation of $250. Your research shows that scooters in Kyrgyzstan are going for about 65,000 Kyrgyzstan soms with a standard deviation of 500 soms. You have to survive on your profit, so you want to estimate what you can expect in your pocket after the sale and subsequent purchase. One U.S. dollar is worth about 43 Kyrgyzstan soms.

Think   Plan   State the problem.

I want to estimate how much money I'd have (in soms) after selling my Isuzu and buying the scooter.

Variables   Define the random variables.

Let A = sale price of my Isuzu (in dollars),
    B = price of a scooter (in soms), and
    D = profit (in soms).

Write an appropriate equation.
Check the conditions.

$D = 43A - B$

✔ **Independence:** The prices are independent.

Show   Mechanics   Find the expected value, using the appropriate rules.

$$E(D) = E(43A - B)$$
$$= 43E(A) - E(B)$$
$$= 43(6,940) - (65,000)$$
$$E(D) = 233,420 \text{ soms}$$

Find the variance, using the appropriate rules. Be sure to check the conditions first!

Since sale and purchase prices are independent,
$$Var(D) = Var(43A - B)$$
$$= (43)^2 Var(A) + Var(B)$$
$$= 1849(250)^2 + (500)^2$$
$$Var(D) = 115,812,500$$

Find the standard deviation.

$$SD(D) = \sqrt{115,812,500} = 10,762 \text{ soms}$$

Tell   Conclusion   Interpret your results in context. (Here that means talking about dollars.)

I can expect to clear about 233,420 soms with a standard deviation of 10,762 soms.

REALITY CHECK ⟩ Given the initial cost estimates, the mean and standard deviation seem reasonable.

In dollars, I calculate a mean profit of about $5428, with a standard deviation of about $250.

Continuous Random Variables

A S **Numeric Outcomes.**
You've seen how to simulate continuous random outcomes. It turns out there's a tool for simulating continuous outcomes, too. See it here.

A company manufactures small stereo systems. At the end of the production line, the stereos are packaged and prepared for shipping. Stage 1 of this process is called "packing." Workers must collect all the system components (a main unit, two speakers, a power cord, an antenna, and some wires), put each in plastic bags, and then place everything inside a protective styrofoam form. The packed form then moves on to Stage 2, called "boxing." There, workers place the form and a packet of instructions in a cardboard box, close it, then seal and label the box for shipping.

The company says that times required for the packing stage can be described by a Normal model with a mean of 9 minutes and standard deviation of 1.5 minutes. The times for the boxing stage can also be modeled as Normal, with a mean of 6 minutes and standard deviation of 1 minute.

This is a common way to model events. Do our rules for random variables apply here? What's different? We no longer have a list of discrete outcomes, with their associated probabilities. Instead, we have **continuous** random variables that can take on any value. Now any single value won't have a probability. We saw this back in Chapter 6 when we first saw the Normal model (although we didn't talk then about "random variables" or "probability"). We know that the probability that $z = 1.5$ doesn't make sense, but we *can* talk about the probability that z lies *between* 0.5 and 1.5. For a Normal random variable (and any other continuous random variable) the probability that it falls within an interval is just the area under the curve over that interval.

A S **Means of Random Variables.** Experiment with continuous random variables to learn how their expected values behave.

As we've seen with Normal models, continuous random variables have means (which we also call *expected values*) and variances. In this book we won't worry about how to calculate them, but we can still work with models for continuous random variables when we're given these parameters.

The good news is that nearly everything we've said about how discrete random variables behave is true of continuous random variables, as well. When two independent continuous random variables have Normal models, so does their sum or difference. This simple fact is a special property of Normal models and is very important. It allows us to apply our knowledge of Normal probabilities to questions about the sum or difference of independent random variables.

Packaging Stereos Step-By-Step

Consider the company that manufactures and ships small stereo systems that we discussed above.

Recall that times required to pack the stereos can be described by a Normal model with a mean of 9 minutes and standard deviation of 1.5 minutes. The times for the boxing stage can also be modeled as Normal, with a mean of 6 minutes and standard deviation of 1 minute.

Two questions:

1. What is the probability that packing two consecutive systems takes over 20 minutes?
2. What percentage of the stereo systems take longer to pack than to box?

Question 1: What is the probability that packing two consecutive systems takes over 20 minutes?

Think

Plan State the problem.

> I want to estimate the probability that packing two consecutive systems takes over 20 minutes.
>
> Let P_1 = time for packing the first system
>
> P_2 = time for packing the second
>
> T = total time to pack two systems

Variables Define your random variables.

Write an appropriate equation.

> $T = P_1 + P_2$

Check the conditions. Sums of independent Normal random variables follow a Normal model. Such simplicity isn't true in general.

> ✔ **Normal models:** We are told that both random variables follow Normal models.
>
> ✔ **Independence:** We can reasonably assume that the two packing times are independent.

Show

Mechanics Find the expected value.

> $E(T) = E(P_1 + P_2)$
>
> $\quad\quad = E(P_1) + E(P_2)$
>
> $\quad\quad = 9 + 9 = 18$ minutes

For sums of independent random variables, variances add. (We don't need the variables to be Normal for this to be true—just independent.)

> Since the times are independent,
>
> $Var(T) = Var(P_1 + P_2)$
>
> $\quad\quad\quad = Var(P_1) + Var(P_2)$
>
> $\quad\quad\quad = 1.5^2 + 1.5^2$
>
> $Var(T) = 4.50$
>
> $SD(T) = \sqrt{4.50} \approx 2.12$ minutes

Find the standard deviation.

Now we use the fact that both random variables follow Normal models to say that their sum is also Normal.

> I'll model T with $N\,(18, 2.12)$.

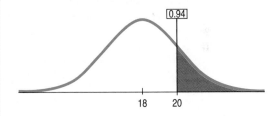

Sketch a picture of the Normal model for the total time, shading the region representing over 20 minutes.

Find the z-score for 20 minutes.

Use technology or a table to find the probability.

> $$z = \frac{20 - 18}{2.12} = 0.94$$
>
> $$P(T > 20) = P(z > 0.94) = 0.1736$$

Tell

Conclusion Interpret your result in context.

> There's a little more than a 17% chance that it will take a total of over 20 minutes to pack two consecutive stereo systems.

Question 2: What percentage of stereo systems take longer to pack than to box?

Think

Plan State the question.

> I want to estimate the percentage of the stereo systems that take longer to pack than to box.

Variables Define your random variables.

> Let P = time for packing a system
>
> B = time for boxing a system
>
> D = difference in times to pack and box a system

Write an appropriate equation.

> $D = P - B$

What are we trying to find?		The probability that it takes longer to pack than to box a system is the probability that the difference $P - B$ is greater than zero.

Don't forget to check the conditions.

✔ **Normal models:** We are told that both random variables follow Normal models.

✔ **Independence:** We can assume that the times it takes to pack and to box a system are independent.

Show **Mechanics** Find the expected value.

$$E(D) = E(P - B)$$
$$= E(P) - E(B)$$
$$= 9 - 6 = 3 \text{ minutes}$$

For the difference of independent random variables, variances add.

Since the times are independent,
$$Var(D) = Var(P - B)$$
$$= Var(P) + Var(B)$$
$$= 1.5^2 + 1^2$$
$$Var(D) = 3.25$$

Find the standard deviation.

$$SD(D) = \sqrt{3.25} \approx 1.80 \text{ minutes}$$

State what model you will use.

I'll model D with $N(3, 1.80)$.

Sketch a picture of the Normal model for the difference in times and shade the region representing a difference greater than zero.

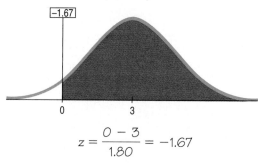

Find the z-score for 0 minutes, then use a table or technology to find the probability.

$$z = \frac{0 - 3}{1.80} = -1.67$$

$$P(D > 0) = P(z > -1.67) = 0.9525$$

Tell **Conclusion** Intepret your result in context.

About 95% of all the stereo systems will require more time for packing than for boxing.

What Can Go Wrong?

- ***Probability models are still just models.*** Models can be useful, but they are not reality. Think about the assumptions behind your models. Are your dice really perfectly fair? (They are probably pretty close.) But when you hear that the probability of a nuclear accident is 1/10,000,000 per year, is that likely to be a precise value? Question probabilities as you would data.
- ***If the model is wrong, so is everything else.*** Before you try to find the mean or standard deviation of a random variable, check to make sure the probability model is reasonable. As a start, the probabilities in your model should add up to 1. If not, you may have calculated a probability incorrectly or left out a value of the random variable. For instance, in the insurance example, the description mentions only death and disability. Good health is by

far the most likely outcome, not to mention the best for both you and the insurance company (who gets to keep your money). Don't overlook that.

To find the expected value of the sum or difference of random variables, we simply add or subtract means. Center is easy; spread is trickier. Watch out for some common traps.

- ***Watch out for variables that aren't independent.*** You can add expected values of *any* two random variables, but you can only add variances of independent random variables. Suppose a survey includes questions about the number of hours of sleep people get each night and also the number of hours they are awake each day. From their answers, we find the mean and standard deviation of hours asleep and hours awake. The expected total must be 24 hours; after all, people are either asleep or awake.[2] The means still add just fine. Since all the totals are exactly 24 hours, however, the standard deviation of the total will be 0. We can't add variances here because the number of hours you're awake depends on the number of hours you're asleep. Be sure to check for independence before adding variances.

- ***Don't forget: Variances of independent random variables add. Standard deviations don't.***

- ***Don't forget: Variances of independent random variables add, even when you're looking at the difference between them.***

- ***Don't write independent instances of a random variable with notation that looks like they are the same variables.*** Make sure you write each instance as a different random variable. Just because each random variable describes a similar situation doesn't mean that each random outcome will be the same. These are *random* variables, not the variables you saw in Algebra. Write $X_1 + X_2 + X_3$ rather than $X + X + X$.

CONNECTIONS

We've seen means, variances, and standard deviations of data. We know that they estimate parameters of models for these data. Now we're looking at the probability models directly. We have only parameters because there are no data to summarize.

It should be no surprise that expected values and standard deviations adjust to shifts and changes of units in the same way as the corresponding data summaries. The fact that we can add variances of independent random quantities is fundamental and will explain why a number of statistical methods work the way they do.

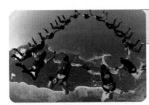

What have we learned?

We've learned to work with random variables. We can use the probability model for a discrete random variable to find its expected value and its standard deviation.

We've learned that the mean of the sum or difference of two random variables, discrete or continuous, is just the sum or difference of their means. And we've learned the

---

[2] Although some students do manage to attain a state of consciousness somewhere between sleeping and wakefulness during Statistics class.

Pythagorean Theorem of Statistics: *For independent random variables,* the variance of their sum or difference is always the *sum* of their variances.

Finally, we've learned that Normal models are once again special. Sums or differences of Normally distributed random variables also follow Normal models.

TERMS

Random variable
A random variable assumes any of several different values as a result of some random event. Random variables are denoted by a capital letter such as X.

Discrete random variable
A random variable that can take one of a finite number of distinct outcomes is called a discrete random variable.

Continuous random variable
A random variable that can take any numeric value within a range of values is called a continuous random variable. The range may be infinite or bounded at either or both ends.

Probability model
The probability model is a function that associates a probability P with each value of a discrete random variable X, denoted $P(X = x)$, or with any interval of values of a continuous random variable.

Expected value
The expected value of a random variable is its theoretical long-run average value, the center of its model. Denoted μ or $E(X)$, it is found (if the random variable is discrete) by summing the products of variable values and probabilities:

$$\mu = E(X) = \sum x \cdot P(X = x).$$

Variance
The variance of a random variable is the expected value of the squared deviation from the mean. For discrete random variables, it can be calculated as:

$$\sigma^2 = Var(X) = \sum (x - \mu)^2 \cdot P(X = x).$$

Standard deviation
The standard deviation of a random variable describes the spread in the model, and is the square root of the variance:

$$\sigma = SD(X) = \sqrt{Var(X)}.$$

Changing a random variable by a constant:
$E(X \pm c) = E(X) \pm c \qquad Var(X \pm c) = Var(X)$
$\quad E(aX) = aE(X) \qquad\qquad Var(aX) = a^2 Var(X)$

Adding or subtracting random variables:
$E(X \pm Y) = E(X) \pm E(Y)$ and *if X and Y are independent,* $Var(X \pm Y) = Var(X) + Var(Y)$ (The Pythagorean Theorem of Statistics).

SKILLS

When you complete this lesson you should:

Think
- Be able to recognize random variables.
- Understand that random variables must be independent in order to determine the variability of their sum or difference by adding variances.

Show
- Be able to find the probability model for a discrete random variable.
- Know how to find the mean (expected value) and the variance of a random variable.

- Always use the proper notation for these population parameters, μ or $E(X)$ for the mean, and σ, $SD(X)$, σ^2, or $Var(X)$ when discussing variability.

- Know how to determine the new mean and standard deviation after adding a constant, multiplying by a constant, or adding or subtracting two independent random variables.

- Be able to interpret the meaning of the expected value and standard deviation of a random variable in the proper context.

Random Variables on the Computer

Statistics packages deal with data, not with random variables. Nevertheless, the calculations needed to find means and standard deviations of random variables are little more than weighted means. Most packages can manage that, but then they are just being overblown calculators. For technological assistance with these calculations, we recommend you pull out your calculator.

TI-83/84 Plus

To calculate the mean and standard deviation of a discrete random variable, enter the probability model in two lists:
- In one list (say, L1) enter the x-values of the variable.
- In a second list (say, L2) enter the associated probabilities $P(X = x)$.
- From the **STAT CALC** menu select **1-VarStats**. L1, L2 and you'll see the mean and standard deviation.

Comments

You can enter the probabilities as fractions; the calculator will change them to decimals for you.
Notice that the calculator knows enough to call the standard deviation σ, but mistakenly uses $\bar{x}$ when it should say μ. Make sure you don't make that mistake!

TI-89

To calculate the mean and standard deviation of a discrete random variable, enter the probability model in two lists:
- In one list (say, list1) enter the x-values of the variable.
- In a second list (say, list2) enter the associated probabilities $P(X = x)$.
- From the **STAT CALC** (F4) menu select **1-VarStats**. Use VAR-LINK to enter the list name list1 in the List box and list2 in the Freq box.

Comments

You can enter the probabilities as fractions; the calculator will change them to decimals for you.
Notice that the calculator knows enough to compute only the standard deviation σ, but mistakenly uses $\bar{x}$ when it should say μ. Make sure you don't make that mistake!

EXERCISES

1. Expected value. Find the expected value of each random variable:

a)

x	10	20	30
P(X = x)	0.3	0.5	0.2

b)

x	2	4	6	8
P(X = x)	0.3	0.4	0.2	0.1

2. Expected value. Find the expected value of each random variable:

a)

x	0	1	2
P(X = x)	0.2	0.4	0.4

b)

x	100	200	300	400
P(X = x)	0.1	0.2	0.5	0.2

3. Pick a card, any card. You draw a card from a deck. If you get a red card, you win nothing. If you get a spade, you win $5. For any club, you win $10 plus an extra $20 for the ace of clubs.
   a) Create a probability model for the amount you win at this game.
   b) Find the expected amount you'll win.
   c) How much would you be willing to pay to play this game?

4. You bet! You roll a die. If it comes up a 6, you win $100. If not, you get to roll again. If you get a 6 the second time, you win $50. If not, you lose.
   a) Create a probability model for the amount you win at this game.
   b) Find the expected amount you'll win.
   c) How much would you be willing to pay to play this game?

5. Kids. A couple plans to have children until they get a girl, but they agree that they will not have more than three children even if all are boys. (Assume boys and girls are equally likely.)
   a) Create a probability model for the number of children they'll have.
   b) Find the expected number of children.
   c) Find the expected number of boys they'll have.

6. Carnival. A carnival game offers a $100 cash prize for anyone who can break a balloon by throwing a dart at it. It costs $5 to play, and you're willing to spend up to $20 trying to win. You estimate that you have about a 10% chance of hitting the balloon on any throw.
   a) Create a probability model for this carnival game.
   b) Find the expected number of darts you'll throw.
   c) Find your expected winnings.

7. Software. A small software company bids on two contracts. It anticipates a profit of $50,000 if it gets the larger contract and a profit of $20,000 on the smaller contract. The company estimates there's a 30% chance it will get the larger contract and a 60% chance it will get the smaller contract. Assuming the contracts will be awarded independently, what's the expected profit?

8. Racehorse. A man buys a racehorse for $20,000, and enters it in two races. He plans to sell the horse afterward, hoping to make a profit. If the horse wins both races, its value will jump to $100,000. If it wins one of the races, it will be worth $50,000. If it loses both races, it will be worth only $10,000. The man believes there's a 20% chance that the horse will win the first race and a 30% chance it will win the second one. Assuming that the two races are independent events, find the man's expected profit.

9. Variation 1. Find the standard deviations of the random variables in Exercise 1.

10. Variation 2. Find the standard deviations of the random variables in Exercise 2.

11. Pick another card. Find the standard deviation of the amount you might win drawing a card in Exercise 3.

12. The die. Find the standard deviation of the amount you might win rolling a die in Exercise 4.

13. Kids. Find the standard deviation of the number of children the couple in Exercise 5 may have.

14. Darts. Find the standard deviation of your winnings throwing darts in Exercise 6.

15. Repairs. The probability model below describes the number of repair calls that an appliance repair shop may receive during an hour.

Repair Calls	0	1	2	3
Probability	0.1	0.3	0.4	0.2

   a) How many calls should the shop expect per hour?
   b) What is the standard deviation?

16. Red lights. A commuter must pass through five traffic lights on her way to work, and will have to stop at each one that is red. She estimates the probability model for the number of red lights she hits, as shown below.

X = # of red	0	1	2	3	4	5
P(X = x)	0.05	0.25	0.35	0.15	0.15	0.05

   a) How many red lights should she expect to hit each day?
   b) What's the standard deviation?

17. Defects. A consumer organization inspecting new cars found that many had appearance defects (dents, scratches, paint chips, etc.). While none had more than three of these defects, 7% had three, 11% two, and 21% one defect. Find the expected number of appearance defects in a new car, and the standard deviation.

18. Insurance. An insurance policy costs $100, and will pay policyholders $10,000 if they suffer a major injury (resulting in hospitalization) or $3000 if they suffer a minor injury (resulting in lost time from work). The company estimates that each year 1 in every 2000 policyholders may have a major injury, and 1 in 500 a minor injury.
   a) Create a probability model for the profit on a policy.
   b) What's the company's expected profit on this policy?
   c) What's the standard deviation?

19. Contest. You play two games against the same opponent. The probability you win the first game is 0.4. If you win the first game, the probability you also win the second is 0.2. If you lose the first game, the probability that you win the second is 0.3.
   a) Are the two games independent? Explain your answer.
   b) What's the probability you lose both games?
   c) What's the probability you win both games?
   d) Let random variable X be the number of games you win. Find the probability model for X.
   e) What are the expected value and standard deviation of X?

20. Contracts. Your company bids for two contracts. You believe the probability you get contract #1 is 0.8. If you get contract #1, the probability you also get contract #2 will be 0.2, and if you do not get #1, the probability you get #2 will be 0.3.
 a) Are the two contracts independent? Explain.
 b) Find the probability you get both contracts.
 c) Find the probability you get no contract.
 d) Let X be the number of contracts you get. Find the probability model for X.
 e) Find the expected value and standard deviation of X.

21. Batteries. In a group of 10 batteries, 3 are dead. You choose 2 batteries at random.
 a) Create a probability model for the number of good batteries you get.
 b) What's the expected number of good ones you get?
 c) What's the standard deviation?

22. Kittens. In a litter of seven kittens, three are female. You pick two kittens at random.
 a) Create a probability model for the number of male kittens you get.
 b) What's the expected number of males?
 c) What's the standard deviation?

23. Random variables. Given independent random variables with means and standard deviations as shown, find the mean and standard deviation of each of these variables:
 a) $3X$
 b) $Y + 6$
 c) $X + Y$
 d) $X - Y$

	Mean	SD
X	10	2
Y	20	5

24. Random variables. Given independent random variables with means and standard deviations as shown, find the mean and standard deviation of each of these variables:
 a) $X - 20$
 b) $0.5Y$
 c) $X + Y$
 d) $X - Y$

	Mean	SD
X	80	12
Y	12	3

25. Random variables. Given independent random variables with means and standard deviations as shown, find the mean and standard deviation of each of these variables:
 a) $0.8Y$
 b) $2X - 100$
 c) $X + 2Y$
 d) $3X - Y$

	Mean	SD
X	120	12
Y	300	16

26. Random variables. Given independent random variables with means and standard deviations as shown, find the mean and standard deviation of each of these variables:
 a) $2Y + 20$
 b) $3X$
 c) $0.25X + Y$
 d) $X - 5Y$

	Mean	SD
X	80	12
Y	12	3

27. Eggs. A grocery supplier believes that in a dozen eggs, the mean number of broken ones is 0.6 with a standard deviation of 0.5 eggs. You buy 3 dozen eggs without checking them.
 a) How many broken eggs do you expect to get?
 b) What's the standard deviation?
 c) What assumptions did you have to make about the eggs in order to answer this question?

28. Garden. A company selling vegetable seeds in packets of 20 estimates that the mean number of seeds that will actually grow is 18, with a standard deviation of 1.2 seeds. You buy 5 different seed packets.
 a) How many bad seeds do you expect to get?
 b) What's the standard deviation?
 c) What assumptions did you make about the seeds? Do you think that assumption is warranted? Explain.

29. Repair calls. Find the mean and standard deviation of the number of repair calls the appliance shop in Exercise 15 should expect during an 8-hour day.

30. Stop! Find the mean and standard deviation of the number of red lights the commuter in Exercise 16 should expect to hit on her way to work during a 5-day work week.

31. Fire! An insurance company estimates that it should make an annual profit of $150 on each homeowner's policy written, with a standard deviation of $6000.
 a) Why is the standard deviation so large?
 b) If it writes only two of these policies, what are the mean and standard deviation of the annual profit?
 c) If it writes 10,000 of these policies, what are the mean and standard deviation of the annual profit?
 d) Do you think the company is likely to be profitable? Explain.
 e) What assumptions underlie your analysis? Can you think of circumstances under which those assumptions might be violated? Explain.

32. Casino. A casino knows that people play the slot machines in hopes of hitting the jackpot, but that most of them lose their dollar. Suppose a certain machine pays out an average of $0.92, with a standard deviation of $120.
 a) Why is the standard deviation so large?
 b) If you play 5 times, what are the mean and standard deviation of the casino's profit?
 c) If gamblers play this machine 1000 times in a day, what are the mean and standard deviation of the casino's profit?
 d) Do you think the casino is likely to be profitable? Explain.

33. Cereal. The amount of cereal that can be poured into a small bowl varies with a mean of 1.5 ounces and a standard deviation of 0.3 ounces. A large bowl holds a mean of 2.5 ounces with a standard deviation of 0.4 ounces. You open a new box of cereal and pour one large and one small bowl.

a) How much more cereal do you expect to be in the large bowl?

b) What's the standard deviation of this difference?

c) If the difference follows a Normal model, what's the probability the small bowl contains more cereal than the large one?

d) What are the mean and standard deviation of the total amount of cereal in the two bowls?

e) If the total follows a Normal model, what's the probability you poured out more than 4.5 ounces of cereal in the two bowls together?

f) The amount of cereal the manufacturer puts in the boxes is a random variable with a mean of 16.3 ounces and a standard deviation of 0.2 ounces. Find the expected amount of cereal left in the box, and the standard deviation.

34. Pets. The American Veterinary Association claims that the annual cost of medical care for dogs averages $100, with a standard deviation of $30, and for cats averages $120, with a standard deviation of $35.

a) What's the expected difference in the cost of medical care for dogs and cats?

b) What's the standard deviation of that difference?

c) If the difference in costs can be described by a Normal model, what's the probability that medical expenses are higher for someone's dog than for her cat?

35. More cereal. In Exercise 33 we poured a large and a small bowl of cereal from a box. Suppose the amount of cereal that the manufacturer puts in the boxes is a random variable with mean 16.2 ounces, and standard deviation 0.1 ounces.

a) Find the expected amount of cereal left in the box.

b) What's the standard deviation?

c) If the weight of the remaining cereal can be described by a Normal model, what's the probability that the box still contains more than 13 ounces?

36. More pets. You're thinking about getting two dogs and a cat. Assume that annual veterinary expenses are independent and have a Normal model with the means and standard deviations described in Exercise 34.

a) Define appropriate variables and express the total annual veterinary costs you may have.

b) Describe the model for this total cost. Be sure to specify its name, expected value, and standard deviation.

c) What's the probability that your total expenses will exceed $400?

37. Medley. In the 4 × 100 medley relay event, four swimmers swim 100 yards, each using a different stroke. A college team preparing for the conference championship looks at the times their swimmers have posted and creates a model based on the following assumptions:
• The swimmers' performances are independent.

• Each swimmer's times follow a Normal model.
• The means and standard deviations of the times (in seconds) are as shown:

Swimmer	Mean	SD
1 (backstroke)	50.72	0.24
2 (breaststroke)	55.51	0.22
3 (butterfly)	49.43	0.25
4 (freestyle)	44.91	0.21

a) What are the mean and standard deviation for the relay team's total time in this event?

b) The team's best time so far this season was 3:19.48. (That's 199.48 seconds.) Do you think the team is likely to swim faster than this at the conference championship? Explain.

38. Bikes. Bicycles arrive at a bike shop in boxes. Before they can be sold, they must be unpacked, assembled, and tuned (lubricated, adjusted, etc.). Based on past experience, the shop manager makes the following assumptions about how long this may take:
• The times for each setup phase are independent.
• The times for each phase follow a Normal model.
• The means and standard deviations of the times (in minutes) are as shown:

Phase	Mean	SD
Unpacking	3.5	0.7
Assembly	21.8	2.4
Tuning	12.3	2.7

a) What are the mean and standard deviation for the total bicycle setup time?

b) A customer decides to buy a bike like one of the display models, but wants a different color. The shop has one, still in the box. The manager says they can have it ready in half an hour. Do you think the bike will be set up and ready to go as promised? Explain.

39. Farmers' market. A farmer has 100 lb of apples and 50 lb of potatoes for sale. The market price for apples (per pound) each day is a random variable with a mean of 0.5 dollars and a standard deviation of 0.2 dollars. Similarly, for a pound of potatoes, the mean price is 0.3 dollars and the standard deviation is 0.1 dollars. It also costs him 2 dollars to bring all the apples and potatoes to the market. The market is busy with eager shoppers, so we can assume that he'll be able to sell all of each type of produce at that day's price.

a) Define your random variables, and use them to express the farmer's net income.

b) Find the mean.

c) Find the standard deviation of the net income.

d) Do you need to make any assumptions in calculating the mean? How about the standard deviation?

40. Bike sale. The bicycle shop in Exercise 38 will be offering 2 specially priced children's models at a sidewalk sale. The basic model will sell for $120 and the deluxe model for $150. Past experience indicates that sales of the basic model will have a mean of 5.4 bikes with a standard deviation of 1.2, and sales of the deluxe model will have a mean of 3.2 bikes with a standard deviation of 0.8 bikes. The cost of setting up for the sidewalk sale is $200.

    a) Define random variables and use them to express the bicycle shop's net income.

    b) What's the mean of the net income?

    c) What's the standard deviation of the net income?

    d) Do you need to make any assumptions in calculating the mean? How about the standard deviation?

---

just checking

Answers

1. a)

Outcome	X = cost	Probability
Recharging works	$60	0.75
Replace control unit	$200	0.25

    b) $60(0.75) + 200(0.25) = \$95$

    c) Car owners with this problem will spend an average of $95 to get it fixed.

2. a) $100 + 100 = 200$ seconds

    b) $\sqrt{50^2 + 50^2} = 70.7$ seconds

    c) The times for the two customers are independent.

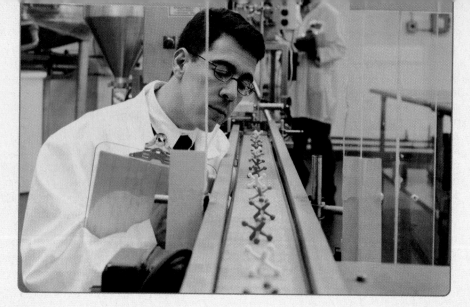

CHAPTER 17

Probability Models

W hen we last saw our heroine, she was . . .
Oops! Wrong book. What we mean to say is that several chapters ago we simulated buying cereal boxes to get sports cards. You remember the setup:

Suppose a cereal manufacturer puts pictures of famous athletes on cards in boxes of cereal, in the hope of increasing sales. The manufacturer announces that 20% of the boxes contain a picture of Tiger Woods, 30% a picture of Lance Armstrong, and the rest a picture of Serena Williams.

Now, instead of simulation, we can use probability.

Searching for Tiger

You've *got* to have the Tiger Woods picture, so you start madly opening boxes of cereal, hoping to find one. Assuming that the pictures are randomly distributed, there's a 20% chance you succeed on any box you open. We call the act of opening a box a "trial," and note that:

- There are only two possible outcomes (called *success* and *failure*) on each trial. Either you get Tiger's picture (success), or you don't (failure).
- The probability of success, denoted p, is the same on every trial. Here $p = 0.20$.
- The trials are independent. Finding Tiger in the first box does not change what might happen when you reach for the next box.

Situations like this occur often, and are called **Bernoulli trials.** Common examples of Bernoulli trials include tossing a coin, looking for defective products rolling off an assembly line, or even shooting free throws in a basketball game.

Tiger Woods.

380

Daniel Bernoulli (1700–1782) was the nephew of Jakob, whom you saw in Chapter 14. He was the first to work out the mathematics for what we now call Bernoulli trials.

A S **Bernoulli Trials.** Guess what! We've been generating Bernoulli trials all along. Look at the Random Simulation Tool in a new way.

Back to Tiger. What's the probability you find his picture in the first box of cereal? It's 20%, of course. We could write $P(\text{\#boxes} = 1) = 0.20$.

How about the probability that you don't find Tiger until the second box? Well, that means you fail on the first trial and then succeed on the second. With the probability of success 20%, the probability of failure, denoted q, is $1 - 0.2 = 80\%$. Since the trials are independent, the probability of getting your first success on the second trial is $P(\text{\#boxes} = 2) = (0.8)(0.2) = 0.16$.

Of course, you could have a run of bad luck. Maybe you won't find Tiger until the fifth box of cereal. What are the chances of that? You'd have to fail 4 straight times and then succeed, so $P(\text{\#boxes} = 5) = (0.8)^4(0.2) = 0.08192$.

How many boxes might you expect to have to open? We could reason that since Tiger's picture is in 20% of the boxes, or 1 in 5, we expect to find his picture, on average, in the fifth box; that is, $\mu = \frac{1}{0.2} = 5$ boxes. That's correct, but not easy to prove.

The Geometric Model

A single Bernoulli trial is usually not all that interesting. We might not care about a particular cereal box. We are more likely to want to know how long it will take us to achieve a success. The model that tells us this probability is called the **Geometric probability model.** Geometric models are completely specified by one parameter, p, the probability of success, and are denoted Geom(p). Since achieving the first success on trial number x requires first experiencing $x - 1$ failures, the probabilities are easily expressed by a formula.

NOTATION ALERT:

Now we have two more reserved letters. Whenever we deal with Bernoulli trials, p represents the probability of success, and q the probability of failure. (Of course, $q = 1 - p$.)

Geometric probability model for Bernoulli trials: Geom(p)

p = probability of success (and $q = 1 - p$ = probability of failure)

X = number of trials until the first success occurs

$$P(X = x) = q^{x-1} p$$

Expected value: $\mu = \dfrac{1}{p}$

Standard deviation: $\sigma = \sqrt{\dfrac{q}{p^2}}$

Independence

One of the important requirements for Bernoulli trials is that the trials be independent. Sometimes that's a reasonable assumption. Is it true for our example? We said that whether we find a Tiger Woods card in one box has no effect on the probabilities in other boxes. This is *almost* true. Technically, if exactly 20% of the boxes have Tiger Woods cards, then when you find one, you've reduced the number of remaining Tiger Woods cards. With a few million boxes of cereal, though, the difference is hardly worth mentioning.

But if you knew there were 2 Tiger Woods cards hiding in the 10 boxes of cereal on the market shelf, then finding one in the first box you try would clearly change your chances of finding Tiger in the next box.

If we had an infinite number of boxes, there wouldn't be a problem. It's selecting from a finite population that causes the probabilities to change, making the trials not independent. Obviously, taking 2 out of 10 boxes changes the probability. Taking even a few hundred out of millions, though, makes very little difference. Fortunately, we have a rule of thumb for the in-between cases. If we've looked at less than 10% of the population, we can pretend that the trials are still independent.

> **The 10% Condition:** Bernoulli trials must be independent. If that assumption is violated, it is still okay to proceed as long as the sample is smaller than 10% of the population.

Working with a Geometric Model   Step-By-Step

People with O-negative blood are called "universal donors" because O-negative blood can be given to anyone else, regardless of the recipient's blood type. Only about 6% of people have O-negative blood. If donors line up at random for a blood drive, how many do you expect to examine before you find someone who has O-negative blood? What's the probability that the first O-negative donor found is one of the first four people in line?

Plan State the questions.

> I want to estimate how many people I'll need to check to find an O-negative donor, and the probability that 1 of the first 4 people is O-negative.

Check to see that these are Bernoulli trials.

> ✔ There are two outcomes:
>   success = O-negative
>   failure = other blood types

> ✔ The probability of success for each person is $p = 0.06$, because they lined up randomly.

> ✔ Trials are not independent because the population is finite, but the donors lined up are fewer than 10% of all possible donors.

Variable Define the random variable.

> Let X = number of donors until one is O-negative.

Model Specify the model.

> I can model X with Geom(0.06).

Show   Mechanics Find the mean.

$$E(X) = \frac{1}{0.06} \approx 16.7$$

Calculate the probability of success on one of the first four trials. That's the probability that $X = 1, 2, 3,$ *or* 4.

$$P(X \le 4) = P(X = 1) + P(X = 2) + P(X = 3) + P(X = 4)$$

$$= (0.06) + (0.94)(0.06) + (0.94)^2(0.06) + (0.94)^3(0.06)$$

$$\approx 0.2193$$

Tell   Conclusion Interpret your results in context.

Blood drives such as this one expect to examine an average of 16.7 people to find a universal donor. About 22% of the time there will be one within the first 4 people in line.

The Binomial Model

A S **The Binomial Distribution.** It's more interesting to combine Bernoulli trials. Simulate this with the Random Tool to get a sense of how Binomial models behave.

Same situation, different question. You buy 5 boxes of cereal. What's the probability you get *exactly* 2 pictures of Tiger Woods? Before, we asked how long it would take until our first success. Now we want to find the probability of getting 2 successes among the 5 trials. We are still talking about Bernoulli trials, but we're asking a different question.

This time we're interested in the *number of successes* in the 5 trials. We want to find $P(\#\text{successes} = 2)$. This is an example of a **Binomial probability.** It takes two parameters to define this **Binomial model:** the number of trials, n, and the probability of success, p. We denote this model $Binom(n, p)$. Here, $n = 5$ trials, and $p = 0.2$, the probability of finding a Tiger Woods card in any trial.

Exactly 2 successes in 5 trials means 2 successes and 3 failures. It seems logical that the probability should be $(0.2)^2(0.8)^3$. Too bad! It's not that easy. That calculation would give you the probability of finding Tiger in the first 2 boxes and not in the next 3—*in that order*. But you could find Tiger in the third and fifth boxes and still have 2 successes. The probability of those outcomes in that particular order is $(0.8)(0.8)(0.2)(0.8)(0.2)$. That's also $(0.2)^2(0.8)^3$. In fact, the probability will always be the same, no matter what order the successes and failures occur in. Anytime we get 2 successes in 5 trials, no matter what the order, the probability will be $(0.2)^2(0.8)^3$. We just need to take account of all the possible orders in which the outcomes can occur.

Fortunately, these possible orders are *disjoint*. (For example, if your successes came on the first two trials, they couldn't come on the last two.) So we can use the Addition Rule and add up the probabilities for all the possible orderings. Since the probabilities are all the same, we only need to know how many orders are possible. For small numbers, we can just make a tree diagram and count the branches. For larger numbers this isn't practical, so we let the computer or calculator do the work.

Each different order in which we can have k successes in n trials is called a "combination." The total number of ways that can happen is written $\binom{n}{k}$ or ${}_nC_k$ and pronounced "*n* choose *k*."

$$_nC_k = \frac{n!}{k!(n-k)!} \text{ where } n! = n \times (n-1) \times \cdots \times 1$$

For 2 successes in 5 trials,

$$_5C_2 = \frac{5!}{2!(5-2)!} = \frac{5 \times 4 \times 3 \times 2 \times 1}{2 \times 1 \times 3 \times 2 \times 1} = \frac{5 \times 4}{2 \times 1} = 10.$$

So there are 10 ways to get 2 Tiger pictures in 5 boxes, and the probability of each is $(0.2)^2(0.8)^3$. Now we can find what we wanted:

$$P(\#successes = 2) = 10(0.2)^2(0.8)^3 = 0.2048$$

In general, the probability of exactly k successes in n trials is $_nC_k\, p^k q^{n-k}$.

Using this formula, we could find the expected value by adding up $x \cdot P(X = x)$ for all values, but it would be a long, hard way to get an answer that you already know intuitively. What's the expected value? If we have 5 boxes, and Tiger's picture is in 20% of them, then we would expect to have $5(0.2) = 1$ success. If we had 100 trials with probability of success 0.2, how many successes would you expect? Can you think of any reason not to say 20? It seems so simple that most people wouldn't even stop to think about it. You just multiply the probability of success by n. In other words, $E(X) = np$.[1]

The standard deviation is less obvious; you can't just rely on your intuition. Fortunately, the formula for the standard deviation also boils down to something simple: $SD(X) = \sqrt{npq}$. In 100 boxes of cereal, we expect to find 20 Tiger Woods cards, with a standard deviation of $\sqrt{100 \times 0.8 \times 0.2} = 4$ pictures.

Time to summarize. A Binomial probability model describes the number of successes in a specified number of trials. It takes two parameters to specify this model: the number of trials n and the probability of success p.

Binomial probability model for Bernoulli trials: Binom(n, p)

n = number of trials
p = probability of success (and $q = 1 - p$ = probability of failure)
X = number of successes in n trials

$$P(X = x) = {}_nC_x\, p^x\, q^{n-x}, \text{ where } {}_nC_x = \frac{n!}{x!(n-x)!}$$

Mean: $\mu = np$
Standard deviation: $\sigma = \sqrt{npq}$

Working with a Binomial Model   Step-By-Step

Suppose 20 donors come to the blood drive. What are the mean and standard deviation of the number of universal donors among them? What is the probability that there are 2 or 3 universal donors?

---

[1] It's an amazing fact that this does work out the long way. That is, $\sum\limits_{x=0}^{n} x \cdot {}_nC_x\, p^x q^{n-x} = np$, something that is not obvious.

Think **Plan** State the question.

> I want to know the mean and standard deviation of the number of universal donors among 20 people, and the probability that there are 2 or 3 of them.

Check to see that these are Bernoulli trials.

> ✔ There are two outcomes:
> success = O-negative
> failure = other blood types
>
> ✔ $p = 0.06$, because people have lined up at random.
>
> ✔ Trials are *not* independent, because the population is finite, but fewer than 10% of all possible donors are lined up.

Variable Define the random variable.

> Let X = number of O-negative donors among $n = 20$ people.

Model Specify the model.

> I can model X with Binom(20, 0.06).

Show **Mechanics** Find the expected value and standard deviation.

> $E(x) = np = 20(0.06) = 1.2$
>
> $SD(X) = \sqrt{npq} = \sqrt{20(0.06)(0.94)} \approx 1.06$

Calculate the probability of 2 or 3 successes using technology.

> $P(X = 2 \text{ or } 3) = P(X = 2) + P(X = 3)$
>
> $= {}_{20}C_2\,(0.06)^2(0.94)^{18}$
>
> $\qquad\qquad + {}_{20}C_3\,(0.06)^3(0.94)^{17}$
>
> $\approx 0.2246 + 0.0860$
>
> $= 0.3106$

Tell **Conclusion** Interpret your results in context.

> In groups of 20 randomly selected blood donors, I expect to find an average of 1.2 universal donors, with a standard deviation of 1.06. About 31% of the time, I'd find 2 or 3 universal donors among the 20 people.

The Normal Model to the Rescue!

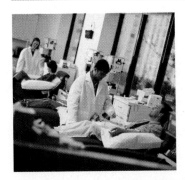

Suppose the Tennessee Red Cross anticipates the need for at least 1850 units of O-negative blood this year. It estimates that it will collect blood from 32,000 donors. How great is the risk that the Tennessee Red Cross will fall short of meeting its need? We've just learned how to calculate such probabilities. We can use the Binomial model with $n = 32,000$ and $p = 0.06$. The probability of getting *exactly* 1850 units of O-negative blood from 32,000 donors is ${}_{32000}C_{1850} \times 0.06^{1850} \times 0.94^{30150}$. No calculator on earth can calculate that first term (it has more than 100,000 digits).[2] And that's just the beginning. The problem said *at least* 1850, so we have to do it again for 1851, for 1852, and all the way up to 32,000. No thanks.

---

[2] If your calculator *can* find Binom(32000, 0.06), then it's smart enough to use an approximation. Read on to see how you can, too.

When we're dealing with a large number of trials like this, making direct calculations of the probabilities becomes tedious (or outright impossible). Here an old friend—the Normal model—comes to the rescue.

The Binomial model has mean np = 1920 and standard deviation $\sqrt{npq}$ ≈ 42.48. We could try approximating its distribution with a Normal model, using the same mean and standard deviation. Remarkably enough, that turns out to be a very good approximation. (We'll see why in the next chapter.) With that approximation, we can find the probability:

$$P(X < 1850) = P\left(z < \frac{1850 - 1920}{42.48}\right) \approx P(z < -1.65) \approx 0.05$$

There seems to be about a 5% chance that this Red Cross chapter will run short of O-negative blood.

Can we always use a Normal model to make estimates of Binomial probabilities? No. Consider the Tiger Woods situation—pictures in 20% of the cereal boxes. If we buy five boxes, the actual Binomial probabilities that we get 0, 1, 2, 3, 4, or 5 pictures of Tiger are 33%, 41%, 20%, 5%, 1%, and 0.03%, respectively. The first histogram shows that this probability model is skewed. That makes it clear that we should not try to estimate these probabilities by using a Normal model.

Now suppose we open 50 boxes of this cereal and count the number of Tiger Woods pictures we find. The second histogram shows this probability model. It is centered at np = 50(0.2) = 10 pictures, as expected, and it appears to be fairly symmetric around that center. Let's have a closer look.

The third histogram again shows Binom(50, 0.2), this time magnified somewhat and centered at the expected value of 10 pictures of Tiger. It looks close to Normal, for sure. With this larger sample size, it appears that a Normal model might be a useful approximation.

A Normal model, then, is a close enough approximation only for a large enough number of trials. And what we mean by "large enough" depends on the probability of success. We'd need a larger sample if the probability of success were very low (or very high). It turns out that a Normal model works pretty well if we expect to see at least 10 successes and 10 failures. That is, we check the **Success/Failure Condition.**

> **The Success/Failure Condition:** A Binomial model is approximately Normal if we expect at least 10 successes and 10 failures:
>
> $$np \geq 10 \text{ and } nq \geq 10$$

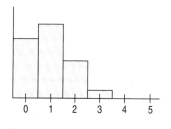

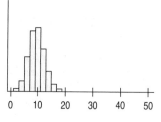

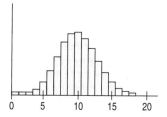

* Why 10? (Well, Actually 9)

It's easy to see where the magic number 10 comes from. You just need to remember how Normal models work. The problem is that a Normal model extends infinitely in both directions, but a Binomial model must have between 0 and n successes. So if we use a Normal to approximate a Binomial, we have to cut off its tails. That's not very important if the center of the Normal model is so far from 0 and n that the lost tails have only a negligible area. More than three standard deviations should do it, because a Normal model has little probability past that.

So the mean needs to be at least 3 standard deviations from 0 and at least 3 standard deviations from n. Let's look at the 0 end.

We require:	$\mu - 3\sigma > 0$
Or, in other words:	$\mu > 3\sigma$
For a Binomial that's:	$np > 3\sqrt{npq}$
Squaring yields:	$n^2p^2 > 9npq$
Now simplify:	$np > 9q$
Since $q \leq 1$, we can require:	$np > 9$

For simplicity we usually demand that np (and nq for the other tail) be at least 10 to use the Normal approximation, the **Success/Failure Condition**.[3]

Continuous Random Variables

There's a problem with approximating a Binomial model with a Normal model. The Binomial is discrete, giving probabilities for specific counts, but the Normal models a **continuous** random variable that can take on *any value*. For continuous random variables we can no longer list all the possible outcomes and their probabilities, as we could for discrete random variables.[4]

As we saw in the previous chapter, models for continuous random variables give probabilities for *intervals* of values. So, when we use the Normal model, we no longer calculate the probability that the random variable equals a *particular* value, but only that it lies *between* two values. We won't calculate the probability of getting exactly 1850 units of blood, but we have no problem approximating the probability of getting 1850 *or more*, which was, after all, what we really wanted.[5]

The Poisson Model

In the early 1990s, a leukemia cluster was identified in the Massachusetts town of Woburn. Many more cases of leukemia, a malignant cancer that originates in a cell in the marrow of bone, appeared in this small town than would be predicted. Was it evidence of a problem in the town, or was it chance? That question led to a famous trial in which the families of eight leukemia victims sued, and became grist for the book and movie *A Civil Action*. Following an 80-day trial, the judge called for a retrial after dismissing the jury's contradictory and confusing findings. Shortly thereafter, the chemical companies and the families settled.

When rare events occur together or in clusters, people often want to know if that happened just by chance, or whether something else is going on. If we

---

[3] Looking at the final step, we see that we need $np > 9$ in the worst case, when q (or p) is near 1, making the Binomial model quite skewed. When q and p are near 0.5—say between 0.4 and 0.6—the Binomial model is nearly symmetric and $np > 5$ ought to be safe enough. Although we'll always check for 10 expected successes and failures, keep in mind that for values of p near 0.5, we can be somewhat more forgiving.

[4] In fact, some people use an adjustment called the "continuity correction" to help with this problem. It's related to the suggestion we made in the previous footnote, and discussed in more advanced textbooks.

[5] If we really had been interested in a single value, we might have found the probability of getting between 1849.5 and 1850.5 units of blood.

assume that the events occur independently, we can use a Binomial model to find the probability that a cluster of events like this occurs. For rare events, p will be quite small, and when n is large it may be difficult to compute the exact probability that a certain size cluster occurs. To see why, let's try to compute the probability that a cluster of cases of size x occurs in Woburn. We'll use the *national average* of leukemia incidence to get a value for p, and the population of Woburn as our value for n. There are about 30,800 new cases of leukemia each year in the United States and about 280,000,000 people in the United States, giving a value for p of about 0.00011. The population of Woburn is about $n = 35,000$. We'd expect $np = 3.85$ new cases of leukemia in Woburn. How unlikely would 8 or more cases be? To answer that, we'll need to calculate the complement, adding the probabilities of no cases, exactly 1 case, etc., up to 7 cases. Each of those probabilities would have the form $_{35000}C_x\, p^x q^{35000-x}$. To find $_{35000}C_x$ we'll need to calculate 35000!, a number with more than 143,800 digits—beyond the reach of most computers. We could use the Normal model to help us approximate the probabilities as long as $\bar{n}p$ is at least 10, but for rare events such as leukemia in Woburn, when p is so small that np falls far below 10, the Normal probabilities won't be accurate.

Simeon Denis Poisson[6] was a French mathematician interested in events with very small probability. He originally derived his model to approximate the Binomial model when the probability of a success, p, is very small and the number of trials, n, is very large. Poisson found a simple approximation to estimate that probability. But the connection to the Binomial isn't obvious from the formula. The Poisson's mean is usually written as λ. To approximate a Binomial model with a Poisson model, just make their means match: $\lambda = np$.

NOTATION ALERT:

Yes, another Greek letter: λ (lambda) represents the mean of a Poisson model.

Poisson probability model for successes: Poisson (λ)

λ = mean number of successes.

X = number of successes.

e is an important mathematical constant (approximately 2.71828).

$$P(X = x) = \frac{e^{-\lambda}\lambda^x}{x!}$$

Expected value: $E(X) = \lambda$

Standard deviation: $SD(X) = \sqrt{\lambda}$

Using Poisson's model, we can easily find the probabilities of a given size cluster of leukemia cases. Setting the mean of the Poisson model, $\lambda = np = 35000 \times 0.00011 = 3.85$ new cases a year, we find that the probability of seeing exactly x cases in a year is $P(X = x) = \frac{e^{-3.85}3.85^x}{x!}$. By adding up these probabilities for $x = 0, 1, \ldots 7$, we'd find that the probability of 8 or more cases with $\lambda = 3.85$ is about 0.043. That's small but not terribly unusual.

● **Where does *e* come from?** You might know that $e = 2.7182818$ (to 7 decimal places), but does it have any real meaning? Yes! In fact, one of the places e originally turned up was in calculating how much more money you'd earn if you could get interest

---

[6] *Poisson* is a French name (meaning "fish"), properly pronounced "pwa sohn" and not with an "oy" sound, as in poison.

compounded more often. Suppose you receive an offer to earn 100% per year simple interest. (Take it!) At the end of the year, you'd have twice as much money as when you started. But suppose that you have the option of having the interest calculated and paid at the end of every month instead. Then each month you'll earn $(100/12)\%$ interest. At the end of the year, you'll wind up with $(1 + 1/12)^{12} = 2.613$ times as much instead of 2. If you could get the interest paid every day, you'd get $(1 + 1/365)^{365} = 2.714$ times as much. If you could convince the bank to compound and pay the interest every second, you'd get $(1 + 1/31536000)^{31536000} = 2.7182818$ times as much at year's end. This is where e shows up. In the limit, if you could get the interest compounded continually, you'd get e times as much. In other words, as n gets large, the limit of $(1 + 1/n)^n = e$. Who discovered this fact? Jakob Bernoulli, in 1683! ●

Although it was originally an approximation to the Binomial, the Poisson model is also used directly to model the probability of the occurrence of events for a variety of phenomena. It's a good model to consider whenever your data consist of counts of occurrences. It requires only that the events be independent and that the mean number of occurrences stays constant for the duration of the data collection (and beyond, if we hope to make predictions). So, for example, we can use the Poisson model to describe arrivals of cars at a tollbooth or emissions of radioactive particles from a chunk of radium.

One nice feature of the Poisson model is that it scales according to the sample size.[7] For example, if we know that the average number of occurrences in a town the size of Woburn, with 35,000 people, is 3.85, we know that the average number of occurrences in a town of only 3500 residents is 0.385. We can use that new value of λ to calculate the probabilities for the smaller town. Using the Poisson formula, we'd find that the probability of 0 occurrences for a town of 3500 is about 0.68.

One of the consequences of the Poisson model is that, as long as the mean rate of occurrences stays constant, the occurrence of past events doesn't change the probability of future events. Just as runs of heads that naturally show up in a random series of coin flips can seem surprising, events that occur according to the Poisson model can *appear* to cluster. Our minds, used to picking out patterns, read more into these apparent clusters than they deserve. We think we see patterns, and start anticipating the next event even though the probability of *another* event occurring is still the same. This is probably one reason why so many people believe in the form of the "Law of Averages" that we argued against in Chapter 14. People think they see patterns in which an athlete, slot machine, or financial market gets "hot." But a run of several successes in a row is not really unusual and is rarely evidence that the mean has changed. Careful studies have been made of "hot streaks," but none has ever demonstrated that they actually exist. When the Poisson model was used to analyze patterns in the bombs dropping over London in World War II (see margin note), the model said that even though several bombs had hit a particular sector, the probability of *another* hit was still the same. So, there was no point in moving people from one sector to another, even after several hits. You can imagine how difficult it must have been trying to convince people of that!

Poisson in History

In his book *Gravity's Rainbow*, Thomas Pynchon describes using the Poisson to model the bombs dropping over London during World War II. "The rockets *are* distributing about London just as Poisson's equation in the textbook predicts. As the data keep coming in, Roger looks more and more like a prophet . . . all I'm doing is plugging numbers into a well-known equation, you can look it up in the book and do it yourself. . . ."

---

[7] Because *Poisson* means "fish" in French, you can remember this fact by the bilingual pun "*Poisson* scales!"

As we noted a few chapters ago, the Pew Research Center reports that they are actually able to contact only 76% of the randomly selected households drawn for a telephone survey.

① Explain why these phone calls can be considered Bernoulli trials.

② Which of the models of this chapter (Geometric, Binomial, Normal, Poisson) would you use to model the number of successful contacts from a list of 1000 sampled households? Explain.

③ Pew further reports that even after they contacted a household, only 38% agree to be interviewed, so the probability of getting a completed interview for a randomly selected household is only 0.29. Which of the models of this chapter would you use to model the number of households Pew has to call before they get the first completed interview?

④ Suppose that in the course of a year, legitimate survey organizations (not folks pretending to take a survey but actually trying to sell you something or change your vote) sample 70,000 of the approximately 107,000,000 households in the United States. You wonder if people from your school are represented in these polls. Counting all the living alumni from your school, there are 10,000 households. How would you model the number of these alumni households you'd expect will be contacted next year by a legitimate poll?

What Can Go Wrong?

- **Be sure you have Bernoulli trials.** Be sure to check the requirements first: two possible outcomes per trial ("success" and "failure"), a constant probability of success, and independence. Remember that the 10% Condition provides a reasonable substitute for independence.
- **Don't confuse Geometric and Binomial models.** Both involve Bernoulli trials, but the issues are different. If you are repeating trials until your first success, that's a Geometric probability. You don't know in advance how many trials you'll need—theoretically, it could take forever. If you are counting the number of successes in a specified number of trials, that's a Binomial probability.
- **Don't use the Normal approximation with small n.** To use a Normal approximation in place of a Binomial model, there must be at least 10 expected successes and 10 expected failures. For large n when np is small, consider using a Poisson model instead.

CONNECTIONS

This chapter builds on what we know about random variables. We now have three more probability models to join the Normal model.

There are a number of "forward" connections from this chapter. We'll see the 10% Condition and the Success/Failure Condition often. And the facts about the Binomial distribution can help explain how proportions behave, as we'll see in the next chapter.

What have we learned?

We've learned that Bernoulli trials show up in lots of places. Depending on the random variable of interest, we can use one of four models to estimate probabilities for Bernoulli trials:

- a Geometric model when we're interested in the number of Bernoulli trials until the next success;
- a Binomial model when we're interested in the number of successes in a certain number of Bernoulli trials;
- a Normal model to approximate a Binomial model when we expect at least 10 successes and 10 failures; or
- a Poisson model to approximate a Binomial model when there are a very large number of trials and the probability of success (or failure) is very small.

TERMS

Bernoulli trials, if . . .
1. there are two possible outcomes.
2. the probability of success is constant.
3. the trials are independent.

Geometric probability model
A Geometric model is appropriate for a random variable that counts the number of Bernoulli trials until the first success.

Binomial probability model
A Binomial model is appropriate for a random variable that counts the number of successes in a fixed number of Bernoulli trials.

Success/Failure Condition
For a Normal model to be a good approximation of a Binomial model, we must expect at least 10 successes and 10 failures. That is, $np \geq 10$ and $nq \geq 10$.

Poisson probability model
A Poisson model can be used to approximate a Binomial model when n is large and p is small. Use $\lambda = np$ as the mean for the Poisson. The Poisson can also be used to model counts of a wide variety of phenomena.

SKILLS

When you complete this lesson you should:

Think
- Know how to tell if a situation involves Bernoulli trials.
- Be able to choose whether to use a Geometric or a Binomial model for a random variable involving Bernoulli trials.
- Be able to identify situations where a Poisson model might be useful.

Show
- Know the appropriate conditions for using a Geometric, Binomial, or Normal model.
- Know how to find the expected value of a Geometric model.
- Be able to calculate Geometric probabilities.
- Know how to find the mean and standard deviation of a Binomial model.
- Be able to calculate Binomial probabilities, perhaps with a Normal or Poisson model.

Tell
- Be able to interpret means, standard deviations, and probabilities in the Bernoulli trial context.

The Binomial, the Geometric, and the Poisson on the Computer

Most statistics packages offer functions that compute Binomial probabilities, and many offer functions for Geometric and Poisson probabilities as well. Some technology solutions automatically use the Normal approximation for the Binomial when the exact calculations become unmanageable.

The only important differences among these functions are in what they are named and the order of their arguments. In these functions, pdf stands for "probability density function"—what we've been calling a probability model. The letters cdf stand for "cumulative distribution function," the technical term when we want to accumulate probabilities over a range of values. These technical terms show up in many of the function names. The term "cumulative" in a function name says that it corresponds to a cdf.

Generically, the four functions are as follows:

Geometric pdf (*prob, x*)	Finds the individual geometric probability of getting the first success on trial *x* when the probability of success is *prob*.	For example, the probability of finding the first Tiger Woods picture in the fifth cereal box is Geometric pdf(0.2, 5)
Geometric cdf (*prob, x*)	Finds the cumulative probability of getting the first success on *or before* trial *x*, when the probability of success is *prob*.	For example, the total probability of finding Tiger's picture in one of the first 4 boxes is Geometric cdf(0.2, 4)
Binomial pdf (*n, prob, x*)	Finds the probability of getting *x* successes in *n* trials *x* when the probability of success is *prob*.	For example, Binomial pdf(5, 0.2, 2) is the probability of finding Tiger's picture exactly twice among 5 boxes of cereal.
Poisson pdf (λ, *x*)	Finds the probability of *x* successes when the mean number of successes is λ.	Best for large *n* and small probability of success.

DATA DESK

BinomDistr(*x, n, prob*) (pdf)
CumBinomDistr(*x, n, prob*) (cdf)
PoisDistr(*x, mean*) (pdf)
CumPoisDistr(*x, mean*) (cdf)

Comments
Data Desk does not compute Geometric probabilities.
These functions work in derived variables or in scratchpads.

EXCEL

Binomdist(*x, n, prob, cumulative*)

Comments
Set cumulative = true for cdf, false for pdf.
Excel's function fails when *x* or *n* is large.
Possibly, it does not use the Normal approximation.
Excel does not compute Geometric or Poisson probabilities.

JMP

Binomial Probability (*prob, n, x*) (pdf)
Binomial Distribution (*prob, n, x*) (cdf)
Poisson Probability (*mean, k*) (pdf)
Poisson Distribution (*mean, k*) (pdf)

Comments
JMP does not compute Geometric probabilities.

MINITAB

Choose **Probability Distributions** from the **Calc** menu.
Choose **Binomial** from the Probability Distributions submenu.
To calculate the probability of getting *x* successes in *n* trials, choose **Probability.**
To calculate the probability of getting *x* or fewer successes among *n* trials, choose **Cumulative Probability.**
For Poisson, choose **Poisson** from the Probability Distribution submenu.

Comments
Minitab does not compute Geometric probabilities.

SPSS

PDF.GEOM(*x, prob*)
CDF.GEOM(*x, prob*)
PDF.Poisson(*x, mean*)

CDF.Poisson(*x, mean*)
PDF.BINOM(*x, n, prob*)
CDF.BINOM(*x, n, prob*)

TI-83/84 Plus

geometpdf(*prob, x*)
geometcdf(*prob, x*)
binompdf(*n, prob, x*)
binomcdf(*n, prob, x*)
poissonpdf(*mean, x*)
poissoncdf(*mean, x*)

Comments
Find these commands in the **2nd DISTR** menu (the calculator refers to models as "distributions").

TI-89

Find the commands under the F5 (Distributions) menu.
- F: **Geometric Pdf** will ask for *p* and *x*. It returns the probability of the first success occurring on the *x*th trial.
- G: **Geometric Cdf** will ask for *p* and the upper and lower values of interest, say *a* and *b*. It returns $P(a \leq X \leq b)$, the probability the first success occurs between the ath and b^{th} trials, inclusive.
- A: **Binomial Pdf** asks for *n*, *p*, and *x*.
- B: **Binomial Cdf** asks for *n*, *p*, and the lower and upper values of interest.
- D: **Poisson Pdf** asks for mean and *x*.
- E: **Poisson Cdf** asks for mean, *x*, and the lower and upper values of interest.

Comments
- For Geometric variables, when finding $P(X \geq a)$ specify an upper value of infinity, 1EE99, or a very large number.
- For Binomial variables, when finding $P(X \geq a)$ the upper value is *n*.

EXERCISES

1. **Bernoulli.** Can we use probability models based on Bernoulli trials to investigate the following situations? Explain.
   a) We roll 50 dice to find the distribution of the number of spots on the faces.
   b) How likely is it that in a group of 120 the majority may have Type A blood, given that Type A is found in 43% of the population?
   c) We deal 5 cards from a deck and get all hearts. How likely is that?
   d) We wish to predict the outcome of a vote on the school budget, and poll 500 of the 3000 likely voters to see how many favor the proposed budget.
   e) A company realizes that about 10% of its packages are not being sealed properly. In a case of 24, is it likely that more than 3 are unsealed?

2. **Bernoulli 2.** Can we use probability models based on Bernoulli trials to investigate the following situations? Explain.
   a) You are rolling 5 dice and need to get at least two 6's to win the game.
   b) We record the eye colors found in a group of 500 people.
   c) A manufacturer recalls a doll because about 3% have buttons that are not properly attached. Customers return 37 of these dolls to the local toy store. Is the manufacturer likely to find any dangerous buttons?
   d) A city council of 11 Republicans and 8 Democrats picks a committee of 4 at random. What's the probability they choose all Democrats?
   e) A 2002 Rutgers University study found that 74% of high-school students have cheated on a test at least once. Your local high-school principal conducts a survey in homerooms and gets responses that admit to cheating from 322 of the 481 students.

3. **Simulating the model.** Think about the Tiger Woods picture search again. You are opening boxes of cereal one at a time looking for his picture, which is in 20% of the boxes. You want to know how many boxes you might have to open in order to find Tiger.
   a) Describe how you would simulate the search for Tiger using random numbers.
   b) Run at least 30 trials.
   c) Based on your simulation, estimate the probabilities that you might find your first picture of Tiger in the first box, the second, etc.
   d) Calculate the actual probability model.
   e) Compare the distribution of outcomes in your simulation to the probability model.

4. **Simulation II.** You are one space short of winning a child's board game, and must roll a 1 on a die to claim victory. You want to know how many rolls it might take.
   a) Describe how you would simulate rolling the die until you get a 1.
   b) Run at least 30 trials.
   c) Based on your simulation, estimate the probabilities that you might win on the first roll, the second, the third, etc.
   d) Calculate the actual probability model.
   e) Compare the distribution of outcomes in your simulation to the probability model.

5. **Tiger again.** Let's take one last look at the Tiger Woods picture search. You know his picture is in 20% of the cereal boxes. You buy five boxes to see how many pictures of Tiger you might get.
   a) Describe how you would simulate the number of pictures of Tiger you might find in five boxes of cereal.
   b) Run at least 30 trials.
   c) Based on your simulation, estimate the probabilities that you get no pictures of Tiger, 1 picture, 2 pictures, etc.
   d) Calculate the actual probability model.
   e) Compare the distribution of outcomes in your simulation to the probability model.

6. **Seatbelts.** Suppose 75% of all drivers always wear their seatbelts. Let's investigate how many of the drivers might be belted among five cars waiting at a traffic light.
   a) Describe how you would simulate the number of seatbelt-wearing drivers among the five cars.
   b) Run at least 30 trials.
   c) Based on your simulation, estimate the probabilities there are no belted drivers, exactly one, two, etc.
   d) Calculate the actual probability model.
   e) Compare the distribution of outcomes in your simulation to the probability model.

7. **Hoops.** A basketball player has made 80% of his foul shots during the season. Assuming the shots are independent, find the probability that in tonight's game he
   a) misses for the first time on his fifth attempt.
   b) makes his first basket on his fourth shot.
   c) makes his first basket on one of his first 3 shots.

8. **Chips.** Suppose a computer chip manufacturer rejects 2% of the chips produced because they fail presale testing.
   a) What's the probability that the fifth chip you test is the first bad one you find?
   b) What's the probability you find a bad one within the first 10 you examine?

9. **More hoops.** For the basketball player in Exercise 7, what's the expected number of shots until he misses?

10. **Chips ahoy.** For the computer chips described in Exercise 8, how many do you expect to test before finding a bad one?

11. Blood. Only 4% of people have Type AB blood.
   a) On average, how many donors must be checked to find someone with Type AB blood?
   b) What's the probability that there is a Type AB donor among the first 5 people checked?
   c) What's the probability that the first Type AB donor will be found among the first 6 people?
   d) What's the probability that we won't find a Type AB donor before the 10th person?

12. Colorblindness. About 8% of males are colorblind. A researcher needs some colorblind subjects for an experiment, and begins checking potential subjects.
   a) On average, how many men should the researcher expect to check to find one who is colorblind?
   b) What's the probability that she won't find anyone colorblind among the first 4 men she checks?
   c) What's the probability that the first colorblind man found will be the sixth person checked?
   d) What's the probability that she finds someone who is colorblind before checking the 10th man?

13. Lefties. Assume that 13% of people are left-handed. If we select 5 people at random, find the probability of each outcome described below.
   a) The first lefty is the fifth person chosen.
   b) There are some lefties among the 5 people.
   c) The first lefty is the second or third person.
   d) There are exactly 3 lefties in the group.
   e) There are at least 3 lefties in the group.
   f) There are no more than 3 lefties in the group.

14. Arrows. An Olympic archer is able to hit the bull's-eye 80% of the time. Assume each shot is independent of the others. If she shoots 6 arrows, what's the probability of each result described below.
   a) Her first bull's-eye comes on the third arrow.
   b) She misses the bull's-eye at least once.
   c) Her first bull's-eye comes on the fourth or fifth arrow.
   d) She gets exactly 4 bull's-eyes.
   e) She gets at least 4 bull's-eyes.
   f) She gets at most 4 bull's-eyes.

15. Lefties redux. Consider our group of 5 people from Exercise 13.
   a) How many lefties do you expect?
   b) With what standard deviation?
   c) If we keep picking people until we find a lefty, how long do you expect it will take?

16. More arrows. Consider our archer from Exercise 14.
   a) How many bull's-eyes do you expect her to get?
   b) With what standard deviation?
   c) If she keeps shooting arrows until she hits the bull's-eye, how long do you expect it will take?

17. Still more lefties. Suppose we choose 12 people instead of the 5 chosen in Exercise 13.
   a) Find the mean and standard deviation of the number of right-handers in the group.
   b) What's the probability that they're not all right-handed?
   c) What's the probability that there are no more than 10 righties?
   d) What's the probability that there are exactly 6 of each?
   e) What's the probability that the majority is right-handed?

18. Still more arrows. Suppose our archer from Exercise 14 shoots 10 arrows.
   a) Find the mean and standard deviation of the number of bull's-eyes she may get.
   b) What's the probability that she never misses?
   c) What's the probability that there are no more than 8 bull's-eyes?
   d) What's the probability that there are exactly 8 bull's-eyes?
   e) What's the probability that she hits the bull's-eye more often than she misses?

19. Tennis, anyone? A certain tennis player makes a successful first serve 70% of the time. Assume that each serve is independent of the others. If she serves 6 times, what's the probability she gets
   a) all 6 serves in?
   b) exactly 4 serves in?
   c) at least 4 serves in?
   d) no more than 4 serves in?

20. Frogs. A wildlife biologist examines frogs for a genetic trait he suspects may be linked to sensitivity to industrial toxins in the environment. Previous research had established that this trait is usually found in 1 of every 8 frogs. He collects and examines a dozen frogs. If the frequency of the trait has not changed, what's the probability he finds the trait in
   a) none of the 12 frogs?
   b) at least 2 frogs?
   c) 3 or 4 frogs?
   d) no more than 4 frogs?

21. And more tennis. Suppose the tennis player in Exercise 19 serves 80 times in a match.
   a) What's the mean and standard deviation of the number of good first serves expected?
   b) Verify that you can use a Normal model to approximate the distribution of the number of good first serves.
   c) Use the 68-95-99.7 Rule to describe this distribution.
   d) What's the probability she makes at least 65 first serves?

22. More arrows. The archer in Exercise 14 will be shooting 200 arrows in a large competition.
   a) What are the mean and standard deviation of the number of bull's-eyes she might get?
   b) Is a Normal model appropriate here? Explain.
   c) Use the 68-95-99.7 Rule to describe the distribution of the number of bull's-eyes she may get.
   d) Would you be surprised if she made only 140 bull's-eyes? Explain.

23. Frogs, part II. Based on concerns raised by his preliminary research, the biologist in Exercise 20 decides to collect and examine 150 frogs.
a) Assuming the frequency of the trait is still 1 in 8, determine the mean and standard deviation of the number of frogs with the trait he should expect to find in his sample.
b) Verify that he can use a Normal model to approximate the distribution of the number of frogs with the trait.
c) He found the trait in 22 of his frogs. Do you think this proves that the trait has become more common? Explain.

24. Apples. An orchard owner knows that he'll have to use about 6% of the apples he harvests for cider because they will have bruises or blemishes. He expects a tree to produce about 300 apples.
a) Describe an appropriate model for the number of cider apples that may come from that tree. Justify your model.
b) Find the probability there will be no more than a dozen cider apples.
c) Is it likely there will be more than 50 cider apples? Explain.

25. Lefties again. A lecture hall has 200 seats with folding arm tablets, 30 of which are designed for left-handers. The average size of classes that meet there is 188, and we can assume that about 13% of students are left-handed. What's the probability that a right-handed student in one of these classes is forced to use a lefty arm tablet?

26. No-shows. An airline, believing that 5% of passengers fail to show up for flights, overbooks (sells more tickets than there are seats). Suppose a plane will hold 265 passengers, and the airline sells 275 seats. What's the probability the airline will not have enough seats so someone gets bumped?

27. Annoying phone calls. A newly hired telemarketer is told he will probably make a sale on about 12% of his phone calls. The first week he called 200 people, but only made 10 sales. Should he suspect he was misled about the true success rate? Explain.

28. The euro. Shortly after the introduction of the euro coin in Belgium, newspapers around the world published articles claiming the coin is biased. The stories were based on reports that someone had spun the coin 250 times and gotten 140 heads—that's 56% heads. Do you think this is evidence that spinning a euro is unfair? Explain.

29. Hurricanes, redux. We first looked at the occurrences of hurricanes in Chapter 4 (Exercise 15) and found that they arrive with a mean of 2.35 per year. Suppose the number of hurricanes can be modeled by a Poisson distribution with this mean.
a) What's the probability of no hurricanes next year?
b) What's the probability that during the next 2 years, there's exactly 1 hurricane?

30. Bank tellers. I am the only bank teller on duty at my local bank. I need to run out for 10 minutes, but I don't want to miss any customers. Suppose the arrival of customers can be modeled by a Poisson distribution with mean 2 customers per hour.
a) What's the probability that no one will arrive in the next 10 minutes?
b) What's the probability that 2 or more people arrive in the next 10 minutes?
c) I've just served 2 customers, who came in one after the other. Is this a better time to run out?

31. TB again. We saw that the probability of contracting TB is small, with p about 0.0005 for a new case in a given year. In a town of 8000 people,
a) what's the expected number of new TB cases?
b) Use the Poisson model to approximate the probability that there will be at least one new case of TB next year.

32. Earthquakes. Suppose the probability of a major earthquake on a given day is 1 out of 10,000.
a) What's the expected number of major earthquakes in the next 1000 days?
b) Use the Poisson model to approximate the probability that there will be at least one major earthquake in the next 1000 days.

33. Seatbelts II. Police estimate that 80% of drivers now wear their seatbelts. They set up a safety roadblock, stopping cars to check for seatbelt use.
a) How many cars do they expect to stop before finding a driver whose seatbelt is not buckled?
b) What's the probability that the first unbelted driver is in the 6th car stopped?
c) What's the probability that the first 10 drivers are all wearing their seatbelts?
d) If they stop 30 cars during the first hour, find the mean and standard deviation of the number of drivers expected to be wearing seatbelts.
e) If they stop 120 cars during this safety check, what's the probability they find at least 20 drivers not wearing their seatbelts?

34. Rickets. Vitamin D is essential for strong, healthy bones. Our bodies produce vitamin D naturally when sunlight falls upon the skin, or it can be taken as a dietary supplement. Although the bone disease rickets was largely eliminated in England during the 1950s, some people there are concerned that this generation of children is at increased risk because they are more likely to watch TV or play computer games than spend time outdoors. Recent research indicated that about 20% of British children are deficient in vitamin D. Suppose doctors test a group of elementary school children.
a) What's the probability that the first vitamin D–deficient child is the 8th one tested?
b) What's the probability that the first 10 children tested are all okay?

c) How many kids do they expect to test before finding one who has this vitamin deficiency?

d) They will test 50 students at the third grade level. Find the mean and standard deviation of the number who may be deficient in vitamin D?

e) If they test 320 children at this school, what's the probability that no more than 50 of them have the vitamin deficiency?

35. ESP. Scientists wish to test the mind-reading ability of a person who claims to "have ESP." They use five cards with different and distinctive symbols (square, circle, triangle, line, squiggle). Someone picks a card at random and thinks about the symbol. The "mind reader" must correctly identify which symbol was on the card. If the test consists of 100 trials, how many would this person need to get right in order to convince you that ESP may actually exist? Explain.

36. True-False. A true-false test consists of 50 questions. How many does a student have to get right to convince you that he is not merely guessing? Explain.

37. Hot hand. A basketball player who ordinarily makes about 55% of his free throw shots has made 4 in a row. Is this evidence that he has a "hot hand" tonight? That is, is this streak so unusual that it means the probability he makes a shot must have changed? Explain.

38. New bow. Our archer in Exercise 14 purchases a new bow, hoping that it will improve her success rate to more than 80% bull's-eyes. She is delighted when she first tests her new bow and hits 6 consecutive bull's-eyes. Do you think this is compelling evidence that the new bow is better? In other words, is a streak like this unusual for her? Explain.

39. Hotter hand. Our basketball player in Exercise 37 has new sneakers, which he thinks improve his game. Over his past 40 shots, he's made 32—much better than the 55% he usually shoots. Do you think his chances of making a shot really increased? In other words, is making at least 32 of 40 shots really unusual for him? (Do you think it's his sneakers?)

40. New bow, again. The archer in Exercise 38 continues shooting arrows, ending up with 45 bull's-eyes in 50 shots. Now are you convinced that the new bow is better? Explain.

---

just checking
Answers

1. There are two outcomes (contact, no contact), the probability of contact is 0.76, and random calls should be independent.

2. Binomial, with $n = 1000$ and $p = 0.76$. For actual calculations, we could approximate using a Normal model with $\mu = np = 1000(0.76) = 760$ and $\sigma = \sqrt{npq} = \sqrt{1000(0.76)(0.24)} \approx 13.5$.

3. Geometric, with $p = 0.29$.

4. Poisson, with $\lambda = np = 10000\left(\dfrac{70000}{107000000}\right) = 6.54$.

REVIEW OF PART **IV** Randomness and Probability

QUICK REVIEW

Here's a brief summary of the key concepts and skills in probability and probability modeling:

▶ The Law of Large Numbers says that the more times we try something, the closer the results will come to theoretical perfection.

  • Don't mistakenly misinterpret the Law of Large Numbers as the "Law of Averages." There's no such thing.

▶ Basic rules of probability can handle most situations:

  • To find the probability that an event OR another event happens, add their probabilities and subtract the probability that both happen.

  • To find the probability that an event AND another independent event both happen, multiply probabilities.

  • Conditional probabilities tell you how likely one event is to happen, knowing that another event has happened.

  • Mutually exclusive events (also called "disjoint") cannot both happen at the same time.

  • Two events are independent if knowing that one happens doesn't change the probability that the other happens.

▶ A probability model for a random variable describes the theoretical distribution of outcomes.

  • The mean of a random variable is its expected value.

  • For sums or differences of independent random variables, variances add.

  • To estimate probabilities involving quantitative variables, you may be able to use a Normal model—but only if the distribution of the variable is unimodal and symmetric.

  • To estimate the probability you'll get your first success on a certain trial, use a Geometric model.

  • To estimate the probability you'll get a certain number of successes in a specified number of independent trials, use a Binomial model.

  • To estimate the probability of the number of occurrences of a relatively rare phenomenon, consider using a Poisson model.

Ready? Here are some opportunities to check your understanding of these ideas.

REVIEW EXERCISES

1. **Quality control.** A consumer organization estimates that 29% of new cars have a cosmetic defect, such as a scratch or a dent, when they are delivered to car dealers. This same organization believes that 7% have a functional defect—something that does not work properly—and that 2% of new cars have both kinds of problems.
   a) If you buy a new car, what's the probability that it has some kind of defect?
   b) What's the probability it has a cosmetic defect but no functional defect?
   c) If you notice a dent on a new car, what's the probability it has a functional defect?
   d) Are the two kinds of defects disjoint events? Explain.
   e) Do you think the two kinds of defects are independent events? Explain.

2. **Workers.** A company's human resources officer reports a breakdown of employees by job type and gender, shown in the table.

		Sex	
		Male	Female
Job Type	Management	7	6
	Supervision	8	12
	Production	45	72

   a) What's the probability that a worker selected at random is
   i) female?
   ii) female or a production worker?
   iii) female, if the person works in production?
   iv) a production worker, if the person is female?
   b) Do these data suggest that job type is independent of gender? Explain.

3. **Airfares.** Each year a company must send 3 officials to a meeting in China and 5 officials to a meeting in France.

Airline ticket prices vary from time to time, but the company purchases all tickets for a country at the same price. Past experience has shown that tickets to China have a mean price of $1000, with a standard deviation of $150, while the mean airfare to France is $500, with a standard deviation of $100.

a) Define random variables and use them to express the total amount the company will have to spend to send these delegations to the two meetings.

b) Find the mean and standard deviation of this total cost.

c) Find the mean and standard deviation of the difference in price of a ticket to China and a ticket to France.

d) Do you need to make any assumptions in calculating these means? How about the standard deviations?

4. Bipolar. Psychiatrists estimate that about 1 in 100 adults suffers from bipolar disorder. What's the probability that in a city of 10,000 there are more than 200 people with this condition? Be sure to verify that a Normal model can be used here.

5. A game. To play a game, you must pay $5 for each play. There is a 10% chance you will win $5, a 40% chance you will win $7, and a 50% chance you will win only $3.

a) What are the mean and standard deviation of your net winnings?

b) You play twice. Assuming the plays are independent events, what are the mean and standard deviation of your total winnings?

6. Emergency switch. Safety engineers must determine whether industrial workers can operate a machine's emergency shutoff device. Among a group of test subjects, 66% were successful with their left hands, 82% with their right hands, and 51% with both hands.

a) What percent of these workers could not operate the switch with either hand?

b) Are success with right and left hands independent events? Explain.

c) Are success with right and left hands mutually exclusive? Explain.

7. Twins. In the United States, the probability of having twins (usually about 1 in 90 births) rises to about 1 in 10 for women who have been taking the fertility drug Clomid. Among a group of 10 pregnant women, what's the probability that

a) at least one will have twins if none were taking a fertility drug?

b) at least one will have twins if all were taking Clomid?

c) at least one will have twins if half were taking Clomid?

8. Deductible. A car owner may buy insurance that will pay the full price of repairing the car after an at-fault accident, or save $12 a year by getting a policy with a $500 deductible. Her insurance company says that about 0.5% of drivers in her area have an at-fault auto accident

during any given year. Based on this information, should she buy the policy with the deductible or not? How does the value of her car influence this decision?

9. More twins. A group of 5 women became pregnant while undergoing fertility treatments with the drug Clomid, discussed in Exercise 7. What's the probability that

a) none will have twins?

b) exactly 1 will have twins?

c) at least 3 will have twins?

10. At fault. The car insurance company in Exercise 8 believes that about 0.5% of drivers have an at-fault accident during a given year. Suppose the company insures 1355 drivers in that city.

a) What are the mean and standard deviation of the number who may have at-fault accidents?

b) Can you describe the distribution of these accidents with a Normal model? Explain.

11. Twins, part III. At a large fertility clinic, 152 women became pregnant while taking Clomid. (See Exercise 7.)

a) What are the mean and standard deviation of the number of twin births we might expect?

b) Can we use a Normal model in this situation? Explain.

c) What's the probability that no more than 10 of the women have twins?

12. Child's play. In a board game you determine the number of spaces you may move by spinning a spinner and rolling a die. The spinner has three regions: Half of the spinner is marked "5," and the other half is equally divided between "10" and "20." The six faces of the die show 0, 0, 1, 2, 3, and 4 spots. When it's your turn, you spin and roll, adding the numbers together to determine how far you may move.

a) Create a probability model for the outcome on the spinner.

b) Find the mean and standard deviation of the spinner results.

c) Create a probability model for the outcome on the die.

d) Find the mean and standard deviation of the die results.

e) Find the mean and standard deviation of the number of spaces you get to move.

13. Language. Neurological research has shown that in about 80% of people, language abilities reside in the brain's left side. Another 10% display right-brain language centers, and the remaining 10% have two-sided language control. (The latter two groups are mainly left-handers; *Science News*, 161 no. 24 [2002])

a) Assume that a freshman composition class contains 25 randomly selected people. What's the probability that no more than 15 of them have left-brain language control?

b) In a randomly assigned group of 5 of these students, what's the probability that no one has two-sided language control?

c) In the entire freshman class of 1200 students, how many would you expect to find of each type?

d) What are the mean and standard deviation of the number of these freshmen who might be right-brained in language abilities?

e) If an assumption of Normality is justified, use the 68-95-99.7 Rule to describe how many students in the freshman class might have right-brain language control.

14. Play again. If you land in a "penalty zone" on the game board described in Exercise 12, your move will be determined by subtracting the roll of the die from the result on the spinner. Now what are the mean and standard deviation of the number of spots you may move?

15. Beanstalks. In some cities tall people who want to meet and socialize with other tall people can join Beanstalk Clubs. To qualify, a man must be over 6'2" tall, and a woman over 5'10." According to the National Health Survey, heights of adults may have a Normal model with mean heights of 69.1" for men and 64.0" for women. The respective standard deviations are 2.8" and 2.5."

a) You're probably not surprised to learn that men are generally taller than women, but what does the greater standard deviation for men's heights indicate?

b) Are men or women more likely to qualify for Beanstalk membership?

c) Beanstalk members believe that height is an important factor when people select their spouses. To investigate, we select at random a married man and, independently, a married woman. Define two random variables, and use them to express how many inches taller the man is than the woman.

d) What's the mean of this difference?

e) What's the standard deviation of this difference?

f) What's the probability that the man is taller than the woman (that the difference in heights is greater than 0)?

g) Suppose a survey of married couples reveals that 92% of the husbands were taller than their wives. Based on your answer to part f, do you believe that people chose spouses independent of height? Explain.

16. Stocks. Since the stock market began in 1872, stock prices have risen in about 73% of the years. Assuming that market performance is independent from year to year, what's the probability that

a) the market will rise for 3 consecutive years?

b) the market will rise 3 years out of the next 5?

c) the market will fall during at least 1 of the next 5 years?

d) the market will rise during a majority of years over the next decade?

17. Multiple choice. A multiple choice test has 50 questions, with 4 answer choices each. You must get at least 30 correct to pass the test, and the questions are very difficult.

a) Are you likely to be able to pass by guessing on every question? Explain.

b) Suppose, after studying for a while, you believe you have raised your chances of getting each question right to 70%. How likely are you to pass now?

c) Assuming you are operating at the 70% level and the instructor arranges questions randomly, what's the probability that the third question is the first one you get right?

18. Stock strategy. Many investment advisors argue that after stocks have declined in value for 2 consecutive years, people should invest heavily because the market rarely declines 3 years in a row.

a) Since the stock market began in 1872, there have been two consecutive losing years eight times. In six of those cases, the market rose during the following year. Does this confirm the advice?

b) Overall, stocks have risen in value during 95 of the 130 years since the market began in 1872. How is this fact relevant in assessing the statistical reasoning of the advisors?

19. Insurance. A 65-year-old woman takes out a $100,000 term life insurance policy. The company charges an annual premium of $520. Estimate the company's expected profit on such policies if mortality tables indicate that only 2.6% of women age 65 die within a year.

20. Teen smoking. The Centers for Disease Control say that about 30% of high-school students smoke tobacco (down from a high of 38% in 1997). Suppose you randomly select high-school students to survey them on their attitudes toward scenes of smoking in the movies. What's the probability that

a) none of the first 4 students you interview is a smoker?

b) the first smoker is the sixth person you choose?

c) there are no more than 2 smokers among 10 people you choose?

21. Passing stats. Molly's college offers two sections of Statistics 101. From what she has heard about the two professors listed, Molly estimates that her chances of passing the course are 0.80 if she gets Professor Scedastic and 0.60 if she gets Professor Kurtosis. The registrar uses a lottery to randomly assign the 120 enrolled students based on the number of available seats in each class. There are 70 seats in Professor Scedastic's class and 50 in Professor Kurtosis's class.

a) What's the probability that Molly will pass Statistics?

b) At the end of the semester, we find out that Molly failed. What's the probability that she got Professor Kurtosis?

22. Teen smoking II. Suppose that, as reported by the Centers for Disease Control, about 30% of high-school students smoke tobacco. You randomly select 120 high-school students to survey them on their attitudes toward scenes of smoking in the movies.
a) What's the expected number of smokers?
b) What's the standard deviation of the number of smokers?
c) The number of smokers among 120 randomly selected students will vary from group to group. Explain why that number can be described with a Normal model.
d) Using the 68-95-99.7 Rule, create and interpret a model for the number of smokers among your group of 120 students.

23. Random variables. Given independent random variables with means and standard deviations as shown, find the mean and standard deviation of each of these variables:
a) $X + 50$
b) $10Y$
c) $X + 0.5Y$
d) $X - Y$

	Mean	SD
X	50	8
Y	100	6

24. Merger. Explain why the facts you know about variances of independent random variables might encourage two small insurance companies to merge. (*Hint:* Think about the expected amount and potential variability in payouts for the separate and the merged companies.)

25. Youth survey. According to a recent Gallup survey, 93% of teens use the Internet, but there are differences in how teen boys and girls say they use computers. The telephone poll found that 77% of boys had played computer games in the past week, compared with 65% of girls. On the other hand, 76% of girls said they had e-mailed friends in the past week, compared with only 65% of boys.
a) For boys, the cited percentages are 77% playing computer games and 65% using e-mail. That total is 142%, so there is obviously a mistake in the report. No? Explain.
b) Based on these results, do you think playing games and using e-mail are mutually exclusive? Explain.
c) Do you think e-mailing friends and gender are independent? Explain.
d) Suppose that in fact 93% of the teens in your area do use the Internet. You want to interview a few who do not, so you start contacting teenagers at random. What is the probability that it takes you 5 interviews until you find the first person who does not use the Internet?

26. Meals. A college student on a seven-day meal plan reports that the amount of money he spends daily on food varies with a mean of $13.50 and a standard deviation of $7.

a) What are the mean and standard deviation of the amount he might spend in two consecutive days?
b) What assumption did you make in order to find that standard deviation? Are there any reasons you might question that assumption?
c) Estimate his average weekly food costs, and the standard deviation.
d) Do you think it likely he might spend less than $50 in a week? Explain, including any assumptions you make in your analysis.

27. Travel to Kyrgyzstan. Your pocket copy of *Kyrgyzstan on 4237±360 Soms a Day* claims that you can expect to spend about 4237 soms each day with a standard deviation of 360 soms. How well can you estimate your expenses for the trip?
a) Your budget allows you to spend 90,000 soms. To the nearest day, how long can you afford to stay in Kyrgyzstan, on average?
b) What's the standard deviation of your expenses for a trip of that duration?
c) You doubt that your total expenses will exceed your expectations by more than two standard deviations. How much extra money should you bring? On average, how much of a "cushion" will you have per day?

28. Picking melons. Two stores sell watermelons. At the first store the melons weigh an average of 22 pounds, with a standard deviation of 2.5 pounds. At the second store the melons are smaller, with a mean of 18 pounds and a standard deviation of 2 pounds. You select a melon at random at each store.
a) What's the mean difference in weights of the melons?
b) What's the standard deviation of the difference in weights?
c) If a Normal model can be used to describe the difference in weights, what's the probability that the melon you got at the first store is heavier?

29. Home sweet home. According to the 2000 Census, 66% of U.S. households own the home they live in. A mayoral candidate conducts a survey of 820 randomly selected homes in your city and finds only 523 owned by the current residents. The candidate then attacks the incumbent mayor, saying that there is an unusually low level of home ownership in the city. Do you agree? Explain.

30. Buying melons. The first store in Exercise 28 sells watermelons for 32 cents a pound. The second store is having a sale on watermelons—only 25 cents a pound. Find the mean and standard deviation of the difference in the price you may pay for melons randomly selected at each store.

31. Who's the boss? The 2000 Census revealed that 26% of all firms in the United States are owned by women. You call some firms doing business locally, assuming that the national percentage is true in your area.
a) What's the probability that the first 3 you call are all owned by women?

b) What's the probability that none of your first 4 calls finds a firm that is owned by a woman?

c) Suppose none of your first 5 calls found a firm owned by a woman. What's the probability that your next call does?

32. Jerseys. A Statistics professor comes home to find that all four of his children got white team shirts from soccer camp this year. He concludes that this year, unlike other years, the camp must not be using a variety of colors. But then he finds out that in each child's age group there are 4 teams, only 1 of which wears white shirts. Each child just happened to get on the white team at random.

a) Why was he so surprised? If each age group uses the same 4 colors, what's the probability that all four kids would get the same color shirt?

b) What's the probability that all 4 would get white shirts?

c) We lied. Actually, in the oldest child's group there are 6 teams instead of the 4 teams in each of the other three groups. How does this change the probability you calculated in part b?

33. When to stop? In Exercise 27 of the Review Exercises for Part III, we posed this question:

You play a game that involves rolling a die. You can roll as many times as you want, and your score is the total for all the rolls. But . . . if you roll a 6, your score is 0 and your turn is over. What might be a good strategy for a game like this?

You attempted to devise a good strategy by simulating several plays to see what might happen. Let's try calculating a strategy.

a) On what roll would you expect to get a 6 for the first time?

b) So, roll *one time less* than that. Assuming all those rolls were not 6's, what's your expected score?

c) What's the probability that you can roll that many times without getting a 6?

34. Plan B. Here's another attempt at developing a good strategy for the dice game in Exercise 33. Instead of stopping after a certain number of rolls, you could decide to stop when your score reaches a certain number of points.

a) How many points would you expect a roll to *add* to your score?

b) In terms of your current score, how many points would you expect a roll to *subtract* from your score?

c) Based on your answers in parts a and b, at what score will another roll "break even"?

d) Describe the strategy this result suggests.

35. Technology on campus. Every 5 years the Conference Board of the Mathematical Sciences surveys college math departments. In 2000 the board reported that 51% of all undergraduates taking Calculus I were in classes that used graphing calculators and 31% were in classes that used computer assignments. Suppose that 16% used both calculators and computers.

a) What percent used neither kind of technology?

b) What percent used calculators but not computers?

c) What percent of the calculator users had computer assignments?

d) Based on this survey, do calculator and computer use appear to be independent events? Explain.

36. Dogs. A census by the county dog control officer found that 18% of homes kept one dog as a pet, 4% had two dogs, and 1% had three or more. If a salesman visits two homes selected at random, what's the probability he encounters

a) no dogs?   c) dogs in each home?

b) some dogs?   d) more than one dog in each home?

37. O-rings. Failures of O-rings on the space shuttle are fairly rare, but often disastrous, events. If we are testing O-rings, suppose that the probability of a failure of any one O-ring is 0.01. Let X be the number of failures in the next 10 O-rings tested.

a) What model might you use to model X?

b) What is the mean number of failures in the next 10 O-rings?

c) What is the probability that there is exactly one failure in the next 10 O-rings?

d) What is the probability that there is at least one failure in the next 10 O-rings?

38. Volcanoes. Almost every year, there is some incidence of volcanic activity on the island of Japan. In 2002 there were 5 volcanic episodes, defined as either eruptions or sizable seismic activity. Suppose the mean number of episodes is 2.4 per year. Let X be the number of episodes in the 2-year period 2004–2005.

a) What model might you use to model X?

b) What is the mean number of episodes in this period?

c) What is the probability that there will be no episodes in this period?

d) What is the probability that there are more than 3 episodes in this period?

39. Socks. In your sock drawer you have 4 blue socks, 5 grey socks, and 3 black ones. Half asleep one morning, you grab 2 socks at random and put them on. Find the probability you end up wearing

a) 2 blue socks.

b) no grey socks.

c) at least 1 black sock.

d) a green sock.

e) matching socks.

40. Coins. A coin is to be tossed 36 times.

a) What are the mean and standard deviation of the number of heads?

b) Suppose the resulting number of heads is unusual, two standard deviations above the mean. How many "extra" heads were observed?

c) If the coin were tossed 100 times, would you still consider the same number of extra heads unusual? Explain.

d) In the 100 tosses, how many extra heads would you need to observe in order to say the results were unusual?

e) Explain how these results refute the "Law of Averages" but confirm the Law of Large Numbers.

41. The Drake equation. In 1961 astronomer Frank Drake developed an equation to try to estimate the number of extraterrestrial civilizations in our galaxy that might be able to communicate with us via radio transmissions. Now largely accepted by the scientific community, the Drake equation has helped spur efforts by radio astronomers to search for extraterrestrial intelligence.

Here is the equation:

$$N_c = N \cdot f_p \cdot n_e \cdot f_l \cdot f_i \cdot f_c \cdot f_L$$

OK, it looks a little messy, but here's what it means:

Factor	What It Represents	Possible Value
N	Number of stars in the Milky Way Galaxy	200–400 billion
f_p	Probability that a star has planets	20%–50%
n_e	Number of planets in a solar system capable of sustaining earth-type life	1? 2?
f_l	Probability that life develops on a planet with a suitable environment	1%–100%
f_i	Probability that life evolves intelligence	50%?
f_c	Probability that intelligent life develops radio communication	10%–20%
f_L	Fraction of the planet's life for which the civilization survives	$\frac{1}{1,000,000}$?
N_c	Number of extraterrestrial civilizations in our galaxy with which we could communicate	?

So, how many ETs are out there? That depends; values chosen for the many factors in the equation depend on ever-evolving scientific knowledge and one's personal guesses. But now, some questions.

a) What quantity is calculated by the first product $N \cdot f_p$?

b) What quantity is calculated by the product $N \cdot f_p \cdot n_e \cdot f_l \cdot f_i$?

c) What probability is calculated by the product $f_l \cdot f_i$?

d) Which of the factors in the formula are conditional probabilities? Restate each in a way that makes the condition clear.

Note: A quick Internet search will find you a site where you can play with the Drake equation yourself.

42. Recalls. In a car rental company's fleet, 70% of the cars are American brands, 20% are Japanese, and the rest are German. The company notes that manufacturers' recalls seem to affect 2% of the American cars, but only 1% of the others.

a) What's the probability that a randomly chosen car is recalled?

b) What's the probability that a recalled car is American?

43. Pregnant? Suppose that 70% of the women who suspect they may be pregnant and purchase an in-home pregnancy test are actually pregnant. Further suppose that the test is 98% accurate. What's the probability that a woman whose test indicates that she is pregnant actually is?

44. Door prize. You are among 100 people attending a charity fundraiser at which a large-screen TV will be given away as a door prize. To determine who wins, 99 white balls and 1 red ball have been placed in a box and thoroughly mixed. The guests will line up and, one at a time, pick a ball from the box. Whoever gets the red ball wins the TV, but if the ball is white, it is returned to the box. If none of the 100 guests gets the red ball, the TV will be auctioned off for additional benefit of the charity.

a) What's the probability that the first person in line wins the TV?

b) You are the third person in line. What's the probability that you win the TV?

c) What's the probability that the charity gets to sell the TV because no one wins?

d) Suppose you get to pick your spot in line. Where would you want to be in order to maximize your chances of winning?

e) After hearing some protest about the plan, the organizers decide to award the prize by not returning the white balls to the box, thus insuring that 1 of the 100 people will draw the red ball and win the TV. Now what position in line would you choose in order to maximize your chances?

CHAPTER 18

Sampling Distribution Models

WHO	U.S. voters
WHAT	Presidential preference
WHEN	Oct. 2004
WHERE	United States
WHY	Election prediction

O n October 4, 2004, less than a month before the presidential election, a Gallup Poll asked 1016 national adults, aged 18 and older, whom they supported for President; 49% said they'd chosen John Kerry. A Rasmussen Poll taken just a few weeks later found 45.9% of 1000 likely voters supporting Kerry. Was one poll "wrong"? Pundits claimed that the race was "too close to call." What's going on? Should we be surprised to find that we could get proportions this different from properly selected random samples drawn from the same population? You're probably used to seeing that observations vary, but how much variability among polls should we expect to see?

Why do sample proportions vary at all? How can surveys conducted at essentially the same time by the same organization asking the same questions get different results? The answer is at the heart of Statistics. It's because each survey is based on a *different* sample of 1000 people. The proportions vary from sample to sample because the samples are composed of different people.

It's actually pretty easy to predict how much a proportion will vary under circumstances like this. When we can understand and predict the variability of our estimates, we've taken the essential step toward seeing past that variability, so we can understand the world.

Modeling the Distribution of Sample Proportions

Imagine

We see only the sample that we actually drew, but by simulating or modeling, we can *imagine* what we might have seen had we drawn other possible random samples.

We've talked about *Think, Show,* and *Tell.* Now we have to add *Imagine.* We want to imagine the results from all the random samples of size 1000 that we didn't take. What would the histogram of all the sample proportions look like?

For Kerry vs. Bush in October of 2004, where do you expect the center of that histogram to be? Of course, we don't *know* the answer to that (and probably never will). But we know that it will be at the true proportion in the population, and we can call that *p*. (See the Notation Alert on the next page.) How about the *shape* of the histogram?

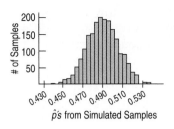

The distribution of observed sample proportions $\hat{p}$ of voters supporting Kerry in samples of 1000 simulated voters with $p = 0.49$. We can see that these sample proportions vary, with a distribution that is centered at the true proportion p.   **Figure 18.1**

A S **Sampling Distribution of a Proportion.** You don't have to imagine—you can simulate. See how here.

A S **Simulating Sampling Distributions.** Watch the Normal model appear from random proportions. Doing it for yourself is more convincing than just reading about it.

We don't have to just imagine. We can simulate. We want to simulate a bunch of those random samples of 1000 that we didn't really draw. To do that, we'll simulate many independent random samples of size 1000, keeping the same probability of success. We chose 0.49 for p as a reasonable value. Here's a histogram of the proportions saying they would vote for Kerry for 2000 independent samples of 1000 voters when the true proportion is $p = 0.49$. It should be no surprise that we don't get the same proportion for each sample we draw, even though the underlying true value is the same for the population from which we've drawn the samples. Each $\hat{p}$ comes from a different independent sample.

Does it surprise you that the histogram is unimodal? Symmetric? That it is centered at p? You probably don't find any of this shocking. Does the shape remind you of any model that we've discussed? It's an amazing and fortunate fact that a Normal model is just the right one for the histogram of sample proportions.

To use a Normal model, we need to specify two parameters: its mean *and* standard deviation. Which Normal model is the right one? The center of the histogram is naturally at p, so we'll put μ, the mean of the Normal, at p.

What about the standard deviation? Usually the mean gives us no information about the standard deviation. Suppose we told you that a batch of bike helmets had a mean diameter of 26 centimeters, and asked what the standard deviation was. If you said, "I have no idea," you'd be exactly right. There's no information about σ from knowing the value of μ.

But there's a special fact about proportions. With proportions we get something for free. Once we know the mean, p, we automatically also know the standard deviation. The standard deviation of the *proportion* of successes, $\hat{p}$, is

$$\sigma(\hat{p}) = SD(\hat{p}) = \sqrt{\frac{p(1-p)}{n}} = \sqrt{\frac{pq}{n}}.$$

When we draw simple random samples of n individuals, the proportions we find will vary from sample to sample. We can model the distribution of these sample proportions with a probability model that is

$$N\left(p, \sqrt{\frac{pq}{n}}\right).$$

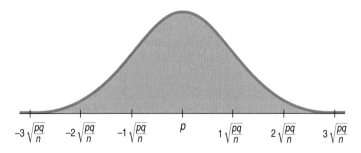

A Normal model centered at p with a standard deviation of $\sqrt{\dfrac{pq}{n}}$ is a good model for a collection of proportions found for many random samples of size n from a population with success probability p.   **Figure 18.2**

For the presidential election of 2004, we know that the true proportion favoring Kerry at the time of the election was 48.48%. Earlier in October, though, the true proportion very well may have been different. We'll never know what that true percentage was at the time the polls were being taken, but let's suppose it was 49%. Here's a picture of the Normal model for our simulation histogram:

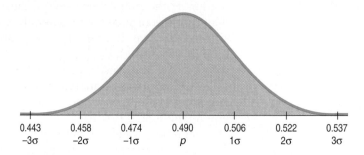

0.443	0.458	0.474	0.490	0.506	0.522	0.537
-3σ	-2σ	-1σ	*p*	1σ	2σ	3σ

Using 0.49 for *p* gives this Normal model for Figure 18.1's histogram of the sample proportions of voters supporting Kerry. **Figure 18.3**

A S **The Standard Deviation of a Proportion.** Do you believe this formula for standard deviation? Don't just take our word for it—prove it for yourself with an experiment.

Once we put the center at *p* = 0.49, the standard deviation becomes

$$\sqrt{\frac{pq}{n}} = \sqrt{\frac{(0.49)(0.51)}{1000}} = 0.0158, \text{ or } 1.58\%.$$

Because we have a Normal model, we can use the 68-95-99.7 Rule or look up other probabilities using a table or technology. For example, we know that 95% of Normally distributed values are within two standard deviations of the mean, so we should not be surprised if 95% of various polls gave results that were near 49% (0.49) but varied above and below that by no more than 3.16% (0.0316). Now we can see the two polls really didn't disagree. The proportions supporting Kerry found in the two polls—49% and 45.9%—are both *consistent* with a true proportion of 49%. This is what we mean by **sampling error.** It's not really an *error* at all, but just *variability* you'd expect to see from one sample to another. A better term would be **sampling variability.**

How Good Is the Normal Model?

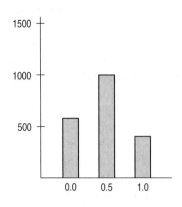

Proportions from samples of size 2 can take on only three possible values. A Normal model does not work well. **Figure 18.4**

Stop and think for a minute about what we've just said. It's a remarkable claim. We've said that if we draw repeated random samples of the same size, *n*, from some population and measure the proportion, $\hat{p}$, we get for each sample, then the collection of these proportions will pile up around the underlying population proportion, *p*, in such a way that a histogram of the sample proportions can be modeled well by a Normal model.

There must be a catch. Suppose the samples were of size 2, for example. Then the only possible proportion values would be 0, 0.5, and 1. There's no way the histogram could ever look like a Normal model with only three possible values for the variable.

Well, there *is* a catch. The claim is only approximately true. (But, that's OK. After all, models are only supposed to be approximately true.) And the model becomes a better and better representation of the distribution of the sample proportions as the sample size gets bigger.[1] Samples of size 1 or 2 just aren't going to work very

---

[1] Formally, we say the claim is true in the limit as *n* grows.

well. But the distributions of proportions of many larger samples do have histograms that are remarkably close to a Normal model.

Assumptions and Conditions

Most models are useful only when specific assumptions are true. In the case of the model for the distribution of sample proportions, there are two assumptions:

1. The sampled values must be independent of each other.
2. The sample size, n, must be large enough.

Of course, assumptions are hard—often impossible—to check. That's why we *assume* them. But, as we discussed in Chapter 8, we should check to see whether the assumptions are reasonable. Fortunately, we can often check *conditions* that provide information about the assumptions. Check the following corresponding conditions before using the Normal to model the distribution of sample proportions:

1. **10% Condition:** If sampling has not been made with replacement (that is, returning each sampled individual to the population before drawing the next individual), then the sample size, n, must be no larger than 10% of the population. For the polls, the total population is very large, so the 1000 that were sampled is a small fraction of the population. There are other ways in which samples can fail to be independent, but the only good protection from such failures is to think carefully about possible reasons for the data to fail to be independent. There are no simple conditions to check that guarantee independence.

> The terms "success" and "failure" for the outcomes that have probability p and q are common in Statistics. But they are completely arbitrary labels. When we say that a disease occurs with probability p, we certainly don't mean that getting sick is a "success" in the normal sense of the word.

2. **Success/Failure Condition:** The sample size has to be big enough so that both np and nq are at least 10.[2] If we call the outcome that has probability p a "success" and the outcome with probability $(1 - p) = q$ a "failure," then this condition says that we need to expect at least 10 successes and at least 10 failures to have enough data for sound conclusions. For the polls, a "success" might be voting for Kerry. With $p = 0.49$, we expect $1000 \times 0.49 = 490$ successes and $1000 \times 0.51 = 510$ failures. Both are at least 10, so there are certainly enough successes and enough failures for the condition to be satisfied.

These two conditions seem to contradict each other. Condition 2 wants a big sample size. How big depends on p. If p is near 0.5, we need a sample of only 20 or so. If p is only 0.01, however, we'd need 1000. But condition 1 says that the sample size can be no larger than 10% of the population. Fortunately, this isn't usually a problem in practice. Often, as in polls that sample from all U.S. adults, or industrial samples from a day's production, the populations are much larger than 10 times the sample size.

A Sampling Distribution Model for a Proportion

We've simulated repeated samples and looked at a histogram of the sample proportions. We modeled that histogram with a Normal model. Why do we bother to model it? Because this model will give us insight into how much the sample

---

[2] Why 10? See page 386.

proportion will vary from sample to sample. We've simulated many of the other random samples we might have gotten. The model is an attempt to show the distribution from *all* the random samples. But how do we know that a Normal model will really work? Is this just an observation based on some simulations that *might* be approximately true some of the time?

It turns out that this model can be justified theoretically with just a little mathematics. We won't bother you with the math because in this instance, it really wouldn't help your understanding.[3] Nevertheless, the fact that we can think of proportions from random samples as random quantities and then say something this specific about their distribution is a fundamental insight—one that we will use in each of the next four chapters.

We have changed our point of view in a very important way. No longer is a proportion something we just compute for a set of data. We now see it as a random quantity that has a distribution. We call that distribution the **sampling distribution model** for the proportion, and we'll make good use of it.

Without the sampling model, the rest of Statistics just wouldn't exist. Sampling models are what makes Statistics work. They inform us about the amount of variation we should expect when we sample. Suppose we spin a coin 100 times in order to decide whether it's fair or not. If we get 52 heads, we're probably not surprised. Although we'd expect 50 heads, 52 doesn't seem particularly unusual for a fair coin. But we would be surprised to see 90 heads; that might really make us doubt that the coin is fair. How about 64 heads? Harder to say. That's a case where we need the sampling distribution model. The sampling model quantifies the variability, telling us how surprising any sample proportion is. And it enables us to make informed decisions about how precise our estimate of the true proportion might be. That's exactly what we'll be doing for the rest of this book.

These sampling distribution models are strange because even though we depend on them, we never actually get to see them. We could take repeated samples to develop the theory and our own intuition about sampling distribution models. But, in practice, we only imagine (or simulate) them. They're important because they act as a bridge from the real world of data to the imaginary model of the statistic and enable us to say something about the population when all we have is data from the real world. This is the huge leap of Statistics. Rather than thinking about the sample proportion as a fixed quantity calculated from our data, we now think of it as a random quantity—just one of many values we might have seen had we chosen a different random sample. By thinking about what *might* happen if we were to draw many, many samples from the same population, we can learn a lot about how *one particular* sample will behave.

> ### The sampling distribution model for a proportion
> Provided that the sampled values are independent and the sample size is large enough, the sampling distribution of $\hat{p}$ is modeled by a Normal model with mean $\mu(\hat{p}) = p$, and standard deviation $SD(\hat{p}) = \sqrt{\dfrac{pq}{n}}$.

Sampling distribution models tame the variation in statistics enough to allow us to measure how close our computed statistic values are likely to be to the underlying parameters they estimate. That's the path to the *margin of error* you hear about in polls and surveys. We'll see how to do that in the next chapter.

---

[3] Actually, we're not sure it helps *our* understanding all that much either.

A S **Simulate the Sampling Distribution Model of a Proportion.** You probably don't want to work through the formal mathematical proof; a simulation is far more convincing!

We have now answered the question raised at the start of the chapter. To know how variable a sample proportion is, we need to know the proportion and the size of the sample. That's all.

Once we think of the idea of a sampling distribution model, it should be clear that we can find a sampling distribution model for any statistic we can compute on sampled data. What isn't at all clear is whether these sampling distribution models will always be simple probability models like the Normal model. The Normal is a good model for proportions, and we'll soon see that it works for means as well. But it won't work for every statistic. Fortunately, many of the statistics we work with do have simple sampling distribution models. Usually we need only a parameter or two to describe them.

Why do we really care about these distributions that we can never see? The answer is that once we have estimates for those parameters, we can use the model to help us understand how the statistic would have been distributed if we had drawn all those samples—but without the work and expense of actually drawing them. This will help us generalize from our data to the world at large. That's just the sort of thing models are good for.

just checking

1. You want to poll a random sample of 100 students on campus to see if they are in favor of the proposed location for the new student center. Of course, you'll get just one number, your sample proportion, $\hat{p}$. But if you imagined all the possible samples of 100 students you could draw and imagined the histogram of all the sample proportions from these samples, what shape would it have?

2. Where would the center of that histogram be?

3. If you think that about half the students are in favor of the plan, what would the standard deviation of the sample proportions be?

Working with Sampling Distribution Models for Proportions Step-By-Step

Suppose that about 13% of the population is left-handed.[4] A 200-seat school auditorium has been built with 15 "lefty seats," seats that have the built-in desk on the left rather than the right arm of the chair. (For the right-handed readers among you, have you ever tried to take notes in a chair with the desk on the left side?) In a class of 90 students, what's the probability that there will not be enough seats for the left-handed students?

Think

Plan State what we want to know.

I want to find the probability that in a group of 90 students, more than 15 will be left-handed. Since 15 out of 90 is 16.7%, I need the probability of finding more than 16.7% left-handed students out of a sample of 90 if the proportion of lefties is 13%.

Model Check the conditions.

✔ **10% Condition:** The 90 students in the class can be thought of as a random sample of

---

[4] Actually, it's quite difficult to get an accurate estimate of the proportion of lefties in the population. Estimates range from 8% to 15%.

students and are surely less than 10% of the population of all students. (Even if the school itself is small, I'm thinking of the population of all *possible* students who could have gone to the school.)

✔ **Success/Failure Condition:**

$$np = 90(0.13) = 11.7 \geq 10$$

$$nq = 90(0.87) = 78.3 \geq 10$$

State the parameters and the sampling distribution model.

The population proportion is $p = 0.13$. The conditions are satisfied, so I'll model the sampling distribution of $\hat{p}$ with a Normal model with mean 0.13 and a standard deviation of

$$SD(\hat{p}) = \sqrt{\frac{pq}{n}} = \sqrt{\frac{(0.13)(0.87)}{90}} \approx 0.035$$

My model for $\hat{p}$ is $N(0.13, 0.035)$.

Show

Plot Make a picture. Sketch the model and shade the area we're interested in, in this case the area to the right of 16.7%.

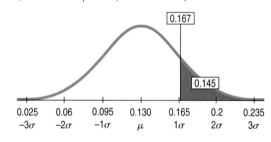

Mechanics Use the standard deviation as a ruler to find the *z*-score of the cutoff proportion. We see that 16.7% lefties would be just over one standard deviation above the mean.

Find the resulting probability from a table of Normal probabilities, a computer program, or a calculator.

$$z = \frac{\hat{p} - p}{SD(\hat{p})} = \frac{0.167 - 0.13}{0.035} = 1.06$$

$$P(\hat{p} > 0.167) = P(z > 1.06) = 0.1446$$

Tell

Conclusion Interpret the probability in the context of the question.

There is about a 14.5% chance that there will not be enough seats for the left-handed students in the class.

What About Quantitative Data?

Proportions summarize categorical variables. And the Normal sampling distribution model looks like it is going to be very useful. But can we do something similar with quantitative data?

Of course we can (or we wouldn't have asked). Even more remarkable, not only can we use all of the same concepts, but almost the same model.

What are the concepts? We know that when we sample at random or randomize an experiment, the results we get will vary from sample-to-sample and from experiment-to-experiment. The Normal model seems an incredibly simple way to summarize all that variation. Could something that simple work for means? We won't keep you in suspense. It turns out that means also have a sampling distribution that we can model with a Normal model. And it turns out that there's a theoretical result that proves it to be so. And, as we did with proportions, we can get some insight from a simulation.

Simulating the Sampling Distribution of a Mean

Here's a simple simulation. Let's start with one fair die. If we toss this die 10,000 times, what should the histogram of the numbers on the face of the die look like? Here are the results of a simulated 10,000 tosses:

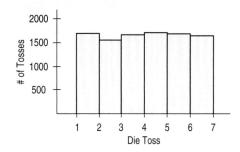

Now let's toss a *pair* of dice and record the average of the two. If we repeat this (or at least simulate repeating it) 10,000 times, recording the average of each pair, what will the histogram of these 10,000 averages look like? Before you look, think a minute. Is getting an average of 1 on *two* dice as likely as getting an average of 3 or 3.5?

Let's see:

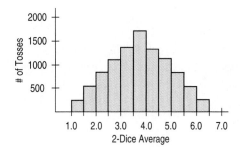

We're much more likely to get an average near 3.5 than we are to get one near 1 or 6. Without calculating those probabilities exactly, it's fairly easy to see that the *only* way to get an average of 1 is to get two 1's. To get a total of 7 (for an average of 3.5), though, there are many more possibilities. This distribution even has a name—the *triangular* distribution.

What if we average 3 dice? We'll simulate 10,000 tosses of 3 dice and take their average:

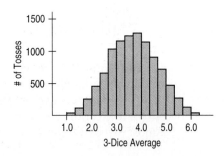

What's happening? First notice that it's getting harder to have averages near the ends. Getting an average of 1 or 6 with 3 dice requires all three to come up 1 or 6, respectively. That's less likely than for 2 dice to come up both 1 or both 6. The distribution is being pushed toward the middle. But what's happening to the shape? (This distribution doesn't have a name, as far as we know.)

Let's continue this simulation to see what happens with larger samples. Here's a histogram of the averages for 10,000 tosses of 5 dice:

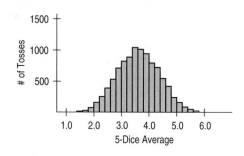

The pattern is becoming clearer. Two things continue to happen. The first fact we knew already from the Law of Large Numbers. It says that as the sample size (number of dice) gets larger, each sample average is more likely to be closer to the population mean. So, we see the shape continuing to tighten around 3.5. But the shape of the distribution is the surprising part. It's becoming bell-shaped. And not just bell-shaped; it's approaching the Normal model.

Are you convinced? Let's skip ahead and try 20 dice. The histogram of averages for 10,000 throws of 20 dice looks like this:

A S **The Sampling Distribution Model for Means.** Don't just sit there reading about the simulation—do it yourself. That's the best way to see the Normal model magically appear in the sampling distribution.

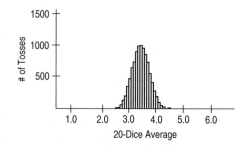

Now we see the Normal shape again (and notice how much smaller the spread is). But can we count on this happening for situations other than dice throws? What kinds of sample means have sampling distributions that we can model with a Normal model? It turns out that Normal models work well amazingly often.

The Fundamental Theorem of Statistics

The dice simulation may look like a special situation. But it turns out that what we saw with dice is true for means of repeated samples for almost every situation. When we looked at the sampling distribution of a proportion, we had to check only a few conditions. For means, the result is even more remarkable. There are almost no conditions at all.

Pierre-Simon Laplace 1749–1827.

"The theory of probabilities is at bottom nothing but common sense reduced to calculus."
—*Laplace, in* Théorie Analytique des Probabilitiés, *1812*

Laplace was one of the greatest scientists and mathematicians of his time. In addition to his contributions to probability and statistics, he published many new results in mathematics, physics, and astronomy (where his nebular theory was one of the first to describe the formation of the solar system in much the way it is understood today). He also played a leading role in establishing the metric system of measurement.

His brilliance, though, sometimes got him into trouble. A visitor to the Académie des Sciences in Paris reported that Laplace let it be known widely that he considered himself the best mathematician in France. The effect of this on his colleagues was not eased by the fact that Laplace was right.

Let's say that again: The sampling distribution of *any* mean becomes Normal as the sample size grows. All we need is for the observations to be independent and collected with randomization. We don't even care about the shape of the population distribution![5] This surprising fact was proved in a fairly general form in 1810 by Pierre-Simon Laplace, and caused quite a stir (at least in mathematics circles) because it is so unintuitive. Laplace's result is called the **Central Limit Theorem**[6] (CLT).

Why should the Normal model be so special as to pop up like this, seemingly out of nowhere and for the sampling distributions of both proportions and means? For once in this text, we're not going to try to persuade you that it is obvious, clear, simple, or straightforward. In fact, the CLT is surprising and a bit weird. Not only does the distribution of means of many random samples get closer and closer to a Normal model as the sample size grows, *this is true regardless of the shape of the population distribution!* Even if we sample from a skewed or bimodal population, the Central Limit Theorem tells us that means of repeated random samples will tend to follow a Normal model as the sample size grows. Of course, you won't be surprised to learn that it works better and faster the closer the population distribution is to a Normal model. And it works better for larger samples. If the data come from a population that's exactly Normal to start with, then the observations themselves are Normal. If we take samples of size 1, their "means" are just the observations—so, of course, they have Normal sampling distribution. But now suppose the population distribution is very skewed (like the CEO data from Chapter 4, for example). The CLT works, although it may take a sample size of dozens or even hundreds of observations for the Normal model to work well.

For example, think about a really bimodal population, one that consists of only 0's and 1's. The CLT says that even means of samples from this population will follow a Normal sampling distribution model. But wait. Suppose we have a categorical variable and we assign a 1 to each individual in the category and a 0 to each individual not in the category. And then we find the mean of these 0's and 1's. That's the same as counting the number of individuals who are in the category and dividing by *n*. That mean will be . . . the *sample proportion*, $\hat{p}$, of individuals who are in the category (a "success"). So maybe it wasn't so surprising after all that proportions, like means, have Normal sampling distribution models; they are actually just a special case of Laplace's remarkable theorem. Of course, for such an extremely bimodal population, we'll need a reasonably large sample size—and that's where the special conditions for proportions come in.

The Central Limit Theorem (CLT)

The mean of a random sample has a sampling distribution whose shape can be approximated by a Normal model. The larger the sample, the better the approximation will be.

---

[5] OK, one technical condition. The data must come from a population with a finite variance. You probably can't imagine a population with an infinite variance, but statisticians can construct such things, so we have to discuss them in footnotes like this that go on and on and really make no difference at all in how you think about the important stuff back in the page. So you can just forget we mentioned it.

[6] The word "central" in the name of the theorem means "fundamental." It doesn't refer to the center of a distribution.

The Real World and the Model World

Be careful. We have been slipping smoothly between the real world, in which we draw random samples of data, and a magical mathematical model world, in which we describe how the sample means and proportions we observe in the real world might behave if we could see the results from every random sample that we might have drawn. Now we have *two* distributions to deal with. The first is the real world distribution of the sample, which we might display with a histogram (for quantitative data) or with a bar chart or table (for categorical data). The second is the math world *sampling distribution* of the statistic, which we model with a Normal model based on the Central Limit Theorem. Don't confuse the two.

For example, don't mistakenly think the CLT says that the *data* are Normally distributed as long as the sample is large enough. In fact, as samples get larger, we expect the distribution of the data to look more and more like the population from which it is drawn—skewed, bimodal, whatever—but not necessarily Normal. You can collect a sample of CEO salaries for the next 1000 years, but the histogram will never look Normal. It will be skewed to the right. The Central Limit Theorem doesn't talk about the distribution of the data from the sample. It talks about the sample *means* and sample *proportions* of many different random samples drawn from the same population. Of course, we never actually draw all those samples, so the CLT is talking about an imaginary distribution—the sampling distribution model.

And the CLT does require that the sample be big enough when the population shape is not unimodal and symmetric. But it is still a very surprising and powerful result.

But Which Normal?

The CLT says that the sampling distribution of any mean or proportion is approximately Normal. But which Normal model? We know that any Normal is specified by its mean and standard deviation. For proportions, the sampling distribution is centered at the population proportion. For means, it's centered at the population mean. What else would we expect?

What about the standard deviations, though? We noticed in our dice simulation that the histograms got narrower as we averaged more and more dice together. This shouldn't be surprising. Means vary less than the individual observations. Think about it for a minute. Which would be more surprising, having *one* person in your Statistics class who is over 6'9" tall or having the *mean* of 100 students taking the course be over 6'9"? The first event is fairly rare.[7] You may have seen somebody this tall in one of your classes sometime. But finding a class of 100 whose mean height is over 6'9" tall just won't happen. Why? Because *means have smaller standard deviations than individuals.*

How much smaller? Well, we have good news and bad news. The good news is that the standard deviation of $\bar{y}$ falls as the sample size grows. The bad news is

A S **The Standard Deviation of Means.** Experiment to see how the variability of the mean changes with the sample size. Here's another surprising result you don't have to trust us for. Find out for yourself.

---

[7] If students are a random sample of adults, fewer than 1 out of 10,000 should be taller than 6'9". Why might college students not really be a random sample with respect to height? Even if they're not a perfectly random sample, a college student over 6'9" tall is still rare.

that it doesn't drop as fast as we might like. It only goes down by the *square root* of the sample size.

That is, the Normal model for the sampling distribution of the mean has a standard deviation equal to

$$SD(\bar{y}) = \frac{\sigma}{\sqrt{n}}$$

"The n's justify the means."

—Apocryphal statistical saying

where σ is the standard deviation of the population. To emphasize that this is a standard deviation *parameter* of the sampling distribution model for the sample mean, $\bar{y}$, we write $SD(\bar{y})$ or $\sigma(\bar{y})$.

A S **The Sampling Distribution of the Mean.** The CLT tells us what to expect. In this activity you can work with the CLT, or simulate it if you prefer.

The sampling distribution model for a mean

When a random sample is drawn from any population with mean μ and standard deviation σ, its sample mean, $\bar{y}$, has a sampling distribution with the same *mean* μ but whose *standard deviation* is $\frac{\sigma}{\sqrt{n}}$ (and we write $\sigma(\bar{y}) = SD(\bar{y}) = \frac{\sigma}{\sqrt{n}}$). No matter what population the random sample comes from, the *shape* of the sampling distribution is approximately Normal as long as the sample size is large enough. The larger the sample used, the more closely the Normal approximates the sampling distribution for the mean.

We now have two closely related sampling distribution models. Which one we use depends on which kind of data we have.

- When we have categorical data, we calculate a sample proportion, $\hat{p}$; its sampling distribution has a Normal model with a mean at the true proportion ("Greek letter") p, and a standard deviation of $SD(\hat{p}) = \sqrt{\frac{pq}{n}} = \frac{\sqrt{pq}}{\sqrt{n}}$. We'll use this model throughout Chapters 19 through 22.

- When we have quantitative data, we calculate a sample mean, $\bar{y}$; its sampling distribution has a Normal model with a mean at the true mean, μ, and a standard deviation of $SD(\bar{y}) = \frac{\sigma}{\sqrt{n}}$. We'll use this model throughout Chapters 23, 24, and 25.

The means of these models are easy to remember, so all you need to be careful about is the standard deviations. Remember that these are standard deviations of the *statistics* $\hat{p}$ and $\bar{y}$. They both have a square root of n in the denominator. That tells us that the larger the sample, the less either statistic will vary. The only difference is in the numerator. If you just start by writing $SD(\bar{y})$ for quantitative data and $SD(\hat{p})$ for categorical data, you'll be able to remember which formula to use.

just checking

4 *Human gestation times have a mean of about 266 days, with a standard deviation of about 16 days. Does this mean that if we record the gestation times of a sample of 100 women, that histogram will be well modeled by a Normal model?*

5 *Suppose we look at the average gestation times for a sample of 100 women. If we imagined all the possible random samples of 100 women we could take and looked at the histogram of all the sample means, what shape would it have?*

6 *Where would the center of that histogram be?*

7 *What would be the standard deviation of that histogram?*

Assumptions and Conditions

A S **The Central Limit Theorem.** Does it really work for samples from non-Normal populations? See for yourself.

The CLT requires remarkably few assumptions, so there are few conditions to check:

1. **Random Sampling Condition:** The data values must be sampled randomly or the concept of a sampling distribution makes no sense.

2. **Independence Assumption:** The sampled values must be mutually independent. There's no way to check in general whether the observations are independent. However, when the sample is drawn without replacement (as is usually the case), you should check the . . .

   10% Condition: The sample size, n, is no more than 10% of the population.

Diminishing Returns

The standard deviation of the sampling distribution declines only with the square root of the sample size and not, for example, with $1/n$. The mean of a random sample of 4 has half $\left(\frac{1}{\sqrt{4}} = \frac{1}{2}\right)$ the standard deviation of an individual data value. To cut it in half again, we'd need a sample of 16, and a sample of 64 to halve it once more. In practice, random sampling works well, and means have smaller standard deviations than the individual data values that were averaged. This is the power of averaging.

If only we could afford a much larger sample, we could get the standard deviation of the sampling distribution *really* under control so that the sample mean could tell us still more about the unknown population mean. As we shall see, that nasty square root limits how much we can make a sample tell about the population. This is an example of something that's known as the Law of Diminishing Returns.

Working with the Sampling Distribution Model for the Mean Step-By-Step

Suppose that mean adult weight is 175 pounds with a standard deviation of 25 pounds. An elevator in our building has a weight limit of 10 persons or 2000 pounds. What's the probability that the 10 people who get on the elevator overload its weight limit?

Think **Plan** State what we want to know.

Asking the probability that the total weight of a sample of 10 people exceeds 2000 pounds is equivalent to asking the probability that their mean weight is greater than 200 pounds.

Model Check the conditions.

✔ **Random Sampling Condition:** I'll assume that the 10 people getting on the elevator are a random sample from the population.

✔ **Independence Assumption:** It's reasonable to think that the weights of 10 randomly sampled people will be mutually independent (but there could be exceptions—for example, if they were all from the same family or if the elevator were in a building with a diet clinic!).

✔ **10% Condition:** 10 people is surely less than 10% of the population of possible elevator riders.

State the parameters and the sampling distribution model. Even though the sample is small, the distribution of population weights is probably close to Normal. (This is not a concern for large samples.)

I've assumed that the model for all weights is roughly mound-shaped, with a mean $\mu = 175$ and standard deviation $\sigma = 25$.

Under these conditions, the CLT says that the sampling distribution of $\bar{y}$ has a Normal model with mean 175 and standard deviation

$$SD(\bar{y}) = \frac{\sigma}{\sqrt{n}} = \frac{25}{\sqrt{10}} \approx 7.91$$

Show   **Plot** Make a picture. Sketch the model and shade the area we're interested in. Here the mean weight of 200 pounds appears to be far out on the right tail of the curve.

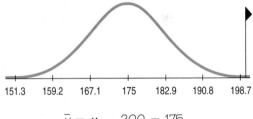

151.3	159.2	167.1	175	182.9	190.8	198.7

Mechanics Use the standard deviation as a ruler to find the z-score of the cutoff mean weight. We see that an average of 200 pounds is more than 3 standard deviations above the mean.

$$z = \frac{\bar{y} - \mu}{SD(\bar{y})} = \frac{200 - 175}{7.91} = 3.16$$

Find the resulting probability from a table of Normal probabilities, such as Table Z, a computer program, or a calculator.

$$P(\bar{y} > 200) = P(z > 3.16) = 0.0008$$

Tell   **Conclusion** Interpret your result in the proper context, being careful to relate it to the original question.

The chance that a random collection of 10 adults will exceed the elevator's weight limit is only 0.0008. So, if they are a random sample, it is quite unlikely that 10 people will exceed the total weight allowed on the elevator.

Standard Error

Both of the sampling distributions we've looked at are Normal. We know for proportions,

$$SD(\hat{p}) = \sqrt{\frac{pq}{n}}$$

and for means,

$$SD(\bar{y}) = \frac{\sigma}{\sqrt{n}}.$$

These are great if we know, or can pretend that we know, p or σ, and sometimes we'll do that.

Often we know only the observed proportion, $\hat{p}$, or the sample standard deviation, s. So of course we just use what we know, and we estimate. That may not seem like a big deal, but it gets a special name. Whenever we estimate the standard deviation of a sampling distribution, we call it a **standard error**.[8]

For a sample proportion, $\hat{p}$, the standard error is

$$SE(\hat{p}) = \sqrt{\frac{\hat{p}\hat{q}}{n}}.$$

For the sample mean, $\bar{y}$, the standard error is

$$SE(\bar{y}) = \frac{s}{\sqrt{n}}.$$

You may see a "standard error" reported by a computer program in a summary or offered by a calculator. It's safe to assume that if no statistic is specified, what was meant is $SE(\bar{y})$, the standard error of the mean.

Sampling Distribution Models

Let's summarize what we've learned about sampling distributions. At the heart is the idea that *the statistic itself is a random quantity*. We can't know what our statistic will be because it comes from a random sample. It's just one instance of something that happened for our particular random sample. A different random sample would have given a different result. This sample to sample variability is what generates the sampling distribution. The sampling distribution shows us the distribution of possible values that the statistic could have had.

We could simulate that distribution by pretending to take lots of samples. Fortunately, for the mean and the proportion, the CLT tells us that we can model their sampling distribution directly with a Normal model.

The two basic truths about sampling distributions are:

1. Sampling distributions arise because samples vary. Each random sample will contain different cases and, so, a different value of the statistic.

2. Although we can always simulate a sampling distribution, the Central Limit Theorem saves us the trouble for means and proportions.

On the next page is a picture showing the process going into the sampling distribution model:

A S **The CLT for Real Data.** Why settle for a picture when you can see it in action?

---

[8] This isn't such a great name because it isn't standard and nobody made an error. But it's much shorter and more convenient than saying, "the estimated standard deviation of the sampling distribution of the sample statistic."

We start with a population model, which can have any shape. It can even be bimodal or skewed (as this one is). We label the mean of this model μ and its standard deviation, σ.

We draw one real sample (solid line) of size n and show its histogram and summary statistics. We *imagine* (or simulate) drawing many other samples (dotted lines), which have their own histograms and summary statistics.

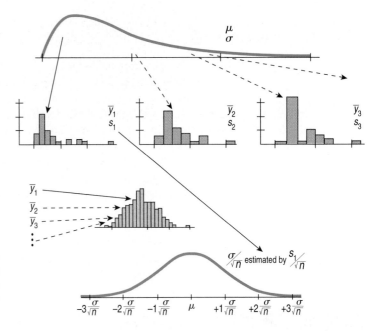

We (imagine) gathering all the means into a histogram.

The CLT tells us we can model the shape of this histogram with a Normal model. The mean of this Normal is μ, and the standard deviation is $SD(\bar{y}) = \frac{\sigma}{\sqrt{n}}$. When we don't know σ, we estimate it with the standard deviation of the one real sample. That gives us the *standard error*, $SE(\bar{y}) = \frac{s}{\sqrt{n}}$. **Figure 18.5**

What Can Go Wrong?

- ***Don't confuse the sampling distribution with the distribution of the sample.*** When you take a sample, you always look at the distribution of the values, usually with a histogram, and you may calculate summary statistics. Examining the distribution of the sample like this is wise. But that's not the sampling distribution. The sampling distribution is an imaginary collection of the values that a statistic *might* have taken for all the random samples—the one you got and the ones that you didn't get. We use the sampling distribution model to make statements about how the statistic varies.

- ***Beware of observations that are not independent.*** The CLT depends crucially on the assumption of independence. If our elevator riders are related, are all from the same school (for example, an elementary school), or in some other way aren't a random sample, then the statements we try to make about the mean are going to be wrong. Unfortunately, this isn't something you can check in your data. You have to think about how the data were gathered. Good sampling practice and well-designed randomized experiments ensure independence.

- ***Watch out for small samples from skewed populations.*** The CLT assures us that the sampling distribution model is Normal if n is large enough. If the population is nearly Normal, even small samples (like our 10 elevator riders) work. If the population is very skewed, then n will have to be large before the Normal model will work well. If we sampled 15 or even 20 CEOs and used $\bar{y}$ to make a statement about the mean of all CEOs' compensation, we'd likely get into trouble because the underlying data distribution is so skewed. Unfortunately, there's no good rule of thumb.[9] It just depends on how skewed the data distribution is. Always plot the data to check.

---

[9] For proportions, of course, there is a rule: the Success/Failure Condition. That works for proportions because the standard deviation of a proportion is linked to its mean.

CONNECTIONS

The concept of a sampling distribution connects to almost everything we have done. The fundamental connection is to the deliberate application of randomness in random sampling and randomized comparative experiments. If we didn't employ randomness to generate unbiased data, then repeating the data collection would just get the same data values again (with perhaps a few new measurement or recording errors). The distribution of statistic values arises directly because different random samples and randomized experiments would generate different statistic values.

The connection to the Normal distribution is obvious. We first introduced the Normal model before because it was "nice." As a unimodal, symmetric distribution with 99.7% of its area within three standard deviations of the mean, the Normal model is easy to work with. Now we see that the Normal holds a special place among distributions because we can use it to model the sampling distributions of the mean and the proportion.

We use simulation to understand sampling distributions. In fact, some important sampling distributions were discovered first by simulation.

What have we learned?

Way back in Chapter 1 we said that Statistics is about variation. We know that no sample fully and exactly describes the population; sample proportions and means will vary from sample to sample. That's sampling error (or, better, sampling variability). We know it will always be present—indeed, the world would be a boring place if variability didn't exist. You might think that sampling variability would prevent us from learning anything reliable about a population by looking at a sample, but that's just not so. The fortunate fact is that sampling variability is not just unavoidable—it's predictable!

We've learned to describe the behavior of sample proportions—shape, center, and spread—as long as certain conditions are met. The sample must be random, of course, and large enough that we expect at least 10 successes and failures. Then:

- The sampling distribution (the imagined histogram of the proportions from all possible samples) is shaped like a Normal model.
- The mean of the sampling model is the true proportion in the population.
- The standard deviation of the sample proportions is $\sqrt{\dfrac{pq}{n}}$.

And we've learned to describe the behavior of sample means as well, based on an amazing result known as the Central Limit Theorem—the Fundamental Theorem of Statistics. Again the sample must be random—no surprise there—and needs to be larger if our data come from a population that's not roughly unimodal and symmetric. Then:

- Regardless of the shape of the original population, the shape of the distribution of the means of all possible samples can be described by a Normal model, provided the samples are large enough.
- The center of the sampling model will be the true mean of the population from which we took the sample.
- The standard deviation of the sample means is the population's standard deviation divided by the square root of the sample size, $\dfrac{\sigma}{\sqrt{n}}$.

TERMS

Sampling distribution model Different random samples give different values for a statistic. The sampling distribution model shows the behavior of the statistic over all the possible samples for the same size *n*.

Sampling distribution model for a proportion If assumptions of independence and random sampling are met, and we expect at least 10 successes and 10 failures, then the sampling distribution of a proportion is modeled by a Normal model with a mean equal to the true proportion value, *p*, and a standard deviation equal to $\sqrt{\dfrac{pq}{n}}$.

Central Limit Theorem The Central Limit Theorem (CLT) states that the sampling distribution model of the sample mean (and proportion) is approximately Normal for large *n*, *regardless of the distribution of the population, as long as the observations are independent.*

Sampling distribution model for a mean If assumptions of independence and random sampling are met, and the sample size is large enough, the sampling distribution of the sample mean is modeled by a Normal model with a mean equal to the population mean, μ, and a standard deviation equal to $\dfrac{\sigma}{\sqrt{n}}$.

Standard error When we estimate the standard deviation of a sampling distribution using statistics found from the data, the estimate is called a standard error.

SKILLS *When you complete this lesson you should:*

Think
- Understand that the variability of a statistic (as measured by the standard deviation of its sampling distribution) depends on the size of the sample. Statistics based on larger samples are less variable.
- Understand that the Central Limit Theorem gives the sampling distribution model of the mean for sufficiently large samples regardless of the underlying population.

Show
- Be able to demonstrate a sampling distribution by simulation.
- Be able to use a sampling distribution model to make simple statements about the distribution of a proportion or mean under repeated sampling.

Tell
- Be able to interpret a sampling distribution model as describing the values taken by a statistic in all possible realizations of a sample or randomized experiment under the same conditions.

EXERCISES

1. Coin tosses. In a large class of introductory Statistics students, the professor has each person toss a coin 16 times and calculate the proportion of his or her tosses that were heads. The students then report their results, and the professor plots a histogram of these several proportions.
  a) What shape would you expect this histogram to be? Why?
  b) Where do you expect the histogram to be centered?
  c) How much variability would you expect among these proportions?
  d) Explain why a Normal model should not be used here.

2. M&M's. The candy company claims that 10% of the M&M's it produces are green. Suppose that the candies are packaged at random in small bags containing about 50 M&M's. A class of elementary school students learning about percents opens several bags, counts the various colors of the candies, and calculates the proportion that are green.
  a) If we plot a histogram showing the proportions of green candies in the various bags, what shape would you expect it to have?
  b) Can that histogram be approximated by a Normal model? Explain.

c) Where should the center of the histogram be?

d) What should the standard deviation of the proportion be?

3. More coins. Suppose the class in Exercise 1 repeats the coin tossing experiment.

a) The students toss the coins 25 times each. Use the 68-95-99.7 Rule to describe the sampling distribution model.

b) Confirm that you can use a Normal model here.

c) They increase the number of tosses to 64 each. Draw and label the appropriate sampling distribution model. Check the appropriate conditions to justify your model.

d) Explain how the sampling distribution model changes as the number of tosses increases.

4. Bigger bag. Suppose the class in Exercise 2 buys bigger bags of candy, with 200 M&M's each. Again the students calculate the proportion of green candies they find.

a) Explain why it's appropriate to use a Normal model to describe the distribution of the proportion of green M&M's they might expect.

b) Use the 68-95-99.7 Rule to describe how this proportion might vary from bag to bag.

c) How would this model change if the bags contained even more candies?

5. Just (un)lucky? One of the students in the introductory Statistics class in Exercise 3 claims to have tossed her coin 200 times and found only 42% heads. What do you think of this claim? Explain.

6. Too many green ones? In a really large bag of M&M's, the students in Exercise 4 found 500 candies, and 12% of them were green. Is this an unusually large proportion of green M&M's? Explain.

7. Speeding. State police believe that 70% of the drivers traveling on a major interstate highway exceed the speed limit. They plan to set up a radar trap and check the speeds of 80 cars.

a) Using the 68-95-99.7 Rule, draw and label the distribution of the proportion of these cars the police will observe speeding.

b) Do you think the appropriate conditions necessary for your analysis are met? Explain.

8. Smoking. Public health statistics indicate that 26.4% of American adults smoke cigarettes. Using the 68-95-99.7 Rule, describe the sampling distribution model for the proportion of smokers among a randomly selected group of 50 adults. Be sure to discuss your assumptions and conditions.

9. Loans. Based on past experience, a bank believes that 7% of the people who receive loans will not make payments on time. The bank has recently approved 200 loans.

a) What are the mean and standard deviation of the proportion of clients in this group who may not make timely payments?

b) What assumptions underlie your model? Are the conditions met? Explain.

c) What's the probability that over 10% of these clients will not make timely payments?

10. Contacts. Assume that 30% of students at a university wear contact lenses.

a) We randomly pick 100 students. Let $\hat{p}$ represent the proportion of students in this sample who wear contacts. What's the appropriate model for the distribution of $\hat{p}$? Specify the name of the distribution, the mean, and the standard deviation. Be sure to verify that the conditions are met.

b) What's the approximate probability that more than one third of this sample wear contacts?

11. Back to school? Best known for its testing program, ACT, Inc., also compiles data on a variety of issues in education. In 2004 the company reported that the national college freshman-to-sophomore retention rate held steady at 74% over the previous four years. Consider colleges with freshman classes of 400 students. Use the 68-95-99.7 Rule to describe the sampling distribution model for the percentage of those students we expect to return to that school for their sophomore years. Do you think the appropriate conditions are met?

12. Binge drinking. As we learned in Chapter 15, a national study found that 44% of college students engage in binge drinking (5 drinks at a sitting for men, 4 for women). Use the 68-95-99.7 Rule to describe the sampling distribution model for the proportion of students in a randomly selected group of 200 college students who engage in binge drinking. Do you think the appropriate conditions are met?

13. Back to school, again. Based on the 74% national retention rate described in Exercise 11, does a college where 522 of their 603 freshman returned the next year as sophomores have a right to brag that it has an unusually high retention rate? Explain.

14. Binge sample. After hearing of the national result that 44% of students engage in binge drinking (5 drinks at a sitting for men, 4 for women), a professor surveyed a random sample of 244 students at his college and found that 96 of them admitted to binge drinking in the past week. Should he be surprised at this result? Explain.

15. Polling. Just before a referendum on a school budget, a local newspaper polls 400 voters in an attempt to predict whether the budget will pass. Suppose that the budget actually has the support of 52% of the voters. What's the probability the newspaper's sample will lead them to predict defeat? Be sure to verify that the assumptions and conditions necessary for your analysis are met.

16. Seeds. Information on a packet of seeds claims that the germination rate is 92%. What's the probability that more than 95% of the 160 seeds in the packet will germinate? Be sure to discuss your assumptions and check the conditions that support your model.

17. Apples. When a truckload of apples arrives at a packing plant, a random sample of 150 is selected and examined for bruises, discoloration, and other defects. The whole truckload will be rejected if more than 5% of the sample is unsatisfactory. Suppose that in fact 8% of the apples on the truck do not meet the desired standard. What's the probability that the shipment will be accepted anyway?

18. Genetic defect. It's believed that 4% of children have a gene that may be linked to juvenile diabetes. Researchers hoping to track 20 of these children for several years test 732 newborns for the presence of this gene. What's the probability that they find enough subjects for their study?

19. Nonsmokers. While some nonsmokers do not mind being seated in a smoking section of a restaurant, about 60% of the customers demand a smoke-free area. A new restaurant with 120 seats is being planned. How many seats should be in the nonsmoking area in order to be very sure of having enough seating there? Comment on the assumptions and conditions that support your model, and explain what "very sure" means to you.

20. Meals. A restaurateur anticipates serving about 180 people on a Friday evening, and believes that about 20% of the patrons will order the chef's steak special. How many of those meals should he plan on serving in order to be pretty sure of having enough steaks on hand to meet customer demand? Justify your answer, including an explanation of what "pretty sure" means to you.

21. Sampling. A sample is chosen randomly from a population that can be described by a Normal model.
 a) What's the sampling distribution model for the sample mean? Describe shape, center, and spread.
 b) If we choose a larger sample, what's the effect on this sampling distribution model?

22. Sampling, part II. A sample is chosen randomly from a population that was strongly skewed to the left.
 a) Describe the sampling distribution model for the sample mean if the sample size is small.
 b) If we make the sample larger, what happens to the sampling distribution model's shape, center, and spread?
 c) As we make the sample larger, what happens to the expected distribution of the data in the sample?

23. GPAs. A college's data about the incoming freshmen indicates that the mean of their high school GPAs was 3.4, with a standard deviation of 0.35; the distribution was roughly mound-shaped and only slightly skewed. The students are randomly assigned to freshman writing seminars in groups of 25. What might the mean GPA of one of these seminar groups be? Describe the appropriate sampling distribution model—shape, center, and spread—with attention to assumptions and conditions. Make a sketch using the 68-95-99.7 Rule.

24. Home values. Assessment records indicate that the value of homes in a small city is skewed right, with a mean of $140,000 and standard deviation of $60,000. To check the accuracy of the assessment data, officials plan to conduct a detailed appraisal of 100 homes selected at random. Using the 68-95-99.7 Rule, draw and label an appropriate sampling model for the mean value of the homes selected.

25. Pregnancy. Assume that the duration of human pregnancies can be described by a Normal model with mean 266 days and standard deviation 16 days.
 a) What percentage of pregnancies should last between 270 and 280 days?
 b) At least how many days should the longest 25% of all pregnancies last?
 c) Suppose a certain obstetrician is currently providing prenatal care to 60 pregnant women. Let $\bar{y}$ represent the mean length of their pregnancies. According to the Central Limit Theorem, what's the distribution of this sample mean, $\bar{y}$? Specify the model, mean, and standard deviation.
 d) What's the probability that the mean duration of these patients' pregnancies will be less than 260 days?

26. Rainfall. Statistics from Cornell's Northeast Regional Climate Center indicate that Ithaca, NY, gets an average of 35.4" of rain each year, with a standard deviation of 4.2". Assume that a Normal model applies.
 a) During what percentage of years does Ithaca get more than 40" of rain?
 b) Less than how much rain falls in the driest 20% of all years?
 c) A Cornell University student is in Ithaca for 4 years. Let $\bar{y}$ represent the mean amount of rain for those 4 years. Describe the sampling distribution model of this sample mean, $\bar{y}$.
 d) What's the probability that those 4 years average less than 30" of rain?

27. Pregnant again. The duration of human pregnancies may not actually follow a Normal model, as described in Exercise 25.
 a) Explain why it may be somewhat skewed to the left.
 b) If the correct model is in fact skewed, does that change your answers to parts a, b, and c of Exercise 25? Explain why or why not for each.

28. At work. Some business analysts estimate that the length of time people work at a job has a mean of 6.2 years and a standard deviation of 4.5 years.
 a) Explain why you suspect this distribution may be skewed to the right.
 b) Explain why you could estimate the probability that 100 people selected at random had worked for their employers an average of 10 years or more, but you could not estimate the probability that an individual had done so.

29. Dice and dollars. You roll a die, winning nothing if the number of spots is odd, $1 for a 2 or a 4, and $10 for a 6.
a) Find the expected value and standard deviation of your prospective winnings.
b) You play twice. Find the mean and standard deviation of your total winnings.
c) You play 40 times. What's the probability that you win at least $100?

30. New game. You pay $10 and roll a die. If you get a 6, you win $50. If not, you get to roll again. If you get a 6 this time, you get your $10 back.
a) Create a probability model for this game.
b) Find the expected value and standard deviation of your prospective winnings.
c) You play this game five times. Find the expected value and standard deviation of your average winnings.
d) 100 people play this game. What's the probability the person running the game makes a profit?

31. AP Stats. The College Board reported the score distribution shown in the table for all students who took the 2004 AP Statistics exam.

Score	Percent of Students
5	12.5
4	22.5
3	24.8
2	19.8
1	20.4

a) Find the mean and standard deviation of the scores.
b) If we select a random sample of 40 AP Statistics students, would you expect their scores to follow a Normal model? Explain.
c) Consider the mean scores of random samples of 40 AP Statistics students. Describe the sampling model for these means (shape, center, and spread).

32. Museum membership. A museum offers several levels of membership, as shown in the table.

Member Category	Amount of Donation ($)	Percent of Members
Individual	50	41
Family	100	37
Sponsor	250	14
Patron	500	7
Benefactor	1000	1

a) Find the mean and standard deviation of the donations.
b) During their annual membership drive, they hope to sign up 50 new members each day. Would you expect the distribution of the donations for a day to follow a Normal model? Explain.

c) Consider the mean donation of the 50 new members each day. Describe the sampling model for these means (shape, center, and spread).

33. AP Stats, again. An AP Statistics teacher had 63 students preparing to take the AP exam discussed in Exercise 31. Though they were obviously not a random sample, he considered his students to be "typical" of all the national students. What's the probability that his students will achieve an average score of at least 3?

34. Joining the museum. One of the museum's phone volunteers sets a personal goal of getting an average donation of at least $100 from the new members she enrolls during the membership drive. If she gets 80 new members and they can be considered a random sample of all the museum's members, what is the probability that she can achieve her goal?

35. Pollution. Carbon monoxide (CO) emissions for a certain kind of car vary with mean 2.9 g/mi and standard deviation 0.4 g/mi. A company has 80 of these cars in its fleet. Let $\bar{y}$ represent the mean CO level for the company's fleet.
a) What's the approximate model for the distribution of $\bar{y}$? Explain.
b) Estimate the probability that $\bar{y}$ is between 3.0 and 3.1 g/mi.
c) There is only a 5% chance that the fleet's mean CO level is greater than what value?

36. Potato chips. The weight of potato chips in a medium-size bag is stated to be 10 ounces. The amount that the packaging machine puts in these bags is believed to have a Normal model with mean 10.2 ounces and standard deviation 0.12 ounces.
a) What fraction of all bags sold are underweight?
b) Some of the chips are sold in "bargain packs" of 3 bags. What's the probability that none of the 3 is underweight?
c) What's the probability that the mean weight of the 3 bags is below the stated amount?
d) What's the probability that the mean weight of a 24-bag case of potato chips is below 10 ounces?

37. Tips. A waiter believes the distribution of his tips has a model that is slightly skewed to the right, with a mean of $9.60 and a standard deviation of $5.40.
a) Explain why you cannot determine the probability that a given party will tip him at least $20.
b) Can you estimate the probability that the next 4 parties will tip an average of at least $15? Explain.
c) Is it likely that his 10 parties today will tip an average of at least $15? Explain.

38. Groceries. Grocery store receipts show that customer purchases have a skewed distribution with a mean of $32 and a standard deviation of $20.
a) Explain why you cannot determine the probability that the next customer will spend at least $40.

b) Can you estimate the probability that the next 10 customers will spend an average of at least $40? Explain.

c) Is it likely that the next 50 customers will spend an average of at least $40? Explain.

39. More tips. The waiter in Exercise 37 usually waits on about 40 parties over a weekend of work.

a) Estimate the probability that he will earn at least $500 in tips.

b) How much does he earn on the best 10% of such weekends?

40. More groceries. Suppose the store in Exercise 38 had 312 customers today.

a) Estimate the probability that the store's revenues were at least $10,000.

b) If in a typical day, the store serves 312 customers, how much does the store take in on the worst 10% of such days?

41. IQs. Suppose that IQs of East State University's students can be described by a Normal model with mean 130 and standard deviation 8 points. Also suppose that IQs of students from West State University can be described by a Normal model with mean 120 and standard deviation 10.

a) We select 1 student at random from East State. Find the probability that this student's IQ is at least 125 points.

b) We select 1 student at random from each school. Find the probability that the East State student's IQ is at least 5 points higher than the West State student's IQ.

c) We select 3 West State students at random. Find the probability that this group's average IQ is at least 125 points.

d) We also select 3 East State students at random. What's the probability that their average IQ is at least 5 points higher than the average for the 3 West Staters?

42. Milk. Although most of us buy milk by the quart or gallon, farmers measure daily production in pounds. Ayrshire cows average 47 pounds of milk a day, with a standard deviation of 6 pounds. For Jersey cows, the mean daily production is 43 pounds, with a standard deviation of 5 pounds. Assume that Normal models describe milk production for these breeds.

a) We select an Ayrshire at random. What's the probability that she averages more than 50 pounds of milk a day?

b) What's the probability that a randomly selected Ayrshire gives more milk than a randomly selected Jersey?

c) A farmer has 20 Jerseys. What's the probability that the average production for this small herd exceeds 45 pounds of milk a day?

d) A neighboring farmer has 10 Ayrshires. What's the probability that his herd average is at least 5 pounds higher than the average for the Jersey herd?

---

just checking
Answers

1. A Normal model (approximately).

2. At the actual proportion of all students who are in favor.

3. $SD(\hat{p}) = \sqrt{\dfrac{(0.5)(0.5)}{100}} = 0.05$

4. No, that histogram will approximate the distribution of human gestation periods; that may not be Normal.

5. A Normal model (approximately).

6. 266 days

7. $\dfrac{16}{\sqrt{100}} = 1.6$ days

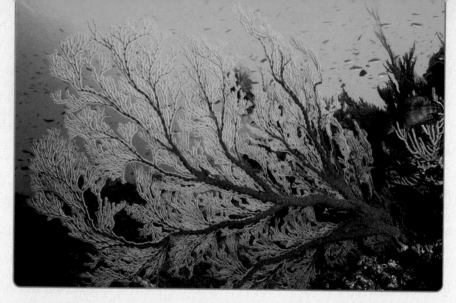

Confidence Intervals for Proportions

oral are in decline worldwide, possibly because of pollution or changes in sea temperature. The consequences of this decline are far-reaching. The death of coral reefs may be an early warning of climate change and could have far-reaching economic impact.

One spectacular kind of coral, the sea fan, looks like a plant growing from the sea floor, but it is actually an animal.[1] A sea fan of the size in the photo above may have taken 40 years to grow, but it can be killed quickly by pollution or disease. Sea fans in the Caribbean Sea have been under attack by the disease *aspergillosis*. In June of 2000, the sea fan disease team from Dr. Drew Harvell's lab randomly sampled some sea fans at the Las Redes Reef in Akumal, Mexico, at a depth of 40 feet. They found that 54 of the 104 sea fans they sampled were infected with the disease. What might this say about the prevalence of this disease among sea fans in general?

We have a sample proportion, which we write as $\hat{p}$, of 54/104, or 51.9%. Our first guess might be that this observed proportion is close to the population proportion, p. But we also know that because of natural sampling variability, if the researchers had drawn a second sample of 104 sea fans at roughly the same time, the proportion infected from that sample probably wouldn't have been exactly 51.9%.

What *can* we say about the population proportion, p? To start to answer this question, think about how different the sample proportion might have been if we'd taken another random sample from the same population. But wait. Remember—we aren't actually going to take more samples. We just want to *imagine* how the sample proportions might vary from sample to sample. In other words, we want to know about the *sampling distribution* of the sample proportion of infected sea fans.

WHO	Sea fans
WHAT	Percent infected
WHEN	June 2000
WHERE	Las Redes Reef, Akumal, Mexico, 40 feet deep
WHY	Research

---

[1] For purists, sea fans are actually colonies of genetically identical animals.

A Confidence Interval

A S **Confidence Intervals and Sampling Distributions.** Simulate the sampling distribution, and see how it gives a confidence interval.

Let's look at our model for the sampling distribution. What do we know about it? We know it's approximately Normal (under certain assumptions, which we should be careful to check) and that its mean is the proportion of all infected sea fans on the Las Redes Reef. Is the infected proportion of *all* sea fans 51.9%? No, that's just $\hat{p}$, our estimate. We don't know the proportion, p, of all the infected sea fans; that's what we're trying to find out. We do know, though, that the sampling distribution model of $\hat{p}$ is centered at p, and we know that the standard deviation of the sampling distribution is $\sqrt{\dfrac{pq}{n}}$.

Since we don't know p, we can't find the true standard deviation of the sampling distribution model. Instead we'll use $\hat{p}$ and find the standard error,

$$SE(\hat{p}) = \sqrt{\frac{\hat{p}\hat{q}}{n}} = \sqrt{\frac{(0.519)(0.481)}{104}} = 0.049 = 4.9\%.$$

NOTATION ALERT:

Remember that $\hat{p}$ is our sample-based estimate of the true proportion p. Recall also that q is just shorthand for $1 - p$, and $\hat{q} = 1 - \hat{p}$.

Now we know that the sampling model for $\hat{p}$ should look like this:

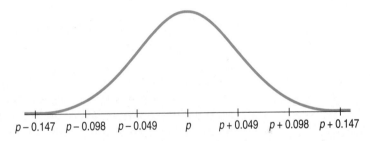

The sampling distribution model for $\hat{p}$ is Normal with a mean of p and a standard deviation we estimate to be 0.049.   **Figure 19.1**

$p - 0.147 \quad p - 0.098 \quad p - 0.049 \quad p \quad p + 0.049 \quad p + 0.098 \quad p + 0.147$

Great. What does that tell us? Well, because it's Normal, it says that about 68% of all samples of 104 sea fans will have $\hat{p}$'s within 1 *SE*, 0.049, of p. And about 95% of all these samples will be within $p \pm 2$ *SE*s. But where is *our* sample proportion in this picture? And what value does p have? We still don't know!

We do know that for 95% of random samples, $\hat{p}$ will be no more than 2 *SE*s away from p. So let's look at this from $\hat{p}$'s point of view. If I'm $\hat{p}$, there's a 95% chance that p is no more than 2 *SE*s away from me. If I reach out 2 *SE*s, or 2×0.049, away from me on both sides, I'm 95% sure that p will be within my grasp. Now, I've got him! Probably. Of course, even if my interval does catch p, I still don't know its true value. The best I can do is an interval, and even then I can't be positive it contains p.

Reaching out 2 *SE*s on either side of $\hat{p}$ makes us 95% confident we'll trap the true proportion, p.   **Figure 19.2**

ACME *p*-trap: Guaranteed* to capture *p*.

*with 95% confidence

$\hat{p} - 2\ SE \qquad\qquad \hat{p} \qquad\qquad \hat{p} + 2\ SE$

So what can we really say about *p*? Here's a list of things we'd like to be able to say, in order of strongest to weakest and the reasons we can't say most of them:

A S **Can We Estimate
a Parameter?** Consider these four
interpretations of a confidence interval by simulating to see whether
they could be right.

1. **"51.9% of *all* sea fans on the Las Redes Reef are infected."** It would be nice to be able to make absolute statements about population values with certainty, but we just don't have enough information to do that. There's no way to be sure that the population proportion is the same as the sample proportion; in fact, it almost certainly isn't. Observations vary. Another sample would yield a different sample proportion.

2. **"It is *probably* true that 51.9% of all sea fans on the Las Redes Reef are infected."** No. In fact we can be pretty sure that whatever the true proportion is, it's not exactly 51.900%. So the statement is not true.

3. **"We don't know exactly what proportion of sea fans on the Las Redes Reef are infected but we *know* that it's within the interval 51.9% ± 2 × 4.9%. That is, it's between 42.1% and 61.7%."** This is getting closer, but we still can't be certain. We can't know *for sure* that the true proportion is in this interval—or in any particular interval.

4. **"We don't know exactly what proportion of sea fans on the Las Redes Reef are infected, but the interval from 42.1% to 61.7% *probably* contains the true proportion."** We've now fudged twice—first by giving an interval and second by admitting that we only think the interval "probably" contains the true value. And this statement is true.

*"Far better an approximate an-
swer to the right question, . . .
than an exact answer to the
wrong question."*
—John W. Tukey

That last statement may be true, but it's a bit wishy-washy. We can tighten it up a bit by quantifying what we mean by "probably." We saw that 95% of the time when we reach out 2 *SE*s from $\hat{p}$ we capture *p*, *so we can be 95% confident that this is one of those times.* After putting a number on the probability that this interval covers the true proportion, we've given our best guess of where the parameter is and how certain we are that it's within some range.

5. **"We are 95% confident that between 42.1% and 61.7% of Las Redes sea fans are infected."** Statements like these are called **confidence intervals.** They're the best we can do.

Each confidence interval discussed in the book has a name. You'll see many different kinds of confidence intervals in the following chapters. Some will be about more than *one* sample, some will be about statistics other than *proportions*, some will use models other than the Normal. The interval calculated and interpreted here is sometimes called a **one-proportion z-interval.**[2]

In the 1992 presidential election Bill Clinton ran against George H. W. Bush and Ross Perot. That June the Gallup organization asked registered voters if there was "Some chance they could vote for other candidates" besides their expressed first choice. At that time, 62% of registered voters said "yes," there was some chance they might switch. In June 2004, Gallup/CNN/USA Today asked 909 registered voters the same question. Only 18% indicated that there was some chance they might switch. The resulting 95% confidence interval is 0.18 ± 0.025 = 15.5% to 20.5%. Which of the following are correct statements about the 2004 presidential election, and why?

---

[2] In fact, this confidence interval is so standard for a single proportion that you may see it simply called a "confidence interval for the proportion."

1. In the sample of 909 registered voters, somewhere between 15.5% and 20.5% of them said there is a chance they might switch votes.

2. We are 95% confident that 18% of all U.S. registered voters had some chance of switching votes.

3. We are 95% confident that between 15.5% and 20.5% of all U.S. registered voters had some chance of switching votes.

4. We know that between 15.5% and 20.5% of all U.S. registered voters had some chance of switching votes.

5. 95% of all U.S. registered voters had some chance of switching votes.

What Does "95% Confidence" Really Mean?

A S **Confidence Intervals for Proportions.** Here's a new interactive tool that makes it easy to construct and experiment with confidence intervals. We'll use this tool for the rest of the course—sure beats calculating by hand!

What do we mean when we say we have 95% confidence that our interval contains the true proportion? Back in Chapter 18 we looked at how proportions vary from sample to sample. If other researchers select their own samples of sea fans, they'll also find some infected by the disease, but each person's sample proportion will almost certainly differ from ours. When they each try to estimate the true rate of infection in the entire population, they'll center *their* confidence intervals at the proportions they observed in their own samples. Each of us will end up with a different interval.

Our interval guessed the true proportion of infected sea fans to be between about 42% and 62%. Another researcher whose sample contained more infected fans than ours did might guess between 46% and 66%. Still another who happened to collect fewer infected fans might estimate the true proportion to be between 23% and 43%. And so on. Every possible sample would produce yet another confidence interval. Although wide intervals like these can't pin down the actual rate of infection very precisely, we expect that most of them should be winners, capturing the true value. Nonetheless, some will be duds, missing the population proportion entirely.

The figure below shows the confidence intervals produced by simulating 20 different random samples. The red dots are the proportions of infected fans in each sample, and the blue segments show the confidence intervals found for each.

The horizontal green line shows the true percentage of all sea fans that are infected. Most of the 20 simulated samples produced confidence intervals that captured the true value, but a few missed.

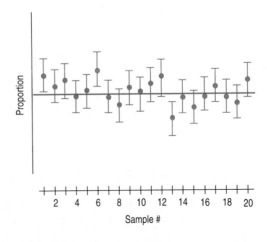

The green line represents the true rate of infection in the population, so you can see that most of the intervals caught it—but a few missed. (And notice again that it is the *intervals* that vary from sample to sample; the green line doesn't move.)

Of course, there's a huge number of possible samples that *could* be drawn, each with its own sample proportion. These are just some of them. Each sample proportion can be used to make a confidence interval. That's a large pile of possible confidence intervals, and ours is just one of those in the pile. Did *our* confidence interval "work"? We can never be sure, because we'll never know the true proportion of all the sea fans that are infected. However, the Central Limit Theorem assures us that 95% of the intervals in the pile are winners, covering the true value, and only 5% are duds. *That's* why we're 95% confident that our interval is a winner!

Margin of Error: Certainty vs. Precision

We've just claimed that with a certain confidence we've captured the true proportion of all infected sea fans. Our confidence interval had the form

$$\hat{p} \pm 2\,SE(\hat{p}).$$

The extent of the interval on either side of $\hat{p}$ is called the **margin of error** *(ME)*. We'll want to use the same approach for many other situations besides estimating proportions. In general, confidence intervals look like this:

$$estimate \pm ME.$$

The margin of error for our 95% confidence interval was 2 *SE*. What if we wanted to be more confident? To be more confident, we'll need to capture p more often, and to do that we'll need to make the interval wider. For example, if we want to be 99.7% confident, the margin of error will have to be 3 *SE*.

Reaching out 3 *SE*s on either side of $\hat{p}$ makes us 99.7% confident we'll trap the true proportion p. Compare with Figure 19.2.   **Figure 19.3**

The more confident we want to be, the larger the margin of error must be. We can be 100% confident that the proportion of infected sea fans is between 0% and 100%, but this isn't likely to be very useful. On the other hand, we could give a confidence interval from 51.8% to 52.0%, but we can't be very confident about a precise statement like this. Every confidence interval is a balance between certainty and precision.

A S **Balancing Precision and Certainty.** What percent of parents expect their kids to pay for college with a student loan? Use the interactive tool to investigate the balance between the precision and the certainty of a confidence interval.

The tension between certainty and precision is always there. Fortunately, in most cases we can be both sufficiently certain and sufficiently precise to make useful statements. There is no simple answer to the conflict. You must choose a confidence level yourself. The data can't do it for you. The choice of confidence level is somewhat arbitrary. The most commonly chosen confidence levels are 90%, 95%, and 99%, but any percentage can be used. (In practice, though, using something like 92.9% or 97.2% is likely to make people think you're up to something.)

Garfield © 1999 Paws, Inc. Reprinted with permission of UNIVERSAL PRESS SYNDICATE. All rights reserved.

Critical Values

In our sea fans example we used 2 *SE* to give us a 95% confidence interval. To change the confidence level, we'd need to change the *number* of SEs so that the size of the margin of error corresponds to the new level. This number of SEs is called the **critical value.** Here it's based on the Normal model, so we denote it z^*. For any confidence level, we can find the corresponding critical value from a computer, a calculator, or a Normal probability table, such as Table Z.

For a 95% confidence interval, you'll find the precise critical value is $z^* = 1.96$. That is, 95% of a Normal model is found within ± 1.96 standard deviations of the mean. We've been using $z^* = 2$ from the 68-95-99.7 Rule because it's easy to remember.

For a 90% confidence interval, the critical value is 1.645, because for a Normal model 90% of the values are within 1.645 standard deviations from the mean. **Figure 19.4**

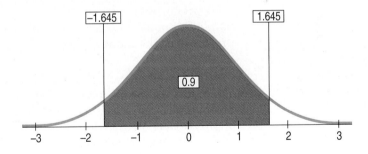

Assumptions and Conditions

We've just made some pretty sweeping statements about sea fans. Those statements were possible because we used a Normal model for the sampling distribution. But is that model appropriate?

As we've seen, all statistical models make **assumptions.** Different models make different assumptions. If those assumptions are not true, the model might be inappropriate and our conclusions based on it may be wrong.

Instead of saying "assume," maybe we should say "pretend." We can never be certain that an assumption is true, but we can decide intelligently whether to pretend it is. We can even have situations in which we know the assumption is technically false, but it is "close enough," so we pretend it is true anyway. When we have data, we can often decide whether an assumption is plausible by checking a related **condition.** For linear regression, we checked the Linearity Assumption by looking at the scatterplot of the data to see if the relationship was *straight enough.* That was a direct way to see whether the linear model was reasonable for our data. Now we need to go further. We want to make a statement about the world at large, not just about the data we collected. So the assumptions we make are not just about how our data look, but about how representative they are.

Here, for example, are the assumptions and the corresponding conditions to check before creating (and believing) a confidence interval about a proportion.

Independence Assumption

A S **Assumptions and Conditions.** Here's an animated review of the assumptions and conditions.

The data values are assumed to be independent from each other. Independence is a fundamental property, but not one we can check in the data. Instead, we think about whether independence is reasonable, and we check three conditions.

Plausible Independence Condition: Is there any reason to believe that the data values somehow affect each other? (For example, might the disease in sea fans be contagious?) This condition depends on your knowledge of the situation. It's not one you can check by looking at the data.

Randomization Condition: Were the data sampled at random or generated from a properly randomized experiment? Proper randomization can help ensure independence.

10% Condition: Samples are almost always drawn without replacement. Usually, of course, we'd like to have as large a sample as we can. But when the population itself is small we have another concern. When we sample from small populations, the probability of success may be different for the last few individuals we draw than it was for the first few. For example, if most of the women have already been sampled, the chance of drawing a woman from the remaining population is lower. If the sample exceeds 10% of the population, the probability of a success changes so much during the sampling that our Normal model may no longer be appropriate. But if less than 10% of the population is sampled, we can pretend to have independence. This is an example of a condition that gives us permission to pretend even though we know the formal assumption is false.

Sample Size Assumption

The model we use for inference is based on the Central Limit Theorem. The sample must be large enough to make the sampling model for the sample proportions approximately Normal. It turns out that we need more data as the proportion gets closer and closer to either extreme (0 or 1). This requirement is easy to check as the following condition:

Success/Failure Condition: We must expect at least 10 "successes" and at least 10 "failures." Recall that by tradition we arbitrarily label one alternative (usually the outcome being counted) as a "success" even if it's something bad (like a sick sea fan). The other alternative is, of course, then a "failure."[3]

A S **A Confidence Interval for p.** View the video story of pollution in Chesapeake Bay, and make a confidence interval for the analysis (with the interactive tool).

> ### One-proportion z-interval
>
> When the conditions are met, we are ready to find the confidence interval for the population proportion, p. The confidence interval is $\hat{p} \pm z^* \times SE(\hat{p})$ where the standard deviation of the proportion is estimated by $SE(\hat{p}) = \sqrt{\dfrac{\hat{p}\hat{q}}{n}}$.

---

[3] We saw where the "10" comes from in Chapter 17.

A Confidence Interval for a Proportion Step-By-Step

WHO	Adults in the United States
WHAT	Response to a question about the death penalty
WHEN	May 2002
WHERE	United States
HOW	537 adults were randomly sampled and asked by the Gallup Poll
WHY	Public opinion research

In May 2002, the Gallup Poll asked 537 randomly sampled adults the question "Generally speaking, do you believe the death penalty is applied fairly or unfairly in this country today?" Of these, 53% answered "Fairly" and 7% said they didn't know. What can we conclude from this survey?

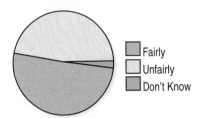

To answer this question, we'll build a confidence interval for the proportion of all U.S. adults who believe the death penalty is applied fairly. There are four steps to building a confidence interval for proportions: Plan, Model, Mechanics, and Conclusion.

Think

Plan State the problem and the W's.

Identify the *parameter* you wish to estimate.

Identify the *population* about which you wish to make statements.

Choose and state a confidence level.

Model Check the conditions. Usually we can't check assumptions. The conditions are the practical aspects of the data that we *can* check to decide whether it is OK to use the Normal model.

State the sampling distribution model for the statistic.
Choose your method.

I want to find an interval that is likely with 95% confidence to contain the true proportion, p, of U.S. adults who think the death penalty is applied fairly. I have a random sample of 537 U.S. adults.

✔ **Plausible Independence Condition:** Gallup phoned a random sample of U.S. adults. It is very unlikely that any of their respondents influenced each other.

✔ **Randomization Condition:** Gallup drew a random sample from all U.S. adults. I don't have details of their randomization, but assume that I can trust it.

✔ **10% Condition:** Although sampling was necessarily without replacement, there are many more U.S. adults than were sampled. The sample is certainly less than 10% of the population.

✔ **Success/Failure Condition:**
$n\hat{p} = 537(53.0\%) = 285 \geq 10$ and
$n\hat{q} = 537(47.0\%) = 252 \geq 10$,
so the sample is large enough.

The conditions are satisfied, so I can use a Normal model to find a **one-proportion z-interval.**

Mechanics Construct the confidence interval.

First find the standard error. (Remember: It's called the "standard error" because we don't know p and have to use $\hat{p}$ instead.)

$n = 537$, $\hat{p} = 0.53$, so

$$SE(\hat{p}) = \sqrt{\frac{\hat{p}\hat{q}}{n}} = \sqrt{\frac{(0.53)(0.47)}{537}} = 0.022$$

Because the sampling model is Normal, for a 95% confidence interval, the critical value $z^* = 1.96$.

Next find the margin of error. We could informally use 2 for our critical value, but 1.96 is more accurate.

The margin of error is

$$ME = z^* \times SE(\hat{p}) = 1.96(0.022) = 0.043$$

Write the confidence interval.

So the 95% confidence interval is

$$0.53 \pm 0.043 \text{ or } (0.487, 0.573).$$

REALITY CHECK The CI is centered at the sample proportion and about as wide as we might expect for a sample of 500.

Conclusion Interpret the confidence interval in the proper context. We're 95% confident that our interval captured the true proportion.

I am 95% confident that between 48.7% and 57.3% of all U.S. adults think that the death penalty is applied fairly.

Think some more about the 95% confidence interval we just created for the proportion of U.S. adults who think the death penalty is applied fairly.

6 If we wanted to be 98% confident, would our confidence interval need to be wider or narrower?

7 Our margin of error was about ±4%. If we wanted to reduce it to ±3%, would our level of confidence be higher or lower?

8 If the Gallup organization had polled more people, would the interval's margin of error have been larger or smaller?

What Can Go Wrong?

Confidence intervals are powerful tools. Not only do they tell what we know about the parameter value, but—more important—they also tell what we *don't* know. In order to use confidence intervals effectively, you must be clear about what you say about them.

Don't Misstate What the Interval Means

- **Don't suggest that the parameter varies.** A statement like "There is a 95% chance that the true proportion is between 42.7% and 51.3%" sounds as though you think the population proportion wanders around and sometimes happens to fall between 42.7% and 51.3%. When you interpret a confidence interval, make it clear that *you* know that the population parameter is fixed and that it is the interval that varies from sample to sample.

What *Can* I Say?

Confidence intervals are based on random samples, so the interval is random, too. The CLT tells us that 95% of the random samples will yield intervals that capture the true value. That's what we mean by being 95% confident.

Technically, we should say, "I am 95% confident that the interval from 42.1% to 61.7% captures the true proportion of infected sea fans." That formal phrasing emphasizes that *our confidence (and our uncertainty) is about the interval, not the true proportion.* But you may choose a more casual phrasing like "I am 95% confident that between 42.1% and 61.7% of the Las Redes fans are infected." Because you've made it clear that the uncertainty is yours and you didn't suggest that the randomness is in the true proportion, this is OK. Keep in mind that it's the interval that's random and is the focus of both our confidence and doubt.

- ***Don't claim that other samples will agree with yours.*** Keep in mind that the confidence interval makes a statement about the true population proportion. An interpretation such as "In 95% of samples of U.S. adults the proportion who think marijuana should be decriminalized will be between 42.7% and 51.3%" is just wrong. The interval isn't about sample proportions, but about the population proportion.
- ***Don't be certain about the parameter.*** Saying "Between 42.1% and 61.7% of sea fans are infected" asserts that the population proportion cannot be outside that interval. Of course, we can't be absolutely certain of that. (Just pretty sure.)
- ***Don't forget: It's the parameter.*** Don't say, "I'm 95% confident that $\hat{p}$ is between 42.1% and 61.7%." Of course you are—in fact, we calculated that $\hat{p} = 51.9\%$ of the fans in our sample were infected. So we already *know* the sample proportion. The confidence interval is about the (unknown) population parameter, p.
- ***Don't claim to know too much.*** Don't say, "I'm 95% confident that between 42.1% and 61.7% of all the sea fans in the world are infected." You didn't sample from all 500 species of sea fans found in coral reefs around the world. Just those of this type on the Las Redes Reef.
- ***Do take responsibility.*** Confidence intervals are about *uncertainty. You* are the one who is uncertain, not the parameter. You have to accept the responsibility and consequences of the fact that not all the intervals you compute will capture the true value. In fact, about 5% of the 95% confidence intervals you find will fail to capture the true value of the parameter. You *can* say, "I am 95% confident that between 42.1% and 61.7% of the sea fans on the Las Redes Reef are infected."[4]

Margin of Error Too Large to Be Useful

We know we can't be exact, but how precise do we need to be? A confidence interval that says that the percentage of infected sea fans is between 10% and 90% wouldn't be of much use. Most likely, you have some sense of how large a margin of error you can tolerate. What can you do?

One way to make the margin of error smaller is to reduce your level of confidence. But that may not be a useful solution. It's a rare study that reports confidence levels lower than 80%. Levels of 95% or 99% are more common.

The time to think about your margin of error to see whether it's small enough to be useful is when you design your study. Don't wait until you compute your confidence interval. To get a narrower interval without giving up confidence, you need to have less variability in your sample proportion. How can you do that? Choose a larger sample.

Choosing Your Sample Size

Suppose a candidate is planning a poll and wants to estimate voter support within 3% with 95% confidence. How large a sample does she need?

---

[4] When we are being very careful we say, "95% of samples of this size will produce confidence intervals that capture the true proportion of infected sea fans on the Las Redes Reef."

Let's look at the margin of error:

$$ME = z^* \sqrt{\frac{\hat{p}\hat{q}}{n}}$$

$$0.03 = 1.96 \sqrt{\frac{\hat{p}\hat{q}}{n}}.$$

We want to find n, the sample size. To find n we need a value for $\hat{p}$. We don't know a $\hat{p}$ because we don't have a sample yet, but we can probably guess a value. The worst case—the value that makes $\hat{p}\hat{q}$ (and therefore n) largest—is 0.50, so if we use that value for $\hat{p}$, we'll certainly be safe. Our candidate probably expects to be near 50% anyway.
Our equation, then, is

$$0.03 = 1.96 \sqrt{\frac{(0.5)(0.5)}{n}}.$$

To extract n from inside the square root, we square both sides of the equation:

$$0.03^2 = 1.96^2 \left(\frac{0.5 \times 0.5}{n} \right).$$

And solve for n:

$$n = \frac{1.96^2 \times 0.5 \times 0.5}{0.03^2} = 1067.1$$

> In general, the sample size needed to produce a confidence interval with a given margin of error at a given confidence level is
> $$n = \frac{z^{*2}\hat{p}\hat{q}}{ME^2}$$
> where z^* is the critical value for the confidence level you specified.

To be safe, we round up and conclude that we need at least 1068 respondents to keep the margin of error as small as 3% with a confidence level of 95%.

Unfortunately, bigger samples cost more money and more effort. Because the standard error declines only with the *square root* of the sample size, to cut the standard error (and thus the ME) in half, we must *quadruple* the sample size.

Here's a table showing sample sizes and the 95% margin of error for an estimated proportion of 0.5 (the safest guess for $\hat{p}$).

n	$SE(\hat{p})$	95% ME
100	0.050	±10%
200	0.035	±7%
400	0.025	±5%
600	0.020	±4%
1,100	0.015	±3%
2,400	0.010	±2%
10,000	0.005	±1%

Generally a margin of error of 5% or less is acceptable, but different circumstances call for different standards. For a pilot study, a margin of error of 10% may be fine, so a sample of 100 will do quite well. In a close election, a polling organization might want to get the margin of error down to 2%. Drawing a large sample to get a smaller ME, however, can run into trouble. It takes time to survey 2400 people, and a survey that extends over a week or more may be trying to hit a target that moves during the time of the survey. An important event can change public opinion in the middle of the survey process.

Public opinion polls often use a sample size of 1000, which gives an ME of 3% when $p = 0.5$. But businesses and non-profit organizations often use much larger samples to estimate the response to a direct mailing campaign. Why? Because the proportion of people who respond to these mailings is very small, often below 5%. So a margin of error of 3% isn't going to be of much use.

Keep in mind that the sample size for a survey is the number of respondents, not the number of people to whom questionnaires were sent or whose phone numbers were dialed. And keep in mind that a low response rate turns any study essentially into a voluntary response study, which is of little value for inferring population values. It's almost always better to spend resources on increasing the response rate than on surveying a larger group. A full or nearly full response by a modest-size sample can yield useful results.

Of course, surveys are not the only place where proportions pop up. Credit card banks sample huge mailing lists to find out what kind of response rate they're likely to get on a new type of credit card. Even pilot studies are likely to be mailed to 50,000 customers or more. In cases like these, the proportions are likely to be quite small, as will the margins of error needed to say anything useful. When p is small, very large samples are usually used. There's a catch-22 here, though. You need to know p to find the necessary sample size, but you need the sample to estimate p. We used $p = 0.50$ for our election poll calculation, but with such a low return rate expected the banks shouldn't do that. If they can make an educated guess for the value of p—suppose they know the response rate is usually around 4%—they will be better off using that value when setting up their equation to find the desired size for their mailing.

Violations of Assumptions

Confidence intervals and margins of error are often reported along with poll results and other analyses. But it's easy to misuse them, and wise to be aware of the ways things can go wrong.

- **Watch out for biased samples.** Don't forget about the potential sources of bias in surveys that we discussed in Chapter 12. Just because we have more statistical machinery now doesn't mean we can forget what we've already learned. A questionnaire that finds that 85% of people enjoy filling out surveys still suffers from nonresponse bias even though now we're able to put confidence intervals around this (biased) estimate.
- **Think about independence.** The assumption that the values in our sample are mutually independent is one that we usually cannot check. It always pays to think about it, though. For example, the disease affecting the sea fans might be contagious, so that fans growing near a diseased fan are more likely themselves to be diseased. Such contagion would violate the Independence Assumption and could severely affect our sample proportion. It could be that the proportion of infected sea fans on the entire reef is actually quite small, and the researchers just happened to find an infected area. To avoid this, the researchers should be careful to sample sites far enough apart to make contagion unlikely.

CONNECTIONS

Now we can see a practical application of sampling distributions. To find a confidence interval, we lay out an interval measured in standard deviations. We're using the standard deviation as a ruler again. But now the standard deviation we need is the standard deviation of the sampling distribution. That's the one that tells how much the proportion varies. (And when we estimate it from the data, we call it a standard error.)

What have we learned?

The first 10 chapters of the book explored graphical and numerical ways of summarizing and presenting sample data. We've learned (at last!) to use the sample we have at hand to say something about the *world at large*. This process, called statistical inference, is based on our understanding of sampling models, and will be our focus for the rest of the book.

As our first step in statistical inference, we've learned to use our sample to make a *confidence interval* that estimates what proportion of a population has a certain characteristic. We've learned that:

- Our best estimate of the true population proportion is the proportion we observed in the sample, so we center our confidence interval there.
- Samples don't represent the population perfectly, so we create our interval with a *margin of error.*
- This method successfully captures the true population proportion most of the time, providing us with a level of confidence in our interval.
- The higher the level of confidence we want, the *wider* our confidence interval becomes.
- The larger the sample size we have, the *narrower* our confidence interval can be.
- When designing a study, we can calculate the sample size we'll need to be able to reach conclusions that have a desired degree of precision and level of confidence.
- There are important assumptions and conditions we must check before using this (or any) statistical inference procedure.

We've learned to interpret a confidence interval by *Telling* what we believe is true in the entire population from which we took our random sample. Of course, we can't be *certain.* We've learned not to overstate or misinterpret what the confidence interval says.

TERMS

Confidence interval A level C confidence interval for a model parameter is an interval of values usually of the form

$$estimate \pm margin\ of\ error$$

found from data in such a way that C% of all random samples will yield intervals that capture the true parameter value.

One-proportion z-interval A confidence interval for the true value of a proportion. The confidence interval is

$$\hat{p} \pm z^*SE(\hat{p})$$

where z^* is a critical value from the Standard Normal model corresponding to the specified confidence level.

Margin of error In a confidence interval, the extent of the interval on either side of the observed statistic value is called the margin of error. A margin of error is typically the product of a critical value from the sampling distribution and a standard error from the data. A small margin of error corresponds to a confidence interval that pins down the parameter precisely. A large margin of error corresponds to a confidence interval that gives relatively little information about the estimated parameter.

Critical value   The number of standard errors to move away from the mean of the sampling distribution to correspond to the specified level of confidence. The critical value, denoted z^*, is usually found from a table or with technology.

Assumptions   Every model depends on assumptions. Although we may be able to think about whether an assumption is plausible, it remains something that we *assume* and cannot *verify*.

Conditions   Although we cannot verify assumptions, often there are conditions about the data that we can check to see whether an assumption is at least reasonable.

SKILLS   *When you complete this lesson you should:*

- Understand confidence intervals as a balance between the precision and the certainty of a statement about a model parameter.
- Understand that the margin of error of a confidence interval for a proportion changes with the sample size and the level of confidence.
- Know how to examine your data for violations of conditions that would make inference about a population proportion unwise or invalid.

- Be able to construct a one-proportion *z*-interval.

Tell

- Be able to interpret a one-proportion *z*-interval in a simple sentence or two. Write such an interpretation so that it does not state or suggest that the parameter of interest is itself random, but rather that the bounds of the confidence interval are the random quantities about which we state our degree of confidence.

Confidence Intervals for Proportions on the Computer

Confidence intervals for proportions are so easy and natural that many statistics packages don't offer special commands for them. Most statistics programs want the "raw data" for computations. For proportions, the raw data are the "success" and "failure" status for each case. Usually, these are given as 1 or 0, but they might be category names like "yes" and "no." Often we just know the proportion of successes, $\hat{p}$, and the total count, *n*. Computer packages don't usually deal with summary data like this easily, but the statistics routines found on many graphing calculators allow you to create confidence intervals from summaries of the data—usually all you need to enter are the number of successes and the sample size.

In some programs you can reconstruct variables of 0's and 1's with the given proportions. But even when you have (or can reconstruct) the raw data values, you may not get *exactly* the same margin of error from a computer package as you would find working by hand. The reason is that some packages make approximations or use other methods. The result is very close, but not exactly the same. Fortunately, Statistics means never having to say you're certain, so the approximate result is good enough.

DATA DESK

Data Desk does not offer built-in methods for inference with proportions.

Comments

For summarized data, open a Scratchpad to compute the standard deviation and margin of error by typing the calculation. Then use **z-interval for individual μs.**

EXCEL

Inference methods for proportions are not part of the standard Excel tool set.

Comments

For summarized data, type the calculation into any cell and evaluate it.

JMP

For a **categorical** variable that holds category labels, the **Distribution** platform includes tests and intervals for proportions.

For summarized data, put the category names in one variable and the frequencies in an adjacent variable. Designate the frequency column to have the **role** of **frequency.** Then use the **Distribution** platform.

Comments

JMP uses slightly different methods for proportion inferences than those discussed in this text. Your answers are likely to be slightly different, especially for small samples.

MINITAB

Choose **Basic Statistics** from the **Stat** menu.
- Choose **1Proportion** from the Basic Statistics submenu.
- If the data are category names in a variable, assign the variable from the variable list box to the **Samples in columns** box. If you have summarized data, click the **Summarized Data** button and fill in the number of trials and the number of successes.
- Click the **Options** button and specify the remaining details.
- If you have a large sample, check **Use test and interval based on normal distribution.**

  Click the **OK** button.

Comments

When working from a variable that names categories, MINITAB treats the last category as the "success" category. You can specify how the categories should be ordered.

SPSS

SPSS does not find confidence intervals for proportions.

TI-83/84 Plus

To calculate a confidence interval for a population proportion:
- Go to the **STATS TESTS** menu and select **A:1-PropZInt.**
- Enter the number of successes observed and the sample size.
- Specify a confidence level.
- Calculate the interval.

Comments

Beware: When you enter the value of *x*, you need the count, not the percentage. The count must be a whole number.

TI-89

To calculate a confidence interval for a population proportion:
- Go to the **Ints** menu (2nd F2) and select **5:1-PropZInt.**
- Enter the number of successes observed and the sample size.
- Specify a confidence level.
- Calculate the interval.

Comments

Beware: When you enter the value of *x*, you need the count, not the percentage. The count must be a whole number. If the number of successes are given as a percentage, you must first multiply *np* and round the result.

EXERCISES

1. **Margin of error.** A TV newsman reports the results of a poll of voters, and then says, "The margin of error is plus or minus 4%." Explain carefully what that means.

2. **Margin of error.** A medical researcher estimates the percentage of children who are exposed to lead-base paint, adding that he believes his estimate has a margin of error of about 3%. Explain what the margin of error means.

3. **Conditions.** Consider each situation described below. Identify the population and the sample, explain what p and $\hat{p}$ represent, and tell whether the methods of this chapter can be used to create a confidence interval.
   a) Police set up an auto checkpoint at which drivers are stopped and their cars inspected for safety problems. They find that 14 of the 134 cars stopped have at least one safety violation. They want to estimate the percentage of all cars that may be unsafe.
   b) A TV talk show asks viewers to register their opinions on prayer in schools by logging on to a Web site. Of the 602 people who voted, 488 favored prayer in schools. We want to estimate the level of support among the general public.
   c) A school is considering requiring students to wear uniforms. The PTA surveys parent opinion by sending a questionnaire home with all 1245 students; 380 surveys are returned, with 228 families in favor of the change.
   d) A college admits 1632 freshmen one year, and four years later 1388 of them graduate on time. The college wants to estimate the percentage of all their freshman enrollees who graduate on time.

4. **More conditions.** Consider each situation described. Identify the population and the sample, explain what p and $\hat{p}$ represent, and tell whether the methods of this chapter can be used to create a confidence interval.
   a) A consumer group hoping to assess customer experiences with auto dealers surveys 167 people who recently bought new cars; 3% of them expressed dissatisfaction with the salesperson.
   b) What percent of college students have cell phones? 2883 students were asked as they entered a football stadium, and 243 indicated they had phones with them.
   c) 240 potato plants in a field in Maine are randomly checked and only 7 show signs of blight. How severe is the blight problem for the U.S. potato industry?
   d) 12 of the 309 employees of a small company suffered an injury on the job last year. What can the company expect in future years?

5. **Conclusions.** A catalog sales company promises to deliver orders placed on the Internet within 3 days. Follow-up calls to a few randomly selected customers show that a 95% confidence interval for the proportion of all orders that arrive on time is 88% ± 6%. What does this mean? Are these conclusions correct? Explain.
   a) Between 82% and 94% of all orders arrive on time.
   b) 95% of all random samples of customers will show that 88% of orders arrive on time.
   c) 95% of all random samples of customers will show that 82% to 94% of orders arrive on time.
   d) We are 95% sure that between 82% and 94% of the orders placed by the customers in this sample arrived on time.
   e) On 95% of the days, between 82% and 94% of the orders will arrive on time.

6. **More conclusions.** In January 2002, two students made worldwide headlines by spinning a Belgian euro 250 times and getting 140 heads—that's 56%. That makes the 90% confidence interval (51%, 61%). What does this mean? Are these conclusions correct? Explain.
   a) Between 51% and 61% of all euros are unfair.
   b) We are 90% sure that in this experiment this euro landed heads on between 51% and 61% of the spins.
   c) We are 90% sure that spun euros will land heads between 51% and 61% of the time.
   d) If you spin a euro many times, you can be 90% sure of getting between 51% and 61% heads.
   e) 90% of all spun euros will land heads between 51% and 61% of the time.

7. **Confidence intervals.** Several factors are involved in the creation of a confidence interval. Among them are the sample size, the level of confidence, and the margin of error. Which statements are true?
   a) For a given sample size, higher confidence means a smaller margin of error.
   b) For a specified confidence level, larger samples provide smaller margins of error.
   c) For a fixed margin of error, larger samples provide greater confidence.
   d) For a given confidence level, halving the margin of error requires a sample twice as large.

8. **Confidence intervals, again.** Several factors are involved in the creation of a confidence interval. Among them are the sample size, the level of confidence, and the margin of error. Which statements are true?
   a) For a given sample size, reducing the margin of error will mean lower confidence.
   b) For a certain confidence level, you can get a smaller margin of error by selecting a bigger sample.
   c) For a fixed margin of error, smaller samples will mean lower confidence.
   d) For a given confidence level, a sample 9 times as large will make a margin of error one third as big.

9. Cars. What fraction of cars are made in Japan? The computer output below summarizes the results of a random sample of 50 autos. Explain carefully what it tells you.

z-Interval for proportion
With 90.00% confidence,
0.29938661 < p(japan) < 0.46984416

10. Parole. A study of 902 decisions made by the Nebraska Board of Parole produced the following computer output. Assuming these cases are representative of all cases that may come before the Board, what can you conclude?

z-Interval for proportion
With 95.00% confidence,
0.56100658 < p(parole) < 0.62524619

11. Ghosts. A May 2000 Gallup Poll found that 38% of a random sample of 1012 adults said that they believe in ghosts.
a) Find the margin of error for this poll if we want 90% confidence in our estimate of the percent of American adults who believe in ghosts.
b) Explain what that margin of error means.
c) If we want to be 99% confident, will the margin of error be larger or smaller? Explain.
d) Find that margin of error.
e) In general, if all other aspects of the situation remain the same, will smaller margins of error involve greater or less confidence in the interval?

12. Cloning. A May 2002 Gallup Poll found that only 8% of a random sample of 1012 adults approved of attempts to clone a human.
a) Find the margin of error for this poll if we want 95% confidence in our estimate of the percent of American adults who approve of cloning humans.
b) Explain what that margin of error means.
c) If we only need to be 90% confident, will the margin of error be larger or smaller? Explain.
d) Find that margin of error.
e) In general, if all other aspects of the situation remain the same, would smaller samples produce smaller or larger margins of error?

13. Teenage drivers. An insurance company checks police records on 582 accidents selected at random and notes that teenagers were at the wheel in 91 of them.
a) Create a 95% confidence interval for the percentage of all auto accidents that involve teenage drivers.
b) Explain what your interval means.
c) Explain what "95% confidence" means.
d) A politician urging tighter restrictions on drivers' licenses issued to teens says, "In one of every five auto accidents, a teenager is behind the wheel." Does your confidence interval support or contradict this statement? Explain.

14. Junk mail. Direct mail advertisers send solicitations (a.k.a. "junk mail") to thousands of potential customers in the hope that some will buy the company's product. The response rate is usually quite low. Suppose a company wants to test the response to a new flyer, and sends it to 1000 people randomly selected from their mailing list of over 200,000 people. They get orders from 123 of the recipients.
a) Create a 90% confidence interval for the percentage of people the company contacts who may buy something.
b) Explain what this interval means.
c) Explain what "90% confidence" means.
d) The company must decide whether to now do a mass mailing. The mailing won't be cost-effective unless it produces at least a 5% return. What does your confidence interval suggest? Explain.

15. Safe food. Some food retailers propose subjecting food to a low level of radiation in order to improve safety, but sale of such "irradiated" food is opposed by many people. Suppose a grocer wants to find out what his customers think. He has cashiers distribute surveys at checkout and ask customers to fill them out and drop them in a box near the front door. He gets responses from 122 customers, of whom 78 oppose the radiation treatments. What can the grocer conclude about the opinions of all his customers?

16. Local news. The mayor of a small city has suggested that the state locate a new prison there, arguing that the construction project and resulting jobs will be good for the local economy. A total of 183 residents show up for a public hearing on the proposal, and a show of hands finds only 31 in favor of the prison project. What can the city council conclude about public support for the mayor's initiative?

17. Death penalty, again. In the survey on the death penalty you read about in the chapter, the Gallup Poll actually split the sample at random, asking 538 respondents the question quoted earlier, "Generally speaking, do you believe the death penalty is applied fairly or unfairly in this country today?" The other half were asked "Generally speaking, do you believe the death penalty is applied unfairly or fairly in this country today?" Seems like the same question, but sometimes the order of the choices matters. Suppose that for the second way of phrasing it, only 44% said they thought the death penalty was fairly applied.
a) Construct a 95% confidence interval for the true proportion of adults who approve of the way the death penalty is currently applied, according to the responses for this second question.
b) Recall that 53% of the respondents in the other random half of the study said that the death penalty is applied fairly. Does a proportion of 0.53 fall inside the confidence interval you just found?

18. Drinking. A national health organization warns that 30% of middle school students nationwide have been drunk. Concerned, a local health agency randomly and anonymously surveys 110 of the 1212 middle school students in its city. Only 21 of them report having been drunk.
   a) What proportion of the sample reported having been drunk?
   b) Does this mean that this city's youth are not drinking as much as the national data would indicate? Explain.
   c) Create a 95% confidence interval for the proportion of the city's middle school students who have been drunk.
   d) Is there any reason to believe that the national level of 30% is not true of the middle school students in this city?

19. Death penalty poll, part III. In the chapter's example and again in Exercise 17 we looked at a Gallup Poll investigating the public's attitude toward the death penalty. In response to one question, 53% thought it was fair, but when the question was phrased differently the proportion in favor dropped to 44%.
   a) What kind of bias may be present here?
   b) Each group consisted of 538 respondents. If we combine them, considering the overall group to be one larger random sample, what is a 95% confidence interval for the proportion of the general public that thinks the death penalty is being fairly applied?
   c) How does the margin of error based on this pooled sample compare with the margins of error from the separate groups? Why?

20. Gambling. A city ballot includes a local initiative that would legalize gambling. The issue is hotly contested, and two groups decide to conduct polls to predict the outcome. The local newspaper finds that 53% of 1200 randomly selected voters plan to vote "yes," while a college Statistics class finds 54% of 450 randomly selected voters in support. Both groups will create 95% confidence intervals.
   a) Without finding the confidence intervals, explain which one will have the larger margin of error.
   b) Find both confidence intervals.
   c) Which group concludes that the outcome is too close to call? Why?

21. Rickets. Vitamin D, whether ingested as a dietary supplement or produced naturally when sunlight falls upon the skin, is essential for strong, healthy bones. The bone disease rickets was largely eliminated in England during the 1950s, but now there is concern that a generation of children more likely to watch TV or play computer games than spend time outdoors is at increased risk. A recent study of 2700 children randomly selected from all parts of England found 20% of them deficient in vitamin D.
   a) Find a 98% confidence interval.
   b) Explain carefully what your interval means.
   c) Explain what "98% confidence" means.

22. Pregnancy. In 1998 a San Diego reproductive clinic reported 49 births to 207 women under the age of 40 who had previously been unable to conceive.
   a) Find a 90% confidence interval for the success rate at this clinic.
   b) Interpret your interval in this context.
   c) Explain what "90% confidence" means.
   d) Would it be misleading for the clinic to advertise a 25% success rate? Explain.

23. Only child. In a random survey of 226 college students, 20 reported being "only" children (with no siblings). Estimate the proportion of students nationwide who are only children.
   a) Check the conditions (to the extent you can) for constructing a confidence interval.
   b) Construct a 95% confidence interval.
   c) Interpret your interval.
   d) Explain what "95% confidence" means in this context.

24. Back to campus. In 2004 ACT, Inc., reported that 74% of 1644 randomly selected college freshmen returned to college the next year. Estimate the national freshman-to-sophomore retention rate.
   a) Verify that the conditions are met.
   b) Construct a 98% confidence interval.
   c) Interpret your interval.
   d) Explain what "98% confidence" means in this context.

25. First lady. A June 2004 Gallup Poll asked Americans who they thought better fit their idea of what a first lady should be, Laura Bush or Hillary Rodham Clinton. More Americans believed Bush fit the bill, 52% to 43%. The remaining 5% felt that both women equally fit their idea of a first lady or neither of them did, or had no opinion. The poll was based on a random sample of 1005 adults aged 18 and older.
   a) Find the 95% confidence interval for the true proportion of all U.S. adults who believe Laura Bush fits their idea of a first lady.
   b) If someone asserts that half of the U.S. adult population thinks Hillary Clinton fits the bill, what would you say?

26. Back to campus again. The ACT, Inc., study described in Exercise 24 was actually stratified by type of college—public or private. The retention rates were 71.9% among 505 students enrolled in public colleges and 74.9% among 1139 students enrolled in private colleges.
   a) Will the 95% confidence interval for the true national retention rate in private colleges be wider or narrower than the 95% confidence interval for the retention rate in public colleges? Explain.
   b) Find the 95% confidence interval for the public college retention rate.

c) Should a public college whose retention rate is 75% proclaim that they do a better job than other public colleges of keeping freshmen in school? Explain.

27. First lady redux. The June 2004 Gallup Poll on first ladies (see Exercise 25) of 1005 U.S. adults split the sample into four age groups: ages 18–29, 30–49, 50–64 and 65+. In the youngest age group, 62% said that Hillary Clinton fit their idea of a first lady, as opposed to 35%, who preferred Laura Bush. The sample included 250 18- to 29-year-olds.
a) Do you expect the 95% confidence interval for the true proportion of all 18- to 29-year-olds who prefer Clinton to be wider or narrower than the 95% confidence interval for the true proportion of all U.S. adults? Explain.
b) Find the 95% confidence interval for the true proportion of all 18 to 29 year olds who would choose Hillary Clinton.

28. Legal Music. A random sample of 168 students were asked how many songs were in their digital music library and what fraction of them were legally purchased. Overall, they reported having a total of 117,079 songs, of which 23.1% were legal. The music industry would like a good estimate of the fraction of songs in students' digital music libraries that are legal.
a) Think carefully. What is the parameter being estimated? What is the population? What is the sample size?
b) Check the conditions for making a confidence interval.
c) Construct a 95% confidence interval for the fraction of legal digital music.
d) Explain what this interval means. Do you believe that you can be this confident about your result? Why or why not?

29. Deer ticks. Wildlife biologists inspect 153 deer taken by hunters and find 32 of them carrying ticks that test positive for Lyme disease.
a) Create a 90% confidence interval for the percentage of deer that may carry such ticks.
b) If the scientists want to cut the margin of error in half, how many deer must they inspect?
c) What concerns do you have about this sample?

30. Pregnancy II. The San Diego reproductive clinic in Exercise 22 wants to publish more definitive information on its success rate.
a) The clinic wants to cut the stated margin of error in half. How many patients' results must be used?
b) Do you have any concerns about this sample? Explain.

31. Graduation. It's believed that as many as 25% of adults over 50 never graduated from high school. We wish to see if this percentage is the same among the 25 to 30 age group.

a) How many of this younger age group must we survey in order to estimate the proportion of non-grads to within 6% with 90% confidence?
b) Suppose we want to cut the margin of error to 4%. What's the necessary sample size?
c) What sample size would produce a margin of error of 3%?

32. Hiring. In preparing a report on the economy, we need to estimate the percentage of businesses that plan to hire additional employees in the next 60 days.
a) How many randomly selected employers must we contact in order to create an estimate in which we are 98% confident with a margin of error of 5%?
b) Suppose we want to reduce the margin of error to 3%. What sample size will suffice?
c) Why might it not be worth the effort to try to get an interval with a margin of error of only 1%?

33. Graduation, again. As in Exercise 31, we hope to estimate the percentage of adults aged 25 to 30 who never graduated from high school. What sample size would allow us to increase our confidence level to 95% while reducing the margin of error to only 2%?

34. Better hiring info. Editors of the business report in Exercise 32 are willing to accept a margin of error of 4%, but want 99% confidence. How many randomly selected employers will they need to contact?

35. Pilot study. A state's environmental agency worries that many cars may be violating clean air emissions standards. The agency hopes to check a sample of vehicles in order to estimate that percentage with a margin of error of 3% and 90% confidence. To gauge the size of the problem, the agency first picks 60 cars and finds 9 with faulty emissions systems. How many should be sampled for a full investigation?

36. Another pilot study. During routine screening, a doctor notices that 22% of her adult patients show higher than normal levels of glucose in their blood—a possible warning signal for diabetes. Hearing this, some medical researchers decide to conduct a large-scale study, hoping to estimate the proportion to within 4% with 98% confidence. How many randomly selected adults must they test?

37. Approval rating. A newspaper reports that the governor's approval rating stands at 65%. The article adds that the poll is based on a random sample of 972 adults and has a margin of error of 2.5%. What level of confidence did the pollsters use?

38. Amendment. A TV news reporter says that a proposed constitutional amendment is likely to win approval in the upcoming election because a poll of 1505 likely voters indicated that 52% would vote in favor. The reporter goes on to say that the margin of error for this poll was 3%.
a) Explain why the poll is actually inconclusive.
b) What confidence level did the pollsters use?

just checking
Answers

1. No. We know that in the sample 18% said "yes"; there's no need for a margin of error.

2. No, we are 95% confident that the percentage falls in some interval, not exactly on a particular value.

3. Yes. That's what the confidence interval means.

4. No. We don't know for sure that's true; we are only 95% confident.

5. No. That's our level of confidence, not the proportion of voters. The sample suggests the proportion is much lower.

6. Wider.

7. Lower.

8. Smaller.

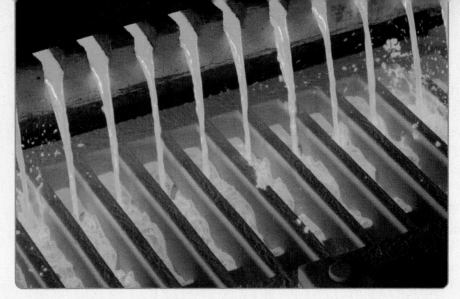

CHAPTER 20

Testing Hypotheses About Proportions

A S **Testing a Claim.** Can we really draw a reasonable conclusion from a random sample? This simulation poses the challenge and lets you try. Run it before you read the chapter and you'll gain a solid sense of what we're doing here.

Ingots are huge pieces of metal, often weighing in excess of 20,000 pounds, made in a giant mold. They must be cast in one large piece for use in fabricating large structural parts for cars and planes. If they crack while being made, the crack may propagate into the zone required for the part, compromising its integrity. Airplane manufacturers insist that metal for their planes be defect-free, so the ingot must be made over if any cracking is detected.

Even though the metal from the cracked ingot is recycled, the scrap cost runs into the tens of thousands of dollars. Metal manufacturers would like to avoid cracking if at all possible. But the casting process is complicated and not everything is completely under control. In one plant, only about 80% of the ingots have been free of cracks. In an attempt to reduce the cracking proportion, the plant engineers and chemists recently tried out some changes in the casting process. Since then, 400 ingots have been cast and only 17% of them have cracked. Should management declare victory? Has the cracking rate really decreased or was 17% just due to luck?

We can treat the 400 ingots cast with the new method as a random sample. We know that each random sample will have a somewhat different proportion of cracked ingots. Is the 17% we observe merely a result of natural sampling variability, or is this lower cracking rate strong enough evidence to assure management that the true cracking rate now is really below 20%?

"Half the money I spend on advertising is wasted; the trouble is I don't know which half."
—John Wanamaker
(attributed)

People want answers to questions like these all the time. Has the president's approval rating changed since last month? Has teenage smoking decreased in the past five years? Is the global temperature increasing? Did the Super Bowl ad we bought actually increase sales? To answer such questions, we test *hypotheses* about models.

Hypotheses

In Statistics, a hypothesis proposes a model for the world. Then we look at the data. If the data are consistent with that model, we have no reason to disbelieve

For hundreds of years it was assumed that the sun revolved around the earth. The fact that the sun appears to rise in the east every morning and set in the west every evening is *consistent* with this hypothesis and *seems* to lend support to it, but it certainly doesn't prove it. It may explain, however, why Copernicus, who believed that the earth revolves around the sun, had such a hard time convincing the world of his (correct) model.

A S The Reasoning of
Hypothesis Testing. Our reasoning is based on a rule of logic that dates back to ancient scholars. Here's a modern discussion of it.

NOTATION ALERT:

We have many P's to keep straight. We use an upper-case P for probabilities, as in $P(\mathbf{A})$, and also for the special probability we care about in hypothesis testing, the P-value.

We use lower-case p to denote our model's underlying proportion parameter, and $\hat{p}$ to denote our observed proportion statistic.

the hypothesis. But if the facts are inconsistent with the model, what then? It depends. If the data are only slightly out of step with the model, we might stick with the model. If the data dramatically contradict the model, though, that's strong evidence that the model is incorrect.

Notice the difference between the two possibilities. If the facts are consistent with the model, they *lend support* to the hypothesis. Does this *prove* the hypothesis is true? No. Lending support is not the same as proving something. Many other models may also be consistent with the same set of facts. Even if a hypothesis is true, we can never prove it.

When a hypothesis is false, however, we might be able to recognize that. When data are glaringly *inconsistent* with the hypothesis, it becomes clear that the hypothesis can't be right. Then we can reject it. This is the logic of traditional scientific thinking and one of the principal differences between scientific thinking and other kinds of reasoning.[1]

This is also the logic of jury trials. To prove someone is guilty, we start by *assuming* they are innocent. We retain that hypothesis until the facts make it unlikely beyond a reasonable doubt. Then, and only then, we reject the hypothesis of innocence and declare the person guilty.

In statistical tests of hypotheses, we use the same logic. We begin by assuming that a hypothesis is true. Next we consider whether the data are consistent with the hypothesis. If they are, all we can do is retain the hypothesis we started with. If they are not, then like a jury, we ask whether they are unlikely beyond a reasonable doubt. The statistical twist is that we can quantify our level of doubt.

If the data are surprising, but we believe they're trustworthy, then we start to doubt the hypothesis. What do we mean by "surprising"? Seeing an event that has a very low probability of occurring is a surprise by definition. But wait! We can use the model proposed by our hypothesis to calculate the probability that the event we've witnessed could happen. And that's just the probability we're looking for. It quantifies exactly how surprised we are. Statisticians are so thrilled with their ability to measure precisely how surprised they are that they give this probability a special name. It's called a **P-value.**[2]

When the data are consistent with the model from the null hypothesis, the P-value is high and we are unable to reject the null hypothesis. In that case, we have to "retain" the null hypothesis we started with. We can't claim to have proved it, but when the data are consistent with the null hypothesis model and in line with what we would expect from natural sampling variability we'll say that we "*fail to reject the null hypothesis*" to highlight the fact that we were unable to say very much. But if the P-value is low enough, then what we have observed would be very unlikely were the null model true, and so we'll *reject the null hypothesis*.

Of course, even when the probability is very low, it may be that the hypothesis is actually true and we just witnessed something extremely unlikely. When it comes to data though, statisticians don't believe in miracles. Instead, they say that if the data are unlikely *enough,* they'll reject the hypothesis.

---

[1] Throughout the book, when we talk about science and scientific thinking, we include the social sciences as well. Indeed, other disciplines, such as History, include works of scientific thinking. When we speak of "the scientist," we don't necessarily mean people in white lab coats; we'll mean any person thinking scientifically.

[2] You'd think if they were so excited, they'd give it a better name, but "P-value" is about as excited as statisticians get.

Testing Hypotheses

How can we state and test a hypothesis about ingot cracking? Remember that we start by *assuming* a hypothesis is true. What should we assume here? The cracking rate has been at 20% for years. To test whether the changes made by the engineers have improved the cracking rate, we assume that they have in fact made no difference. Our starting hypothesis, called the null hypothesis, is that the proportion of cracks is still 20%.

The **null hypothesis,** which we denote H_0, specifies a population model parameter of interest and proposes a value for that parameter. We usually write down the null hypothesis in the form H_0: *parameter = hypothesized value*. This is a concise way to specify the two things we need most: the identity of the parameter we hope to learn about and a specific hypothesized value for that parameter. (We need a hypothesized value so we can compare our observed statistic value to it. That's how we find the P-value.) Which value to take is often obvious from the context of the problem itself, from the *Who* and *What* of the data. But sometimes it takes a bit of thinking to translate the question we hope to answer into a hypothesis about a parameter of a particular model. Here we can write H_0: $p = 0.20$.

What would convince you that the cracking rate had actually gone down? Probably only if you observed a cracking rate *much lower* than 20% in your sample. If only 3 out of the next 400 ingots crack (0.75%), you'll probably be convinced the changes helped. If the sample cracking rate is only slightly lower than 20%, though, you might still be skeptical. After all, observations do vary, so we should not be surprised that natural sampling variability will result in small differences. How big must the difference be before we are convinced that the cracking rate has changed?

Whenever we ask how big a difference is, we think of the standard deviation. (You *did* think of the standard deviation, didn't you?) Let's start by finding the standard deviation of the sample cracking rate. The sample cracking rate is a proportion, so we already know how to find its standard deviation.

Since the change in process, 400 new ingots have been cast. The sample size of 400 is big enough to satisfy the Success/Failure Condition. (We expect $0.20 \times 400 = 80$ ingots to crack.) We have no reason to think the ingots are not independent, so the Normal sampling distribution model should work well. The standard deviation of the sampling model is

> **NOTATION ALERT:**
>
> Capital H is the standard letter for hypotheses, and H_0 always labels the null hypothesis.

> To remind us that the parameter value comes from the null hypothesis, it is sometimes written as p_0 and the standard deviation as
>
> $$SD(\hat{p}) = \sqrt{\frac{p_0 q_0}{n}}.$$

$$SD(\hat{p}) = \sqrt{\frac{pq}{n}} = \sqrt{\frac{(0.20)(0.80)}{400}} = 0.02$$

● **Why is this a standard deviation and not a standard error?** Because we haven't estimated anything. Once we assume that the null hypothesis is true, it gives us a value for the model parameter p. With proportions, if we know p then we also automatically know its standard deviation. And because we find the standard deviation from the model parameter, this is a standard deviation and not a standard error. When we found a confidence interval for p, we could not assume that we knew its value, so we estimated the standard deviation from the sample value $\hat{p}$. ●

Now we know both parameters of the Normal sampling distribution model: $p = 0.20$ and $SD(\hat{p}) = 0.02$. So, we can find out how likely it would be to see the observed value of $\hat{p} = 17\%$. Since we are using a Normal model, we find the z-score:

$$z = \frac{0.17 - 0.20}{0.02} = -1.5$$

and we ask, "How likely is it to observe a value at least 1.5 standard deviations below the mean of a Normal model?" The answer (from a calculator, computer program, or the Normal table) is about 0.067. This is the probability of observing a cracking rate of 17% or less if the null model is true. Management now must decide whether an event that would happen 6.7% of the time by chance is strong enough evidence to conclude that the true cracking proportion has decreased.

Beyond a Reasonable Doubt
We ask whether the data were unlikely beyond a reasonable doubt. We've just calculated that probability. The probability that the observed statistic value (or an even more extreme value) could occur if the null model were true—in this case, 0.067—is the P-value.

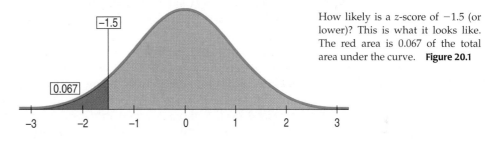

How likely is a z-score of −1.5 (or lower)? This is what it looks like. The red area is 0.067 of the total area under the curve. **Figure 20.1**

A Trial as a Hypothesis Test

We've said that hypothesis testing is very much like a court trial. To understand the logic better, let's push the analogy a little further. In our imagined trial, suppose the defendant has been accused of robbery. We wonder whether he is guilty or not. In British common law and those systems derived from it (including U.S. law), the null hypothesis is that the defendant is innocent. Instructions to juries are quite explicit about this.

We then marshal the evidence. For us, this means collecting data. In the trial the prosecutor presents evidence. This evidence takes the form of facts that seem to contradict the presumption of innocence. ("If the defendant were innocent, wouldn't it be remarkable that the police found him at the scene of the crime with a bag full of money in his hand, a mask on his face, and a getaway car parked outside?")

The next step is to judge the evidence. Evaluating the evidence is the responsibility of the jury in a trial, but it falls on your shoulders in hypothesis testing. The jury considers the evidence in light of the *presumption* of innocence and judges whether the evidence against the defendant would be plausible *if the defendant were in fact innocent*.

Like the jury, we ask, "Could these data plausibly have happened by chance if the null hypothesis were true?" If they were very unlikely to have occurred, then the evidence raises more than a reasonable doubt in our minds about the null hypothesis.

Ultimately, we must make a decision. The standard of "beyond a reasonable doubt" is wonderfully ambiguous because it leaves the jury to decide the degree to which the evidence contradicts the presumption of innocence. We have to ask the same question. How unlikely is unlikely? Some people advocate setting rigid standards, like 1 time out of 20 (0.05) or 1 time out of 100 (0.01). But if *you* have to make the decision, you must judge for yourself in any particular situation whether the probability is small enough to constitute "reasonable doubt."

What to Do with an "Innocent" Defendant

If the evidence is not strong enough to reject the defendant's presumption of innocence, what verdict does the jury return? They say "not guilty." Notice that they do not say that the defendant is innocent. Nor do they employ a standard of "reasonable certainty" of guilt in place of the standard of "reasonable doubt" of innocence. All they say is that they have not seen sufficient evidence to convict, to reject innocence. The defendant may, in fact, be innocent, but the jury has no way to be sure.

Said statistically, the jury's null hypothesis is H_0: innocent defendant. If the evidence is too unlikely given this assumption, the jury rejects the null hypothesis and finds the defendant guilty. But—and this is an important distinction—if there is *insufficient evidence* to convict the defendant, the jury does not decide that H_0 is true and declare the defendant innocent. Juries can only *fail to reject* the null hypothesis and declare the defendant "not guilty."

In the same way, if the data are not particularly unlikely under the assumption that the null hypothesis is true, then the most we can do is to "fail to reject" our null hypothesis. We never declare the null hypothesis to be true, because we simply do not know whether it's true or not. (After all, more evidence may come along later.) Sometimes in this case we say that the *null hypothesis has been retained.*

In the trial, the burden of proof is on the prosecution. In a hypothesis test, the burden of proof is on the unusual claim. The null hypothesis is the ordinary state of affairs, so it's the alternative to the null hypothesis that we consider unusual and for which we must marshal evidence.

Imagine a clinical trial testing the effectiveness of a new headache remedy. We will most likely test it against a placebo, and the null hypothesis will be that the new drug is no more effective than the placebo. If we use only six people to test the drug, what are we likely to conclude? The results are likely *not to be clear* with only six people in the trial. We will be unable to reject the hypothesis. Does this mean the drug doesn't work? Of course not. It simply means that we don't have enough evidence to reject our assumption. That's why we don't start by assuming that the drug *is more effective.* If we were to do so, then we could test just a few people, find that the results aren't clear, and claim that we've been unable to reject our original assumption. The FDA is unlikely to be impressed by such an argument.

> "If the People fail to satisfy their burden of proof, you must find the defendant not guilty."
> —NY state jury instructions

1. A research team wants to know if aspirin helps to thin blood. The null hypothesis says that it doesn't. They test 12 patients, observe the proportion with thinner blood, and get a P-value of 0.32. They proclaim that aspirin doesn't work. What would you say?

2. An allergy drug has been tested and found to give relief to 75% of the patients in a large clinical trial. Now the scientists want to see if the new improved version works even better. What would the null hypothesis be?

3. The new drug is tested and the P-value is 0.0001. What would you conclude about the new drug?

The Reasoning of Hypothesis Testing

"The null hypothesis is never proved or established, but is possibly disproved, in the course of experimentation. Every experiment may be said to exist only in order to give the facts a chance of disproving the null hypothesis."
—*Sir Ronald Fisher,* The Design of Experiments

Hypothesis tests follow a carefully structured path. To avoid getting lost as we navigate down it, we divide that path into four distinct sections.

1. Hypotheses

First we state the null hypothesis. That's usually the skeptical claim that nothing's different. Are we considering a (New! Improved!) possibly better method? The null hypothesis says "Oh yeah? Convince me!" To convert a skeptic, we must pile up enough evidence against the null hypothesis that we can reasonably reject it.

But what should we believe if we reject the null? That's called the alternative hypothesis. The alternative hypothesis usually gives an interval of outcomes. The null hypothesis gives a specific value for the parameter. Even when we reject the null hypothesis, we won't know the true value of the parameter. But we need to state the alternative clearly to understand what we are testing.

In statistical hypothesis testing, hypotheses are almost always about model parameters. To assess how unlikely our data may be, we need a null model. The null hypothesis specifies a particular parameter value to use in our model. The alternative hypothesis usually gives a *range* of other possible values.

The null hypothesis. To perform a hypothesis test, we must first translate our question of interest into a statement about model parameters. Suppose we want to see if people prefer Coke or Pepsi. How do we translate this into a null hypothesis we can test? We need to specify a parameter and a value for it. If we let p be the proportion of people who prefer Coke to Pepsi, we could let the null hypothesis be "$H_0: p = 0.50$." Now we can collect data and test this hypothesis. The key step is to take your question and translate it into a statement about the parameter of a model so that we can test it. In the usual shorthand, write H_0: *parameter = hypothesized value*.

The alternative hypothesis. The **alternative hypothesis,** H_A, contains the values of the parameter we accept if we reject the null. If we reject the null hypothesis, then we accept the alternative. In the Coke vs. Pepsi example, our null hypothesis is that $p = 0.50$. What's the alternative? We would be interested in learning that either cola was preferred. In terms of the parameter, we can write $H_A: p \neq 0.50$. If the data convince us that we should reject the null hypothesis, we would accept the alternative.

NOTATION ALERT:

H_A always labels the alternative hypothesis.

Some folks pronounce the hypothesis labels "Ho!" and "Ha!" (but it makes them seem overexcitable). Others pronounce H_0 "H naught" (as in "all is for naught"). Nobody, it seems, says "H zero" or "H-Oh," so you probably shouldn't either.

2. Model

To plan a statistical hypothesis test, specify the *model* you will use to test the null hypothesis and the parameter of interest. Of course, all models require assumptions, so you will need to state them and check any corresponding conditions.

Your Model step should end with a statement such as

> *Because the conditions are satisfied, I can model the sampling distribution of the proportion with a Normal model.*

Watch out, though. Your Model step could end with *Because the conditions are not satisfied, I can't proceed with the test.* If that's the case, stop and reconsider.

Each test we discuss in the book has a name that you should include in your report. We'll see many tests in the following chapters. Some will be about more than one sample, some will involve statistics other than proportions, some will use

When the Conditions Fail . . .
You might proceed with caution, explicitly stating your concerns. Or you may need to do the analysis with and without an outlier, or on different subgroups, or after re-expressing the response variable. Or you may not be able to proceed at all.

models other than the Normal (and so will not use *z*-scores). The test about proportions is called a **one-proportion *z*-test**.[3]

A S **Was the Observed Outcome Unlikely?** Complete the test you started in the first activity for this chapter. The narration explains the steps of the hypothesis test.

One-proportion z-test

The conditions for the one-proportion *z*-test are the same as for the one-proportion *z*-interval. We test the hypothesis $H_0: p = p_0$ using the statistic

$z = \dfrac{(\hat{p} - p_0)}{SD(\hat{p})}$. We use the hypothesized proportion to find the standard

deviation, $SD(\hat{p}) = \sqrt{\dfrac{p_0 q_0}{n}}$.

When the conditions are met and the null hypothesis is true, this statistic follows the standard Normal model, so we can use that model to obtain a P-value.

---

3. Mechanics

Conditional Probability
Did you notice that a P-value is a conditional probability? It's the probability that the observed results could have happened *if the null hypothesis is true*.

Under "mechanics" we place the actual calculation of our test statistic from the data. Different tests we encounter will have different formulas and different test statistics. Usually, the mechanics are handled by a statistics program or calculator, but it's good to have the formulas recorded for reference and to know what's being computed. The ultimate goal of the calculation is to obtain a **P-value**—the probability that the observed statistic value (or an even more extreme value) could occur if the null model were correct. If the P-value is small enough, we'll reject the null hypothesis.

4. Conclusion

". . . They make things admirably plain,
But one hard question will remain:
If one hypothesis you lose,
Another in its place you choose . . ."
—James Russell Lowell,
Credidimus Jovem Regnare

The conclusion in a hypothesis test is always a statement about the null hypothesis. The conclusion must state either that we reject or that we fail to reject the null hypothesis. And, as always, the conclusion should be stated in context.

Your conclusion about the null hypothesis should never be the end of a testing procedure. Often there are actions to take or policies to change. In our ingot example, management must decide whether to continue the changes proposed by the engineers. The decision always includes the practical consideration of whether the new method is worth the cost. Suppose management decides to reject the null hypothesis of 20% cracking in favor of the alternative that it has been reduced. They must still evaluate how much the cracking rate has been reduced and how much it cost to accomplish it. The *size of the effect* is always a concern when we test hypotheses. A good way to look at the effect size is to examine a confidence interval.

● **How much does it cost?** Formal tests of a null hypothesis base the decision of whether to reject the null hypothesis solely on the size of the P-value. But in real life, we want to evaluate the costs of our decisions as well. How much would you be willing to pay for a faster computer? Shouldn't your decision depend on how much faster? And on how much more it costs? Costs are not just monetary, either. Would you use the same standard of proof for testing the safety of an airplane as for the speed of your new computer? ●

---

[3] It's also called the "one sample test for a proportion."

Alternative Alternatives

In the Coke vs. Pepsi example, we were equally interested in proportions that deviate from 50% in *either* direction. So we wrote our alternative hypothesis as $H_A: p \neq 0.50$.

Such an alternative hypothesis is known as a **two-sided alternative** because we are equally interested in deviations on either side of the null hypothesis value. For two-sided alternatives, the P-value is the probability of deviating in *either* direction from the null hypothesis value.

The P-value for a two-sided alternative adds the probabilities in both tails of the sampling distribution model outside the value that corresponds to the test statistic.

Figure 20.2

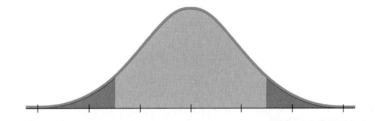

In the ingot example, management is particularly interested in *lowering* the cracking rate below 20%. Management probably doesn't want to pursue changes in the process that make the cracking rate worse. The only alternative of interest is that the cracking rate *decreases*. We would write our alternative hypothesis, then, as $H_A: p < 0.20$. An alternative hypothesis that focuses on deviations from the null hypothesis value in only one direction is called a **one-sided alternative.**

The P-value for a one-sided alternative considers only the probability of values beyond the test statistic value in the specified direction.

Figure 20.3

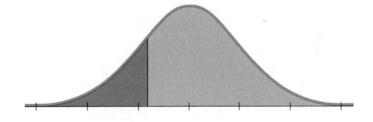

For a hypothesis test with a one-sided alternative, the P-value is the probability of deviating *only in the direction of the alternative* away from the null hypothesis value.

● **Aren't we overlooking some values?** When we use a one-sided alternative, it may look like we're ignoring some possibilities. What about the other tail? We're not really ignoring it; we're just being honest. We're admitting up front that we really don't care about those possibilities. For example, in the case of the ingots the company is really interested only in finding out whether the cracking rate has *improved*. If it has stayed the same, they have not solved the problem. And if it became *worse*—well, they really aren't interested. They care only if it got better.

Our null hypothesis then, is that the cracking rate is still 20%, and our alternative is that the cracking rate has improved—that *p* is now less than 20%. We use a model with the proposed parameter value, *p* = 20%, to see how unusual our observed results are. (The P-value measures that.) That helps us decide whether the observed cracking rate is convincing evidence of an improvement (to something *less* than 20%). If so, we reject the null hypothesis. Should we happen to observe a proportion that's in the other tail—a $\hat{p}$ that's greater than 20%—we'd hardly think that's evidence of *improvement* in the cracking rate, so we certainly wouldn't reject the null. While the higher proportion may suggest that the cracking rate has gotten worse, that's not the issue we care about. You may occasionally see a null hypothesis written as if it included all those possible parameter values on the other side, like $p \geq 20\%$. Just ignore the extra values. The null's job is to propose a model parameter we can use to find the P-value. To do that, we need a specific value for p (the value on the edge of no-man's land), not all the other possibilities. ●

How do we know whether we want a one- or two-sided alternative? To answer this, we return to the W's, and specifically to the *Why* of the study. In the ingot example, the data were collected to see whether the changes *improved* the cracking rate. When we test a hypothesis, we must specify both the null and the alternative. But the alternative is often the hypothesis that motivated the study.

In some social sciences, alternative hypotheses are called "research hypotheses" and are stated clearly at the beginning of research reports. Ask "Why was this study done?" and you are likely to hear about the alternative hypothesis. Keep in mind, though, that you can't start by assuming that the alternative is true. To be able to do the test, you must specify and assume the null. The burden of proof is on the claim made by the research hypothesis. Without sufficient evidence, we won't reject the null hypothesis.

A S **Practice with Testing Hypotheses About Proportions.** Here's an interactive tool that makes it easy to see what's going on in a hypothesis test. Learn to use it here (just in time).

Testing a Hypothesis   Step-By-Step

Advances in medical care, such as prenatal ultrasound examination, now make it possible to determine the child's sex early in a pregnancy. Because some cultures value male children more highly than female children, there's a fear that some parents may not carry pregnancies of girls to term. A study in Punjab, India (E. E. Booth, M. Verma, R. S. Beri, "Fetal Sex Determination in Infants in Punjab, India: Correlations and Implications," *BMJ* 309[12 November 1994]:1259–1261) reports that in 1993 in one hospital, 56.9% of the 550 live births that year were boys. It's a medical fact that male babies are slightly more common than female babies. The authors report a baseline for this region of 51.7% male live births. Is the sample proportion of 56.9% evidence of a higher proportion of male births?

After stating what we want to find out, we'll follow the four main steps of performing a hypothesis test: Hypotheses, Model, Mechanics, and Conclusion.

Think	**Plan** State what we want to know. Define the variables and discuss the W's. **Hypotheses** The null hypothesis makes the claim of no difference from the baseline. We are interested only in an increase in male births, so the alternative hypothesis is one-sided.	*I want to know whether the proportion of male births has increased from the established baseline of 51.7%. The data are the recorded sexes of the 550 live births from a hospital in Punjab, India, in 1993, collected for a study on fetal sex determination. The parameter of interest, p, is the proportion of male births.* $H_O: p = 0.517$ $H_A: p > 0.517$

Model

Check the appropriate conditions.

For testing proportions, the conditions are the same ones we had for making confidence intervals, except that we check the Success/Failure Condition with the *hypothesized* proportions rather than with the *observed* proportions.

✔ **Independence Assumption:** There is no reason to think that the sex of one baby can affect the sex of other babies, so births can reasonably be assumed to be independent with regard to the sex of the child.

✔ **Random Sampling Condition:** The 550 live births are not a random sample. However, I hope that this is a representative year, and think that the hospital may be typical of hospitals in this area of India.

✔ **10% Condition:** I would like to be able to make statements about births at this hospital or at similar hospitals in India. These 550 births are fewer than 10% of all of those births.

✔ **Success/Failure Condition:** Both $np_O = 550(0.517) = 284.35$ and $nq_O = 550(0.483) = 265.65$ are greater than 10; I expect the births of at least 10 boys and at least 10 girls, so the sample is large enough.

Specify the sampling distribution model.

Tell what test you plan to use.

The conditions are satisfied, so I can use a Normal model and perform a **one-proportion z-test.**

Show

Mechanics

The null model gives us the mean, and (because we are working with proportions) the mean gives us the standard deviation.

We find the z-score for the observed proportion to find out how many standard deviations it is from the hypothesized proportion.

From the z-score we can find the P-value, which tells us the probability of observing a value that extreme (or more).

The probability of observing a value 2.44 standard deviations above the mean of a Normal model or higher can be found by computer, calculator, or table (see the table on p. 458) to be 0.0073.

The null model is a normal distribution with a mean of 0.517 and a standard deviation of

$$\sqrt{\frac{0.517 \times (1 - 0.517)}{550}} = 0.0213$$

The observed proportion, $\hat{p}$, is 0.569,

so the z-value is

$$z = \frac{0.569 - 0.517}{0.0213} = 2.44$$

and the sample proportion lies 2.44 standard deviations above the mean.

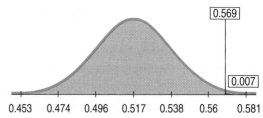

The corresponding P-value is 0.0073.

Tell

Conclusion State your conclusion in context.

This P-value is roughly 1 time in 150. That's clearly significant, but don't jump to other conclusions. We can't be sure how this deviation

The P-value of 0.0073 says that if the true proportion of male babies were still at 51.7%, then an observed proportion of 56.9% male babies would occur at random only about 7 times

came about. For instance, we don't know whether this hospital is typical, or whether the time period studied was selected at random.

in 1000. With a P-value this small, I reject H_O. This is strong evidence that the birth ratio of boys to girls is not equal to its natural level, but rather has increased.

Here's a portion of a Normal table that includes the part that gives the probability we needed for the test.

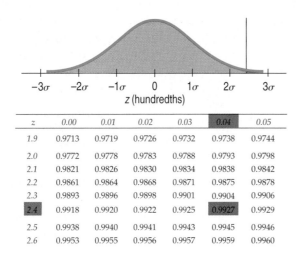

A portion of a Normal table with the look-up for z = 2.44 indicated. The P-value we want is 1 − 0.9927 = 0.0073.   **Figure 20.4**

z	0.00	0.01	0.02	0.03	0.04	0.05
1.9	0.9713	0.9719	0.9726	0.9732	0.9738	0.9744
2.0	0.9772	0.9778	0.9783	0.9788	0.9793	0.9798
2.1	0.9821	0.9826	0.9830	0.9834	0.9838	0.9842
2.2	0.9861	0.9864	0.9868	0.9871	0.9875	0.9878
2.3	0.9893	0.9896	0.9898	0.9901	0.9904	0.9906
2.4	0.9918	0.9920	0.9922	0.9925	0.9927	0.9929
2.5	0.9938	0.9940	0.9941	0.9943	0.9945	0.9946
2.6	0.9953	0.9955	0.9956	0.9957	0.9959	0.9960

P-Values and Decisions: What to Tell About a Hypothesis Test

MORE

Hypothesis tests are particularly useful when we must make a decision. Is the defendant to be considered guilty or not? Should we choose print advertising or television? Questions like these cannot always be answered with the margins of error of confidence intervals. The absolute nature of the hypothesis test decision, however, makes some people (including the authors) uneasy. If possible, it's often a good idea to report a confidence interval for the parameter of interest as well.

How small should the P-value be in order for you to reject the null hypothesis? A jury needs enough evidence to show the defendant guilty "beyond a reasonable doubt." How does that translate to P-values? The answer is that it's highly context-dependent. When we're screening for a disease and want to be sure we treat all those who are sick, we may be willing to reject the null hypothesis of no disease with a P-value as large as 0.10. We would rather treat the occasional healthy person than fail to treat someone who was really sick. But a longstanding hypothesis, believed by many to be true, needs stronger evidence (and a correspondingly small P-value) to reject it.

See if you require the same P-value to reject each of the following null hypotheses:

- A renowned musicologist claims that she can distinguish between the works of Mozart and Haydn simply by hearing a randomly selected 20 seconds of music from any work by either composer. What's the null hypothesis? If she's just guessing, she'll get 50% of the pieces correct on average. So our null hypothesis is that p is 50%. If she's for real, she'll get more than 50% correct. Now, we present her with 10 pieces of Mozart or Haydn chosen at random. She gets 9 out of 10 correct. It turns out that the P-value associated with that result is 0.011. (In other words, if you tried to just guess, you'd get at least 9 out of 10 correct only about 1% of the time.) What would *you* conclude? Most people would probably reject the null hypothesis and be convinced that she has some ability to do as she claims. Why? Because the P-value is small, and we don't have any particular reason to doubt the alternative.

"*Extraordinary claims require extraordinary proof.*"
—*Carl Sagan*

- On the other hand, imagine a student who bets that he can make a flipped coin land the way he wants just by thinking hard. To test him, we flip a fair coin 10 times. Suppose he gets 9 out of 10 right. This also has a P-value of 0.011. Are you willing now to reject this null hypothesis? Are you convinced that he's not just lucky? What amount of evidence *would* convince you? We require more evidence when rejecting the null hypotheses would contradict longstanding beliefs or other scientific results. Of course, with sufficient evidence we would revise our opinions (and scientific theories). That's how science makes progress.

Another factor in choosing a P-value is the importance of the issue being tested. Consider the following two tests:

- A researcher claims that the proportion of college students who hold part-time jobs now is higher than the proportion known to hold such jobs a decade ago. You might be willing to believe the claim (and reject the null hypothesis of no change) with a P-value of 10%.
- An engineer claims that the proportion of rivets holding the wing on an airplane that are likely to fail is below the proportion at which the wing would fall off. What P-value would be small enough to get you to fly on that plane?

A S **Hypothesis Tests for Proportions.** You've probably noticed that the tools for confidence intervals and for hypothesis tests are similar. See just how tests and intervals for proportions are related—and an important way in which they differ.

Your conclusion about any null hypothesis should be accompanied by the P-value of the test. When you have the actual data, you should also include a confidence interval for the parameter of interest. Don't just declare the null hypothesis rejected or not rejected. Report the P-value to show the strength of the evidence against the hypothesis. This will let each reader decide whether or not to reject the null hypothesis.

Tests and Intervals Step-By-Step

Anyone who plays or watches sports has heard of the "home field advantage." Teams tend to win more often when they play at home. Or do they? In the 2003 major league baseball season, 2429 regular season games were played.

Let's consider two related questions. First, we wonder whether there actually is a home field advantage. Every team plays roughly half of their games at home, and half away. We'd expect, on average, that the home team should win half the games if there was no home field advantage. In the 2003 major league baseball season, the home team won 1335 of the 2429 games, or 54.96% of them. Could this deviation from 50% be explained just from natural sampling variability, or is this evidence to suggest that there really is a home field advantage, at least in professional baseball?

We might also want to know how large the advantage might be. We might be able to discern a statistically significant difference from 50% for the entire league, but it might amount to less than a one game difference for any particular team—no big deal. So we'll also want a confidence interval.

Tests and intervals are closely related. But for proportions, they are not quite identical calculations. Seeing them together step by step helps to show their similarities and differences.

Think

Plan

State what we want to know.

Define the variables and discuss the W's.

Hypotheses

The null hypothesis makes the claim of no difference from the baseline. Here, that means no home field advantage.

We are interested only in a home field *advantage*, so the alternative hypothesis is one-sided.

Model

Check the appropriate conditions.

I want to know whether the home team in professional baseball is more likely to win. The data are all 2429 games from the 2003 major league baseball season. The variable is whether or not the home team won. The parameter of interest is the proportion of home team wins. If there's no advantage, I'd expect that proportion to be 0.50.

$$H_0: p = 0.50$$
$$H_A: p > 0.50$$

✔ **Independence Assumption:** Generally, the outcome of one game has no effect on the outcome of another game. But this may not always be strictly true. For example, if a key player is injured, the probability that the team will win in the next couple of games may decrease slightly, but independence is still roughly true. The data come from one entire season, but I expect other seasons to be similar.

✔ **Random Sampling Condition:** I have results for all 2429 games of the 2003 season. But I'm not just interested in 2003, and those games, while not randomly selected, are a reasonable representative sample of all recent professional baseball games.

✔ **10% Condition:** These 2429 games are fewer than 10% of all games played over the years.

✔ **Success/Failure Condition:** Both $np_0 = 2429(0.50) = 1214.5$ and $nq_0 = 2429(0.50) = 1214.5$ are greater than 10.

Specify the sampling distribution model.

Tell what test you plan to use.

Because the conditions are satisfied, I'll use a Normal model for the sampling distribution of the proportion and do a **one-proportion z-test**.

Show

Mechanics

The null model gives us the mean, and (because we are working with proportions) the mean gives us the standard deviation.

The null model is a Normal distribution with a mean of 0.50 and a standard deviation of

$$SD(\hat{p}) = \sqrt{\frac{0.5 \times (1 - 0.5)}{2429}} = 0.01015$$

We next find the z-score for the observed proportion to find out how many standard deviations it is from the hypothesized proportion.

From the z-score we can find the P-value, which tells us the probability of observing a value that extreme (or more).

The probability of observing a value 4.89 standard deviations above the mean of a Normal model or higher can be found by computer, calculator, or table to be < 0.0001.

The observed proportion, $\hat{p}$, is 0.5496.

So the z-value is

$$z = \frac{0.5496 - 0.5}{0.01015} = 4.89$$

The sample proportion lies 4.89 standard deviations above the mean.

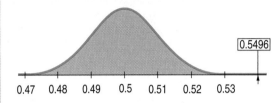

The corresponding P-value is < 0.0001.

Tell

Conclusion State your conclusion about the parameter—in context, of course!

The P-value of <0.0001 says that if the true proportion of home team wins were 0.50, then the observed value of 0.5496 (or larger) would occur less than 1 out of 10,000 times. With a P-value that small I reject H_0. I conclude that the true proportion of home team wins is not 50% and that there is a home field advantage.

Think

AGAIN

How big a difference are we talking about? Let's find a confidence interval for the home field advantage.

Model

Check the conditions.

The conditions are identical to those for the hypothesis test, with one difference. Now we are not given a hypothesized proportion, p_0, so we must instead work with the observed proportion $\hat{p}$.

Specify the sampling distribution model.

Tell what method you plan to use.

✔ **Success/Failure Condition:** $n\hat{p} = 1335$ and $n\hat{q} = 1094$ are both at least 10.

The conditions are satisfied, so I can model the sampling distribution of the proportion with a Normal model and find a **one-proportion z-interval.**

Show

MORE

Mechanics

We can't find the sampling model standard deviation from the null model proportion. (In fact, we've just rejected it.) Instead, we find the standard error of $\hat{p}$ from the *observed* proportions. Other than that substitution, the calculation looks the same as for the hypothesis test.

$$SE(\hat{p}) = \sqrt{\frac{0.5496 \times (1 - 0.5496)}{2429}}$$

$$= 0.0101$$

The sampling model is Normal, so for a 95% confidence interval, the critical value $z^* = 1.96$.

With this large a sample size, the difference is negligible, but in smaller samples, it could make a bigger difference.

The margin of error is

$$ME = z^* \times SE(\hat{p}) = 1.96 \times 0.0101 = 0.0198.$$

So the 95% confidence interval is

$$0.5496 \pm 0.0198 \text{ or } (0.5298, 0.5694).$$

Tell

All

Conclusion

Confidence intervals help us think about the size of the effect. Here we can see that the home field advantage may affect enough games to make a real difference.

I am 95% confident that, in professional baseball, home teams win between 53.0% and 56.9% of the games.

In a season of 162 games, the low end of this interval, 53.0% of the 81 home games, is about two extra home victories. The upper end, 56.9%, is nearly six extra wins.

just checking

4. A bank is testing a new method for getting delinquent customers to pay their past-due credit card bills. The standard way was to send a letter (costing about $0.40 each) asking the customer to pay. That worked 30% of the time. They want to test a new method that involves sending a video tape to the customer encouraging them to contact the bank and set up a payment plan. Developing and sending the video costs about $10.00 per customer. What is the parameter of interest? What are the null and alternative hypotheses?

5. The bank sets up an experiment to test the effectiveness of the video tape. They mail it out to several randomly selected delinquent customers and keep track of how many actually do contact the bank to arrange payments. The bank's statistician calculates a P-value of 0.003. What does this P-value suggest about the video tape?

6. The statistician tells the bank's management that the results are clear and that they should switch to the video tape method. Do you agree? What else might you want to know?

*A Better Confidence Interval for Proportions

Confidence intervals and hypothesis tests for proportions use different standard deviation formulas because of the special property that links a proportion's value with its standard deviation. It turns out, however, that we can improve on the confidence interval for a proportion with a simple adjustment. The adjustment brings the standard deviations for the test and confidence interval into closer agreement.

The improved method is a little strange. It takes the original counts and adds four *phony* observations, two to the successes, two to the failures. So instead of the proportion $\hat{p} = \dfrac{y}{n}$, we use the adjusted proportion $\tilde{p} = \dfrac{y + 2}{n + 4}$ and, for convenience, we write $\tilde{n} = n + 4$. We modify the interval by using these adjusted values for both the center of the interval *and* the margin of error. Now the adjusted interval is

$$\tilde{p} \pm z^* \sqrt{\frac{\tilde{p}(1 - \tilde{p})}{\tilde{n}}}.$$

This adjusted form gives better performance overall[4] and works much better for proportions near 0 or 1. It has the additional advantage that we no longer need to check the Success/Failure Condition that $n\hat{p}$ and $n\hat{q}$ are greater than 10.

What Can Go Wrong?

Hypothesis tests are so widely used—and so widely misused—that we've devoted all of the next chapter to discussing the pitfalls involved, but there are a few issues that we can talk about already.

- ***Don't base your null hypotheses on what you see in the data.*** You are not allowed to look at the data first and then adjust your null hypothesis so that it will be rejected. When your sample value turns out to be $\hat{p} = 51.8\%$, with a standard deviation of 1%, don't form a null hypothesis like $H_0: p = 49.8\%$, knowing that you can reject it. You should always *Think* about the situation you are investigating and make your null hypothesis describe the "nothing interesting" or "nothing has changed" scenario. No peeking at the data!

A S **The Alternative Hypotheses.** This interactive tool provides easy ways to visualize how one- and two-tailed alternative hypotheses work.

- ***Don't base your alternative hypothesis on the data, either.*** Again, you need to *Think* about the situation. Are you interested only in knowing whether something has *increased*? Then write a one-tail (upper tail) alternative. Or would you be equally interested in a change in either direction? Then you want a two-tailed alternative. You should decide whether to do a one- or two-tailed test based on what results would be of interest to you, not what you might see in the data.

- ***Don't make your null hypothesis what you want to show to be true.*** Remember, the null hypothesis is the status quo, the nothing-is-strange-here position a skeptic would take. You wonder whether the data cast doubt on that. You can reject the null hypothesis, but you can never "accept" or "prove" the null.

- ***Don't forget to check the conditions.*** The reasoning of inference depends on randomization. No amount of care in calculating a test result can recover from a biased sample. The probabilities we compute depend on the independence assumption. And our sample must be large enough to justify our use of a Normal model.

CONNECTIONS

Hypothesis tests and confidence intervals share many of the same concepts. Both rely on sampling distribution models, and because the models are the same and require the same assumptions, both check the same conditions. They also calculate many of the same statistics. Like confidence intervals, hypothesis tests use the standard deviation of the sampling distribution as a ruler, as we first saw in Chapter 6.

For testing, we find ourselves looking once again at *z*-scores, and we compute the P-value by finding the distance of our test statistic from the center of the null model. P-values are conditional

---

[4] By "better performance" we mean that a 95% confidence interval has more nearly a 95% chance of covering the true population proportion. Simulation studies have shown that our original, simpler confidence interval in fact is less likely than 95% to cover the true population proportion until the sample size gets much larger. (A. Agresti and B. A. Coull, "Approximate Is Better Than 'Exact' for Interval Estimation of Binomial Proportions," *The American Statistician,* 52 [1998]: 119–126.)

probabilities. They give the probability of observing the result we have seen (or one even more extreme) *given* that the null hypothesis is true.

The Standard Normal model is here again as our connection between *z*-score values and probabilities.

What have we learned?

We've learned to use what we see in a random sample to test a particular hypothesis about the world. This is our second step in statistical inference, complementing our use of confidence intervals.

We've learned that testing a hypothesis involves proposing a model, then seeing whether the data we observe are consistent with that model or so unusual that we must reject it. We do this by finding a P-value—the probability that data like ours could have occurred if the model is correct.

We've learned that:

- We start with a null hypothesis specifying the parameter of a model we'll test using our data.
- Our alternative hypothesis can be one- or two-sided, depending on what we want to learn.
- We must check the appropriate assumptions and conditions before proceeding with our test.
- If the data are out of line with the null hypothesis model, the P-value will be small and we will reject the null hypothesis.
- If the data are consistent with the null hypothesis model, the P-value will be large and we will not reject the null hypothesis.
- We must always state our conclusion in the context of the original question.

And we've learned that confidence intervals and hypothesis tests go hand in hand in helping us think about models. A hypothesis test makes a yes/no decision about the plausibility of a parameter value. The confidence interval shows us the range of plausible values for the parameter.

TERMS

Null hypothesis   The claim being assessed in a hypothesis test is called the null hypothesis. Usually, the null hypothesis is a statement of "no change from the traditional value," "no effect," "no difference," or "no relationship." For a claim to be a testable null hypothesis, it must specify a value for some population parameter that can form the basis for assuming a sampling distribution for a test statistic.

Alternative hypothesis   The alternative hypothesis proposes what we should conclude if we find the null hypothesis to be unlikely.

P-value   The probability of observing a value for a test statistic at least as far from the hypothesized value as the statistic value actually observed if the null hypothesis is true. A small P-value indicates either that the observation is improbable or that the probability calculation was based on incorrect assumptions. The assumed truth of the null hypothesis is the assumption under suspicion.

One-proportion z-test   A test of the null hypothesis that the proportion of a single sample equals a specified value

$$(\text{H}_0: p = p_0) \text{ by referring the statistic } z = \frac{\hat{p} - p_0}{SD(\hat{p})} \text{ to a Standard Normal model.}$$

Two-sided alternative   An alternative hypothesis is two-sided ($H_A: p \neq p_0$) when we are interested in deviations in *either* direction away from the hypothesized parameter value.

One-sided alternative   An alternative hypothesis is one-sided (e.g., $H_A: p > p_0$ or $H_A: p < p_0$) when we are interested in deviations in *only one* direction away from the hypothesized parameter value.

S K I L L S   *When you complete this lesson you should:*

Think
- Be able to state the null and alternative hypotheses for a one-proportion z-test.
- Know the conditions that must be true for a one-proportion z-test to be appropriate and how to examine your data for violations of those conditions.
- Be able to identify and use the alternative hypothesis when testing hypotheses. Understand how to choose between a one-sided and two-sided alternative hypothesis and know how to defend the choice of a one-sided alternative.

Show
- Be able to perform a one-proportion z-test.

Tell
- Be able to write a sentence interpreting the results of a one-proportion z-test.
- Know how to interpret the meaning of a P-value in nontechnical language, making clear that the probability claim is made about computed values under the assumption that the null model is true and not about the population parameter of interest.

Hypothesis Tests for Proportions on the Computer

Hypothesis tests for proportions are so easy and natural that many statistics packages don't offer special commands for them. Most statistics programs want to know the "success" and "failure" status for each case. Usually these are given as 1 or 0, but they might be category names like "yes" and "no." Often we just know the proportion of successes, $\hat{p}$, and the total count, n. Computer packages don't usually deal naturally with summary data like this, but the statistics routines found on many graphing calculators do. These calculators allow you to test hypotheses from summaries of the data—usually all you need to enter are the number of successes and the sample size.

In some programs you can reconstruct the original values. But even when you have reconstructed (or can reconstruct) the raw data values, often you won't get *exactly* the same test statistic from a computer package as you would find working by hand. The reason is that when the packages treat the proportion as a mean, they make some approximations. The result is very close, but not exactly the same.

DATA DESK

Data Desk does not offer built-in methods for inference with proportions. The **Replicate Y by X** command in the **Manip** menu will "reconstruct" summarized count data so you can display it.

Comments

For summarized data, open a Scratchpad to compute the standard deviation and margin of error by typing the calculation. Then perform the test with the **z-test for individual μs** found in the Test command.

EXCEL

Inference methods for proportions are not part of the standard Excel tool set.

Comments

For summarized data, type the calculation into any cell and evaluate it.

JMP

For a **categorical** variable that holds category labels, the **Distribution** platform includes tests and intervals of proportions. For summarized data, put the category names in one variable and the frequencies in an adjacent variable. Designate the frequency column to have the **role** of **frequency**. Then use the **Distribution** platform.

Comments

JMP uses slightly different methods for proportion inferences than those discussed in this text. Your answers are likely to be slightly different.

MINITAB

Choose **Basic Statistics** from the **Stat** menu.

* Choose **1Proportion** from the Basic Statistics submenu.
* If the data are category names in a variable, assign the variable from the variable list box to the **Samples in columns** box.
* If you have summarized data, click the **Summarized Data** button and fill in the number of trials and the number of successes.
* Click the **Options** button and specify the remaining details.
* If you have a large sample, check **Use test and interval based on Normal distribution.**
* Click the **OK** button.

Comments

When working from a variable that names categories, MINITAB treats the last category as the "success" category. You can specify how the categories should be ordered.

SPSS

SPSS does not find hypothesis tests for proportions.

TI-83/84 Plus

To do the mechanics of a hypothesis test for a proportion,

* Select **5:1-PropZTest** from the **STAT TESTS** menu.
* Specify the hypothesized proportion.
* Enter the observed value of x.
* Specify the sample size.
* Indicate what kind of test you want: one-tail lower tail, two-tail, or one-tail upper tail.
* Calculate the result.

Comments

Beware: When you enter the value of x, you need the *count,* not the percentage. The count must be a whole number.

TI-89

To do the mechanics of a hypothesis test for a proportion,

* Select **5:1-PropZTest** from the **STAT TESTS** 2nd F1 menu.
* Specify the hypothesized proportion.
* Enter the observed value of x.
* Specify the sample size.
* Indicate what kind of test you want: one-tail lower tail, two-tail, or one-tail upper tail.
* Specify whether to calculate the result or draw the result (a normal curve with p-value area shaded.)

Comments

Beware: When you enter the value of x, you need the *count,* not the percentage. The count must be a whole number. If the number of successes are given as a percent, you must first multiply np and round the result to obtain x.

EXERCISES

1. Hypotheses. Write the null and alternative hypotheses you would use to test each of the following situations.
 a) A governor is concerned about his "negatives"—the percentage of state residents who express disapproval of his job performance. His political committee pays for a series of TV ads, hoping that they can keep the negatives below 30%. They will use follow-up polling to assess the ads' effectiveness.
 b) Is a coin fair?
 c) Only about 20% of people who try to quit smoking succeed. Sellers of a motivational tape claim that listening to the recorded messages can help people quit.

2. More hypotheses. Write the null and alternative hypotheses you would use to test each of the following situations.
 a) In the 1950s only about 40% of high school graduates went on to college. Has the percentage changed?
 b) 20% of cars of a certain model have needed costly transmission work after being driven between 50,000 and 100,000 miles. The manufacturer hopes that redesign of a transmission component has solved this problem.
 c) We field test a new flavor soft drink, planning to market it only if we are sure that over 60% of the people like the flavor.

3. Negatives. After the political ad campaign described in Exercise 1a, pollsters check the governor's negatives. They test the hypothesis that the ads produced no change against the alternative that the negatives are now below 30%, and find a P-value of 0.22. Which conclusion is appropriate?
 a) There's a 22% chance that the ads worked.
 b) There's a 78% chance that the ads worked.
 c) There's a 22% chance that the poll they conducted is correct.
 d) There's a 22% chance that poll results could be just natural sampling variation rather than a real change in public opinion.

4. Dice. The seller of a loaded die claims that it will favor the outcome 6. We don't believe that claim, and roll the die 200 times to test an appropriate hypothesis. Our P-value turns out to be 0.03. Which conclusion is appropriate?
 a) There's a 3% chance that the die is fair.
 b) There's a 97% chance that the die is fair.
 c) There's a 3% chance that a loaded die could randomly produce the results we observed, so it's reasonable to conclude that the die is fair.
 d) There's a 3% chance that a fair die could randomly produce the results we observed, so it's reasonable to conclude that the die is loaded.

5. Relief. A company's old antacid formula provided relief for 70% of the people who used it. The company tests a new formula to see if it is better, and gets a P-value of 0.27. Is it reasonable to conclude that the new formula and the old one are equally effective? Explain.

6. Cars. A survey investigating whether the proportion of today's high school seniors who own their own cars is higher than it was a decade ago finds a P-value of 0.017. Is it reasonable to conclude that more high schoolers have cars? Explain.

7. He cheats! A friend of yours claims that when he tosses a coin he can control the outcome. You are skeptical and want him to prove it. He tosses the coin, and you call heads; it's tails. You try again, and lose again.
 a) Do two losses in a row convince you that he really can control the toss? Explain.
 b) You try a third time, and again you lose. What's the probability of losing three tosses in a row if the process is fair?
 c) Would three losses in a row convince you that your friend cheats? Explain.
 d) How many times in a row would you have to lose in order to be pretty sure that this friend really can control the toss? Justify your answer by calculating a probability and explaining what it means.

8. Candy. Someone hands you a box of a dozen chocolate-covered candies, telling you that half are vanilla creams, and the other half peanut butter. You pick candies at random and discover that the first three you eat are all vanilla.
 a) If there really were 6 vanilla and 6 peanut butter candies in the box, what is the probability you would have picked three vanillas in a row?
 b) Do you think there really might have been 6 of each? Explain.
 c) Would you continue to believe it if the fourth one you try is also vanilla? Explain.

9. Blinking timers. Many people have trouble programming their VCRs, so a company has developed what it hopes will be easier instructions. The goal is to have at least 96% of customers succeed. The company tests the new system on 200 people, of whom 188 were successful. Is this strong evidence that the new system fails to meet the company's goal? A student's test of this hypothesis is shown below. How many mistakes can you find?

$H_0: \hat{p} = 0.96$

$H_A: \hat{p} \neq 0.96$

SRS, $0.96(200) > 10$

$\dfrac{188}{200} = 0.94;$ $SD(\hat{p}) = \sqrt{\dfrac{(0.94)(0.06)}{200}} = 0.017$

$z = \dfrac{0.96 - 0.94}{0.017} = 1.18$

$P = P(z > 1.18) = 0.12$

There is strong evidence that the new system does not work.

10. Got milk? In November 2001, the *Ag Globe Trotter* newsletter reported that 90% of adults drink milk. A regional farmers' organization planning a new marketing campaign across its multi-county area polls a random sample of 750 adults living there. In this sample, 657 people said that they drink milk. Do these responses provide strong evidence that the 90% figure is not accurate for this region? Correct the mistakes you find in a student's attempt to test an appropriate hypothesis.

$H_0: \hat{p} = 0.9$

$H_A: \hat{p} < 0.9$

SRS, $750 > 10$

$\dfrac{657}{750} = 0.876;$ $SD(\hat{p}) = \sqrt{\dfrac{(0.88)(0.12)}{750}} = 0.012$

$z = \dfrac{0.876 - 0.94}{0.012} = -2$

$P = P(z > -2) = 0.977$

There is more than a 97% chance that the stated percentage is correct for this region.

11. Dowsing. In a rural area only about 30% of the wells that are drilled find adequate water at a depth of 100 feet or less. A local man claims to be able to find water by "dowsing"—using a forked stick to indicate where the well should be drilled. You check with 80 of his customers and find that 27 have wells less than 100 feet deep. What do you conclude about his claim? (We consider a P-value of around 5% to represent strong evidence.)
a) Write appropriate hypotheses.
b) Check the necessary assumptions.
c) Perform the mechanics of the test. What is the P-value?
d) Explain carefully what the P-value means in this context.
e) What's your conclusion?

12. Abnormalities. In the 1980s it was generally believed that congenital abnormalities affected about 5% of the nation's children. Some people believe that the increase in the number of chemicals in the environment has led to an increase in the incidence of abnormalities. A recent study examined 384 children and found that 46 of them showed signs of an abnormality. Is this strong evidence that the risk has increased? (We consider a P-value of around 5% to represent strong evidence.)
a) Write appropriate hypotheses.

b) Check the necessary assumptions.
c) Perform the mechanics of the test. What is the P-value?
d) Explain carefully what the P-value means in this context.
e) What's your conclusion?
f) Do environmental chemicals cause congenital abnormalities?

13. Absentees. The National Center for Education Statistics monitors many aspects of elementary and secondary education nationwide. Their 1996 numbers are often used as a baseline to assess changes. In 1996 34% of students had not been absent from school even once during the previous month. In the 2000 survey, responses from 8302 students showed that this figure had slipped to 33%. Officials would, of course, be concerned if student attendance were declining. Do these figures give evidence of a decrease in student attendance?
a) Write appropriate hypotheses.
b) Check the assumptions and conditions.
c) Perform the test and find the P-value.
d) State your conclusion.
e) Do you think this difference is meaningful? Explain.

14. Educated mothers. The National Center for Education Statistics monitors many aspects of elementary and secondary education nationwide. Their 1996 numbers are often used as a baseline to assess changes. In 1996 31% of students reported that their mothers had graduated from college. In 2000, responses from 8368 students found that this figure had grown to 32%. Is this evidence of a change in education level among mothers?
a) Write appropriate hypotheses.
b) Check the assumptions and conditions.
c) Perform the test and find the P-value.
d) State your conclusion.
e) Do you think this difference is meaningful? Explain.

15. Smoking. National data in the 1960s showed that about 44% of the adult population had never smoked cigarettes. In 1995 a national health survey interviewed a random sample of 881 adults and found that 52% had never been smokers.
a) Create a 95% confidence interval for the proportion of adults (in 1995) who had never been smokers.
b) Does this provide evidence of a change in behavior among Americans? Using your confidence interval, test an appropriate hypothesis and state your conclusion.

16. Satisfaction. A company hopes to improve customer satisfaction, setting as a goal no more than 5% negative comments. A random survey of 350 customers found only 10 with complaints.
a) Create a 95% confidence interval for the true level of dissatisfaction among customers.
b) Does this provide evidence that the company has reached its goal? Using your confidence interval, test an appropriate hypothesis and state your conclusion.

17. Pollution. A company with a fleet of 150 cars found that the emissions systems of 7 out of the 22 they tested failed to meet pollution control guidelines. Is this strong evidence that more than 20% of the fleet might be out of compliance? Test an appropriate hypothesis and state your conclusion. Be sure the appropriate assumptions and conditions are satisfied before you proceed.

18. Scratch and dent. An appliance manufacturer stockpiles washers and dryers in a large warehouse for shipment to retail stores. Sometimes in handling them the appliances get damaged. Even though the damage may be minor, the company must sell those machines at drastically reduced prices. The company goal is to keep the level of damaged machines below 2%. One day an inspector randomly checks 60 washers and finds that 5 of them have scratches or dents. Is this strong evidence that the warehouse is failing to meet the company goal? Test an appropriate hypothesis and state your conclusion. Be sure the appropriate assumptions and conditions are satisfied before you proceed.

19. Twins. In 2001 a national vital statistics report indicated that about 3% of all births produced twins. Data from a large city hospital found only 7 sets of twins were born to 469 teenage girls. Does that suggest that mothers under age 20 may be less likely to have twins? Test an appropriate hypothesis and state your conclusion. Be sure the appropriate assumptions and conditions are satisfied before you proceed.

20. Football. During the 2000 season, the home team won 138 of the 240 regular season National Football League games. Is this strong evidence of a home field advantage in professional football? Test an appropriate hypothesis and state your conclusion. Be sure the appropriate assumptions and conditions are satisfied before you proceed.

21. WebZine. A magazine is considering the launch of an online edition. The magazine plans to go ahead only if it's convinced that more than 25% of current readers would subscribe. The magazine contacts a simple random sample of 500 current subscribers, and 137 of those surveyed expressed interest. What should the company do? Test an appropriate hypothesis and state your conclusion. Be sure the appropriate assumptions and conditions are satisfied before you proceed.

22. Seeds. A garden center wants to store leftover packets of vegetable seeds for sale the following spring, but the center is concerned that the seeds may not germinate at the same rate a year later. The manager finds a packet of last year's green bean seeds and plants them as a test. Although the packet claims a germination rate of 92%, only 171 of 200 test seeds sprout. Is this evidence that the seeds have lost viability during a year in storage? Test an appropriate hypothesis and state your conclusion. Be sure the appropriate assumptions and conditions are satisfied before you proceed.

23. Women executives. A company is criticized because only 13 of 43 people in executive-level positions are women. The company explains that although this proportion is lower than it might wish, it's not surprising given that only 40% of all their employees are women. What do you think? Test an appropriate hypothesis and state your conclusion. Be sure the appropriate assumptions and conditions are satisfied before you proceed.

24. Jury. Census data for a certain county shows that 19% of the adult residents are Hispanic. Suppose 72 people are called for jury duty, and only 9 of them are Hispanic. Does this apparent underrepresentation of Hispanics call into question the fairness of the jury selection system? Explain.

25. Dropouts. Some people are concerned that new tougher standards and high-stakes tests adopted in many states may drive up the high school dropout rate. The National Center for Education Statistics reported that the high school dropout rate for the year 2000 was 10.9%. One school district, whose dropout rate has always been very close to the national average, reports that 210 of their 1782 students dropped out last year. Is their experience evidence that the dropout rate may be increasing? Explain.

26. Acid rain. A study of the effects of acid rain on trees in the Hopkins Forest shows that of 100 trees sampled, 25 of them exhibited some sort of damage from acid rain. This rate seemed to be higher than the 15% quoted in a recent *Environmetrics* article on the average proportion of damaged trees in the Northeast. Does the sample suggest that trees in the Hopkins Forest are more susceptible than the rest of the region? Comment, and write up your own conclusions based on an appropriate confidence interval as well as a hypothesis test. Include any assumptions you made about the data.

27. Lost luggage. An airline's public relations department says that the airline rarely loses passengers' luggage. It further claims that on those occasions when luggage is lost, 90% is recovered and delivered to its owner within 24 hours. A consumer group who surveyed a large number of air travelers found that only 103 of 122 people who lost luggage on that airline were reunited with the missing items by the next day. Does this cast doubt on the airline's claim? Explain.

28. TV ads. A start-up company is about to market a new computer printer. It decides to gamble by running commercials during the Super Bowl. The company hopes that name recognition will be worth the high cost of the ads. The goal of the company is that at least 40% of the public recognize its brand name and associate it with computer equipment. The day after the game, a pollster contacts 420 randomly chosen adults, and finds that 181 of them know that this company manufactures printers. Would you recommend that the company continue to advertise during Super Bowls? Explain.

29. John Wayne. Like a lot of other Americans, John Wayne died of cancer. But is there more to this story? In 1955

Wayne was in Utah shooting the film *The Conqueror.* Across the state line, in Nevada, the United States military was testing atomic bombs. Radioactive fallout from those tests drifted across the film location. A total of 46 of the 220 people working on the film eventually died of cancer. Cancer experts estimate that one would expect only about 30 cancer deaths in a group this size.

a) Is the death rate observed in the movie crew unusually high?

b) Does this prove that exposure to radiation may increase the risk of cancer?

30. AP Stats. The College Board reported that 57% of all students who took the 2002 AP Statistics exam earned scores of 3 or higher. One teacher was particularly pleased that 65% of her 54 students achieved scores of 3 or better. She considered that year's group typical of the students who will take AP Stats at her high school. Should she brag that her school's students outperform the national results? Explain.

just checking

Answers

1. You can't conclude that the null hypothesis is true. You can conclude only that the experiment was unable to reject the null hypothesis. They were unable, on the basis of 12 patients, to show that aspirin was effective.

2. The null hypothesis is $H_0: p = 0.75$.

3. With a P-value of 0.0001, this is very strong evidence against the null hypothesis. We can reject H_0 and conclude that the improved version of the drug gives relief to a higher proportion of patients.

4. The parameter of interest is the proportion, p, of all delinquent customers who will pay their bills. $H_0: p = 0.30$ and $H_A: p > 0.30$.

5. The very low P-value leads us to reject the null hypothesis. There is strong evidence that the video tape is more effective in getting people to start paying their debts than just sending a letter had been.

6. All we know is that there is strong evidence to suggest that $p > 0.30$. We don't know how much higher than 30% the new proportion is. We'd like to see a confidence interval to see if the new method is worth the cost.

CHAPTER 21

More About Tests

WHO	Trials of TT practitioners attempting to detect a HEF
WHAT	Success/failure
WHEN	1998 publication date
WHY	Test claim of HEF/debunk TT?

A S **Is There Evidence for Therapeutic Touch?** This video shows the experiment and tells the story.

Therapeutic touch (TT), taught in many schools of nursing, is a therapy in which the practitioner moves her hands near, but not touching, a patient in an attempt to manipulate a "human energy field" (HEF). Therapeutic touch practitioners believe that by adjusting this field they can promote healing. However, no instrument has ever detected a human energy field and no experiment has ever shown that TT practitioners can detect such a field.

In 1998 the *Journal of the American Medical Association* published a paper reporting work by a then 9-year-old girl (L. Rosa, E. Rosa, L. Sarner, S. Barrett, "A Close Look at Therapeutic Touch," *JAMA* 279(13) [1 April 1998]:1005–1010). She had performed a simple experiment in which she challenged 15 TT practitioners to detect whether her unseen hand was hovering over their left or right hand (selected by a coin flip). The practitioners "warmed up" with a period during which they could see the experimenter's hand, and each said that they could detect the girl's human energy field. Then a screen was placed so that the practitioners could not see the girl's hand, and they attempted 10 trials each. Overall, of 150 trials, the TT practitioners were successful 70 times, for a success proportion of 46.7%. Is there evidence from this experiment that TT practitioners can successfully detect a "human energy field"?

Zero In on the Null

Null hypotheses have special requirements. In order to perform a statistical test of the hypothesis, the null must be a statement about the value of a parameter for a model. We use this value to compute the probability that the observed sample statistic—or something even farther from the null value—would occur.

How do we choose the null hypothesis? The appropriate null arises directly from the context of the problem. It is not dictated by the data, but instead by the situation.

One good way to identify both the null and alternative hypotheses is to think about the *Why* of the situation. A pharmaceutical company wanting to develop

471

and market a new drug needs to show that the new drug is effective. The Food and Drug Administration (FDA) will not allow a new drug to be sold in the United States that has not been proven effective in a double-blind, randomized clinical trial. (Most countries have similar restrictions on the introduction of new medications.) The typical null hypothesis in this case is that the proportion of patients recovering after receiving the new drug is the same as we would expect of patients receiving a placebo. The alternative hypothesis is that the new drug cures a higher proportion. But a physician seeking the best treatment for her patients would probably prefer to see the new drug tested against a currently available treatment.

To write a null hypothesis, you can't just choose any parameter value you like. The null must relate to the question at hand. Even though the null usually means no difference or no change, you can't automatically interpret "null" to mean zero. A claim that the TT practitioners should guess correctly *none of the time* would be absurd. The null hypothesis in this case is that the proportion of correct responses is 0.5—what you'd expect from random guessing. You need to find the value for the parameter in the null hypothesis from the context of the problem.

● **ESP?** In the 1930s, a series of experiments was performed at Duke University in an attempt to see whether humans were capable of extrasensory perception, or ESP. A set of cards with 5 symbols, made famous in the movie *Ghostbusters*, was used:

$$\bigcirc \quad \square \quad \bigstar \quad + \quad \wr\wr\wr$$

In the experiment, the "sender" selects one of the 5 cards at random from a deck and then concentrates on it. The "receiver" tries to determine which card it is. If we let p be the proportion of correct responses, what's the null hypothesis? The null hypothesis is that ESP makes no difference. Without ESP, the receiver would just be guessing, and since there are 5 possible responses, there would be a 20% chance of guessing each card correctly. So, H_0 is $p = 0.20$. What's the alternative? It seems that it should be $p > 0.20$, a one-sided alternative. But some ESP researchers have expressed the claim that if the proportion guessed were much *lower* than expected, that would show an "interference" and should be considered evidence for ESP as well. So they argue for a two-sided alternative. ●

A S **Testing Therapeutic Touch.** Perform the one-proportion z-test using *ActivStats* technology. Compare this approach to the Step-by-Step below. The test in *ActivStats* is two-sided. Which do you think is the more appropriate choice?

There is a temptation to state your *claim* as the null hypothesis. As we have seen, however, you cannot prove a null hypothesis true any more than you can prove a defendant innocent. So, it makes more sense to use what you want to show as the *alternative*. This way, when you reject the null, you are left with what you want to show.

Another One-Proportion z-Test Step-By-Step

Let's look at the therapeutic touch experiment with a one-proportion z-test.

Think

Plan State the problem and discuss the variables and the W's.

Hypotheses

The null hypothesis is that the TT practitioners are just guessing. Because there are two

I want to know whether the TT practitioners could detect a "human energy field" (HEF). The data are from a designed experiment where 15 practitioners were asked in 10 trials apiece to identify which of their hands was close to the experimenter's hand. For each trial, the experi-

choices, guessing should succeed about half the time.

Here it seems natural to test a one-sided alternative hypothesis. We wouldn't be interested in below-chance performance.

menter recorded whether the practitioner correctly identified the hand or not. The parameter of interest is the proportion of successful identifications.

$H_0: p = 0.50$

$H_A: p > 0.50$

Model

Check the conditions.

If some TT practitioners had been successful, their trials might not be independent, but the assumption only needs to hold if the null hypothesis is true. Under the null hypothesis that there is no HEF, each of these trials is an independent randomized guess.

✔ **Independence Assumption:** The choice of hand was randomized with a coin flip. Under the null hypothesis, the trials should be independent.

✔ **Randomization:** The experiment was randomized by flipping a coin.

✔ **10% Condition:** The experiment observes some of what could be an infinite number of trials.

✔ **Success/Failure Condition:** Both $np_0 = 150(0.5) = 75$ and $nq_0 = 150(0.5) = 75$ are greater than 10. I expect at least 10 successes and at least 10 failures, so enough trials were run.

Specify the sampling distribution model.

Name the test.

The conditions are satisfied, so I can use a Normal model and perform a **one-proportion z-test.**

Show ⬮ **Mechanics**

Find the standard deviation of the sampling model using the hypothesized proportion, p_0.

The null model has a mean of 0.5 and a standard deviation of

$$SD(\hat{p}) = \sqrt{\frac{p_0 q_0}{n}} = \sqrt{\frac{(0.5)(0.5)}{150}} \approx 0.041$$

The observed proportion $\hat{p}$ is 0.467.

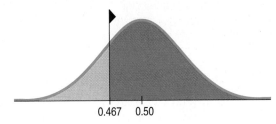

Make a picture. Sketch a Normal model centered at the hypothesized success rate of 50%. This is an upper tail test, so shade the region to the right of the observed success rate.

Find the z-score for the observed proportion.

Find the P-value by computer, calculator, or table (shown below). The P-value is the probability of observing a sample proportion at least as large as 0.467 if the true proportion were 0.5.

$$z = \frac{\hat{p} - p_0}{SD(\hat{p})} = \frac{0.467 - 0.5}{0.041} = -0.805$$

The observed success rate is 0.805 standard deviations below the hypothesized mean.

$$P = P(z > -0.805) = 0.788$$

Tell ⬮ **Conclusion** Link the P-value to your decision about the null hypothesis, then state your conclusion in context.

The P-value of 0.788 says that if the true proportion of successful detections of a human energy field is 50%, then an observed proportion

If possible, propose a course of action. In the case of this study, the editor of the *Journal of the American Medical Association* suggested that insurance companies reconsider paying for therapeutic touch treatments. This, of course, raised a storm of protest letters from those who practice TT.

of 46.7% successes or more would occur at random about 8 times in 10. That's not a rare event, so I don't reject the null hypothesis.

There is insufficient evidence to conclude that the TT practitioners can perform better than they would if they were just guessing.

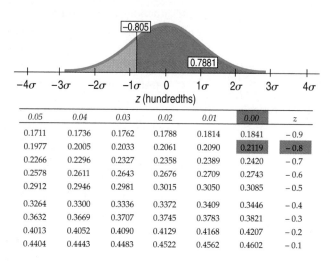

0.05	0.04	0.03	0.02	0.01	0.00	z
0.1711	0.1736	0.1762	0.1788	0.1814	0.1841	− 0.9
0.1977	0.2005	0.2033	0.2061	0.2090	0.2119	− 0.8
0.2266	0.2296	0.2327	0.2358	0.2389	0.2420	− 0.7
0.2578	0.2611	0.2643	0.2676	0.2709	0.2743	− 0.6
0.2912	0.2946	0.2981	0.3015	0.3050	0.3085	− 0.5
0.3264	0.3300	0.3336	0.3372	0.3409	0.3446	− 0.4
0.3632	0.3669	0.3707	0.3745	0.3783	0.3821	− 0.3
0.4013	0.4052	0.4090	0.4129	0.4168	0.4207	− 0.2
0.4404	0.4443	0.4483	0.4522	0.4562	0.4602	− 0.1

A portion of the z-table corresponding to the test. The value -0.805 falls between -0.80 and -0.81. We'd have to interpolate to get from the table the accuracy we can get from technology. The P-value we want is the area of the right tail, so it's $1 - 0.2119 = 0.7881$.   **Figure 21.1**

How to Think About P-values

Which Conditional?

Suppose that as a political science major you are offered the chance to be a White House intern. There would be a very high probability that next summer you'll be in Washington, D.C. That is, $P(\text{Washington}\,|\,\text{Intern})$ would be high. But if we find a student in Washington, D.C., is it likely that he's a White House intern? Almost surely not; $P(\text{Intern}\,|\,\text{Washington})$ is low. You can't switch around conditional probabilities. The P-value is $P(\text{data}\,|\,H_0)$. We might like to report $P(H_0\,|\,\text{data})$, but these two quantities are NOT the same.

A P-value actually is a conditional probability. It tells us the probability of the observed statistic *given* that the null hypothesis is true. To show this more clearly, we can write P-value = $P(\text{observed statistic value [or even more extreme]}\,|\,H_0)$.

Writing the P-value this way helps to make clear that the P-value is *not* the probability that the null hypothesis is true. It is a probability about the data. Let's say that again:

The P-value is not the probability that the null hypothesis is true.

The P-value is not even the conditional probability that the null hypothesis is true given the data. We would write that probability as $P(H_0\,|\,\text{observed statistic value})$. This is the conditional probability for the P-value, in reverse. It would be nice to know this, but it's impossible to calculate without making additional assumptions. As we saw in Chapter 15, reversing the order in a conditional probability is difficult, and the results can be counterintuitive.

● **How guilty is the suspect?** We might like to know $P(H_0\,|\,\text{data})$, but when you think about it, we can't talk about the probability that the null hypothesis is true. The null is not a random event, so either it is true or it isn't. The data, however, are random in the sense that if we were to repeat a randomized experiment or draw another random sample, we'd get different data and expect to find a different statistic value. So we can talk about the probability of the data given the null hypothesis, and that's the P-value.

But it does make sense that the smaller the P-value, the more confident we can be in declaring that we doubt the null hypothesis. Think again about the jury trial. Our null hypothesis is that the defendant is innocent. Then the evidence starts rolling in. A car the same color as his was parked in front of the bank. Well, there are lots of cars that color. The probability of that happening (given his innocence) is pretty high, so we're not persuaded that he's guilty. The bank's security camera showed the robber was male and about his height and weight. Hmmm. Could that be a coincidence? If he's innocent, then it's a little less likely that the car and description would *both* match, so our P-value goes down. We're starting to question his innocence a little. Witnesses said the robber wore a blue jacket just like the one the police found in a garbage can behind the defendant's house. Well, if he's innocent, then that doesn't seem very likely, does it? If he's really innocent, the probability that all of these could have happened is getting pretty low. Now our P-value may be small enough to be called "beyond a reasonable doubt" and lead to a conviction. Each new piece of evidence strains our skepticism a bit more. The more compelling the evidence—the more *unlikely* it would be were he innocent—the more convinced we become that he's guilty.

But even though it may make *us* more confident in declaring him guilty, additional evidence does not make *him* any guiltier. Either he robbed the bank or he didn't. Additional evidence (like the teller picking him out of a police lineup) just makes us more confident that we did the right thing when we convicted him. The lower the P-value, the more comfortable we feel about our decision to reject the null hypothesis, but the null hypothesis doesn't get any more false. ●

We can find the P-value, P(observed statistic value $\mid H_0$), because H_0 gives the parameter values that we need to find the required probability. But there's no direct way to find $P(H_0 \mid$ observed statistic value).[1] As tempting as it may be to say that a P-value of 0.03 means there's a 3% chance that the null hypothesis is true, that just isn't right. All we can say is that given the null hypothesis there's a 3% chance of observing the statistic value that we have actually observed (or one more unlike the null value).

Alpha Levels

Sometimes we need to make a firm decision about whether or not to reject the null hypothesis. A jury must *decide* whether the evidence reaches the level of "beyond a reasonable doubt." A business must *select* a Web design. You need to decide which section of Statistics to enroll in.

When the P-value is small, it tells us that our data are rare *given the null hypothesis*. As humans, we are suspicious of rare events. If the data are "rare enough," we just don't think that could have happened by chance. Since the data *did* happen, something must be wrong. All we can do now is to reject the null hypothesis.

But how rare is "rare"?

We can define "rare event" arbitrarily by setting a threshold for our P-value. If our P-value falls below that point, we'll reject the null hypothesis. We call such results **statistically significant.** The threshold is called an **alpha level.** Not surprisingly, it's labeled with the Greek letter α. Common α levels are 0.10, 0.05, and

---

[1] The approach to statistical inference known as Bayesian Statistics addresses the question in just this way, but it requires more advanced mathematics and more assumptions.

Sir Ronald Fisher (1890–1962) was one of the founders of modern Statistics.

NOTATION ALERT:

The first Greek letter, α, is used in Statistics for the threshold value of a hypothesis test. You'll hear it referred to as the alpha level. Common values are 0.10, 0.05, 0.01, and 0.001.

Of course, if the null hypothesis *is* true, no matter what alpha level you choose, you still have a probability α of rejecting the null hypothesis by mistake. This is the rare event we want to protect ourselves against. When we do reject the null hypothesis, no one ever thinks that *this* is one of those rare times. As statistician Stu Hunter notes: *The statistician says "rare events do happen—but not to me!"*

0.01. You have the option—almost the *obligation*—to consider your alpha level carefully and choose an appropriate one for the situation. If you're assessing the safety of airbags, you'll want a low alpha level; even 0.01 might not be low enough. If you're just wondering whether folks prefer their pizza with or without pepperoni, you might be happy with $\alpha = 0.10$. It can be hard to justify your choice of α, though, so often we arbitrarily choose 0.05.

 Where did the value 0.05 come from? In 1931 in a famous book called *The Design of Experiments*, Sir Ronald Fisher discussed the amount of evidence needed to reject a null hypothesis. He said that it was *situation dependent*, but remarked, somewhat casually, that for many scientific applications, 1 out of 20 *might be* a reasonable value. Since then, some people—indeed some entire disciplines—have acted as if the number 0.05 is sacrosanct.

The alpha level is also called the **significance level.** When we reject the null hypothesis, we say that the test is "significant at that level." For example, we might say that we reject the null hypothesis "at the 5% level of significance." You must select the alpha level *before* you look at the data. Otherwise you can be accused of cheating by tuning your alpha level to suit the data.

What can you say if the P-value does not fall below α?

When you have not found sufficient evidence to reject the null according to the standard you have established, you should say that "The data have failed to provide sufficient evidence to reject the null hypothesis." Don't say that you "accept the null hypothesis." You certainly haven't proven or established it; it was assumed to begin with. You *could* say that you have *retained* the null hypothesis, but it's better to say that you've failed to reject it.

Look again at the Step-by-Step example for therapeutic touch. The P-value was 0.788. This is so much larger than any reasonable alpha level that we can't reject H_0. We concluded: "We fail to reject the null hypothesis. There is insufficient evidence to conclude that the practitioners are performing better than they would if they were just guessing."

The automatic nature of the reject/fail-to-reject decision when we use an alpha level may make you uncomfortable. If your P-value falls just slightly above your alpha level, you're not allowed to reject the null. Yet a P-value just barely below the alpha level leads to rejection. If this bothers you, you're in good company. Many statisticians think it better to report the P-value than to choose an alpha level and carry the decision through to a final reject/fail-to-reject verdict. So when you decide to declare a verdict, it's always a good idea to report the P-value as an indication of the strength of the evidence.

 It's in the stars Some disciplines carry the idea further and code P-values by their size. In this scheme, a P-value between 0.05 and 0.01 gets highlighted by *. A P-value between 0.01 and 0.001 gets **, and a P-value less than 0.001 gets ***. This can be a convenient summary of the weight of evidence against the null hypothesis if it's not taken too literally. But we warn you against taking the distinctions too seriously and against making a black-and-white decision near the boundaries. The boundaries are a matter of tradition, not science; there is nothing special about 0.05. A P-value of 0.051 should be looked at very seriously and not casually thrown away just because it's larger than 0.05, and one that's 0.009 is not very different from one that's 0.011.

Sometimes it's best to report that the conclusion is not yet clear and to suggest that more data be gathered. (In a trial, a jury may "hang" and be unable to return a verdict.) In these cases, it's an especially good idea to report the P-value, since it's the best summary we have of what the data say or fail to say about the null hypothesis.

What Not to Say About Significance

Practical vs. Statistical Significance

A large insurance company mined its data and found a statistically significant ($P = 0.04$) difference between the mean value of policies sold in 2001 and 2002. The difference in the mean values was $9.83. Even though it was statistically significant, management did not see this as an important difference when a typical policy sold for more than $1000. On the other hand, even a clinically important improvement of 10% in cure rate with a new treatment is not likely to be statistically significant in a study of fewer than 225 patients. A small clinical trial would probably not be conclusive.

What do we mean when we say that a test is statistically significant? All we mean is that the test statistic had a P-value lower than our alpha level. Don't be lulled into thinking that statistical significance carries with it any sense of practical importance or impact.

For large samples, even small, unimportant ("insignificant") deviations from the null hypothesis can be statistically significant. On the other hand, if the sample is not large enough, even large, financially or scientifically "significant" differences may not be statistically significant.

It's good practice to report the magnitude of the difference between the observed statistic value and the null hypothesis value (in the data units) along with the P-value on which we base statistical significance.

Critical Values Again

If you need to make a decision on the fly with no technology, remember "2." That's our old friend from the 68-95-99.7 Rule. It's roughly the critical value for testing a hypothesis against a two-sided alternative at $\alpha = 0.05$. The exact critical value for z is 1.96, but 2 is close enough for most decisions.

When making a confidence interval, we've found a **critical value**, z^*, to correspond to our selected confidence level. Critical values can also be used as a shortcut for hypothesis tests. Before computers and calculators were common, P-values were hard to find. It was easier to select a few common alpha levels (0.05, 0.01, 0.001, for example) and learn the corresponding critical values for the Normal model (that is, the critical values corresponding to confidence levels 0.95, 0.99, and 0.999, respectively). Rather than looking up the probability corresponding to your z-score in the table, you'd just check your z-score directly against these z^* values. Any z-score larger in magnitude (that is, more extreme) than a particular critical value has to be less likely, so it will have a P-value smaller than the corresponding alpha. If we are willing to settle for a flat reject/fail-to-reject decision, comparing an observed z-score with the critical value for a specified alpha level gives a shortcut path to that decision. For the therapeutic touch example, if we choose $\alpha = 0.05$, then in order to reject H_0 our z-score has to be larger than the one-sided critical value of 1.645. Our calculated z-score is -0.805, so we clearly fail to reject the null hypothesis. This is perfectly correct, and does give us a yes/no decision, but it gives us less information about the hypothesis because we don't have the P-value to think about. With technology, P-values are easy to find. And since they give more information about the strength of the evidence, you should always report them.

Here are the traditional critical values from the Normal model[2]:

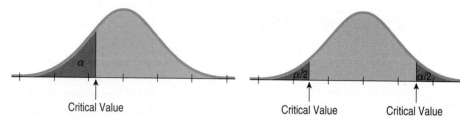

α	1-sided	2-sided
0.05	1.645	1.96
0.01	2.33	2.576
0.001	3.09	3.29

When the alternative is one-sided, the critical value puts all of α on one side.   **Figure 21.2**

When the alternative is two-sided, the critical value splits α equally into two tails.   **Figure 21.3**

Confidence Intervals and Hypothesis Tests

For the therapeutic touch example, a 90% confidence interval would give $0.47 \pm 1.65 \times 0.04 = (0.404, 0.536)$, or 40.4% to 53.6%. We could not reject H_0: $p = 0.50$ because 50% is a plausible value in the confidence interval. Indeed, *any* hypothesized value for the true proportion of successes in this interval is consistent with the data. For example, a success rate of 53% is a plausible value for the practitioners' true success rate. But we'd be pretty skeptical of a claim that TT practitioners can detect a human energy field 75% of the time. Any value outside the confidence interval would make a null hypothesis that we would reject, but we'd feel more strongly about values far outside the interval.

Confidence intervals and hypothesis tests are built from the same calculations.[3] They have the same assumptions and conditions. As we have just seen, you can approximate a hypothesis test by examining the confidence interval. Just ask whether the null hypothesis value is consistent with a confidence interval for the parameter at the corresponding confidence level. Because confidence intervals are naturally two-sided, they correspond to two-sided tests. For example, a 95% confidence interval corresponds to a two-sided hypothesis test at $\alpha = 5\%$. In general, a confidence interval with a confidence level of $C\%$ corresponds to a two-sided hypothesis test with an α level of $100 - C\%$.

---

[2] In a sense, these are the flip side of the 68-95-99.7 Rule. There we chose simple statistical distances from the mean and recalled the areas of the tails. Here we select convenient tail areas (0.05, 0.01, and 0.001, either on one side or adding both together) and record the corresponding statistical distances.

[3] As we saw in Chapter 20, this is not *exactly* true for proportions. For a confidence interval, we estimate the standard deviation of $\hat{p}$ from $\hat{p}$ itself. Because we estimate it from the data, we have a *standard error*. For the corresponding hypothesis test, we use the model's standard deviation for $\hat{p}$, based on the null hypothesis value p_0. When $\hat{p}$ and p_0 are close, these calculations give very similar results. When they differ, you're likely to reject H_0 (because the observed proportion is far from your hypothesized value). In that case, you're better off building your confidence interval with a standard error estimated from the data.

The relationship between confidence intervals and one-sided hypothesis tests is a little more complicated. For a one-sided test with $\alpha = 5\%$, the corresponding confidence interval has a confidence level of 90%—that's 5% in each tail. In general, a confidence interval with a confidence level of C% corresponds to a one-sided hypothesis test with an α level of $\frac{1}{2}(100 - C)\%$.

just checking

1. An experiment to test the fairness of a roulette wheel gives a z-score of 0.62. What would you conclude?

2. In the last chapter we encountered a bank that wondered if it could get more customers to make payments on delinquent balances by sending them a video tape urging them to set up a payment plan. Well, the bank just got back the results on their test of the video tape strategy. A 90% confidence interval for the success rate is (0.29, 0.45). Their old send-a-letter method had worked 30% of the time. Can you reject the null hypothesis that the proportion is still 30% at $\alpha = 0.05$? Explain.

3. Given the confidence interval the bank found in their trial of video tapes, what would you recommend that they do? Should they scrap the video tape strategy?

Wear that Seatbelt! Step-By-Step

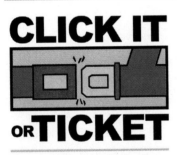

CLICK IT
OR TICKET

Massachusetts is Serious About Saving Lives

Teens are at the greatest risk of being killed or injured in traffic crashes. According to the National Highway Traffic Safety Administration, 65% of young people killed were not wearing a safety belt. In 2001, a total of 3322 teens were killed in motor vehicle crashes, an average of 9 teenagers a day. Because many of these deaths could easily be prevented by the use of safety belts, several states have begun "Click It or Ticket" campaigns in which increased enforcement and publicity have resulted in significantly higher seat belt use. Overall use in Massachusetts increased from 51% to 62% from 2002 to 2003. The goal for 2004 was 65%. A local newspaper reports that a roadblock resulted in 42 tickets to drivers who were unbelted out of 134 stopped for inspection. Does this provide evidence that the goal of 65% compliance was met?

Let's use a confidence interval to test this hypothesis.

Think

Plan State the problem and discuss the variables and the W's.

Hypotheses

The null hypothesis is that the compliance rate is 65%. The alternative is that it has increased. It's clearly a one-sided test, so if we use a confidence interval, we'll have to be careful about what level we use.

Model

Check the conditions.

The data come from a local newspaper report that tells the number of tickets issued and number of drivers stopped at a recent roadblock. I want to know whether the rate of compliance with the seatbelt law is greater than 65%.

$$H_0: p = 0.65$$
$$H_A: p > 0.65$$

✔ **Independence Assumption:** Drivers are not likely to influence one another when it comes to wearing a seatbelt.

✔ **Random Sampling Condition:** This wasn't a random sample, but I assume these drivers are representative of the driving public.

We are finding a confidence interval, so we work from the data rather than the null model.

✔ **10% Condition:** The police stopped fewer than 10% of all drivers.

✔ **Success/Failure Condition:** There were 92 successes and 42 failures, both greater than 10. The sample is large enough.

Under these conditions, the sampling model is Normal. I'll create a **one-proportion z-interval.**

State your method.

Show   **Mechanics** Write down the given information, and determine the sample proportion.

$n = 134$, so

$$\hat{p} = \frac{92}{134} = 0.687 \text{ and}$$

To use a confidence interval, we need a confidence level that corresponds to the alpha level of the test. If we use $\alpha = 0.05$, we should construct a 90% confidence interval, because this is a one-sided test. That will leave 5% on *each* side of the observed proportion. Determine the standard error of the sample proportion and the margin of error. The critical value is $z^* = 1.645$.

$$SE(\hat{p}) = \sqrt{\frac{\hat{p}\hat{q}}{n}} = \sqrt{\frac{(0.687)(0.313)}{134}} = 0.040$$

$$ME = z^* \times SE(\hat{p})$$
$$= 1.645(0.040) = 0.066$$

The confidence interval is

estimate ± margin of error.

The 90% confidence interval is:

$$0.687 \pm 0.066 \text{ or}$$
$$(0.621, 0.753).$$

Tell   **Conclusion**

Link the confidence interval to your decision about the null hypothesis, then state your conclusion in context.

I am 90% confident that between 62.1% and 75.3% of all drivers wear their seatbelts. Because the hypothesized rate of 65% is within this interval, I do not reject the null hypothesis. There is insufficient evidence to conclude that more than 65% of all drivers are now wearing seatbelts.

The upper limit of the confidence interval shows it's possible that the program is quite successful, but the small sample size makes the interval too wide to be very specific.

Making Errors

A S   **Type I and Type II Errors.** View an animated exploration of Type I and Type II errors—not quite a first-run movie, but good backup for the reading in this section.

Nobody's perfect. Even with lots of evidence, we can still make the wrong decision. In fact, when we perform a hypothesis test, we can make mistakes in *two* ways:

  I. The null hypothesis is true, but we mistakenly reject it.

  II. The null hypothesis is false, but we fail to reject it.

These two types of errors are known as **Type I** and **Type II errors.** One way to keep the names straight is to remember that we start by assuming the null hypothesis is true, so a Type I error is the first kind of error we could make.

Some false-positive results mean no more than an unnecessary chest X-ray. But for a drug test or a disease like AIDS, a false-positive result that is not kept confidential could have serious consequences.

In medical disease testing, the null hypothesis is usually the assumption that a person is healthy. The alternative is that he or she has the disease we're testing for. So a Type I error is a *false positive*; a healthy person is diagnosed with the disease. A Type II error, in which an infected person is diagnosed as disease free, is a *false negative*. These errors have other names, depending on the particular discipline and context.

Which type of error is more serious depends on the situation. In the jury trial, a Type I error occurs if the jury convicts an innocent person. A Type II error occurs if the jury fails to convict a guilty person. Which seems more serious? In medical diagnosis, a false negative could mean that a sick patient goes untreated. A false positive might mean that the person must undergo further tests. In a Statistics final exam (with H_0: the student has learned only 60% of the material), a Type I error would be passing a student who in fact learned less than 60% of the material, while a Type II error would be failing a student who knew enough to pass. Which of these errors seems more serious? It depends on the situation, the cost, and your point of view.

Here's an illustration of the situations:

A S **Hypothesis Tests Are Random.** Simulate hypothesis tests and watch Type I errors occur. When you conduct real hypothesis tests you'll never know, but simulation makes you omniscient so you can tell when you've made an error.

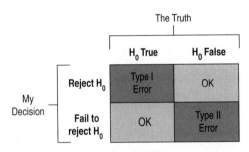

How often will a Type I error occur? It happens when the null hypothesis is true but we've had the bad luck to draw an unusual sample. To reject H_0, the P-value must fall below α. When H_0 is true, that happens *exactly* with probability α. So when you choose level α, you're setting the probability of a Type I error to α.

What if H_0 is not true? Then we can't possibly make a Type I error. You can't get a false positive from a sick person. A Type I error can happen only when H_0 is true.

When H_0 is false and we reject it, we have done the right thing. A test's ability to detect a false hypothesis is called the **power** of the test. In a jury trial, power is the ability of the criminal justice system to convict people who are guilty—a good thing! We'll have a lot more to say about power soon.

In Statistics, α is almost always saved for the alpha level. But β has already been used for the parameters of a linear model. Fortunately, it's usually clear whether we're talking about a Type II error probability or the slope or intercept of a regression model.

When H_0 is false but we fail to reject it, we have made a Type II error. We assign the letter β to the probability of this mistake. What's the value of β? That's harder to assess than α because we don't know what the value of the parameter really is. When H_0 is true, it specifies a single parameter value. But when H_0 is false, we don't have a specific one; we have many possible values. We can compute the probability β for any parameter value in H_A. But which one should we choose?

The null hypothesis specifies a single value for the parameter. So it's easy to calculate the probability of a Type I error. But the alternative gives a whole range of possible values, and we may want to find a β for several of them.

There is no single value for β. In fact, we can think of a whole collection of β's, one for each incorrect parameter value. Then we can display them as a curve, plotting each β against the corresponding parameter value in H_A. One way to focus our attention is by thinking about the *effect size*. That is, ask *"How big a difference would matter?"* Suppose a charity wants to test whether placing personalized address labels in the envelope along with a request for a donation increases the response rate above the baseline of 5%. If the minimum response that would pay for the address labels is 6%, they would calculate β for the alternative $p = 0.06$.

We have seen ways to find a sample size by specifying the margin of error. Choosing the sample size to achieve a specified β (for a particular alternative value) is sometimes more appropriate, but the calculation is more complex and lies beyond the scope of this book.

Of course, we could reduce β for *all* alternative parameter values by increasing α. By making it easier to reject the null, we'd be more likely to reject it whether it's true or not. So we'd reduce β, the chance that we fail to reject a false null—but we'd make more Type I errors. This tension between Type I and Type II errors is inevitable. In the political arena, think of the ongoing debate between those who favor provisions to reduce Type I errors in the courts (Miranda rights, warrants required for wire taps, legal representation paid by the state if needed, etc.) and those who are concerned that Type II errors have become too common (admitting into evidence confessions made when no lawyer is present, eavesdropping on conferences with lawyers, restricting paths of appeal, etc.).

The only way to reduce *both* types of error is to collect more evidence or, in statistical terms, to collect more data. Otherwise, we just wind up trading off one kind of error against the other. Whenever you design a survey or experiment, it's a good idea to calculate β (for a reasonable α level). Use a parameter value in the alternative that corresponds to an effect size that you want to be able to detect. Too often, studies fail because their sample sizes are too small to detect the change they are looking for.

Power

When we failed to reject the null hypothesis about TT practitioners, did we prove that they were just guessing? No, it could be that they actually *can* discern a human energy field but we just couldn't tell. For example, suppose they really have the ability to get 53% of the trials right but just happened to get only 47% in our experiment. Our confidence interval shows that with these data we wouldn't have rejected the null. And if we retained the null even though the true proportion was actually greater than 50%, we would have made a Type II error because we failed to detect their ability.

Remember, we can never prove a null hypothesis true. We can only fail to reject it. But when we fail to reject a null hypothesis, it's natural to wonder whether we looked hard enough. Might the null hypothesis actually be false and our test too weak to tell?

When the null hypothesis actually *is* false, we hope our test is strong enough to reject it. We'd like to know how likely we are to succeed. The power of the test gives us a way to think about that. The **power** of a test is the probability that it correctly rejects a false null hypothesis. When the power is high, we can be confident that we've looked hard enough. We know that β is the probability that a test *fails* to reject a false null hypothesis, so the power of the test is the complement, $1 - \beta$. We might have just written $1 - \beta$, but power is such an important concept that it gets its own name.

Whenever a study fails to reject its null hypothesis, the test's power comes into question. Was the sample size big enough to detect an effect, had there been one? Might we have missed an effect large enough to be interesting just because we failed to gather sufficient data or because there was too much variability in the data we could gather? The therapeutic touch experiment failed to reject the null hypothesis that the TT practitioners were just guessing. Might the problem be that the experiment simply lacked adequate power to detect their ability?

When we calculate power, we imagine that the null hypothesis is false. The value of the power depends on how far the truth lies from the null hypothesis

A S **The Power of a Test.**
Power is a concept that's much easier to understand when you can visualize what's happening. This interactive animation will help.

value. We call the distance between the null hypothesis value, p_0, and the truth, p, the **effect size.** The power depends directly on the effect size. It's easier to see larger effects, so the farther p_0 is from p, the greater the power. If the therapeutic touch practitioners were in fact able to detect human energy fields 90% of the time, it should be easy to see that they aren't guessing. With an effect size this large, we'd have a powerful test. If their true success rate were only 53%, however, we'd need a larger sample size to have a good chance of noticing that (and rejecting H_0).

How can we decide what power we need? Choice of power is more a financial or scientific decision than a statistical one because to calculate the power, we need to specify the "true" parameter value we're interested in. In other words, power is calculated for a particular effect size, and it changes depending on the size of the effect we want to detect. For example, do you think that health insurance companies should pay for therapeutic touch if practitioners could detect a human energy field only 53% of the time—just slightly better than chance? That doesn't seem clinically useful.[4] How about 75% of the time? No therapy works all the time, and insurers might be quite willing to pay for such a success rate. Let's take 75% as a reasonably interesting effect size (keeping in mind that 50% is the level of guessing). With 150 trials, the TT experiment would have been able to detect such an ability with a power of 99.99%. So power was not an issue in this study. There is only a very small chance that the study would have failed to detect a practitioner's ability, had it existed. The sample size was clearly big enough.

just checking

4. Remember our bank that's sending out video tapes to try to get customers to make payments on delinquent loans? It is looking for evidence that the costlier video tape strategy produces a higher success rate than the letters it has been sending. Explain what a Type I error is in this context, and what the consequences would be to the bank.

5. What's a Type II error in the bank experiment context, and what would the consequences be?

6. If the video tape strategy *really* works well, actually getting 60% of the people to pay off their balances after receiving the tape, would the power of the test be higher or lower compared to a 32% pay off rate? Explain briefly.

A Picture Worth $\dfrac{1}{P(z > 3.09)}$ Words

It makes intuitive sense that the larger the effect size, the easier it should be to see it. Obtaining a larger sample size decreases the probability of a Type II error, so it increases the power. It also makes sense that the more we're willing to accept a Type I error, the less likely we will be to make a Type II error.

---

[4] On the other hand, a scientist might be interested in anything clearly different from the 50% guessing rate because that might suggest an entirely new Physics at work. In fact, it could lead to a Nobel prize.

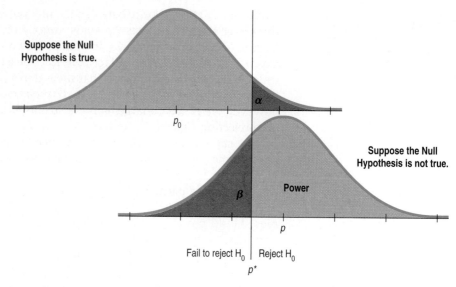

The power of a test is the probability that it rejects a false null hypothesis. The upper figure shows the null hypothesis model. We'd reject the null in a one-sided test if we observed a value of $\hat{p}$ in the red region to the right of the critical value, p^*. The lower figure shows the true model. If the true value of p is greater than p_0, then we're more likely to observe a value that exceeds the critical value, and make the correct decision to reject the null hypothesis. The power of the test is the blue region on the right of the lower figure. Of course, even drawing samples whose observed proportions are distributed around p, we'll sometimes get a value in the red region on the left and make a Type II error of failing to reject the null. **Figure 21.4**

Figure 21.4 shows a good way to visualize the relationships among these concepts. Suppose we are testing H_0: $p = p_0$ against the alternative H_A: $p > p_0$. We'll reject the null if the observed proportion, $\hat{p}$, is big enough. By big enough, we mean $\hat{p} > p^*$ for some critical value, p^* (shown as the red region in the right tail of the upper curve). For example, we might be willing to believe the ability of therapeutic touch practitioners if they were successful in 65% of our trials. This is what the upper model shows. It's a picture of the sampling distribution model for the proportion when the null hypothesis is true. If the null were true, then this would be a picture of that truth. We'd make a Type I error whenever the sample gave us $\hat{p} > p^*$ because we would reject the (true) null hypothesis. And unusual samples like that would happen only with probability α.

In reality, though, the null hypothesis is rarely *exactly* true. The lower probability model supposes that H_0 is not true. In particular, it supposes that the true value is p, not p_0. (Perhaps the TT practitioner really can detect the human energy field 72% of the time.) It shows a distribution of possible observed $\hat{p}$ values around this true value. Because of sampling variability, sometimes $\hat{p} < p^*$ and we fail to reject the (false) null hypothesis. Suppose a TT practitioner with a true ability level of 72% is actually successful on fewer than 65% of our tests. Then we'd make a Type II error. The area under the curve to the left of p^* in the bottom model represents how often this happens. The probability is β. In this picture, β is less than half, so most of the time we *do* make the right decision. The *power* of the test—the probability that we make the right decision—is shown as the region to the right of p^*. It's $1 - \beta$.

We calculate p^* based on the upper model because p^* depends only on the null model and the alpha level. No matter what the true proportion, no matter whether the practitioners can detect a human energy field 90%, 53%, or 2% of the time, p^* doesn't change. After all, we don't *know* the truth, so we can't use it to determine the critical value. But we always reject H_0 when $\hat{p} > p^*$.

How often we reject H_0 when it's *false* depends on the effect size. We can see from the picture that if the true proportion were farther from the hypothesized value, the bottom curve would shift to the right, making the power greater.

We can see several important relationships from this figure:

- Power = $1 - \beta$.
- Reducing α to lower Type I error will move the critical value, p^*, to the right (in this example), and have the effect of increasing β, the probability of a Type II error, and correspondingly reducing the power.
- The larger the real difference between the hypothesized value, p_0, and the true population value, p, the smaller the chance of making a Type II error and the greater the power of the test. If the two proportions are very far apart, the two models will barely overlap, and we would not be likely to make any Type II errors at all—but then, we are unlikely to really need a formal hypothesis testing procedure to see such an obvious difference. If the TT practitioners were successful almost all the time, we'd be able to see that with even a small experiment.

Reducing Both Type I and Type II Error

Figure 21.4 seems to show that if we reduce Type I error, we automatically must increase Type II error. But there is a way to reduce both. Can you think of it?

If we can make both curves narrower, as shown in Figure 21.5, then both the probability of Type I errors and the probability of Type II errors will decrease, and the power of the test will increase.

A S **Power and Sample Size.** Investigate how the power of a test changes with the sample size. The interactive tool is really the only way you can see this easily.

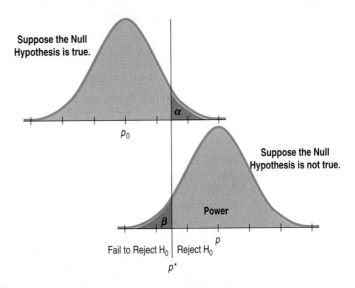

Making the standard deviations smaller increases the power without changing the alpha level or corresponding critical value. The means are just as far apart as in Figure 21.4, but the error rates are reduced   **Figure 21.5**

How can we accomplish that? The only way is to reduce the standard deviations by increasing the sample size. (Remember, these are pictures of sampling distribution models, not of data.) Increasing the sample size works regardless of the true population parameters. But recall the curse of diminishing returns. The standard deviation of the sampling distribution model decreases only as the *square root* of the sample size, so to halve the standard deviations we must *quadruple* the sample size.

What Can Go Wrong?

- ***Don't interpret the P-value as the probability that H₀ is true.*** The P-value is about the data, not the hypothesis. It's the probability of the data *given* that H_0 is true, not the other way around.
- ***Don't believe too strongly in arbitrary alpha levels.*** There's not really much difference between a P-value of 0.051 and a P-value of 0.049, but sometimes it's regarded as the difference between night (having to retain H_0) and day (being able to shout to the world that your results are "statistically significant"). It may just be better to report the P-value and a confidence interval and let the world decide along with you.
- ***Don't confuse practical and statistical significance.*** A large sample size can make it easy to discern even a trivial change from the null hypothesis value. On the other hand, an important difference can be missed if your test lacks sufficient power.
- ***Don't forget that in spite of all your care, you might make a wrong decision.*** We can never reduce the probability of a Type I error (α) or of a Type II error (β) to zero (but increasing the sample size helps).

CONNECTIONS

All of the hypothesis tests we'll see boil down to the same question: "Is the difference between two quantities large?" We always measure "how large" by finding a ratio of this difference to the standard deviation of the sampling distribution of the statistic. Using the standard deviation as our ruler for inference is one of the core ideas of statistical thinking.

We've discussed the close relationship between hypothesis tests and confidence intervals. They are two sides of the same coin.

This chapter also has natural links to the discussion of probability, to the Normal model, and to the two previous chapters on inference.

What have we learned?

We've learned that there's a lot more to hypothesis testing than a simple yes/no decision.

- We've learned that the P-value can indicate evidence against the null hypothesis when it's small, but does not tell us the probability that the null hypothesis is true.
- We've learned that the alpha level of the test establishes the level of proof we'll require. That determines the critical value of *z* that will lead us to reject the null hypothesis.
- We've also learned more about the connection between hypothesis tests and confidence intervals; they're really two ways of looking at the same question. The hypothesis test gives us the answer to a decision about a parameter; the confidence interval tells us the plausible values of that parameter.

We've learned about the two kinds of errors we might make and we've seen why in the end we're never sure we've made the right decision.

- If the null hypothesis is really true and we reject it, that's a Type I error; the alpha level of the test is the probability that this happens.
- If the null hypothesis is really false, but we fail to reject it, that's a Type II error.
- The power of the test is the probability that we reject the null hypothesis when it's false. The larger the size of the effect we're testing for, the greater the power of the test in detecting it.
- We've seen that tests with greater likelihood of Type I error have more power and less chance of a Type II error. We can increase power while reducing the chances of both kinds of error by increasing the sample size.

TERMS

Statistically significant When the P-value falls below the alpha level, we say that the test is "statistically significant" at that alpha level.

Alpha level The threshold P-value that determines when we reject a null hypothesis. If we observe a statistic whose P-value based on the null hypothesis is less than α, we reject that null hypothesis.

Significance level The alpha level is also called the significance level, most often in a phrase such as a conclusion that a particular test is "significant at the 5% significance level."

Critical value The value in the sampling distribution model of the statistic whose P-value is equal to the alpha level. Any statistic value farther from the null hypothesis value than the critical value will have a smaller P-value than α and will lead to rejecting the null hypothesis. The critical value is often denoted with an asterisk, as z^*, for example.

Type I error The error of rejecting a null hypothesis when in fact it is true (also called a "false positive"). The probability of a Type I error is α.

Type II error The error of failing to reject a null hypothesis when in fact it is false (also called a "false negative"). The probability of a Type II error is commonly denoted β, and depends on the effect size.

Power The probability that a hypothesis test will correctly reject a false null hypothesis is the power of the test. To find power, we must specify a particular alternative parameter value as the "true" value. For any specific value in the alternative, the power is $1 - \beta$.

Effect size The difference between the null hypothesis value and true value of a model parameter is called the effect size.

SKILLS

Think

When you complete this lesson you should:

- Understand that statistical significance does not measure the importance or magnitude of an effect. Recognize when others misinterpret statistical significance as proof of practical importance.
- Understand the close relationship between hypothesis tests and confidence intervals.
- Be able to identify and use the alternative hypothesis when testing hypotheses. Understand how to choose between a one-sided and two-sided alternative hypothesis, and know how to defend the choice of a one-sided alternative.
- Understand how the critical value for a test is related to the specified alpha level.
- Understand that the power of a test gives the probability that it correctly rejects a false null hypothesis when a specified alternative is true.

- Understand that the power of a test depends in part on the sample size. Large sample sizes lead to greater power (and thus fewer Type II errors).

Show

- Know how to complete a hypothesis test for a population proportion.

Tell

- Be able to interpret the meaning of a P-value in nontechnical language.
- Understand that the P-value of a test does not give the probability that the null hypothesis is correct.
- Know that we do not "accept" a null hypothesis if we cannot reject it, but rather that we can only "fail to reject" the hypothesis for lack of evidence against it.

Hypothesis Tests on the Computer

Reports about hypothesis tests generated by technologies don't follow a standard form. Most will name the test, provide the test statistic value, its standard deviation, and the P-value. But these elements may not be labeled clearly. For example, the expression "Prob $> |z|$" means the probability (the "Prob") of observing a test statistic whose magnitude (the absolute value tells us this) is larger than that of the one (the "z") found in the data (which, because it is written as "z," we know follows a Normal model). That is a fancy (and not very clear) way of saying P-value. In some packages you can specify that the test be one-sided. Others might report three P-values, covering the ground for both one-sided tests and the two-sided test.

Sometimes a confidence interval and hypothesis test are automatically given together. The CI ought to be for the corresponding confidence level: $1.0 - \alpha$.

Often, the standard deviation of the statistic is called the "standard error," and usually that's appropriate because we've had to estimate its value from the data. That's not the case for proportions, however. We get the standard deviation for a proportion from the null hypothesis value. Nevertheless, you may see the standard deviation called a "standard error" even for tests with proportions.

It's common for statistics packages and calculators to report more digits of "precision" than could possibly have been found from the data. You can safely ignore them. Round values such as the standard deviation to one digit more than the number of digits reported in your data.

Here are the kind of results you might see. This is not from any program or calculator we know of, but it shows some of the things you might see in typical computer output.

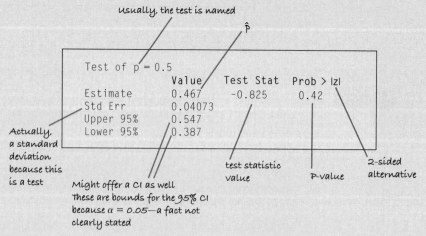

For information on hypothesis testing with particular statistics packages, see the table in Chapter 20 on pages 465–466.

EXERCISES

1. **P-value.** A medical researcher has tested a new treatment for poison ivy against the traditional ointment. With a P-value of 0.047, he concludes the new treatment is more effective. Explain what the P-value means in this context.

2. **Another P-value.** Have harsher penalties and ad campaigns increased seatbelt use among drivers and passengers? Observations of commuter traffic failed to find evidence of a significant change compared with three years ago. Explain what the study's P-value of 0.17 means in this context.

3. **Alpha.** A researcher developing scanners to search for hidden weapons at airports has concluded that a new device is significantly better than the current scanner. He made this decision based on a test using $\alpha = 0.05$. Would he have made the same decision at $\alpha = 0.10$? How about $\alpha = 0.01$? Explain.

4. **Alpha again.** Environmentalists concerned about the impact of high-frequency radio transmissions on birds found that there was no evidence of a higher mortality rate among hatchlings in nests near cell towers. They based this conclusion on a test using $\alpha = 0.05$. Would they have made the same decision at $\alpha = 0.10$? How about $\alpha = 0.01$? Explain.

5. **Significant?** Public health officials believe that 90% of children have been vaccinated against measles. A random survey of medical records at many schools across the country found that among more than 13,000 children only 89.4% had been vaccinated. A statistician would reject the 90% hypothesis with a P-value of $P = 0.011$.
   a) Explain what the P-value means in this context.
   b) The result is statistically significant, but is it important? Comment.

6. **Significant again?** A new reading program may reduce the number of elementary students who read below grade level. The company that developed this program supplied materials and teacher training for a large-scale test involving nearly 8500 children in several different school districts. Statistical analysis of the results showed that the percentage of students who did not attain the grade level standard was reduced from 15.9% to 15.1%. The hypothesis that the new reading program produced no improvement was rejected with a P-value of 0.023.
   a) Explain what the P-value means in this context.
   b) Even though this reading method has been shown to be significantly better, why might you not recommend that your local school adopt it?

7. **Success.** In August 2004, *Time* magazine reported the results of a random telephone poll commissioned by the Spike network. Of the 1302 men who responded, only 39 said that their most important measure of success was their work.
   a) Estimate the percentage of all American males who measure success primarily from their work. Use a 98% confidence interval. Don't forget to check the conditions first.
   b) Some believe that few contemporary men judge their success primarily by their work. Suppose we wished to conduct a hypothesis test to see if the fraction has fallen below the 5% mark. What does your confidence interval indicate? Explain.
   c) What is the level of significance of this test? Explain.

8. **Is the Euro fair?** Soon after the Euro was introduced as currency in Europe, it was widely reported that someone had spun a Euro coin 250 times and gotten heads 140 times. We wish to test a hypothesis about the fairness of spinning the coin.
   a) Estimate the true proportion of heads. Use a 95% confidence interval. Don't forget to check the conditions first.
   b) Does your confidence interval provide evidence that the coin is unfair when spun? Explain.
   c) What is the significance level of this test? Explain.

9. **Approval.** In July 2004, George W. Bush's approval rating stood at 49% according to a CNN/*USA Today*/Gallup poll of 1000 randomly selected adults.
   a) Make a 95% confidence interval for his approval rating by all U.S. adults.
   b) Based on the confidence interval, test the null hypothesis that half of the country approved of the way he was handling his job at that time.
   c) How might Democrats and Republicans try to spin the confidence interval differently?

10. **Superdads.** The Spike network commissioned a telephone poll of randomly sampled U.S. men. Of the 712 respondents who had children, 22% said "yes" to the question "Are you a stay-at-home dad?" [*Time*, August 23, 2004]
   a) To help them market commercial time, Spike wants an accurate estimate of the true percentage of stay-at-home dads. Construct a 95% confidence interval.
   b) An advertiser of baby-carrying slings for dads will buy commercial time if at least 25% of men are stay-at-home dads. Use your confidence interval to test an appropriate hypothesis and make a recommendation to the advertiser.
   c) Could Spike claim to the advertiser that it is possible that 25% of men with young children are stay-at-home dads? What is wrong with the reasoning?

11. Homeowners. In 2003 the Department of Commerce reported that 68.3% of American families owned their homes. In one small city, census data reveal that the ownership rate is much lower. The City Council is debating a plan to offer tax breaks to first-time home buyers in order to encourage people to become homeowners. They decide to adopt the plan on a 2-year trial basis and use the data they collect to make a decision about continuing the tax breaks. Since this plan costs the city tax revenues, they will continue to use it only if there is strong evidence that the rate of home ownership is increasing.
a) In words, what will their hypotheses be?
b) What would a Type I error be?
c) What would a Type II error be?
d) For each type of error, tell who would be harmed.
e) What would the power of the test represent in this context?

12. Alzheimer's. Testing for Alzheimer's disease can be a long and expensive process, consisting of lengthy tests and medical diagnosis. Recently, a group of researchers (Solomon *et al.*, 1998) devised a 7-minute test to serve as a quick screen for the disease for use in the general population of senior citizens. A patient who tested positive would then go through the more expensive battery of tests and medical diagnosis. The authors reported a false positive rate of 4% and a false negative rate of 8%.
a) Put this in the context of a hypothesis test. What are the null and alternative hypotheses?
b) What would a Type I error mean?
c) What would a Type II error mean?
d) Which is worse here, a Type I or Type II error? Explain.
e) What is the power of this test?

13. Testing cars. A clean air standard requires that vehicle exhaust emissions not exceed specified limits for various pollutants. Many states require that cars be tested annually to be sure they meet these standards. Suppose state regulators double check a random sample of cars that a suspect repair shop has certified as okay. They will revoke the shop's license if they find significant evidence that the shop is certifying vehicles that do not meet standards.
a) In this context, what is a Type I error?
b) In this context, what is a Type II error?
c) Which type of error would the shop's owner consider more serious?
d) Which type of error might environmentalists consider more serious?

14. Quality control. Production managers on an assembly line must monitor the output to be sure that the level of defective products remains small. They periodically inspect a random sample of the items produced. If they find a significant increase in the proportion of items that must be rejected, they will halt the assembly process until the problem can be identified and repaired.

a) In this context, what is a Type I error?
b) In this context, what is a Type II error?
c) Which type of error would the factory owner consider more serious?
d) Which type of error might customers consider more serious?

15. Cars again. As in Exercise 13, state regulators are checking up on repair shops to see if they are certifying vehicles that do not meet pollution standards.
a) In this context, what is meant by the power of the test the regulators are conducting?
b) Will the power be greater if they test 20 or 40 cars? Why?
c) Will the power be greater if they use a 5% or a 10% level of significance? Why?
d) Will the power be greater if the repair shop's inspectors are only a little out of compliance or a lot? Why?

16. Production. Consider again the task of the quality control inspectors in Exercise 14.
a) In this context, what is meant by the power of the test the inspectors conduct?
b) They are currently testing 5 items each hour. Someone has proposed they test 10 each hour instead. What are the advantages and disadvantages of such a change?
c) Their test currently uses a 5% level of significance. What are the advantages and disadvantages of changing to an alpha level of 1%?
d) Suppose that as a day passes one of the machines on the assembly line produces more and more items that are defective. How will this affect the power of the test?

17. Equal opportunity? A company is sued for job discrimination because only 19% of the newly hired candidates were minorities when 27% of all applicants were minorities. Is this strong evidence that the company's hiring practices are discriminatory?
a) Is this a one-tailed or a two-tailed test? Why?
b) In this context, what would a Type I error be?
c) In this context, what would a Type II error be?
d) In this context, describe what is meant by the power of the test.
e) If the hypothesis is tested at the 5% level of significance instead of 1%, how will this affect the power of the test?
f) The lawsuit is based on the hiring of 37 employees. Is the power of the test higher than, lower than, or the same as it would be if it were based on 87 hires?

18. Stop signs. Highway safety engineers test new road signs, hoping that increased reflectivity will make them more visible to drivers. Volunteers drive through a test course with several of the new and old style signs and rate which kind shows up the best.
a) Is this a one-tailed or a two-tailed test? Why?
b) In this context, what would a Type I error be?

c) In this context, what would a Type II error be?

d) In this context, describe what is meant by the power of the test.

e) If the hypothesis is tested at the 1% level of significance instead of 5%, how will this affect the power of the test?

f) The engineers hoped to base their decision on the reactions of 50 drivers, but time and budget constraints may force them to cut back to 20. How would this affect the power of the test? Explain.

19. Dropouts. A Statistics professor has observed that for several years about 13% of the students who initially enroll in his Introductory Statistics course withdraw before the end of the semester. A salesman suggests that he try a statistics software package that gets students more involved with computers, predicting that it will cut the dropout rate. The software is expensive, and the salesman offers to let the professor use it for a semester to see if the dropout rate goes down significantly. The professor will have to pay for the software only if he chooses to continue using it.

a) Is this a one-tailed or two-tailed test? Explain.

b) Write the null and alternative hypotheses.

c) In this context, explain what would happen if the professor makes a Type I error.

d) In this context, explain what would happen if the professor makes a Type II error.

e) What is meant by the power of this test?

20. Ads. A company is willing to renew its advertising contract with a local radio station only if the station can prove that more than 20% of the residents of the city have heard the ad and recognize the company's product. The radio station conducts a random phone survey of 400 people.

a) What are the hypotheses?

b) The station plans to conduct this test using a 10% level of significance, but the company wants the significance level lowered to 5%. Why?

c) What is meant by the power of this test?

d) For which level of significance will the power of this test be higher? Why?

e) They finally agree to use $\alpha = 0.05$, but the company proposes that the station call 600 people instead of the 400 initially proposed. Will that make the risk of Type II error higher or lower? Explain.

21. Dropouts, part II. Initially, 203 students signed up for the Stats course in Exercise 19. They used the software suggested by the salesman, and only 11 dropped out of the course.

a) Should the professor spend the money for this software? Support your recommendation with an appropriate test.

b) Explain carefully what your P-value means in this context.

22. Testing the ads. The company in Exercise 20 contacts 600 people selected at random, and only 133 remember the ad.

a) Should the company renew the contract? Support your recommendation with an appropriate test.

b) Explain carefully what your P-value means in this context.

23. Hoops. A basketball player with a poor foul-shot record practices intensively during the off-season. He tells the coach that he has raised his proficiency from 60% to 80%. Dubious, the coach asks him to take 10 shots, and is surprised when the player hits 9 out of 10. Did the player prove that he has improved?

a) Suppose the player really is no better than before— still a 60% shooter. What's the probability he could hit at least 9 of 10 shots anyway? (*Hint:* Use a Binomial model.)

b) If that is what happened, now the coach thinks the player has improved, when he has not. Which type of error is that?

c) If the player really can hit 80% now, and it takes at least 9 out of 10 successful shots to convince the coach, what's the power of the test?

d) List two ways the coach and player could increase the power to detect any improvement.

24. Pottery. An artist experimenting with clay to create pottery with a special texture has been experiencing difficulty with these special pieces. About 40% break in the kiln during firing. Hoping to solve this problem, she buys some more expensive clay from another supplier. She plans to make and fire 10 pieces, and will decide to use the new clay if at most one of them breaks.

a) Suppose the new expensive clay really is no better than her usual clay. What's the probability that this test convinces her to use it anyway? (*Hint:* Use a Binomial model.)

b) If she decides to switch to the new clay and it is no better, what kind of error did she commit?

c) If the new clay really could reduce breakage to only 20%, what's the probability that her test will not detect the improvement?

d) How can she improve the power of her test? Offer at least two suggestions.

25. Survey. A company has surveyed a stratified sample of its employees to find out how many might take advantage of a program to help people stop smoking. You may assume that the sampling strategy was properly random and that the data-gathering methodology avoided biases. The table on the next page shows the results. For which of the groups could we use our methods of inference to determine a 95% confidence interval for the proportion of employees who would participate in the stop smoking program?

Group	Number of Employees	Number Surveyed	Percent to Participate
Laborers	6235	300	9%
Clerical	1520	200	7%
Management	342	25	8%

26. Fire safety. A city law requires all buildings to have a fire safety inspection at least once every three years. Concerned that some of these inspections are not being done, the mayor orders a survey to see what fraction of the city's buildings may be out of compliance. Results for the sample appear in the table below. For which of the building classifications could you use our methods of inference to create 95% confidence intervals for the proportion of buildings lacking the required inspection?

Class	Number of Buildings	Number Surveyed	Percent Not Inspected
Single-family home	7742	200	7%
Apartment building	205	20	10%
Commercial	407	70	16%

27. Little League. In a 1999–2000 longitudinal study of youth baseball, researchers found that 26% of 298 young pitchers complained of elbow pain after pitching.
   a) Create a 90% confidence interval for the percentage of young players who may develop elbow pain after pitching.
   b) A coach claims that only about 1 player in 5 is at risk of arm injury from pitching. Is this claim consistent with your confidence interval?

28. News sources. In May of 2000 the Pew Research Foundation sampled 1593 respondents and asked how they obtain news. The foundation reports that 33% now say they obtain news from the Internet at least once a week. Pew reports a margin of error of ±3% for this result. It had generally been assumed, based on earlier polls, that only 25% got news from the Internet. Does the Pew result provide strong evidence that the percentage has increased? Use the confidence interval to test an appropriate hypothesis, and state your conclusion.

just checking

Answers

1. With a z-score of 0.62, you can't reject the null hypothesis. The experiment shows no evidence that the wheel is not fair.

2. At $\alpha = 0.05$, you can't reject the null hypothesis because 0.30 is contained in the 90% confidence interval—it's plausible that sending the video tapes is no more effective than just sending letters.

3. The confidence interval is from 29% to 45%. The video tape strategy is more expensive, and may not be worth it. We can't distinguish the success rate from 30% given the results of this experiment, but 45% would represent a large improvement. The bank should consider another trial, increasing their sample size to get a narrower confidence interval.

4. A Type I error would mean deciding that the video tape success rate is higher than 30%, when it really isn't. They would adopt a more expensive method for collecting payments that's no better than the less expensive strategy.

5. A Type II error would mean deciding that there's not enough evidence to say the video tape strategy works when in fact it does. The bank would fail to discover an effective method for increasing their revenue from delinquent accounts.

6. Higher; the larger the effect size, the greater the power. It's easier to detect an improvement to a 60% success rate than to a 32% rate.

CHAPTER 22

Comparing Two Proportions

WHO	1026 U.S. adults
WHAT	Opinions on traits
WHY	Polling by Gallup Poll
WHEN	2001
WHERE	United States

Who do you think are more intelligent, men or women? To find out what people think, the Gallup Poll selected a random sample of 520 women and 506 men. The pollsters showed them a list of personal attributes and asked them to indicate whether each attribute was "generally more true of men or of women." When asked about intelligence, 28% of the men thought men were generally more intelligent, but only 14% of the women agreed. Is there a gender gap in opinions about which sex is smarter? This is only a random sample. What would we estimate the true size of that gap to be?

Comparisons between two percentages are much more common than questions about isolated percentages. And they are more interesting. We often want to know how two groups differ, whether a treatment is better than a placebo control, or whether this year's results are better than last year's.

Another Ruler

We know the *difference* between the proportions of our two random samples. It's 14%. That sounds like a lot, but is it the sort of difference that might just be due to random sampling? To decide about that, we need a new ruler—the standard deviation of the sampling distribution model for the difference in the proportions. Now we have two proportions, and each will vary from sample to sample. We are interested in the difference between them. So what is the correct standard deviation?

The answer is simple, but not obvious. (You may have seen it in Chapter 16.) It makes sense that the difference between two random quantities should vary more than either one of them. After all, there are two varying quantities. But it isn't obvious how much more variation to expect. It turns out that we add their *variances* (not the standard deviations) and take a square root to get back to natural units:

The variance of the sum or difference of two independent random variables is the sum of their variances.

493

For independent random variables, **variances add.**

A S **Compare Two Proportions.** Does a preschool program help disadvantaged children later in life? Test whether the proportions of success are different using this intereactive tool.

This is such an important (and powerful) idea in Statistics that it's worth pausing a moment to think about it. Here's some intuition about why variation increases even when we subtract two random quantities.

Grab a full box of cereal. The box claims to contain 16 ounces of cereal. We know that's not exact: There's some small variation from box to box. Now pour a bowl of cereal. Of course, your 2-ounce serving will not be exactly 2 ounces. There'll be some variation there, too. How much cereal would you guess was left in the box? Do you think your guess will be as close as your guess for the full box? *After* you pour your bowl, the amount of cereal in the box is still a random quantity (with a smaller mean than before), but it is even *more variable* because of the additional variation in the amount you poured.

According to our rule, the variance of the amount of cereal left in the box would now be the *sum* of the two *variances.*

We want a standard deviation, not a variance, but that's just a square root away. We can write symbolically what we've just said:

$$Var(X - Y) = Var(X) + Var(Y), \text{ so}$$
$$SD(X - Y) = \sqrt{SD^2(X) + SD^2(Y)} = \sqrt{Var(X) + Var(Y)}.$$

That formula may look sort of familiar. Think of a right triangle. Remember the Pythagorean Theorem?[1] It says that the length of the longest side of a right triangle is the square root of the sum of the squares of the other two sides. Standard deviations add in exactly the same way. (You can think of this as the Pythagorean Theorem of Statistics.)

Just as the Pythagorean Theorem works only for right triangles, our formula works only for independent random variables. Always check for independence before using it. Be careful, though—this simple formula applies only when X and Y are independent.

The Standard Deviation of the Difference Between Two Proportions

Fortunately, proportions observed in independent random samples *are* independent, so we can put the two proportions in for X and Y and add their variances. We just need to use careful notation to keep things straight.

When we have two samples, each can have a different size and proportion value, so we keep them straight with subscripts. Often we choose subscripts that remind us of the groups. For our example, we might use "$_M$" and "$_F$", but generically we'll just use "$_1$" and "$_2$". We will represent the two sample proportions as $\hat{p}_1$ and $\hat{p}_2$, and the two sample sizes as n_1 and n_2.

Combining independent random quantities always *increases* the overall variation, so even for *differences* of independent random variables, **variances add.**

The standard deviations of the sample proportions are $SD(\hat{p}_1) = \sqrt{\dfrac{p_1 q_1}{n_1}}$ and $SD(\hat{p}_2) = \sqrt{\dfrac{p_2 q_2}{n_2}}$, so the variance of the difference in the proportions is

$$Var(\hat{p}_1 - \hat{p}_2) = \left(\sqrt{\dfrac{p_1 q_1}{n_1}}\right)^2 + \left(\sqrt{\dfrac{p_2 q_2}{n_2}}\right)^2 = \dfrac{p_1 q_1}{n_1} + \dfrac{p_2 q_2}{n_2}.$$

---

[1] If you don't remember, don't rely on the Scarecrow from *The Wizard of Oz*. He may have a brain, and his Th. D., but he gets the formula wrong.

The standard deviation is the square root of that variance:

$$SD(\hat{p}_1 - \hat{p}_2) = \sqrt{\frac{p_1 q_1}{n_1} + \frac{p_2 q_2}{n_2}}.$$

We usually don't know the true values of p_1 and p_2. When we have the sample proportions in hand from the data, we use them to estimate the variances. So the standard error is

$$SE(\hat{p}_1 - \hat{p}_2) = \sqrt{\frac{\hat{p}_1 \hat{q}_1}{n_1} + \frac{\hat{p}_2 \hat{q}_2}{n_2}}.$$

Before we look at our example, we need to check assumptions and conditions.

1 A June 2004 public opinion poll asked 1000 randomly selected adults whether the United States should decrease the amount of immigration allowed; 49% of those responding said "yes." In June of 1995, a random sample of 1000 had found that 65% of adults thought immigration should be curtailed. To see if that percentage has decreased, why can't we just use a one-proportion z-test of H_0: $p = 0.65$ and see what the P-value of $\hat{p} = 0.49$ is?

2 For opinion polls like this, which has more variability—the percentage of respondents answering "yes" in either year, or the difference in the percentages between the two years?

Assumptions and Conditions

Independence Assumptions

Within each group the data should be based on results for independent individuals. We can't check that for certain, but we *can* check the following:

Randomization Condition: The data in each group should be drawn independently and at random from a homogeneous population or generated by a randomized comparative experiment.

The 10% Condition: If the data are sampled without replacement, the sample should not exceed 10% of the population.

Because we are comparing two groups in this way, we need an additional Independence Assumption. In fact, this is the most important of these assumptions. If it is violated, these methods just won't work. We check it with the Independent Groups Condition.

Independent Groups Assumption: The two groups we're comparing must also be independent *of each other*. Usually, the independence of the groups from each other is evident from the way the data were collected.

Why is the Independent Groups Assumption so important? If we compare husbands with their wives, or a group of subjects before and after some treatment, we can't just add the variances. Subjects' performance before a treatment might very well be related to their performance after the treatment. So the proportions are not independent and the Pythagorean-style variance formula does not hold. We'll see a way to compare a common kind of nonindependent samples in a later chapter.

Sample Size Condition

Each of the groups must be big enough. As with individual proportions, we need larger groups to estimate proportions that are near 0% or 100%. We usually check the Success/Failure Condition for each group.

Success/Failure Condition: Both groups are big enough that at least 10 successes and at least 10 failures have been observed in each.

The Sampling Distribution

We're almost there. We just need one more fact about proportions. We know already that for large enough samples, each of our proportions has an approximately Normal sampling distribution. The same is true of their difference.

The sampling distribution model for a difference between two independent proportions

Provided that the sampled values are independent, the samples are independent, and the sample sizes are large enough, the sampling distribution of $\hat{p}_1 - \hat{p}_2$ is modeled by a Normal model with mean $\mu = p_1 - p_2$ and standard deviation

$$SD(\hat{p}_1 - \hat{p}_2) = \sqrt{\frac{p_1 q_1}{n_1} + \frac{p_2 q_2}{n_2}}.$$

The sampling distribution model and the standard deviation give us all we need to find a margin of error for the difference in proportions—or at least they would if we knew the true proportions, p_1 and p_2. However, we don't know the true values, so we'll work with the observed proportions, $\hat{p}_1$ and $\hat{p}_2$, and use $SE(\hat{p}_1 - \hat{p}_2)$ to estimate the standard deviation. The rest is just like a one-proportion z-interval.

A two-proportion z-interval

When the conditions are met, we are ready to find the confidence interval for the difference of two proportions, $p_1 - p_2$. The confidence interval is

$$(\hat{p}_1 - \hat{p}_2) \pm z^* \times SE(\hat{p}_1 - \hat{p}_2)$$

where we find the standard error of the difference, $SE(\hat{p}_1 - \hat{p}_2) = \sqrt{\frac{\hat{p}_1 \hat{q}_1}{n_1} + \frac{\hat{p}_2 \hat{q}_2}{n_2}}$, from the observed proportions.

The critical value z^* depends on the particular confidence level, C, that you specify.

A Two-Proportion z-Interval Step-By-Step

Now we are ready to answer the question of how big a gap there is between the sexes in their opinions of whether men are more intelligent. To estimate the size of the gap we need a confidence interval, so let's follow the four confidence interval steps. The method is called the **two-proportion z-interval.**

Think

Plan   State what you want to know. Discuss the variables and the W's.

Identify the parameter you wish to estimate. (It usually doesn't matter in which direction we subtract, so for convenience we usually choose the direction with a positive difference.)

Choose and state a confidence level.

Model   Check the conditions.

The Success/Failure Condition must hold for each group.

State the sampling distribution model for the statistic.

Choose your method.

Show

Mechanics   Construct the confidence interval.

As often happens, the key step in finding the confidence interval is estimating the standard deviation of the sampling distribution model of the statistic. Here the statistic is the difference in the proportions of men and women who think that men are intelligent. Substitute the data values into the formula.

I want to know the true difference in the population proportion, p_M, of American men who think that men are more intelligent and the proportion, p_F, of American women who think so. The data are from a random sample of 1026 U.S. adults taken by the Gallup Poll in 2001. Pollsters asked respondents whether they thought that intelligence (among other attributes) was generally more true of men or women. The parameter of interest is the difference, $p_M - p_F$.

I will find a 95% confidence interval for this parameter.

✔ **Randomization Condition:** Gallup drew a random sample of U.S. adults.

✔ **10% Condition:** Although sampling was necessarily without replacement, there are many more U.S. women and U.S. men than were sampled in each of the groups polled.

✔ **Independent Groups Assumption:** The sample of women and the sample of men are independent of each other.

✔ **Success/Failure Condition:**

$n\hat{p}_M = 506 \times 28.0\% = 142 \geq 10$

$n\hat{q}_M = 506 \times 72.0\% = 364 \geq 10$

$n\hat{p}_F = 520 \times 14.0\% = 73 \geq 10$

$n\hat{q}_F = 520 \times 86.0\% = 447 \geq 10$

Both samples exceed the minimum size.

Under these conditions, the sampling distribution of the difference between the sample proportions is approximately Normal, so I'll find a **two-proportion z-interval.**

I know:
$$n_M = 506, n_F = 520.$$

The observed sample proportions are
$$\hat{p}_M = 0.28, \hat{p}_F = 0.14$$

Estimate $SD(\hat{p}_M - \hat{p}_F)$ as

$$SE(\hat{p}_M - \hat{p}_F) = \sqrt{\frac{\hat{p}_M\hat{q}_M}{n_M} + \frac{\hat{p}_F\hat{q}_F}{n_F}}$$

$$= \sqrt{\frac{0.28(1 - 0.28)}{506} + \frac{0.14(1 - 0.14)}{520}}$$

$$= 0.025$$

The sampling distribution is Normal, so the critical value for a 95% confidence interval, z^*, is 1.96. The margin of error is the critical value times the SE.

$$ME = z^* \times SE(\hat{p}_M - \hat{p}_F)$$
$$= 1.96(0.025) = 0.049$$

The confidence interval is the statistic ± ME.

The observed difference in proportions is $(\hat{p}_M - \hat{p}_F) = 0.28 - 0.14 = 0.14$, so the 95% confidence interval is

$$0.14 \pm 0.049,$$
$$\text{or 9\% to 19\%.}$$

Conclusion Interpret your confidence interval in the proper context. (Remember: We're 95% confident that our interval captured the true difference.)

I am 95% confident that the proportion of American men who think that the attribute "intelligent" applies more to men than to women is between 9% and 19% more than the proportion of American women who think that.

Will I Snore When I'm 64?

WHO	Randomly selected U.S. adults over age 18
WHAT	Proportion who snore, categorized by age (less than 30, 30 or older)
WHEN	2001
WHERE	United States
WHY	To study sleep behaviors of U.S. adults

The National Sleep Foundation asked a random sample of 1010 U.S. adults questions about their sleep habits. The sample was selected in the fall of 2001 from random telephone numbers, stratified by region and sex, guaranteeing that an equal number of men and women were interviewed (2002 Sleep in America Poll, National Sleep Foundation, Washington, D.C.).

One of the questions asked about snoring. Of the 995 respondents, 37% of adults reported that they snored at least a few nights a week during the past year. Would you expect that percentage to be the same for all age groups? Split into two age categories, 26% of the 184 people under 30 snored, compared with 39% of the 811 in the older group. Is this difference of 13% real, or due only to natural fluctuations in the sample we've chosen?

The question calls for a hypothesis test. Now the parameter of interest is the true *difference* between the (reported) snoring rates of the two age groups.

What's the appropriate null hypothesis? That's easy here. We hypothesize that there is no difference in the proportions. This is such a natural null hypothesis that we rarely consider any other. But instead of writing $H_0: p_1 = p_2$, we usually express it in a slightly different way. To make it relate directly to the *difference*, we hypothesize that the difference in proportions is zero:

$$H_0: p_1 - p_2 = 0.$$

Everyone into the Pool

Our hypothesis is about a new parameter—the *difference* in proportions. We'll need a standard error for that. Wait—don't we know that already? Yes and no. We know that the standard error of the difference in proportions is

$$SE(\hat{p}_1 - \hat{p}_2) = \sqrt{\frac{\hat{p}_1\hat{q}_1}{n_1} + \frac{\hat{p}_2\hat{q}_2}{n_2}}$$

and we could just plug in the numbers, but we can do even better. The secret is that proportions and their standard deviations are linked. There are two proportions in the standard error formula—but look at the null hypothesis. It says that these proportions are equal. To do a hypothesis test, we *assume* that the null hypothesis is true. So there should be just a single value of $\hat{p}$ in the SE formula (and, of course, $\hat{q}$ is just $1 - \hat{p}$).

How would we do this for the snoring example? If the null hypothesis is true, then among all adults the two groups have the same proportion. Overall, we saw $48 + 318 = 366$ snorers out of a total of $184 + 811 = 995$ adults who responded to this question. The overall proportion of snorers was $366/995 = 0.3678$.

Combining the counts like this to get an overall proportion is called **pooling.** (Whenever we have data from different sources or different groups but we believe that they really came from the same underlying population, we pool them to get better estimates.)

When we have counts for each group, we can find the pooled proportion as

$$\hat{p}_{\text{pooled}} = \frac{Success_1 + Success_2}{n_1 + n_2}$$

where $Success_1$ is the number of successes in group 1 and $Success_2$ is the number of successes in group 2. That's the overall proportion of success.

When we have only proportions and not the counts, as in the snoring example, we have to reconstruct the number of successes by multiplying the sample sizes by the proportions:

$$Success_1 = n_1\hat{p}_1 \text{ and } Success_2 = n_2\hat{p}_2.$$

If these calculations don't come out to whole numbers, round them first. There must have been a whole number of successes, after all. (This is the *only* time you should round values in the middle of a calculation.)

We then put this pooled value into the formula, substituting it for *both* sample proportions in the standard error formula:

$$SE_{\text{pooled}}(\hat{p}_1 - \hat{p}_2) = \sqrt{\frac{\hat{p}_{\text{pooled}}\,\hat{q}_{\text{pooled}}}{n_1} + \frac{\hat{p}_{\text{pooled}}\,\hat{q}_{\text{pooled}}}{n_2}}$$

$$= \sqrt{\frac{0.3678 \times (1 - 0.3678)}{184} + \frac{0.3678 \times (1 - 0.3678)}{811}}$$

> When finding the number of successes, round the values to integers. For example, the 48 snorers among the 184 under-30 respondents are actually 26.1% of 184. We round back to the nearest whole number to find the count that could have yielded the rounded percent we were given.

which comes out to 0.039375.

Compared to What?

Naturally, we'll reject our null hypothesis if we see a large enough difference in the two proportions. How can we decide whether the difference we see, $\hat{p}_1 - \hat{p}_2$, is large? The answer is the same as always: We just compare it with its standard deviation.

Unlike previous hypothesis testing situations, the null hypothesis doesn't provide a standard deviation, so we'll use a standard error (here, pooled). Since the sampling distribution is Normal, we can divide the observed difference by its standard error to get a z-score. The z-score will tell us how many standard errors the observed difference is away from 0. We can then use the 68-95-99.7 Rule to decide whether this is large, or some technology to get an exact P-value. The result is a **two-proportion z-test.**

Two-proportion z-test

The conditions for the two-proportion z-test are the same as for the two-proportion z-interval. We are testing the hypothesis

$$H_0: p_1 - p_2 = 0.$$

Because we hypothesize that the proportions are equal, we pool the groups to find

$$\hat{p}_{pooled} = \frac{Success_1 + Success_2}{n_1 + n_2}$$

and use that pooled value to estimate the standard error.

$$SE_{pooled}(\hat{p}_1 - \hat{p}_2) = \sqrt{\frac{\hat{p}_{pooled}\,\hat{q}_{pooled}}{n_1} + \frac{\hat{p}_{pooled}\,\hat{q}_{pooled}}{n_2}}$$

Now we find the test statistic using the statistic,

$$z = \frac{\hat{p}_1 - \hat{p}_2}{SE_{pooled}(\hat{p}_1 - \hat{p}_2)}.$$

When the conditions are met and the null hypothesis is true, this statistic follows the standard Normal model, so we can use that model to obtain a P-value.

A S **Test for a Difference Between Two Proportions.** Is premium-brand chicken less likely to be contaminated than store-brand chicken? Use the interactive tool to perform a two-proportion z-test. And you can play with the tool to see how much difference pooling makes.

A Two-Proportion z-Test Step-By-Step

Let's look at the snoring rates of the two groups.

Think

Plan State what you want to know. Discuss the variables and the W's.

I want to know whether snoring rates differ for those under and over 30 years old. The data are from a random sample of 1010 U.S. adults surveyed in the 2002 Sleep in America Poll. Of these, 995 responded to the question about snoring, indicating whether or not they had snored at least a few nights a week in the past year.

Hypotheses The study simply broke down the responses by age, so there is no sense that either alternative was of interest. A two-sided alternative hypothesis is appropriate.

H_0: There is *no difference* in snoring rates in the two age groups:

$$p_{old} - p_{young} = 0.$$

H_A: The rates are different: $p_{old} - p_{young} \neq 0.$

Model Check the conditions.

✔ **Randomization Condition:** The patients were randomly selected by telephone number and stratified by sex and region.

✔ **10% Condition:** The number of adults surveyed is certainly far less than 10% of the population.

✔ **Independent Groups Assumption:** *The two groups are independent of each other because the sample was selected at random.*

✔ **Success/Failure Condition:**[2] *In the younger age group, 48 snored and 136 didn't. In the older group, 318 snored and 493 didn't. The observed number of both successes and failures is much more than 10 for both groups.*[3]

State the null model.

Choose your method.

*Because the conditions are satisfied, I'll use a Normal model and perform a **two-proportion z-test.***

Show   **Mechanics**

$$n_{young} = 184, n_{old} = 811$$

$$\hat{p}_{young} = 0.261, \hat{p}_{old} = 0.392$$

The hypothesis is that the proportions are equal, so pool the sample data.

Use the pooled SE to estimate $SD(p_{old} - p_{young})$.

$$\hat{p}_{pooled} = \frac{y_{old} + y_{young}}{n_{old} + n_{young}} = \frac{318 + 48}{811 + 184} = 0.3678$$

$$SE_{pooled}(\hat{p}_{old} - \hat{p}_{young})$$

$$= \sqrt{\frac{\hat{p}_{pooled}\,\hat{q}_{pooled}}{n_{old}} + \frac{\hat{p}_{pooled}\,\hat{q}_{pooled}}{n_{young}}}$$

$$= \sqrt{\frac{(0.3678)(0.6322)}{811} + \frac{(0.3678)(0.6322)}{184}}$$

$$\approx 0.039375$$

The observed difference in sample proportions is:
$$\hat{p}_{old} - \hat{p}_{young} = 0.392 - 0.261 = 0.131.$$

Make a picture. Sketch a Normal model centered at the hypothesized difference of 0. Shade the region to the right of the observed difference, and because this is a two-tailed test, also shade the corresponding region in the other tail.

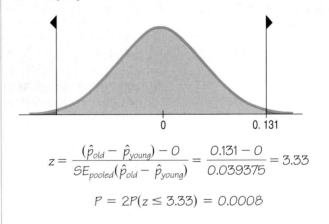

Find the *z*-score for the observed difference in proportions, 0.131.

$$z = \frac{(\hat{p}_{old} - \hat{p}_{young}) - 0}{SE_{pooled}(\hat{p}_{old} - \hat{p}_{young})} = \frac{0.131 - 0}{0.039375} = 3.33$$

Find the P-value using Table Z or technology. Because this is a two-tailed test, we must *double* the probability we find in the upper tail.

$$P = 2P(z \le 3.33) = 0.0008$$

---

[2] Technically, this is the first time we've seen that the observed counts of successes and failures can be different from those we'd expect. We expect $n\hat{p}_{pooled}$ but observe $n\hat{p}$. Checking the observed counts is easy. But if your data just barely miss (say, you see 9 successes), check $n\hat{p}_{pooled}$ as well. If it passes the test, you're OK.

[3] This is one of those situations in which the traditional term "success" seems a bit weird. A success here could be that a person snores. "Success" and "failure" are arbitrary labels left over from studies of gambling games.

Tell

Conclusion

Link the P-value to your decision about the null hypothesis, and state your conclusion in context.

The P-value of 0.0008 says that if there really were no difference in (reported) snoring rates between the two age groups, then the difference observed in this study would happen only 8 times in 10,000. This is so small that I reject the null hypothesis of no difference and conclude that there is a difference in the rate of snoring between older adults and younger adults. It appears that older adults are more likely to snore.

A public broadcasting station plans to launch a special appeal for additional contributions from current members. Unsure of the most effective way to contact people, they run an experiment. They randomly select two groups of current members. They send the same request for donations to everyone, but it goes to one group by e-mail and to the other group by regular mail. The station was successful in getting contributions from 26% of the members they e-mailed, but only from 15% of those who received the request by regular mail. A 90% confidence interval estimated the difference in donation rates to be 11% ± 7%.

③ Based on this confidence interval, what conclusion would we reach if we tested the hypothesis that there's no difference in the response rates to the two methods of fund raising? Explain.

④ Interpret the confidence interval in this context.

What Can Go Wrong?

- ***Don't use two-sample proportion methods when the samples aren't independent.*** These methods give wrong answers when this assumption of independence is violated. Good random sampling is usually the best insurance of independent groups. Make sure there is no relationship between the two groups. For example, you can't compare the proportion of respondents who own SUVs with the proportion of those same respondents who think the tax on gas should be eliminated. The responses are not independent because you've asked many of the same people. To use these methods to estimate or test the difference, you'd need to survey two different groups of people.

  Alternatively, if you have a random sample, you can split your respondents according to their answers to one question and treat the two resulting groups as independent samples. So, you could test whether the proportion of SUV owners who favored eliminating the gas tax was the same as the corresponding proportion among non–SUV owners.

- ***Don't apply inference methods where there was no randomization.*** If the data do not come from representative random samples or from a properly randomized experiment, then the inference about the differences in proportions will be wrong.

- ***Don't interpret a significant difference in proportions causally.*** It turns out that people with higher incomes are more likely to snore. Does that mean money affects sleep patterns? Probably not. We have seen that older people are more likely to snore, and they are also likely to earn more. In a prospective or retrospective study, there is always the danger that other lurking variables not accounted for are the real reason for an observed difference. Be careful not to jump to conclusions about causality.

CONNECTIONS

In Chapter 3 we looked at contingency tables for two categorical variables. Differences in proportions are just 2 × 2 contingency tables. You'll often see data presented in this way. For example, the snoring data could be shown as

	18–29	30 and over	Total
Snore	48	318	366
Don't snore	136	493	629
Total	184	811	995

We tested whether the column percentages of snorers were the same for the two age groups.

This chapter gives the first examples we've seen of inference methods for a parameter other than a simple proportion. Although we have a different standard error, the step-by-step procedures are almost identical. In particular, once again we divide the statistic (the difference in proportions) by its standard error and get a z-score. You should feel right at home.

What have we learned?

In the last few chapters we began our exploration of statistical inference; we learned how to create confidence intervals and test hypotheses about a proportion. Now we've looked at inference for the difference in two proportions. In doing so, perhaps the most important thing we've learned is that the concepts and interpretations are essentially the same—only the mechanics have changed slightly.

We've learned that hypothesis tests and confidence intervals for the difference in two proportions are based on Normal models. Both require us to find the standard error of the difference in two proportions. We do that by adding the variances of the two sample proportions, assuming our two groups are independent. When we test a hypothesis that the two proportions are equal, we pool the sample data; for confidence intervals we don't pool.

TERMS

Variances of independent random variables add
The variance of a sum or difference of independent random variables is the sum of the variances of those variables.

Sampling distribution of the difference between two proportions
The sampling distribution of $\hat{p}_1 - \hat{p}_2$ is, under appropriate assumptions, modeled by a Normal model with mean $\mu = p_1 - p_2$ and standard deviation $SD(\hat{p}_1 - \hat{p}_2) = \sqrt{\dfrac{p_1 q_1}{n_1} + \dfrac{p_2 q_2}{n_2}}$.

Two-proportion z-interval
A two-proportion z-interval gives a confidence interval for the true difference in proportions, $p_1 - p_2$, in two independent groups.

The confidence interval is $(\hat{p}_1 - \hat{p}_2) \pm z^* \times SE(\hat{p}_1 - \hat{p}_2)$, where z^* is a critical value from the standard Normal model corresponding to the specified confidence level.

Pooling
When we have data from different sources that we believe are homogeneous, we can get a better estimate of the common proportion and its standard deviation. We can combine, or pool,

the data into a single group for the purpose of estimating the common proportion. The resulting pooled standard error is based on more data and is thus more reliable (if the null hypothesis is true and the groups are truly homogeneous).

Two-proportion z-test   Test the null hypothesis $H_0: p_1 = p_2$ by referring the statistic

$$z = \frac{\hat{p}_1 - \hat{p}_2}{SE_{pooled}(\hat{p}_1 - \hat{p}_2)}$$

to a standard Normal model.

S K I L L S   *When you complete this lesson you should:*

Think

- Be able to state the null and alternative hypotheses for testing the difference between two population proportions.

- Know how to examine your data for violations of conditions that would make inference about the difference between two population proportions unwise or invalid.

- Understand that the formula for the standard error of the difference between two independent sample proportions is based on the principle that when finding the sum or difference of two independent random variables, their variances add.

Show

- Know how to find a confidence interval for the difference between two proportions.
- Be able to perform a significance test of the natural null hypothesis that two population proportions are equal.

Tell

- Know how to write a sentence describing what is said about the difference between two population proportions by a confidence interval.

- Know how to write a sentence interpreting the results of a significance test of the null hypothesis that two population proportions are equal.

- Be able to interpret the meaning of a P-value in nontechnical language, making clear that the probability claim is made about computed values and not about the population parameter of interest.

- Know that we do not "accept" a null hypothesis if we fail to reject it.

Inferences for the Difference Between Two Proportions on the Computer

It is so common to test against the null hypothesis of no difference between the two true proportions that most statistics programs simply assume this null hypothesis. And most will automatically use the pooled standard deviation. If you wish to test a different null (say that the true difference is 0.3), you may have to search for a way to do it.

Many statistics packages don't offer special commands for inference for differences between proportions. As with inference for single proportions, most statistics programs want the "success" and "failure" status for each case. Usually these are given as 1 or 0, but they might be category names like "yes" and "no." Often we just know the proportions of successes, $\hat{p}_1$ and $\hat{p}_2$, and the counts, n_1 and n_2. Computer packages don't usually deal with summary data like these easily. Calculators typically do a better job.

In some programs you can reconstruct the original values. But even when you have (or can reconstruct) the raw data values, often you won't get *exactly* the same test statistic from a computer package as you would find working by hand. The reason is that when the packages treat the proportion as a mean, they make some approximations. The result is very close, but not exactly the same.

DATA DESK

Data Desk does not offer built-in methods for inference with proportions. Use **Replicate Y by X** to construct data corresponding to given proportions and totals.

Comments

For summarized data, open a Scratchpad to compute the standard deviations and margin of error by typing the calculation.

EXCEL

Inference methods for proportions are not part of the standard Excel tool set.

Comments

For summarized data, type the calculation into any cell and evaluate it.

JMP

For a **categorical** variable that holds category labels, the **Distribution** platform includes tests and intervals of proportions.

For summarized data, put the category names in one variable and the frequencies in an adjacent variable. Designate the frequency column to have the **role** of **frequency.** Then use the **Distribution** platform.

Comments

JMP uses slightly different methods for proportion inferences than those discussed in this text. Your answers are likely to be slightly different.

MINITAB

To find a hypothesis test for a proportion, Choose **Basic Statistics** from the **Stat** menu.

Choose **2Proportions. . .** from the Basic Statistics submenu.

If the data are organized as category names in one column and case IDs in another, assign the variables from the variable list box to the **Samples in one column** box. If the data are organized as two separate columns of responses, click on **Samples in different columns:** and assign the variables from the variable list box. If you have summarized data, click the **Summarized Data** button and fill in the number of trials and the number of successes for each group.

Click the **Options** button and specify the remaining details. Remember to click the **Use pooled estimate of *p* for test** box when testing the null hypothesis of no difference between proportions.

Click the **OK** button.

Comments

When working from a variable that names categories, MINITAB treats the last category as the "success" category. You can specify how the categories should be ordered

SPSS

SPSS does not find hypothesis tests for proportions.

TI-83/84 Plus

To calculate a confidence interval for the difference between two population proportions:

- Select **B:2-PropZInt** from the **STAT TESTS** menu.
- Enter the observed counts and the sample sizes for both samples.
- Specify a confidence level.
- Calculate the interval.

To do the mechanics of a hypothesis test for equality of population proportions:

- Select **6:2-PropZTest** from the **STAT TESTS** menu.
- Enter the observed counts and sample sizes.
- Indicate what kind of test you want: one-tail upper tail, lower tail, or two-tail.
- Calculate the result.

Comments

Beware: When you enter the value of *x*, you need the *count*, not the percentage. The count must be a whole number.

TI-89

To calculate a confidence interval for the difference between two population proportions:

- Select **6:2-PropZInt** from the **STAT Ints** menu.
- Enter the observed counts and the sample sizes for both samples.
- Specify a confidence level.
- Calculate the interval.

To do the mechanics of a hypothesis test for equality of population proportions:

- Select **6:2-PropZTest** from the **STAT Tests** menu.
- Enter the observed counts and sample sizes.
- Indicate what kind of test you want: one-tail upper tail, lower tail, or two-tail.
- Specify if results should simply be calculated or displayed with the area corresponding to the *p*-value of the test shaded.

Comments

Beware: When you enter the value of *x*, you need the *count*, not the percentage. The count must be a whole number. If the number of successes is given as a percent, you must first multiply *np* and round the result to obtain *x*.

EXERCISES

1. **Gender gap.** A presidential candidate fears he has a problem with women voters. His campaign staff plans to run a poll to assess the situation. They'll randomly sample 300 men and 300 women, asking if they have a favorable impression of the candidate. Obviously, the staff can't know this, but suppose the candidate has a positive image with 59% of males, but with only 53% of females.
   a) What kind of sampling design is his staff planning to use?
   b) What difference would you expect the poll to show?
   c) Of course, sampling error means the poll won't reflect the difference perfectly. What's the standard error for the difference in the proportions?
   d) Sketch a sampling model for the size difference in proportions of men and women with favorable im-

   pressions of this candidate that might appear in a poll like this.
   e) Could the campaign be misled by the poll, concluding that there really is no gender gap? Explain.

2. **Buy it again?** A consumer magazine plans to poll car owners to see if they are happy enough with their vehicles that they would purchase the same model again. They'll randomly select 450 owners of American-made cars and 450 owners of Japanese models. Obviously the actual opinions of the entire population couldn't be known, but suppose 76% of owners of American cars and 78% of owners of Japanese cars would purchase another.
   a) What kind of sampling design is the magazine planning to use?
   b) What difference would you expect their poll to show?

c) Of course, sampling error means the poll won't reflect the difference perfectly. What's the standard error for the difference in the proportions?

d) Sketch a sampling model for the difference in proportions that might appear in a poll like this.

e) Could the magazine be misled by the poll, concluding that owners of American cars are much happier with their vehicles than owners of Japanese cars? Explain.

3. Arthritis. The Centers for Disease Control report a survey of randomly selected Americans age 65 and older, which found that 411 of 1012 men and 535 of 1062 women suffered from some form of arthritis.

a) Are the assumptions and conditions necessary for inference satisfied? Explain.

b) Create a 95% confidence interval for the difference in the proportions of senior men and women who have this disease.

c) Interpret your interval in this context.

d) Does this confidence interval suggest that arthritis is more likely to afflict women than men? Explain.

4. Graduation. In October 2000 the U.S. Department of Commerce reported the results of a large-scale survey on high school graduation. Researchers contacted more than 25,000 Americans aged 24 years to see if they had finished high school; 84.9% of the 12,460 males and 88.1% of the 12,678 females indicated that they had high school diplomas.

a) Are the assumptions and conditions necessary for inference satisfied? Explain.

b) Create a 95% confidence interval for the difference in graduation rates between males and females.

c) Interpret your confidence interval.

d) Does this provide strong evidence that girls are more likely than boys to complete high school? Explain.

5. Pets. In 1991, researchers at the National Cancer Institute released the results of a study that investigated the effect of weed-killing herbicides on house pets. They examined 827 dogs from homes where an herbicide was used on a regular basis, diagnosing malignant lymphoma in 473 of them. Of the 130 dogs from homes where no herbicides were used, only 19 were found to have lymphoma.

a) What's the standard error of the difference in the two proportions?

b) Construct a 95% confidence interval for this difference.

c) State an appropriate conclusion.

6. Carpal tunnel. The painful wrist condition called carpal tunnel syndrome can be treated with surgery or less invasive wrist splints. In September 2002, *Time* magazine reported on a study of 176 patients. Among the half that had surgery, 80% showed improvement after three months, but only 54% of those who used the wrist splints improved.

a) What's the standard error of the difference in the two proportions?

b) Construct a 95% confidence interval for this difference.

c) State an appropriate conclusion.

7. Prostate cancer. There has been debate among doctors over whether surgery can prolong life among men suffering from prostate cancer, a type of cancer that typically develops and spreads very slowly. In the summer of 2003, *The New England Journal of Medicine* published results of some Scandinavian research. Men diagnosed with prostate cancer were randomly assigned to either undergo surgery or not. Among the 347 men who had surgery, 16 eventually died of prostate cancer, compared with 31 of the 348 men who did not have surgery.

a) Was this an experiment or an observational study? Explain.

b) Create a 95% confidence interval for the difference in rates of death for the two groups of men.

c) Based on your confidence interval, is there evidence that surgery may be effective in preventing death from prostate cancer? Explain.

8. Race and smoking. In 1995, 24.8% of 550 white adults surveyed reported that they smoked cigarettes, while 25.7% of the 550 black adults surveyed were smokers.

a) Create a 90% confidence interval for the difference in the percentages of smokers among black and white American adults.

b) Does this survey indicate a race-based difference in smoking among American adults? Explain, using your confidence interval to test an appropriate hypothesis.

9. Politics. A poll checking on the level of public support for proposed antiterrorist legislation reported that 68% of the respondents were in favor. The pollsters reported a sampling error of ±3%. When the responses were broken down by party affiliation, support was 2% higher among Republican respondents than Democrats. The pollsters said the margin of error for this difference was ±4%.

a) Why is the margin of error larger for the difference in support between the parties than for the overall level of support?

b) Based on these results, can we conclude that support is significantly higher among Republicans? Explain.

10. War. In September 2002, a Gallup Poll found major differences of opinion based on political affiliation over whether Congress should give President Bush authority to take military action in Iraq. Overall, about 50% of the 1010 respondents were in favor, with a margin of error of ±3%. Opinion differed greatly among Republicans, Democrats, and Independents. (See the bar graph, next page.)

a) How large did this poll estimate the difference in support between Republicans and Democrats to be?

b) Was the margin of error for that difference equal to, greater than, or less than 3%? Explain.

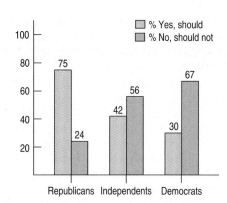

11. Teen smoking, part I. A Vermont study published in December 2001 by the American Academy of Pediatrics examined parental influence on teenagers' decisions to smoke. A group of students who had never smoked were questioned about their parents' attitudes toward smoking. These students were questioned again two years later to see if they had started smoking. The researchers found that among the 284 students who indicated that their parents disapproved of kids smoking, 54 had become established smokers. Among the 41 students who initially said their parents were lenient about smoking, 11 became smokers. Do these data provide strong evidence that parental attitude influences teenagers' decisions about smoking?
   a) What kind of design did the researchers use?
   b) Write appropriate hypotheses.
   c) Are the assumptions and conditions necessary for inference satisfied?
   d) Test the hypothesis and state your conclusion.
   e) Explain in this context what your P-value means.
   f) If that conclusion is actually wrong, which type of error did you commit?

12. Depression. A study published in the *Archives of General Psychiatry* in March 2001 examined the impact of depression on a patient's ability to survive cardiac disease. Researchers identified 450 people with cardiac disease, evaluated them for depression, and followed the group for 4 years. Of the 361 patients with no depression, 67 died. Of the 89 patients with minor or major depression, 26 died. Among people who suffer from cardiac disease, are depressed patients more likely to die than nondepressed ones?
   a) What kind of design was used to collect these data?
   b) Write appropriate hypotheses.
   c) Are the assumptions and conditions necessary for inference satisfied?
   d) Test the hypothesis and state your conclusion.
   e) Explain in this context what your P-value means.
   f) If your conclusion is actually incorrect, which type of error did you commit?

13. Teen smoking, part II. Consider again the Vermont study discussed in Exercise 11.

a) Create a 95% confidence interval for the difference in proportion of children who may smoke and have approving parents and those who may smoke and have disapproving parents.
   b) Interpret your interval in this context.
   c) Carefully explain what "95% confidence" means.

14. Depression revisited. Consider again the study of the association between depression and cardiac disease survivability in Exercise 12.
   a) Create a 95% confidence interval for the difference in survival rates.
   b) Interpret your interval in this context.
   c) Carefully explain what "95% confidence" means.

15. Pregnancy. In 1998, a San Diego reproductive clinic reported 42 live births to 157 women under the age of 38, but only 7 live births for 89 clients aged 38 and older. Is this strong evidence of a difference in the effectiveness of the clinic's methods for older women?
   a) Test an appropriate hypothesis and state your conclusion.
   b) If you concluded there was a difference, estimate that difference with a confidence interval and interpret your interval in context.

16. Suicide. The August 2001 issue of *Pediatrics* reported on a study of adolescent suicide attempts. Questionnaires were given to 6577 middle and high school students, 214 of whom were adopted. Of the 6577, 213 youngsters said they had attempted suicide within the last year—16 of those who were adopted and 197 of those who were not. Does this indicate a significantly different rate of suicide among adopted teens?
   a) Test an appropriate hypothesis and state your conclusion.
   b) If you concluded there was a difference, estimate that difference with a confidence interval and interpret your interval in context.

17. Politics and sex. One month before the election, a poll of 630 randomly selected voters showed 54% planning to vote for a certain candidate. A week later it became known that he had had an extramarital affair, and a new poll showed only 51% of 1010 voters supporting him. Do these results indicate a decrease in voter support for his candidacy?
   a) Test an appropriate hypothesis and state your conclusion.
   b) If you concluded there was a difference, estimate that difference with a confidence interval and interpret your interval in context.

18. Retirement. The Employee Benefit Research Institute reports that 27% of males anticipate having enough money to live comfortably in retirement, but only 18% of females express that confidence. If these results were based on samples of 250 people of each gender, would you consider this strong evidence that men and women have different outlooks?

a) Test an appropriate hypothesis and state your conclusion.

b) If you concluded there was a difference, estimate that difference with a confidence interval and interpret your interval in context.

19. Twins. In 2001, one county reported that among 3132 white women who had babies, 94 were multiple births. There were also 20 multiple births to 606 black women. Does this indicate any racial difference in the likelihood of multiple births?

a) Test an appropriate hypothesis and state your conclusion.

b) If your conclusion is incorrect, which type of error did you commit?

20. Shopping. A survey of 430 randomly chosen adults found that 21% of the 222 men and 18% of the 208 women had purchased books online.

a) Is there evidence that men are more likely than women to make online purchases of books? Test an appropriate hypothesis and state your conclusion in context.

b) If your conclusion in fact proves to be wrong, did you make a Type I or Type II error?

21. Mammograms. It's widely believed that regular mammogram screening may detect breast cancer early, resulting in fewer deaths from that disease. One study that investigated this issue over a period of 18 years was published during the 1970s. Among 30,565 women who had never had mammograms, 196 died of breast cancer, while only 153 of 30,131 who had undergone screening died of breast cancer.

a) Do these results suggest that mammograms may be an effective screening tool to reduce breast cancer deaths?

b) If your conclusion is incorrect, what type of error have you committed?

22. Mammograms redux. In 2001, the conclusion of the study outlined in Exercise 21 was questioned. A new 9-year study was conducted in Sweden, comparing 21,088 women who had mammograms with 21,195 who did not. Of the women who underwent screening, 63 died of breast cancer, compared with 66 deaths among the control group. (*New York Times,* Dec 9, 2001)

a) Do these results support the effectiveness of regular mammograms in preventing deaths from breast cancer?

b) If your conclusion is incorrect, what kind of error have you committed?

23. Pain. Researchers comparing the effectiveness of two pain medications randomly selected a group of patients who had been complaining of a certain kind of joint pain. They randomly divided these people into two groups, then administered the pain killers. Of the 112 people in the group who received medication A, 84 said this pain reliever was effective. Of the 108 people in the other group, 66 reported that pain reliever B was effective.

a) Write a 95% confidence interval for the percent of people who may get relief from this kind of joint pain by using medication A. Interpret your interval.

b) Write a 95% confidence interval for the percent of people who may get relief by using medication B. Interpret your interval.

c) Do the intervals for A and B overlap? What do you think this means about the comparative effectiveness of these medications?

d) Find a 95% confidence interval for the difference in the proportions of people who may find these medications effective. Interpret your interval.

e) Does this interval contain zero? What does that mean?

f) Why do the results in parts c and e seem contradictory? If we want to compare the effectiveness of these two pain relievers, which is the correct approach? Why?

24. Gender gap. Candidates for political office realize that different levels of support among men and women may be a crucial factor in determining the outcome of the election. One candidate finds that 52% of 473 men polled say they will vote for him, but only 45% of the 522 women in the poll express support.

a) Write a 95% confidence interval for the percent of male voters who may vote for this candidate. Interpret your interval.

b) Write a 95% confidence interval for the percent of female voters who may vote for him. Interpret your interval.

c) Do the intervals for males and females overlap? What do you think this means about the gender gap?

d) Find a 95% confidence interval for the difference in the proportions of males and females who will vote for this candidate. Interpret your interval.

e) Does this interval contain zero? What does that mean?

f) Why do the results in parts c and e seem contradictory? If we want to see if there is a gender gap among voters with respect to this candidate, which is the correct approach? Why?

25. Convention bounce. Political pundits talk about the "bounce" that a presidential candidate gets after his party's convention. In the last forty years, it has averaged about 6 percentage points. Just before the 2004 Democratic convention, Rasmussen Reports polled 1500 likely voters at random and found that 47% favored John Kerry. Just afterward they took another random sample of 1500 likely voters and found that 49% favored Kerry. That's a two percentage point increase, but the pollsters claimed that there was no bounce. Explain.

26. Stay-at-home dads. In August 2004, a *Time* magazine article about a survey of men's attitudes reported that 11 of 161 black respondents and 20 of 358 Latino respondents responded "Yes" to the question "Are you a stay-at-home dad?" How big is the difference in proportions in the two populations?

a) Construct and interpret an appropriate confidence interval.
b) Overall, the survey contacted 1302 men and claims a margin of error of ±2.9%. Why is the margin of error different for your confidence interval?

27. Sensitive men. In August 2004, *Time* magazine, reporting on a survey of men's attitudes, noted that "Young men are more comfortable than older men talking about their problems." The survey reported that 80 of 129 surveyed 18- to 24-year-old men and 98 of 184 25- to 34-year-old men said they were comfortable. What do you think? Is *Time's* interpretation justified by these numbers?

28. Carbs. In July of 2004, the Gallup Poll asked 1005 U.S. adults if they actively try to avoid carbohydrates in their diet. That number increased to 27% from 20% in a similar 2002 poll. Is this a statistically significant increase? Explain.

29. Intentional walk. During the 2004 baseball season, San Francisco Giants' slugger Barry Bonds was such a dangerous hitter that many teams simply chose to walk him rather than throw him a pitch he could hit. Just before a series of games in New York, an analyst advised the Mets that they should pitch to Bonds. As evidence, he reported that thus far in the season the Giants had scored in 37 of 79 innings when Bonds was walked intentionally, but in only 107 of 298 innings when the opponents did not walk him. Does this provide evidence that teams should not intentionally walk Barry Bonds?

30. Retention rates. In 2004 the testing company ACT, Inc., reported on the percentage of first-year students at 4-year colleges who return for a second year. Their sample of 1139 students in private colleges showed a 74.9% retention rate, while the rate was 71.9% for the sample of 505 students at public colleges. Does this provide evidence that there's a difference in retention rates of first-year students at public and private colleges?

just checking
Answers

1. The proportion from the sample in 1995 has variability, too. If we do a one-proportion *z*-test we won't take that variability into account and our P-value will be incorrect.

2. The difference in the proportions between the two years has more variability than either individual proportion. The variance of the difference is the sum of the two variances.

3. Since a difference of 0 is not in the confidence interval, we'd reject the null hypothesis. There is evidence that more members will donate if contacted by e-mail.

4. We're 90% confident that if members are contacted by e-mail, the donation rate will be between 4 and 18 percentage points higher than if they received regular mail.

QUICK REVIEW

What do samples really tell us about the populations from which they are drawn? Are the results of an experiment meaningful, or are they just sampling error? Statistical inference based on our understanding of sampling models can help answer these questions. Here's a brief summary of the key concepts and skills:

▶ Sampling models describe the variability of sample statistics using a remarkable result called the Central Limit Theorem.

- When the number of trials is sufficiently large, proportions found in different samples vary according to an approximately Normal model.
- When samples are sufficiently large, the means of different samples vary, with an approximately Normal model.
- The variability of sample statistics decreases as sample size increases.
- Statistical inference procedures are based on the Central Limit Theorem.
- No inference procedure is valid unless the underlying assumptions are true. Always check the conditions before proceeding.

▶ A confidence interval uses a sample statistic (such as a proportion) to estimate a range of plausible values for the parameter of a population model.

- All confidence intervals involve an estimate of the parameter, a margin of error, and a level of confidence.
- For confidence intervals based on a given sample, the greater the margin of error the higher the confidence.
- At a given level of confidence, the larger the sample the smaller the margin of error.

▶ A hypothesis test proposes a model for the population, then examines the observed statistics to see if that model is plausible.

- A null hypothesis suggests a parameter value for the population model. Usually, we assume there is nothing interesting, unusual, or different about the sample results.
- The alternative hypothesis states what we will believe if the sample results turn out to be inconsistent with our null model.
- We compare the difference between statistic and hypothesized value with the standard deviation of the statistic. It's the sampling distribution of this ratio that gives us a P-value.
- The P-value of the test is the probability that the null model could produce results at least as extreme as those observed in the sample or the experiment just as a result of sampling error.
- A low P-value indicates evidence against the null model. If it is sufficiently low, we reject the null model.
- A high P-value indicates that the sample results are not inconsistent with the null model, so we cannot reject it. However, this does not prove the null model is true.
- Sometimes we will mistakenly reject the null hypothesis even though it's actually true—that's called a Type I error. If we fail to reject a false null hypothesis, we commit a Type II error.
- The power of a test measures its ability to detect a false null hypothesis.
- You can lower the risk of a Type I error by requiring a higher standard of proof (lower P-value) before rejecting the null hypothesis. But this will raise the risk of a Type II error and decrease the power of the test.
- The only way to increase the power of a test while decreasing the chance of committing either error is to design a study based on a larger sample.

And now for some opportunities to review these concepts and skills . . .

REVIEW EXERCISES

1. Herbal cancer. A report in the *New England Journal of Medicine* (June 6, 2000) notes growing evidence that the herb *Aristolochia fangchi* can cause urinary tract cancer in those who take it. Suppose you are asked to design an experiment to study this claim. Imagine that you have data on urinary tract cancers in subjects who have used this herb and similar subjects who have not used it and that you can measure incidences of cancer and precancerous lesions in these subjects. State the null and alternative hypotheses you would use in your study.

2. Colorblind. Medical literature says that about 8% of males are colorblind. A university's introductory psychology course is taught in a large lecture hall. Among the students there are 325 males. Each semester when the professor discusses visual perception, he shows the class a test for colorblindness. The percentage of males who are colorblind varies from semester to semester.
a) Is the sampling distribution model for the sample proportion likely to be Normal? Explain.
b) What are the mean and standard deviation of this sampling distribution model?
c) Sketch the sampling model using the 68-95-99.7 Rule.
d) Write a few sentences explaining what the model says about this professor's class.

3. Birth days. During a 2-month period in 2002, 72 babies were born at the Tompkins Community Hospital in upstate New York. The table shows how many babies were born on each day of the week.

a) If births are uniformly distributed across all days of the week, how many would you expect on each day?
b) Only 7 births occurred on a Monday. Does this indicate that women might be less likely to give birth on a Monday? Explain.

Day	Births
Mon.	7
Tues.	17
Wed.	8
Thurs.	12
Fri.	9
Sat.	10
Sun.	9

c) Are the 17 births on Tuesdays unusually high? Explain.
d) Can you think of any reasons why births may not occur completely at random?

4. Polling. In the 1992 U.S. presidential election, Bill Clinton received 43% of the vote compared with 38% for George Bush and 19% for Ross Perot. Suppose we had taken a random sample of 100 voters in an exit poll and asked them for whom they had voted.
a) Would you always get 43 votes for Clinton, 38 for Bush, and 19 for Perot in a sample of 100? Why or why not?
b) In 95% of such polls, our sample proportion of voters for Clinton should be between what two values?
c) In 95% of such polls, the sample proportion of Perot votes should be between what two numbers?
d) Would you expect the sample proportion of Perot votes to vary more, less, or about the same as the sample proportion of Bush votes? Why?

5. Leaky gas tanks. Nationwide it is estimated that 40% of service stations have gas tanks that leak to some extent. A new program in California is designed to lessen the prevalence of these leaks. We want to assess the effectiveness of the program by seeing if the percentage of service stations whose tanks leak has decreased. To do this, we randomly sample 27 service stations in California and determine whether there is any evidence of leakage. In our sample, only 7 of the stations exhibit any leakage. Is there evidence that the new program is effective?
a) What are the null and alternative hypotheses?
b) Check the assumptions necessary for inference.
c) Test the null hypothesis.
d) What do you conclude (in plain English)?
e) If the program actually works, have you made an error? What kind?
f) What two things could you do to decrease the probability of making this kind of error?
g) What are the advantages and disadvantages of taking those two courses of action?

6. Surgery and germs. Joseph Lister (for whom Listerine is named!) was a British physician who was interested in the role of bacteria in human infections. He suspected that germs were involved in transmitting infection, and so he tried using carbolic acid as an operating room disinfectant. In 75 amputations, he used carbolic acid 40 times. Of the 40 amputations using carbolic acid, 34 of the patients lived. In the 35 amputations without carbolic acid, 19 lived. The question of interest is whether carbolic acid is effective in increasing the chances of surviving an amputation.
a) What kind of a study is this?
b) What do you conclude? Support your conclusion by testing an appropriate hypothesis.
c) What reservations do you have about the design of the study?

7. Scrabble. Using a computer to play many simulated games of Scrabble, researcher Charles Robinove found that the letter "A" occurred in 54% of the hands. This study had a margin of error of ± 10%. (*Chance*, 15, no. 1 [2002])
a) Explain what the margin of error means in this context.
b) Why might the margin of error be so large?
c) Probability theory predicts that the letter "A" should appear in 63% of the hands. Does this make you concerned that the simulation might be faulty? Explain.

8. Dice. When one die is rolled, the number of spots showing has a mean of 3.5 and a standard deviation of 1.7. Suppose you roll 10 dice. What's the approximate probability that your total is between 30 and 40 (that is, the average for the 10 dice is between 3 and 4)? Specify the model you use and the assumptions and conditions that justify your approach.

9. News sources. In May of 2000, the Pew Research Foundation sampled 1593 respondents and asked how they obtain news. In Pew's report, 33% of respondents say that they now obtain news from the Internet at least once a week.
   a) Pew reports a margin of error of ±3% for this result. Explain what the margin of error means.
   b) Pew also asked about investment information, and 21% of respondents reported that the Internet is their main source of this information. When limited to the 780 respondents who identified themselves as investors, the percent who rely on the Internet rose to 28%. How would you expect the margin of error for this statistic to change in comparison with the margin of error for the percentage of all respondents?
   c) When restricted to the 239 active traders in the sample, Pew reports that 45% rely on the Internet for investment information. Find a confidence interval for this statistic.
   d) How does the margin of error for your confidence interval compare with the values in parts a and b? Explain why.

10. Death penalty. In May of 2002, the Gallup Organization asked a random sample of 537 American adults this question:

> *If you could choose between the following two approaches, which do you think is the better penalty for murder, the death penalty or life imprisonment, with absolutely no possibility of parole?*

Of those polled, 52% chose the death penalty.
   a) Create a 95% confidence interval for the percentage of all American adults who favor the death penalty.
   b) Based on your confidence interval, is it clear that the death penalty has majority support? Explain.
   c) If pollsters wanted to follow up on this poll with another survey that could determine the level of support for the death penalty to within 2% with 98% confidence, how many people should they poll?

11. Bimodal. We are sampling randomly from a distribution known to be bimodal.
   a) As our sample size increases, what's the expected shape of the sample's distribution?
   b) What's the expected value of our sample's mean? Does the size of the sample matter?
   c) How is the variability of this sample's mean related to the standard deviation of the population? Does the size of the sample matter?
   d) How is the shape of the sampling distribution model affected by the sample size?

12. Vitamin D. In July 2002 the *American Journal of Clinical Nutrition* reported that 42% of 1546 African-American women studied had vitamin D deficiency. The data came from a national nutrition study conducted by the Centers for Disease Control in Atlanta.
   a) Do these data meet the assumptions necessary for inference? What would you like to know that you don't?
   b) Create a 95% confidence interval.
   c) Interpret the interval in this context.
   d) Explain in this context what "95% confidence" means.

13. Archery. A champion archer can generally hit the bull's-eye 80% of the time. Suppose she shoots 200 arrows during competition. Let $\hat{p}$ represent the percentage of bull's-eyes she gets (the sample proportion).
   a) What are the mean and standard deviation of the sampling distribution model for $\hat{p}$?
   b) Is a Normal model appropriate here? Explain.
   c) Sketch the sampling model using the 68-95-99.7 Rule.
   d) What's the probability she gets at least 85% bull's-eyes?

14. Free throws. During the 2000–2001 NBA season, San Antonio Spurs player Tim Duncan made 409 of 662 free throws he attempted. During the 2001–2002 season, he made 460 of his 568 free throws.
   a) Write a 95% confidence interval for the increase in percent of foul shots he can make.
   b) Based on your confidence interval, is there strong evidence that he has become a better foul shooter? Explain.

15. Twins. There is some indication in medical literature that doctors may have become more aggressive in inducing labor or doing preterm cesarean sections when a woman is carrying twins. Records at a large hospital show that of the 43 sets of twins born in 1990, 20 were delivered before the 37th week of pregnancy. In 2000, 26 of 48 sets of twins were born preterm. Does this indicate an increase in the incidence of early births of twins? Test an appropriate hypothesis and state your conclusion.

16. Eclampsia. It's estimated that 50,000 pregnant women worldwide die each year of eclampsia, a condition involving elevated blood pressure and seizures. A research team from 175 hospitals in 33 countries investigated the effectiveness of magnesium sulfate in preventing the occurrence of eclampsia in at-risk patients. Results are summarized below. (*Lancet*, June 1, 2002)

	Total Subjects	Reported side effects	Developed eclampsia	Deaths
Treatment Magnesium sulfide	4999	1201	40	11
Placebo	4993	228	96	20

   a) Write a 95% confidence interval for the increase in the proportion of women who may develop side effects from this treatment. Interpret your interval.

b) Is there evidence that the treatment may be effective in preventing the development of eclampsia? Test an appropriate hypothesis and state your conclusion.

17. Eclampsia. Refer again to the research summarized in Exercise 16. Is there any evidence that when eclampsia does occur, the magnesium sulfide treatment may help prevent the woman's death?
a) Write an appropriate hypothesis.
b) Check the assumptions and conditions.
c) Find the P-value of the test.
d) What do you conclude about the magnesium sulfide treatment?
e) If your conclusion is wrong, which type of error have you made?
f) Name two things you could do to increase the power of this test.
g) What are the advantages and disadvantages of those two options?

18. Eggs. The ISA Babcock Company supplies poultry farmers with hens, advertising that a mature B300 Layer produces eggs with a mean weight of 60.7 grams. Suppose that egg weights follow a Normal model with standard deviation 3.1 grams.
a) What fraction of the eggs produced by these hens weigh more than 62 grams?
b) What's the probability that a dozen randomly selected eggs average more than 62 grams?
c) Using the 68-95-99.7 Rule, sketch a model of the total weights of a dozen eggs.

19. Polling disclaimer. A newspaper article that reported the results of an election poll included the following explanation.

> *The Associated Press poll on the 2000 presidential campaign is based on telephone interviews with 798 randomly selected registered voters from all states except Alaska and Hawaii. The interviews were conducted June 21–25 by ICR of Media, Pa.*
>
> *The results were weighted to represent the population by demographic factors such as age, sex, region, and education.*
>
> *No more than 1 time in 20 should chance variations in the sample cause the results to vary by more than 4 percentage points from the answers that would be obtained if all Americans were polled.*
>
> *The margin of sampling error is larger for responses of subgroups, such as income categories or those in political parties. There are other sources of potential error in polls, including the wording and order of questions.*

a) Did they describe the 5 W's well?
b) What kind of sampling design could take into account the several demographic factors listed?
c) What was the margin of error of this poll?
d) What was the confidence level?
e) Why is the margin of error larger for subgroups?

f) Which kinds of potential bias did they caution readers about?

20. Enough eggs? One of the important issues for poultry farmers is the production rate—the percentage of days on which a given hen actually lays an egg. Ideally, that would be 100% (an egg every day), but realistically hens tend to lay eggs on about 3 of every 4 days. ISA Babcock wants to advertise the production rate for the B300 Layer (see Exercise 18) as a 95% confidence interval with a margin of error of ±2%. How many hens must they collect data on?

21. Teen deaths. Traffic accidents are the leading cause of death among people aged 15 to 20. In May 2002, the National Highway Traffic Safety Administration reported that even though only 6.8% of licensed drivers are between 15 and 20 years old, they were involved in 14.3% of all fatal crashes. Insurance companies have long known that teenage boys were high risks, but what about teenage girls? One insurance company found that the driver was a teenage girl in 44 of the 388 fatal accidents they investigated. Is this strong evidence that the accident rate is lower for girls than for teens in general?
a) Test an appropriate hypothesis and state your conclusion.
b) Explain what your P-value means in this context.

22. Perfect pitch. A recent study on perfect pitch tested 2700 students in American music conservatories. It found that 7% of non-Asian and 32% of Asian students have perfect pitch. A test of the difference in proportions resulted in a P-value of < 0.0001.
a) What are the researchers' null and alternative hypotheses?
b) State your conclusion.
c) Explain in this context what the P-value means.
d) The researchers claimed that the data prove that genetic differences between the two populations cause a difference in the frequency of occurrence of perfect pitch. Do you agree? Why or why not?

23. Largemouth bass. Organizers of a fishing tournament believe that the lake holds a sizable population of largemouth bass. They assume that the weights of these fish have a model that is skewed to the right with a mean of 3.5 pounds and a standard deviation of 2.2 pounds.
a) Explain why a skewed model makes sense here.
b) Explain why you cannot determine the probability that a largemouth bass randomly selected ("caught") from the lake weighs over 3 pounds.
c) Each fisherman in the contest catches 5 fish each day. Can you determine the probability that someone's catch averages over 3 pounds? Explain.
d) The 12 fishermen competing each caught the limit of 5 fish. What's the probability that the total catch of 60 fish averaged more than 3 pounds?

24. Cheating. A Rutgers University study released in 2002 found that many high-school students cheat on tests. The researchers surveyed a random sample of 4500 high school students nationwide; 74% of them said they had cheated at least once.
a) Create a 90% confidence interval for the level of cheating among high-school students. Don't forget to check the appropriate conditions.
b) Interpret your interval.
c) Explain what "90% confidence" means.
d) Would a 95% confidence interval be wider or narrower? Explain without actually calculating the interval.

25. Language. Neurological research has shown that in about 80% of people, language abilities reside in the brain's left side. Another 10% display right-brain language centers, and the remaining 10% have two-sided language control. (The latter two groups are mainly left-handers.) (*Science News*, 161, no. 24 [2002])
a) We select 60 people at random. Is it reasonable to use a Normal model to describe the possible distribution of the proportion of the group that has left-brain language control? Explain.
b) What's the probability that our group has at least 75% left-brainers?
c) If the group had consisted of 100 people, would that probability be higher, lower, or about the same? Explain why, without actually calculating the probability.
d) How large a group would almost certainly guarantee at least 75% left-brainers? Explain.

26. Cigarettes. In 1999 the Centers for Disease Control estimated that about 34.8% of high-school students smoked cigarettes. They established a national health goal of reducing that figure to 16% by the year 2010. To that end, they hoped to achieve a reduction to 30% by the end of 2001. Early in 2002 they released a research study in which only 28.5% of a random sample of 10,204 high-school students said they were current smokers. Is this evidence that progress toward the goal is on track?
a) Write appropriate hypotheses.
b) Verify that the appropriate assumptions are satisfied.
c) Find the P-value of this test.
d) Explain what the P-value means in this context.
e) State an appropriate conclusion.
f) Of course, your conclusion may be incorrect. If so, which kind of error did you commit?

27. Crohn's disease. In 2002 the medical journal *Lancet* reported that 335 of 573 patients suffering from Crohn's disease responded positively to injections of the arthritis-fighting drug infliximab.
a) Create a 95% confidence interval for the effectiveness of this drug.
b) Interpret your interval in context.
c) Explain carefully what "95% confidence" means in this context.

28. Teen smoking. The Centers for Disease Control say that about 30% of teenagers smoke tobacco (down from a high of 38% in 1997). A college has 522 students in its freshmen class. Is it likely that more than 40% of them may be smokers? Explain.

29. Alcohol abuse. Growing concern about binge drinking among college students has prompted one large state university to conduct a survey to assess the size of the problem on its campus. The university plans to randomly select students and ask how many have been drunk during the past week. If the school hopes to estimate the true proportion among all its students with 90% confidence and a margin of error of ±4%, how many students must be surveyed?

30. Errors. An auto parts company advertises that its special oil additive will make the engine "run smoother, cleaner, longer, with fewer repairs." An independent laboratory decides to test part of this claim. It arranges to use a taxicab company's fleet of cars. The cars are randomly divided into two groups. The company's mechanics will use the additive in one group of cars but not in the other. At the end of a year the laboratory will compare the percentage of cars in each group that required engine repairs.
a) What kind of a study is this?
b) Will they do a one-tailed or a two-tailed test?
c) Explain in this context what a Type I error would be.
d) Explain in this context what a Type II error would be.
e) Which type of error would the additive manufacturer consider more serious?
f) If the cabs with the additive do indeed run significantly better, can the company conclude it is an effect of the additive? Can they generalize this result and recommend the additive for all cars? Explain.

31. Preemies. Among 242 Cleveland-area children born prematurely at low birth weights between 1977 and 1979, only 74% graduated from high school. Among a comparison group of 233 children of normal birth weight, 83% were high school graduates. ("Outcomes in Young Adulthood for Very-Low-Birth-Weight Infants," *New England Journal of Medicine*, 346, no. 3 [2002])
a) Create a 95% confidence interval for the difference in graduation rates between children of normal and very low birth weights. Be sure to check the appropriate assumptions and conditions.
b) Does this provide evidence that premature birth may be a risk factor for not finishing high school? Use your confidence interval to test an appropriate hypothesis.
c) Suppose your conclusion is incorrect. Which type of error did you make?

32. Safety. Observers in Texas watched children at play in eight communities. Of the 814 children seen biking, roller skating, or skateboarding, only 14% wore a helmet.

a) Create and interpret a 95% confidence interval.

b) What concerns do you have about this study that might make your confidence interval unreliable?

c) Suppose we want to do this study again, picking various communities and locations at random, and hope to end up with a 98% confidence interval having a margin of error of ±4%. How many children must we observe?

33. Fried PCs. A computer company recently experienced a disastrous fire that ruined some of its inventory. Unfortunately, during the panic of the fire, some of the damaged computers were sent to another warehouse where they were mixed with undamaged computers. The engineer responsible for quality control would like to check out each computer in order to decide if it's undamaged or damaged. Each computer undergoes a series of 100 tests. The number of tests it fails will be used to make the decision. If it fails more than a certain number, it will be classified as damaged, and then scrapped. From past history, the distribution of the number of tests failed is known for both undamaged and damaged computers. The probabilities associated with each outcome are listed in the table below:

Number of tests failed	0	1	2	3	4	5	>5
Undamaged (%)	80	13	2	4	1	0	0
Damaged (%)	0	10	70	5	4	1	10

The table indicates, for example, that 80% of the undamaged computers have no failures, while 70% of the damaged computers have 2 failures.

a) To the engineers, this is a hypothesis-testing situation. State the null and alternative hypotheses.

b) Someone suggests classifying a computer as damaged if it fails any of the tests. Discuss the advantages and disadvantages of this test plan.

c) What number of tests would a computer have to fail in order to be classified as damaged if the engineers want to have the probability of a Type I error equal to 5%?

d) What's the power of the test plan in part c?

e) A colleague points out that by increasing α just 2%, the power can be increased substantially. Explain.

34. Power. We are replicating an experiment. How will each of the following changes affect the power of our test? Indicate whether it will increase, decrease, or remain the same, assuming all other aspects of the situation remain unchanged.

a) We increase the number of subjects from 40 to 100.

b) We require a higher standard of proof, changing from $\alpha = 0.05$ to $\alpha = 0.01$.

35. Approval. Of all presidents since World War II, Jimmy Carter suffered the worst overall approval rating at 45%. George W. Bush's approval rating varied widely. It reached a high of 90% just after 9/11, but in mid-July 2004, it stood at 49% as measured by a CNN/*USA Today*/Gallup poll of 1000 randomly selected adults. Is there evidence to suggest that Bush's July 2004 rating reflects higher public approval than the 45% that Jimmy Carter averaged over his presidency?

36. Grade inflation. In 1996, 20% of the students at a major university had an overall grade point average of 3.5 or higher (on a scale of 4.0). In 2000 a random sample of 1100 student records found that 25% had a GPA of 3.5 or higher. Is this evidence of grade inflation?

PART VI

Learning About the World

CHAPTER 23

Inferences About Means

WHO	Vehicles on Triphammer Road
WHAT	Speed
UNITS	Miles per hour
WHEN	April 11, 2000, 1 p.m.
WHERE	A small town in the northeastern United States
WHY	Concern over impact on residential neighborhood

Motor vehicle crashes are the leading cause of death for people between 4 and 33 years old. In the year 2000, motor vehicle accidents claimed the lives of 41,821 people in the United States, up from 41,717 the year before. This means that, on average, motor vehicle crashes resulted in 115 deaths each day, or 1 death every 13 minutes.

Speeding is a contributing factor in 31% of all fatal accidents according to the National Highway Traffic Safety Administration. Not only were 13,713 lives lost in speeding-related crashes during 2002 (up from 12,350 in 2000), but the economic cost of such crashes is estimated to be about $40.4 billion per year.

Triphammer Road is a busy street that passes through a residential neighborhood. Residents there are concerned that vehicles traveling on Triphammer often exceed the posted speed limit of 30 miles per hour. The local police sometimes place a radar speed detector by the side of the road; as a vehicle approaches, this detector displays the vehicle's speed to its driver. The police hope that drivers will slow down if they are reminded of their speed.

The local residents are not convinced that such a passive method is helping the problem. They wish to persuade the village to add extra police patrols to encourage drivers to observe the speed limit. To help their case, a resident stood where he could see the detector and recorded the speed of vehicles passing it during a

The speeds of cars on Triphammer Road seem to be unimodal and symmetric, at least at this scale.

Figure 23.1

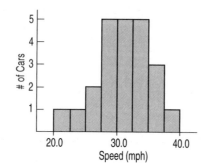

Speed		
29	29	24
34	34	34
34	32	36
28	31	31
30	27	34
29	37	36
38	29	21
31	26	

15-minute period one day. When clusters of vehicles went by, he noted only the speed of the front vehicle. The data and the histogram appear in Figure 23.1.

We're interested both in estimating the true mean speed and in testing whether it exceeds the posted speed limit. Although the sample of vehicles is a convenience sample, not a truly random sample, there's no compelling reason to believe that vehicles at one time of day are driving faster or slower than vehicles at another time of day,[1] so we can take the sample to be representative.

These data differ from data on proportions in one important way. Proportions are usually reported as summaries. After all, individual responses are just "success" and "failure" or "1" and "0." Quantitative data, though, usually report a value for each individual. When you have a value for each individual, you should remember the three rules of data analysis and plot the data, as we have done here.

We have quantitative data, so we summarize with means and standard deviations. Because we want to do inference, we'll think about sampling distributions, too, and we already know most of the facts we need.

Getting Started

You've learned how to create confidence intervals and test hypotheses about proportions (if you haven't, this would be a really good time for a quick review). Now we want to do the same thing for means. Just as we did before, we will base both our confidence interval and our hypothesis test on the sampling distribution model. The Central Limit Theorem told us (back in Chapter 18) that the sampling distribution model for means is Normal:

> ### The Central Limit Theorem
>
> When a random sample is drawn from any population with mean μ and standard deviation σ, its sample mean, $\bar{y}$, has a sampling distribution with the same *mean μ* but whose *standard deviation* is $\dfrac{\sigma}{\sqrt{n}}$ (and we write $\sigma(\bar{y}) = SD(\bar{y}) = \dfrac{\sigma}{\sqrt{n}}$).
>
> No matter what population the random sample comes from, the *shape* of the sampling distribution is approximately Normal as long as the sample size is large enough. The larger the sample used, the more closely the Normal approximates the sampling distribution for the mean.

This is an amazing result. It says the shape of the mean's sampling distribution doesn't depend on the shape of the distribution the data come from. And unless the shape of this underlying distribution is far from unimodal and symmetric, this Normal model works well even for small samples. All we need is a random sample of quantitative data.

And the true population standard deviation, σ.

Uh oh. That could be a problem. How are we supposed to know σ? Proportions have a link between the proportion value and the standard deviation of the sample proportion: $SD(\hat{p}) = \sqrt{\dfrac{pq}{n}}$. And there was an obvious way to estimate it from the data: $SE(\hat{p}) = \sqrt{\dfrac{\hat{p}\hat{q}}{n}}$. But for means, $SD(\bar{y}) = \dfrac{\sigma}{\sqrt{n}}$. So knowing $\bar{y}$ doesn't tell us

> When we estimate the standard deviation of the sampling distribution model from the data, it's called the *standard error*. We'll use that term and the $SE(\bar{y})$ notation. Remember, though, that it's just the estimated standard deviation of the sampling distribution model for means.

---

[1] Except, perhaps, at rush hour. But at that time, traffic is slowed. Our concern is with ordinary traffic during the day.

anything about $SD(\bar{y})$. We know n, the sample size, but the population standard deviation, σ, could be *anything*. So what should we do? We do what any sensible person would do: We estimate the population parameter σ with s, the sample standard deviation based on the data. Then, the resulting standard error is $SE(\bar{y}) = \dfrac{s}{\sqrt{n}}$.

Unfortunately, we may not be home free yet. The sample standard deviation, s, is a statistic. It varies from sample to sample just like any statistic. So our estimated standard error has additional variation in it. How should we allow for that extra variation?

A century ago, people used this standard error with the Normal model, assuming it would work. And for large sample sizes it *did* work pretty well. But they began to notice problems with smaller samples. The extra variation in the standard error was messing up the P-values and margins of error.

William S. Gosset is the man who first investigated this fact. He realized that not only do we need to allow for the extra variation with larger margins of error and P-values, but we even need a new sampling distribution model. In fact, we need a whole *family* of models, depending on the sample size, n. These models are unimodal, symmetric, bell-shaped models, but the smaller our sample, the more we must stretch out the tails. Gosset's work transformed Statistics, but most people who use his work don't even know his name.

Gosset's *t*

To find the sampling distribution of $\dfrac{\bar{y}}{s/\sqrt{n}}$, Gosset simulated it *by hand*. He drew paper slips of small samples from a hat *hundreds of times* and computed the means and standard deviations with a mechanically cranked calculator. Today you could repeat in seconds on a computer the experiment that took him over a year. Gosset's work was so meticulous that not only did he get the shape of the new histogram approximately right, but he even figured out the exact *formula* for it from his sample. The formula was not confirmed mathematically until years later by Sir R. A. Fisher.

Gosset had a job that made him the envy of many. He was the quality control engineer for the Guinness Brewery in Dublin, Ireland. His job was to make sure that the stout (a thick, dark beer) leaving the brewery was of high enough quality to meet the demands of the brewery's many discerning customers. It's easy to imagine, when testing stout, why a large sample with many observations might be undesirable, not to mention dangerous to one's health. So Gosset often used small samples of 3 or 4. But he noticed that with samples of this size, his tests for quality weren't quite right. He knew this because when the batches that he rejected were sent back to the laboratory for more extensive testing, too often they turned out to be OK.

Gosset checked the stout's quality by performing hypothesis tests. He knew that the test would make some Type I errors and reject about 5% of the *good* batches of stout. However, the lab told him that he was in fact rejecting about 15% of the good batches. Gosset knew something was wrong, and it bugged him.

Gosset took time off to study the problem (and earn a graduate degree in the emerging field of Statistics). He figured out that when he used the estimated standard error, $\dfrac{s}{\sqrt{n}}$, the shape of the sampling model changed. He even figured out what the new model should be and called it a *t*-distribution.

The Guinness Company didn't give Gosset a lot of support for his work. In fact, it had a policy against publishing results. Gosset had to convince the company that he was not publishing an industrial secret, and (as part of getting permission to publish) had to use a pseudonym. The pseudonym he chose was "Student," and ever since, the model he found has been known as **Student's *t*.**

The shape of Gosset's model changes with different sample sizes. So the Student's *t*-models form a whole *family* of related distributions that depend on a parameter known as **degrees of freedom.** We often denote degrees of freedom as df, and the model as t_{df}, with the degrees of freedom as a subscript.

What Does This Mean for Means?

To make confidence intervals or test hypotheses for means, we need to use Gosset's model. Which one? Well, for means, it turns out the right value for degrees of freedom is $df = n - 1$.

NOTATION ALERT:

Ever since Gosset, *t* has been reserved in Statistics for his distribution.

A practical sampling distribution model for means

When the conditions are met, the standardized sample mean,

$$t = \frac{\bar{y} - \mu}{SE(\bar{y})},$$

follows a Student's *t*-model with $n - 1$ degrees of freedom. We estimate the standard error with

$$SE(\bar{y}) = \frac{s}{\sqrt{n}}.$$

A S **Student's *t* Distributions.**
Interact with Gosset's family of *t*-models. Watch the shape of the model change as you slide the degrees of freedom up and down—much more fun than just looking at Figure 23.2.

When Gosset corrected the model for the extra uncertainty, the margin of error got bigger, as you might have guessed. When you use Gosset's model instead of the Normal model, your confidence intervals will be just a bit wider and your P-values just a bit larger. That's the correction you need. By using the *t*-model, you've compensated for the extra variability in precisely the right way.

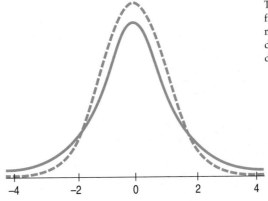

The *t*-model (solid curve) on 2 degrees of freedom has fatter tails than the Normal model (dashed curve). So the 68-95-99.7 Rule doesn't work for *t*-models with only a few degrees of freedom. **Figure 23.2**

Student's *t*-models are unimodal, symmetric, and bell shaped, just like the Normal. But *t*-models with only a few degrees of freedom have much fatter tails than the Normal. (That's what makes the margin of error bigger.) As the degrees of freedom increase, the *t*-models look more and more like the Normal. In fact, the *t*-model with infinite degrees of freedom is exactly Normal.[2] This is great news if you happen to have an infinite number of data values. Unfortunately, that's not practical. Fortunately, above a few hundred degrees of freedom it's very hard to tell the difference. Of course, in the rare situation that we *know* σ, it would be foolish not to use that information. And if we don't have to estimate *s*, we can use the Normal model.

z or t?

If you know σ, use z.
(That's rare!)
Whenever you use *s*
to estimate σ, use *t*.

---

[2] Formally, in the limit as *n* goes to infinity.

● **When σ is known** Administrators of a hospital were concerned about the prenatal care given to mothers in their part of the city. To study this, they examined the gestation times of babies born there. They drew a sample of 25 babies born in their hospital in the previous 6 months. Human gestation times for healthy pregnancies are known to be well modeled by a Normal with a mean of 266 days and a standard deviation of 16 days. The hospital administrators wanted to test the mean gestation time of their sample of babies against the known standard. For this test, they should use the established value for the standard deviation, 16 days, rather than estimating the standard deviation from their sample. Because they use the model parameter value for σ, they should base their test on the Normal model rather than Student's *t*. ●

Finding *t*-Values by Hand

Two tail probability	0.20	0.10	0.05
One tail probability	0.10	0.05	0.025
Table T df			
Values of t_α 1	3.078	6.314	12.706
2	1.886	2.920	4.303
3	1.638	2.353	3.182
4	1.533	2.132	2.776
5	1.476	2.015	2.571
6	1.440	1.943	2.447
7	1.415	1.895	2.365
8	1.397	1.860	2.306
9	1.383	1.833	2.262
10	1.372	1.812	2.228
11	1.363	1.796	2.201
12	1.356	1.782	2.179
13	1.350	1.771	2.160
14	1.345	1.761	2.145
15	1.341	1.753	2.131
16	1.337	1.746	2.120
17	1.333	1.740	2.110
18	1.330	1.734	2.101
19	1.328	1.729	2.093

Part of Table T.

A S **Building *t*-Intervals with the *t*-Table.** Interact with an animated version of Table T.

> As degrees of freedom increase, the shape of Student's *t*-models changes more gradually. Table T at the back of the book includes degrees of freedom between 100 and 1000 selected so that you can pin down the P-value for just about any df. If your df's aren't listed, take the cautious approach by using the next lower value, or use technology.

The Student's *t*-model is different for each value of degrees of freedom. We might print a table like Table Z for each degrees of freedom value, but that's a lot of pages, and not likely to be a bestseller. One way to shorten the book is to admit that while it might be nice to be able to get a critical value for a 93.4% confidence interval, in practice we usually limit ourselves to 90%, 95%, 99%, and 99.9% confidence levels. So Statistics books usually have one table of *t*-model critical values for selected confidence levels. (This one does, too; see Table T.) The tables run down the page for as many degrees of freedom as can fit, and they are actually much easier to use than the Normal tables.

Then they get to the bottom of the page and run out of room. Of course, for *enough* degrees of freedom, the *t*-model gets closer and closer to the Normal, so the tables give a final row with the critical values from the Normal model and label it "∞ df."

Or we could use technology. The Appendix of *ActivStats* on the CD, any graphing calculator or statistics program can give critical values for a *t*-model for any number of degrees of freedom and for any confidence level you please. And they can go straight to P-values when we test a hypothesis. You can also find tables on the Internet. (Search for terms like "statistical tables z t".)

Assumptions and Conditions

Gosset found the *t*-model by simulation. Years later, when Sir Ronald A. Fisher showed mathematically that Gosset was right, he needed to make some assumptions to make the proof work. These are the assumptions we need to use the Student's *t*-models.

Independence Assumption

The data values should be independent. There's really no way to check independence of the data by looking at the sample, but we should think about whether the assumption is reasonable.

A S **Assumptions and Conditions for Student's *t*.** A narrated review.

Randomization Condition: The data arise from a random sample or suitably randomized experiment. Randomly sampled data—and especially data from a Simple Random Sample—are ideal.

We Don't *Want* to Stop
We check conditions hoping that we can make a meaningful analysis of our data. The conditions serve as *disqualifiers*—we keep going unless there's a serious problem. If we find minor issues, we note them and express caution about our results. If the sample is not an SRS, but we believe it's representative of some populations, we limit our conclusions accordingly. If there are outliers, rather than stop, we perform the analysis both with and without them. If the sample looks bimodal, we try to analyze subgroups separately. Only when there's major trouble—like a strongly skewed small sample or an obviously non-representative sample—are we unable to proceed at all.

When a sample is drawn without replacement, technically we ought to confirm that we haven't sampled a large fraction of the population, which would threaten the independence of our selections. We check the

10% Condition: The sample is no more than 10% of the population. When we made inferences about proportions, this condition was crucial. For means, though, it is rarely a problem unless we are sampling from a small population. We can estimate means more reliably than proportions, so we almost never need to take that large a fraction of the population as our sample. We usually won't mention it for means—but remember to check if you have a small population, or a very large sample.

Normal Population Assumption

To use a Student's *t*-model, we are formally required to assume that the data are from a population that follows a Normal model. Practically speaking, there's no way to be certain this is true.

And it's almost certainly *not* true. Models are idealized; real data are, well, real. The good news, however, is that even for small samples, it's sufficient to check the

Nearly Normal Condition: The data come from a distribution that is unimodal and symmetric. Check this condition by making a histogram or Normal probability plot. The importance of Normality for Student's *t* depends on the sample size. Just our luck; it matters most when it's hardest to check.[3]

For very small samples ($n < 15$ or so), the data should follow a Normal model pretty closely. Of course, with so little data, it's rather hard to tell. But if you do find outliers or strong skewness, don't use these methods.

For moderate sample sizes (n between 15 and 40 or so), the *t* methods will work well as long as the data are unimodal and reasonably symmetric. Make a histogram.

When the sample size is larger than 40 or 50, the *t* methods are safe to use even if the data are skewed. Make a histogram anyway. If you find outliers in the data, it's always a good idea to perform the analysis twice, once with and once without the outliers, even for large samples. They may well hold additional information about the data that deserves special attention. If you find multiple modes, you may well have different groups that should be analyzed and understood separately.

NOTATION ALERT:
When we found critical values from a Normal model we called them z^*. When we use a Student's *t*-model, we'll denote the critical values t^*.

 **Student's *t* in Practice.** Use a statistics package to find a *t*-based confidence interval; that's how it's almost always done.

One-sample *t*-interval

When the conditions are met, we are ready to find the confidence interval for the population mean, μ. The confidence interval is

$$\bar{y} \pm t^*_{n-1} \times SE(\bar{y})$$

where the standard error of the mean, $SE(\bar{y}) = \dfrac{s}{\sqrt{n}}$.

The critical value t^*_{n-1} depends on the particular confidence level, C, that you specify and on the number of degrees of freedom, $n - 1$, which we get from the sample size.

---

 just checking

Every 10 years, the United States takes a census. The census tries to count every resident. There are two forms, known as the "short form," answered by most people, and the "long form," slogged through by about one in six or seven households chosen at random. According

---

[3] There are formal tests of Normality, but they don't really help. When we have a small sample—just when we really care about checking Normality—these tests have very little power. So it doesn't make much sense to use them in deciding whether to perform a *t*-test. We don't recommend that you use them.

to the Census Bureau (http://factfinder.census.gov), "... each estimate based on the long form responses has an associated confidence interval."

1 Why does the Census Bureau need a confidence interval for long form information, but not for the questions that appear on both the long and short forms?

2 Why must the Census Bureau base these confidence intervals on *t*-models?

The Census Bureau goes on to say, "These confidence intervals are wider ... for geographic areas with smaller populations and for characteristics that occur less frequently in the area being examined (such as the proportion of people in poverty in a middle-income neighborhood)."

3 Why is this so? For example, why should a confidence interval for the mean amount families spend monthly on housing be wider for a sparsely populated area of farms in the Midwest than for a densely populated area of an urban center? How does the formula show this will happen?

To deal with this problem the Census Bureau reports long-form data only for "... geographic areas from which about two hundred or more long forms were completed—which are large enough to produce good quality estimates. If smaller weighting areas had been used, the confidence intervals around the estimates would have been significantly wider, rendering many estimates less useful ..."

4 Suppose the Census Bureau decided to report on areas from which only 50 long forms were completed. What effect would that have on a 95% confidence interval for, say, the mean cost of housing? Specifically, which values used in the formula for the margin of error would change? Which would change a lot and which would change only slightly?

5 Approximately how much wider would that confidence interval based on 50 forms be than the one based on 200 forms?

A One-Sample *t*-Interval for the Mean Step-By-Step

Let's build a 90% confidence interval for the mean speed of all vehicles traveling on Triphammer Road. The interval that we'll make is called the **one-sample *t*-interval.**

Think

Plan State what we want to know. Identify the parameter of interest.

Identify the variables and review the W's.

Make a picture. Check the distribution shape and look for skewness, multiple modes, and outliers.

REALITY CHECK The histogram centers around 30 mph, and the data lie between 20 and 40 mph. We'd expect a confidence interval to place the population mean within a few mph of 30.

I wish to find a 90% confidence interval for the mean speed, μ, of vehicles driving on Triphammer Road. I have data on the speeds of 23 cars there, sampled on April 11, 2000.

Here's a histogram of the 23 observed speeds.

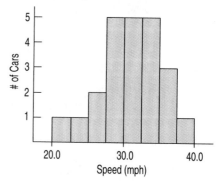

Model Check the conditions.

✔ **Randomization Condition:** Not really met. This is a convenience sample, but I have reason to believe that it is representative.

✔ **Nearly Normal Condition:** The histogram of the speeds is unimodal and symmetric.

State the sampling distribution model for the statistic.

The conditions are satisfied so I will use a Student's *t*-model with

$$(n - 1) = 22 \text{ degrees of freedom}$$

and find a **one-sample *t*-interval for the mean.**

Choose your method.

Show   **Mechanics** Construct the confidence interval.

Be sure to include the units along with the statistics.

Calculating from the data (see p. 518):

$$n = 23 \text{ cars}$$
$$\bar{y} = 31.0 \text{ mph}$$
$$s = 4.25 \text{ mph}$$

The standard error of $\bar{y}$ is:

$$SE(\bar{y}) = \frac{s}{\sqrt{n}} = \frac{4.25}{\sqrt{23}} = 0.886 \text{ mph}.$$

The 90% critical value is $t^*_{22} = 1.717$, so

the margin of error is

$$ME = t^*_{22} \times SE(\bar{y})$$
$$= 1.717(0.886)$$
$$= 1.521 \text{ mph}.$$

The critical value we need to make a 90% interval comes from a Student's *t* table, a computer program, or a calculator. We have $23 - 1 = 22$ degrees of freedom. The selected confidence level says that we want 90% of the probability to be caught in the middle, so we exclude 5% in *each* tail, for a total of 10%. The degrees of freedom and 5% tail probability are all we need to know to find the critical value.

The 90% confidence interval for the mean speed is 31.0 ± 1.5 mph.

REALITY CHECK The result looks plausible and in line with what we thought.

Tell   **Conclusion** Interpret the confidence interval in the proper context.

I am 90% confident that the interval from 29.5 mph to 32.5 mph contains the true mean speed of all vehicles on Triphammer Road.

When we construct confidence intervals in this way, we expect 90% of them to cover the true mean and 10% to miss the true value. That's what "90% confident" means.

Caveat: This was not a random sample of vehicles. It was a convenience sample taken at one time on one day. And the participants were not blinded. Drivers could see the police device, and some may have slowed down. I'm reluctant to extend this inference to other situations.

Using Table T to look up the critical value t^* for a 90% confidence level with 22 degrees of freedom.

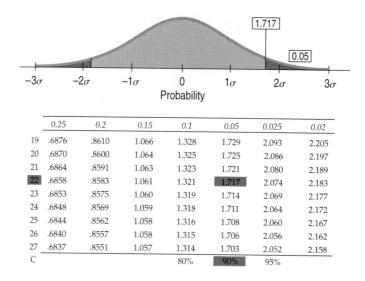

	0.25	0.2	0.15	0.1	0.05	0.025	0.02
19	.6876	.8610	1.066	1.328	1.729	2.093	2.205
20	.6870	.8600	1.064	1.325	1.725	2.086	2.197
21	.6864	.8591	1.063	1.323	1.721	2.080	2.189
22	.6858	.8583	1.061	1.321	1.717	2.074	2.183
23	.6853	.8575	1.060	1.319	1.714	2.069	2.177
24	.6848	.8569	1.059	1.318	1.711	2.064	2.172
25	.6844	.8562	1.058	1.316	1.708	2.060	2.167
26	.6840	.8557	1.058	1.315	1.706	2.056	2.162
27	.6837	.8551	1.057	1.314	1.703	2.052	2.158
C				80%	90%	95%	

A S **The Real Effect of Small Sample Size.** We know that smaller sample sizes lead to wider confidence intervals, but is that just because they have fewer degrees of freedom? Use the interactive confidence interval tool to discover the surprising answer.

Here's the part of the Student's t table that gives the critical value we needed. This table is part of the interactive software on the CD-ROM included with the book. You can find similar technology on a graphing calculator. Table T in the back of the book is a static t table that looks much the same.

To find a critical value, locate the row of the table corresponding to the degrees of freedom and the column corresponding to the probability you want. Our 90% confidence interval leaves 5% of the values on either side, so look for 0.05 at the top of the column or 90% at the bottom. The value in the table at that intersection is the critical value we need: 1.717.

More Cautions About Interpreting Confidence Intervals

So What *Should* We Say?
Since 90% of random samples yield an interval that captures the true mean, we should say "I am 90% confident that the interval from 29.5 to 32.5 mph contains the mean speed of all the vehicles on Triphammer Road." It's also okay to say something less formal: "I am 95% confident that the average speed of all vehicles on Triphammer Road is between 29.5 and 32.5 mph." Remember: *Our uncertainty is about the interval, not the true mean.* The interval varies randomly. The true mean speed is neither variable nor random—just unknown.

Confidence intervals for means offer new tempting wrong interpretations. Here are some things you *shouldn't* say:

- ***Don't say,*** "*90% of all the vehicles* on Triphammer Road drive at a speed between 29.5 and 32.5 mph." The confidence interval is about the *mean* speed, not about the speeds of *individual* vehicles.
- ***Don't say,*** "We are 90% confident that *a randomly selected vehicle* will have a speed between 29.5 and 32.5 mph." This false interpretation is also about individual vehicles rather than about the *mean* of the speeds. We are 90% confident that the *mean* speed of all vehicles on Triphammer Road is between 29.5 and 32.5 mph.
- ***Don't say,*** "The mean speed of the vehicles is 31.0 mph 90% *of the time.*" That's about means, but still wrong. It implies that the true mean varies, when in fact it is the confidence interval that would have been different had we gotten a different sample.
- Finally, ***don't say,*** "90% *of all samples* will have mean speeds between 29.5 and 32.5 mph." That statement suggests that *this* interval somehow sets a standard for every other interval. In fact, this interval is no more (or less) likely to be correct than any other. You could say that 90% of all possible samples will produce intervals that actually do contain the true mean speed. (The problem is that because we'll never know where the true mean speed really is we can't know if our sample was one of those 90%.)

Make a Picture, Make a Picture, Make a Picture

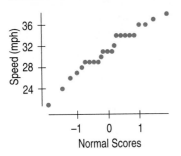

A Normal probability plot of speeds looks reasonably straight.   **Figure 23.3**

The only reasonable way to check the Nearly Normal Condition is with graphs of the data. Make a histogram of the data and verify that its distribution is unimodal and symmetric and that it has no outliers. You may also want to make a Normal probability plot to see that it's reasonably straight. You'll be able to spot deviations from the Normal model more easily with a Normal probability plot, but it's easier to understand the particular nature of the deviations from a histogram.

If you have a computer or graphing calculator doing the work, there's no excuse not to look at *both* displays as part of checking the Nearly Normal Condition.

A Test for the Mean

The residents along Triphammer Road have a more specific concern. It appears that the mean speed along the road is higher than it ought to be. To get the police to patrol more frequently, though, they'll need to show that the true mean speed is *in fact greater* than the 30 mph speed limit. This calls for a hypothesis test called the **one-sample *t*-test for the mean.**

You already know enough to construct this test. The test statistic looks just like the others we've seen. It compares the difference between the observed statistic and a hypothesized value to the standard error of the observed statistic. We already know that for means, the appropriate probability model to use for P-values is Student's *t* with $n-1$ degrees of freedom.

We're ready to go:

> ### One-sample *t*-test for the mean
>
> The conditions for the one-sample *t*-test for the mean are the same as for the one-sample *t*-interval. We test the hypothesis $H_0: \mu = \mu_0$ using the statistic
>
> $$t_{n-1} = \frac{\bar{y} - \mu_0}{SE(\bar{y})}.$$
>
> The standard error of $\bar{y}$ is $SE(\bar{y}) = \dfrac{s}{\sqrt{n}}.$
>
> When the conditions are met and the null hypothesis is true, this statistic follows a Student's *t*-model with $n-1$ degrees of freedom. We use that model to obtain a P-value.

A One-Sample *t*-Test for the Mean Step-By-Step

Let's apply the one-sample *t*-test to the Triphammer Road car speeds. The residents would like to know whether the mean speed exceeds the posted speed limit. The speed limit is 30 mph, so we'll use that as the null hypothesis value.

Think

Plan State what we want to know. Make clear what the population and parameter are.

Identify the variables and review the W's.

Hypotheses The null hypothesis is that the true mean speed is equal to the limit. Because we're interested in whether the vehicles are speeding, the alternative is one sided.

Make a picture. Check the distribution for skewness, multiple modes, and outliers.

REALITY CHECK

The histogram of the observed speeds is clustered around 30, so we'd be surprised to find that the mean was much higher than that. (The fact that 30 is within the confidence interval that we've just found confirms this suspicion.)

Model Check the conditions.

State the sampling distribution model.

Choose your method.

I want to know whether the mean speed of vehicles on Triphammer Road exceeds the posted speed limit of 30 mph. I have a sample of 23 car speeds on April 11, 2000.

H_0: Mean speed, $\mu = 30$ mph
H_A: Mean speed, $\mu > 30$ mph

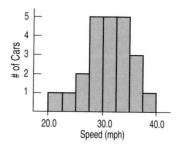

✔ **Randomization Condition:** Although I have a convenience sample, I have reason to believe that it is a representative sample.

✔ **Nearly Normal Condition:** The histogram of the speeds is unimodal and reasonably symmetric.

The conditions are satisfied, so I'll use a Student's *t*-model with $(n - 1) = 22$ degrees of freedom to do a **one-sample *t*-test for the mean.**

Show

Mechanics Be sure to include the units when you write down what you know from the data.

We use the null model to find the P-value. Make a picture of the *t*-model centered at $\mu = 30$. Since this is an upper-tail test, shade the region to the right of the observed mean speed.

The *t*-statistic calculation is just a standardized value, like *z*. We subtract the hypothesized mean and divide by the standard error.

The P-value is the probability of observing a sample mean as large as 31.0 (or larger) *if* the true mean were 30.0, as the null hypothesis states. We can find this P-value from a table, calculator, or computer program.

REALITY CHECK

We're not surprised that the difference isn't statistically significant.

From the data:

$$n = 23 \text{ cars}$$
$$\bar{y} = 31.0 \text{ mph}$$
$$s = 4.25 \text{ mph}$$

$$SE(\bar{y}) = \frac{s}{\sqrt{n}} = \frac{4.25}{\sqrt{23}} = 0.886 \text{ mph}.$$

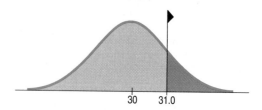

$$t = \frac{\bar{y} - \mu_0}{SE(\bar{y})} = \frac{31.0 - 30.0}{0.886} = 1.13$$

(The observed mean is 1.13 standard errors above the hypothesized value.)

$$\text{P-value} = P(t_{22} > 1.13) = 0.136$$

Tell **Conclusion** Link the P-value to your decision about H_0, and state your conclusion in context.

Unfortunately for the residents, there is no course of action associated with failing to reject this particular null hypothesis.

> The P-value of 0.136 says that if the true mean speed of vehicles on Triphammer Road were 30 mph, samples of 23 vehicles can be expected to have an observed mean of at least 31.0 mph 13.6% of the time. That P-value is not small enough for me to reject the hypothesis that the true mean is 30 mph at any reasonable alpha level. I conclude that there is not enough evidence to say the average speed is too high.

Significance and Importance

Recall that "statistically significant" does not mean "actually important" or "meaningful," even though it sort of sounds that way. In this example, it does seem that speeds may be a bit above 30 miles per hour. Possibly a larger sample would show statistical significance.

But would that be the right decision? The difference between 31 miles per hour and 30 miles per hour doesn't seem meaningful, and rejecting the null hypothesis wouldn't change that. Even with a statistically significant result, it would be hard to convince the police that vehicles on Triphammer Road were driving at dangerously fast speeds. It would probably also be difficult to persuade the town that spending more money to lower the average speed on Triphammer Road would be a good use of the town's resources. Looking at the confidence interval, we can say that, with 90% confidence, the mean speed is somewhere between 29.5 and 32.5 mph. Even in the worst case, if the mean speed is 32.5 mph, would this be a bad enough situation to convince the town to spend more money? Probably not. It's always a good idea when we test a hypothesis to also check the confidence interval and think about the likely values for the mean.

 just checking

In discussing estimates based on the long-form samples, the Census Bureau notes, "The disadvantage . . . is that . . . estimates of characteristics that are also reported on the short form will not match the [long-form estimates]."

The short-form estimates are values from a complete census, so they are the "true" values—something we don't usually have when we do inference.

6. Suppose we use long-form data to make 95% confidence intervals for the mean age of residents for each of 100 of the Census-defined areas. How many of these 100 intervals should we expect will *fail* to include the true mean age (as determined from the complete short-form Census data)?

7. Based on the long-form sample, we might test the null hypothesis that the mean household income in a region was the same as in the previous census. Would the P-value for such a test increase or decrease if we used an area with more long forms?

Intervals and Tests

The 90% confidence interval for the mean speed was 31.0 mph ± 1.5, or (29.5 mph, 32.5 mph). If someone hypothesized that the mean speed was really 30 mph, how would you feel about it? How about 35 mph?

Because the confidence interval included the speed limit of 30 mph, it certainly looked like 30 mph might be a plausible value for the true mean speed of the vehicles on Triphammer Road. In fact, 30 mph gave a P-value of 0.136—too large to reject the null hypothesis. We should have seen this coming. The hypothesized mean of 30 mph lies *within the confidence interval*. It's one of the reasonable values for the mean.

Confidence intervals and significance tests are built from the same calculations. In fact, they are really complementary ways of looking at the same question. Here's the connection: The confidence level contains all the null hypothesis values we can't reject with these data.

More precisely, a level C confidence interval contains *all* of the possible null hypothesis values that would not be rejected by a two-sided hypothesis test at alpha level 1 − C. So a 95% confidence interval matches a 1 − 0.95 = 0.05 level test for these data.

Confidence intervals are naturally two sided, so they match exactly with two-sided hypothesis tests. When, as in our example, the hypothesis is one sided, the corresponding alpha level is (1 − C)/2.

⬤ **Fail to reject** Our 90% confidence interval was 29.5 to 32.5 mph. If any of these values had been the null hypothesis for the mean, then the corresponding hypothesis test at $\alpha = 0.05$ (because $\dfrac{1 - 0.90}{2} = 0.05$) would not have been able to reject the null. That is, the corresponding one-sided P-value for our observed mean of 31 mph would be greater than 0.05. So, we would retain any hypothesized value between 29.5 and 32.5 mph. ⬤

Sample Size

How large a sample do we need? The simple answer is "more." But more data cost money, effort, and time, so how much is enough? Suppose your computer just took 2 hours to download a movie you wanted to watch. You're not happy. You hear about a program that claims to download movies in less than an hour. You're interested enough to spend $29.95 for it, but only if it really delivers. So you get the free evaluation copy and test it by downloading that movie 10 different times. Of course, the mean download time is not exactly 1 hour as claimed. Observations vary. If the margin of error were 8 minutes, though, you'd probably be able to decide whether the software is worth the money. Doubling the sample size would require another 10 hours of testing and would reduce your margin of error to a bit under 6 minutes. You'll need to decide whether that's worth the effort.

As we make plans to collect data, we should have some idea of how large a margin of error we need to be able to draw a conclusion or detect a difference we want to see. If the size of the effect we're studying is large, then we may be able to tolerate a larger *ME*. If we need great precision, however, we'll want a smaller *ME*, and, of course, that means a larger sample size.

Armed with the *ME* and confidence level, we can find the sample size we'll need. Almost.

We know that for a mean, $ME = t^*_{n-1} \times SE(\bar{y})$ and that $SE(\bar{y}) = \dfrac{s}{\sqrt{n}}$, so we can determine the sample size by solving this equation for *n*:

$$ME = t^*_{n-1} \frac{s}{\sqrt{n}}.$$

<table>
<tr><td>

Doing sample size calculations by hand

Let's give the sample size formula a spin. Suppose we want an *ME* of 8 minutes and we think the standard deviation of download times is about 10 minutes. Using a 95% confidence interval, and $z^* = 1.96$, we solve:

$$8 = 1.96\frac{10}{\sqrt{n}}$$

$$\sqrt{n} = \frac{1.96 \times 10}{8} = 2.45$$

$$n = (2.45)^2 = 6.0025$$

That's a small sample size, so we use $(6 - 1) = 5$ degrees of freedom to substitute an appropriate t^* value. At 95%, $t_5^* = 2.571$. Now solve the equation one more time:

$$8 = 2.571\frac{10}{\sqrt{n}}$$

$$\sqrt{n} = \frac{2.571 \times 10}{8} \approx 3.214$$

$$n = (3.214)^2 \approx 10.33$$

To make sure the *ME* is no larger than you want, you should always round *up*, which gives $n = 11$ runs. So, to get an *ME* of 8 minutes, we should find the downloading times for $n = 11$ movies.

</td></tr>
</table>

The good news is that we have an equation; the bad news is that we won't know most of the values we need to solve it. When we thought about sample size for proportions back in Chapter 19, we ran into a similar problem. There we had to guess a working value for p to compute a sample size. Here, we need to know s. We don't know s until we get some data, but we want to calculate the sample size *before* collecting the data. We might be able to make a good guess, and that is often good enough for this purpose. If we have no idea what the standard deviation might be, or if the sample size really matters (for example, because each additional individual is very expensive to sample or experiment on), it might be a good idea to run a small *pilot study* to get some feeling for the standard deviation.

That's not all. Without knowing n, we don't know the degrees of freedom and we can't find the critical value, t_{n-1}^*. One common approach is to use the corresponding z^* value from the Normal model. If you've chosen a 95% confidence level, then just use 2, following the 68-95-99.7 Rule. If your estimated sample size is, say, 60 or more, it's probably okay—z^* was a good guess. If it's smaller than that, you may want to add a step, using z^* at first, finding n, and then replacing z^* with the corresponding t_{n-1}^* and calculating the sample size once more.

Sample size calculations are *never* exact. The margin of error you find *after* collecting the data won't match exactly the one you used to find n. The sample size formula depends on quantities that you won't have until you collect the data, but using it is an important first step. Before you collect data, it's always a good idea to know whether the sample size is large enough to give you a good chance of being able to tell you what you want to know.

Degrees of Freedom

Some calculators offer an alternative button for standard deviation that divides by n instead of $n - 1$. Why don't you stick a wad of gum over the "n" button so you won't be tempted to use it? Use $n - 1$.

The number of degrees of freedom, $(n - 1)$, might have reminded you of the value we divide by to find the standard deviation of the data (since, in fact, it's the same number). We promised back when we introduced that formula to say a bit more about why we divide by $n - 1$ rather than by n. The reason is closely tied to the reasoning of the *t*-distribution.

If only we knew the true population mean, μ, we would find the sample standard deviation as

$$s = \sqrt{\frac{\sum(y - \mu)^2}{n}} \qquad \text{(Equation 23.1)[4]}$$

We use $\bar{y}$ instead of μ, though, and that causes a problem. For any sample, the data values will generally be closer to their own sample mean than to the true population mean, μ. Why is that? Imagine that we take a random sample of 10 high-school seniors. The mean SAT verbal score is 500 in the United States. But

---

[4] Statistics textbooks usually have equation numbers so they can talk about equations by name. We haven't needed equation numbers yet, but we admit it's useful here, so this is our first.

the sample mean, $\bar{y}$, for *these* 10 seniors won't be exactly 500. Are the 10 seniors' scores closer to 500 or $\bar{y}$? They'll always be closer to their own average $\bar{y}$. If we used $\sum(y - \bar{y})^2$ instead of $\sum(y - \mu)^2$ in Equation 23.1 to calculate s, our standard deviation estimate would be too small. How can we fix it? The amazing mathematical fact is that we can compensate for the smaller sum exactly by dividing by $n - 1$ instead of by n. So that's all the $n - 1$ is doing in the denominator of s. And we call $n - 1$ the degrees of freedom.

*The Sign Test—Back to Yes and No

Another, and perhaps simpler, way to look at the Triphammer Road data would be to ignore the actual speed and just ask, "Is the car speeding?" Rather than record the speed, we might have recorded a "yes" (or "1") for cars going over 30 mph and a "no" (or "0") for the cars whose speed is below the posted limit (we'll ignore cars going exactly 30 mph).

What null hypothesis can we use? Well, if drivers were really trying to maintain a 30 mph speed, they might miss randomly above and below that target. We'd expect the number of cars driving faster than 30 mph to be about the same as the number driving slower than 30. In that case 30 mph would be the *median* speed, and so our null hypothesis says that the median is 30. If that null hypothesis were true, we'd expect the proportion of cars driving faster than 30 mph to be 0.50. On the other hand, if the median speed were greater than 30 mph, we'd expect to see more cars driving faster than 30.

What we've done is to turn the quantitative data about car speeds into a set of yes/no values (Bernoulli trials from Chapter 17). And we've turned a question about the median car speed into a test of a proportion (Is the proportion of cars that are going faster than the speed limit 0.50?). We already know how to do a test of proportions, so this isn't a new situation at all.

When we test a median by counting the number of values above and below that value, it's called a **sign test.** The sign test is a *distribution-free* method, so called because there are no *distributional* assumptions or conditions on the data. Specifically, because we no longer have quantitative data, we're not requiring the Nearly Normal Condition.

We already know all we need to do the test Step-by-Step:

*A Sign Test Step-By-Step

 Think

Plan State what we want to know.

Identify the parameter of interest. Here it's the population median.

Identify the variables and review the W's.

Hypotheses Write the null and alternative hypotheses.

I want to know whether the median speed of cars on Triphammer Road is 30 mph.

I have 22 car speeds (one car measured at exactly 30 mph was omitted) and have recorded whether their speed exceeded the 30 mph limit or not.

H_0: The median speed of cars on Triphammer Road is 30 mph. Equivalently, the proportion of cars exceeding 30 mph is 50%.

There is not a great need to plot the data. Medians are resistant to the effects of skewness or outliers.

$H_0: p_0 = 0.50.$

$H_A:$ The true proportion of speeders is more than 0.50. $p_0 > 0.50.$

Model Check the conditions. The sign test doesn't require the Nearly Normal Condition.

✔ **Random Sampling Condition:** The data are a convenience sample, not drawn with randomization, but they are likely to be representative.

✔ **10% Condition:** The data are from what could be a very large number of cars.

✔ **Success/Failure Condition:** Both $np_0 = 22(0.5) = 11$ and $nq_0 = 22(0.5) = 11$ are greater than 10, showing that I expect at least 10 successes and at least 10 failures.

Choose your method. (The sign test is just a one-proportion z-test for $p_0 = 0.5$.)

Because the conditions are satisfied, I'll do a **sign test.**

Show   **Mechanics** We use the null model to find the P-value, the probability of observing a proportion as far from the hypothesized proportion as the one we did observe, or even farther.

$$SD(\hat{p}) = \sqrt{\frac{0.5 \times 0.5}{22}} = 0.107.$$

Of the 22 cars, 13 had speeds over 30 mph (one of the original 23 was going 30 mph), so the observed proportion, $\hat{p}$, is 0.591.

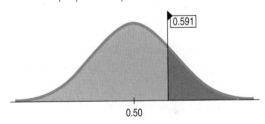

The P-value is the conditional probability of observing a sample proportion as large as 0.591 (or larger) if the null hypothesis is true:

$$P = P(\hat{p} \geq 0.591 \,|\, p_0 = 0.50).$$

The probability of observing a value 0.850 standard deviations or more above the mean of a Normal model can be found by computer, calculator, or table to be 0.197.

The observed value has a z-score of

$z = \frac{0.591 - 0.5}{0.107} = 0.850$, so it is 0.85 standard deviations above the hypothesized proportion.

The P-value is 0.197.

Tell   **Conclusion** Link the P-value to your decision, then state your conclusion in the proper context.

The P-value of 0.197 is not very small, so I fail to reject the null hypothesis. There is insufficient evidence to suggest that cars are speeding on Triphammer Road.

The sign test *is* simpler than the *t*-test, and it requires fewer assumptions. We need only yes/no data. We still should check the Randomization Condition, but we no longer need the Nearly Normal Condition. When the data satisfy all the assumptions and conditions for a *t*-test on the mean, we usually prefer the *t*-test because it is more powerful than the sign test; for the same data, the P-value from the *t*-test would be smaller than the P-value from the sign test. (In fact, the P-value for the *t*-test was 0.136.) That's because the *t*-test uses the actual *values*, which contain much more information than just knowing whether those values are over 30. The more information we use, the more potential our conclusions have to be significant.

On the other hand, the sign test works even when the data have outliers or a skewed distribution—problems that can distort the results of the *t*-test and reduce its power. When we have doubts whether the conditions for the *t*-test are satisfied, it's a good idea to perform a sign test.[5]

What Can Go Wrong?

Student's *t* methods work only when the Normality Assumption is true. Naturally, many of the ways things can go wrong turn out to be different ways that the Normality Assumption can fail. It's always a good idea to look for the most common kinds of failure. It turns out that you can even fix some of them.

- **Beware of multimodality.** The Nearly Normal Condition clearly fails if a histogram of the data has two or more modes. When you see this, look for the possibility that your data come from two groups. If so, your best bet is to try to separate the data into different groups. (Use the variables to help distinguish the modes, if possible. For example, if the modes seem to be composed mostly of men in one and women in the other, split the data according to sex.) Then you could analyze each group separately.

- **Beware of skewed data.** Make a Normal probability plot and a histogram of the data. If the data are very skewed, you might try re-expressing the variable. Re-expressing may yield a distribution that is unimodal and symmetric, more appropriate for the inference methods for means. Re-expression cannot help if the sample distribution is not unimodal. Some people may object to re-expressing the data, but unless your sample is very large, you just can't use the methods of this chapter on skewed data.

> As tempting as it is to get rid of annoying values, you can't just throw away outliers and not discuss them. It isn't appropriate to lop off the highest or lowest values just to improve your results.

- **Set outliers aside.** The Nearly Normal Condition also fails if the data have outliers. If you find outliers in the data, you may want to set them aside before using Student's *t* methods. Once they are singled out, you should look at them carefully. Sometimes, it's obvious that a data value is wrong and the justification for removing or correcting it is clear. When there's no clear justification for removing outliers, you might want to run the analysis both with and without the outliers and note any differences in your conclusions. Any time data values are set aside, you *must* report on them individually. Often they will turn out to be the most informative part of your report on the data.[6]

- **Watch out for bias.** Measurements of all kinds can be biased. If your observations differ from the true mean in a systematic way, your confidence interval may not capture the true mean. And there is no sample size that will save you. A bathroom scale that's 5 pounds off will be 5 pounds off even if

---

[5] It's probably a good idea to routinely compute both. If they agree, then the inference is clear. If they differ, it may be interesting and important to see why.

[6] This suggestion may be controversial in some disciplines. Setting aside outliers is seen by some as "cheating" because the result is likely to be a narrower confidence interval or a smaller P-value. But an analysis of data with outliers left in place is *always* wrong. The outliers violate the Nearly Normal Condition and also the implicit assumption of a homogeneous population, so they invalidate inference procedures. An analysis of the nonoutlying points, along with a separate discussion of the outliers, is often much more informative, and can reveal important aspects of the data.

you weigh yourself 100 times and take the average. We've seen several sources of bias in surveys, but measurements can be biased, too. Be sure to think about possible sources of bias in your measurements.

- **Make sure data are independent.** Student's *t* methods also require the sampled values to be mutually independent. We check for random sampling and the 10% Condition. You should also think hard about whether there are likely violations of independence in the data collection method. If there are, be very cautious about using these methods.

- **Make sure that data are from an appropriately randomized sample.** Ideally, all data that we analyze are drawn from a simple random sample or generated by a randomized experiment. When they're not, be careful about making inferences from them. You may still compute a confidence interval correctly, or get the mechanics of the P-value right, but this might not save you from making a serious mistake in inference.

- **Interpret your confidence interval correctly.** Many statements that sound tempting are, in fact, misinterpretations of a confidence interval for a mean. You might want to have another look at some of the common mistakes, explained on page 526. Keep in mind that a confidence interval is about the mean of the population, not about the means of samples, individuals in samples, or individuals in the population.

CONNECTIONS

The steps for finding a confidence interval or hypothesis test for means are just like the corresponding steps for proportions. Even the form of the calculations is similar. As the *z*-statistic did for proportions, the *t*-statistic tells us how many standard errors our sample mean is from the hypothesized mean. For means, though, we have to estimate the standard error separately. This added uncertainty changes the model for the sampling distribution from *z* to *t*.

As with all of our inference methods, the randomization applied in drawing a random sample or in randomizing a comparative experiment is what generates the sampling distribution. Randomization is what makes inference in this way possible at all.

The new concept of degrees of freedom connects back to the denominator of the sample standard deviation calculation, as shown earlier.

There's just no escaping histograms and Normal probability plots. The Nearly Normal Condition required to use Student's *t* can be checked best by making appropriate displays of the data. Back when we first used histograms, we looked at their shape and in particular checked whether they were unimodal and symmetric and whether they showed any outliers. Those are just the features we check for here. The Normal probability plot zeros in on the Normal model a little closer.

What have we learned?

We first learned to create confidence intervals and test hypotheses about proportions. Now we've turned our attention to means, and learned that statistical inference for means relies on the same concepts; only the mechanics and our model have changed.

- We've learned that what we can say about a population mean is inferred from data, using the mean of a representative random sample.

- We've learned to describe the sampling distribution of sample means using a new model we select from the Student's *t* family based on our degrees of freedom.
- We've learned that our ruler for measuring the variability in sample means is the standard error $SE(\bar{y}) = \dfrac{s}{\sqrt{n}}$.
- We've learned to find the margin of error for a confidence interval using that ruler and critical values based on a Student's *t*-model.
- And we've also learned to use that ruler to test hypotheses about the population mean.

Above all, we've learned that the reasoning of inference, the need to verify that the appropriate assumptions are met, and the proper interpretation of confidence intervals and P-values all remain the same regardless of whether we are investigating means or proportions.

TERMS

Student's *t*
Degrees of freedom (df)
A family of distributions indexed by its degrees of freedom. The *t*-models are unimodal, symmetric, and bell shaped, but generally have fatter tails and a narrower center than the Normal model. As the degrees of freedom increase, *t*-distributions approach the Normal.

One-sample *t*-interval
A one-sample *t*-interval for the population mean is $\bar{y} \pm t^{*}_{n-1} \times SE(\bar{y})$ where

$$SE(\bar{y}) = \frac{s}{\sqrt{n}}.$$

The critical value t^{*}_{n-1} depends on the particular confidence level, C, that you specify and on the number of degrees of freedom, $n - 1$.

One-sample *t*-test for the mean
The one-sample *t*-test for the mean tests the hypothesis $H_0: \mu = \mu_0$ using the statistic

$$t_{n-1} = \frac{\bar{y} - \mu_0}{SE(\bar{y})}.$$

The standard error of $\bar{y}$ is

$$SE(\bar{y}) = \frac{s}{\sqrt{n}}.$$

***Sign test**
A **distribution-free** method to analyze quantitative data by using yes/no values and testing the hypothesis $H_0: p_0 = 0.50$.

SKILLS   *When you complete this lesson you should:*

Think
- Know the assumptions required for *t*-tests and *t*-based confidence intervals.
- Know how to examine your data for violations of conditions that would make inference about the population mean unwise or invalid.
- Understand that a confidence interval and a hypothesis test are essentially equivalent. You can do a two-tailed hypothesis test at level of significance α with a $1 - \alpha$ confidence interval, or a one-tailed test with a $1 - 2\alpha$ confidence interval.

Show
- Be able to compute and interpret a *t*-test for the population mean using a statistics package or working from summary statistics for a sample.
- Be able to compute and interpret a *t*-based confidence interval for the population mean using a statistics package or working from summary statistics for a sample.

Tell

- Be able to explain the meaning of a confidence interval for a population mean. Make clear that the randomness associated with the confidence level is a statement about the interval bounds and not about the population parameter value.

- Understand that a 95% confidence interval does not trap 95% of the sample values.

- Be able to interpret the result of a test of a hypothesis about a population mean.

- Know that we do not "accept" a null hypothesis if we cannot reject it. We say that we fail to reject it.

- Understand that the P-value of a test does not give the probability that the null hypothesis is correct.

Inference for Means on the Computer

Statistics packages offer convenient ways to make histograms of the data. Even better for assessing near-Normality is a Normal probability plot. When you work on a computer, there is simply no excuse for skipping the step of plotting the data to check that it is nearly Normal. *Beware:* Statistics packages don't agree on whether to place the Normal scores on the *x*-axis (as we have done) or the *y*-axis. Read the axis labels.

Any standard statistics package can compute a hypothesis test. Here's what the package output might look like in general (although no package we know gives the results in exactly this form)[7]:

A S Student's *t* in Practice.
We almost always use technology to do inference with Student's *t*. Here's a chance to do that as you investigate several questions.

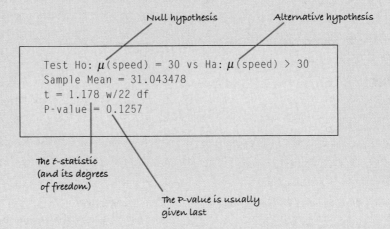

Null hypothesis   *Alternative hypothesis*

```
Test Ho: μ(speed) = 30 vs Ha: μ(speed) > 30
Sample Mean = 31.043478
t = 1.178 w/22 df
P-value = 0.1257
```

The t-statistic (and its degrees of freedom)

The P-value is usually given last

The package computes the sample mean and sample standard deviation of the variable and finds the P-value from the *t*-distribution based on the appropriate number of degrees of freedom. All modern statistics packages report P-values. The package may also provide additional information such as the sample mean, sample standard deviation, *t*-statistic value, and degrees of freedom. These are useful for interpreting the resulting P-value and telling the difference between a meaningful result and one that is merely statistically significant. Statistics packages that report the estimated standard deviation of the sampling distribution usually label it "standard error" or "SE."

---

[7] Many statistics packages keep as many as 16 digits for all intermediate calculations. If we had kept as many, our results in the Step-by-Step section would have been closer to these.

Inference results are also sometimes reported in a table. You may have to read carefully to find the values you need. Often, test results and the corresponding confidence interval bounds are given together. And often you must read carefully to find the alternative hypotheses. Here's an example of that kind of output:

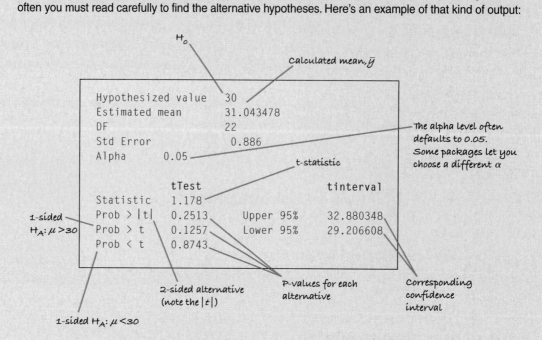

H_0

Calculated mean, $\bar{y}$

Hypothesized value	30
Estimated mean	31.043478
DF	22
Std Error	0.886
Alpha	0.05

The alpha level often defaults to 0.05. Some packages let you choose a different α

t-statistic

	tTest		tinterval
Statistic	1.178		
Prob > |t|	0.2513	Upper 95%	32.880348
Prob > t	0.1257	Lower 95%	29.206608
Prob < t	0.8743		

1-sided $H_A: \mu > 30$

1-sided $H_A: \mu < 30$

2-sided alternative (note the $|t|$)

P-values for each alternative

Corresponding confidence interval

The commands to do inference for means on common statistics programs and calculators are not always obvious. (By contrast, the resulting output is usually clearly labeled and easy to read.) The guides for each program can help you start navigating.

DATA DESK

Select variables.
From the **Calc** menu, choose **Estimate** for confidence intervals or **Test** for hypothesis tests. Select the interval or test from the drop-down menu and make other choices in the dialog.

EXCEL

Specify formulas. Find t^* with the TINV(alpha, df) function.

Comments
Not really automatic. There's no easy way to find P-values in Excel.

JMP

From the **Analyze** menu select **Distribution**. For a confidence interval, scroll down to the "Moments" section to find the interval limits. For a hypothesis test, click the red triangle next to the variable's name and choose **Test Mean** from the menu, then fill in the resulting dialog.

Comments
"Moment" is a fancy statistical term for means, standard deviations, and other related statistics.

MINITAB

From the **Stat** menu choose the **Basic Statistics** submenu. From that menu, choose **1-sample t . . .** Then fill in the dialog.

Comments

The dialog offers a clear choice between confidence interval and test.

SPSS

From the **Analyze** menu choose the **Compare Means** submenu. From that, choose the **One-Sample t-test** command.

Comments

The commands suggest neither a single mean nor an interval. But the results provide both a test and an interval.

TI-83/84 Plus

Finding a confidence interval:
In the **STAT TESTS** menu choose **8:TInterval.** You may specify that you are using data stored in a list, or you may enter the mean, standard deviation, and sample size. You must also specify the desired level of confidence.

Testing a hypothesis:
In the **STAT TESTS** menu choose **2:T-Test.** You may specify that you are using data stored in a list or you may enter the mean, standard deviation, and size of your sample. You must also specify the hypothesized model mean and whether the test is to be two-tail, lower-tail, or upper-tail.

TI-89

Finding a confidence interval:
In the **STAT Ints** menu choose **2:TInterval.** Specify whether you are using data stored in a list or if you will enter the mean, standard deviation, and sample size. You must also specify the desired level of confidence.

Testing a hypothesis:
In the **STAT Tests** menu choose **2:T-Test.** You must specify whether you are using data stored in a list or if you will enter the mean, standard deviation, and size of your sample. You must also specify the hypothesized model mean and whether the test is to be two-tail, lower-tail, or upper-tail. Select whether the test is to be simply computed or whether to display the distribution curve and highlight the area corresponding to the P-value of the test.

EXERCISES

1. *t-models, part I.* Using the *t* tables, software, or a calculator, estimate
   a) the critical value of *t* for a 90% confidence interval with df = 17.
   b) the critical value of *t* for a 98% confidence interval with df = 88.
   c) the P-value for $t \geq 2.09$ with 4 degrees of freedom.
   d) the P-value for $|t| > 1.78$ with 22 degrees of freedom.

2. *t-models, part II.* Using the *t* tables, software, or a calculator, estimate
   a) the critical value of *t* for a 95% confidence interval with df = 7.
   b) the critical value of *t* for a 99% confidence interval with df = 102.
   c) the P-value for $t \leq 2.19$ with 41 degrees of freedom.
   d) the P-value for $|t| > 2.33$ with 12 degrees of freedom.

3. *t-models, part III.* Describe how the shape, center, and spread of *t*-models change as the number of degrees of freedom increases.

4. *t-models, part IV (last one!).* Describe how the critical value of *t* for a 95% confidence interval changes as the number of degrees of freedom increases.

5. **Cattle.** Livestock are given a special feed supplement to see if it will promote weight gain. The researchers report that the 77 cows studied gained an average of 56 pounds, and that a 95% confidence interval for the mean weight gain this supplement produces has a margin of error of ±11 pounds. Some students wrote the following conclusions. Did anyone interpret the interval correctly? Explain any misinterpretations.
   a) 95% of the cows studied gained between 45 and 67 pounds.

b) We're 95% sure that a cow fed this supplement will gain between 45 and 67 pounds.

c) We're 95% sure that the average weight gain among the cows in this study was between 45 and 67 pounds.

d) The average weight gain of cows fed this supplement will be between 45 and 67 pounds 95% of the time.

e) If this supplement is tested on another sample of cows, there is a 95% chance that their average weight gain will be between 45 and 67 pounds.

6. Teachers. Software analysis of the salaries of a random sample of 288 Nevada teachers produced the confidence interval shown below. Which conclusion is correct? What's wrong with the others?

t-Interval for μ: with 90.00% Confidence,
$38944 < \mu(\text{TchPay}) < 42893$

a) If we took many random samples of Nevada teachers, about 9 out of 10 of them would produce this confidence interval.

b) If we took many random samples of Nevada teachers, about 9 out of 10 of them would produce a confidence interval that contained the mean salary of all Nevada teachers.

c) About 9 out of 10 Nevada teachers earn between $38,944 and $42,893.

d) About 9 out of 10 of the teachers surveyed earn between $38,944 and $42,893.

e) We are 90% confident that the average teacher salary in the United States is between $38,944 and $42,893.

7. Meal plan. After surveying students at Dartmouth College, a campus organization calculated that a 95% confidence interval for the mean cost of food for one term (of three in the Dartmouth trimester calendar) is ($780, $920). Now the organization is trying to write its report, and considering the following interpretations. Comment on each.

a) 95% of all students pay between $780 and $920 for food.

b) 95% of the sampled students paid between $780 and $920.

c) We're 95% sure that students in this sample averaged between $780 and $920 for food.

d) 95% of all samples of students will have average food costs between $780 and $920.

e) We're 95% sure that the average amount all students pay is between $780 and $920.

8. Rain. Based on meteorological data for the past century, a local TV weatherman estimates that the region's average winter snowfall is 23", with a margin of error of ±2 inches. Assuming he used a 95% confidence interval, how should viewers interpret this news? Comment on each of these statements.

a) During 95 of the last 100 winters, the region got between 21" and 25" of snow.

b) There's a 95% chance the region will get between 21" and 25" of snow this winter.

c) There will be between 21" and 25" of snow on the ground for 95% of the winter days.

d) Residents can be 95% sure that the area's average snowfall is between 21" and 25".

e) Residents can be 95% confident that the average snowfall during the last century was between 21" and 25" per winter.

T 9. Pulse rates. A medical researcher measured the pulse rates (beats per minute) of a sample of randomly selected adults and found the following Student's *t*-based confidence interval:

With 95.00% Confidence,
$70.887604 < \mu(\text{Pulse}) < 74.497011$

a) Explain carefully what the software output means.

b) What's the margin of error for this interval?

c) If the researcher had calculated a 99% confidence interval, would the margin of error be larger or smaller? Explain.

10. Crawling. Data collected by child development scientists produced this confidence interval for the average age (in weeks) at which babies begin to crawl.

t-Interval for μ
(95.00% Confidence): $29.202 < \mu(\text{age}) < 31.844$

a) Explain carefully what the software output means.

b) What is the margin of error for this interval?

c) If the researcher had calculated a 90% confidence interval, would the margin of error be larger or smaller? Explain.

T 11. Normal temperature. The researcher described in Exercise 9 also measured the body temperatures of that randomly selected group of adults. The data he collected are summarized below. We wish to estimate the average (or "normal") temperature among the adult population.

Summary	Temperature
Count	52
Mean	98.285
Median	98.200
MidRange	98.600
StdDev	0.6824
Range	2.800
IntQRange	1.050

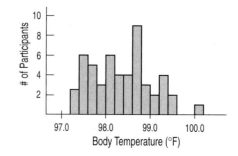

a) Are the necessary conditions for a *t*-interval satisfied? Explain.

b) Find a 98% confidence interval for mean body temperature.

c) Explain the meaning of that interval.

d) Explain what "98% confidence" means in this context.

e) 98.6°F is commonly assumed to be "normal." Do these data suggest otherwise? Explain.

*f) Of the 52 observations, 15 are above 98.6°F, 31 are below, and 6 are recorded as equal to 98.6°F. Perform a sign test to test whether the median temperature is 98.6°F. Does your conclusion agree with what you found in part e?

12. Parking. Hoping to lure more shoppers downtown, a city builds a new public parking garage in the central business district. The city plans to pay for the structure through parking fees. During a two-month period (44 weekdays), daily fees collected averaged $126, with a standard deviation of $15.

a) What assumptions must you make in order to use these statistics for inference?

b) Write a 90% confidence interval for the mean daily income this parking garage will generate.

c) Explain in context what this confidence interval means.

d) Explain what "90% confidence" means in this context.

e) The consultant who advised the city on this project predicted that parking revenues would average $130 per day. Based on your confidence interval, do you think the consultant could have been correct? Why?

Ⓣ 13. Normal temperatures, part II. Consider again the statistics about human body temperature in Exercise 11.

a) Would a 90% confidence interval be wider or narrower than the 98% confidence interval you calculated before? Explain. (You should not need to compute the new interval.)

b) What are the advantages and disadvantages of the 98% confidence interval?

c) If we conduct further research, this time using a sample of 500 adults, how would you expect the 98% confidence interval to change? Explain.

d) How large a sample would you need to estimate the mean body temperature to within 0.1 degrees with 98% confidence?

14. Parking II. Suppose that for budget planning purposes the city in Exercise 12 needs a better estimate of the mean daily income from parking fees.

a) Someone suggests that the city use its data to create a 95% confidence interval instead of the 90% interval first created. How would this interval be better for the city? (You need not actually create the new interval.)

b) How would the 95% interval be worse for the planners?

c) How could they achieve an interval estimate that would better serve their planning needs?

d) How many days' worth of data must they collect to have 95% confidence of estimating the true mean to within $3?

15. Hot dogs. A nutrition laboratory tests 40 "reduced sodium" hot dogs, finding that the mean sodium content is 310 mg, with a standard deviation of 36 mg.

a) Find a 95% confidence interval for the mean sodium content of this brand of hot dog.

b) What assumptions have you made in this inference? Are the appropriate conditions satisfied?

c) Explain clearly what your interval means.

16. Speed of light. In 1882 Michelson measured the speed of light (usually denoted "*c*" as in the famous statement $e = mc^2$). His values are in km/sec and have 299,000 subtracted from them. He reported the results of 23 trials with a mean of 756.22 and a standard deviation of 107.12.

a) Find a 95% confidence interval for the true speed of light from these statistics.

b) State in words what this interval means. Keep in mind that the speed of light is a physical constant that, as far as we know, has a value that is true throughout the universe.

c) What assumptions must you make in order to use your method?

17. Second dog. The nutrition lab in Exercise 15 tests the hot dogs again, this time using a sample of 60 "reduced sodium" frankfurters. The new sample produces a mean of 318 mg of sodium, and the standard deviation is 32 mg.

a) Should this larger sample of hot dogs produce a more accurate estimate of the mean sodium content? Explain.

b) What is the standard error of the mean sodium content?

c) Find and interpret a 95% confidence interval.

d) Food labeling regulations require that any food identified as "reduced sodium" must have at least 30% less sodium than its regular counterpart. If regular franks average 465 mg of sodium, should this brand be labeled "reduced"? Explain, using your confidence interval.

18. Better light. After his first attempt to determine the speed of light (described in Exercise 16), Michelson conducted an "improved" experiment. In 1897 he reported results of 100 trials with a mean of 852.4 and a standard deviation of 79.0.

a) What is the standard error of the mean for these data?

b) Without computing it, how would you expect a 95% confidence interval for the second experiment to differ from the confidence interval for the first? Note at least three specific reasons why they might differ, and indicate the ways in which these differences would change the interval.

c) According to Stigler (who reports these values), the true speed of light is 299,710.5 km/sec, corresponding to a value of 710.5 for Michelson's 1897 measurements. What does this indicate about Michelson's two experiments? Explain, using your confidence interval.

19. Pizza. A researcher tests whether the mean cholesterol level among those who eat frozen pizza exceeds the value considered to indicate a health risk. She gets a P-value of 0.07. Explain in this context what the "7%" represents.

20. Golf balls. The United States Golf Association sets performance standards for golf balls. For example, the initial velocity of the ball may not exceed 250 feet per second when measured by an apparatus approved by the USGA. Suppose a manufacturer introduces a new kind of ball, and provides a sample for testing. Based on the mean speed in the test, the USGA comes up with a P-value of 0.34. Explain in this context what the "34%" represents.

21. TV safety. The manufacturer of a metal stand for home TV sets must be sure that its product will not fail under the weight of the TV. Since some larger sets weigh nearly 300 pounds, the company's safety inspectors have set a standard of ensuring that the stands can support an average of over 500 pounds. Their inspectors regularly subject a random sample of the stands to increasing weight until they fail. They test the hypothesis $H_0: \mu = 500$ against $H_A: \mu > 500$, using the level of significance $\alpha = 0.01$. If the sample of stands fail to pass this safety test, the inspectors will not certify the product for sale to the general public.
a) Is this an upper-tail or lower-tail test? In the context of the problem, why do you think this is important?
b) Explain what will happen if the inspectors commit a Type I error.
c) Explain what will happen if the inspectors commit a Type II error.

22. Catheters. During an angiogram, heart problems can be examined via a small tube (a catheter) threaded into the heart from a vein in the patient's leg. It's important that the company who manufactures the catheter maintain a diameter of 2.00 mm. (The standard deviation is quite small.) Each day, quality control personnel make several measurements to test $H_0: \mu = 2.00$ against $H_A: \mu \neq 2.00$ at a significance level of $\alpha = 0.05$. If they discover a problem, they will stop the manufacturing process until it is corrected.
a) Is this a one-sided or two-sided test? In the context of the problem, why do you think this is important?
b) Explain in this context what happens if the quality control people commit a Type I error.
c) Explain in this context what happens if the quality control people commit a Type II error.

23. TV safety revisited. The manufacturer of the metal TV stands in Exercise 21 is thinking of revising its safety test.
a) If the company's lawyers are worried about being sued for selling an unsafe product, should they increase or decrease the value of α? Explain.

b) In this context, what is meant by the power of the test?
c) If the company wants to increase the power of the test, what options does it have? Explain the advantages and disadvantages of each option.

24. Catheters again. The catheter company in Exercise 22 is reviewing its testing procedure.
a) Suppose the significance level is changed to $\alpha = 0.01$. Will the probability of Type II error increase, decrease, or remain the same?
b) What is meant by the power of the test the company conducts?
c) Suppose the manufacturing process is slipping out of proper adjustment. As the actual mean diameter of the catheters produced gets farther and farther above the desired 2.00 mm, will the power of the quality control test increase, decrease, or remain the same?
d) What could they do to improve the power of the test?

25. Marriage. In 1960, census results indicated that the age at which American men first married had a mean of 23.3 years. It is widely suspected that young people today are waiting longer to get married. We want to find out if the mean age of first marriage has increased during the past 40 years.
a) Write appropriate hypotheses.
b) We plan to test our hypothesis by selecting a random sample of 40 men who married for the first time last year. Do you think the necessary assumptions for inference are satisfied? Explain.
c) Describe the approximate sampling distribution model for the mean age in such samples.
d) The men in our sample married at an average age of 24.2 years with a standard deviation of 5.3 years. What's the P-value for this result?
e) Explain (in context) what this P-value means.
f) What's your conclusion?

26. Fuel economy. A company with a large fleet of cars hopes to keep gasoline costs down, and sets a goal of attaining a fleet average of at least 26 miles per gallon. To see if the goal is being met, they check the gasoline usage for 50 company trips chosen at random, finding a mean of 25.02 mpg and a standard deviation of 4.83 mpg. Is this strong evidence that they have failed to attain their fuel economy goal?
a) Write appropriate hypotheses.
b) Are the necessary assumptions to perform inference satisfied?
c) Describe the sampling distribution model of mean fuel economy for samples like this.
d) Find the P-value.
e) Explain what the P-value means in this context.
f) State an appropriate conclusion.

T 27. Ruffles. Students investigating the packaging of potato chips purchased 6 bags of Lay's Ruffles marked with a net weight of 28.3 grams. They carefully weighed the contents of each bag, recording the following weights (in grams): 29.3, 28.2, 29.1, 28.7, 28.9, 28.5.

a) Do these data satisfy the assumptions for inference? Explain.
b) Find the mean and standard deviation of the observed weights.
c) Create a 95% confidence interval for the mean weight of such bags of chips.
d) Explain in context what your interval means.
e) Comment on the company's stated net weight of 28.3 grams.

Ⓣ 28. Doritos. Some students checked 6 bags of Doritos marked with a net weight of 28.3 grams. They carefully weighed the contents of each bag, recording the following weights (in grams): 29.2, 28.5, 28.7, 28.9, 29.1, 29.5.

a) Do these data satisfy the assumptions for inference? Explain.
b) Find the mean and standard deviation of the observed weights.
c) Create a 95% confidence interval for the mean weight of such bags of chips.
d) Explain in context what your interval means.
e) Comment on the company's stated net weight of 28.3 grams.

Ⓣ 29. Popcorn. Yvon Hopps ran an experiment to test optimum power and time settings for microwave popcorn. His goal was to find a combination of power and time that would deliver high-quality popcorn with only 10% of the kernels left unpopped, on average. After experimenting with several bags, he determined that power 9 at 4 minutes was the best combination.

a) He concluded that this popping method achieved the 10% goal. If it really does not work that well, what kind of error did Hopps make?
b) To be sure that the method was successful, he popped 8 more bags of popcorn (selected at random) at this setting. All were of high quality, with the following percentages of uncooked popcorn: 7, 13.2, 10, 6, 7.8, 2.8, 2.2, 5.2. Does this provide evidence that he met his goal of an average of no more than 10% uncooked kernels? Explain.

Ⓣ 30. Ski wax. Bjork Larsen was trying to decide whether to use a new racing wax for cross-country skis. He decided that the wax would be worth the price if he could average less than 55 seconds on a course he knew well, so he planned to test the wax by racing on the course 8 times.

a) Suppose that he eventually decides not to buy the wax, but it really would lower his average time to below 55 seconds. What kind of error would he have made?
b) His 8 race times were: 56.3, 65.9, 50.5, 52.4, 46.5, 57.8, 52.2, and 43.2 seconds. Should he buy the wax? Explain.

31. Cars. One of the important factors in auto safety is the weight of the vehicle. Insurance companies are inter-

ested in knowing the average weight of cars currently licensed in the United States; they believe it is 3000 pounds. To see if that estimate is correct, they checked a random sample of 91 cars. For that group the mean weight was 2919 pounds, with a standard deviation of 531.5 pounds. Is this strong evidence that the mean weight of all cars is not 3000 pounds?

32. Portable phones. A manufacturer claims that a new design for a portable phone has increased the range to 150 feet, allowing many customers to use the phone throughout their homes and yards. An independent testing laboratory found that a random sample of 44 of these phones worked over an average distance of 142 feet, with a standard deviation of 12 feet. Is there evidence that the manufacturer's claim is false?

Ⓣ 33. Chips Ahoy. In 1998, as an advertising campaign, the Nabisco Company announced a "1000 Chips Challenge," claiming that every 18-ounce bag of their Chips Ahoy cookies contained at least 1000 chocolate chips. Dedicated Statistics students at the Air Force Academy (no kidding) purchased some randomly selected bags of cookies, and counted the chocolate chips. Some of their data are given below. (*Chance*, 12, no. 1 [1999])

1219	1214	1087	1200	1419	1121	1325	1345
1244	1258	1356	1132	1191	1270	1295	1135

a) Check the assumptions and conditions for inference. Comment on any concerns you have.
b) Create a 95% confidence interval for the average number of chips in bags of Chips Ahoy cookies.
c) What does this evidence say about Nabisco's claim? Use your confidence interval to test an appropriate hypothesis and state your conclusion.

Ⓣ 34. Yogurt. *Consumer Reports* tested 14 brands of vanilla yogurt and found the following numbers of calories per serving:

160	200	220	230	120	180	140
130	170	190	80	120	100	170

a) Check the assumptions and conditions for inference.
b) Create a 95% confidence interval for the average calorie content of vanilla yogurt.
c) A diet guide claims that you will get 120 calories from a serving of vanilla yogurt. What does this evidence indicate? Use your confidence interval to test an appropriate hypothesis and state your conclusion.
*d) Perform a sign test to test the hypothesis that the median number of calories is 120. Does your conclusion agree with what you found in part c?

Ⓣ 35. Maze. Psychology experiments sometimes involve testing the ability of rats to navigate mazes. The mazes are classified according to difficulty, as measured by the mean length of time it takes rats to find the food at the end. One researcher needs a maze that will take rats an average of about one minute to solve. He tests one maze on several rats, collecting the data that follow.

a) Plot the data. Do you think the conditions for inference are satisfied? Explain.

b) Test the hypothesis that the mean completion time for this maze is 60 seconds. What is your conclusion?

c) Eliminate the outlier, and test the hypothesis again. What is your conclusion?

d) Do you think this maze meets the "one-minute average" requirement? Explain.

*e) Perform a sign test to see if the median time is one minute or less, keeping the outlier in the data set. Does your conclusion change from the one you arrived at in part d?

Time (sec)	
38.4	57.6
46.2	55.5
62.5	49.5
38.0	40.9
62.8	44.3
33.9	93.8
50.4	47.9
35.0	69.2
52.8	46.2
60.1	56.3
55.1	

36. Braking. A tire manufacturer is considering a newly designed tread pattern for its all-weather tires. Tests have indicated that these tires will provide better gas mileage and longer treadlife. The last remaining test is for braking effectiveness. The company hopes the tire will allow a car traveling at 60 mph to come to a complete stop within an average of 125 feet after the brakes are applied. They will adopt the new tread pattern unless there is strong evidence that the tires do not meet this objective. The distances (in feet) for 10 stops on a test track were 129, 128, 130, 132, 135, 123, 102, 125, 128, and 130. Should the company adopt the new tread pattern? Test an appropriate hypothesis and state your conclusion. Explain how you dealt with the outlier, and why you made the recommendation you did.

Comparing Means

WHO	AA alkaline batteries
WHAT	Length of battery life while playing a CD continuously
UNITS	Minutes
WHY	Class project
WHEN	1998

Should you buy generic rather than brand-name batteries? A Statistics student designed an experiment to test battery life. He wanted to know whether there was any real difference between brand-name batteries and a generic brand. To estimate the difference in mean lifetimes, he kept a battery-powered CD player continuously playing the same CD, with the volume control fixed at 5, and measured the time until no more music was heard through the headphones. (He ran an initial trial to find out approximately how long that would take, so that he didn't have to spend the first 3 hours of each run listening to the same CD.) For the experiment, he used six sets of AA alkaline batteries from two major battery manufacturers: a well-known brand name and a generic brand. In the language of experiments, battery brand was his factor. The brand names were the two levels of his factor. His response variable was the time in minutes until the sound stopped. To account for changes in the CD player's performance over time, he randomized the run order by choosing sets of batteries at random.

Here are his data (times in minutes):

Brand Name	Generic
194.0	190.7
205.5	203.5
199.2	203.5
172.4	206.5
184.0	222.5
169.5	209.4

A S Can Diet Prolong Life?
Watch a video that tells the story of an experiment. We'll analyze the data later in this chapter.

Experiments that compare two groups are common throughout both science and industry. We might want to compare the effects of a new drug with the traditional therapy, the fuel efficiency of two car engine designs, or the sales of new products in two different test cities. In fact, battery manufacturers perform such experiments themselves.

Plot the Data

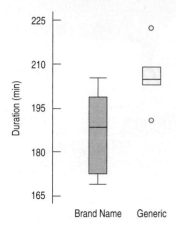

Boxplots comparing the brand name and generic data suggest a difference in duration. **Figure 24.1**

The natural display for comparing two groups is boxplots of the data for the two groups, placed side by side. Although we can't make a confidence interval or test a hypothesis from the boxplots themselves, you should always start with boxplots when comparing groups. Let's look at the boxplots of the battery test data.

It sure looks like the generic batteries lasted longer. And we can see that they were also more consistent. But is the difference large enough to change our battery-buying behavior? Can we be confident that the difference is more than just random fluctuation? That's why we need statistical inference.

The boxplot for the generic data identifies two possible outliers. That's interesting, but with only six measurements in each group, the outlier nomination rule is not very reliable. Both of the extreme values are plausible results and the range of the generic values is smaller than the range of the brand-name values, even with the outliers. So we're probably better off just leaving these values in the data.

Comparing Two Means

Comparing two means is not very different from comparing two proportions. In fact, it's not different in concept from any of the methods we've seen. Now, the population model parameter of interest is the difference between the *mean* battery lifetimes of the two brands, $\mu_1 - \mu_2$.

The rest is the same as before. The statistic of interest is the difference in the two observed means, $\bar{y}_1 - \bar{y}_2$. We'll start with this statistic to build our confidence interval, but we'll need to know its standard deviation and its sampling model. Then we can build confidence intervals and find P-values for hypothesis tests.

We know that for independent random quantities, variances add. It worked for proportions; it works for means. To find the standard deviation of the difference between the two sample means, we add their variances and then take a square root:

$$SD(\bar{y}_1 - \bar{y}_2) = \sqrt{Var(\bar{y}_1) + Var(\bar{y}_2)}$$

$$= \sqrt{\left(\frac{\sigma_1}{\sqrt{n_1}}\right)^2 + \left(\frac{\sigma_2}{\sqrt{n_2}}\right)^2}$$

$$= \sqrt{\frac{\sigma_1^2}{n_1} + \frac{\sigma_2^2}{n_2}}.$$

Of course, we still don't know the true standard deviations of the two groups, σ_1 and σ_2, so as usual, we'll use the estimates, s_1 and s_2. Using the estimates gives us the *standard error,*

$$SE(\bar{y}_1 - \bar{y}_2) = \sqrt{\frac{s_1^2}{n_1} + \frac{s_2^2}{n_2}}.$$

We'll use the standard error to see how big the difference really is. Because we are working with means and estimating the standard error of their difference using the data, we shouldn't be surprised that the sampling model is a Student's *t*.

The confidence interval we build is called a **two-sample *t*-interval** (for the difference in means). The corresponding hypothesis test is called a **two-sample *t*-test.** The interval looks just like all the others we've seen—the statistic plus or minus an estimated margin of error:

$$(\bar{y}_1 - \bar{y}_2) \pm ME$$

$$\text{where } ME = t^* \times SE(\bar{y}_1 - \bar{y}_2).$$

An Easier Rule?
The formula for the degrees of freedom of the sampling distribution of the difference between two means is long, so some books teach an easier rule: The number of degrees of freedom is always at *least* the smaller of the two *n*'s, minus 1 and at most $n_1 + n_2 - 2$. The problem is that if you need to use a two-sample *t* and don't have the formula at hand to find the correct degrees of freedom, you have to be conservative and use the lower value. And *that* approximation can be a poor choice because it can give less than *half* the degrees of freedom you're entitled to from the correct formula.

Compare this formula with the one for the confidence interval for the difference of two proportions we saw in Chapter 22 (p. 496). The formulas are almost the same. It's just that here we use a Student's *t*-model instead of a Normal model to find the appropriate critical t^* value corresponding to our chosen confidence.

What are we missing? Only the degrees of freedom for the Student's *t*-model. Unfortunately, *that* formula is strange.

The deep, dark secret is that the sampling model isn't *really* Student's *t*, but only something close. The trick is that by using a special, adjusted degrees of freedom value, we can make it so close to a Student's *t*-model that nobody can tell the difference. The adjustment formula is straightforward but doesn't help our understanding much, so we leave it to the computer or calculator. (If you are curious and really want to see the formula, look in the footnote.[1])

A sampling distribution for the difference between two means

When the conditions are met, the standardized sample difference between the means of two independent groups,

$$t = \frac{(\bar{y}_1 - \bar{y}_2) - (\mu_1 - \mu_2)}{SE(\bar{y}_1 - \bar{y}_2)},$$

can be modeled by a Student's *t*-model with a number of degrees of freedom found with a special formula. We estimate the standard error with

$$SE(\bar{y}_1 - \bar{y}_2) = \sqrt{\frac{s_1^2}{n_1} + \frac{s_2^2}{n_2}}.$$

Assumptions and Conditions

Now we've got everything we need. Before we can make a two-sample *t*-interval or perform a two-sample *t*-test, though, we have to check the assumptions and conditions.

Independence Assumption

The data in each group must be drawn independently and at random from a homogeneous population, or generated by a randomized comparative experiment.

---

[1]
$$df = \frac{\left(\dfrac{s_1^2}{n_1} + \dfrac{s_2^2}{n_2}\right)^2}{\dfrac{1}{n_1 - 1}\left(\dfrac{s_1^2}{n_1}\right)^2 + \dfrac{1}{n_2 - 1}\left(\dfrac{s_2^2}{n_2}\right)^2}$$

Are you sorry you looked? This formula usually doesn't even give a whole number. If you are using a table, you'll need a whole number, so round down to be safe. If you are using technology, it's even easier. The approximation formulas that computers and calculators use for the Student's *t*-distribution can deal with fractional degrees of freedom, so it's OK to pass the fractional degrees of freedom from this formula through to those formulas.

We can't expect that the data, taken as one big group, come from a homogeneous population, because that's what we're trying to test. But without randomization of some sort, there are no sampling distribution models and no inference. We can check two conditions:

   Randomization Condition: Were the data collected with suitable randomization? For surveys, are they a representative random sample? For experiments, was the experiment randomized?

   10% Condition: We usually don't check this condition for differences of means. It was important for proportions, but we'll check it for means only if we have a very small population or an extremely large sample.

Normal Population Assumption

As we did before with Student's *t*-models, we should check the assumption that the underlying populations are *each* Normally distributed. We check the

   Nearly Normal Condition: We must check this for *both* groups; a violation by either one violates the condition. As we saw for single sample means, the Normality Assumption matters most when sample sizes are small. For samples of $n < 15$ in either group, you should not use these methods if the histogram or Normal probability plot shows severe skewness. For groups of $n < 40$, a mildly skewed histogram is OK, but you should remark on any outliers you find and not work with severely skewed data. When both groups are bigger than 40, the Central Limit Theorem starts to kick in no matter how the data are distributed, so the Nearly Normal Condition for the data matters less. Even in large samples, however, you should still be on the lookout for outliers, extreme skewness, and multiple modes.

Independent Groups Assumption

A S **Does Restricting Diet Prolong Life?** There may be evidence that eating a severely calorie-restricted diet could prolong your life by up to 20%. This activity lets you construct a confidence interval to compare lifespans of rats fed two different diets.

To use the two-sample *t* methods, the two groups we are comparing must be independent of each other. In fact, this test is sometimes called the two *independent samples t*-test. No statistical test can verify this assumption. You have to think about how the data were collected. The assumption would be violated, for example, if one group were comprised of husbands and the other group, their wives. Whatever we measure on couples might naturally be related. Similarly, if we compared subjects' performances before some treatment with their performances afterward, we'd expect a relationship of each "before" measurement with its corresponding "after" measurement. In cases such as these, where the observational units in the two groups are related or matched, *the two-sample methods of this chapter can't be applied.* When this happens, we need a different procedure that we'll see in the next chapter.

Two-sample t-interval

When the conditions are met, we are ready to find the confidence interval for the difference between means of two independent groups, $\mu_1 - \mu_2$. The confidence interval is

$$(\bar{y}_1 - \bar{y}_2) \pm t^*_{df} \times SE(\bar{y}_1 - \bar{y}_2)$$

where the standard error of the difference of the means, $SE(\bar{y}_1 - \bar{y}_2) = \sqrt{\dfrac{s_1^2}{n_1} + \dfrac{s_2^2}{n_2}}$.

The critical value t^*_{df} depends on the particular confidence level, C, that you specify and on the number of degrees of freedom, which we get from the sample sizes and a special formula.

A Two-Sample *t*-Interval Step-By-Step

Judging from the boxplot, the generic batteries seem to have lasted about 20 minutes longer than the brand-name batteries. Before we change our buying habits, what should we expect to happen with the next batteries we buy? How much longer might the generics last? Let's make a confidence interval for the differences of the means.

Think

Plan State what we want to know.

Identify the *parameter* you wish to estimate. Here our parameter is the difference in the means, not the individual group means.

Identify the *population(s)* about which you wish to make statements. We hope to make decisions about purchasing batteries, so we're interested in all the AA batteries of these two brands.

Identify the variables and review the W's.

Make a picture. Boxplots are the display of choice for comparing groups. We'll also want to check the distribution of each group. Histograms or Normal probability plots do a better job there.

REALITY CHECK From the boxplots, it appears our confidence interval should be centered near a difference of 20 minutes. We don't have a lot of intuition about how far the interval should extend on either side of 20.

I have measurements of the lifetimes (in minutes) of 6 generic and 6 brand name AA batteries from a randomized experiment. I want to find an interval that is likely with 95% confidence to contain the true difference $\mu_G - \mu_B$ between the mean lifetime of the generic brand AA batteries and the mean lifetime of the brand-name batteries.

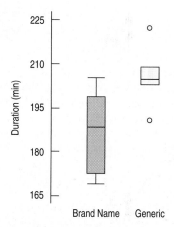

Model

Check the appropriate assumptions and conditions to be sure that a student's *t*-model for the sampling distribution is appropriate.

✔ **Independent Groups Assumption:** Batteries manufactured by two different companies and purchased in separate packages should be independent.

✔ **Randomization Condition:** The batteries were selected at random from those available for sale. Not exactly an SRS, but a reasonably representative random sample. The batteries do come in packs, however, so they may not be independent. For example, a storage problem might affect all the batteries in the same pack. Repeating the experiment for several different packs of batteries would make it stronger.

✔ **Nearly Normal Condition:** The samples are small, but histograms look unimodal and symmetric:

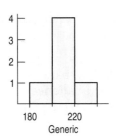
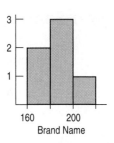

State the sampling distribution model for the statistic. Here the degrees of freedom come from that messy approximation formula.

Under these conditions, it's okay to use a Student's t-model.

Specify your method.

*I'll use a **two-sample t-interval**.*

Show **Mechanics** Construct the confidence interval.

I know:

Be sure to include the units along with the statistics. Use meaningful subscripts to identify the groups.

$$n_G = 6 \qquad\qquad n_B = 6$$
$$\bar{y}_G = 206.0 \text{ min} \quad \bar{y}_B = 187.4 \text{ min}$$
$$s_G = 10.3 \text{ min} \qquad s_B = 14.6 \text{ min}$$

Use the sample standard deviations to find the standard error of the sampling distribution.

The groups are independent, so:

$$SE(\bar{y}_G - \bar{y}_B) = \sqrt{SE^2(\bar{y}_G) + SE^2(\bar{y}_B)}$$
$$= \sqrt{\frac{s_G^2}{n_G} + \frac{s_B^2}{n_B}}$$
$$= \sqrt{\frac{10.3^2}{6} + \frac{14.6^2}{6}}$$
$$= \sqrt{\frac{106.09}{6} + \frac{213.16}{6}}$$
$$= \sqrt{53.208}$$
$$= 7.29 \text{ min.}$$

We have three choices for degrees of freedom. The best alternative is to let the computer or calculator use the approximation formula for df. This gives a fractional degree of freedom (here df = 8.98), and technology can find a corresponding critical value. In this case, it is $t^* = 2.263$.

The approximation formula calls for 8.98 degrees of freedom.[2]

Or we could round the approximation formula's df value down to an integer so we can use a *t* table. That gives 8 df and a critical value, $t^* = 2.306$.

The corresponding critical value for a 95% confidence level from a Student's t-model with 8.98 df is $t^* = 2.263$.

The easy rule says to use only $6 - 1 = 5$ df. That gives a critical value $t^* = 2.571$. The corresponding confidence interval is about 14% wider—a high price to pay for a small savings in effort.

So the margin of error is:

$$ME = t^* \times SE(\bar{y}_G - \bar{y}_B)$$
$$= 2.263(7.29)$$
$$= 16.50 \text{ min.}$$

---

[2] If you try to find the degrees of freedom with that messy approximation formula (We dare you! It's in the footnote on page 547.) using the values above, you'll get 8.99. The minor discrepancy is because we rounded the standard deviations.

The 95% confidence interval is:

$$(206.0 - 187.4) \pm 16.5 \text{ minutes}$$

$$\text{or } 18.6 \pm 16.5 \text{ minutes}$$

$$= (2.1, 35.1) \text{ minutes}.$$

Tell

Conclusion Interpret the confidence interval in the proper context.

Less formally, you could say, "I'm 95% confident that generic batteries last an average of 2.1 to 35.1 minutes longer than brand-name batteries."

I am 95% confident that the interval from 2.1 minutes to 35.1 minutes captures the mean amount of time that generic batteries outlast brand-name batteries for this task. If generic batteries are cheaper, there seems little reason not to use them. If it is more trouble or costs more to buy them, then I'd consider whether the additional performance is worth it.

Another One Just Like the Other Ones?

A S **Find Two-Sample *t*-Intervals.** Who wants to deal with that ugly df formula? We usually find these intervals with a statistics package. Learn how here.

Yes. That's been our point all along. Once again we see a statistic plus or minus the margin of error. And the ME is just a critical value times the standard error. Just look out for that crazy degrees of freedom formula.

Carpal tunnel syndrome (CTS) causes pain and tingling in the hand. It can be bad enough to keep sufferers awake at night and restrict their daily activities. Researchers studied the effectiveness of two alternative surgical treatments for CTS (Mackenzie, Hainer, and Wheatley, *Annals of Plastic Surgery*). Patients were randomly assigned to have endoscopic or open-incision surgery. Four weeks later the endoscopic surgery patients demonstrated a mean pinch strength of 9.1 kg compared to 7.6 kg for the open-incision patients.

1. Why is the randomization of the patients into the two treatments important?

2. A 95% confidence interval for the difference in mean strength is about (0.04 kg, 2.96 kg). Explain what this interval means.

3. Why might we want to examine such a confidence interval in deciding between these two surgical procedures?

4. Why might you want to see the data before believing the confidence interval?

Testing the Difference Between Two Means

If you bought a used camera in good condition from a friend, would you pay the same as you would if you bought the same item from a stranger? A researcher at Cornell University (J. J. Halpern, "The Transaction Index: A Method for Standardizing Comparisons of Transaction Characteristics Across Different Contexts," *Group Decision and Negotiation*, 6, no. 6[1997]: 557–572) wanted to know how friendship might affect simple sales such as this. She randomly divided subjects into two groups and gave each group descriptions of items they might want to buy. One group was told to imagine buying from a friend whom they expected to see again. The other group was told to imagine buying from a stranger.

Here are the prices they offered for a used camera in good condition:

WHO	University students
WHAT	Prices offered for a used camera ($)
WHY	Study of the effects of friendship on transactions
WHEN	1990s
WHERE	Cornell University

PRICE OFFERED FOR A USED CAMERA ($)	
Buying from a Friend	**Buying from a Stranger**
275	260
300	250
260	175
300	130
255	200
275	225
290	240
300	

The researcher who designed this study had a specific concern. Previous theories had doubted that friendship had a measurable effect on pricing. She hoped to find an effect of friendship. This calls for a hypothesis test—in this case a **two-sample t-test for means.**

A Test for the Difference Between Two Means

A S **The Two-Sample *t*-Test.** How different are beef hot dogs and chicken dogs? Test whether measured differences are statistically significant.

You already know enough to construct this test. The test statistic looks just like the others we've seen. It finds the difference between the observed group means and compares this with a hypothesized value for that difference. We'll call that hypothesized difference Δ_0 ("delta naught"). It's so common for that hypothesized difference to be zero that we often just assume $\Delta_0 = 0$. We then compare the difference in the means with the standard error of that difference. We already know that for a difference between independent means, we can find P-values from a Student's *t*-model on that same special number of degrees of freedom.

We're ready to go:

Two-sample t-test

The conditions for the two-sample *t*-test for the difference between the means of two independent groups are the same as for the two-sample *t*-interval. We test the hypothesis

$$H_0: \mu_1 - \mu_2 = \Delta_0,$$

where the hypothesized difference is almost always 0, using the statistic

$$t = \frac{(\bar{y}_1 - \bar{y}_2) - \Delta_0}{SE(\bar{y}_1 - \bar{y}_2)}.$$

The standard error of $\bar{y}_1 - \bar{y}_2$ is

$$SE(\bar{y}_1 - \bar{y}_2) = \sqrt{\frac{s_1^2}{n_1} + \frac{s_2^2}{n_2}}.$$

When the conditions are met and the null hypothesis is true, this statistic can be closely modeled by a Student's *t*-model with a number of degrees of freedom given by a special formula. We use that model to obtain a P-value.

A Two-Sample *t*-test for the Difference   Step-By-Step
Between Two Means

The usual null hypothesis is that there's no difference in means. That's just the right null hypothesis for the camera purchase prices.

Think

Plan State what we want to know.

Identify the *parameter* you wish to estimate. Here our parameter is the difference in the means, not the individual group means.

Identify the variables and check the W's.

I want to know whether people are likely to offer a different amount for a used camera when buying from a friend than when buying from a stranger. I wonder whether the difference between mean amounts is zero. I have bid prices from 8 subjects buying from a friend and 7 buying from a stranger, found in a randomized experiment.

H_O: The difference in mean price offered to friends and the mean price offered to strangers is zero:

$$\mu_F - \mu_S = 0.$$

Hypotheses State the null and alternative hypotheses.

The research claim is that friendship changes what people are willing to pay.[3] The natural null hypothesis is that friendship makes no difference.

We didn't start with any knowledge of whether friendship might increase or decrease the price, so we choose a two-sided alternative.

H_A: The difference in mean prices is not zero:

$$\mu_F - \mu_S \neq 0.$$

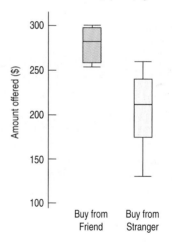

Make a picture. Boxplots are the display of choice for comparing groups. We'll also want to check the distribution of each group. Histograms or Normal probability plots do a better job there.

REALITY CHECK Looks like the prices are higher if you buy from a friend, but it's hard to be sure. The two ranges barely overlap, so we'd be pretty surprised if we don't reject the null hypothesis.

Model Check the conditions.

✔ **Independent Groups Assumption:** Randomizing the experiment gives independent groups.

✔ **Randomization Condition:** The experiment was randomized. Subjects were assigned to treatment groups at random.

✔ **Nearly Normal Condition:** Histograms of the two sets of prices are unimodal and symmetric:

---

[3] This claim is a good example of what is called a "research hypothesis" in many social sciences. The only way to check it is to deny that it's true and see where the resulting null hypothesis leads us.

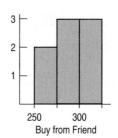
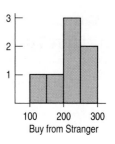

State the sampling distribution model.

Specify your method.

The conditions are okay, so I'll use a Student's *t*-model to perform a **two-sample t-test.**

Show **Mechanics** List the summary statistics. Be sure to use proper notation.

From the data:

$$n_F = 8 \qquad n_S = 7$$
$$\bar{y}_F = \$281.88 \qquad \bar{y}_S = \$211.43$$
$$s_F = \$18.31 \qquad s_S = \$46.43$$

Use the null model to find the P-value. First determine the standard error of the difference between sample means.

For independent groups:

$$SE(\bar{y}_F - \bar{y}_S) = \sqrt{SE^2(\bar{y}_F) + SE^2(\bar{y}_S)}$$

$$= \sqrt{\frac{s_F^2}{n_F} + \frac{s_S^2}{n_S}}$$

$$= \sqrt{\frac{18.31^2}{8} + \frac{46.43^2}{7}}$$

$$= \sqrt{349.87}$$

$$= 18.70$$

The observed difference is

$$(\bar{y}_F - \bar{y}_S) = 281.88 - 211.43 = \$70.45.$$

Make a picture. Sketch the *t*-model centered at the hypothesized difference of zero. Because this is a two-tailed test, shade the region to the right of the observed difference and the corresponding region in the other tail.

Find the *t*-value.

A statistics program or graphing calculator can find the P-value using the fractional degrees of freedom from the approximation formula. If you are doing a test like this without technology, you could use the smaller sample size to determine degrees of freedom. In this case $n_2 - 1 = 6$.

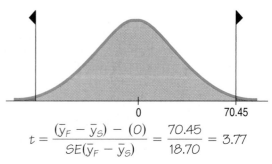

$$t = \frac{(\bar{y}_F - \bar{y}_S) - (0)}{SE(\bar{y}_F - \bar{y}_S)} = \frac{70.45}{18.70} = 3.77$$

The approximation formula gives 7.62 degrees of freedom.[4]

$$\text{P-value} = 2P(t_{7.62} > 3.77) = 0.006$$

---

[4] If you were crazy enough to calculate that messy degrees of freedom formula by hand with the values given here, you'd get about 7.74. Computers work with more precision for the standard deviations than we gave in our example. Many computer programs will round the final result down to 7 degrees of freedom. All give about the same result for the P-value, so it doesn't really matter—the conclusion would be the same.

Tell

Conclusion Link the P-value to your decision about the null hypothesis and state the conclusion in context.

Be cautious about generalizing to items whose prices are outside the range of those in this study.

If possible, propose a course of action.

If there were no difference in the mean prices, a difference this large would occur only 6 times in 1000. That's too rare to believe, so I reject the null hypothesis and conclude that people are likely to offer a friend a different amount than they'd offer a stranger for a used camera (and possibly for other similar items).

I may want to take special care not to pay too much when buying an item such as this from a friend.

Recall the experiment comparing patients 4 weeks after surgery for carpal tunnel syndrome. The patients who had endoscopic surgery demonstrated a mean pinch strength of 9.1 kg compared to 7.6 kg for the open-incision patients.

5 What hypotheses would you test?

6 The P-value of the test was less than 0.05. State a brief conclusion.

7 The study reports work on 36 "hands," but there were only 26 patients. In fact, 7 of the endoscopic surgery patients had both hands operated on, as did 3 of the open-incision group. Does this alter your thinking about any of the assumptions? Explain.

Back into the Pool

Remember that when we know a proportion, we know its standard deviation. When we tested the null hypothesis that two proportions were equal, that link meant we could assume their variances were equal as well. This led us to pool our data to estimate a standard error for the hypothesis test.

For means, there is also a pooled *t*-test. Like the two-proportions *z*-test, this test assumes that the variances in the two groups are equal. But be careful. Knowing the mean of some data doesn't tell you anything about their variance. And knowing that two means are equal doesn't say anything about whether their variances are equal. If we were willing to *assume* that their variances are equal, we could pool the data from two groups to estimate the common variance. We'd estimate this pooled variance from the data, so we'd still use a Student's *t*-model. This test is called a **pooled *t*-test.**

Pooled *t*-tests have a couple of advantages. They often have a few more degrees of freedom than the corresponding two-sample test and a much simpler degrees of freedom formula. But these advantages come at a price. You have to pool the variances and add another assumption. The assumption of equal variances is a strong one, is often not true, and is difficult to check. For these reasons, we recommend that you use a two-sample *t*-test instead.

The pooled *t*-test is the theoretically correct method only when we have a good reason to believe that the variances are equal. And (as we will see shortly) there are times when this makes sense. Keep in mind, however, that it's never wrong *not* to pool.

*The Pooled *t*-Test*

In order to use the pooled *t*-test, we must make the **Equal Variance Assumption** that the variances of the two populations from which the samples have been drawn are equal. That is, $\sigma_1^2 = \sigma_2^2$. (Of course, we can think about the standard deviations being equal instead.) The corresponding **Nearly Equal Spreads Condition** really just consists of looking at the boxplots to check that the spreads are not wildly different. We were going to make boxplots anyway, so there's really nothing new here.

Once we decide to pool, we estimate the common variance by combining numbers we already have:

$$s_{\text{pooled}}^2 = \frac{(n_1 - 1)s_1^2 + (n_2 - 1)s_2^2}{(n_1 - 1) + (n_2 - 1)}.$$

(If the two sample sizes are equal, this is just the average of the two variances.)

Now we just substitute this pooled variance in place of each of the variances in the standard error formula. Remember, the standard error formula for the difference of two independent means is:

$$SE(\bar{y}_1 - \bar{y}_2) = \sqrt{\frac{s_1^2}{n_1} + \frac{s_2^2}{n_2}}.$$

We just substitute the common pooled variance for each of the two variances in this formula and simplify. That makes the pooled standard error formula much simpler:

$$SE_{\text{pooled}}(\bar{y}_1 - \bar{y}_2) = \sqrt{\frac{s_{\text{pooled}}^2}{n_1} + \frac{s_{\text{pooled}}^2}{n_2}} = s_{\text{pooled}}\sqrt{\frac{1}{n_1} + \frac{1}{n_2}}.$$

The formula for degrees of freedom for the Student's *t*-model is simpler, too. It was so complicated for the two-sample *t* that we stuck it in a footnote.[5] Now it's just df $= n_1 + n_2 - 2$.

Substitute the pooled-*t* estimate of the standard error and its degrees of freedom into the steps of the confidence interval or hypothesis test and you'll be using the pooled-*t* method. Of course, if you decide to use a pooled-*t* method, you must defend your assumption that the variances of the two groups are equal.

How can we defend an assumption like that? We're testing whether the means are equal, so we admit that we don't *know* whether they are equal. Doesn't it seem a bit much to just *assume* that the variances are equal? Well, yes—but there are some special cases to consider.

A S **The Pooled *t*-Test.**
It's those hot dogs again. The same interactive tool can handle a pooled *t*-test, too. Take it for a spin here.

---

[5] See footnote 1, page 547.

Pooled *t*-test and confidence interval for means

The conditions for the pooled *t*-test for the difference between the means of two independent groups (commonly called a "pooled *t*-test") are the same as for the two-sample *t*-test with the additional assumption that the variances of the two groups are the same. We test the hypothesis

$$H_0: \mu_1 - \mu_2 = \Delta_0,$$

where the hypothesized difference, Δ_0, is almost always 0, using the statistic

$$t = \frac{(\bar{y}_1 - \bar{y}_2) - \Delta_0}{SE_{\text{pooled}}(\bar{y}_1 - \bar{y}_2)}.$$

The standard error of $\bar{y}_1 - \bar{y}_2$ is

$$SE_{\text{pooled}}(\bar{y}_1 - \bar{y}_2) = \sqrt{\frac{s_{\text{pooled}}^2}{n_1} + \frac{s_{\text{pooled}}^2}{n_2}} = s_{\text{pooled}}\sqrt{\frac{1}{n_1} + \frac{1}{n_2}},$$

where the pooled variance is

$$s_{\text{pooled}}^2 = \frac{(n_1 - 1)s_1^2 + (n_2 - 1)s_2^2}{(n_1 - 1) + (n_2 - 1)}.$$

When the conditions are met and the null hypothesis is true, we can model this statistic's sampling distribution with a Student's *t*-model with $(n_1 - 1) + (n_2 - 1)$ degrees of freedom. We use that model to obtain a P-value for a test or a margin of error for a confidence interval.

The corresponding confidence interval is

$$(\bar{y}_1 - \bar{y}_2) \pm t_{\text{df}}^* \times SE_{\text{pooled}}(\bar{y}_1 - \bar{y}_1),$$

where the critical value t^* depends on the confidence level and is found with $(n_1 - 1) + (n_2 - 1)$ degrees of freedom.

Is the Pool All Wet?

So when *should* you use pooled-*t* methods rather than two-sample *t* methods?

Never.

What, never?

Well, hardly ever.

You see, when the variances of the two groups are in fact equal, the two methods give pretty much the same result. Pooled methods have a small advantage (slightly narrower confidence intervals, slightly more powerful tests) mostly because they usually have a few more degrees of freedom, but the advantage is slight.

When the variances are *not* equal, the pooled methods are just not valid, and can give poor results. You have to use the two-sample methods instead.

As the sample sizes get bigger, the advantages that come from a few more degrees of freedom make less and less difference. So the advantage (such as it is) of the pooled method is greatest when the samples are small—just when it's hardest to check the conditions. And the difference in the degrees of freedom is greatest when the variances are not equal—just when you can't use the pooled method anyway. Our advice is to use the two-sample *t* methods to compare means.

> Because the advantages of pooling are small, and you are allowed to pool only rarely (when the equal variances assumption is met), **don't.**
>
> It's never wrong *not* to pool.

Why did we devote a whole section to a method that we don't recommend using? Good question. The answer is that pooled methods are actually very important in Statistics. We'll see important pooled methods in coming chapters. It's just that the simplest of the pooled methods—those for comparing two means—have good alternatives in the two-sample methods that don't require the extra assumption. Lacking the burden of the Equal Variances Assumption, the two-sample methods apply to more situations and are safer to use.

Why Not Test the Assumption That the Variances Are Equal?

There is a hypothesis test that would do this. However, it is very sensitive to failures of the assumptions and works poorly for small sample sizes—just the situation in which we might care about a difference in the methods. When the choice between two-sample t and pooled t methods makes a difference (that is, when the sample sizes are small), the test for whether the variances are equal hardly works at all.

Is There Ever a Time When Assuming Equal Variances Makes Sense?

Yes. In a randomized comparative experiment, we start by assigning our experimental units to treatments at random. We know that at the start of the experiment each treatment group is a random sample from the same population,[6] so each treatment group begins with the same population variance. In this case, assuming the variances are equal after we apply the treatment is the same as assuming that the treatment doesn't change the variance. When we test whether the true means are equal, we may be willing to go a bit farther and say that the treatments made no difference *at all*. For example, we might suspect that the treatment is no different from the placebo offered as a control. Then it's not much of a stretch to assume that the variances have remained equal. It's still an assumption and there are conditions that need to be checked (make the boxplots, make the boxplots, make the boxplots), but at least it's a plausible assumption.

This line of reasoning is important. The methods used to analyze comparative experiments *do* pool variances in exactly this way and defend the pooling with a version of this argument.

*Tukey's Quick Test

The famous statistician John Tukey[7] was once challenged to come up with a simpler alternative to the two-sample t-test that, like the 68-95-99.7 Rule, had critical values that could be remembered easily. The test he came up with asks you only to count and to remember three numbers: 7, 10, and 13.

---

[6] That is, the population of experimental subjects. Remember that to be valid, experiments do not need a representative sample drawn from a population because we are not trying to estimate a population model parameter.

[7] You know some of his other work. Tukey originated the stem-and-leaf display and the boxplot in the form we use here, and published theoretical work on data re-expression that justifies the recommendations of Chapter 10. He also coined the term "bit" for binary digit.

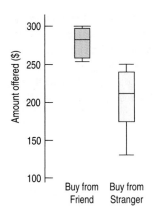

Buying from a Friend	Buying from a Stranger
$275	$260
300	250
260	175
300	130
255	200
275	225
290	240
300	

When you first looked at the boxplots of the price data, you might have noticed that they didn't overlap very much. That's the basis for Tukey's test.

To use Tukey's test, one group must have the highest value and the other, the lowest. We just count how many values in the high group are higher than *all* the values of the lower group. Add to this the number of values in the low group that are lower than *all* the values of the higher group. (You can count ties as 1/2.) Now if this total is 7 or more, we can reject the null hypothesis of equal means at $\alpha = 0.05$. The "critical values" of 10 and 13 give us α's of 0.01 and 0.001.

Let's try it. The "Friend" group has the highest value ($300) and the "Stranger" group has the lowest ($130). Six of the values in the Friend group are higher than the highest value of the Stranger group ($260) and one is a tie. Six of the Stranger values are lower than the lowest value for Friends. That's a total of 12 1/2 values that leak over. That's more than 10, but less than 13. So the P-value is between 0.01 and 0.001—just what we found with the two-sample *t*.

This is another example of a distribution-free test and a remarkably good test. The only assumption it requires is that the two samples be independent. It's so simple to do that there's no reason not to do one to check your two-sample *t* results. If they disagree, check the assumptions. Tukey's quick test, however, is not as widely known or accepted as the two-sample *t*-test, so you still need to know and use the two-sample *t*.

What Can Go Wrong?

- **Watch out for paired data.** The Independent Groups Assumption deserves special attention. Some researchers *deliberately* violate the Independent Groups Assumption. For example, suppose you wanted to test a diet program. You select 10 people at random to take part in your diet. You measure their weights at the beginning of the diet and after 10 weeks of the diet. So, you have two columns of weights, one for *before* and one for *after*. Can you use these methods to test whether the mean has gone down? No! The data are related; each "after" weight goes naturally with the "before" weight for the *same* person. If the samples are not independent, you can't use two-sample methods. This is probably the main thing that can go wrong when using these two-sample methods. Certainly, someone's weight before and after the 10 weeks will be related (whether the diet works or not). The methods of this chapter can be used *only* if the observations in the two groups are *independent*. Matched-pairs designs in which the observations are deliberately related arise often and are important. The next chapter deals with them.

- **Look at the plots.** The usual (by now) cautions about checking for outliers and non-Normal distributions apply, of course. The simple defense is to make and examine boxplots. You may be surprised how often this simple step saves you from the wrong or even absurd conclusions that can be generated by a single undetected outlier. You don't want to conclude that two methods have very different means just because one observation is atypical. 🚫

● **Do what we say, not what we do . . .** Precision machines used in industry often have a bewildering number of parameters that have to be set, so experiments are performed in an attempt to try to find the best settings. Such was the case for a hole-punching machine used by a well-known computer manufacturer to make printed circuit boards. The data were analyzed by one of the authors, but because he was in a hurry, he didn't look at the boxplots first and just performed *t*-tests on the experimental factors. When he found extremely small P-values even for factors that made no sense, he plotted the data. Sure enough, there was one observation 1,000,000 times bigger than the others. It turns out that it had been recorded in microns (millionths of an inch), while all the rest were in inches.

CONNECTIONS

The structure and reasoning of inference methods for comparing two means are very similar to what we used for comparing two proportions. Here we must estimate the standard errors independent of the means, so we use Student's *t*-models rather than the Normal.

We first learned about side-by-side boxplots in Chapter 5. There we made general statements about the shape, center, and spread of each group. When we compared groups, we asked whether their centers looked different compared with how spread out the distributions were. Here we've made that kind of thinking precise with confidence intervals for the difference and tests of whether the means are the same.

We use Student's *t* as we did for single sample means, and for the same reasons. We are using standard errors from the data to estimate the standard deviation of the sample statistic. As before, to work with Student's *t*-models we need to check the Nearly Normal Condition. Histograms and Normal probability plots are the best methods for such checks.

As always, we've decided whether a statistic is large by comparing it with its standard error. In this case, our statistic is the difference in means.

We pooled data to find a standard deviation when we tested the hypothesis of equal proportions. For that test, the assumption of equal variances was a consequence of the null hypothesis that the proportions were equal, so it didn't require an extra assumption. When two proportions are equal, so are their variances. But means don't have a linkage with their corresponding variances, so to use pooled-*t* methods, we must make the additional assumption of equal variances. When we can make the assumption, the pooled variance calculations are very similar to those for proportions, combining the squared deviations of each group from its own mean to find a common variance.

What have we learned?

Are the means of two groups the same? If not, how different are they? We've learned to use statistical inference to compare the means of two independent groups.

- We've seen that confidence intervals and hypothesis tests about the difference between two means, like those for an individual mean, use *t*-models.
- Once again we've seen the importance of checking assumptions that tell us whether our method will work.
- We've seen that, as when comparing proportions, finding the standard error for the difference in sample means depends on believing that our data come from independent groups. Unlike proportions, however, pooling is usually not the best choice here.
- And we've seen once again that we can add variances of independent random variables to find the standard deviation of the difference in two independent means.
- Finally we've learned that the reasoning of statistical inference remains the same; only the mechanics change.

TERMS

Two-sample *t* methods

Two-sample *t* methods allow us to draw conclusions about the difference between the means of two independent groups. The two-sample methods make relatively few assumptions about the underlying populations, so they are usually the method of choice for comparing two sample means. However, the Student's *t*-models are only approximations for their true sampling distribution. To make that approximation work well, the two-sample *t* methods have a special rule for estimating degrees of freedom.

Two-sample *t*-interval

A confidence interval for the difference in the means of two independent groups found as

$$(\bar{y}_1 - \bar{y}_2) \pm t^*_{df} \times SE(\bar{y}_1 - \bar{y}_2), \text{ where}$$

$$SE(\bar{y}_1 - \bar{y}_2) = \sqrt{\frac{s_1^2}{n_1} + \frac{s_2^2}{n_2}}$$

and the number of degrees of freedom is given by a special formula (see footnote 1 on page 547 if you really want the formula).

Two-sample *t*-test

A hypothesis test for the difference in the means of two independent groups. It tests the null hypothesis

$$H_0: \mu_1 - \mu_2 = \Delta_0,$$

where the hypothesized difference, Δ_0, is almost always 0, using the statistic

$$t_{df} = \frac{(\bar{y}_1 - \bar{y}_2) - \Delta_0}{SE(\bar{y}_1 - \bar{y}_2)}$$

with the number of degrees of freedom given by the special formula.

Pooling

Data from two or more populations may sometimes be combined, or *pooled,* to estimate a statistic (typically a pooled variance) when we are willing to assume that the estimated value is the same in both populations. The resulting larger sample size may lead to an estimate with lower sample variance. However, pooled estimates are appropriate only when the required assumptions are true.

Pooled-*t* methods

Pooled-*t* methods provide inference about the difference between the means of two independent populations under the assumption that both populations have the same standard deviation. When the assumption is justified, pooled-*t* methods generally produce slightly narrower confidence intervals and more powerful significance tests than two-sample *t* methods. When the assumption is not justified, they generally produce worse results—sometimes substantially worse.

We recommend that you use two-sample *t* methods instead.

*Pooled *t*-interval

A confidence interval for the difference in the means of two independent groups used when we are willing and able to make the additional assumption that the variances of the groups are equal. It is found as

$$(\bar{y}_1 - \bar{y}_2) \pm t^*_{df} \times SE_{pooled}(\bar{y}_1 - \bar{y}_2),$$

$$\text{where } SE_{pooled}(\bar{y}_1 - \bar{y}_2) = \sqrt{\frac{s_{pooled}^2}{n_1} + \frac{s_{pooled}^2}{n_2}} = s_{pooled}\sqrt{\frac{1}{n_1} + \frac{1}{n_2}},$$

and pooled variance is

$$s_{pooled}^2 = \frac{(n_1 - 1)s_1^2 + (n_2 - 1)s_2^2}{(n_1 - 1) + (n_2 - 1)}$$

and the number of degrees of freedom is $(n_1 - 1) + (n_2 - 1)$.

*Pooled *t*-test A hypothesis test for the difference in the means of two independent groups when we are willing and able to assume that the variances of the groups are equal. It tests the null hypothesis

$$H_0: \mu_1 - \mu_2 = \Delta_0,$$

where the hypothesized difference, Δ_0, is almost always 0, using the statistic

$$t_{df} = \frac{(\bar{y}_1 - \bar{y}_2) - \Delta_0}{SE_{pooled}(\bar{y}_1 - \bar{y}_2)},$$

where the pooled standard error is defined as for the pooled interval and the degrees of freedom is $(n_1 - 1) + (n_2 - 1)$.

S K I L L S   *When you complete this lesson you should:*

Think
- Be able to recognize situations in which we want to do inference on the difference between the means of two independent groups.

- Know how to examine your data for violations of conditions that would make inference about the difference between two population means unwise or invalid.

- Be able to recognize when a pooled-*t* procedure might be appropriate and be able to explain why you decided to use a two-sample method anyway.

Show
- Be able to perform a two-sample *t*-test using a statistics package or calculator (at least for finding the degrees of freedom).

Tell
- Be able to interpret a test of the null hypothesis that the means of two independent groups are equal. (If the test is a pooled *t*-test, your interpretation should include a defense of your assumption of equal variances.)

Two-Sample Methods on the Computer

Here's some typical computer package output with comments:

May just say "difference of means"

Test Statistic

```
2-Sample t-Test of μ1-μ2 = 0 vs ≠ 0

Difference Between Means = 0.99145299 t-Statistic = 1.540
w/196 df
Fail to reject Ho at Alpha = 0.05
P = 0.1251
```

Some programs will draw a conclusion about the test. Others just give the P-value and let you decide for yourself.

df found from approximation formula and rounded down. The unrounded value may be given, or may be used to find the P-value.

Many programs give far too many digits. Ignore the excess ones.

Most statistics packages compute the test statistic for you and report a P-value corresponding to that statistic. And, of course, statistics packages make it easy to examine the boxplots of the two groups, so you have no excuse for skipping this important check.

Some statistics software automatically tries to test whether the variances of the two groups are equal. Some automatically offer both the two-sample-*t* and pooled-*t* results. Ignore the test for the variances; it has little power in any situation in which its results could matter. If the pooled and two-sample methods differ in any important way, you should stick with the two-sample method. Most likely, the Equal Variance Assumption needed for the pooled method has failed.

The degrees of freedom approximation usually gives a fractional value. Most packages seem to round the approximate value down to the next smallest integer (although they may actually compute the P-value with the fractional value, gaining a tiny amount of power).

There are two ways to organize data when we want to compare two independent groups. The data can be in two lists, as in the table at the start of this chapter. Each list can be thought of as a variable. In this method, the variables in the batteries example would be *brand name* and *generic*. Graphing calculators usually prefer this form, and some computer programs can use it as well.

There's another way to think about the data. What is the response variable for the battery life experiment? It's the time until the music stopped. But the values of this variable are in both columns, and actually there's an experiment factor here, too—namely, the brand of the battery. So, we could put the data into two different columns, one with the *times* in it and one with the *brand*. Then the data would look like this:

Time	Brand
194.0	Brand name
205.5	Brand name
199.2	Brand name
172.4	Brand name
184.0	Brand name
169.5	Brand name
190.7	Generic
203.5	Generic
203.5	Generic
206.5	Generic
222.5	Generic
209.4	Generic

This way of organizing the data makes sense as well. Now the factor and the response variables are clearly visible. You'll have to see which method your program requires. Some packages even allow you to structure the data either way.

The commands to do inference for two independent groups on common statistics technology are not always found in obvious places. Here are some starting guidelines.

DATA DESK

Select variables.
From the **Calc** menu, choose **Estimate** for confidence intervals or **Test** for hypothesis tests. Select the interval or test from the drop-down menu and make other choices in the dialog.

Comments

Data Desk expects the two groups to be in separate variables.

EXCEL

From the **Tools** menu, choose **Data Analysis.** From the **Data Analysis** menu, choose **t-test: two-sample assuming unequal variances.**

Fill in the cell ranges for the two groups, the hypothesized difference, and the alpha level.

Comments

Excel expects the two groups to be in separate cell ranges. Notice that, contrary to Excel's wording, we do not need to assume that the variances are *not* equal; we simply choose not to assume that they *are* equal.

JMP

From the **Analyze** menu select **Fit y by x.** Select variables: a **Y, Response** variable that holds the data and an **X, Factor** variable that holds the group names. JMP will make a dotplot. Click the **red triangle** in the dotplot title, and choose **Unequal variances.** The *t*-test is at the bottom of the resulting table. Find the P-value from the Prob>F section of the table (they are the same).

Comments

JMP expects data in one variable and category names in the other. Don't be misled: There is no need for the variances to be unequal to use two-sample *t* methods.

MINITAB

From the **Stat** menu choose the **Basic Statistics** submenu. From that menu, choose **2-sample t...** Then fill in the dialog.

Comments

The dialog offers a choice of data in two variables or data in one variable and category names in the other.

SPSS

From the **Analyze** menu choose the **Compare Means** submenu. From that, choose the **Independent-Samples t-test** command. Specify the data variable and "group variable." Then type in the labels used in the group variable. SPSS offers both the two-sample and pooled-*t* results in the same table.

Comments

SPSS expects the data in one variable and group names in the other. If there are more than two group names in the group variable, only the two that are named in the dialog box will be compared.

TI-83/84 Plus

For a confidence interval:

In the **STAT TESTS** menu choose **0:2-SampTInt.** You may specify that you are using data stored in two lists or you may enter the means, standard deviations, and sizes of both samples. You must also indicate whether to pool the variances (when in doubt, say no) and specify the desired level of confidence.

To test a hypothesis:

In the **STAT TESTS** menu choose **4:2-SampTTest.** You may specify if you are using data stored in two lists or you may enter the means, standard deviations, and sizes of both samples. You must also indicate whether to pool the variances (when in doubt, say no) and specify whether the test is to be two-tail, lower-tail, or upper-tail.

TI-89

For a confidence interval:

In the **STAT Ints** menu choose **4:2-SampTInt.** You must specify if you are using data stored in two lists or if you will enter the means, standard deviations, and sizes of both samples. You must also indicate whether to pool the variances (when in doubt, say no) and specify the desired level of confidence.

To test a hypothesis:

In the **STAT TESTS** menu choose **4:2-SampTTest.** You must specify if you are using data stored in two lists or if you will enter the means, standard deviations, and sizes of both samples. You must also indicate whether to pool the variances (when in doubt, say no) and specify whether the test is to be two-tail, lower-tail, or upper-tail.

EXERCISES

1. **Learning math.** The Core Plus Mathematics Project (CPMP) is an innovative approach to teaching Mathematics that engages students in group investigations and mathematical modeling. After field tests in 36 high schools over a three-year period, researchers compared the performances of CPMP students with those taught using a traditional curriculum. In one test, students had to solve applied Algebra problems using calculators. Scores for 320 CPMP students were compared with those of a control group of 273 students in a traditional Math program. Computer software was used to create a confidence interval for the difference in mean scores. (*Journal for Research in Mathematics Education*, 31, no. 3 [2000])

   Conf level: 95%   Variable: Mu(CPMP) − Mu(Ctrl)
   Interval: (5.573, 11.427)

   a) What's the margin of error for this confidence interval?
   b) If we had created a 98% CI, would the margin of error be larger or smaller?
   c) Explain what the calculated interval means in this context.
   d) Does this result suggest that students who learn Mathematics with CPMP will have significantly higher mean scores in Algebra than those in traditional programs? Explain.

2. **Stereograms.** Stereograms appear to be composed entirely of random dots. However, they contain separate images that a viewer can "fuse" into a three-dimensional (3D) image by staring at the dots while defocusing the eyes. An experiment was performed to determine whether knowledge of the form of the embedded image affected the time required for subjects to fuse the images. One group of subjects (group NV) received no information or just verbal information about the shape of the embedded object. A second group (group VV) received both verbal information and visual information (specifically, a drawing of the object). The experimenters measured how many seconds it took for the subject to report that he or she saw the 3D image.

   2-Sample t-Interval for $\mu 1 - \mu 2$ df = 70
   Conf level = 90%
   $\mu(NV) - \mu(VV)$ interval: (0.55, 5.47)

   a) Interpret your interval in context.
   b) Does it appear that viewing a picture of the image helps people "see" the 3D image in a stereogram?
   c) What's the margin of error for this interval?
   d) Explain carefully what the 90% confidence level means.
   e) Would you expect a 99% confidence level to be wider or narrower? Explain.
   f) Might that change your conclusion in part b? Explain.

3. **CPMP, again.** During the study described in Exercise 1, students in both CPMP and traditional classes took another Algebra test that did not allow them to use calculators. The table below shows the results. Are the mean scores of the two groups significantly different? Test an appropriate hypothesis and state your conclusion.

Math Program	n	Mean	SD
CPMP	312	29.0	18.8
Traditional	265	38.4	16.2

   Performance on Algebraic Symbolic Manipulation Without Use of Calculators

   a) Write an appropriate hypothesis.
   b) Do you think the assumptions for inference are satisfied? Explain.
   c) Here is computer output for this hypothesis test. Explain what the P-value means in this context.

   2-Sample t-Test of $\mu 1 - \mu 2 \neq 0$
   t-Statistic = −6.451 w/574.8761 df
   P < 0.0001

   d) State a conclusion about the CPMP program.

4. **CPMP and word problems.** The study of the new CPMP Mathematics methodology described in Exercise 1 also tested students' abilities to solve word problems. This table shows how the CPMP and traditional groups performed. What do you conclude?

Math Program	n	Mean	SD
CPMP	320	57.4	32.1
Traditional	273	53.9	28.5

5. **Commuting.** A man who moves to a new city sees that there are two routes he could take to work. A neighbor who has lived there a long time tells him Route A will average 5 minutes faster than Route B. The man decides to experiment. Each day he flips a coin to determine which way to go, driving each route 20 days. He finds that Route A takes an average of 40 minutes, with standard deviation 3 minutes, and Route B takes an average of 43 minutes, with standard deviation 2 minutes. Histograms of travel times for the routes are roughly symmetric and show no outliers.

   a) Find a 95% confidence interval for the difference in average commuting time for the two routes.
   b) Should the man believe the old-timer's claim that he can save an average of 5 minutes a day by always driving Route A? Explain.

6. Pulse rates. A researcher wanted to see whether there is a significant difference in resting pulse rates for men and women. The data she collected are displayed in the boxplots and summarized below.

Sex	Male	Female
Count	28	24
Mean	72.75	72.625
Median	73	73
StdDev	5.37225	7.69987
Range	20	29
IQR	9	12.5

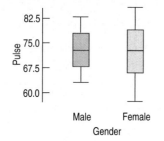

a) What do the boxplots suggest about any gender differences in pulse rates?
b) Is it appropriate to analyze these data using the methods of inference discussed in this chapter? Explain.
c) Create a 90% confidence interval for the difference in mean pulse rates.
d) Does the confidence interval confirm your answer to part a? Explain.

7. Cereal. The data below show the sugar content (as a percentage of weight) of several national brands of children's and adults' cereals. Create and interpret a 95% confidence interval for the difference in mean sugar content. Be sure to check the necessary assumptions and conditions.

Children's cereals: 40.3, 55, 45.7, 43.3, 50.3, 45.9, 53.5, 43, 44.2, 44, 47.4, 44, 33.6, 55.1, 48.8, 50.4, 37.8, 60.3, 46.6

Adults' cereals: 20, 30.2, 2.2, 7.5, 4.4, 22.2, 16.6, 14.5, 21.4, 3.3, 6.6, 7.8, 10.6, 16.2, 14.5, 4.1, 15.8, 4.1, 2.4, 3.5, 8.5, 10, 1, 4.4, 1.3, 8.1, 4.7, 18.4

8. Egyptians. Some archaeologists theorize that ancient Egyptians interbred with several different immigrant populations over thousands of years. To see if there is any indication of changes in body structure that might have resulted, they measured 30 skulls of male Egyptians dated from 4000 B.C.E. and 30 others dated from 200 B.C.E. (A. Thomson and R. Randall-Maciver, *Ancient Races of the Thebaid*, Oxford: Oxford University Press, 1905.)

a) Are these data appropriate for inference? Explain.
b) Create a 95% confidence interval for the difference in mean skull breadth between these two eras.
c) Do these data provide evidence that the mean breadth of males' skulls changed over this time period? Explain.
*d) Perform Tukey's test for the difference. Do your conclusions of part c change?

Maximum Skull Breadth (mm)			
4000 B.C.E.	4000 B.C.E.	200 B.C.E.	200 B.C.E.
131	131	141	131
125	135	141	129
131	132	135	136
119	139	133	131
136	132	131	139
138	126	140	144
139	135	139	141
125	134	140	130
131	128	138	133
134	130	132	138
129	138	134	131
134	128	135	136
126	127	133	132
132	131	136	135
141	124	134	141

9. Reading. An educator believes that new reading activities for elementary school children will improve reading comprehension scores. She randomly assigns third graders to an eight-week program in which some will use these activities and others will experience traditional teaching methods. At the end of the experiment, both groups take a reading comprehension exam. Their scores are shown in the back-to-back stem-and-leaf display. Do these results suggest that the new activities are better? Test an appropriate hypothesis and state your conclusion. (*Would Tukey's test be appropriate here? Explain.

New Activities		Control
	1	07
4	2	068
3	3	377
96333	4	12222238
9876432	5	355
721	6	02
1	7	
	8	5

10. Streams. Researchers collected samples of water from streams in the Adirondack Mountains to investigate the effects of acid rain. They measured the pH (acidity) of the water and classified the streams with respect to the kind of substrate (type of rock over which they flow). A lower pH means the water is more acidic. Here is a plot of the pH of the streams by substrate (limestone, mixed, or shale):

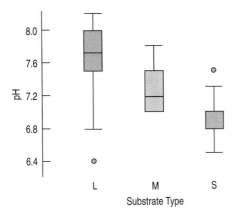

Here are selected parts of a software analysis comparing the pH of streams with limestone and shale substrates:

> 2-Sample t-Test of $\mu 1 - \mu 2$
> Difference Between Means = 0.735
> t-Statistic = 16.30 w/133 df
> p ≤ 0.0001

a) State the null and alternative hypotheses for this test.
b) From the information you have, do the assumptions and conditions appear to be met?
c) What conclusion would you draw?

T 11. Hurricanes. The data below show the number of hurricanes recorded annually before and after 1970. Create an appropriate visual display and determine whether these data are appropriate for testing whether there has been a change in the frequency of hurricanes.

1944–1969	**1970–2000**
3, 2, 1, 2, 4, 3, 7, 2, 3, 3, 2, 5, 2,	2, 1, 0, 1, 2, 3, 2, 1, 2, 2, 2, 3, 1, 1,1, 3,
2, 4, 2, 2, 6, 0, 2, 5, 1, 3, 1, 0, 3	0, 1, 3, 2, 1, 2, 1, 1, 0, 5, 6, 1, 3, 5, 3

T 12. Memory. Does ginkgo biloba enhance memory? In an experiment to find out, subjects were assigned randomly to take ginkgo biloba supplements or a placebo. Their memory was tested to see whether it improved. Here are boxplots comparing the two groups and some computer output from a two-sample *t*-test computed for the data.

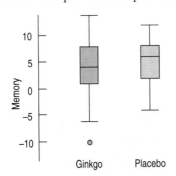

> 2-Sample t-Test of $\mu_G - \mu_P > 0$
> Difference Between Means = −0.9914
> t-Statistic = −1.540 w/196 df
> P = 0.9374

a) Explain in this context what the P-value means.
b) State your conclusion about the effectiveness of ginkgo biloba.
c) Proponents of ginkgo biloba continue to insist that it works. What type of error do they claim your conclusion makes? Explain.

T 13. Baseball. American League baseball teams play their games with the designated hitter rule, meaning that pitchers do not bat. The league believes that replacing the pitcher, traditionally a weak hitter, with another player in the batting order produces more runs and generates more interest among fans. Below are the average numbers of runs scored in American League and National League stadiums for the first half of the 2001 season.

American				National			
11.1	10.8	10.8	10.3	14.0	11.6	10.4	10.3
10.3	10.1	10.0	9.5	10.2	9.5	9.5	9.5
9.4	9.3	9.2	9.2	9.5	9.1	8.8	8.4
	9.0	8.3		8.3	8.2	8.1	7.9

a) Create an appropriate display of these data. What do you see?
b) With a 95% confidence interval, estimate the mean number of runs scored in American League games.
c) What concerns do you have about making a similar confidence interval for National League games? What could you do?
d) Coors Field, in Denver, stands a mile above sea level, an altitude far greater than that of any other major league ball park. Some believe that the thinner air makes it harder for pitchers to throw curve balls and easier for batters to hit the ball a long way. Do you think the 14 runs scored per game at Coors is unusual? Explain.
e) Explain why you should not use two separate confidence intervals to decide whether the two leagues differ in average number of runs scored.

14. Handy. A factory hiring people to work on an assembly line gives job applicants a test of manual agility. This test counts how many strangely shaped pegs the applicant can fit into matching holes in a one-minute period. The table below summarizes the data by gender of the job applicant. Assume that all conditions necessary for inference are met.

	Male	Female
Number of subjects	50	50
Pegs placed:		
Mean	19.39	17.91
SD	2.52	3.39

a) Find 95% confidence intervals for the average number of pegs that males and females can each place.
b) Those intervals overlap. What does this suggest about any gender-based difference in manual agility?
c) Find a 95% confidence interval for the difference in the mean number of pegs that could be placed by men and women.
d) What does this interval suggest about any gender-based difference in manual agility?
e) The two results seem contradictory. Which method is correct: doing two-sample inference, or doing one-sample inference twice?
f) Why don't the results agree?

T 15. Double header. Do the data in Exercise 13 suggest that the American League's designated hitter rule may lead to more runs?
a) Using a 95% confidence interval, estimate the difference between the mean number of runs scored in American and National League games.
b) Interpret your interval.
c) Does that interval suggest that the two leagues may differ in average number of runs scored per game?
d) Does omitting the 14 runs scored per game at Coors Field affect your decision?

T 16. Hard water. In an investigation of environmental causes of disease, data were collected on the annual mortality rate (deaths per 100,000) for males in 61 large towns in England and Wales. In addition, the water hardness was recorded as the calcium concentration (parts per million, ppm) in the drinking water. The data set also notes for each town whether it was south or north of Derby. Is there a significant difference in mortality rates in the two regions? Here are the summary statistics.

Summary of: **mortality**
For categories in: **Derby**

Group	Count	Mean	Median	StdDev
North	34	1631.59	1631	138.470
South	27	1388.85	1369	151.114

a) Test appropriate hypotheses and state your conclusion.
b) The boxplots of the two distributions show an outlier among the data north of Derby. What effect might that have had on your test?

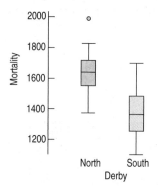

T 17. Job satisfaction. A company institutes an exercise break for its workers to see if this will improve job satisfaction, as measured by a questionnaire that assesses workers' satisfaction. Scores for 10 randomly selected workers before and after implementation of the exercise program are shown. The company wants to assess the effectiveness of the exercise program. Explain why you can't use the methods discussed in this chapter to do that. (Don't worry, we'll give you another chance to do this the right way.)

Worker Number	Job Satisfaction Index	
	Before	After
1	34	33
2	28	36
3	29	50
4	45	41
5	26	37
6	27	41
7	24	39
8	15	21
9	15	20
10	27	37

18. Summer school. Having done poorly on their math final exams in June, six students repeat the course in summer school, then take another exam in August. If we consider these students representative of all students who might attend this summer school in other years, do these results provide evidence that the program is worthwhile?

June	54	49	68	66	62	62
Aug.	50	65	74	64	68	72

19. Sex and violence. In June 2002, the *Journal of Applied Psychology* reported on a study that examined whether the content of TV shows influenced the ability of viewers to recall brand names of items featured in the commercials. The researchers randomly assigned volunteers to watch one of three programs, each containing the same nine commercials. One of the programs had violent content, another sexual content, and the third neutral content. After the shows ended, the subjects were asked to recall the brands of products that were advertised. Results are summarized below.

	Program Type		
	Violent	Sexual	Neutral
No. of subjects	108	108	108
Brands recalled			
Mean	2.08	1.71	3.17
SD	1.87	1.76	1.77

a) Do these results indicate that viewer memory for ads may differ depending on program content? A test of the hypothesis that there is no difference in ad memory between programs with sexual content and those with violent content has a P-value of 0.136. State your conclusion.

b) Is there evidence that viewer memory for ads may differ between programs with sexual content and those with neutral content? Test an appropriate hypothesis and state your conclusion.

20. Ad campaign. You are a consultant to the marketing department of a business preparing to launch an ad campaign for a new product. The company can afford to run ads during one TV show, and has decided not to sponsor a show with sexual content. You read the study described in Exercise 19, then use a computer to create a confidence interval for the difference in mean number of brand names remembered between the groups watching violent shows and those watching neutral shows.

TWO-SAMPLE T
95% CI FOR MU viol – MU neut: (−1.578, −0.602)

a) At the meeting of the marketing staff, you have to explain what this output means. What will you say?

b) What advice would you give the company about the upcoming ad campaign?

21. Sex and violence II. In the study described in Exercise 19, the researchers also contacted the subjects again, 24 hours later, and asked them to recall the brands advertised. Results are summarized below.

	Program Type		
	Violent	**Sexual**	**Neutral**
No. of subjects	101	106	103
Brands recalled			
Mean	3.02	2.72	4.65
SD	1.61	1.85	1.62

a) Is there a significant difference in viewers' abilities to remember brands advertised in shows with violent vs. neutral content?

b) Find a 95% confidence interval for the difference in mean number of brand names remembered between the groups watching shows with sexual content and those watching neutral shows. Interpret your interval in this context.

22. Ad recall. In Exercises 19 and 21, we see the number of advertised brand names people recalled immediately after watching TV shows and 24 hours later. Strangely enough, it appears that they remembered more about the ads the next day. Should we conclude this is true in general about people's memory of TV ads?

a) Suppose one analyst conducts a two-sample hypothesis test to see if memory of brands advertised during violent TV shows is higher 24 hours later. The

P-value of his test is 0.00013. What might he conclude?

b) Explain why his procedure was inappropriate. Which of the assumptions for inference was violated?

c) How might the design of this experiment have tainted these results?

d) Suggest a design that could compare immediate brand name recall with recall one day later.

23. Lower scores? Newspaper headlines recently announced a decline in science scores among high school seniors. In 2000, 15,109 seniors tested by The National Assessment in Education Program (NAEP) scored a mean of 147 points. Four years earlier, 7537 seniors had averaged 150 points. The standard error of the difference in the mean scores for the two groups was 1.22.

a) Have the science scores declined significantly? Cite appropriate statistical evidence to support your conclusion.

b) The sample size in 2000 was almost double that in 1996. Does this make the results more convincing, or less? Explain.

24. The Internet. The NAEP report described in Exercise 23 compared science scores for students who had home Internet access with the scores of those who did not, as shown in the graph. They report that the differences are statistically significant.

a) Explain what "statistically significant" means in this context.

b) If their conclusion is incorrect, which type of error did the researchers commit?

c) Does this prove that using the Internet at home can improve a student's performance in science?

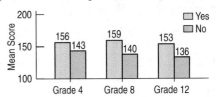

25. Crossing Ontario. Between 1954 and 2003, swimmers have crossed Lake Ontario 43 times. Both women and men have made the crossing. Here are some plots (we've omitted a crossing by Vikki Kieth, who swam a round trip—North to South to North—in 3390 minutes):

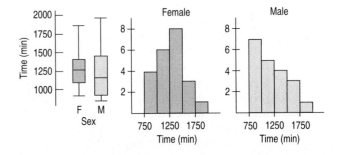

Summary of Time (min)			
Group	Count	Mean	StdDev
F	22	1271.59	261.111
M	20	1196.75	304.369

How much difference is there between the mean amount of time (in minutes) it would take female and male swimmers to swim the lake?

a) Construct and interpret a 95% confidence interval for the difference between female and male crossing times.

b) Comment on the assumptions and conditions.

26. Still swimming. Here's some additional information about the Ontario crossing times presented in Exercise 25. It is generally thought to be harder to swim across the lake from north to south. Indeed, this has been done only 5 times. Every one of those crossings was by a woman. If we omit those 5 crossings, the boxplots look like this:

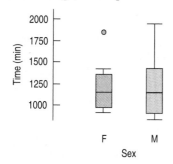

The female outlier is Vikki Kieth, who crossed the lake swimming only butterfly stroke. Omitting that extraordinary swim gives the following summary statistics:

Summary of Time (min)			
Group	Count	Mean	StdDev
F	16	1175.25	177.794
M	20	1196.75	304.369

a) Do women or men appear to be faster at swimming across the lake? Support your answer by interpreting a confidence interval.

b) Vikki Kieth was responsible for two of the more remarkable crossings, but she also swam Lake Ontario two other times. In fact, of the 36 crossings in this analysis, 7 were repeat crossings by a swimmer who'd crossed the lake before. How does this fact affect your thoughts about the confidence interval?

27. Olympic heats. In Olympic running events, preliminary heats are determined by random draw, so we should expect that the ability level of runners in the various heats to be about the same, on average. Here are the times (in seconds) for the 400-m women's run in the 2004 Olympics in Athens for preliminary heats 2 and 5. Is there any evidence that the mean time to finish is different

for randomized heats? Explain. Be sure to include a discussion of assumption and conditions for your analysis.

Country	Name	Heat	Time
USA	HENNAGAN Monique	2	51.02
BUL	DIMITROVA Mariyana	2	51.29
CHA	NADJINA Kaltouma	2	51.50
JAM	DAVY Nadia	2	52.04
BRA	ALMIRAO Maria Laura	2	52.10
FIN	MYKKANEN Kirsi	2	52.53
CHN	BO Fanfang	2	56.01
BAH	WILLIAMS-DARLING Tonique	5	51.20
BLR	USOVICH Svetlana	5	51.37
UKR	YEFREMOVA Antonina	5	51.53
CMR	NGUIMGO Mireille	5	51.90
JAM	BECKFORD Allison	5	52.85
TOG	THIEBAUD-KANGNI Sandrine	5	52.87
SRI	DHARSHA K V Damayanthi	5	54.58

T 28. Swimming heats. In Exercise 27 we looked at the times in two different heats for the 400-m women's run from the 2004 Olympics. Unlike track events, swimming heats are *not* determined at random. Instead, swimmers are seeded so that better swimmers are placed in later heats. Here are the times (in seconds) for the women's 400-m freestyle from heats 2 and 5. Do these results suggest that the mean times of seeded heats are not equal? Explain. Include a discussion of assumption and conditions for your analysis.

Country	Name	Heat	Time
ARG	BIAGIOLI Cecilia Elizabeth	2	256.42
SLO	CARMAN Anja	2	257.79
CHI	KOBRICH Kristel	2	258.68
MKD	STOJANOVSKA Vesna	2	259.39
JAM	ATKINSON Janelle	2	260.00
NZL	LINTON Rebecca	2	261.58
KOR	HA Eun-Ju	2	261.65
UKR	BERESNYEVA Olga	2	266.30
FRA	MANAUDOU Laure	5	246.76
JPN	YAMADA Sachiko	5	249.10
ROM	PADURARU Simona	5	250.39
GER	STOCKBAUER Hannah	5	250.46
AUS	GRAHAM Elka	5	251.67
CHN	PANG Jiaying	5	251.81
CAN	REIMER Brittany	5	252.33
BRA	FERREIRA Monique	5	253.75

29. Tees. Does it matter what kind of tee a golfer places the ball on? The company that manufactures "Stinger" tees claims that the thinner shaft and smaller head will lessen resistance and drag, reducing spin and allowing the ball to travel farther. In August 2003, Golf Laboratories, Inc., compared the distance traveled by golf balls hit off regular wooden tees to those hit off Stinger tees. All the balls were struck by the same golf club using a robotic device set to swing the club head at approximately 95 miles per hour. Summary statistics from the test are shown in the table. Assume that 6 balls were hit off each tee and that the data were suitable for inference.

		Total Distance (yards)	Ball Velocity (mph)	Club Velocity (mph)
Regular tee	Avg.	227.17	127.00	96.17
	SD	2.14	0.89	0.41
Stinger tee	Avg.	241.00	128.83	96.17
	SD	2.76	0.41	0.52

Is there evidence that balls hit off the Stinger tees would have a higher initial velocity?

30. Golf again. Given the test results on golf tees described in Exercise 29, is there evidence that balls hit off Stinger tees would travel farther? Again assume that 6 balls were hit off each tee and that the data were suitable for inference.

31. Statistics journals. When a professional statistician has information to share with colleagues, he or she will submit an article to one of several Statistics journals for publication. This can be a lengthy process; typically, the article must be circulated for "peer review" and perhaps edited before being accepted for publication. Then the article must wait in line with other articles before actually appearing in print. In the Winter 1998 issue of *Chance* magazine, Eric Bradlow and Howard Wainer reported on this delay for several journals between 1990 and 1994. For 288 articles published in *The American Statistician*, the mean length of time between initial submission and publication was 21 months, with a standard deviation of 8 months. For 209 *Applied Statistics* articles, the mean time to publication was 31 months, with a standard deviation of 12 months. Create and interpret a 90% confidence interval for the difference in mean delay, and comment on the assumptions that underlie your analysis.

32. Music and memory. Is it a good idea to listen to music when studying for a big test? In a study conducted by some Statistics students, 62 people were randomly assigned to listen to rap music, music by Mozart, or no music while attempting to memorize objects pictured on a page. They were then asked to list all the objects they could remember. Here are summary statistics for each group:

	Rap	Mozart	No Music
Count	29	20	13
Mean	10.72	10.00	12.77
SD	3.99	3.19	4.73

a) Does it appear that it is better to study while listening to Mozart than to rap music? Test an appropriate hypothesis and state your conclusion.
b) Create a 90% confidence interval for the mean difference in memory score between students who study to Mozart and those who listen to no music at all. Interpret your interval.

33. Rap. Using the results of the experiment described in Exercise 32, does it matter whether one listens to rap music while studying, or is it better to study without music at all?
a) Test an appropriate hypothesis and state your conclusion.
b) If you concluded there is a difference, estimate the size of that difference with a confidence interval and explain what your interval means.

T 34. Cuckoos. Cuckoos lay their eggs in the nests of other (host) birds. The eggs are then adopted and hatched by the host birds. But the potential host birds lay eggs of different sizes. Does the cuckoo change the size of her eggs for different foster species? These are lengths (in mm) of cuckoo eggs found in nests of three different species of other birds. The data are drawn from the work of O. M. Latter in 1902 and were used in a fundamental textbook on statistical quality control by L. H. C. Tippett (1902–1985), one of the pioneers in that field.

CUCKOO EGG LENGTH (MM)		
Foster Parent Species		
Sparrow	Robin	Wagtail
20.85	21.05	21.05
21.65	21.85	21.85
22.05	22.05	21.85
22.85	22.05	21.85
23.05	22.05	22.05
23.05	22.25	22.45
23.05	22.45	22.65
23.05	22.45	23.05
23.45	22.65	23.05
23.85	23.05	23.25
23.85	23.05	23.45
23.85	23.05	24.05
24.05	23.05	24.05
25.05	23.05	24.05
	23.25	24.85
	23.85	

Investigate the question of whether the mean length of cuckoo eggs is the same for different species and state your conclusion. (*Use Tukey's test to compare the mean egg length of each pair of species. Do your conclusions change?)

just checking

Answers

1. Randomization balances unknown sources of variability in the two groups of patients and helps us believe the two groups are independent.

2. We can be 95% confident that after 4 weeks endoscopic surgery patients will have a mean pinch strength between 0.04 kg and 2.96 kg higher than open-incision patients.

3. The lower bound of this interval is close to 0, so the difference may not be great enough that patients could actually notice the difference. We may want to consider other issues such as cost or risk in making a recommendation about the two surgical procedures.

4. Without data, we can't check the Nearly Normal Condition.

5. H_0: Mean pinch strength is the same after both surgeries. H_A: Mean pinch strength is different after the two surgeries.

6. With a P-value this low, we reject the null hypothesis. We can conclude that mean pinch strength differs after 4 weeks in patients who undergo endoscopic surgery and open-incision surgery. Results suggest that the endoscopic surgery patients may be stronger on average.

7. If some patients contributed two hands to the study, then the groups may not be internally independent. It is reasonable to assume that two hands from the same patient might respond in similar ways to similar treatments.

CHAPTER 25

Paired Samples and Blocks

WHO	11 Healthcare workers
WHAT	Annual mileage driven
WHEN	1993–1994
WHERE	Illinois
WHY	Experiment to see whether change in work week increased efficiency

Do flexible schedules reduce the demand for resources? The Lake County, IL, Health Department experimented with a flexible four-day work week. For a year, the department recorded the mileage driven by 11 field workers on an ordinary five-day work week. Then it changed to a flexible four-day work week and recorded mileage for another year (Charles S. Catlin, "Four-day Work Week Improves Environment," *Journal of Environmental Health*, Denver, 59 [March 1997]:7).

Here are the data:

Name	5-Day mileage	4-Day mileage
Jeff	2798	2914
Betty	7724	6112
Roger	7505	6177
Tom	838	1102
Aimee	4592	3281
Greg	8107	4997
Larry G.	1228	1695
Tad	8718	6606
Larry M.	1097	1063
Leslie	8089	6392
Lee	3807	3362

Using boxplots to compare 5-day-week driving with 4-day-week driving shows little because it ignores the natural pairing. Those who drive a lot continue to do so.

Figure 25.1

Boxplots of the mileages don't show much. We can't tell whether the 4-day work week was successful in reducing miles driven. A two-sample *t*-test wouldn't help much either. (It would find a P-value of 0.4, clearly not evidence of a difference.) But wait a minute. The two-sample *t*-test is not valid here. We have 11 individuals, and we know their mileage *before and after* the change in schedule. These data are not independent.

Paired Data

Data such as these are called **paired.** We have each worker's mileage both before and after the work schedule change. The two groups represent the *same people* at different times, so they can't be independent. And since they're not independent, we can't use the two-sample *t* methods. Instead, we can focus on the *changes* in driving mileage.

Paired data arise in a number of ways. Perhaps the most common is to compare subjects with themselves before and after a treatment. When pairs arise from an experiment, the pairing is a type of *blocking*. When they arise from an observational study, it is a form of *matching*.

Pairing isn't a problem; it's an opportunity. If you know the data are paired, you can take advantage of that fact—in fact, you *must* take advantage of it. You *may not* use the two-sample and pooled methods of the previous chapter when the data are paired. You must decide whether the data are paired from understanding how they were collected and what they mean (check the W's). There is no test to determine whether the data are paired.

Once we recognize that the mileage data are matched pairs, it makes sense to consider the change in annual miles driven for each worker as the work schedule changed. So we look at the collection of *pairwise* differences:

A S **Differences in Means of Paired Groups.** Are married couples typically the same age, or do wives tend to be younger than their husbands, on average? This is a paired situation, and the tool in *ActivStats* can handle it. Learn how here.

Name	5-Day mileage	4-Day mileage	Difference
Jeff	2798	2914	−116
Betty	7724	6112	1612
Roger	7505	6177	1328
Tom	838	1102	−264
Aimee	4592	3281	1311
Greg	8107	4997	3110
Larry G.	1228	1695	−467
Tad	8718	6606	2112
Larry M.	1097	1063	34
Leslie	8089	6392	1697
Lee	3807	3362	445

Jeff actually increased his mileage under the 4-day plan, so his difference is negative. But Greg (who started out driving a far greater distance) reduced his mileage substantially, saving over 3100 miles under the new plan. Because it is the *differences* we care about, we'll treat them as if *they* were the data, ignoring the original two columns. Now that we have only one column of values to consider, we can use a simple one-sample *t*-test. Mechanically, a **paired *t*-test** is just a one-sample *t*-test for the means of these pairwise differences. The sample size is the number of pairs.

So much for *Show.*

Assumptions and Conditions

Paired Data Assumption

The data must be paired. You can't just decide to pair data when in fact the samples are independent. When you have two groups with the same number of observations, it may be tempting to match them up.

Don't, unless you are prepared to justify your claim that the data are paired.

On the other hand, be sure to recognize paired data when you have it. Remember, two-sample *t* methods aren't valid without independent groups, and paired groups aren't independent.

Independence Assumption

If the data are paired, the *groups* are not independent. For these methods, it's the *differences* that must be independent of each other. In our example, the fact that Jeff drove farther doesn't affect how far Betty drove. (If Jeff and Betty carpooled for part of the year, this wouldn't be true and the assumption would be violated.)

Randomization Condition: Randomness can arise in many ways. The pairs may be a random sample. In an experiment, the order of the two treatments may be randomly assigned, or the treatments may be randomly assigned to one member of each pair. In a before-and-after study, we may believe that the observed differences are a representative sample from a population of interest. If we have any doubts, we'll need to include a control group to be able to draw conclusions. What we want to know usually focuses our attention on where the randomness should be.

In our example, our main concern is testing whether a 4-day work week reduces mileage. The randomness comes from all the many random driving events that make up the annual totals. If we were to repeat the experiment in two other years, we would find different mileage totals, but (we hope) the same overall pattern of change.

If we also wanted to claim that this change would be typical of all state employees, *then* we might be concerned with how the workers were selected. Are they a random sample of Illinois workers? Was this department selected randomly? If not, our inference may be restricted only to these 11 individuals.

Normal Population Assumption

We need to assume that the population of *differences* follows a Normal model. We don't need to check the individual groups.

Nearly Normal Condition: This condition can be checked with a histogram or Normal probability plot of the differences. As with the one-sample *t*-methods, this assumption matters less as we have more pairs to consider. You may be pleasantly surprised when you check this condition. Even if your original measurements are skewed or bimodal, the *differences* may be nearly Normal. After all, the individual who was way out in the tail on an initial measurement is likely to still be out there on the second one, giving a perfectly ordinary difference.

The paired *t*-test

When the conditions are met, we are ready to test whether the mean of paired differences is significantly different from zero. We test the hypothesis

$$H_0: \mu_d = \Delta_0,$$

where the *d*'s are the pairwise differences and Δ_0 is almost always 0.
We use the statistic,

$$t_{n-1} = \frac{\bar{d} - \Delta_0}{SE(\bar{d})},$$

where $\bar{d}$ is the mean of the pairwise differences, *n* is the number of *pairs,* and

$$SE(\bar{d}) = \frac{s_d}{\sqrt{n}}.$$

$SE(\bar{d})$ is the ordinary standard error for the mean applied to the differences.
When the conditions are met and the null hypothesis is true, we can model the sampling distribution of this statistic with a Student's *t*-model with $n - 1$ degrees of freedom, and use that model to obtain a P-value.

A Paired *t*-Test   Step-By-Step

The steps of testing a hypothesis for paired differences are very much like the steps for a one-sample *t*-test for a mean. Only now we have to check that the data are paired. Then we take the difference of each pair and work with them as our data values.

Think

Plan State what we want to know.

Identify the *parameter* we wish to estimate. Here our parameter is the mean difference in mileage driven.

Identify the variables and check the W's.

Hypotheses State the null and alternative hypotheses.

Although we hope for a reduction in miles driven, we have no reason to suppose that the difference must be in that direction, so we'd better test a two-sided alternative.

REALITY CHECK The individual differences are all in the hundreds to low thousands of miles. We should expect the mean difference to be comparable in magnitude.

Model Check the conditions.
State why you think the data are paired. Simply having the same number of individuals in each group, displaying them in side-by-side columns, doesn't make them paired.

I want to know whether the change in schedule changed the mean number of miles driven. I have the miles driven for a year by 11 employees under one schedule and then for a second year under the alternative schedule.

H_0: The change in the health department workers' schedules didn't change the mean mileage driven; the mean difference is zero: $\mu_d = 0$.

H_A: The mean difference is different from zero: $\mu_d \neq 0$.

✔ **Paired Data Assumption:** The data are paired because they are measurements on the same individuals before and after a change in work schedule.

Think about what we hope to learn and where the randomization comes from. Here, the randomization comes from the random events that happen to each driver during the study.

Make a picture—just one. Don't plot separate distributions of the two groups—that entirely misses the pairing. For paired data, it's the Normality of the differences that we care about. Treat those paired differences as you would a single variable, and check the Nearly Normal Condition with a histogram or a Normal probability plot.

✔ **Independence Assumption:** The behavior of any individual is independent of the behavior of the others, so the differences are mutually independent.

✔ **Randomization Condition:** The measured values are the sums of individual trips, each of which experienced random events that arose while driving. Repeating the experiment in two new years would give randomly different values.

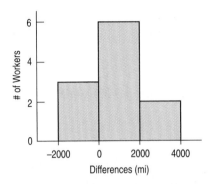

✔ **Nearly Normal Condition:** The histogram of the differences is unimodal and symmetric.

Specify the sampling distribution model.

Choose the method.

The conditions are met, so I'll use a Student's t-model with $(n - 1) = 10$ degrees of freedom, and perform a **paired t-test.**

Show

Mechanics

n is the number of *pairs*; in this case, the number of workers.

$\bar{d}$ is the mean difference.

s_d is the standard deviation of the differences.

Make a picture. Sketch a *t*-model centered at the hypothesized mean of 0. Because this is a two-tail test, shade both the region to the right of the observed mean difference of 982 miles and the corresponding region in the lower tail.

Find the standard error and the *t*-score of the observed mean difference. There is nothing new in the mechanics of the paired-*t* methods. These are the mechanics of the *t*-test for a mean applied to the differences. Find the P-value, using technology.

The data give:

$$n = 11 \text{ pairs}$$
$$\bar{d} = 982 \text{ miles}$$
$$s_d = 1139.6 \text{ miles.}$$

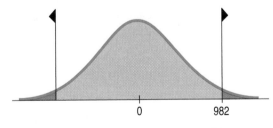

I estimate the standard deviation of $\bar{d}$ using:

$$SE(\bar{d}) = \frac{s_d}{\sqrt{n}} = \frac{1139.6}{\sqrt{11}} = 343.6$$

$$\text{So } t_{10} = \frac{\bar{d} - 0}{SE(\bar{d})} = \frac{982.0}{343.6} = 2.86$$

$$\text{P-value} = 2P(t_{10} > 2.86) = 0.017$$

The P-value is 0.017.

REALITY CHECK The mean difference is 982 miles. That's more than twice the SE of 344 miles, so it looks like the change is real. The *t*-value of 2.86 and P-value of 0.017 are consistent with that.

Conclusion Link the P-value to your decision about H_0, and state your conclusion in context.

> The P-value is small, so I reject the null hypothesis and conclude that the change in work week did lead to a change in driving mileage.

It appears that the change in work schedule reduced the mileage driven by workers. We should propose a course of action, but it's hard to tell from the hypothesis test whether the reduction matters. Is the difference in mileage important in the sense of reducing air pollution or costs, or is it merely statistically significant? To help make that decision, we should look at a confidence interval. If the difference in mileage proves to be large in a practical sense, then we might recommend a change in schedule for the rest of the department.

Confidence Intervals for Matched Pairs

In developed countries, the average age of women is generally higher than that of men. After all, women tend to live longer. But if we look at *married couples*, husbands tend to be slightly older than wives. How much older on average are husbands? We have data from a random sample of 200 British couples, the first 16 of which are shown below. Only 170 couples provided ages for both husband and wife, so we can work only with that many pairs. Let's form a confidence interval for the mean difference of husband's and wife's age for these 170 couples. Here are the first 16 pairs:

WHO	170 Randomly sampled couples
WHAT	Ages (years)
WHEN	Recently
WHERE	Britain

Wife's Age	Husband's Age	Difference (husband–wife)
43	49	6
28	25	−3
30	40	10
57	52	−5
52	58	6
27	32	5
52	43	−9
43	47	4
23	31	8
25	26	1
39	40	1
32	35	3
35	35	0
33	35	2
43	47	4
35	38	3
⋮	⋮	⋮

Clearly, these data are paired. The survey selected *couples* at random, not individuals. We're interested in the mean age difference within couples. How would we construct a confidence interval for the true mean difference in ages?

> ### Paired *t*-interval
>
> When the conditions are met, we are ready to find the confidence interval for the mean of the paired differences. The confidence interval is
>
> $$\bar{d} \pm t^*_{n-1} \times SE(\bar{d}),$$
>
> where the standard error of the mean difference is $SE(\bar{d}) = \frac{s_d}{\sqrt{n}}$.
>
> The critical value t^* from the Student's *t*-model depends on the particular confidence level, *C*, that you specify and on the degrees of freedom, $n-1$, which is based on the number of pairs, *n*.

A Paired *t*-Interval Step-By-Step

Making confidence intervals for matched pairs follows exactly the steps for a one-sample *t*-interval.

Think

Plan State what we want to know.

Identify the variables and check the W's.

Identify the parameter you wish to estimate. For a paired analysis, the parameter of interest is the mean of the differences. The population of interest is the population of differences.

Model Check the conditions.

Make a picture. We focus on the differences, so a histogram or Normal probability plot is best here.

REALITY CHECK The histogram shows husbands are often older than wives (because most of the differences are greater than 0). The mean difference seen here of about 2 years is reasonable.

I want to estimate the mean difference in age between husbands and wives. I have a random sample of 200 British couples, 170 of whom provided both ages.

✔ **Paired Data Assumption:** The data are paired because they are on members of married couples.

✔ **Randomization Condition:** These couples were randomly sampled.

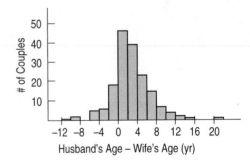

Husband's Age – Wife's Age (yr)

✔ **Nearly Normal Condition:** The histogram of the husband–wife differences is unimodal and symmetric.

The conditions are met, so I can use a Student's *t*-model with $(n-1) = 169$ degrees of freedom and find a **paired *t*-interval.**

State the sampling distribution model.

Choose your method.

Show   Mechanics

n is the number of *pairs,* here, the number of couples.

$\bar{d}$ is the mean difference.

s_d is the standard deviation of the differences.

Be sure to include the units along with the statistics.

The critical value we need to make a 95% interval comes from a Student's *t* table, a computer program, or a calculator.

$n = 170$ couples

$\bar{d} = 2.2$ years

$s_d = 4.1$ years

I estimate the standard error of $\bar{d}$ as:

$$SE(\bar{d}) = \frac{s_d}{\sqrt{n}} = \frac{4.1}{\sqrt{170}} = 0.31 \text{ years.}$$

The df for the *t*-model is $n - 1 = 169$.

The 95% critical value for t_{169} (from the table) is 1.97.

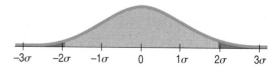

The margin of error is

$$ME = t^*_{169} \times SE(\bar{d}) = 1.97(0.31) = 0.61$$

So the 95% confidence interval is:

$$2.2 \pm 0.6 \text{ years,}$$

or an interval of (1.6, 2.8) years.

REALITY CHECK This result makes sense. Our everyday experience confirms that an average age difference of about 2 years is reasonable.

Tell   Conclusion Interpret the confidence interval in context.

I am 95% confident that British husbands are, on average, 1.6 to 2.8 years older than their wives.

Blocking

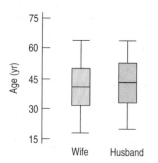

This display is worthless. It does no good to compare all the wives as a group with all the husbands. We care about the paired differences.

Figure 25.2

Because the sample includes both older and younger couples, there's a lot of variation in the ages of the men and in the ages of the women. In fact, that variation is so great that a boxplot of the two groups would show little difference. But that would be the wrong plot. It's the difference we care about. Pairing removes the extra variation and allows us to focus on the individual differences. In experiments, we block to remove the variability between identifiable groups of subjects, allowing us to better see variability among treatment groups due to their response to the treatment. A paired design is an example of blocking.

When we pair, we have roughly half the degrees of freedom of a two-sample test. You may see discussions that suggest that in "choosing" a paired analysis we "give up" these degrees of freedom. This isn't really true, though. If the data are paired, then there never were additional degrees of freedom, and we have no "choice." The fact of the pairing determines how many degrees of freedom are available.

Matching pairs generally removes so much extra variation that it more than compensates for having only half the degrees of freedom. Of course, inappropriate matching when the groups are in fact independent (say, by matching on the first letter of the last name of subjects) would cost degrees of freedom without the benefit of reducing the variance. When you design a study or experiment, you should consider using a paired design if possible.

just checking

Think about each of the situations described below.

- Would you use a two-sample *t* or paired-*t* method (or neither)? Why?
- Would you perform a hypothesis test or find a confidence interval?

1. Random samples of 50 men and 50 women are asked to imagine buying a birthday present for their best friend. We want to estimate the difference in how much they are willing to spend.

2. Mothers of twins were surveyed and asked how often in the past month strangers had asked whether the twins were identical.

3. Are parents equally strict with boys and girls? In a random sample of families, researchers asked a brother and sister from each family to rate how strict their parents were.

4. Forty-eight overweight subjects are randomly assigned to either aerobic or stretching exercise programs. They are weighed at the beginning and at the end of the experiment to see how much weight they lost.
   a) We want to estimate the mean amount of weight lost by those doing aerobic exercise.
   b) We want to know which program is more effective at reducing weight.

5. Couples at a dance club were separated and each person was asked to rate the band. Do men or women like this group more?

*The Sign Test Again?

Because we have paired data, we've been using a simple *t*-test for the paired differences. This suggests that if we want a distribution-free method, it would be natural to compute a sign test on the paired differences and test whether the *median* of the differences is 0. That's exactly what we do. The test is very simple. We record a 0 for every paired difference that's negative and a 1 for each positive difference, ignoring pairs for which the difference is exactly 0. We test the associated proportion $p = 0.5$ using a *z*-test if the number of pairs is at least 20 (so that we expect at least 10 successes and failures), or compute the exact Binomial probabilities if $n < 20$ (as discussed in Chapter 17).

Let's try it on the married couples data. Of the 170 couples, there are 119 with the husband older, 32 with the wife older, and 19 that have the same age. So we test $p = 0.5$, with a sample proportion of $119/151 = 0.788$. The Success/Failure Condition is easily satisfied. Applying the one-proportion *z*-test to the differences gives a *z*-score of 7.08, with a P-value < 0.00001. We can be pretty confident that the median is not 0. (Just for comparison, the *t*-statistic for testing $\mu_d = 0$ is 7.152 with 169 df—almost the identical result.)

As with other distribution-free tests, the advantage of the ***sign test for matched pairs*** is that we don't require the Nearly Normal Condition for the paired differences. Because it looks only at the direction of the difference, the sign test isn't affected by outliers—extraordinarily large differences—which can be an advantage. On the other hand, when the assumptions of the *paired t-test* are met, the paired *t*-test is more powerful than the sign test.

What Can Go Wrong?

- **Don't use a two-sample t-test for paired data.** See the What Can Go Wrong? discussion in Chapter 24.
- **Don't use a paired-t method when the samples aren't paired.** When two groups don't have the same number of values, it's pretty easy to see that they can't be paired. But just because two groups have the same number of observations doesn't mean they can be paired, even if they are shown side-by-side in a table. We might have 25 men and 25 women in our study, but they might be completely independent of one another. If they were siblings or spouses, we might consider them paired. Remember that you cannot *choose* which method to use based on your preferences. If the data are from two independent samples, use two-sample *t* methods. If the data are from an experiment in which observations were paired, you must use a paired method. If the data are from an observational study, you must be able to defend your decision to use matched pairs or independent groups.
- **Don't forget outliers.** The outliers we care about now are in the differences. A subject who is extraordinary both before and after a treatment may still have a perfectly typical difference. But one outlying difference can completely distort your conclusions. Be sure to plot the differences (even if you also plot the data).
- **Don't look for the difference in side-by-side boxplots.** The point of the paired analysis is to remove extra variation. The boxplots of each group still contain that variation. Comparing them is likely to be misleading. A scatterplot of the two variables can sometimes be helpful.

CONNECTIONS

The most important connection is to the concept of blocking that we first discussed when we considered designed experiments in Chapter 13. Pairing is a basic and very effective form of blocking.

Of course, the details of the mechanics for paired *t*-tests and intervals are identical to those for the one-sample *t* methods. Everything we know about those methods applies here.

The connection to the two-sample and pooled methods of the previous chapter is that when the data are naturally paired, those methods are not appropriate because paired data fail the required condition of independence.

What have we learned?

When we looked at various ways to design experiments, back in Chapter 13, we saw that pairing can be a very effective strategy. Because pairing can help control variability between individual subjects, paired methods are usually more powerful than methods that compare independent groups. Now we've learned that analyzing data from matched pairs requires different inference procedures.

- We've learned that paired *t*-methods look at pairwise differences. Based on these differences we test hypotheses and generate confidence intervals. Our procedures are mechanically identical to the one-sample *t* methods we saw in Chapter 23.

- We've also learned to *Think* about the design of the study that collected the data before we proceed with inference. We must be careful to recognize pairing when it is present, but not assume it when it is not. Making the correct decision about whether to use independent *t*-procedures or paired *t*-methods is the first critical step in analyzing the data.

TERMS

Paired data   Data are paired when the observations are collected in pairs or the observations in one group are naturally related to observations in the other. The simplest form of pairing is to measure each subject twice—often before and after a treatment is applied. More sophisticated forms of pairing in experiments are a form of blocking and arise in other contexts. Pairing in observational and survey data is a form of matching.

Paired *t*-test   A hypothesis test for the mean of the pairwise differences of two groups. It tests the null hypothesis

$$H_0: \mu_d = \Delta_0,$$

where the hypothesized difference is almost always 0, using the statistic

$$t = \frac{\bar{d} - \Delta_0}{SE(\bar{d})}$$

with $n - 1$ degrees of freedom, where $SE(\bar{d}) = \dfrac{s_d}{\sqrt{n}}$, and n is the number of pairs.

Paired-*t* confidence interval   A confidence interval for the mean of the pairwise differences between independent groups found as

$$\bar{d} \pm t^*_{n-1} \times SE(\bar{d}), \text{ where } SE(\bar{d}) = \frac{s_d}{\sqrt{n}}, \text{ and } n \text{ is the number of pairs.}$$

SKILLS   *When you complete this lesson you should:*

Think
- Be able to recognize whether a design that compares two groups is paired or not.

Show
- Be able to find a paired confidence interval, recognizing that it is mechanically equivalent to doing a one-sample *t*-interval applied to the differences.
- Be able to perform a paired *t*-test, recognizing that it is mechanically equivalent to a one-sample *t*-test applied to the differences.

Tell
- Be able to interpret a paired *t*-test, recognizing that the hypothesis tested is about the mean of the differences between paired values rather than about the differences between the means of two independent groups.
- Be able to interpret a paired *t*-interval, recognizing that it gives an interval for the mean difference in the pairs.

Paired *t* on the Computer

Most statistics programs can compute paired-*t* analyses. Some may want you to find the differences yourself and use the one-sample *t* methods. Those that perform the entire procedure will need to know the two variables to compare. The computer, of course, cannot verify that the variables are naturally paired. Most programs will check whether the two variables have the same number of observations, but some stop there, and that can cause trouble. (See the comments on individual packages for details.) Most programs will automatically omit any pair that is missing a value for either variable (as we did with the British couples). You must look carefully to see whether that has happened.

Computers make it easy to examine the boxplots of the two groups[1] and the histogram of the differences—both important steps. Some programs offer a scatterplot of the two variables. That can be helpful. In terms of the scatterplot, a paired *t*-test is about whether the points tend to be above or below the 45° line $y = x$. (Note, though, that pairing says nothing about whether the scatterplot should be straight. That doesn't matter for our *t*-methods.)

As we've seen with other inference results, some packages stack a lot of information into a simple table, but you must locate what you want for yourself. Here's a generic example with comments:

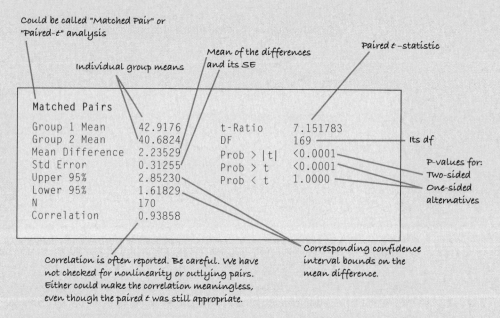

Other packages try to be more descriptive. It may be easier to find the results, but you may get less information from the output table:

---

[1] Even though our concern is with the differences of the pairs, you should also examine the boxplots of the two groups separately so you are aware of any outliers that might be of interest.

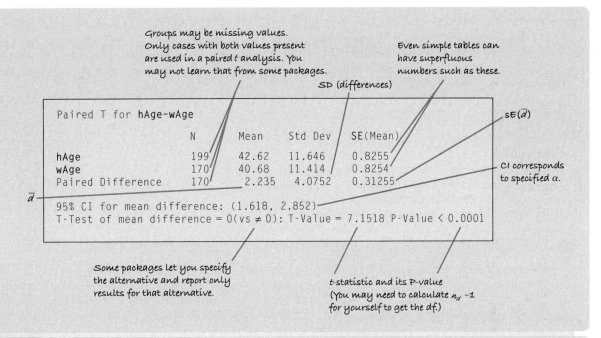

Groups may be missing values. Only cases with both values present are used in a paired *t* analysis. You may not learn that from some packages.

Even simple tables can have superfluous numbers such as these.

SD (differences)

$SE(\bar{d})$

```
Paired T for hAge-wAge

                  N      Mean    Std Dev   SE(Mean)
hAge             199    42.62    11.646    0.8255
wAge             170    40.68    11.414    0.8254
Paired Difference 170    2.235   4.0752    0.31255

95% CI for mean difference: (1.618, 2.852)
T-Test of mean difference = 0(vs ≠ 0): T-Value = 7.1518 P-Value < 0.0001
```

$\bar{d}$

CI corresponds to specified α.

Some packages let you specify the alternative and report only results for that alternative.

t-statistic and its P-value (You may need to calculate $n_d - 1$ for yourself to get the df.)

DATA DESK

Select variables.

From the **Calc** menu, choose **Estimate** for confidence intervals or **Test** for hypothesis tests. Select the interval or test from the drop-down menu and make other choices in the dialog.

Comments

Data Desk expects the two groups to be in separate variables and in the same "Relation"—that is, about the same cases.

EXCEL

From the **Tools** menu, choose **Data Analysis**. From the **Data Analysis** menu, choose **t-test: paired two-sample for Means.** Fill in the cell ranges for the two groups, the hypothesized difference, and the alpha level.

Comments

Excel expects the two groups to be in separate cell ranges.
Warning: Do not compute this test in Excel without checking for missing values. If there are any missing values (empty cells), Excel will usually give a wrong answer. Excel compacts each list, pushing values up to cover the missing cells and then checks only that it has the same number of values in each list. The result is mismatched pairs and an entirely wrong analysis.

JMP

From the **Analyze** menu select **Matched Pairs.** Specify the columns holding the two groups in the **Y Paired Response** Dialog. Click **OK.**

MINITAB

From the **Stat** menu choose the **Basic Statistics** submenu. From that menu, choose **Paired t...** Then fill in the dialog.

Comments

Minitab takes "First sample" minus "Second sample."

SPSS

From the **Analyze** menu, choose the **Compare Means** submenu. From that, choose the **Paired-Samples t-test** command. Select pairs of variables to compare and click the arrow to add them to the selection box.

Comments

You can compare several pairs of variables at once. Options include the choice to exclude cases missing in any pair from all tests.

TI-83/84 Plus

If the data are stored in two lists, say, L1 and L2, create a list of the differences:

L1–L2 → L3. (The arrow is the STO button.) Since inference for paired differences uses one-sample *t*-procedures, select **2:T-Test** or **8:TInterval** from the **STAT TESTS** menu. Specify as your data the list of differences you just created and apply the procedure.

TI-89

If the data are stored in two lists, say, list1 and list2, create a list of the differences: Move the cursor to the name of an empty list, then use VAR-LINK to enter the command list1-list2. Press ENTER to perform the subtraction.

Since inference for paired differences uses one-sample *t*-procedures, select **2:T-Test** or **2:TInterval** from the **STAT Tests** or **Ints** menu. Specify as your data the list of differences you just created and apply the procedure.

EXERCISES

1. More eggs? Can a food additive increase egg production? Agricultural researchers want to design an experiment to find out. They have 100 hens available. They have two kinds of feed—the regular feed and the new feed with the additive. They plan to run their experiment for a month, recording the number of eggs each hen produces.

a) Design an experiment that will require a two-sample *t* procedure to analyze the results.

b) Design an experiment that will require a matched-pairs *t* procedure to analyze the results.

c) Which experiment would you consider the stronger design? Why?

2. MTV. Some students do homework with the TV on. (Anyone come to mind?) Some researchers want to see if people can work as effectively with as without distraction. The researchers will time some volunteers to see how long it takes them to complete some relatively easy crossword puzzles. During some of the trials, the room will be quiet; during other trials in the same room, a TV will be on, tuned to MTV.

a) Design an experiment that will require a two-sample *t* procedure to analyze the results.

b) Design an experiment that will require a matched-pairs *t* procedure to analyze the results.

c) Which experiment would you consider the stronger design? Why?

3. Sex sells? Ads for many products use sexual images to try to attract attention to the product. But do these ads bring people's attention to the item that was being advertised? We want to design an experiment to see if the presence of sexual images in an advertisement affects people's ability to remember the product.

a) Describe an experimental design that would require a matched-pairs *t* procedure to analyze the results.

b) Describe an experimental design that would require an independent sample procedure to analyze the results.

4. Freshman 15? Many people believe that students gain weight as freshmen. Suppose we plan to conduct a study to see if this is true.

a) Describe a study design that would require a matched-pairs *t* procedure to analyze the results.

b) Describe a study design that would require a two-sample *t* procedure to analyze the results.

5. Women. Values for the labor force participation rate of women (LFPR) are published by the U.S. Bureau of Labor Statistics. We are interested in whether there was a difference between female participation in 1968 and 1972, a time of rapid change for women. We check LFPR values for 19 randomly selected cities for 1968 and 1972. Shown below is software output for two possible tests.

Paired t-Test of $\mu(1 - 2)$
Test Ho: $\mu(1972\text{-}1968) = 0$ vs Ha: $\mu(1972\text{-}1968) \neq 0$
Mean of Paired Differences = 0.0337
t-Statistic = 2.458 w/18 df
p = 0.0244

2-Sample t-Test of $\mu1 - \mu2$
Ho: $\mu1 - \mu2 = 0$ Ha: $\mu1 - \mu2 \neq 0$
Test Ho: $\mu(1972) - \mu(1968) = 0$ vs
Ha: $\mu(1972) - \mu(1968) \neq 0$
Difference Between Means = 0.0337
t-Statistic = 1.496 w/35 df
p = 0.1434

a) Which of these tests is appropriate for these data? Explain.
b) Using the test you selected, state your conclusion.

T 6. Rain. Simpson, Alsen, and Eden (*Technometrics* 1975) report the results of trials in which clouds were seeded and the amount of rainfall recorded. The authors report on 26 seeded and 26 unseeded clouds in order of the amount of rainfall, largest amount first. Here are two possible tests to study the question of whether cloud seeding works. Which test is appropriate for these data? Explain your choice. Using the test you select, state your conclusion.

Paired t-Test of $\mu(1 - 2)$
Mean of Paired Differences = −277.39615
t-Statistic = −3.641 w/25 df
p = 0.0012

2-Sample t-Test of $\mu1 - \mu2$
Difference Between Means = −277.4
t-Statistic = −1.998 w/33 df
p = 0.0538

a) Which of these tests is appropriate for these data? Explain.
b) Using the test you selected, state your conclusion.

T 7. Friday the 13th, I. In 1993 the *British Medical Journal* published an article titled, "Is Friday the 13th Bad for Your Health?" Researchers in Britain examined how Friday the 13th affects human behavior. One question was whether people tend to stay at home more on Friday the 13th. The data below are the number of cars passing Junctions 9 and 10 on the M25 motorway for consecutive Fridays (the 6th and 13th) for five different time periods.

Year	Month	6th	13th
1990	July	134012	132908
1991	September	133732	131843
1991	December	121139	118723
1992	March	124631	120249
1992	November	117584	117263

Here are summaries of two possible analyses:
Paired t-Test of mu1 = mu2 vs. mu1 > mu2
Mean of Paired Differences: 2022.4
t-Statistic = 2.9377 w/4 df
P = 0.0212

2-Sample t-Test of mu1 = mu2 vs. mu1 > mu2
Difference Between Means: 2022.4
t-Statistic = 0.4273 w/7.998 df
P = 0.3402

a) Which of the tests is appropriate for these data? Explain.
b) Using the test you selected, state your conclusion.
c) Are the assumptions and conditions for inference met?

T 8. Friday the 13th, II: The researchers in Exercise 7 also examined the number of people admitted to emergency rooms for vehicular accidents on 12 Friday evenings (6 each on the 6th and 13th).

Year	Month	6th	13th
1989	October	9	13
1990	July	6	12
1991	September	11	14
1991	December	11	10
1992	March	3	4
1992	November	5	12

Based on these data, is there evidence that more people are admitted on average on Friday the 13th? Here are two possible analyses of the data:

Paired t-Test of mu1 = mu2 vs. mu1 < mu2
Mean of Paired Differences = 3.333
t-Statistic = 2.7116 w/5 df
P = 0.0211

2-Sample t-Test of mu1 = mu2 vs. mu1 < mu2
Difference Between Means = 3.333
t-Statistic = 1.6644 w/9.940 df
P = 0.0636

a) Which of these tests is appropriate for these data? Explain.
b) Using the test you selected, state your conclusion.
c) Are the assumptions and conditions for inference met?

9. Wheelchair marathon. The Boston Marathon has had a wheelchair division since 1977. Who do you think is typically faster, the men's marathon winner on foot or the women's wheelchair marathon winner? Because the conditions differ year to year, and speeds have improved over the years, it seems best to treat these as paired measurements. Here are summary statistics for the pairwise differences in finishing time (in minutes):

Summary of	wheelchrF − runM
N =	28
Mean =	0.089286
SD =	37.7241

a) Comment on the assumptions and conditions.
b) Assuming that these times are representative of such races and the differences appeared acceptable for inference, construct and interpret a 95% confidence interval for the mean difference in finishing times.
c) Would a hypothesis test at $\alpha = 0.05$ reject the null hypothesis of no difference? What conclusion would you draw?

10. Boston startup years. When we considered the Boston Marathon in Exercise 9, we were unable to check the Nearly Normal Condition. Here's a histogram of the differences:

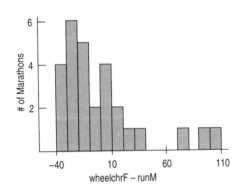

Those three large differences are the first three years of wheelchair competition, 1977, 1978, and 1979. Often the startup years of new events are different; later on more athletes train and compete. If we omit those three years, the summary statistics change as follows:

Summary of	wheelchrF − runM
N	25
Mean	−10.9093
SD	20.3071

a) Comment on the assumptions and conditions.
b) Assuming that these times are representative of such races, construct and interpret a 95% confidence interval for the mean difference in finishing time.
c) Would a hypothesis test at $\alpha = 0.05$ reject the null hypothesis of no difference? What conclusion would you draw?

11. BST. Many dairy cows now receive injections of BST, a hormone intended to spur greater milk production. After the first injection, a test herd of 60 Ayrshire cows increased their mean daily production from 47 pounds to 61 pounds of milk. The standard deviation of the increases was 5.2 pounds. We want to estimate the mean increase a farmer could expect in his own cows.
a) Check the assumptions and conditions for inference.
b) Write a 95% confidence interval.
c) Explain what your interval means in this context.
d) Given the cost of BST, a farmer believes he cannot afford to use it unless he is sure of attaining at least a 25% increase in milk production. Based on your confidence interval, what advice would you give him?

12. BST II. In the experiment about hormone injections in cows described in Exercise 11, a group of 52 Jersey cows increased average milk production from 43 pounds to 52 pounds per day, with a standard deviation of 4.8 pounds. Is this evidence that the hormone may be more effective in one breed than the other? Test an appropriate hypothesis and state your conclusion. Be sure to discuss any assumptions you make.

13. Temperatures. The table below gives the average high temperatures in January and July for several European cities. Write a 90% confidence interval for the mean temperature difference between summer and winter in Europe. Be sure to check conditions for inference, and clearly explain what your interval means.

	Mean High Temperatures (°F)	
City	**Jan.**	**July**
Vienna	34	75
Copenhagen	36	72
Paris	42	76
Berlin	35	74
Athens	54	90
Rome	54	88
Amsterdam	40	69
Madrid	47	87
London	44	73
Edinburgh	43	65
Moscow	21	76
Belgrade	37	84

14. Marathons. Shown are the winning times (in minutes) for men and women in the New York City Marathon between 1978 and 1998. Assuming that performances in the Big Apple resemble performances elsewhere, we can think of these data as a sample of performance in marathon competitions. Create a 90% confidence interval for the mean difference in winning times for male and female marathon competitors. (*Chance*, 12, no. 4, [1999])

Year	Men	Women
1978	132.2	152.5
1979	131.7	147.6
1980	129.7	145.7
1981	128.2	145.5
1982	129.5	147.2
1983	129.0	147.0
1984	134.9	149.5
1985	131.6	148.6
1986	131.1	148.1
1987	131.0	150.3
1988	128.3	148.1
1989	128.0	145.5
1990	132.7	150.8
1991	129.5	147.5
1992	129.5	144.7
1993	130.1	146.4
1994	131.4	147.6
1995	131.0	148.1
1996	129.9	148.3
1997	128.2	148.7
1998	128.8	145.3

15. Push-ups. Every year the students at Gossett High School take a physical fitness test during their gym classes. One component of the test asks them to do as many push-ups as they can. Results for one class are shown below, according to gender. Assuming that students at Gossett are assigned to gym classes at random, create a 90% confidence interval for how many more pushups boys can do than girls, on average, at that high school.

Boys	17	27	31	17	25	32	28	23	25	16	11	34
Girls	24	7	14	16	2	15	19	25	10	27	31	8

16. Exercise. An August 2001 article in the journal *Medicine and Science in Sports and Exercise* compared how long it would take men and women to burn 200 calories during light or heavy workouts on various kinds of exercise equipment. The results summarized in the table are the average times for a group of physically active young men and women whose performances were measured on each type of equipment.

a) On average, how many minutes longer than a man must a woman exercise at a light exertion rate in order to burn 200 calories? Give a 95% confidence interval.

b) Estimate the average number of minutes longer a woman must work out at light exertion than at heavy exertion to get the same benefit. Give a 95% confidence interval.

c) These data are actually averages rather than individual times. How might this affect the margins of error in these confidence intervals?

	AVERAGE MINUTES TO BURN 200 CALORIES			
	Hard Exertion		**Light Exertion**	
	Men	**Women**	**Men**	**Women**
Treadmill	12	17	14	22
X-C skier	12	16	16	23
Stair climber	13	18	20	37
Rowing machine	14	16	21	25
Exercise rider	22	24	27	36
Exercise bike	16	20	29	44

(*Machine Type* label appears vertically at left of the table rows.)

17. Job satisfaction. (When you first read about this exercise break plan in Chapter 24, you did not have an inference method that would work. Try again now.) A company institutes an exercise break for its workers to see if this will improve job satisfaction, as measured by a questionnaire that assesses workers' satisfaction. Scores for 10 randomly selected workers before and after the implementation of the exercise program are shown.

a) Identify the procedure you would use to assess the effectiveness of the exercise program, and check to see if the conditions allow use of that procedure.

b) Test an appropriate hypothesis and state your conclusion.

*c) Use a matched-pairs sign test to test the appropriate hypothesis. Do your conclusions change from those in part b?

Worker	Job Satisfaction Index	
Number	Before	After
1	34	33
2	28	36
3	29	50
4	45	41
5	26	37
6	27	41
7	24	39
8	15	21
9	15	20
10	27	37

T 18. Summer school. (When you first read about the summer school issue in Chapter 24 you did not have an inference method that would work. Try again now.) Having done poorly on their Math final exams in June, six students repeat the course in summer school and take another exam in August. If we consider these students to be representative of all students who might attend this summer school in other years, do these results provide evidence that the program is worthwhile?

June	54	49	68	66	62	62
Aug.	50	65	74	64	68	72

19. Sleep. W. S. Gosset (Student) refers to data recording the number of hours of additional sleep gained by 10 patients from the use of *laevohyoscyamine hydrobromide*. We want to see if there is strong evidence that the herb can help people get more sleep.
a) State the null and alternative hypotheses clearly.
b) A *t*-test of the null hypothesis of no gain has a *t*-statistic of 3.680 with 9 degrees of freedom. Find the P-value.
c) Interpret this result by explaining the meaning of the P-value.
d) State your conclusion regarding the hypotheses.
e) This conclusion, of course, may be incorrect. If so, which type of error was made?

T 20. Gasoline. Many drivers of cars that can run on regular gas actually buy premium in the belief that they will get better gas mileage. To test that belief, we use 10 cars in a company fleet in which all the cars run on regular gas. Each car is filled first with either regular or premium gasoline, decided by a coin toss, and the mileage for that tankful recorded. Then the mileage is recorded again for the same cars for a tankful of the other kind of gasoline. We don't let the drivers know about this experiment. Here are the results (miles per gallon):

Car #	1	2	3	4	5	6	7	8	9	10
Regular	16	20	21	22	23	22	27	25	27	28
Premium	19	22	24	24	25	25	26	26	28	32

a) Is there evidence that cars get significantly better fuel economy with premium gasoline?
b) How big might that difference be? Check a 90% confidence interval.
c) Even if the difference is significant, why might the company choose to stick with regular gasoline?
d) Suppose you had done a "bad thing." (We're sure you didn't.) Suppose you had mistakenly treated these data as two independent samples instead of matched pairs. What would the significance test have found? Carefully explain why the results are so different.
*e) Use a matched-pairs sign test to test the appropriate hypothesis. Do your conclusions change from those in part a?

T 21. Yogurt. Do these data suggest that there is a significant difference in calories between servings of strawberry and vanilla yogurt? Test an appropriate hypothesis and state your conclusion. Don't forget to check assumptions and conditions!

Brand	Calories per Serving	
	Strawberry	Vanilla
America's Choice	210	200
Breyer's Lowfat	220	220
Columbo	220	180
Dannon Light 'n Fit	120	120
Dannon Lowfat	210	230
Dannon laCreme	140	140
Great Value	180	80
La Yogurt	170	160
Mountain High	200	170
Stonyfield Farm	100	120
Yoplait Custard	190	190
Yoplait Light	100	100

22. Caffeine. A student experiment investigating the potential impact of caffeine on studying for a test involved 30 subjects, randomly divided into two groups. Each group took a memory test. The subjects then each drank two cups of regular (caffeinated) cola or caffeine-free cola. Thirty minutes later they each took another version of the memory test, and the changes in their scores were noted. Among the 15 subjects who drank caffeine, scores fell an average of -0.933 points, with a standard deviation of 2.988 points. Among the no-caffeine group, scores went up an average of 1.429 points with a standard deviation of 2.441 points. Assumptions of Normality were deemed reasonable based on histograms of differences in scores.
a) Did scores change significantly for the group who drank caffeine? Test an appropriate hypothesis and state your conclusion.
b) Did scores change significantly for the no-caffeine group? Test an appropriate hypothesis and state your conclusion.
c) Does this indicate that some mystery substance in noncaffeinated soda may aid memory? What other explanation is plausible?

T 23. Braking. In a test of braking performance, a tire manufacturer measured the stopping distance for one of its tire models. On a test track, a car made repeated stops from 60 miles per hour. The test was run on both dry and wet pavement, with results as shown in the table. (Note that actual *braking distance*, which takes into account the driver's reaction time, is much longer, typically nearly 300 feet at 60 mph!)
a) Write a 95% confidence interval for the mean dry pavement stopping distance. Be sure to check the ap-

propriate assumptions and conditions, and explain what your interval means.

b) Write a 95% confidence interval for the mean increase in stopping distance on wet pavement. Be sure to check the appropriate assumptions and conditions, and explain what your interval means.

Stopping Distance (ft)	
Dry Pavement	Wet Pavement
145	211
152	191
141	220
143	207
131	198
148	208
126	206
140	177
135	183
133	223

T 24. Brain waves. An experiment was performed to see whether sensory deprivation over an extended period of time has any effect on the alpha-wave patterns produced by the brain. To determine this, 20 subjects, inmates in a Canadian prison, were randomly split into two groups. Members of one group were placed in solitary confinement. Those in the other group were allowed to remain in their own cells. Seven days later, alpha-wave frequencies were measured for all subjects, as shown in the following table (P. Gendreau et al., "Changes in EEG Alpha Frequency and Evoked Response Latency During Solitary Confinement," *Journal of Abnormal Psychology* 79 [1972]: 54–59):

Nonconfined	Confined
10.7	9.6
10.7	10.4
10.4	9.7
10.9	10.3
10.5	9.2
10.3	9.3
9.6	9.9
11.1	9.5
11.2	9.0
10.4	10.9

a) What are the null and alternative hypotheses? Be sure to define all the terms and symbols you use.
b) Are the assumptions necessary for inference met?
c) Perform the appropriate test, indicating the formula you used, the calculated value of the test statistic, and the P-value.
d) State your conclusion.

T 25. Braking, test 2. For another test of the tires in Exercise 23, the company tried them on 10 different cars, recording the stopping distance for each car on both wet and dry pavement. Results are shown in the table.

	Stopping Distance (ft)	
Car #	Dry Pavement	Wet Pavement
1	150	201
2	147	220
3	136	192
4	134	146
5	130	182
6	134	173
7	134	202
8	128	180
9	136	192
10	158	206

a) Write a 95% confidence interval for the mean dry pavement stopping distance. Be sure to check the appropriate assumptions and conditions, and explain what your interval means.
b) Write a 95% confidence interval for the mean increase in stopping distance on wet pavement. Be sure to check the appropriate assumptions and conditions, and explain what your interval means.

T 26. Tuition. How much more do public colleges and universities charge out-of-state students for tuition per semester? A random sample of 19 public colleges and universities listed in the *Information Please Almanac* found the data shown below. Tuition figures per semester are rounded to the nearest hundred dollars.

Institution	Resident	Nonresident
U Akron (OH)	3200	7900
Athens State (GA)	3800	7500
Ball State (IN)	3000	7800
Bloomsburg U (PA)	3700	8500
UC Irvine (CA)	4300	12000
Central State (OH)	2900	6400
Clarion U (PA)	3900	8600
Dakota State	2500	4600
Fairmont State (WV)	1800	4500
Johnson State (VT)	4000	8600
Lock Haven U (PA)	7400	12100
New College of S. Fla.	2000	7900
Oakland U (MI)	3100	8700
U Pittsburgh	5400	11200
Savannah State (GA)	2000	5200
SW Louisiana	1900	4900
W Liberty State (WV)	1900	4500
W Texas State	1600	6000
Worcester State (MA)	2700	6200

a) Create a 90% confidence interval for the mean difference in cost. Be sure to justify the procedure you use.
b) Interpret your interval in context.
c) A national magazine claims that public institutions charge state residents an average of $4000 less for tuition each semester. What does your confidence interval indicate about the validity of this assertion?

T 27. Sex sells, part II. In Exercise 3 you considered the question of whether sexual images in ads affected people's abilities to remember the item being advertised. To investigate, a group of Statistics students cut ads out of magazines. They were careful to find two ads for each of ten similar items, one with a sexual image and one without. They arranged the ads in random order and had 39 subjects look at them for one minute. Then they asked the subjects to list as many of the products as they could remember. Their data are shown in the table. Is there evidence that the sexual images mattered?

Subject Number	Ads Remembered Sexual Image	Ads Remembered No Sex	Subject Number	Ads Remembered Sexual Image	Ads Remembered No Sex
1	2	2	21	2	3
2	6	7	22	4	2
3	3	1	23	3	3
4	6	5	24	5	3
5	1	0	25	4	5
6	3	3	26	2	4
7	3	5	27	2	2
8	7	4	28	2	4
9	3	7	29	7	6
10	5	4	30	6	7
11	1	3	31	4	3
12	3	2	32	4	5
13	6	3	33	3	0
14	7	4	34	4	3
15	3	2	35	2	3
16	7	4	36	3	3
17	4	4	37	5	5
18	1	3	38	3	4
19	5	5	39	4	3
20	2	2			

T 28. Freshman 15, revisited. In Exercise 4 you thought about how to design a study to see if it's true that students tend to gain weight during their first year in college. Well, Cornell Professor of Nutrition David Levitsky did just that. He recruited students from two large sections of an introductory health course. Although they were volunteers, they appeared to match the rest of the freshman class in terms of demographic variables such as gender and ethnicity. The students were weighed during the first week of the semester then again 12 weeks later. Based on Professor Levitsky's data estimate the mean weight gain in first-semester freshmen, and comment on the "freshman 15." (Weights are in pounds.)

Subject Number	Initial Weight	Terminal Weight	Subject Number	Initial Weight	Terminal Weight
1	171	168	35	148	150
2	110	111	36	164	165
3	134	136	37	137	138
4	115	119	38	198	201
5	150	155	39	122	124
6	104	106	40	146	146
7	142	148	41	150	151
8	120	124	42	187	192
9	144	148	43	94	96
10	156	154	44	105	105
11	114	114	45	127	130
12	121	123	46	142	144
13	122	126	47	140	143
14	120	115	48	107	107
15	115	118	49	104	105
16	110	113	50	111	112
17	142	146	51	160	162
18	127	127	52	134	134
19	102	105	53	151	151
20	125	125	54	127	130
21	157	158	55	106	108
22	119	126	56	185	188
23	113	114	57	125	128
24	120	128	58	125	126
25	135	139	59	155	158
26	148	150	60	118	120
27	110	112	61	149	150
28	160	163	62	149	149
29	220	224	63	122	121
30	132	133	64	155	158
31	145	147	65	160	161
32	141	141	66	115	119
33	158	160	67	167	170
34	135	134	68	131	131

T 29. Strikes. Advertisements for an instructional video claim that the techniques will improve the ability of Little League pitchers to throw strikes, and that, after undergoing the training, players will be able to throw strikes on at least 60% of their pitches. To test this claim, we have 20 Little Leaguers throw 50 pitches each, and we record the number of strikes. After the players participate in the training program, we repeat the test. The table shows the number of strikes each player threw before and after the training.

a) Is there evidence that after training players can throw strikes more than 60% of the time?

b) Is there evidence that the training is effective in improving a player's ability to throw strikes?

*c) Use a matched-pairs sign test to test an appropriate hypothesis. Do your conclusions change from those in part a?

Number of Strikes (out of 50)		Number of Strikes (out of 50)	
Before	After	Before	After
28	35	33	33
29	36	33	35
30	32	34	32
32	28	34	30
32	30	34	33
32	31	35	34
32	32	36	37
32	34	36	33
32	35	37	35
33	36	37	32

T 30. Uninsured. During the economic recession of 2000–2001, unemployment increased and job benefits were scaled back. The table shows the percentage of people in each state who lacked health insurance during the '99–'00 and '00–'01, as reported by the U.S. Census Bureau. Is there any evidence that more people were without health insurance during the recession? (*Use a matched-pairs sign test as well. Do the conclusions change?)

State	'99–'00	'00–'01	State	'99–'00	'00–'01
AL	13.3	13.2	MT	17.3	15.2
AK	18.6	17.3	NE	17.7	16.5
AZ	18.6	17.3	NV	17.7	16.5
AR	14.4	15.2	NH	8.7	8.9
CA	19.0	19.0	NJ	12.1	12.6
CO	14.9	14.9	NM	24.4	22.4
CT	9.4	10.0	NY	15.9	15.9
DE	9.6	9.2	NC	14.0	14.0
DC	14.1	13.4	ND	11.5	10.5
FL	17.6	17.9	OH	10.7	11.2
GA	14.7	15.5	OK	17.7	18.6
HI	9.8	9.5	OR	13.3	12.7
ID	16.8	15.7	PA	8.5	9.0
IL	13.6	13.7	RI	6.9	7.6
IN	10.3	11.5	SC	13.8	12.2
IA	8.2	8.2	SD	10.9	10.2
KS	11.4	11.1	TN	10.6	11.1
KY	13.4	13.0	TX	22.7	23.2
LA	19.9	18.7	UT	13.0	13.7
ME	10.8	10.6	VT	9.8	9.1
MD	10.8	11.3	VA	12.4	11.3
MA	9.0	8.5	WA	13.7	13.3
MI	9.7	9.8	WV	14.7	13.6
MN	7.8	8.1	WI	8.9	7.6
MS	14.6	15.0	WY	15.4	15.8
MO	8.1	9.9			

just checking
Answers

1. These are independent groups sampled at random, so use a two-sample t confidence interval to estimate the size of the difference.

2. There is only one sample. Use a one-sample t-interval.

3. A brother and sister from the same family represent a matched pair. The question calls for a paired t-test.

4. a) A before-and-after study calls for paired t-methods. To estimate the loss, find a confidence interval for the before–after differences.
   b) The two treatment groups were assigned randomly, so they are independent. Use a two-sample t-test to assess whether the mean weight losses differ.

5. Sometimes it just isn't clear. Most likely, couples would discuss the band or even decide to go to the club because they both like a particular band. If we think that's likely, then these data are paired. But maybe not. If we asked them their opinions of, say, the decor or furnishings at the club, the fact that they were couples might not affect the independence of their answers.

REVIEW OF PART VI — Learning About the World

QUICK REVIEW

We continue to explore how to answer questions about the statistics we get from samples and experiments. In this part, those questions have been about means— means of one sample, two independent samples, or matched pairs. Here's a brief summary of the key concepts and skills:

▶ A confidence interval uses a sample statistic to estimate a range of possible values for a parameter of interest.

▶ A hypothesis test proposes a model, then examines the plausibility of that model by seeing how surprising our observed data would be if the model were true.

▶ Statistical inference procedures for proportions are based on the Central Limit Theorem. We can make inferences about a single proportion or the difference of two proportions using Normal models.

▶ Statistical inference procedures for means are also based on the Central Limit Theorem, but we don't usually know the population standard deviation. Student's *t*-models take into account the additional uncertainty of independently estimating the standard deviation.

- We can make inferences about one mean, the difference of two independent means, or the mean of paired differences using *t*-models.

- No inference procedure is valid unless the underlying assumptions are true. Always check the conditions before proceeding.

- Because *t*-models assume that samples are drawn from Normal populations, data in the sample should appear to be nearly Normal. Skewness and outliers are particularly problematic, especially for small samples.

- When there are two variables, you must think carefully about how the data were collected. You may use two-sample *t* procedures only if the groups are independent.

- Unless there is some obvious reason to suspect that two independent populations have the same standard deviation, you should not pool the variances. It is never wrong to use unpooled *t* procedures.

- If the two groups are somehow paired, the data are *not* from independent groups. You must use matched-pairs *t* procedures.

Now for some opportunities to review these concepts. Be careful. You have a lot of thinking to do. These review exercises mix questions about proportions and means. You have to determine which of our inference procedures is appropriate in each situation. Then you have to check the proper assumptions and conditions. Keeping track of those can be difficult, so first we summarize the many procedures with their corresponding assumptions and conditions on the next page. Look them over carefully . . . then, on to the Exercises!

Quick Guide to Inference

Think				Show				Tell?
Inference about?	One group or two?	Procedure	Model	Parameter	Estimate	SE		Chapter
Proportions	One sample	1-Proportion z-Interval	z	p	$\hat{p}$	$\sqrt{\dfrac{\hat{p}\hat{q}}{n}}$		19
		1-Proportion z-Test				$\sqrt{\dfrac{p_0 q_0}{n}}$		20, 21
	Two independent groups	2-Proportion z-Interval	z	$p_1 - p_2$	$\hat{p}_1 - \hat{p}_2$	$\sqrt{\dfrac{\hat{p}_1\hat{q}_1}{n_1} + \dfrac{\hat{p}_2\hat{q}_2}{n_2}}$		22
		2-Proportion z-Test				$\sqrt{\dfrac{\hat{p}\hat{q}}{n_1} + \dfrac{\hat{p}\hat{q}}{n_2}}$,	$\hat{p} = \dfrac{y_1 + y_2}{n_1 + n_2}$	22
Means	One sample	t-Interval / t-Test	t df = $n - 1$	μ	$\bar{y}$	$\dfrac{s}{\sqrt{n}}$		23
	Two independent groups	2-Sample t-Test / 2-Sample t-Interval	t df from technology	$\mu_1 - \mu_2$	$\bar{y}_1 - \bar{y}_2$	$\sqrt{\dfrac{s_1^2}{n_1} + \dfrac{s_2^2}{n_2}}$		24
	Matched pairs	Paired t-Test / Paired t-Interval	t df = $n - 1$	μ_d	$\bar{d}$	$\dfrac{s_d}{\sqrt{n}}$		25

Assumptions for Inference | And the Conditions That Support or Override Them

Proportions (z)
- **One sample**
  1. Individuals are independent.
  2. Sample is sufficiently large.

  1. SRS and $n < 10\%$ of the population.
  2. Successes and failures each ≥ 10.

- **Two groups**
  1. Groups are independent.
  2. Data in each group are independent.

  3. Both groups are sufficiently large.

  1. (Think about how the data were collected.)
  2. Both are SRSs and $n < 10\%$ of populations OR random allocation.
  3. Successes and failures each ≥ 10 for both groups.

Means (t)
- **One sample** (df = $n - 1$)
  1. Individuals are independent.
  2. Population has a Normal model.

  1. SRS and $n < 10\%$ of the population.
  2. Histogram is unimodal and symmetric.*

- **Matched pairs** (df = $n - 1$)
  1. Data are matched.
  2. Individuals are independent.
  3. Population of differences is Normal.

  1. (Think about the design.)
  2. SRS and $n < 10\%$ OR random allocation.
  3. Histogram of differences is unimodal and symmetric.*

- **Two independent groups** (df from technology)
  1. Groups are independent.
  2. Data in each group are independent.
  3. Both populations are Normal.

  1. (Think about the design.)
  2. SRSs and $n < 10\%$ OR random allocation.
  3. Both histograms are unimodal and symmetric.*

(*less critical as n increases)

REVIEW EXERCISES

1. Crawling. A study published in 1993 found that babies born at different times of the year may develop the ability to crawl at different ages! The author of the study suggested that these differences may be related to the temperature at the time the infant is 6 months old. (Benson and Janette, *Infant Behavior and Development* [1993])

a) The study found that 32 babies born in January crawled at an average age of 29.84 weeks, with a standard deviation of 7.08 weeks. Among 21 July babies, crawling ages averaged 33.64 weeks with a standard deviation of 6.91 weeks. Is this difference significant?

b) For 26 babies born in April, the mean and standard deviation were 31.84 and 6.21 weeks, while for 44 October babies the mean and standard deviation of crawling ages were 33.35 and 7.29 weeks. Is this difference significant?

c) Are these results consistent with the researcher's claim? (We'll examine these data in more detail in a later chapter.)

2. Mazes and smells. Can pleasant smells improve learning? Researchers timed 21 subjects as they tried to complete paper-and-pencil mazes. Each subject attempted a maze both with and without the presence of a floral aroma. Subjects were randomized with respect to whether they did the scented trial first or second. Some of the data collected are shown in the table. Is there any evidence that the floral scent improved the subjects' ability to complete the mazes? (A. R. Hirsch and L. H. Johnston, "Odors and Learning." Chicago: Smell and Taste Treatment and Research Foundation)

Time to Complete the Maze (sec)	
Unscented	Scented
25.7	30.2
41.9	56.7
51.9	42.4
32.2	34.4
64.7	44.8
31.4	42.9
40.1	42.7
43.2	24.8
33.9	25.1
40.4	59.2
58.0	42.2
61.5	48.4
44.6	32.0
35.3	48.1
37.2	33.7
39.4	42.6
77.4	54.9
52.8	64.5
63.6	43.1
56.6	52.8
58.9	44.3

3. Women. The U.S. Census Bureau reports that 26% of all U.S. businesses are owned by women. A Colorado consulting firm surveys a random sample of 410 businesses in the Denver area and finds that 115 of them have women owners. Should the firm conclude that its area is unusual? Test an appropriate hypothesis and state your conclusion.

4. Drugs. In a full-page ad that ran in many U.S. newspapers in August 2002, a Canadian discount pharmacy listed costs of drugs that could be ordered from a Web site in Canada. The table compares prices (in US$) for commonly prescribed drugs.

	Cost per 100 Pills		
Drug Name	United States	Canada	Percent savings
Cardizem	131	83	37
Celebrex	136	72	47
Cipro	374	219	41
Pravachol	370	166	55
Premarin	61	17	72
Prevacid	252	214	15
Prozac	263	112	57
Tamoxifen	349	50	86
Vioxx	243	134	45
Zantac	166	42	75
Zocor	365	200	45
Zoloft	216	105	51

a) Give a 95% confidence interval for the average savings in dollars.

b) Give a 95% confidence interval for the average savings in percent.

c) Which analysis do you think is more appropriate? Why?

d) In small print the newspaper ad says, "Complete list of all 1500 drugs available on request." How does this comment affect your conclusions above?

5. Pottery. Archaeologists can use the chemical composition of clay found in pottery artifacts to determine whether different sites were populated by the same ancient people. They collected five samples of Romano-British pottery from each of two sites in Great Britain—the Ashley Rails site and the New Forest site—and measured the percentage of aluminum oxide in each. Based on these data, do you think the same people used these two kiln sites? Base your conclusion on a 95% confidence interval for the difference in aluminum oxide content of pottery made at the sites. (A. Tubb, A. J. Parker, and G. Nickless, "The Analysis of Romano-British Pottery by Atomic Absorption Spectrophotometry." *Archaeometry*, 22[1980]:153–171)

Ashley Rails	19.1	14.8	16.7	18.3	17.7
New Forest	20.8	18.0	18.0	15.8	18.3

6. Streams. Researchers in the Adirondack Mountains collect data on a random sample of streams each year. One of the variables recorded is the substrate of the stream—the type of soil and rock over which they flow. The researchers found that 69 of the 172 sampled streams had a substrate of shale. Construct a 95% confidence interval for the proportion of Adirondack streams with a shale substrate. Clearly interpret your interval in this context.

7. Gehrig. Ever since Lou Gehrig developed amyotrophic lateral sclerosis (ALS), this deadly condition has been commonly known as Lou Gehrig's disease. Some believe that ALS is more likely to strike athletes or the very fit. Columbia University neurologist Lewis P. Rowland recorded personal histories of 431 patients he examined between 1992 and 2002. He diagnosed 280 as having ALS; 38% of them had been varsity athletes. The other 151 had other neurological disorders, and only 26% of them had been varsity athletes. (*Science News*, Sept. 28 [2002])
a) Is there evidence that ALS is more common among athletes?
b) What kind of study is this? How does that affect the inference you made in part a?

T 8. Teen drinking. A study of the health behavior of school-aged children asked a sample of 15-year-olds in several different countries if they had been drunk at least twice.

Country	Percent of 15-Year-Olds Drunk at Least Twice	
	Female	Male
Denmark	63	71
Wales	63	72
Greenland	59	58
England	62	51
Finland	58	52
Scotland	56	53
No. Ireland	44	53
Slovakia	31	49
Austria	36	49
Canada	42	42
Sweden	40	40
Norway	41	37
Ireland	29	42
Germany	31	36
Latvia	23	47
Estonia	23	44
Hungary	22	43
Poland	21	39
USA	29	34
Czech Rep.	22	36
Belgium	22	36
Russia	25	32
Lithuania	20	32
France	20	29
Greece	21	24
Switzerland	16	25
Israel	10	18

The results are shown in the table, by gender. Give a 95% confidence interval for the difference in the rates for males and females. Be sure to check the assumptions that support your chosen procedure, and explain what your interval means. (*Health and Health Behavior Among Young People.* Copenhagen: World Health Organization, 2000)

9. Babies. The National Perinatal Statistics Unit of the Sydney Children's Hospital reports that the mean birth weight of all babies born in Australia in 1999 was 3360 grams—about 7.41 pounds. A Missouri hospital reports that the average weight of 112 babies born there last year was 7.68 pounds, with a standard deviation of 1.31 pounds. If we believe the Missouri babies fairly represent American newborns, is there any evidence that U.S. babies and Australian babies do not weigh the same amount at birth?

10. Petitions. To get a voter initiative on a state ballot, petitions that contain at least 250,000 valid voter signatures must be filed with the Elections Commission. The board then has 60 days to certify the petitions. A group wanting to create a statewide system of universal health insurance has just filed petitions with a total of 304,266 signatures. As a first step in the process, the Board selects an SRS of 2000 signatures and checks them against local voter lists. Only 1772 of them turn out to be valid.
a) What percent of the sample signatures were valid?
b) What percent of the petition signatures submitted must be valid in order to have the initiative certified by the Elections Commission?
c) What will happen if the Elections Commission commits a Type I error?
d) What will happen if the Elections Commission commits a Type II error?
e) Does the sample provide evidence in support of certification? Explain.
f) What could the Elections Commission do to increase the power of the test?

11. Feeding fish. In the midwestern United States, a large aquaculture industry raises largemouth bass. Researchers wanted to know whether the fish would grow better if fed a natural diet of fathead minnows or an artificial diet of food pellets. They stocked six ponds with bass fingerlings weighing about 8 grams. For one year, the fish in three of the ponds were fed minnows, and the others were fed the commercially prepared pellets. The fish were then harvested, weighed, and measured. The bass fed a natural food source had a higher average length (19.6 cm) and weight (95.9 g) than those fed the commercial fish food (17.3 cm and 72.0 g, respectively). The researchers reported P-values for both measurements to be less than 0.001.
a) Explain to someone who has not studied Statistics what the P-values mean here.
b) What advice should the researchers give the people who raise largemouth bass?
c) If that advice turns out to be incorrect, what type of error occurred?

⊤ 12. Risk. A study of auto safety determined the number of driver deaths per million vehicle sales for the model years 1995–1999, classified by type of vehicle. The data below are for 6 midsize models and 6 SUVs. We wonder if there is evidence that drivers of SUVs are safer, hoping to create a 95% confidence interval for the difference in driver death rates for the two types of vehicles. Are these data appropriate for this inference? Explain. (Ross and Wenzel, *An Analysis of Traffic Deaths by Vehicle Type and Model,* March 2002)

Midsize	47	54	64	76	88	97
SUV	55	60	62	76	91	109

13. Age. In a study of how depression may impact one's ability to survive a heart attack, the researchers reported the ages of the two groups they examined. The mean age of 2397 patients without cardiac disease was 69.8 years (SD = 8.7 years), while for the 450 patients with cardiac disease the mean and standard deviation of the ages were 74.0 and 7.9, respectively.
a) Create a 95% confidence interval for the difference in mean ages of the two groups.
b) How might an age difference confound these research findings about the relationship between depression and ability to survive a heart attack?

14. Smoking. In the depression and heart attack research described in Exercise 13, 32% of the diseased group were smokers, compared with only 23.7% of those free of heart disease.
a) Create a 95% confidence interval for the difference in the proportions of smokers in the two groups.
b) Is this evidence that the two groups in the study were different? Explain.
c) Could this be a problem in analyzing the results of the study? Explain.

15. Computer use. A Gallup telephone poll of 1240 teens conducted in 2001 found that boys were more likely than girls to play computer games, by a margin of 77% to 65%. An equal number of boys and girls were surveyed.
a) What kind of sampling design was used?
b) Give a 95% confidence interval for the difference in game playing by gender.
c) Does your confidence interval suggest that among all teens a higher percentage of boys than girls play computer games?

16. Recruiting. In September 2002, CNN reported on a method of grad student recruiting by the Haas School of Business at U.C.-Berkeley. The school notifies applicants by formal letter that they have been admitted, and also e-mails the accepted students a link to a Web site that greets them with personalized balloons, cheering, and applause. The director of admissions says this extra effort at recruiting has really worked well. The school accepts 500 applicants each year, and the percentage who actually choose to enroll at Berkeley has increased from 52% the year before the Web greeting to 54% this year.
a) Create a 95% confidence interval for the change in enrollment rates.
b) Based on your confidence interval, are you convinced that this new form of recruiting has been effective? Explain.

⊤ 17. Hearing. Fitting someone for a hearing aid requires assessing the patient's hearing ability. In one method of assessment, the patient listens to a tape of 50 English words. The tape is played at low volume, and the patient is asked to repeat the words. The patient's hearing ability score is the number of words perceived correctly. Four tapes of equivalent difficulty are available so that each ear can be tested with more than one hearing aid. These lists were created to be equally difficult to perceive in silence, but hearing aids must work in the presence of background noise. Researchers had 24 subjects with normal hearing compare two of the tapes when a background noise was present, with the order of the tapes randomized. Is it reasonable to assume that the two lists are still equivalent for purposes of the hearing test when there is background noise? Base your decision on a confidence interval for the mean difference in the number of words people might misunderstand. (Faith Loven, *A Study of the Interlist Equivalency of the CID W-22 Word List Presented in Quiet and in Noise.* University of Iowa [1981])

Subject	List A	List B
1	24	26
2	32	24
3	20	22
4	14	18
5	32	24
6	22	30
7	20	22
8	26	28
9	26	30
10	38	16
11	30	18
12	16	34
13	36	32
14	32	34
15	38	32
16	14	18
17	26	20
18	14	20
19	38	40
20	20	26
21	14	14
22	18	14
23	22	30
24	34	42

18. Cesareans. Some people fear that differences in insurance coverage can affect healthcare decisions. A survey of several randomly selected hospitals found that 16.6% of 223 recent births in Vermont involved cesarean deliveries, compared with 18.8% of 186 births in New Hampshire. Is this evidence that the rate of cesarean births in the two states is different?

⊤ 19. Newspapers. Who reads the newspaper more, men or women? Eurostat, an agency of the European Union (EU), conducts surveys on several aspects of daily life in EU countries. Recently, the agency asked samples of 1000 respondents in each of 14 European countries whether they read the newspaper on a daily basis. Below are the data by country and gender.

	% Reading a Newspaper Daily	
Country	Men	Women
Belgium	56.3	45.5
Denmark	76.8	70.3
Germany	79.9	76.8
Greece	22.5	17.2
Spain	46.2	24.8
Ireland	58.0	54.0
Italy	50.2	29.8
Luxembourg	71.0	67.0
Netherlands	71.3	63.0
Austria	78.2	74.1
Portugal	58.3	24.1
Finland	93.0	90.0
Sweden	89.0	88.0
UK	32.6	30.4

a) Examine the differences in the percentages for each country. Which of these countries seem to be outliers? What do they have in common?

b) After eliminating the outliers, is there evidence that in Europe men are more likely than women to read the newspaper?

T 20. Meals. A college student is on a "meal program." His budget allows him to spend an average of $10 per day for the semester. He keeps track of his daily food expenses for 2 weeks; the data are given in the table. Is there strong evidence that he will overspend his food allowance? Explain.

Date	Cost ($)
7/29	15.20
7/30	23.20
7/31	3.20
8/1	9.80
8/2	19.53
8/3	6.25
8/4	0
8/5	8.55
8/6	20.05
8/7	14.95
8/8	23.45
8/9	6.75
8/10	0
8/11	9.01

21. Wall Street. In September of 2000, the Harris Poll organization asked 1002 randomly sampled American adults whether they agreed or disagreed with the following statement:

Most people on Wall Street would be willing to break the law if they believed they could make a lot of money and get away with it.

Of those asked, 60% said they agreed with this statement. We know that if we could ask the entire population of American adults, we would not find that exactly 60% think that Wall Street workers would be willing to break the law to make money. Construct a 95% confidence interval for the true percentage of American adults who agree with the statement.

22. Teach for America. Several programs attempt to address the shortage of qualified teachers by placing uncertified instructors in schools with acute needs—often in inner cities. A 1999–2000 study compared students taught by certified teachers with others taught by undercertified teachers in the same schools. Reading scores of the students of certified teachers averaged 35.62 points with standard deviation 9.31. The scores of students instructed by undercertified teachers had mean 32.48 points with standard deviation 9.43 points on the same test. There were 44 students in each group. The appropriate t procedure has 86 degrees of freedom. Is there evidence of lower scores with uncertified teachers? Discuss. (*The Effectiveness of "Teach for America" and Other Under-certified Teachers on Student Academic Achievement: A Case of Harmful Public Policy.* Education Policy Analysis Archives [2002])

T 23. Legionnaires' disease. In 1974, the Bellevue-Stratford Hotel in Philadelphia was the scene of an outbreak of what later became known as legionnaires' disease. The cause of the disease was finally discovered to be bacteria that thrived in the air-conditioning units of the hotel. Owners of the Rip Van Winkle Motel, hearing about the Bellevue-Stratford, replace their air-conditioning system. The following data are the bacteria counts, in the air of eight rooms, before and after a new air-conditioning system was installed (measured in colonies per cubic foot of air). The objective is to find out whether the new system has succeeded in lowering the bacterial count. You are the statistician assigned to report to the hotel whether the strategy has worked. Base your analysis on a confidence interval. Be sure to list all your assumptions, methods, and conclusions.

Room Number	Before	After
121	11.8	10.1
163	8.2	7.2
125	7.1	3.8
264	14	12
233	10.8	8.3
218	10.1	10.5
324	14.6	12.1
325	14	13.7

24. Teach for America, Part II. The study described in Exercise 22 also looked at scores in mathematics and language. Here are software outputs for the appropriate tests. Explain what they show.

Mathematics

T-TEST OF Mu(1) − Mu(2) = 0
Mu(Cert) − Mu(NoCert) = 4.53 t (86) = 2.95 p = 0.002

Language

T-TEST OF Mu(1) − Mu(2) = 0
Mu(Cert) − Mu(NoCert) = 2.13 t (84) = 1.71 p = 0.045

25. Bipolar kids. The June 2002 *American Journal of Psychiatry* reported that researchers used medication and psychotherapy to treat children aged 7 to 16 who exhibit bipolar symptoms. After 2 years, symptoms had cleared up in only 26 of the 89 children involved in the study.
a) Write a 95% confidence interval, and interpret it in this context.
b) If researchers subsequently hope to produce an estimate of treatment effectiveness for bipolar disorder that has a margin of error of only 6%, how many patients should they study?

T 26. Online testing. The Educational Testing Service is now administering several of its standardized tests online, the CLEP and GMAT exams, for example. Since taking a test on a computer is different from taking a test with pencil and paper, one wonders if the scores will be the same. To investigate this question, researchers created two versions of an SAT-type test and got 20 volunteers to participate in an experiment. Each volunteer took both versions of the test, one with pencil and paper and the other online. Subjects were randomized with respect to the order in which they sat for the tests (online/paper) and which form they took (Test A, Test B) in which environment. The scores (out of a possible 20) are summarized in the table.

Subject	Paper	Online
	Test A	Test B
1	14	13
2	10	13
3	16	8
4	15	14
5	17	16
6	14	11
7	9	12
8	12	12
9	16	16
10	7	14
	Test B	Test A
11	8	13
12	11	13
13	15	17
14	11	13
15	13	14
16	9	9
17	15	9
18	14	15
19	16	12
20	8	10

a) Were the two forms (A/B) of the test equivalent in terms of difficulty? Test an appropriate hypothesis and state your conclusion.
b) Is there evidence that testing environment (paper/on-line) matters? Test an appropriate hypothesis and state your conclusion.

27. Bread. Clarksburg Bakery is trying to predict how many loaves of bread to bake. In the last 100 days, the bakery has sold between 95 and 140 loaves per day. Here are a histogram and the summary statistics for the number of loaves sold for the last 100 days.

Summary of Sales	
Mean	103
Median	100
SD	9.000
Min	95
Max	140
Lower 25th %tile	97
Upper 25th %tile	105.5

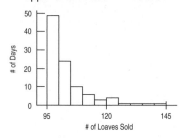

a) Can you use these data to estimate the number of loaves sold on the busiest 10% of all days? Explain.
b) Explain why you can use these data to construct a 95% confidence interval for the mean number of loaves sold per day.
c) Calculate a 95% confidence interval and carefully interpret what that confidence interval means.
d) If the bakery would have been satisfied with a confidence interval whose margin of error was twice as wide, how many days' data could they have used?
e) When the bakery opened, the owners estimated that they would sell an average of 100 loaves per day. Does your confidence interval provide strong evidence that this estimate was incorrect? Explain.

28. Irises. Can measurements of the petal length of flowers be of value when you need to determine the species of a certain flower? Here are the summary statistics from measurements of the petals of two species of irises. (R. A. Fisher, "The Use of Multiple Measurements in Axonomic Problems." *Annals of Eugenics* 7 [1936]:179–188)

	Species	
	Versicolor	Virginica
Count	50	50
Mean	55.52	43.22
Median	55.50	44.00
SD	5.519	5.362
Min	45	30
Max	69	56
Lower Quartile	51	40
Upper Quartile	59	47

a) Make parallel boxplots of petal lengths for the two species.

b) Describe the differences seen in the boxplots.

c) Write a 95% confidence interval for this difference.

d) Explain what your interval means.

e) Based on your confidence interval, is there evidence of a difference in petal length? Explain.

29. Insulin and diet. A study published in the *Journal of the American Medical Association* examined people to see if they showed any signs of IRS (insulin resistance syndrome) involving major risk factors for Type 2 diabetes and heart disease. Among 102 subjects who consumed dairy products more than 35 times per week, 24 were identified with IRS. In comparison, IRS was identified in 85 of 190 individuals with the lowest dairy consumption, fewer than 10 times per week.

a) Is this strong evidence that IRS risk is different in people who frequently consume dairy products than in those who do not?

b) Does this prove that dairy consumption influences the development of IRS? Explain.

30. Speeding. A newspaper report in August 2002 raised the issue of racial bias in the issuance of speeding tickets. The following facts were noted:

- 16% of drivers registered in New Jersey are black.
- Of the 324 speeding tickets issued in one month on a 65-mph section of the New Jersey Turnpike, 25% went to black drivers.

a) Is the percentage of speeding tickets issued to blacks unusually high compared with the state's registration information?

b) Does this prove that racial profiling may be present?

c) What other statistics would you like to know about this situation?

31. Rainmakers? In an experiment to determine whether seeding clouds with silver iodide increases rainfall, 52 clouds were randomly assigned to be seeded or not. The amount of rain they generated was then measured (in acre-feet). Create a 95% confidence interval for the average amount of additional rain created by seeding clouds. Explain what your interval means.

	Unseeded Clouds	Seeded Clouds
Count	26	26
Mean	164.588	441.985
Median	44.200	221.600
SD	278.426	650.787
IntQRange	138.600	337.600
25 %ile	24.400	92.400
75 %ile	163	430

32. Fritos. As a project for an introductory Statistics course, students checked 6 bags of Fritos marked with a net weight of 35.4 grams. They carefully weighed the contents of each bag, recording the following weights (in grams): 35.5, 35.3, 35.1, 36.4, 35.4, 35.5. Is there evidence that the mean weight of bags of Fritos is less than advertised?

a) Write appropriate hypotheses.

b) Do these data satisfy the assumptions for inference? Explain.

c) Test your hypothesis using all 6 weights.

d) Retest your hypothesis with the one unusually high weight removed.

e) What would you conclude about the stated net weight?

T 33. Color or text? In an experiment, 32 volunteer subjects are briefly shown seven cards, each displaying the name of a color printed in a different color (example: red, blue, and so on). The subject is asked perform one of two tasks: memorize the order of the words or memorize the order of the colors. Researchers record the number of cards remembered correctly. Then the cards are shuffled and the subject is asked to perform the other task. The table displays the results for each subject. Is there any evidence that either the color or the written word dominates perception?

a) What role does randomization play in this experiment?

b) State appropriate hypotheses.

c) Are the assumptions necessary for inference reasonable here?

d) Perform the test.

e) State your conclusion.

Subject	Color	Word	Subject	Color	Word
1	4	7	17	4	3
2	1	4	18	7	4
3	5	6	19	4	3
4	1	6	20	0	6
5	6	4	21	3	3
6	4	5	22	3	5
7	7	3	23	7	3
8	2	5	24	3	7
9	7	5	25	5	6
10	4	3	26	3	4
11	2	0	27	3	5
12	5	4	28	1	4
13	6	7	29	2	3
14	3	6	30	5	3
15	4	6	31	3	4
16	4	7	32	6	7

34. And it means? Every statement about a confidence interval contains two parts—the level of confidence and the interval. Suppose that an insurance agent estimating the mean loss claimed by clients after home burglaries created the 95% confidence interval ($1644, $2391).
a) What's the margin of error for this estimate?
b) Carefully explain what the interval means.
c) Carefully explain what the 95% confidence level means.

35. Batteries. We work for the "Watchdog for the Consumer" consumer advocacy group. We've been asked to look at a battery company that claims its batteries last an average of 100 hours under normal use. There have been several complaints that the batteries don't last that long, so we decide to test them. To do this we select 16 batteries and run them until they die. They lasted a mean of 97 hours, with a standard deviation of 12 hours.
a) One of the editors of our newsletter (who does not know statistics) says that 97 hours is a lot less than the advertised 100 hours, so we should reject the company's claim. Explain to him the problem with doing that.
b) What are the null and alternative hypotheses?
c) What assumptions must we make in order to proceed with inference?
d) At a 5% level of significance, what do you conclude?
e) Suppose that, in fact, the average life of the company's batteries is only 98 hours. Has an error been made in part d? If so, what kind?

36. Hamsters. How large are hamster litters? Among 47 golden hamster litters recorded, there were an average of 7.72 baby hamsters, with a standard deviation of 2.5 hamsters per litter.
a) Create and interpret a 90% confidence interval.
b) Would a 98% confidence interval have a larger or smaller margin of error? Explain.
c) How many litters must be used to estimate the average litter size to within 1 baby hamster with 95% confidence?

CHAPTER 26

Comparing Counts

oes your zodiac sign determine how successful you will be in later life? *Fortune* magazine collected the zodiac signs of 256 heads of the largest 400 companies. Here are the number of births for each sign:

Births	Sign
23	Aries
20	Taurus
18	Gemini
23	Cancer
20	Leo
19	Virgo
18	Libra
21	Scorpio
19	Sagittarius
22	Capricorn
24	Aquarius
29	Pisces

Birth totals by sign
for 256 Fortune 400
executives.

WHO	Executives of Fortune 400 companies
WHAT	Zodiac birth sign
WHY	Maybe the researcher was a Gemini and naturally curious?

A S **Children at Risk.** See how a contingency table helps us understand the different risks to which an incident exposed children.

We can see some variation in the number of births per sign and there *are* more Pisces, but is it enough to claim that successful people are more likely to be born under some signs than others?

Goodness-of-Fit

"All creatures have their determined time for giving birth and carrying fetus, only a man is born all year long, not in determined time, one in the seventh month, the other in the eighth, and so on till the beginning of the eleventh month."
—*Aristotle*

If births were distributed uniformly across the year, we would expect about 1/12 of them to occur under each sign of the zodiac. That suggests 256/12, or about 21.3 births per sign. How closely do the observed numbers of births per sign fit this simple "null" model?

A hypothesis test to address this question is called a test of "**goodness-of-fit.**" The name suggests a certain badness-of-grammar, but it is quite standard. After all, we are asking whether the model that births are uniformly distributed over the signs fits the data good, . . . er, well. Goodness-of-fit involves testing a hypothesis. We have specified a model for the distribution and want to know whether it fits. There is no single parameter to estimate, so a confidence interval wouldn't make much sense.

If the question were about only one astrological sign (for example, "Are executives more likely to be Pisces?"[1]), we could use a one-proportion z-test and ask if the true proportion of executives with that sign is equal to 1/12. However, here we have 12 hypothesized proportions, one for each sign. We need a test that considers all of them together and gives an overall idea of whether the observed distribution differs from the hypothesized one.

Assumptions and Conditions

These data are organized in tables as we saw in Chapter 3, and the assumptions and conditions reflect that. Rather than having an observation for each individual, we typically work with summary counts in categories. In our example, we don't know the birth signs of each of the 256 executives, only the totals for each sign.

Counted Data Condition

The first condition is just to check that the data are *counts* for the categories of a categorical variable. This might seem a simplistic, even silly condition. But many kinds of values can be assigned to categories, and it is unfortunately common to find the methods of this chapter applied incorrectly to proportions or amounts just because they happen to be organized in a two-way table. So check to be sure you really have counts.

Independence Assumption

The easiest case is when the individuals who are counted in the cells are sampled independently from some population. That's what we'd like to have if we want to draw conclusions about that population. Randomness can arise in other ways, though. For example, if we care only about Fortune 400 executives, we might still think that the birth date of a particular executive is randomly distributed through the year. If we want to generalize to a large population, we should check the Randomization Condition.

Randomization Condition: The individuals who have been counted and whose counts are available for analysis should be a random sample from some population.

---

[1] A question actually asked us by someone who was undoubtedly a Pisces.

Sample Size Assumption

We must have enough data for the methods to work. That turns out not to be a simple question. We usually check the following:

Expected Cell Frequency Condition: We should expect to see at least 5 individuals in each cell.

The Expected Cell Frequency Condition sounds like—and is, in fact, quite similar to—the condition that np and nq be at least 10 when we tested proportions.

Calculations

We have observed a count in each category from the data, and have an expected count for each category from the hypothesized proportions. We wonder, are the differences just natural sampling variability, or are they so large that they indicate something important? Naturally enough, we look at the *differences* between these observed and expected counts, denoted $(Obs - Exp)$. We'd like to think about the total of the differences, but just adding them won't work because some differences are positive, others negative. We've been in this predicament before—once when we looked at deviations from the mean and again when we dealt with residuals. In fact, these *are* residuals. They're just the differences between the observed data and the counts given by the (null) model. We'll handle these residuals in essentially the same way we did in regression: We square them. That gives us positive values and focuses attention on any cells with large differences from what we expected. Because the differences between observed and expected counts generally get larger the more data we have, we also need to get an idea of the *relative* sizes of the differences. To do that, we divide each squared difference by the expected count for that cell.

> **NOTATION ALERT:**
>
> We compare the counts *observed* in each cell with the counts we *expect* to find. The usual notation uses O's and E's or abbreviations such as those we've used here. The method for finding the expected counts depends on the model.

The test statistic, called the **chi-square** (or chi-squared) **statistic,** is found by adding up the sum of the squares of the deviations between the observed and expected counts divided by the expected counts:

$$\chi^2 = \sum_{all\ cells} \frac{(Obs - Exp)^2}{Exp}.$$

The chi-square statistic is denoted χ^2, where χ is the Greek letter chi (pronounced "ky" as in "sky"). It refers to a family of sampling distribution models we have not seen before called (remarkably enough) the **chi-square models.**

This family of models, like the Student's t-models, differ only in the number of degrees of freedom. The number of degrees of freedom for a goodness-of-fit test is $n - 1$. Here, however, n is *not* the sample size, but instead is the number of categories. For the zodiac example, we have 12 signs, so our χ^2 statistic has 11 degrees of freedom.

> **NOTATION ALERT:**
>
> The only use of the Greek letter χ in Statistics is to represent this statistic and the associated sampling distribution. This is another violation of our "rule" that Greek letters represent population parameters. Here we are using a Greek letter simply to name a family of distribution models and a statistic.

One-Sided or Two-Sided?

The chi-square statistic is used only for testing hypotheses, not for constructing confidence intervals. If the observed counts don't match the expected, the statistic will be large. It can't be "too small." That would just mean that our model *really* fit the data well. So the chi-square test is always one-sided. If the calculated statistic value is large enough, we'll reject the null hypothesis. What could be simpler?

Even though its mechanics work like a one-sided test, the interpretation of a chi-square test is in some sense *many*-sided. With more than two proportions, there are many ways the null hypothesis can be wrong. By squaring the differences, we made all the deviations positive, whether our observed counts were higher or lower than expected. There's no direction to the rejection of the null model. All we know is that it doesn't fit.

A Chi-Square Test for Goodness-of-Fit Step-By-Step

We have counts of 256 executives in 12 zodiac sign categories. The natural null hypothesis is that birth dates of executives are divided equally among all the zodiac signs. The test statistic looks at how closely the observed data match this idealized situation.

Think

Plan State what you want to know.

Identify the variables and check the W's.

Hypotheses State the null and alternative hypotheses. For χ^2 tests it's usually easier to do that in words than in symbols.

I want to know whether births of successful people are *uniformly* distributed across the signs of the zodiac. I have counts of 256 Fortune 400 executives, categorized by their birth sign.

H_0: Births are uniformly distributed over zodiac signs.[2]

H_A: Births are not uniformly distributed over zodiac signs.

Model Check the conditions.

✔ **Counted Data Condition:** I have counts of the number of executives in 12 categories.

✔ **Randomization Condition:** This is a convenience sample of executives, but there's no reason to suspect bias.

✔ **Expected Cell Frequency Condition:** The null hypothesis expects that 1/12 of the 256 births, or 21.333, should occur in each sign. These expected values are all at least 5, so the condition is satisfied.

Specify the sampling distribution model.

Name the test you will use.

The conditions are satisfied, so I'll use a χ^2 model with $12 - 1 = 11$ degrees of freedom and do a **chi-square goodness-of-fit test.**

Show

Mechanics Each cell contributes an $\dfrac{(Obs - Exp)^2}{Exp}$ value to the chi-square sum.

The expected value for each zodiac sign is 21.333.

---

[2] It may seem that we have broken our rule of thumb that null hypotheses should specify parameter values. If you want to get formal about it, the null hypothesis is that

$$p_{Aries} = p_{Taurus} = \cdots = p_{Pisces}.$$

That is, we hypothesize that the true proportions of births of CEOs under each sign are equal. The role of the null hypothesis is to specify the model so that we can compute the test statistic. That's what this one does.

We add up these components for each zodiac sign. If you do it by hand, it can be helpful to arrange the calculation in a table. We show that after this Step-by-Step.

The P-value is the area in the upper tail of the χ^2 model above the computed χ^2 value.

The χ^2 models are skewed to the high end, and change shape depending on the degrees of freedom. The P-value considers only the right tail. Large χ^2 statistic values correspond to small P-values, which lead us to reject the null hypothesis.

$$\chi^2 = \sum \frac{(Obs - Exp)^2}{Exp} = \frac{(23 - 21.333)^2}{21.333} + $$
$$\frac{(20 - 21.333)^2}{21.333} + \cdots$$
$$= 5.094 \text{ for all 12 signs.}$$

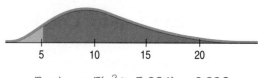

$$\text{P-value} = P(\chi^2 > 5.094) = 0.926$$

Tell   **Conclusion** Link the P-value to your decision. Remember to state your conclusion in terms of what the data mean rather than just making a statement about the distribution of counts.

The P-value of 0.926 says that if the zodiac signs of executives were in fact distributed uniformly, an observed chi-square value of 5.09 or higher would occur about 93% of the time. This certainly isn't unusual, so I fail to reject the null hypothesis, and conclude that these data show virtually no evidence of nonuniform distribution of zodiac signs among executives.

The Chi-Square Calculation

Let's make the chi-square procedure very clear. Here are the steps:

1. **Find the expected values.** These come from the null hypothesis model. Every model gives a hypothesized proportion for each cell. The expected value is the product of the total number of observations times this proportion.

   For our example, the null model hypothesizes *equal* proportions. With 12 signs, 1/12 of the 256 executives should be in each category. The expected number for each sign is 21.333.

A S **Calculating Standardized Residuals.** Women were at risk, too. Standardized residuals help us understand the relative risks. Here the calculation is animated to help you follow it.

2. **Compute the residuals.** Once you have expected values for each cell, find the residuals, *Observed − Expected*.

3. **Square the residuals.**

4. **Compute the components.** Now find the component, $\frac{(Observed - Expected)^2}{Expected}$, for each cell.

5. **Find the sum of the components.** That's the chi-square statistic.

6. **Find the degrees of freedom.** It's equal to the number of cells minus one. For the zodiac signs, that's $12 - 1 = 11$ degrees of freedom.

A S **The Chi-Square Test.** This animation completes the calculation of the chi-square statistic and the hypothesis test based on it.

7. **Test the hypothesis.** Large chi-square values mean lots of deviation from the hypothesized model, so they give small P-values. Look up the critical value from a table of chi-square values, or use technology to find the P-value directly.

The steps of the chi-square calculations are often laid out in tables. Use one row for each category, and columns for observed counts, expected counts, residuals, squared residuals, and the contributions to the chi-square total like this:

Sign	Observed	Expected	Residual = $(Obs - Exp)$	$(Obs - Exp)^2$	Component = $\dfrac{(Obs - Exp)^2}{Exp}$
Aries	23	21.333	1.667	2.778889	0.130262
Taurus	20	21.333	−1.333	1.776889	0.083293
Gemini	18	21.333	−3.333	11.108889	0.520737
Cancer	23	21.333	1.667	2.778889	0.130262
Leo	20	21.333	−1.333	1.776889	0.083293
Virgo	19	21.333	−2.333	5.442889	0.255139
Libra	18	21.333	−3.333	11.108889	0.520737
Scorpio	21	21.333	−0.333	0.110889	0.005198
Sagittarius	19	21.333	−2.333	5.442889	0.255139
Capricorn	22	21.333	0.667	0.444889	0.020854
Aquarius	24	21.333	2.667	7.112889	0.333422
Pisces	29	21.333	7.667	58.782889	2.755491
					$\sum = 5.094$

● **How big is big?** When we calculated χ^2 for the zodiac sign example, we got 5.094. That value would have been big for z or t, leading us to reject the null hypothesis. Not here. Were you surprised that $\chi^2 = 5.094$ had a huge P-value of 0.926? What *is* big for a χ^2 statistic, anyway?

Think about how χ^2 is calculated. In every cell any deviation from the expected count contributes to the sum. Large deviations generally contribute more, but if there are a lot of cells even small deviations can add up, making the χ^2 value larger. So the more cells there are, the higher the value of χ^2 has to get before it becomes noteworthy. For χ^2, then, the decision about how big is big depends on the number of degrees of freedom.

Unlike the Normal and t families, χ^2 models are skewed. Curves in the χ^2 family change both shape and center as the number of degrees of freedom grows. Here, for example, are the χ^2 curves for 5 and 9 degrees of freedom.

A S **The Chi-Square Family of Curves.** (Not an activity like the others, but there's no better way to see how χ^2 changes with more df.) Click on the Lesson Book's Resources tab and open the chi-square table. Watch the curve at the top as you scroll down the degrees of freedom column.

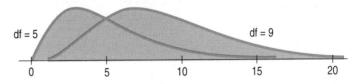

Notice that the value $\chi^2 = 10$ might seem somewhat extreme when there are 5 degrees of freedom, but appears to be rather ordinary for 9 degrees of freedom. Here are two simple facts to help you think about χ^2 models:

- The mode is at $\chi^2 = df - 2$. (Look at the curves; their peaks are at 3 and 7, see?)
- The expected value (mean) of a χ^2 model is its number of degrees of freedom. That's a bit to the right of the mode—as we would expect for a skewed distribution.

Our test for zodiac birthdays had 11 df, so the relevant χ^2 curve peaks at 9 and has a mean of 11. Knowing that, we might have easily guessed that the calculated χ^2 value of 5.094 wasn't going to be significant. ●

But I Believe the Model . . .

Goodness-of-fit tests are likely to be performed by people who have a theory of what the proportions *should* be in each category, and who believe their theory to be true. Unfortunately, the only *null* hypothesis available for a goodness-of-fit test is that the theory is true. And as we know, the hypothesis testing procedure allows us only to *reject* the null or *fail to reject* it. We can never confirm that a theory is in fact true, which is often what people want to do.

Unfortunately, they're stuck. At best, we can point out that the data are consistent with the proposed theory. But this doesn't *prove* the theory. The data *could* be consistent with the model even if the theory were wrong. In that case, we fail to reject the null hypothesis, but can't conclude anything for sure about whether the theory is true.

And we can't fix the problem by turning things around. Suppose we try to make our favored hypothesis the alternative. Then it is impossible to pick a single null. For example, suppose, as a doubter of astrology, you want to prove that the distribution of executive births is uniform. If you choose uniform as the null hypothesis, you can only *fail* to reject it. So you'd like uniformity to be your alternative hypothesis. Which particular violation of equally distributed births would you choose as your null? The problem is that the model can be wrong in many, many ways. There's no way to frame a null hypothesis the other way around. There's just no way to prove that a favored model is true.

● **Why can't we prove the null?** A biologist wants to show that her inheritance theory about fruit flies is valid. It says that 10% of the flies should be type 1, 70% type 2, and 20% type 3. After her students collected data on 100 flies, she did a goodness-of-fit test and found a P-value of 0.07. She started celebrating, since her null hypothesis wasn't rejected—that is, until her students collected data on 100 more flies. With 200 flies, the P-value dropped to 0.02. Although she knew the answer was probably no, she asked the statistician somewhat hopefully if she could just ignore half the data and stick with the original 100. By this reasoning we could always "prove the null" just by not collecting much data. With only a little data, the chances are good that they'll be consistent with almost anything. But they also have little chance of disproving anything either. In this case, the test has no power. Don't let yourself be lured into this scientist's reasoning. With data, more is always better. But you can't ever prove that your null hypothesis is true. ●

Comparing Observed Distributions

Many high schools survey graduating classes to determine the plans of the graduates. We might wonder whether the plans of students have stayed roughly the same over past decades or whether they have changed. At the top of the next page is a summary table from one high school. Each **cell** of the table shows how many students from a particular graduating class made a certain choice.

We might wonder about changes in the choice of military service as a post–high-school activity. The numbers for 1980 and 1990 look similar, until you notice the class sizes are quite different. The 18 seniors in 1980 who chose military service were only about 4% of the graduating class. In 1990, the 19 seniors making the same

		WHO	Graduates from an up-state NY high school

	Graduates from an up-state NY high school
WHO	Graduates from an up-state NY high school
WHAT	Postgraduation activities
WHEN	1980, 1990, 2000
WHY	Regular survey for general information

	1980	1990	2000	Total
College/Post-HS education	320	245	288	853
Employment	98	24	17	139
Military	18	19	5	42
Travel	17	2	5	24
Total	453	290	315	1058

Activities of graduates of a high school reported a year after graduation, by graduation year.   **Table 26.1**

choice represented 6.6% of the class. Because class sizes are so different, we see changes better by examining the proportions for each class rather than the counts:

A S **The Incident.** You may have guessed which famous incident put women and children at risk. Here you can view the story complete with rare film footage.

	1980	1990	2000	Total
College	70.64	84.48	91.43	**80.62**
Employment	21.63	8.28	5.40	**13.14**
Military	3.97	6.55	1.59	**3.97**
Travel	3.75	0.69	1.59	**2.27**
Total	100	100	100	100

Activities of graduates of a high school reported a year after graduation, as percentage of the class.
Table 26.2

We already know how to test whether *two* proportions are the same. For example, we could consider whether the proportion of students choosing military service was the same between 1980 and 1990 with a two-proportion z-test. But now we have more than two groups. We want to test whether the students' choices are the same across all three graduating classes. The z-test for two proportions generalizes to a **chi-square test of homogeneity.**

Chi-square again? It turns out that the mechanics of this test are *identical* to the chi-square test for goodness-of-fit that we just saw. (How similar can you get?) Why a different name, then? The goodness-of-fit test compared counts with a theoretical model. But here we're asking whether choices have changed, so we find the expected counts for each category directly from the data. As a result, we count the degrees of freedom slightly differently as well.

The term "homogeneity" means that things are the same. Here, we ask whether the post–high-school choices made by students are the *same* for these three graduating classes. The homogeneity test comes with a built-in null hypothesis. We hypothesize that the distribution does not change from group to group. The test looks for differences large enough to step beyond what we might expect from random sample-to-sample variation. It can reveal a large deviation in a single category or small but persistent differences over all the categories—or anything in between.

Assumptions and Conditions

The assumptions and conditions are the same as for the chi-square test for goodness-of-fit. The **Counted Data Condition** says that these data must be counts. You can't do a test of homogeneity on proportions, so we have to work with the counts of graduates given in the first table. Also, you can't do a chi-square test on measurements. For example, if we had recorded GPAs for these same groups over the same time span, we wouldn't be able to determine whether the mean GPAs had changed using this test.[3]

Often when we test for homogeneity, we aren't interested in some larger population, so we don't really need a random sample. (We would need one if we wanted to draw a more general conclusion—say, about the choices made by all high-school students in these graduation years.) Don't we need *some* randomness, though? Fortunately, the null hypothesis can be thought of as a model in which the counts in the table are distributed as if each student chose a plan randomly according to the overall proportions of the choices, regardless of the student's class. As long as we don't want to generalize, we don't have to check the **Randomization Condition.**

We still must be sure we have enough data for this method to work. The **Expected Cell Frequency Condition** says that the expected count in each cell must be at least 5. We'll confirm that as we do the calculations.

Calculations

The null hypothesis says that the proportions of graduates choosing each alternative should be the same for all three classes. So for each choice the expected proportion is just the overall proportion of students making that choice. The expected counts are those proportions applied to the number of students in each graduating class.

For example, overall, 139, or about 13.14%, of the 1058 students covered by these three surveys were employed. If the distributions are homogeneous (as the null hypothesis asserts), then 13.14% of the 453 students in the class of 1980 (or about 59.52 students) should be employed. Similarly, 13.14% of the 290 students in the class of 1990 (or about 38.1) should be employed.

Working in this way, we (or, more likely, the computer) can fill in expected values for each cell. Because these are theoretical values, they don't have to be integers. The expected values look like this:

	1980	1990	2000	Total
College	365.2259	233.8091	253.9650	853
Employment	59.5151	38.1002	41.3847	139
Military	17.9830	11.5123	12.5047	42
Travel	10.2760	6.5784	7.1456	24
Total	453	290	315	1058

Expected values for the high-school graduates.   **Table 26.3**

---

[3] To do that, you'd use a method called Analysis of Variance. (See Chapter 28 on the CD.)

Now check the **Expected Cell Frequency Condition.** Indeed, there are at least 5 individuals expected in each cell.

Following the pattern of the goodness-of-fit test, we compute the component for each cell of the table. For the highlighted cell, employed students graduating in 1980, that's

$$\frac{(Obs - Exp)^2}{Exp} = \frac{(98 - 59.5151)^2}{59.5151} = 24.886$$

Summing these components across all cells gives

$$\chi^2 = \sum_{all\ cells} \frac{(Obs - Exp)^2}{Exp} = 72.77$$

How about the degrees of freedom? We don't really need to calculate all the expected values in the table. We know there is a total of 139 employed students, so once we find the expected values for 1980 and 1990, we can determine the expected number of employed students for 2000 by just subtracting. Similarly, we know how many students graduated in each class, so after filling in three rows, we can find the expected values for the remaining row by subtracting. To fill out the table, we need to know the counts in only $R - 1$ rows and $C - 1$ columns. So the table has $(R - 1)(C - 1)$ degrees of freedom.

In our example, we need to calculate only 3 choices in each column and counts for 2 classes, for a total of $3 \times 2 = 6$ degrees of freedom. We'll need the degrees of freedom to find a P-value for the chi-square statistic.

> **NOTATION ALERT:**
>
> For a contingency table, R represents the number of rows and C the number of columns.

A Chi-Square Test for Homogeneity Step-By-Step

We have counts of 1058 students observed in three different years a decade apart and categorized according to their postgraduation activities.

Think

Plan State what you want to know.

Identify the variables and check the W's.

> I want to know whether the choices made by high-school graduates in what they do after high school have changed across three graduating classes. I have a table of counts classifying students according to their postgraduation plans.

Hypotheses State the null and alternative hypotheses.

> H_0: The post–high-school choices made by the classes of 1980, 1990, and 2000 have the same distribution (are homogeneous).
>
> H_A: The post–high-school choices made by the classes of 1980, 1990, and 2000 do not have the same distribution.

Model Check the conditions.

> ✔ **Counted Data Condition:** I have counts of the number of students in categories.
>
> ✔ **Randomization Condition:** I don't want to draw inferences to other high schools or other classes, so there is no need to check for a random sample.

State the sampling distribution model.

Name the test you will use.

✔ **Expected Cell Frequency Condition:** The expected values (shown below) are all at least 5.

The conditions seem to be met, so I can use a χ^2 model with $(4 - 1) \times (3 - 1) = 6$ degrees of freedom and do a **chi-square test of homogeneity.**

Show

Mechanics Show the expected counts for each cell of the data table. You could make separate tables for the observed and expected counts, or put both counts in each cell as shown here. While observed counts must be whole numbers, expected counts rarely are—don't be tempted to round those off.

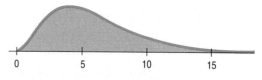

	1980	1990	2000
Coll.	320 / 365.23	245 / 233.81	288 / 253.97
Emp.	98 / 59.52	24 / 38.10	17 / 41.38
Mil.	18 / 17.98	19 / 11.51	5 / 12.50
Trav.	17 / 10.28	2 / 6.58	5 / 7.15

Calculate χ^2.

$$\chi^2 = \sum_{all\ cells} \frac{(Obs - Exp)^2}{Exp} = 72.77$$

The shape of a χ^2 model depends on the degrees of freedom. A χ^2 model with 6 df is skewed to the high end.

The P-value considers only the right tail. Here, the calculated value of the χ^2 statistic is off the scale, so the P-value is quite small.

P-value $= P(\chi^2 > 72.77) < 0.0001$

Tell

Conclusion State your conclusion in the context of the data. You should specifically talk about whether the distributions for the groups appear to be different.

The P-value is very small, so I reject the null hypothesis and conclude that the choices made by high-school graduates have changed over the two decades examined.

If you find that simply rejecting the hypothesis of homogeneity is a bit unsatisfying, you're in good company. It's hardly a shock that high-school students made different choices in 2000 than in 1980. What we'd really like to know is how big the differences were, where they were greatest, and where they were smallest. The test for homogeneity doesn't answer these interesting questions, but it does provide some evidence that can help us.

just checking

Tiny black potato flea beetles can damage potato plants in a vegetable garden. These pests chew holes in the leaves, causing the plants to wither or die. They can be killed with an insecticide, but a canola oil spray has been suggested as a non-chemical "natural" method of controlling the beetles. To conduct an experiment to test the effectiveness of the natural spray, we gather 500 beetles and place them in three Plexiglas containers. Two hundred beetles go in the first container, where we spray them with the canola oil mixture. Another 200 beetles go in the second container; we spray them with the insecticide. The remaining 100 beetles in the last container serve as a control group; we simply spray them with water. Then we wait 6 hours and count the number of surviving beetles in each container.

1 Why do we need the control group?

2 What would our null hypothesis be?

3 After the experiment is over we could summarize the results in a table like this. How many degrees of freedom does our χ^2 test have?

	Natural spray	Insecticide	Water	Total
Survived				
Died				
Total	200	200	100	500

4 Suppose that all together 125 beetles survived. (That's the first row total.) What's the expected count in the first cell—survivors among those sprayed with the natural spray?

5 If it turns out that only 40 of the beetles in the first container survived, what's the calculated component of χ^2 for that cell?

6 If the total calculated value of χ^2 for this table turns out to be around 10, would you expect the P-value of our test to be large or small? Explain.

Examining the Residuals

Whenever we reject the null hypothesis, it's a good idea to examine residuals. (We don't need to do that when we fail to reject because when the χ^2 value is small, all of its components must have been small.) For chi-square tests, we want to compare residuals for cells that may have very different counts. So we're better off standardizing the residuals. We know the mean residual is zero,[4] but we need to know each residual's standard deviation. When we tested proportions, we saw a link between the expected proportion and its standard deviation. For counts, there's a similar link. To standardize a cell's residual, we just divide by the square root of its expected value:

$$c = \frac{(Obs - Exp)}{\sqrt{Exp}}.$$

Notice that these **standardized residuals** are just the square roots of the **components** we calculated for each cell, with the + or the − sign indicating whether we observed more cases than we expected, or fewer.

The standardized residuals give us a chance to think about the underlying patterns and to consider the ways in which the distribution may have changed from class to class. Now that we've subtracted the mean (zero) and divided by their standard deviations, these are z-scores. If the null hypothesis were true, we could even appeal to the Central Limit Theorem, think of the Normal model, and use the 68-95-99.7 Rule to judge how extraordinary the large ones are.

---

[4] Residual = observed − expected. Because the total of the expected values is set to be the same as the observed total, the residuals must sum to zero.

Here are the standardized residuals for the high-school data:

	1980	1990	2000
College	−2.366	0.732	2.136
Employment	4.989	−2.284	−3.791
Military	0.004	2.207	−2.122
Travel	2.098	−1.785	−0.803

Standardized residuals can help show how the table differs from the null hypothesis pattern. **Table 26.4**

The row for Employment immediately attracts our attention. It holds the largest (in magnitude) two standardized residuals, and they're both big. It looks like employment was a more common choice for the class of 1980 and a less common choice for the classes of 1990 and 2000. This trend may have been influenced by changes in the economy or by a growing sense that college might be a better choice. We can see an opposite but less dramatic trend in the College row, from moderate negative residual to moderate positive residual across the three classes, indicating that going to college has become more common.

Independence

A study from the University of Texas Southwestern Medical Center examined whether the risk of hepatitis C was related to whether people had tattoos and to where they got their tattoos. Hepatitis C causes about 10,000 deaths each year in the United States, but often lies undetected for years after infection.

The data from this study can be summarized in a two-way table, as follows:

	Hepatitis C	No Hepatitis C	Total
Tattoo, parlor	17	35	52
Tattoo, elsewhere	8	53	61
None	22	491	513
Total	47	579	626

Counts of patients classified by their hepatitis C test status according to whether they had a tattoo from a tattoo parlor or from another source, or had no tattoo. **Table 26.5**

WHO   Patients being treated for non–blood-related disorders

WHAT   Tattoo status and hepatitis C status

WHEN   1991, 1992

WHERE   Texas

These data differ from the kinds of data we've considered before in this chapter because they categorize subjects from a single group on two categorical variables rather than on only one. The categorical variables here are *hepatitis C status* ("Hepatitis C" or "No Hepatitis C") and *tattoo status* ("Parlor," "Elsewhere," "None"). We've seen counts classified by two categorical variables displayed like this in Chapter 3, so we know such tables are called contingency tables. **Contingency tables** categorize counts on two (or more) variables so that we can see whether the distribution of counts on one variable is contingent on the other.

The natural question to ask of these data is whether the chance of having hepatitis C is *independent* of tattoo status. Recall that for events **A** and **B** to be independent, $P(\mathbf{A})$ must equal $P(\mathbf{A} \mid \mathbf{B})$. Here, this means the probability that a randomly selected patient has hepatitis C should not change when we learn the patient's tattoo status. We examined the question of independence in just this way back in Chapter 15, but we lacked a way to test it. The rules for independent events are much too precise and absolute to work well with real data. A **chi-square test for independence** is the test called for here.

If *hepatitis status* is independent of tattoos, we'd expect the proportion of people testing positive for hepatitis to be the same for the three levels of *tattoo status*. This sounds a lot like the test of homogeneity. In fact, the mechanics of the calculation are identical.

The difference is that now we have two categorical variables measured on a single population. For the homogeneity test, we had a single categorical variable measured independently on two or more populations. But now we ask a different question: "Are the variables independent?" rather than "Are the groups homogeneous?" These are subtle differences, but they are important when we draw conclusions.

Assumptions and Conditions

Of course, we still need counts and enough data so that the expected values are at least 5 in each cell.

If we're interested in the independence of variables, we usually want to generalize from the data to some population. In that case, we'll need to check that the data are a representative random sample from that population.

A Chi-Square Test for Independence Step-By-Step

We have counts of 626 individuals categorized according to their "tattoo status" and their "hepatitis status."

Think

Plan State what you want to know.

Identify the variables and check the W's.

Hypotheses State the null and alternative hypotheses.

We perform a test of independence when we suspect the variables may not be independent. We are on the familiar ground of making a claim (in this case, that knowing *tattoo status* will change probabilities for *hepatitis C status*), and testing the null hypothesis that it is *not* true.

I want to know whether the categorical variables *tattoo status* and *hepatitis status* are statistically independent. I have a contingency table of 626 Texas patients with an unrelated disease.

H_0: *Tattoo status and hepatitis status are independent.*[5]

H_A: *Tattoo status and hepatitis status are not independent.*

---

[5] Once again, parameters are hard to express. The hypothesis of independence itself tells us how to find expected values for each cell of the contingency table. That's all we need.

Model Check the conditions.

✔ **Counted Data Condition:** I have counts of individuals categorized on two categorical variables.

✔ **Randomization Condition:** These data are from a retrospective study of patients being treated for something unrelated to hepatitis. Although they are not an SRS, they were selected to avoid biases.

✘ **Expected Cell Frequency Condition:** The expected values do not meet the condition that all are at least 5.

Warning: Be very wary of proceeding when there are small expected counts. If we see expected counts that fall far short of 5, or if many cells violate the condition, we should not use χ^2. (We will soon discuss ways you can fix the problem.) If you do continue, always check the residuals to be sure those cells did not have a major influence on your result.

This table shows both the observed and expected counts for each cell. Notice that the expected counts are calculated exactly as they were for a test of homogeneity.

	Hepatitis C	No Hepatitis C	Total
Tattoo, parlor	17 3.904	35 48.096	52
Tattoo, elsewhere	8 4.580	53 56.420	61
None	22 38.516	491 474.484	513
Total	47	579	626

Specify the model.

Name the test you will use.

Although the Expected Cell Frequency Condition is not satisfied, the values are close to 5. I'll go ahead, but I'll check the residuals carefully. I'll use a χ^2 model with $(3 - 1) \times (2 - 1) = 2$ df and do a **chi-square test of independence.**

Show Mechanics Calculate χ^2.

$$\chi^2 = \sum_{all\ cells} \frac{(Obs - Exp)^2}{Exp} = 57.91$$

The shape of a chi-square model depends on its degrees of freedom. With 2 df, the model looks quite different, as you can see here. We still care only about the right tail.

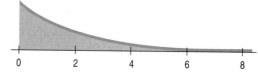

P-value = $P(\chi^2 > 57.91) < 0.0001$

Tell Conclusion Link the P-value to your decision. State your conclusion about the independence of the two variables.

(We should be wary of this conclusion because of the small expected counts. A complete solution must include the additional analysis, recalculation, and final conclusion discussed in the following section.)

The P-value is very small, so I reject the null hypothesis and conclude that *hepatitis status is not independent of tattoo status.* Because the Expected Cell Frequency Condition was violated, I need to check that the two cells with small expected counts did not influence this result too greatly.

Examine the Residuals

Each cell of the contingency table contributes a term to the chi-square sum. As we did earlier, we should examine the residuals because we have rejected the null hypothesis. In this instance, we have an additional concern that the cells with small expected frequencies not be the ones that make the chi-square statistic large.

Our interest in the data arises from the potential for improving public health. If patients with tattoos are more likely to test positive for hepatitis C, perhaps physicians should be advised to suggest blood tests for such patients.

The standardized residuals look like this:

Think
AGAIN

	Hepatitis C	No Hepatitis C
Tattoo, parlor	6.628	−1.888
Tattoo, elsewhere	1.598	−0.455
None	−2.661	0.758

Standardized residuals for the hepatitis and tattoos data. Are any of them particularly large in magnitude?

Table 26.6

Show
MORE

The chi-square value of 57.91 is the sum of the squares of these six values. The cell for people with tattoos obtained in a tattoo parlor who have hepatitis C is large and positive, indicating there are more people in that cell than the null hypothesis of independence would predict. This result suggests that a principal source of infection may be tattoo parlors.

The second largest component is a negative value for those with no tattoos who test positive for hepatitis C. A negative value says that there are fewer people in this cell than independence would expect. That is, those who have no tattoos are less likely to be infected with hepatitis C than we might expect if the two variables are independent.

What about the cells with small expected counts? The formula for the chi-square standardized residuals divides each residual by the square root of the expected frequency. Too small an expected frequency can arbitrarily inflate the residual, and lead to an inflated chi-square statistic. Any expected count close to the arbitrary minimum of 5 calls for checking that cell's standardized residual to be sure it is not particularly large. In this case, the standardized residual for the "Hepatitis C and Tattoo, elsewhere" cell is not particularly large, but the standardized residual for the "Hepatitis C and Tattoo, parlor" cell is large.

We might choose not to report the results because of concern with the small expected frequency. Alternatively, we could include a warning along with our report of the results. Yet another approach is to combine categories to get a larger sample size and correspondingly larger expected frequencies, if there are some categories that can be appropriately combined. Here, we might naturally combine the two rows for tattoos, obtaining a 2 × 2 table:

	Hepatitis C	No Hepatitis C	Total
Tattoo	25	88	113
None	22	491	513
Total	47	579	626

Combining the two tattoo categories gives a table with all expected counts greater than 5. **Table 26.7**

Tell

ALL

This table has expected values of at least 5 in every cell, and a chi-square value of 42.42 on 1 degree of freedom. The corresponding P-value is < 0.0001.

We conclude that tattoo status and hepatitis C status are not independent. The data *suggest* that tattoo parlors may be a particular problem, but we haven't enough data to draw that conclusion.

just checking

Which of the three chi-square tests would you use in each of the following situations—goodness-of-fit, homogeneity, or independence?

7. A restaurant manager wonders whether customers who dine on Friday nights have the same preferences among the four chef's special entrées as those who dine on Saturday nights. One weekend he has the wait staff record which entrées were ordered each night. Assuming these customers to be typical of all weekend diners, he'll compare the distributions of meals chosen Friday and Saturday.

8. Company policy calls for parking spaces to be assigned to everyone at random, but you suspect that may not be so. There are three lots of equal size: lot A, next to the building, lot B, a bit farther away, and lot C on the other side of the highway. You gather data about employees at middle management level and above to see how many were assigned parking in each lot.

9. Is a student's social life affected by where the student lives? A campus survey asked a random sample of students whether they lived in a dormitory, in off-campus housing, or at home, and whether they had been out on a date 0, 1–2, 3–4, or 5 or more times in the past two weeks.

Chi-Square and Causation

Chi-square tests are common. Tests for independence are especially widespread. Unfortunately, many people interpret a small P-value as proof of causation. We know better. Just as correlation between quantitative variables does not demonstrate causation, a failure of independence between two categorical variables does not show a cause-and-effect relationship between them, nor should we say that one variable *depends* on the other.

The chi-square test for independence treats the two variables symmetrically. There is no way to differentiate the direction of any possible causation from one variable to the other. In our example, it is unlikely that having hepatitis causes one to crave a tattoo, but other examples are not so clear.

Of course, there's never any way to eliminate the possibility that a lurking variable is responsible for the observed lack of independence. For example, it might be that people who have body piercings or those who inject drugs are both more likely to get tattooed and more likely to contract hepatitis C. Even a small subpopulation of drug users among those with tattoos might be enough to create the observed result.

In some sense, a failure of independence between two categorical variables is less impressive than a strong, consistent, linear association between quantitative variables. Two categorical variables can fail the test of independence in many ways, including ways that show no consistent pattern of failure. Examination of the chi-square standardized residuals can help you think about the underlying patterns.

What Can Go Wrong?

 Sample Size and Chi-Square. Chi-square statistics have a peculiar problem. They don't respond to increasing the sample size in quite the same way you might expect. This simulation shows what does happen.

- **Don't use chi-square methods unless you have counts.** All three of the chi-square tests apply only to counts. Other kinds of data can be arrayed in two-way tables. Just because numbers are in a two-way table doesn't make them suitable for chi-square analysis. Data reported as proportions or percentages can be suitable for chi-square procedures, *but only after they are converted to counts.* If you try to do the calculations without first finding the counts, your results will be wrong.

- **Beware large samples.** Beware *large* samples?! That's not the advice you're used to hearing. The chi-square tests, however, are unusual. You should be wary of chi-square tests performed on very large samples. No hypothesized distribution fits perfectly, no two groups are exactly homogeneous, and two variables are rarely perfectly independent. The degrees of freedom for chi-square tests don't grow with the sample size. With a sufficiently large sample size, a chi-square test can always reject the null hypothesis. But we have no measure of how far the data are from the null model. There are no confidence intervals to help us judge the effect size.

- **Don't say that one variable "depends" on the other just because they're not independent.** Dependence suggests a model or a pattern, but variables can fail to be independent in many different ways. When variables fail the test for independence, you might just say they are "associated."

CONNECTIONS

Chi-square methods relate naturally to inference methods for proportions. We can think of a test of homogeneity as stepping from a comparison of two proportions to a question of whether three or more proportions are equal. The standard deviations of the residuals in each cell are linked to the expected counts much like the standard deviations we found for proportions.

Independence is, of course, a fundamental concept in Statistics. But chi-square tests do not offer a general way to check on independence for all those times when we have had to assume it.

Stacked bar charts or side-by-side pie charts can help us think about patterns in two-way tables. A histogram or boxplot of the standardized residuals can help locate extraordinary values.

What have we learned?

We've learned how to test hypotheses about categorical variables. We use one of three related methods. All look at counts of data in categories, and all rely on chi-square models, a new family indexed by degrees of freedom.

- Goodness-of-fit tests compare the observed distribution of a single categorical variable to an expected distribution based on a theory or model.
- Tests of homogeneity compare the distribution of several groups for the same categorical variable.
- Tests of independence examine counts from a single group for evidence of an association between two categorical variables.

We've seen that mechanically these tests are almost identical. Although the tests appear to be one-sided, we've learned that conceptually they are many-sided, because there are many ways that a table of counts can deviate significantly from what we hypothesized. When that happens and we reject the null hypothesis, we've learned to examine standardized residuals in order to better understand the patterns as in the table.

TERMS

Goodness-of-fit

A test of whether the distribution of counts in one categorical variable matches the distribution predicted by a model is called a test of goodness-of-fit. A chi-square test of goodness-of-fit finds

$$\chi^2 = \sum_{all\ cells} \frac{(Obs - Exp)^2}{Exp},$$

where the expected counts come from the predicting model. It finds a P-value from a chi-square model with $n - 1$ degrees of freedom, where n is the number of categories in the categorical variable.

Chi-square statistic

The chi-square statistic is found by summing the chi-square components. Chi-square tests can be used to test goodness-of-fit, homogeneity, or independence.

Cell

A cell of a two-way table is one element of the table corresponding to a specific row and a specific column. Table cells can hold counts, percentages, or measurements on other variables. Or they can hold several values.

Homogeneity

A test comparing the distribution of counts for two or more groups on the same categorical variable is called a test of *homogeneity*. A chi-square test of homogeneity finds

$$\chi^2 = \sum_{all\ cells} \frac{(Obs - Exp)^2}{Exp},$$

where the expected counts are based on the overall frequencies, adjusted for the totals in each group. We find a P-value from a chi-square distribution with $(\#Rows - 1) \times (\#Cols - 1)$ degrees of freedom, where $\#Rows$ gives the number of categories and $\#Cols$ gives the number of independent groups.

Standardized residual

In each cell of a two-way table, a standardized residual is the square root of the chi-square component for that cell with the sign of the *Observed − Expected* difference:

$$\frac{(Obs - Exp)}{\sqrt{Exp}}.$$

When we reject a chi-square test, an examination of the standardized residuals can sometimes reveal more about how the data deviate from the null model.

Contingency table A two-way table that classifies individuals according to two categorical variables is called a *contingency table*.

Independence A test of whether two categorical variables are independent examines the distribution of counts for one group of individuals classified according to both variables. A chi-square test of *independence* uses the same calculation as a test of homogeneity. We find a P-value from a chi-square distribution with (#*Rows* − 1) × (#*Cols* − 1) degrees of freedom, where #*Rows* gives the number of categories in one variable and #*Cols* gives the number of categories in the other.

Two-way table Each *cell* of a two-way table shows counts of individuals. One way classifies a sample according to a categorical variable. The other way can classify different groups of individuals according to the same variable or classify the same individuals according to a different categorical variable.

Chi-square models Chi-square models are skewed to the right. They are parameterized by their degrees of freedom, and become less skewed with increasing degrees of freedom.

Chi-square component The components of a chi-square calculation are

$$\frac{(Observed - Expected)^2}{Expected},$$

found for each cell of the table.

SKILLS

When you complete this lesson you should:

Think
- Be able to recognize when a test of goodness-of-fit, a test of homogeneity, or a test of independence would be appropriate for a table of counts.
- Understand that the degrees of freedom for a chi-square test depend on the dimensions of the table and not on the sample size. Understand that this means that increasing the sample size increases the ability of chi-square procedures to reject the null hypothesis.

Show
- Be able to display and interpret counts in a two-way table.
- Know how to use the chi-square tables to perform chi-square tests.
- Know how to compute a chi-square test using your statistics software or calculator.
- Be able to examine the standardized residuals to explain the nature of the deviations from the null hypothesis.

Tell
- Know how to interpret chi-square as a test of goodness-of-fit in a few sentences.
- Know how to interpret chi-square as a test of homogeneity in a few sentences.
- Know how to interpret chi-square as a test of independence in a few sentences.

Chi-Square on the Computer

Most statistics packages associate chi-square tests with contingency tables. Often chi-square is available as an option only when you make a contingency table. This organization can make it hard to locate the chi-square test and may confuse the three different roles that the chi-square test can take. In particular, chi-square tests for goodness-of-fit may be hard to find or missing entirely. Chi-square tests for homogeneity are computationally the same as chi-square tests for independence, so you may have to perform the mechanics as if they were tests of independence and interpret them afterwards as tests of homogeneity.

Most statistics packages work with data on individuals rather than with the summary counts. If the only information you have is the table of counts, you may find it more difficult to get a statistics package to compute chi-square. Some packages offer a way to reconstruct the data from the summary counts so that they can then be passed back through the chi-square calculation, finding the cell counts again. Many packages offer chi-square standardized residuals (although they may be called something else).

DATA DESK

Select variables.

From the **Calc** menu, choose **Contingency Table.** From the table's HyperView menu choose **Table Options.** (Or Choose **Calc** > **Calculation Options** > **Table Options.**) In the dialog, check the boxes for **Chi Square** and for **Standardized Residuals.** Data Desk will display the chi-square and its P-value below the table, and the standardized residuals within the table.

Comments

Data Desk automatically treats variables selected for this command as categorical variables even if their elements are numerals.

The **Compute Counts** command in the table's HyperView menu will make variables that hold the table contents (as selected in the Table Options dialog) including the standardized residuals.

EXCEL

Excel offers the function

CHITEST(actual_range, expected_range), which computes a chi-square value for homogeneity. Both ranges are of the form UpperleftCell:LowerRightCell, specifying two rectangular tables that must hold counts (although Excel will not check for integer values). The two tables must be of the same size and shape.

Comments

Excel's documentation claims this is a test for independence and labels the input ranges accordingly, but Excel offers no way to find expected counts, so the function is not particularly useful for testing independence. You can only use this function if you already know both tables of counts or are willing to program additional calculations.

JMP

From the **Analyze** menu select **Fit Y by X.** Select variables: a Y, Response variable that holds responses for one variable, and an X, Factor variable that holds responses for the other. Both selected variables must be Nominal or Ordinal. JMP will make a plot and a contingency table. Below the contingency table, JMP offers a **Tests** panel. In that panel the Chi Square for independence is called a **Pearson ChiSquare.** The table also offers the P-value.

Click on the Contingency Table title bar to drop down a menu that offers to include a **Deviation** and **Cell Chi square** in each cell of the table.

Comments

JMP will choose a chi-square analysis for a **Fit Y by X** if both variables are nominal or ordinal (marked with an N or O), but not otherwise. Be sure the variables have the right type.

Deviations are the observed − expected differences in counts.

Cell chi-squares are the squares of the standardized residuals. Refer to the deviations for the sign of the difference.

Look under **Distributions** in the **Analyze** menu to find a chi-square test for goodness-of-fit.

MINITAB

From the **Stat** menu choose the **Tables** submenu. From that menu, choose **Chi Square Test . . . .** In the dialog identify the columns that make up the table. Minitab will display the table and print the chi-square value and its P-value.

Comments

Alternatively, select the **Cross Tabulation . . .** command to see more options for the table, including expected counts and standardized residuals.

SPSS

From the **Analyze** menu choose the **Descriptive Statistics** submenu. From that submenu, choose **Crosstabs . . .** In the Crosstabs dialog, assign the row and column variables from the variable list. Both variables must be categorical. Click the **Cells** button to specify that standardized residuals should be displayed. Click the **Statistics** button to specify a chi-square test.

Comments

SPSS offers only variables that it knows to be categorical in the variable list for the Crosstabs dialog. If the variables you want are missing, check that they have the right type.

TI-83/84 Plus

The TI-83 does not have a routine for the chi-square goodness-of-fit test.

To test hypothesis of homogeneity or independence, you need to enter the data as a matrix. Push the **MATRIX** button, and choose to **EDIT** matrix. Specify the dimensions of the table, rows × columns, then enter the appropriate counts. To do the test, choose **C: χ^2-Test** from the **STAT TESTS** menu. Note that the calculator automatically stores the expected counts in a matrix you specify.

TI-89

To test goodness-of-fit, enter the observed counts in a list and the expected counts in another list. Expected counts can be entered as n*p and the calculator will compute them for you.

From the **STAT Tests** menu select **7:Chi2 GOF.** Enter the list names using VAR-LINK and the degrees of freedom, k − 1, where k is the number of categories. Select whether to simply calculate or display the result with the area corresponding to the P-value highlighted.

To test a hypothesis of homogeneity or independence, you need to enter the data as a matrix. From the home screen, press APPS and select **6:Data/Matrix Editor,** then select **3:New.** Specify type as Matrix and name the matrix in the **Variable** box. Specify the number of rows and columns. Type the entries, pressing ENTER after each. Press 2nd ESC to leave the editor.

To do the test, choose **8:Chi2 2-way** from the **STAT Tests** menu.

EXERCISES

1. Which test? For each of the following situations, state whether you'd use a chi-square goodness-of-fit test, a chi-square test of homogeneity, a chi-square test of independence, or some other statistical test.

a) A brokerage firm wants to see whether the type of account a customer has (Silver, Gold, or Platinum) affects the type of trades that customer makes (in person, by phone, or on the Internet). It collects a random sample of trades made for its customers over the past year and performs a test.

b) That brokerage firm also wants to know if the type of account affects the size of the account (in dollars). It performs a test to see if the mean size of the account is the same for the three account types.

c) The academic research office at a large community college wants to see whether the distribution of courses chosen (Humanities, Social Science, or Science) is different for its residential and nonresidential students. It assembles last semester's data and performs a test.

2. Which test again? For each of the following situations, state whether you'd use a chi-square goodness-of-fit test, a chi-square test of homogeneity, a chi-square test of independence, or some other statistical test.

a) Is the quality of a car affected by what day it was built? A car manufacturer examines a random sample of the warranty claims filed over the past two years to test whether defects are randomly distributed across days of the work week.

b) A medical researcher wants to know if blood cholesterol level is related to heart disease. She examines a database of 10,000 patients, testing whether the cholesterol level (in milligrams) is related to whether a person has heart disease or not.

c) A student wants to find out whether political leaning (liberal, moderate, or conservative) is related to choice of major. He surveys 500 randomly chosen students and performs a test.

3. Dice. After getting trounced by your little brother in a children's game, you suspect the die he gave you to roll may be unfair. To check, you roll it 60 times, recording the number of times each face appears. Do these results cast doubt on the die's fairness?

a) If the die is fair, how many times would you expect each face to show?

b) To see if these results are unusual, will you test goodness-of-fit, homogeneity, or independence?

c) State your hypotheses.

d) Check the conditions.

e) How many degrees of freedom are there?

f) Find χ^2 and the P-value.

g) State your conclusion.

Face	Count
1	11
2	7
3	9
4	15
5	12
6	6

4. M&M's. As noted in an earlier chapter, the Masterfoods Company says that until very recently yellow candies made up 20% of its milk chocolate M&M's, red another 20%, and orange, blue, and green 10% each. The rest are brown. On his way home from work the day he was writing these exercises, one of the authors bought a bag of plain M&M's. He got 29 yellow ones, 23 red, 12 orange, 14 blue, 8 green, and 20 brown. Is this sample consistent with the company's stated proportions? Test an appropriate hypothesis and state your conclusion.

a) If the M&M's are packaged in the stated proportions, how many of each color should the author have expected to get in his bag?

b) To see if his bag was unusual, should he test goodness-of-fit, homogeneity, or independence?

c) State the hypotheses.

d) Check the conditions.

e) How many degrees of freedom are there?

f) Find χ^2 and the P-value.

g) State a conclusion.

5. Nuts. A company says its premium mixture of nuts contains 10% Brazil nuts, 20% cashews, 20% almonds, and 10% hazelnuts, and the rest are peanuts. You buy a large can and separate the various kinds of nuts. Upon weighing them, you find there are 112 grams of Brazil nuts, 183 grams of cashews, 207 grams of almonds, 71 grams of hazelnuts, and 446 grams of peanuts. You wonder whether your mix is significantly different from what the company advertises.

a) Explain why the chi-square goodness-of-fit test is not an appropriate way to find out.

b) What might you do instead of weighing the nuts in order to use a χ^2 test?

6. Mileage. A salesman who is on the road visiting clients thinks that on average he drives the same distance each day of the week. He keeps track of his mileage for several weeks and discovers that he averages 122 miles on Mondays, 203 miles on Tuesdays, 176 miles on Wednesdays, 181 miles on Thursdays, and 108 miles on Fridays. He wonders if this evidence contradicts his belief in a uniform distribution of miles across the days of the week. Explain why it is not appropriate to test his hypothesis using the chi-square goodness-of-fit test.

7. NYPD and race. Census data for New York City indicate that 29.2% of the under-18 population is white, 28.2% black, 31.5% Latino, 9.1% Asian, and 2% other ethnicities. The New York Civil Liberties Union points out that of 26,181 police officers, 64.8% are white, 14.5% black, 19.1% Hispanic, and 1.4% Asian. Do the police officers reflect the ethnic composition of the city's youth? Test an appropriate hypothesis and state your conclusion.

8. Violence against women. In its study *When Men Murder Women*, the Violence Policy Center reported that 2129 women were murdered by men in 1996. Of these victims, a weapon could be identified for 2013 of them. Of those for whom a weapon could be identified, 1139 were killed by guns, 372 by knives or other cutting instruments, 158 by other weapons, and 344 by personal attack (battery, strangulation, etc.). The FBI's Uniform Crime Report says that among all murders nationwide the weapon use rates were as follows: guns 63.4%, knives 13.1%, other weapons, 16.8%, personal attack 6.7%. Is there evidence that violence against women involves different weapons than other violent attacks in the United States?

9. Fruit flies. Offspring of certain fruit flies may have yellow or ebony bodies and normal wings or short wings. Genetic theory predicts that these traits will appear in the ratio 9:3:3:1 (9 yellow, normal: 3 yellow, short: 3 ebony, normal: 1 ebony, short). A researcher checks 100 such flies and finds the distribution of the traits to be 59, 20, 11, and 10, respectively.

a) Are the results this researcher observed consistent with the theoretical distribution predicted by the genetic model?

b) If the researcher had examined 200 flies and counted exactly twice as many in each category—118, 40, 22, 20—what conclusion would he have reached?

c) Why is there a discrepancy between the two conclusions?

T 10. Pi. Many people know the mathematical constant π is approximately 3.14. But that's not exact. To be more precise, here are 20 decimal places: 3.14159265358979323846. Still not exact, though. In fact, the actual value is irrational, a decimal that goes on forever without any repeating pattern. But notice that there are no 0's and only one 7 in the 20 decimal places above. Does that pattern persist, or do all the digits show up with equal frequency? The table shows the number of times each digit appears in the first million digits. Test the hypothesis that the digits 0 through 9 are uniformly distributed in the decimal representation of π.

The first million digits of π	
Digit	Count
0	99959
1	99758
2	100026
3	100229
4	100230
5	100359
6	99548
7	99800
8	99985
9	100106

T 11. *Titanic.* Here is a table we first saw in Chapter 3 showing who survived the sinking of the *Titanic* based on whether they were crew members, or passengers booked in first-, second-, or third-class staterooms:

	Crew	First	Second	Third	Total
Alive	212	202	118	178	**710**
Dead	673	123	167	528	**1491**
Total	**885**	**325**	**285**	**706**	**2201**

a) If we draw an individual at random from this table, what's the probability that we will draw a member of the crew?

b) What's the probability of randomly selecting a third-class passenger who survived?

c) What's the probability of a randomly selected passenger surviving, given that the passenger was a first-class passenger?

d) If someone's chances of surviving were the same regardless of their status on the ship, how many members of the crew would you expect to have lived?

e) State the null and alternative hypotheses we would test here.

f) Give the degrees of freedom for the test.

g) The chi-square value for the table is 187.8, and the corresponding P-value is barely greater than 0. State your conclusions about the hypotheses.

T 12. NYPD and gender. The table in the next column shows the rank attained by male and female officers in the New York City Police Department. Do these data indicate that men and women are equitably represented at all levels of the department?

Rank	Male	Female
Officer	21,900	4,281
Detective	4,058	806
Sergeant	3,898	415
Lieutenant	1,333	89
Captain	359	12
Higher ranks	218	10

a) What's the probability that a person selected at random from the NYPD is a female?

b) What's the probability that a person selected at random from the NYPD is a detective?

c) Assuming no bias in promotions, how many female detectives would you expect the NYPD to have?

d) To see if there is evidence of differences in ranks attained by males and females, will you test goodness-of-fit, homogeneity, or independence?

e) State the hypotheses.

f) Test the conditions.

g) How many degrees of freedom are there?

h) Find χ^2 and the P-value.

i) State your conclusion.

j) If you concluded that the distributions are not the same, analyze the differences using the standardized residuals of your calculations.

13. Birth order and college choice. Students in an Introductory Statistics class at a large university were classified by birth order and by the college they attend.

College	Birth Order (1 = oldest or only child)				
	1	2	3	4 or more	Total
Arts and Sciences	34	14	6	3	57
Agriculture	52	27	5	9	93
Social Science	15	17	8	3	43
Professional	13	11	1	6	31
Total	**114**	**69**	**20**	**21**	**224**

College	Expected Values Birth Order (1 = oldest or only child)			
	1	2	3	4 or more
Arts and Sciences	29.0089	17.5580	5.0893	5.3438
Agriculture	47.3304	28.6473	8.3036	8.7188
Social Science	21.8839	13.2455	3.8393	4.0313
Professional	15.7768	9.5491	2.7679	2.9063

a) What kind of chi-square test is appropriate—goodness-of-fit, homogeneity, or independence?

b) State your hypotheses.

c) State and check the conditions.
d) How many degrees of freedom are there?
e) The calculation yields $\chi^2 = 17.78$, with P = 0.0378. State your conclusion.
f) Examine and comment on the standardized residuals. Do they challenge your conclusion? Explain.

Standardized Residuals
Birth Order
(1 = oldest or only child)

College	1	2	3	4 or more
Arts and Sciences	0.92667	−0.84913	0.40370	−1.01388
Agriculture	0.67876	−0.30778	−1.14640	0.09525
Social Science	−1.47155	1.03160	2.12350	−0.51362
Professional	−0.69909	0.46952	−1.06261	1.81476

14. Binging and politics. Students in a large Statistics class at a university in the Northeast were asked to describe their political position and whether they had engaged in binge drinking (5 drinks at a sitting for a man; 4 for a woman).

Political Position

Binge Drinking	Conservative	Moderate	Liberal	Total
In the past 3 days	7	19	14	40
In the past week	9	19	28	56
In the past month	3	8	23	34
Some other time	8	15	22	45
Never	9	22	20	51
Total	36	83	107	226

Expected Values

	Conservative	Moderate	Liberal
In the past 3 days	6.37168	14.6903	18.9381
In the past week	8.92035	20.5664	26.5133
In the past month	5.41593	12.4867	16.0973
Some other time	7.16814	16.5265	21.3053
Never	8.12389	18.7301	24.1460

a) What kind of chi-square test is appropriate—goodness-of-fit, homogeneity, or independence?
b) State your hypotheses.
c) State and check the conditions.
d) How many degrees of freedom are there?
e) The calculation yields $\chi^2 = 10.10$, with P = 0.2578. State your conclusion.
f) Would you expect that a larger sample might find statistical significance? Explain.

T 15. Cranberry juice. It's common folk wisdom that drinking cranberry juice can help prevent urinary tract infections in women. In 2001 the *British Medical Journal* reported the results of a Finnish study in which three groups of 50 women were monitored for these infections over 6 months. One group drank cranberry juice daily, another group drank a lactobacillus drink, and the third drank neither of those beverages, serving as a control group. In the control group, 18 women developed at least one infection compared with 20 of those who consumed the lactobacillus drink and only 8 of those who drank cranberry juice. Does this study provide supporting evidence for the value of cranberry juice in warding off urinary tract infections?

a) Is this a survey, a retrospective study, a prospective study, or an experiment? Explain.
b) Will you test goodness-of-fit, homogeneity, or independence?
c) State the hypotheses.
d) Test the conditions.
e) How many degrees of freedom are there?
f) Find χ^2 and the P-value.
g) State your conclusion.
h) If you concluded that the groups are not the same, analyze the differences using the standardized residuals of your calculations.

16. Cars. A random survey of autos parked in student and staff lots at a large university classified the brands by country of origin, as seen in the table. Are there differences in the national origins of cars driven by students and staff?

Driver

Origin	Student	Staff
American	107	105
European	33	12
Asian	55	47

a) Is this a test of independence or homogeneity?
b) Write appropriate hypotheses.
c) Check the necessary assumptions and conditions.
d) Find the P-value of your test.
e) State your conclusion and analysis.

T 17. Montana. A 1992 poll conducted by the University of Montana classified respondents by gender and political party, as shown in the table. We wonder if there is evidence of an association between gender and party affiliation.

	Democrat	Republican	Independent
Male	36	45	24
Female	48	33	16

a) Is this a test of homogeneity or independence?
b) Write an appropriate hypothesis.
c) Are the conditions for inference satisfied?
d) Find the P-value for your test.
e) State a complete conclusion.

T 18. Fish diet. Medical researchers followed 6272 Swedish men for 30 years to see if there was any association between the amount of fish in their diet and prostate cancer. ("Fatty Fish Consumption and Risk of Prostate Cancer," *Lancet* [June 2001])

Fish Consumption	Total Subjects	Prostate Cancers
Never/seldom	124	14
Small part of diet	2621	201
Moderate part	2978	209
Large part	549	42

a) Is this a survey, a retrospective study, a prospective study, or an experiment? Explain.
b) Is this a test of homogeneity or independence?
c) Do you see evidence of an association between the amount of fish in a man's diet and his risk of developing prostate cancer?
d) Does this study prove that eating fish does not prevent prostate cancer? Explain.

T 19. Montana revisited. The poll described in Exercise 17 also investigated the respondents' party affiliations based on what area of the state they lived in. Test an appropriate hypothesis about this table, and state your conclusions.

	Democrat	Republican	Independent
West	39	17	12
Northeast	15	30	12
Southeast	30	31	16

T 20. Working parents. In July 1991 and again in April 2001, the Gallup Poll asked random samples of 1015 adults about their opinions on working parents. The table summarizes responses to the question "Considering the needs of both parents and children, which of the following do you see as the ideal family in today's society?"

	1991	2001
Both work full time	142	131
One works full time, other part time	274	244
One works, other works at home	152	173
One works, other stays home for kids	396	416
No opinion	51	51

a) Is this a survey, a retrospective study, a prospective study, or an experiment? Explain.
b) Will you test goodness-of-fit, homogeneity, or independence?
c) Based on these results, do you think there was a change in people's attitudes during the 10 years between these polls?

T 21. Grades. Two different professors teach an introductory Statistics course. The table shows the distribution of final grades they reported. We wonder whether one of these professors is an "easier" grader.

	Prof. Alpha	Prof. Beta
A	3	9
B	11	12
C	14	8
D	9	2
F	3	1

a) Will you test goodness-of-fit, homogeneity, or independence?
b) Write appropriate null hypotheses.
c) Find the expected counts for each cell, and explain why the chi-square procedures are not appropriate for this table.

T 22. Full moon. Some people believe that a full moon elicits unusual behavior in people. The table shows the number of arrests made in a small town during weeks of six full moons and six other randomly selected weeks during the same year. We wonder if there is evidence of a difference in the types of illegal activity that take place.

	Full Moon	Not Full
Violent (murder, assault, rape, etc.)	2	3
Property (burglary, vandalism, etc.)	17	21
Drugs/Alcohol	27	19
Domestic abuse	11	14
Other offenses	9	6

a) Will you test goodness-of-fit, homogeneity, or independence?
b) Write appropriate null hypotheses.
c) Find the expected counts for each cell, and explain why the chi-square procedures are not appropriate for this table.

T 23. Grades again. In some situations where the expected cell counts are too small, as in the case of the grades given by Professors Alpha and Beta in Exercise 21, we can complete an analysis anyway. We can often proceed after combining cells in some way that makes sense and also produces a table in which the conditions are satisfied. Here we create a new table displaying the same data, but calling D's and F's "Below C", as shown.

	Prof. Alpha	Prof. Beta
A	3	9
B	11	12
C	14	8
Below C	12	3

a) Find the expected counts for each cell in this new table, and explain why a chi-square procedure is now appropriate.
b) With this change in the table, what has happened to the number of degrees of freedom?
c) Test your hypothesis about the two professors, and state an appropriate conclusion.

T 24. Full moon, next phase. In Exercise 22 you found that the expected cell counts failed to satisfy the conditions for inference.
a) Find a sensible way to combine some cells that will make the expected counts acceptable.
b) Test a hypothesis about the full moon and state your conclusion.

T 25. Racial steering. A subtle form of racial discrimination in housing is "racial steering." Racial steering occurs when real estate agents show prospective buyers only homes in neighborhoods already dominated by that family's race. This violates the Fair Housing Act of 1968. According to an article in *Chance* magazine (Vol. 14, no. 2 [2001]), tenants at a large apartment complex recently filed a lawsuit alleging racial steering. The complex is divided into two parts, Section A and Section B. The plaintiffs claimed that white potential renters were steered to Section A, while African Americans were steered to Section B. The table displays the data that were presented in court to show the locations of recently rented apartments. Do you think there is evidence of racial steering?

New Renters			
	White	Black	Total
Section A	87	8	95
Section B	83	34	117
Total	170	42	212

26. Titanic, redux. Newspaper headlines at the time, and traditional wisdom in the succeeding decades, have held that women and children escaped the *Titanic* in greater proportion than men. Here's a table with data by gender. Do you think that survival was independent of gender? Defend your conclusion.

	Female	Male	Total
Alive	343	367	710
Dead	127	1364	1491
Total	470	1731	2201

27. Steering revisited. You could have checked the data in Exercise 25 for evidence of racial steering using two-proportion z procedures.

a) Find the z–value for this approach, and show that when you square your z-value you get the value of χ^2 you calculated in Exercise 25.
b) Show that the resulting P-values are the same.

28. Survival and gender, one more time. In Exercise 26 you could have checked for a difference in the chances of survival for men and women using two-proportion z procedures.
a) Find the z–value for this approach.
b) Show that the square of your calculated value of z is the value of χ^2 you calculated in Exercise 26.
c) Show that the resulting P-values are the same.

T 29. Race and education. Data from the U.S. Census Bureau show levels of education attained by age 30 for a sample of U.S. residents.

	Not HS Grad	HS Diploma	College Grad	Adv. Degree
White	810	6429	4725	1127
Black	263	1598	549	117
Hispanic	1031	1269	412	99
Other	66	341	305	197

Do these data highlight significant differences in education levels attained by these groups?

T 30. Pregnancies. Most pregnancies result in live births, but some end in miscarriages or stillbirths. A June 2001 National Vital Statistics Report examined those outcomes in the United States during 1997, broken down by the age of the mother. The table shows counts consistent with that report. Is there evidence that the distribution of outcomes is not the same for these age groups?

		Live Births	Fetal Losses
Age of Mother	Under 20	49	13
	20–29	201	41
	30–34	88	21
	35 or over	49	21

T 31. Race and education, part 2. Consider only the people who have graduated from high school from Exercise 29. Do these data suggest there are significant differences in opportunities for black and Hispanic Americans who have completed high school to pursue college or advanced degrees?

	HS Diploma	College Grad	Adv. Degree
Black	1598	549	117
Hispanic	1269	412	99

T 32. Education by age. Use the survey results in the table to investigate differences in education level attained among different age groups in the United States.

	Age Group				
Education Level		25–34	35–44	45–54	55–64	≥ 65
Not HS grad	27	50	52	71	101
HS	82	19	88	83	59
1–3 years college	43	56	26	20	20
≥ 4 years college	48	75	34	26	20

33. Ranking universities. In 2004 the Institute of Higher Education at Shanghai's Jiao Tong University evaluated the world's universities. Among their criteria were the size of the institution, the number of Nobel Prizes and Fields Medals won by faculty and alumni, and the faculty's research output. This ranking of the top 502 universities included 200 in North or Latin America, 209 in Europe, and 93 in the rest of the world (Asia/Pacific/Africa). A closer examination of the top 100 showed 55 in the Americas, 37 in Europe, and 8 elsewhere. Is there anything unusual about the geographical distribution of the world's top 100 universities?

34. Ranking universities, redux. In Exercise 33 you read about the world's top universities, as ranked by the Institute of Higher Education. The table shows the geographical distribution of these universities by groups of 100. (Not all groups have exactly 100 because of ties in the rankings.) Do these institutions appear to be randomly distributed around the world, or does there appear to be an association between ranking and location?

		Ranking				
Location		1–100	101–200	201–300	301–400	401–500
	North/Latin America	55	46	37	26	36
	Europe	37	42	46	46	38
	Asia/Africa/Pacific	8	13	17	30	25

just checking

Answers

1. We need to know how well beetles can survive 6 hours in a Plexiglas box so that we have a baseline to compare the treatments.

2. There's no difference in survival rate in the three groups.

3. $(2 - 1)(3 - 1) = 2$ *df*

4. 50

5. 2

6. The mean value for a χ^2 with 2 df is 2, so 10 seems pretty large. The P-value is probably small.

7. This is a test of homogeneity. The clue is that the question asks whether the distributions are alike.

8. This is a test of goodness-of-fit. We want to test the model of equal assignment to all lots against what actually happened.

9. This is a test of independence. We have responses on two variables for the same individuals.

CHAPTER 27

Inferences for Regression

WHO	250 Male subjects
WHAT	Body fat and waist size
UNITS	%Body fat and inches
WHEN	1990s
WHERE	United States
WHY	Scientific research

Three percent of a man's body is essential fat. (For a woman, the percentage is closer to 12.5%.) As the name implies, essential fat is necessary for a normal, healthy body. Fat is stored in small amounts throughout your body. Too much body fat, however, can be dangerous to your health. For men between 18 and 39 years old, a healthy percent body fat ranges from 8% to 19%. (For women of the same age, it's 21% to 32%.)

Measuring body fat can be tedious and expensive. The "standard reference" measurement is by dual-energy X-ray absorptiometry (DEXA), which involves two low-dose X-ray generators and takes from 10 to 20 minutes.

Percent body fat vs. *waist size* for 250 men of various ages. The scatterplot shows a strong, positive, linear relationship. **Figure 27.1**

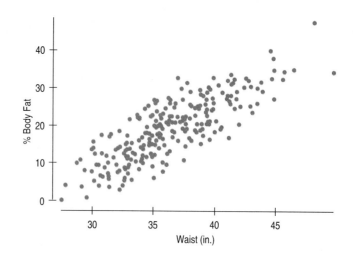

How close can we get to a useable prediction of body fat from easily measurable variables such as *height*, *weight*, or *waist* size? Figure 27.1 shows a scatterplot of the *%body fat* plotted against *waist* size for a sample of 250 males of various ages.

Back in Chapter 8 we modeled relationships like this by fitting a least squares line. The plot is clearly straight, so we can find that line. The equation of the least squares line for these data is:

$$\widehat{\%body\ fat} = -42.7 + 1.7\ waist.$$

The slope says that, on average, *%body fat* is greater by 1.7 percent for each additional inch around the waist.

How useful is this model? When we fit linear models before, we used them to describe the relationship between the variables and we interpreted the slope and intercept as descriptions of the data. Now we'd like to know what the regression model can tell us beyond the 250 men in this study. To do that, we'll want to make confidence intervals and test hypotheses about the slope and intercept of the regression line.

The Population and the Sample

When we found a confidence interval for a mean, we could imagine a single, true underlying value for the mean. When we tested whether two means or two proportions were equal, we imagined a true underlying difference. But what does it mean to do inference for regression? We know better than to think that even if we knew every population value, the data would line up perfectly on a straight line. After all, even in our sample, not all men who have 38-inch waists have the same *%body fat*. In fact, there's a whole distribution of *%body fat* for these men:

The distribution of *%body fat* for men with a *waist* size of 38 inches is unimodal and symmetric.
Figure 27.2

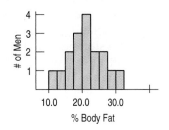

This is true at each *waist* size. In fact, we could depict the distribution of *%body fat* at different *waist* sizes like this:

There's a distribution of *%body fat* for *each* value of *waist* size. We'd like the means of these distributions to line up.   **Figure 27.3**

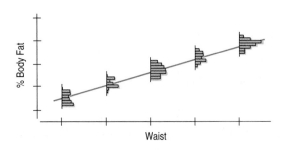

But we want to *model* the relationship between *%body fat* and *waist* size for all men. To do that, we imagine an idealized regression line. The model assumes that the *means* of the distributions of *%body fat* for each *waist* size fall along the line even though the individuals are scattered around it. We know that this model is not a perfect description of how the variables are associated, but it may be useful for predicting *%body fat* and for understanding how it's related to *waist* size.

If only we had all the values in the population, we could find the slope and intercept of this *idealized regression line* explicitly by using least squares. Following our usual conventions, we write the idealized line with Greek letters and consider the coefficients (the slope and intercept) to be *parameters*: β_0 is the intercept and β_1 is the slope. Corresponding to our fitted line of $\hat{y} = b_0 + b_1x$, we write

$$\mu_y = \beta_0 + \beta_1 x.$$

Why μ_y insead of $\hat{y}$? Because this is a model. There is a distribution of *%body fat* for each *waist* size. The model places the *means* of the distributions of *%body fat* for each *waist* size on the same straight line. And, since this is a model, we use Greek letters for the parameters.

Of course, not all the individual y's are at these means. (In fact, the line will miss most—and quite possibly all—of the plotted points.) Some individuals lie above and some below the line, so, like all models, this one makes **errors.** Lots of them. In fact, one at each point. These errors can be positive or negative, depending on which side of the line the data fall (or even zero if the line happens to pass right through the point). We denote the errors by ε and write $\varepsilon = y - \mu_y$ for each data point (x, y).

When we put the errors into the equation, we can account for each individual y:

$$y = \beta_0 + \beta_1 x + \varepsilon.$$

This equation is now true for each data point (since there is an ε to soak up the deviation), so the model gives a value of y for any value of x.

For the body fat data, an idealized model such as this provides a summary of the relationship between *%body fat* and *waist* size. Like all models, it simplifies the real situation. We know there is more to predicting body fat than waist size alone. But the advantage of a model is that the simplification might help us to think about the situation and assess how well *%body fat* can be predicted from simpler measurements.

We estimate the β's by finding a regression line, $\hat{y} = b_0 + b_1x$, as we did in Chapter 8. The residuals, $e = y - \hat{y}$, are the sample-based versions of the errors, ε. We'll use them to help us assess the regression model.

We know that least squares regression will give reasonable estimates of the parameters of this model from a random sample of data. Our challenge is to account for our uncertainty in how well they do. For that, we need to make some assumptions about the model and the errors.

Assumptions and Conditions

Back in Chapter 8 when we fit lines to data, we needed to check only the Straight Enough Condition. Now, when we want to make inferences about the coefficients of the line, we'll have to make more assumptions. Fortunately, we can check conditions to help us judge whether these assumptions are reasonable for our data.

And as we've done before, we'll make some checks *after* we find the regression equation.

Also, we need to be careful about the order in which we check conditions. If our initial assumptions are not true, it makes no sense to check the later ones. So now we number the assumptions to keep them in order.

1. Linearity Assumption

If the true relationship is far from linear and we use a straight line to fit the data, our entire analysis will be useless, so we always check this first.

The **Straight Enough Condition** is satisfied if a scatterplot looks straight. It's generally not a good idea to draw a line through the scatterplot when checking. That can fool your eye into seeing the plot as more straight. Sometimes it's easier to see violations of the Straight Enough Condition by looking at a scatterplot of the residuals against x or against the predicted values, $\hat{y}$. That plot will have a horizontal direction and should have no pattern if the condition is satisfied.

If the scatterplot is straight enough, we can go on to some assumptions about the errors. If not, stop here, or consider re-expressing the data (see Chapter 10) to make the scatterplot more nearly linear. For the *%body fat* data, the scatterplot is beautifully linear.

2. Independence Assumption

The errors in the true underlying regression model (the ε's) must be mutually independent. As usual, there's no way to be sure that the Independence Assumption is true.

Usually when we care about inference for the regression parameters it's because we think our regression model might apply to a larger population. In such cases, we can check a **Randomization Condition,** that the individuals are a representative sample from that population.

We can also check displays of the regression residuals for evidence of patterns, trends, or clumping, any of which would suggest a failure of independence. In the special case when the x-variable is related to time, a common violation of the Independence Assumption is for the errors to be correlated. (The error our model makes today may be similar to the one it made for yesterday.) This violation can be checked by plotting the residuals against the x-variable and looking for patterns.

The *%body fat* data were collected on a sample of men taken to be representative. The subjects were not related in any way, so we can be pretty sure that their measurements are independent. The residuals plot shows no pattern.

3. Equal Variance Assumption

The variability of y should be about the same for all values of x. In Chapter 8 we looked at the standard deviation of the residuals (s_e) to measure the size of the scatter. Now we'll need this standard deviation to build confidence intervals and test hypotheses. The standard deviation of the residuals is the building block for the standard errors of all the regression parameters. But it only makes sense if the scatter of the residuals is the same everywhere. In effect, the standard deviation of the residuals "pools" information across all of the individual distributions at each x-value, and pooled estimates are appropriate only when they combine information for groups with the same variance.

Practically, what we can check is the **Does The Plot Thicken? Condition.** A scatterplot of y against x offers a visual check. Fortunately, we've already made

Check the scatterplot.
The shape must be linear or we can't use regression at all.

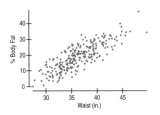

Check the residuals plot (1).
The residuals should appear to be randomly scattered.

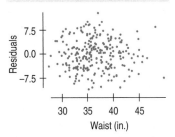

The residuals show only random scatter when plotted against waist size. **Figure 27.4**

Check the residuals plot (2).
The vertical spread of the residuals should be roughly the same everywhere.

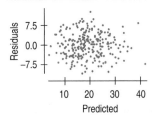

A scatterplot of residuals against predicted values can help check for plot thickening. Note that this plot looks identical to the plot of residuals against waist size. For a regression of one response variable on one predictor, these plots differ only in the labels on the x-axis. **Figure 27.5**

one. Make sure the spread around the line is nearly constant. Be alert for a "fan" shape or other tendency for the variation to grow or shrink in one part of the scatterplot. Often it is better to look at the residuals plotted against the predicted values, $\hat{y}$. With the slope of the line removed, it's easier to see patterns left behind. For the body fat data, the spread of *%body fat* around the line is remarkably constant across *waist* sizes from 30 inches to about 45 inches.

If the plot is straight enough, the data are independent, and the plot doesn't thicken, you can now move on to the final assumption.

4. Normal Population Assumption

We assume the errors around the idealized regression line at each value of *x* follow a Normal model. We need this assumption so that we can use a Student's *t*-model for inference.

As with other times when we've used Student's *t*, we'll settle for the residuals satisfying the **Nearly Normal Condition.** Look at a histogram or Normal probability plot of the residuals.[1]

The histogram of residuals in the *%body fat* regression certainly looks nearly Normal. As we have noted before, the Normality Assumption becomes less important as the sample size grows because the model is about means and the Central Limit Theorem takes over.

If all four assumptions were true, the idealized regression model would look like this:

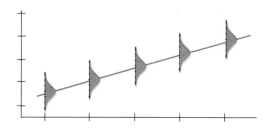

At each value of *x* there is a distribution of *y*-values that follows a Normal model, and each of these Normal models is centered on the line and has the same standard deviation. Of course, we don't expect the assumptions to be exactly true, and we know that all models are wrong, but the linear model is often close enough to be very useful.

> **Check a histogram of the residuals.**
>
> The distribution of the residuals should be unimodal and symmetric.

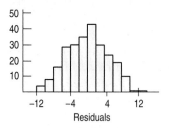

A histogram of the residuals is one way to check whether they are nearly Normal. Alternatively, we can look at a Normal probability plot. **Figure 27.6**

The regression model has a distribution of *y*-values for each *x*-value. These distributions follow a Normal model with means lined up along the line and the same standard deviations. **Figure 27.7**

Which Come First: the Conditions or the Residuals?

> *"Truth will emerge more readily from error than from confusion."*
> —*Francis Bacon*
> *(1561–1626)*

In regression, there's a little catch. The best way to check many of the conditions is with the residuals, but we get the residuals only *after* we compute the regression. Before we compute the regression, however, we should check at least one of the conditions.

---

[1] *This* is why we have to check the conditions in order. We have to check that the residuals are independent and that the variation is the same for all *x*'s so that we can lump all the residuals together for a single check of the Nearly Normal Condition.

So we work in this order:

1. Make a scatterplot of the data to check the Straight Enough Condition. (If the relationship is curved, try re-expressing the data. Or stop.)

2. If the data are straight enough, fit a regression and find the residuals, e, and predicted values, $\hat{y}$.

3. Make a scatterplot of the residuals against x or the predicted values. This plot should have no pattern. Check in particular for any bend (which would suggest that the data weren't all that straight after all) for any thickening (or thinning) and, of course, for any outliers. (If there are outliers, and you can correct them or justify removing them, do so and go back to step 1, or consider performing two regressions—one with and one without the outliers.)

4. If the data are measured over time, plot the residuals against time to check for evidence of patterns that might suggest they are not independent.

5. If the scatterplots look OK, then make a histogram and Normal probability plot of the residuals to check the Nearly Normal Condition.

6. If all the conditions seem to be reasonably satisfied, go ahead with inference.

Regression Inference Step-By-Step

If our data can jump through all these hoops, we're ready to do regression inference. Let's try one on the body fat data, gathered at Brigham Young University.

  Plan Specify the question of interest.

Name the variables and report the W's.

Identify the parameters you want to estimate.

Model Check the conditions.

Make pictures. For regression inference you'll need a scatterplot, a residuals plot, and either a histogram or a Normal probability plot of the residuals.

I have body measurements on 250 adult males from the BYU Human Performance Research Center. I want to find the slope of the relationship between %body fat and waist size.

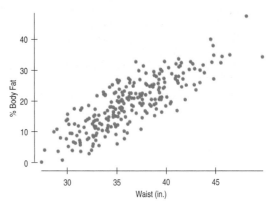

✔ **Straight Enough Condition:** There's no obvious bend in the original scatterplot of the data or in the plot of residuals against predicted values.

✔ **Independence Assumption:** These data are not collected over time and there's no reason to think that the %body fat of one man influences the %body fat of another.

(We've seen plots of the residuals already. See Figures 27.5 and 27.6.)

✔ **Does the Plot Thicken? Condition:** Neither the original scatterplot nor the residual scatterplot shows any changes in the spread about the line.

✔ **Nearly Normal Condition:** A histogram of the residuals is unimodal and symmetric. The Normal probability plot of the residuals is quite straight, indicating that the Normal model is reasonable for the errors.

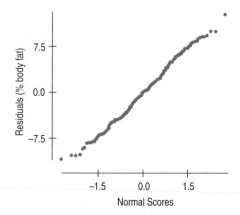

Under these conditions the regression model is appropriate.

I will fit a **regression model.**

Choose your method.

 Show

Mechanics Let's just "push the button" and see what the regression looks like.

The formula for the regression equation can be found in Chapter 8, and the standard error formulas will be shown a bit later, but regressions are almost always computed with a computer program or calculator.

Write the regression equation.

Tell **Conclusion**

Interpret your results in context.

More Interpretation We haven't worked it out in detail yet, but the output gives us numbers labeled as *t*-statistics and

Here's the computer output for this regression:

Dependent variable is: %BF

R-squared = 67.8%
s = 4.713 with 250 − 2 = 248 degrees of freedom

Variable	Coeff	SE(Coeff)	t-ratio	P-value
Intercept	−42.734	2.717	−15.7	<0.0001
Waist	1.70	0.0743	22.9	<0.0001

The estimated regression equation is

$$\widehat{\%body\ fat} = -42.73 + 1.70\ waist.$$

The R^2 for the regression is 67.8%. Waist size seems to account for about 2/3 of the %body fat variation in men. The slope of the regression says that %body fat increases by about 1.7 percentage points per inch of waist size, on average.

The standard error of 0.07 for the slope is much smaller than the slope itself, so it looks like the estimate is reasonably precise. And there are a couple of t-ratios and P-values given. Because the P-values are small, it appears that some null hypotheses can be rejected.

corresponding P-values, and we have a general idea of what those mean.

(It's time to learn more about regression inference so we can figure out what the rest of the output means.)

Intuition About Regression Inference

A S **Simulate the Sampling Distribution of a Regression Slope.** Draw samples repeatedly to see for yourself how slope can vary from sample to sample. This lets you build up a histogram to see the sampling distribution.

Wait a minute! We've just pulled a fast one. We've pushed the "regression button" on our computer or calculator, but haven't discussed where the standard errors for the slope or intercept come from. We know that if we had collected similar data on a different random sample of men, the slope and intercept would be different. Each sample would have produced its own regression line, with slightly different b_0's and b_1's. This sample-to-sample variation is what generates the sampling distributions for the coefficients.

There's only one regression model; each sample regression is trying to estimate the same parameters, β_0 and β_1. We expect any sample to produce a b_1 whose expected value is the true slope, β_1. What about its standard deviation? What aspects of the data affect how much the slope (and intercept) vary from sample to sample?

- **Spread around the line.** Here are two situations in which we might do regression. Which situation would yield the more consistent slope? That is, if we were to sample over and over from these two underlying populations that these samples come from and compute all the slopes, which group of slopes would vary less?

Which of these scatterplots shows a situation that would give the more consistent regression slope estimate if we were to sample repeatedly from their underlying population?

Figure 27.8

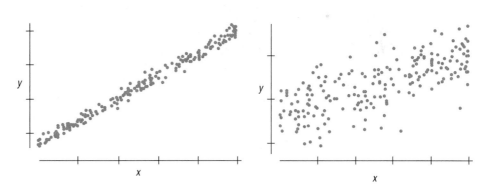

Clearly, data like those in the left plot give more consistent slopes.

Less scatter around the line means the slope will be more consistent from sample to sample. The spread around the line is measured with the residual standard deviation, s_e. You can always find s_e in the regression output, often just labeled s. You're probably not going to calculate the error standard deviation by hand. As we noted when we first saw this formula in Chapter 8, it looks a lot like the standard deviation of y, only now subtracting the predicted values rather than the mean and dividing by $n - 2$ instead of $n - 1$:

$$s_e = \sqrt{\frac{\sum(y - \hat{y})^2}{n - 2}}.$$

n − 2?
For standard deviation (in Chapter 5), we divided by *n* − 1 because we didn't know the true mean and had to estimate it. Now it's later in the course and there's even more we don't know. Here we don't know *two* things: the slope and the intercept. If we knew them both, we'd divide by *n* and have *n* degrees of freedom. When we estimate both, however, we adjust by subtracting 2, so we divide by *n* − 2 and (as we will see soon) have 2 fewer degrees of freedom.

The less scatter around the line, the smaller the error standard deviation and the stronger the relationship between *x* and *y*.

Some people prefer to assess the strength of a regression by looking at s_e rather than R^2. After all, s_e has the same units as *y* and because it's the standard deviation of the errors around the line, it tells you how close the data are to our model. By contrast, R^2 is the proportion of the variation of *y* accounted for by *x*. We say, why not look at both?

- **Spread of the *x*'s:** Here are two more situations. Which of these would yield more consistent slopes?

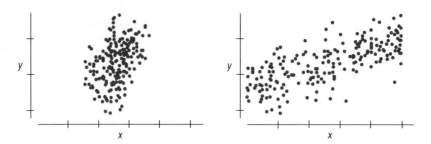

Which of these scatterplots shows a situation that would give the more consistent regression slope estimate if we were to sample repeatedly from the underlying population?
Figure 27.9

A plot like the one on the right has a broader range of *x*-values, so it gives a more stable base for the slope. We'd expect the slopes of samples from situations like that to vary less from sample to sample. A large standard deviation of *x*, s_x, provides a more stable regression.

- **Sample size.** Here we go again. What about these two?

Which of these scatterplots shows a situation that would give the more consistent regression slope estimate if we were to sample repeatedly from the underlying population?
Figure 27.10

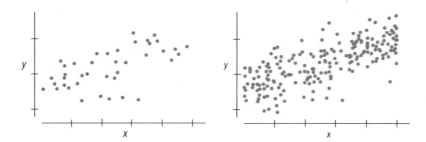

It shouldn't be a surprise that having a larger sample size, *n*, gives more consistent estimates from sample to sample.

Standard Error for the Slope

Three aspects of the scatterplot, then, affect the standard error of the regression slope:

- Spread around the line: s_e
- Spread of *x* values: s_x
- Sample size: *n*

These are in fact the *only* things that affect the standard error of the slope. Although you'll probably never have to calculate it by hand, the formula for the standard error is:

$$SE(b_1) = \frac{s_e}{\sqrt{n-1}\, s_x}.$$

A S **Regression Slope Standard Error.** See how $SE(b_1)$ is constructed and where the values used in the formula are found in the regression output table.

The error standard deviation, s_e, is in the *numerator*, since spread around the line *increases* the slope's standard error. The denominator has both a sample size term ($\sqrt{n-1}$) and s_x because increasing either of these *decreases* the slope's standard error.

A S **x-Variance and Slope Variance.** You don't have to just imagine how the variability of the slope depends on the spread of the x's. This simulation lets you see for yourself.

We know the b_1's vary from sample to sample. As you'd expect, their sampling distribution model is centered at β_1, the slope of the idealized regression line. Now we can estimate its standard deviation with $SE(b_1)$. What about its shape? Here the Central Limit Theorem and "Wild Bill" Gosset come to the rescue again. When we standardize the slopes by subtracting the model mean and dividing by their standard error, we get a Student's t-model, this time with $n-2$ degrees of freedom:

$$\frac{b_1 - \beta_1}{SE(b_1)} \sim t_{n-2}.$$

A sampling distribution for regression slopes

When the conditions are met, the standardized estimated regression slope,

$$t = \frac{b_1 - \beta_1}{SE(b_1)},$$

follows a Student's t-model with $n-2$ degrees of freedom. We estimate the standard error with

$$SE(b_1) = \frac{s_e}{\sqrt{n-1}\, s_x}, \text{ where } s_e = \sqrt{\frac{\sum(y - \hat{y})^2}{n-2}}, \, n \text{ is the number of data values, and } s_x$$

is the ordinary standard deviation of the x-values.

What About the Intercept?

The same reasoning applies for the intercept. We could write

$$\frac{b_0 - \beta_0}{SE(b_0)} \sim t_{n-2}$$

and use it to construct confidence intervals and test hypotheses, but often the value of the intercept isn't something we care about. The intercept usually isn't interesting. Most hypothesis tests and confidence intervals for regression are about the slope. But in case you really want to see the formula for the standard error of the intercept, we've parked it in a footnote.[2]

---

[2] $SE(b_0) = s_e\sqrt{\dfrac{1}{n} + \dfrac{\bar{x}^2}{\sum(x - \bar{x})^2}}$

Regression Inference

What if the Slope Were 0?

If $b_1 = 0$, our prediction is $\hat{y} = b_0 + 0x$. The equation collapses to just $\hat{y} = b_0$. Now x is nowhere in sight, so y doesn't depend on x at all.

And b_0 would turn out to be $\bar{y}$. Why? We know that $b_0 = \bar{y} - b_1\bar{x}$, but when $b_1 = 0$ that becomes simply $b_0 = \bar{y}$. It turns out, then, that when the slope is 0 the equation is just $\hat{y} = \bar{y}$; at every value of x we always predict the mean value for y.

Now that we have the standard error of the slope and its sampling distribution, we can test a hypothesis about it and make confidence intervals. The usual null hypothesis about the slope is that it's equal to 0. Why? Well, a slope of zero would say that y doesn't tend to change linearly when x changes—in other words, that there is no linear association between the two variables. If the slope were zero, there wouldn't be much left of our regression equation.

So a null hypothesis of a zero slope questions the entire claim of a linear relationship between the two variables—and often that's just what we want to know. In fact, every software package or calculator that does regression simply assumes that you want to test the null hypothesis that the slope is really zero.

To test $H_0: \beta_1 = 0$ we find

$$t = \frac{b_1 - 0}{SE(b_1)}.$$

This is just like every t-test we've seen: a difference between the statistic and its hypothesized value divided by its standard error.

For our body fat data, the computer found the slope (1.7), its standard error (0.0743), and the ratio of the two: $\dfrac{1.7 - 0}{0.0743} = 22.9$ (see p. 638). Nearly 23 standard errors from the hypothesized value certainly seems big. The P-value (<0.0001) confirms that a t-ratio this large would be very unlikely to occur if the true slope were zero.

Maybe the standard null hypothesis isn't all that interesting here. Did you have any doubts that *%body fat* is related to *waist* size? A more sensible use of these same values might be to make a confidence interval for the slope instead.

We can build a confidence interval in the usual way, as an estimate plus or minus a margin of error. As always, the margin of error is just the product of the standard error and a critical value. Here the critical value comes from the t distribution with $n - 2$ degrees of freedom, so a 95% confidence interval for β is

$$b_1 \pm t^*_{n-2} \times SE(b_1).$$

For the body fat data, $t^*_{248} = 1.970$, so that comes to $1.7 \pm 1.97 \times 0.074$, or an interval from 1.55 to 1.85 *%body fat* per inch of *waist* size.

just checking

Researchers in Food Science studied how big people's mouths tend to be. They measured mouth volume by pouring water into the mouth of subjects who lay on their backs. Unless this is your idea of a good time, it would be helpful to have a model to estimate mouth volume more simply. Fortunately, mouth volume is related to height. (Mouth volume is measured in cubic centimeters and height in meters.)

The data were checked and deemed suitable for regression. Here is some computer output:

Summary of	Mouth Volume
Mean	60.2704
StdDev	16.8777

Dependent variable is: Mouth Volume

R squared = 15.3%

s = 15.66 with 61 − 2 = 59 degrees of freedom

Variable	Coefficient	SE(coeff)	t-ratio	P-value
Intercept	−44.7113	32.16	−1.39	0.1697
Height	61.3787	18.77	3.27	0.0018

1. What does the *t*-ratio of 3.27 tell us about this relationship? How does the P-value help our understanding?

2. Would you say that measuring a person's height could reliably be used as a substitute for the wetter method of determining how big a person's mouth is? What numbers in the output helped you reach that conclusion?

3. What does the value of *s* add to this discussion?

Another Example

Every spring, Nenana, Alaska, hosts a contest in which participants try to guess the exact minute that a wooden tripod placed on the frozen Tanana River will fall through the breaking ice. The contest started in 1917 as a diversion for railroad engineers, with a jackpot of $800 for the closest guess. It has grown into an event in which hundreds of thousands of entrants enter their guesses on the Internet[3] and vie for more than $300,000.

Because so much money and interest depends on the time of breakup, it has been recorded to the nearest minute with great accuracy ever since 1917. And because a standard measure of breakup has been used throughout this time, the data are consistent. An article in *Science* ("Climate Change in Nontraditional Data Sets." *Science* 294 [26 October 2001]: 811) used the data to investigate global warming—whether greenhouse gasses and other human actions have been making the planet warmer. Others might just want to make a good prediction of next year's breakup time.

Of course, we can't use regression to tell the *causes* of any change. But we can estimate the *rate* of change (if any) and use it to make better predictions.

Here are some of the data:

WHO	Years
WHAT	Year, day, and hour of ice breakup
UNITS	*x* is in years since 1900. *y* is in days after midnight Dec. 31.
WHEN	1917 – present
WHERE	Nenana, Alaska
WHY	Wagering, but proposed to look at global warming

Year (since 1900)	Breakup Date (days after Jan. 1)	Year (since 1900)	Breakup Date (days after Jan. 1)
17	119.4792	30	127.7938
18	130.3979	31	129.3910
19	122.6063	32	121.4271
20	131.4479	33	127.8125
21	130.2792	34	119.5882
22	131.5556	35	134.5639
23	128.0833	36	120.5403
24	131.6319	37	131.8361
25	126.7722	38	125.8431
26	115.6688	39	118.5597
27	131.2375	40	110.6437
28	126.6840	41	122.0764
29	124.6535	⋮	⋮

---

[3] Search for "Nenana Ice Classic."

Has the *date* of ice breakup on the Tanana River changed since 1917, when record keeping began? Earlier breakup dates might be a sign of global warming. **Figure 27.11**

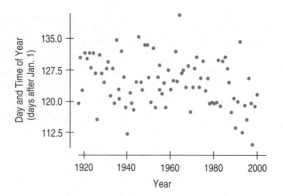

A S **A Hypothesis Test for the Regression Slope.** View an animated discussion of testing the standard null hypothesis for slope.

A Regression Slope *t*-Test   Step-By-Step

The slope of the regression gives the change in breakup date per year. Let's test the hypothesis that the slope is zero.

Think

Plan State what you want to know.

Identify the *parameter* you wish to estimate. Here our parameter is the slope.

Identify the variables and review the W's.

Hypotheses Write your null and alternative hypotheses.

Model Check the assumptions and conditions.

Make pictures. We saw a scatterplot of the data earlier. Because it seems straight enough, we can find and plot the residuals.

Usually, we check for suggestions that the Independence Assumption fails by plotting the residuals against the predicted values. Patterns and clusters in that plot raise our suspicions. But when the data are measured over time, it is always a good idea to plot residuals against time to look for trends and oscillations.

I wonder whether the *date* of ice breakup has changed over time. The slope of that change might indicate climate change. I have the date of ice breakup annually since 1917, recorded as the number of days and fractions of a day until the ice breakup.

H_0: There is no change in the *date* of ice breakup: $\beta_1 = 0$

H_A: Yes, there is: $\beta_1 \neq 0$

✔ **Straight Enough Condition:** There is no obvious bend in the scatterplot.

✔ **Independence Assumption:** These data are a time series, which raises my suspicions that they may not be independent. To check, here's a plot of the residuals against time, the x-variable of the regression:

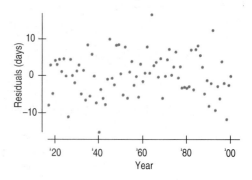

I see a hint that the data oscillate up and down, which suggests some failure of independence, but not so strongly that I can't proceed with the analysis. These data are not a random sample, so I'm reluctant to extend my conclusions beyond this river and these years.

✔ **Does the Plot Thicken? Condition:** The residuals plot shows no obvious trends in the spread.

✔ **Nearly Normal Condition:** A histogram of the residuals is unimodal and symmetric.

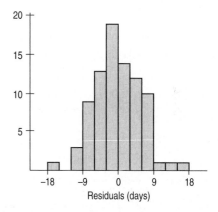

Residuals (days)

A S **Conditions for Regression Inference.** View an illustrated discussion of the conditions for regression inference.

State the sampling distribution model.

Under these conditions, the sampling distribution of the regression slope can be modeled by a Student's *t*-model with $(n - 2) = 82$ degrees of freedom.

Choose your method.

I'll do a **regression slope *t*-test.**

Show   **Mechanics**

Here's the computer output for this regression:

The regression equation can be found from the formulas in Chapter 8, but regressions are almost always found from a computer program or calculator.

Dependent variable is: date

R-squared = 8.6%
s = 5.747 with 84 − 2 = 82 degrees of freedom

The P-values given in the regression output table are from the Student's *t* distribution on $(n - 1) = 82$ degrees of freedom. They are appropriate for two-sided alternatives.

Variable	Coeff	SE(Coeff)	t-ratio	P-value
Intercept	128.753	1.638	78.6	<0.0001
Year	−0.07196	0.0259	−2.78	0.0067

The estimated regression equation is

$$\widehat{date} = 128.8 - 0.07 \text{ year.}$$

Tell   **Conclusion**

Link the P-value to your decision and state your conclusion in the proper context.

The P-value of 0.0067 means that the association we see in the data is unlikely to have occurred by chance. I reject the null hypothesis, and conclude that there is strong evidence that, on average, the ice breakup is occurring

Show

MORE

Create a confidence interval
for the true slope

Tell

MORE

Interpret the interval

Simply rejecting the standard null hypothesis
doesn't guarantee that the size of the effect is
large enough to be important. Whether we
want to know the breakup time to the nearest
minute or are interested in global warming, a
change measured in hours each year is big
enough to be interesting.

earlier each year. But the oscillation pattern in
the residuals raises concerns.

A 95% confidence interval for β_1 is:

$$b_1 \pm t^*_{82} \times SE(b_1)$$
$$-0.072 \pm (1.99)(0.0259)$$

$$\text{or } (-0.12, -0.02) \text{ days.}$$

I am 95% confident that the ice has been breaking up, on average, between 0.02 days (about a
half-hour) and 0.12 days (about 3 hours) earlier
each year since 1900.

(Technically: I am 95% confident that the interval from 0.02 to 0.12 days per year captures
the true rate at which the time of ice breakup
has been getting earlier since 1900.)

● **But is it global warming?** So the ice is breaking up earlier. Temperatures are
higher. Must be global warming, right?

Maybe.

An article challenging the original analysis of the Nenana data proposed a possible confounding variable. It noted that the city of Fairbanks is upstream from Nenana and suggested that the growth of Fairbanks could have warmed the river. So maybe it's not global
warming.

Or maybe global warming is a lurking variable, leading more people to move to a now
balmier Fairbanks and also leading to generally earlier ice breakup in Nenana.

Or maybe there's some other variable or combination of variables at work. We can't set
up an experiment, so we may never really know.

Only one thing is for sure. When you try to explain an association by claiming cause and
effect, you're bound to be on thin ice.[4] ●

Standard Errors for Predicted Values

Once we have a useful regression, how can we indulge our natural desire to predict, without being irresponsible? We know how to compute predicted values of y
for any value of x. We first did that in Chapter 8. This predicted value would be
our best estimate, but it's still just an informed guess.

Now, however, we have standard errors. We can use those to construct a confidence interval for the predictions, smudging the results in the right way to report
our uncertainty honestly.

---

[4] How *do* scientists sort out such messy situations? Even though they can't conduct an experiment,
they *can* look for replications elsewhere. A number of studies of ice on other bodies of water have also
shown earlier ice breakup times in recent years. That suggests they need an explanation that's more
comprehensive than just Fairbanks and Nenana.

From our model of *%body fat* and *waist* size, we might want to use *waist* size to get a reasonable estimate of *%body fat*. A confidence interval can tell us how precise that prediction will be. The precision depends on the question we ask, however, and there are two questions. Do we want to know the mean *%body fat* for *all* men with a *waist* size of, say, 38 inches? Or do we want to estimate the *%body fat* for a particular man with a 38-inch *waist* without making him climb onto the X-ray table?

What's the difference between the two questions? The predicted *%body fat* is the same, but one question leads to an answer much more precise than the other. We can predict the *mean %body fat* for *all* men whose *waist* size is 38 inches with a lot more precision than we can predict the *%body fat* of a *particular individual* whose *waist* size happens to be 38 inches. Both are interesting questions.

We start with the same prediction in both cases. We are predicting the value for a new individual, one that was not part of the original data set. To emphasize this, we'll call his *x*-value "*x* sub new" and write it x_ν.[5] Here, x_ν is 38 inches. The regression equation predicts *%body fat* as $\hat{y}_\nu = b_0 + b_1 x_\nu$.

Now that we have the predicted value, we construct both intervals around this same number. Both intervals take the form

$$\hat{y}_\nu \pm t^*_{n-2} \times SE.$$

Even the *t** value is the same for both. It's the critical value (from Table T or technology) for $n - 2$ degrees of freedom and the specified confidence level. The difference between the two intervals is all in the standard errors. Our choice of ruler depends on which interval we want.

The standard errors for prediction depend on the same kinds of things as the coefficients' standard errors. If there is more spread around the line, we'll be less certain when we try to predict the response. Of course, if we're less certain of the slope, we'll be less certain of our prediction. If we have more data, our estimate will be more precise. And there's one more piece. If we're farther from the center of our data, our prediction will be less precise. The last factor is new, but makes intuitive sense. It's a lot easier to predict a data point near the middle of the data set than to extrapolate far from the center.

Each of these factors contributes uncertainty—that is, variability—to the estimate. Because the factors are independent of each other, we can add their variances to find the total variability. The resulting formula for the standard error of the predicted *mean* value explicitly takes into account each of the factors:

$$SE(\hat{\mu}_\nu) = \sqrt{SE^2(b_1) \times (x_\nu - \bar{x})^2 + \frac{s_e^2}{n}}.$$

Individual values vary more than means, so the standard error for a single predicted value has to be larger than the standard error for the mean. In fact, the standard error of a single predicted value has an *extra* source of variability: the variation of individuals around the predicted mean. That appears as the extra variance term, s_e^2, at the end under the square root:

$$SE(\hat{y}_\nu) = \sqrt{SE^2(b_1) \times (x_\nu - \bar{x})^2 + \frac{s_e^2}{n} + s_e^2}.$$

For the Nenana Ice Classic, someone who planned to place a bet would want to predict this year's breakup time. By contrast, scientists studying global warming are likely to be more interested in the mean breakup time. Unfortunately if you want to gamble, the variability is greater for predicting for a single year.

---

[5] Yes, this is a bilingual pun. The Greek letter ν is called "nu." Don't blame me; my co-author suggested this.

Remember to keep this distinction between the two kinds of standard errors when looking at computer output. The smaller one is for the predicted *mean* value, and the larger one is for an predicted *individual* value.[6]

Confidence Intervals for Predicted Values

Now that we have standard errors, we can ask how well our analysis can predict the mean *%body fat* for men with 38-inch *waists*. The regression output table provides most of the numbers we need:

$$s_e = 4.713$$
$$n = 250$$
$$SE(b_1) = 0.074, \text{ and from the data we need to know that}$$
$$\bar{x} = 36.3$$

The regression model gives a predicted value at $x_\nu = 38$ of

$$\hat{y}_\nu = -42.7 + 1.7 \times 38 = 21.9\%.$$

Let's find the 95% confidence interval for the mean *%body fat* for all men with 38-inch *waists*. We find the standard error from the formula:

$$SE(\hat{\mu}_\nu) = \sqrt{0.074^2 \times (38 - 36.3)^2 + \frac{4.713^2}{250}} = 0.32\%.$$

The t^* value that excludes 2.5% in either tail with $250 - 2 = 248$ df is (according to the tables) 1.97.
Putting it all together, we find the margin of error as

$$ME = 1.97 \times 0.32 = 0.63\%.$$

So, we are 95% confident that the mean *%body fat* for men with 38-inch *waists* is

$$21.9\% \pm 0.63\%.$$

Suppose, instead, we want to predict the *%body fat* for an individual man with a 38-inch *waist*. We need the larger standard error:

$$SE(\hat{y}_\nu) = \sqrt{0.074^2 \times (38 - 36.3)^2 + \frac{4.713^2}{250} + 4.713^2} = 4.72\%.$$

The corresponding margin of error is

$$ME = 1.97 \times 4.72 = 9.30\%,$$

so the prediction interval is:

$$21.9\% \pm 9.30\%.$$

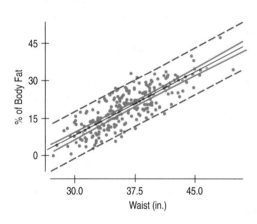

A scatterplot of *%body fat* vs. *waist* size with a least squares regression line. The solid green lines near the regression line show the extent of the 95% confidence intervals for *mean %body fat* at each *waist* size. The dashed red lines show the prediction intervals. Most of the points are contained within the prediction intervals, but not within the confidence intervals.

Figure 27.12

---

[6] You may see the standard error expressions written in other equivalent ways. The most common alternatives are

$$SE(\hat{\mu}_\nu) = s_e \sqrt{\frac{1}{n} + \frac{(x_\nu - \bar{x})^2}{\sum(x - \bar{x})^2}} \quad \text{and} \quad SE(\hat{y}_\nu) = s_e \sqrt{1 + \frac{1}{n} + \frac{(x_\nu - \bar{x})^2}{\sum(x - \bar{x})^2}}.$$

We can think of this interval as having a 95% chance of capturing the true *%body fat* of a randomly selected man whose *waist* is 38 inches.[7] Notice how much wider this interval is than the previous one. As we've known since Chapter 18, the mean is much less variable than a randomly selected individual value.

Keep in mind this distinction between the two kinds of confidence intervals. The narrower interval is a **confidence interval for the predicted mean value** at x_v, and the wider interval is a **prediction interval for an individual** with that *x*-value.

What Can Go Wrong?	In this chapter we've added inference to the regression explorations that we did in Chapters 8 and 9. Everything covered in those chapters that could go wrong with regression can still go wrong. It's probably a good time to review Chapter 9. We'll let you do that.

With inference, we've put numbers on our estimates and predictions, but these numbers are only as good as the model. Here are the main things to watch out for:

- ***Don't fit a linear regression to data that aren't straight.*** This is the most fundamental assumption. If the relationship between *x* and *y* isn't approximately linear, there's no sense in fitting a straight line to it.
- ***Watch out for the plot thickening.*** The common part of confidence and prediction intervals is the estimate of the error standard deviation, the spread around the line. If it changes with *x*, the estimate won't make sense. Imagine making a prediction interval for these data.

  When *x* is small, we can predict *y* precisely, but as *x* gets larger, it's much harder to pin *y* down. Unfortunately, if the spread changes, the single value of s_e won't pick that up. The prediction interval will use the average spread around the line, with the result that we'll be too pessimistic about our precision for low *x*-values and too optimistic for high *x*-values. A re-expression of *y* is often a good fix for changing spread.
- ***Make sure the errors are Normal.*** When we make a prediction interval for an individual, the Central Limit Theorem can't come to our rescue. For us to believe the prediction interval, the errors must be from the Normal model. Check the histogram and Normal probability plot of the residuals to see if this assumption looks reasonable.
- ***Watch out for extrapolation.*** It's tempting to think that because we have prediction *intervals*, they'll take care of all of our uncertainty so we don't have to worry about extrapolating. Wrong. The interval is only as good as the model. The uncertainty our intervals predict is only correct if our model is true. There's no way to adjust for wrong models. That's why it's always dangerous to predict for *x*-values that lie far from the center of the data.
- ***Watch out for high-influence points and outliers.*** We always have to be on the lookout for a few points that have undue influence on our estimated model—and regression is certainly no exception.
- ***Watch out for one-tailed tests.*** Because tests of hypotheses about regression coefficients are usually two-tailed, software packages report two-tailed P-values. If you are using software to conduct a one-tailed test about slope, you'll need to divide the reported P-value in half.

---
[7] Technically, it's a little more complicated, but it's very close to this.

CONNECTIONS

We would never consider a regression analysis without first making a scatterplot. And the aspects of scatterplots that we always look for relate directly to regression. We can't use regression methods unless the *form* of the relationship is linear. The *direction* of the relationship gives the sign of the regression slope. The *strength* in the plot is measured by the R^2 statistic, which gives the fraction of the variability of y accounted for by the regression model and is the square of the correlation. It's also reflected in the standard deviation of the residuals—which plays a central role in the inference calculations.

Regression inference is connected to just about every inference method we have seen for measured data. The assumption that the spread of data about the line is constant is essentially the same as the assumption of equal variances required for the pooled-t methods. Our use of all the residuals together to estimate their standard deviation is a form of pooling.

Inference for regression is closely related to inference for means, so your understanding of means transfers pretty directly to your understanding of regression. Here's a table that displays the similarities:

	Means	**Regression Slope**
Parameter	μ	β_1
Statistic	$\bar{y}$	b_1
Population spread estimate	$s_y = \sqrt{\dfrac{\sum(y - \bar{y})^2}{n - 1}}$	$s_e = \sqrt{\dfrac{\sum(y - \hat{y})^2}{n - 2}}$
Standard error of the statistic	$SE(\bar{y}) = \dfrac{s_y}{\sqrt{n}}$	$SE(b_1) = \dfrac{s_e}{s_x\sqrt{n - 1}}$
Test statistic	$\dfrac{\bar{y} - \mu_0}{SE(\bar{y})} \sim t_{n-1}$	$\dfrac{b_1 - \beta_1}{SE(b_1)} \sim t_{n-2}$
Margin of error	$ME = t^*_{n-1} \times SE(\bar{y})$	$ME = t^*_{n-2} \times SE(b_1)$

What have we learned?

In Chapters 7, 8, and 9, we learned to examine the relationship between two quantitative variables in a scatterplot, to summarize its strength with correlation, and to fit linear relationships by least squares regression. And we saw that these methods are particularly powerful and effective for modeling, predicting, and understanding these relationships.

Now we have completed our study of inference methods by applying them to these regression models. We've found that the same methods we used for means—Student's t-models—work for regression in much the same way as they did for means. And we've seen that although this makes the mechanics familiar, there are new conditions to check and a need for care in describing the hypotheses we test and the confidence intervals we construct.

- We've learned that under certain assumptions, the sampling distribution for the slope of a regression line can be modeled by a Student's t-model with $n - 2$ degrees of freedom.
- We've learned to check four conditions to verify those assumptions before we proceed with inference. We've learned the importance of checking these conditions in order, and we've seen that most of the checks can be made by graphing the data and the residuals with the methods we learned in Chapters 4, 5, and 8.

- We've learned to use the appropriate *t*-model to test a hypothesis about the slope. If the slope of our regression line is significantly different from zero, we have strong evidence that there is an association between the two variables.
- We've also learned to create and interpret a confidence interval for the true slope.
- And we've been reminded yet again never to mistake the presence of an association for proof of causation.

TERMS

Conditions for inference in regression (and checks for some of them)

- Straight Enough Condition for linearity. (Check that the scatterplot of *y* against *x* has linear form and that the scatterplot of residuals against predicted values has no obvious pattern.)
- Independence Assumption. (Think about the nature of the data. Check a residuals plot.)
- Does the Plot Thicken? Condition for constant variance? (Check that the scatterplot shows consistent spread across the range of the *x*-variable, and that the residuals plot has constant variance too. A common problem is increasing spread with increasing predicted values—the *plot thickens!*)
- Nearly Normal Condition for Normality of the residuals. (Check a histogram of the residuals.)

Residual standard deviation

The spread of the data around the regression line is measured with the residual standard deviation, s_e:

$$s_e = \sqrt{\frac{\sum(y - \hat{y})^2}{n - 2}} = \sqrt{\frac{\sum e^2}{n - 2}}.$$

t-test for the regression slope

When the assumptions are satisfied, we can perform a test for the slope coefficient. We usually test the null hypothesis that the true value of the slope is zero against the alternative that it is not. A zero slope would indicate a complete lack of linear relationship between *y* and *x*.

To test $H_0: \beta_1 = 0$ we find:

$$t = \frac{b_1 - 0}{SE(b_1)}$$

where

$$SE(b_1) = \frac{s_e}{\sqrt{n - 1}\, s_x}, \quad s_e = \sqrt{\frac{\sum(y - \hat{y})^2}{n - 2}},$$

n is the number of cases, and s_x is the standard deviation of the *x*-values. We find the P-value from the Student's *t*-model with $n - 2$ degrees of freedom.

Confidence interval for the regression slope

When the assumptions are satisfied, we can find a confidence interval for the slope parameter from $b_1 \pm t^*_{n-2} \times SE(b_1)$. The critical value, t^*_{n-2}, depends on the confidence interval specified and on Student's *t*-model with $n - 2$ degrees of freedom.

Confidence interval for a predicted mean value

Different samples will give different estimates of the regression model and, so, different predicted values for the same value of *x*. We find a confidence interval for the mean of these predicted values at a specified *x*-value, x_ν, as $\hat{y}_\nu \pm t^*_{n-2} \times SE(\hat{\mu}_\nu)$ where

$$SE(\hat{\mu}_\nu) = \sqrt{SE^2(b_1) \times (x_\nu - \bar{x})^2 + \frac{s_e^2}{n}}.$$

The critical value, t^*_{n-2}, depends on the specified confidence level and the Student's *t*-model with $n - 2$ degrees of freedom.

Prediction interval for an individual

Different samples will give different estimates of the regression model and, so, different predicted values for the same value of *x*. We can make a confidence interval to capture a certain percentage of the entire distribution of predicted values. This makes it much wider than the corresponding confidence interval for the mean. The confidence interval takes the form $\hat{y}_\nu \pm t^*_{n-2} \times SE(\hat{y}_\nu)$, where

$$SE(\hat{y}_\nu) = \sqrt{SE^2(b_1) \times (x_\nu - \bar{x})^2 + \frac{s_e^2}{n} + s_e^2}.$$

The critical value, t^*_{n-2}, depends on the specified confidence level and the Student's t-model with $n - 2$ degrees of freedom.

SKILLS

When you complete this lesson you should:

Think

- Understand that the "true" regression line does not fit the population data perfectly, but rather is an idealized summary of that data.

- Know how to examine your data and a scatterplot of y vs. x for violations of assumptions that would make inference for regression unwise or invalid.

- Know how to examine displays of the residuals from a regression to double-check that the conditions required for regression have been met. In particular, know how to judge linearity and constant variance from a scatterplot of residuals against predicted values. Know how to judge Normality from a histogram and Normal probability plot.

- Remember to be especially careful to check for failures of the Independence Assumption when working with data recorded over time. To search for patterns, examine scatterplots both of x against time and of the residuals against time.

Show

- Know how to test the standard hypothesis that the true regression slope is zero. Be able to state the null and alternative hypotheses. Know where to find the relevant numbers in standard computer regression output.

- Be able to find a confidence interval for the predicted mean of the y-values and a prediction interval for a particular y-value based on the summary statistics for x and the values reported in a standard regression output table. Know how to interpret these intervals in terms of the regression.

Tell

- Be able to summarize a regression in words. In particular, be able to state the meaning of the true regression slope, the standard error of the estimated slope, and the standard deviation of the errors.

- Be able to interpret the P-value of the t-statistic for the slope to test the standard null hypothesis and to interpret a confidence interval for the slope.

- Be able to interpret a prediction interval for a predicted y-value for a given x-value and a confidence interval for the mean of y-values for a given x-value.

Regression Analysis on the Computer

All statistics packages make a table of results for a regression. These tables differ slightly from one package to another, but all are essentially the same. We've seen two examples of such tables already.

All packages offer analyses of the residuals. With some, you must request plots of the residuals as you request the regression. Others let you find the regression first and then analyze the residuals afterward. Either way, your analysis is not complete if you don't check the residuals with a histogram or Normal probability plot and a scatterplot of the residuals against *x* or the predicted values.

You should, of course, always look at the scatterplot of your two variables (make a picture, make a picture, . . .) before computing a regression. In Chapter 8 we showed how to go conveniently from a scatterplot to regression for many packages, but some of those regressions gave only the regression equation and not the *t*-statistics or P-values.

Here we show the direct path from the data to a regression table. Some packages, however, don't generate the plots you need automatically when you just ask for a regression.

Can we trust you to ask for the scatterplots yourself? (We knew we could.)

Regressions are almost always found with a computer or calculator. The calculations are too long to do conveniently by hand for data sets of any reasonable size. No matter how the regression is computed, the results are usually presented in a table that has a standard form. Here's a portion of a typical regression results table along with annotations showing where the numbers come from:

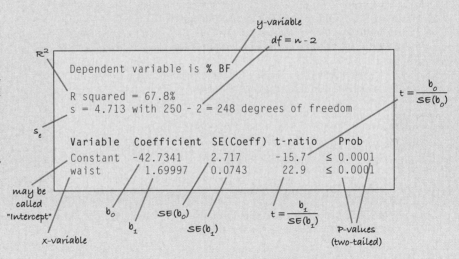

The regression table gives the coefficients (once you find them in the middle of all this other information), so we can see that the regression equation is

$$\widehat{\%BF} = -42.73 + 1.7 \ waist$$

and that the R^2 for the regression is 67.8%. (Is accounting for 68% of the variation in *%body fat* good enough to be useful? Is a prediction *ME* of more than 9% good enough? Health professionals might not be satisfied.)

The column of *t*-ratios gives the test statistics for the respective null hypotheses that the true values of the coefficients are zero. The corresponding P-values are also usually reported.

DATA DESK

- Select *Y*- and *X*-variables.
- From the **Calc** menu, choose **Regression.**
- Data Desk displays the regression table.
- Select plots of residuals from the Regression table's HyperView menu.

Comments

You can change the regression by dragging the icon of another variable over either the *Y*- or *X*-variable name in the table and dropping it there. The regression will re-compute automatically.

EXCEL

- From the **Tools** menu, select Data Analysis.
- Select Regression from the **Analysis Tools** list.
- Click the **OK** button.
- Enter the data range holding the Y-variable in the box labeled "Y-range".
- Enter the range of cells holding the X-variable in the box labeled "X-range."
- Select the **New Worksheet Ply** option.
  Select **Residuals** options. Click the **OK** button.

Comments

The Y and X ranges do not need to be in the same rows of the spreadsheet, although they must cover the same number of cells. But it is a good idea to arrange your data in parallel columns as in a data table.

Although the dialog offers a Normal probability plot of the residuals, the data analysis add-in does not make a correct probability plot, so don't use this option.

JMP

- From the **Analyze** menu select **Fit Y by X.**
- Select variables: a Y, Response variable, and an X, Factor variable. Both must be continuous (quantitative).
- JMP makes a scatterplot.
- Click on the red triangle beside the heading labeled **Bivariate Fit...** and choose **Fit Line.** JMP draws the least squares regression line on the scatterplot and displays the results of the regression in tables below the plot.
- The portion of the table labeled "Parameter Estimates" gives the coefficients, their standard errors, *t*-ratios, and P-values.

Comments

JMP chooses a regression analysis when both variables are "Continuous." If you get a different analysis, check the variable types.

The Parameter table does not include the residual standard deviation s_e. You can find that as Root Mean Square Error in the Summary of Fit panel of the output.

MINITAB

- Choose **Regression** from the **Stat** menu.
- Choose **Regression...** from the **Regression** submenu.
- In the Regression dialog, assign the Y-variable to the Response box and assign the X-variable to the Predictors box.
- Click the **Graphs** button.
- In the Regression-Graphs dialog, select **Standardized residuals,** and check **Normal plot of residuals** and **Residuals versus fits.**
- Click the **OK** button to return to the Regression dialog.
- Click the **OK** button to compute the regression.

Comments

You can also start by choosing a Fitted Line plot from the **Regression** submenu to see the scatterplot first—usually good practice.

SPSS

- Choose **Regression** from the **Analyze** menu.
- Choose **Linear** from the **Regression** submenu.
- In the Linear Regression dialog that appears, select the Y-variable and move it to the dependent target. Then move the X-variable to the independent target.
- Click the **Plots** button.

- In the Linear Regression Plots dialog, choose to plot the *SRESIDs against the *ZPRED values.
- Click the **Continue** button to return to the Linear Regression dialog.
- Click the **OK** button to compute the regression.

TI-83/84 Plus

Under **STAT TESTS** choose **E.LinRegTTest.** Specify the two lists where the data are stored and (usually) choose the two-tail option. In addition to reporting the calculated value of *t* and the P-value, the calculator will tell you the coefficients of the regression equation (*a* and *b*), the values of r^2 and *r*, and value of *s* you need for confidence or prediction intervals.

TI-89

Under **STAT Tests** choose **A:LinRegTTest.** Specify the two lists where the data are stored and (usually) choose the two-tail option. Select an equation name to store the resulting line. In addition to reporting the calculated value of *t* and the P-value, the calculator will tell you the coefficients of the regression equation (*a* and *b*), the values of r^2 and *r*, the value of *s* used in prediction and confidence intervals, and the standard error of the slope. For 95% prediction and confidence intervals, choose **7:LinRegTint** from the **STAT Ints** menu. Specify the two lists where the data are stored and select an equation name to store the resulting line. Select for an interval for the slope or for a response. If for a response, enter the *x*-value.

EXERCISES

T 1. Ms. President? In Chapter 7 we saw data collected by the Gallup organization. They have, over six decades, periodically asked the following question:

If your party nominated a generally well-qualified person for president who happened to be a woman, would you vote for that person?

Here is a scatterplot of the percentage answering "yes" vs. the year of the century (37 = 1937):

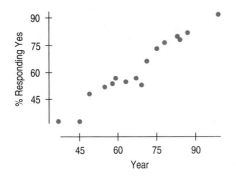

In Chapter 7 we could describe the relationship only in general terms. Now we can learn more. Here is the regression analysis:

Dependent variable is: Yes
R-squared = 94.2%
s = 4.274 with 16 − 2 = 14 degrees of freedom

Variable	Coefficient	SE(Coeff)	t-ratio	P-value
Intercept	−5.58269	4.582	−1.22	0.2432
Year	0.999373	0.0661	15.1	<0.0001

a) Explain in words and numbers what the regression says.
b) State the hypothesis about the slope (both numerically and in words) that describes how voters' thoughts have changed about voting for a woman.
c) Assuming that the assumptions for inference are satisfied, perform the hypothesis test and state your conclusion. Be sure to state it in terms of voters' opinions.
d) Explain what the *R*-squared in this regression means.

T 2. Drug use. The *European School Study Project on Alcohol and Other Drugs,* published in 1995, investigated the use of marijuana and other drugs. Data from 11 countries are summarized in the scatterplot and regression analysis below. They show the association between the percentage of a country's ninth graders who report having smoked marijuana and who have used other drugs such as LSD, amphetamines, and cocaine.

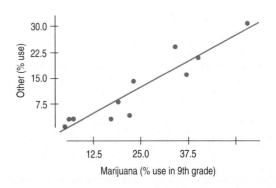

Dependent variable is: Other
R-squared = 87.3%
s = 3.853 with 11 − 2 = 9 degrees of freedom

Variable	Coefficient	SE(Coeff)	t-ratio	P-value
Intercept	−3.06780	2.204	−1.39	0.1974
Marijuana	0.615003	0.0784	7.85	<0.0001

a) Explain in words and numbers what the regression says.
b) State the hypothesis about the slope (both numerically and in words) that describes how use of marijuana is associated with other drugs.
c) Assuming that the assumptions for inference are satisfied, perform the hypothesis test and state your conclusion in context.
d) Explain what the *R*-squared in this regression means.
e) Do these results indicate that marijuana use leads to the use of harder drugs? Explain.

T 3. No opinion. Here's a regression of the percentage of respondents whose response to the question about voting for a woman president was "no opinion." We wonder if the percentage of the public who have no opinion on this issue has changed over the years. Assume that the conditions for inference are satisfied.

Dependent variable is: No Opinion
R-squared = 9.5%
s = 2.280 with 16 − 2 = 14 degrees of freedom

Variable	Coefficient	SE(Coeff)	t-ratio	P-value
Intercept	7.69262	2.445	3.15	0.0071
Year	−0.042708	0.0353	−1.21	0.2458

a) State the appropriate hypothesis for the slope.
b) Test your hypothesis and state your conclusion in the proper context.
c) Below is the scatterplot corresponding to the regression for No Opinion. How does the scatterplot change your opinion of the trend in "no opinion" responses? Do you think the true slope is negative? Does this change the conclusion of your hypothesis test of part b? Explain.

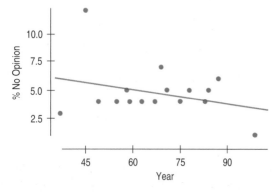

T 4. Cholesterol. Does a person's cholesterol level tend to increase with age? Data collected in Framingham, MA,

from 294 adults aged 45 to 62 produced the regression analysis shown. Assuming that the data satisfy the conditions for inference, examine the association between age and cholesterol level.

Dependent variable is: Chol

Variable	Coefficient	SE(Coeff)	t-ratio	P-value
Intercept	196.619	33.21	5.92	≤0.0001
Age	0.745779	0.6075	1.23	0.2206

a) State the appropriate hypothesis for the slope.
b) Test your hypothesis and state your conclusion in the proper context.

T 5. Marriage age. The scatterplot suggests a decrease in the difference in ages at first marriage for men and women since 1975. We want to examine the regression to see if this decrease is significant.

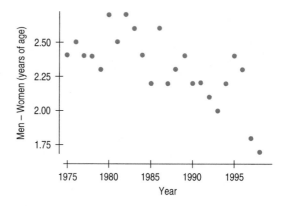

Dependent variable is: Men − Women
R-squared = 46.3%
s = 0.1866 with 24 − 2 = 22 degrees of freedom

Variable	Coefficient	SE(Coeff)	t-ratio	P-value
Intercept	49.9021	10.93	4.56	0.0002
Year	−0.023957	0.0055	−4.35	0.0003

a) Write appropriate hypotheses.
b) Here are the residuals plot and a histogram of the residuals. Do you think the conditions for inference are satisfied? Explain.

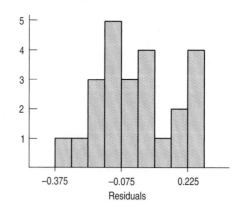

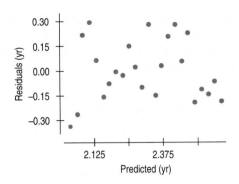

c) Test the hypothesis and state your conclusion about the trend in age at first marriage.

T 6. Used cars. Classified ads in a newspaper offered several used Toyota Corollas for sale. Listed below are the ages of the cars and the advertised prices.

Age (yr)	Prices Advertised ($)
1	12,995; 10,950
2	10,495
3	10,995; 10,995
4	6,995; 7,990
5	8,700; 6,995
6	5,990; 4,995
9	3,200; 2,250; 3,995
11	2,900; 2,995
13	1,750

a) Make a scatterplot for these data.
b) Do you think a linear model is appropriate? Explain.
c) Find the equation of the regression line.
d) Check the residuals to see if the conditions for inference are met.

T 7. Marriage age, again. Based on the analysis of marriage ages since 1975 given in Exercise 5, give a 95% confidence interval for the rate at which the age gap is closing. Clearly explain what your confidence interval means.

T 8. Used cars, again. Based on the analysis of used car prices you did for Exercise 6, create a 95% confidence interval for the slope of the regression line, and explain what your interval means in context.

T 9. Fuel economy. A consumer organization has reported test data for 50 car models. We will examine the association between the weight of the car (in thousands of pounds) and the fuel efficiency (in miles per gallon). Shown are the summary statistics, scatterplot, and regression analysis:

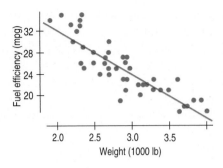

Variable	Count	Mean	StdDev
MPG	50	25.0200	4.83394
wt/1000	50	2.88780	0.511656

Dependent variable is: MPG
R-squared = 75.6%
s = 2.413 with 50 − 2 = 48 df

Variable	Coefficient	SE(Coeff)	t-ratio	P-value
Intercept	48.7393	1.976	24.7	≤0.0001
Weight	−8.21362	0.6738	−12.2	≤0.0001

a) Is there strong evidence of an association between the weight of a car and its gas mileage? Write an appropriate hypothesis.
b) Are the assumptions for regression satisfied?

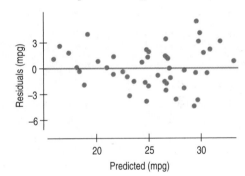

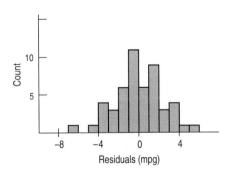

c) Test your hypothesis and state your conclusion.

T 10. SAT scores. How strong is the association between student scores on the Math and Verbal sections of the SAT? Scores on this exam range from 200 to 800, and are widely used by college admissions offices. Here are summaries and plots of the scores for a recent graduating class at Ithaca High School.

Variable	Count	Mean	Median	StdDev	Range	IntQRange
Verbal	162	596.296	610	99.5199	490	140
Math	162	612.099	630	98.1343	440	150

Dependent variable is: Math
R-squared = 46.9%
s = 71.75 with 162 − 2 = 160 df

Variable	Coefficient	SE(Coeff)	t-ratio	P-value
Intercept	209.554	34.35	6.10	≤0.0001
Verbal	0.675075	0.0568	11.9	≤0.0001

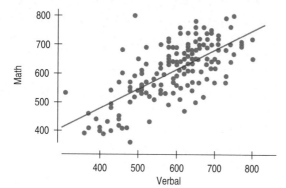

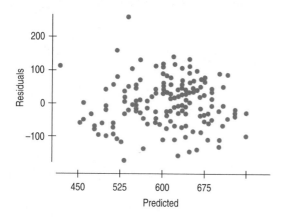

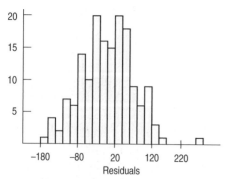

a) Is there evidence of an association between Math and Verbal scores? Write an appropriate hypothesis.
b) Discuss the assumptions for inference.

c) Test your hypothesis and state an appropriate conclusion.

11. Fuel economy, part II. Consider again the data in Exercise 9 about the gas mileage and weights of cars.
   a) Create a 95% confidence interval for the slope of the regression line.
   b) Explain in this context what your confidence interval means.

12. SATs, part II. Consider the high school SAT scores data from Exercise 10.
   a) Find a 90% confidence interval for the slope of the true line describing the association between Math and Verbal scores.
   b) Explain in this context what your confidence interval means.

13. MPG revisited. Consider again the data in Exercise 9 about the gas mileage and weights of cars.
   a) Create a 95% confidence interval for the average fuel efficiency among cars weighing 2500 pounds, and explain what your interval means.
   b) Create a 95% prediction interval for the gas mileage you might get driving your new 3450-pound SUV, and explain what that interval means.

14. SATs again. Consider the high-school SAT scores data from Exercise 10 once more.
   a) Find a 90% confidence interval for the mean SAT-Math score for all students with an SAT-Verbal score of 500.
   b) Find a 90% prediction interval for the Math score of the senior class president, if you know she scored 710 on the Verbal section.

15. Cereal. A healthy cereal should be low in both calories and sodium. Data for 77 cereals were examined and judged acceptable for inference. The 77 cereals had between 50 and 160 calories per serving and between 0 and 320 mg of sodium per serving. The regression analysis is shown.

Dependent variable is: Sodium
R-squared = 9.0%
s = 80.49 with 77 − 2 = 75 degrees of freedom

Variable	Coefficient	SE(Coeff)	t-ratio	P-value
Intercept	21.4143	51.47	0.416	0.6786
Calories	1.29357	0.4738	2.73	0.0079

a) Is there an association between the number of calories and the sodium content of cereals? Explain.
b) Do you think this association is strong enough to be useful? Explain.

16. Brain size. Does your IQ depend on the size of your brain? A group of female college students took a test that measured their verbal IQs and also underwent an MRI scan to measure the size of their brains (in 1000s of pixels). The scatterplot and regression analysis are shown, and the assumptions for inference were satisfied.

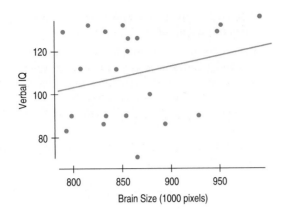

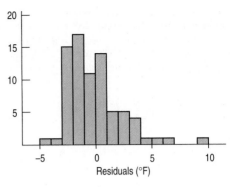

Dependent variable is: IQ_Verbal
R-squared = 6.5%

Variable	Coefficient	SE(Coeff)
Intercept	24.1835	76.38
Size	0.098842	0.0884

a) Test an appropriate hypothesis about the association between brain size and IQ.
b) State your conclusion about the strength of this association.

T 17. Another bowl. Further analysis of the data for the breakfast cereals in Exercise 15 looked for an association between *fiber* content and *calories* by attempting to construct a linear model. Several graphs are shown. Which of the assumptions for inference are violated? Explain.

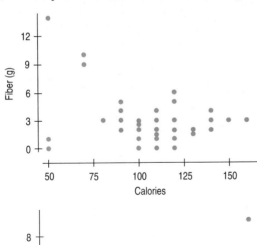

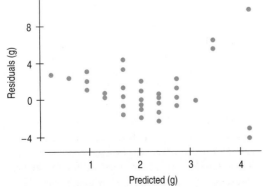

T 18. Winter. The output shows an attempt to model the association between average *January temperature* (in degrees Fahrenheit) and *latitude* (in degrees north of the equator) for 59 U.S. cities. Which of the assumptions for inference do you think are violated? Explain.

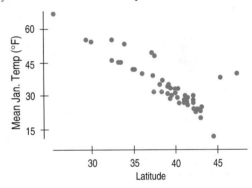

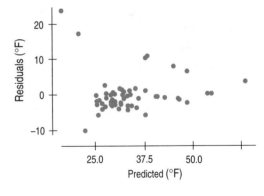

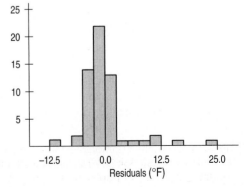

T 19. Acid rain. Biologists studying the effects of acid rain on wildlife collected data from 163 streams in the Adirondack Mountains. They recorded the *pH* (acidity) of the water and the *BCI*, a measure of biological diversity. Following is a scatterplot of *BCI* against *pH*.

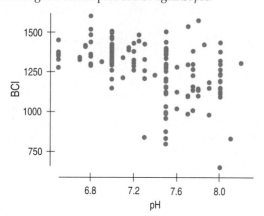

And here is part of the regression analysis:

Dependent variable is: BCI
R-squared = 27.1%
s = 140.4 with 163 − 2 = 161 degrees of freedom

Variable	Coefficient	SE(Coeff)
Intercept	2733.37	187.9
pH	−197.694	25.57

a) State the null and alternative hypotheses under investigation.
b) Assuming that the assumptions for regression inference are reasonable, find the *t* and P-value for the test.
c) State your conclusion.

T 20. El Niño. Concern over the weather associated with El Niño has increased interest in the possibility that the climate on earth is getting warmer. The most common theory relates an increase in atmospheric levels of carbon dioxide (CO_2), a greenhouse gas, to increases in temperature. Here is part of a regression analysis of the mean annual CO_2 concentration in the atmosphere measured in parts per million (ppm) at the top of Mauna Loa in Hawaii and the mean annual air temperature over both land and sea across the globe, in degrees Celsius. The scatterplots and residuals plots indicated that the data were appropriate for inference.

Dependent variable is: Temp
R-squared = 33.4%
s = 0.0809 with 37 − 2 = 35 degrees of freedom

Variable	Coefficient	SE(Coeff)
Intercept	15.3066	0.3139
CO2	0.004	0.0009

a) Write the equation of the regression line.
b) Is there evidence of an association between CO_2 level and global temperature?

c) Do you think predictions made by this regression will be very accurate? Explain.

21. Ozone. The Environmental Protection Agency is examining the relationship between the ozone level (in parts per million) and the population (in millions) of U.S. cities. Part of the regression analysis is shown.

Dependent variable is: Ozone
R-squared = 84.4%
s = 5.454 with 16 − 2 = 14 df

Variable	Coefficient	SE(Coeff)
Intercept	18.892	2.395
Pop	6.650	1.910

a) We suspect that the greater the population of a city the higher its ozone level. Is the relationship significant? Assuming the conditions for inference are satisfied, test an appropriate hypothesis and state your conclusion in context.
b) Do you think that the population of a city is a useful predictor of ozone level? Use the values of both R^2 and *s* in your explanation.

T 22. Sales and profits. A business analyst was interested in the relationship between a company's sales and its profits. She collected data (in millions of dollars) from a random sample of Fortune 500 companies, and created the regression analysis and summary statistics shown. The assumptions for regression inference appeared to be satisfied.

	Profits	Sales	Dependent variable is: Profits
Count	79	79	R-squared = 66.2% s = 466.2
Mean	209.839	4178.29	**Variable Coefficient SE(Coeff)**
Variance	635,172	49,163,000	Intercept −176.644 61.16
Std Dev	796.977	7011.63	Sales 0.092498 0.0075

a) Is there a significant association between sales and profits? Test an appropriate hypothesis and state your conclusion in context.
b) Do you think that a company's sales serve as a useful predictor of their profits? Use the values of both R^2 and *s* in your explanation.

23. Ozone, again. Consider again the relationship between the population and ozone level of U.S. cities that you analyzed in Exercise 21.
a) Give a 90% confidence interval for the approximate increase in ozone level associated with each additional million city inhabitants.
b) For the cities studied, the mean population was 1.7 million people. The population of Boston is approximately 0.6 million people. Predict the mean ozone level for cities of that size with an interval in which you have 90% confidence.

T 24. More sales and profits. Consider again the relationship between the sales and profits of Fortune 500 companies that you analyzed in Exercise 22.
a) Find a 95% confidence interval for the slope of the regression line. Interpret your interval in context.

b) Last year the drug manufacturer Eli Lilly, Inc. reported gross sales of $9 billion (that's $9,000 million). Create a 95% prediction interval for the company's profits, and interpret your interval in context.

Ⓣ 25. Start the car! In October 2002, *Consumer Reports* listed the price (in dollars) and power (in cold cranking amps) of auto batteries. We want to know if more expensive batteries are generally better in terms of starting power. Here are several software displays.

Dependent variable is: Power
R-squared = 25.2%
s = 116.0 with 33 − 2 = 31 degrees of freedom

Variable	Coefficient	SE(Coeff)	t-ratio	P-value
Intercept	384.594	93.55	4.11	0.0003
Cost	4.14649	1.282	3.23	0.0029

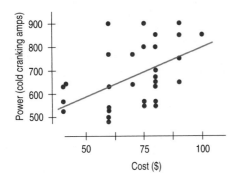

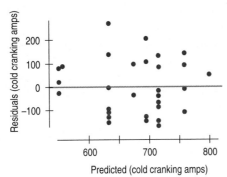

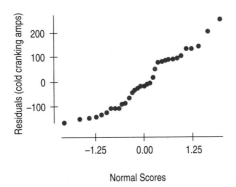

a) How many batteries were tested?
b) Are the conditions for inference satisfied? Explain.

c) Is there evidence of an association between the cost and cranking power of auto batteries? Test an appropriate hypothesis and state your conclusion.
d) Is the association strong? Explain.
e) What is the equation of the regression line?
f) Create a 90% confidence interval for the slope of the true line.
g) Interpret your interval in this context.

Ⓣ 26. Crawling. Researchers at the University of Denver Infant Study Center wondered if temperature might influence the age at which babies learn to crawl. Perhaps the extra clothing that babies wear in cold weather would restrict movement and delay the age at which they started crawling. Data were collected on 208 boys and 206 girls. Parents reported the month of the baby's birth and the age (in weeks) at which their child first crawled. The table gives the average *temperature* (°F) when the babies were 6 months old and average *crawling age* (in weeks) for each month of the year. Make the plots and compute the analyses necessary to answer the following questions.

Birth Month	6-Month Temperature	Average Crawling Age
Jan.	66	29.84
Feb.	73	30.52
Mar.	72	29.70
April	63	31.84
May	52	28.58
June	39	31.44
July	33	33.64
Aug.	30	32.82
Sept.	33	33.83
Oct.	37	33.35
Nov.	48	33.38
Dec.	57	32.32

a) Would this association appear to be weaker, stronger, or the same if data had been plotted for individual babies instead of using monthly averages? Explain.
b) Is there evidence of an association between *temperature* and *crawling age*? Test an appropriate hypothesis and state your conclusion. Don't forget to check the assumptions.
c) Create and interpret a 95% confidence interval for the slope of the true relationship.

Ⓣ 27. Printers. In March 2002, *Consumer Reports* reviewed several models of inkjet printers. Shown are the speed of the printer (in pages per minute) and the cost per page printed. Is there evidence of an association between *speed* and *cost*? Test an appropriate hypothesis and state your conclusion.

Speed (ppm)	Cost (cents/page)
4.6	12.0
5.5	8.5
4.5	6.2
3.8	3.4
4.6	2.6
3.7	4.0
4.7	5.8
4.7	8.1
4.0	9.4
3.1	14.9
1.9	2.6
2.2	4.3
1.8	4.6
2.0	14.8
2.0	4.4

Waist (in.)	Weight (lb)	Body Fat (%)	Waist (in.)	Weight (lb)	Body Fat (%)
32	175	6	33	188	10
36	181	21	40	240	20
38	200	15	36	175	22
33	159	6	32	168	9
39	196	22	44	246	38
40	192	31	33	160	10
41	205	32	41	215	27
35	173	21	34	159	12
38	187	25	34	146	10
38	188	30	44	219	28

T 30. Body fat, again. Use these data to examine the association between weight and *%body fat*.
a) Find a 90% confidence interval for the slope of the regression line of *%body fat* on weight.
b) Interpret your interval in context.
c) Give a 95% prediction interval for the *%body fat* of an individual who weighs 165 pounds.

T 31. Grades. The data set below shows midterm scores from an Introductory Statistics course.

T 28. Strike two. Remember the Little League instructional video discussed in Chapter 25? Ads claimed that the techniques would improve the performances of Little League pitchers. To test this claim, 20 Little Leaguers threw 50 pitches each, and we recorded the number of strikes. After the players participated in the training program, we repeated the test. The table shows the number of strikes each player threw before and after the training. A test of paired differences failed to show that this training was effective in improving a player's ability to throw strikes. Is there any evidence that the effectiveness of the video depends on the player's initial ability to throw strikes? Test an appropriate hypothesis and state your conclusion.

Number of Strikes (out of 50)

Before	After	Before	After
28	35	33	33
29	36	33	35
30	32	34	32
32	28	34	30
32	30	34	33
32	31	35	34
32	32	36	37
32	34	36	33
32	35	37	35
33	36	37	32

T 29. Body fat. Do the data shown in the table indicate an association between *waist* size and *%body fat*?
a) Test an appropriate hypothesis and state your conclusion.
b) Give a 95% confidence interval for the mean *%body fat* found in people with 40-inch *waists*.

First Name	Midterm 1	Midterm 2	Homework
Timothy	82	30	61
Karen	96	68	72
Verena	57	82	69
Jonathan	89	92	84
Elizabeth	88	86	84
Patrick	93	81	71
Julia	90	83	79
Thomas	83	21	51
Marshall	59	62	58
Justin	89	57	79
Alexandra	83	86	78
Christopher	95	75	77
Justin	81	66	66
Miguel	86	63	74
Brian	81	86	76
Gregory	81	87	75
Kristina	98	96	84
Timothy	50	27	20
Jason	91	83	71
Whitney	87	89	85
Alexis	90	91	68
Nicholas	95	82	68
Amandeep	91	37	54
Irena	93	81	82
Yvon	88	66	82
Sara	99	90	77
Annie	89	92	68

First Name	Midterm 1	Midterm 2	Homework
Benjamin	87	62	72
David	92	66	78
Josef	62	43	56
Rebecca	93	87	80
Joshua	95	93	87
Ian	93	65	66
Katharine	92	98	77
Emily	91	95	83
Brian	92	80	82
Shad	61	58	65
Michael	55	65	51
Israel	76	88	67
Iris	63	62	67
Mark	89	66	72
Peter	91	42	66
Catherine	90	85	78
Christina	75	62	72
Enrique	75	46	72
Sarah	91	65	77
Thomas	84	70	70
Sonya	94	92	81
Michael	93	78	72
Wesley	91	58	66
Mark	91	61	79
Adam	89	86	62
Jared	98	92	83
Michael	96	51	83
Kathryn	95	95	87
Nicole	98	89	77
Wayne	89	79	44
Elizabeth	93	89	73
John	74	64	72
Valentin	97	96	80
David	94	90	88
Marc	81	89	62
Samuel	94	85	76
Brooke	92	90	86

a) Fit a model predicting the second midterm score from the first.
b) Comment on the model you found, including a discussion of the assumptions and conditions for regression. Is the coefficient for the slope statistically significant?
c) A student comments that because the P-value for the slope is very small, Midterm 2 is very well predicted from Midterm 1. So, he reasons, next term the professor can give just one midterm. What do you think?

T 32. Grades? The professor teaching the Introductory Statistics class discussed in Exercise 31 wonders whether performance on homework can accurately predict midterm scores.

a) To investigate it, she fits a regression of the sum of the two midterms scores on homework scores. Fit the regression model.
b) Comment on the model including a discussion of the assumptions and conditions for regression. Is the coefficient for the slope "statistically significant"?
c) Do you think she can accurately judge a student's performance without giving the midterms? Explain.

T 33. Swimming Ontario. Between 1974 and 2003, swimmers have crossed Lake Ontario 37 times. Have swimmers been getting faster or slower, or has there been no change in swimming speed? A regression predicting time for the swim (in minutes) against year of the swim looks like this:

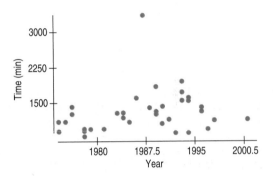

Dependent variable is: Time
R squared = 7.0%
s = 447.3 with 37 − 2 = 35 degrees of freedom

Variable	Coefficient	SE(Coeff)	t-ratio	P-value
Intercept	−29161.9	18830	−1.55	0.1305
Year	15.3323	9.479	1.62	0.1147

a) What do you hope to learn from these data? Define the variables and state the null and alternative hypotheses about the slope.
b) Check assumptions and conditions.
c) If they are satisfied, complete the analysis.
d) The longest time is for a swim in which Vicki Kieth swam round trip, crossing the lake twice. If we remove it from the data, the new analysis is shown. Check the conditions again.

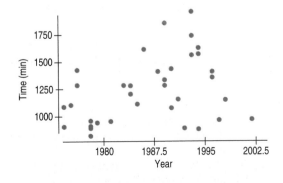

Dependent variable is: Time
R squared = 15.9%
s = 277.2 with 36 − 2 = 34 degrees of freedom

Variable	Coefficient	SE(Coeff)	t-ratio	P-value
Constant	−28399.9	11671	−2.43	0.0204
Year	14.9198	5.875	2.54	0.0158

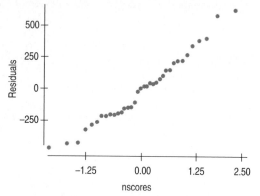

e) If the conditions are satisfied, complete the analysis.
f) The slope itself didn't change very much when we removed the outlier, but the P-value did. Consider the formula for the *t*-ratio and explain what changed.

T 34. All the efficiency money can buy. A sample of 84 model 2004 cars from an online information service was examined to see how fuel efficiency (as highway mpg) relates to the cost (Manufacturer's Suggested Retail Price in dollars) of cars. Here are displays and computer output:

Dependent variable is: Highway MPG
R squared = 30.1%
s = 5.298 with 84 − 2 = 82 degrees of freedom

Variable	Coefficient	SE (Coeff)	t-ratio	P-value
Constant	33.0581	1.299	25.5	≤0.0001
MSRP	−2.16543e-4	0.0000	−5.95	≤0.0001

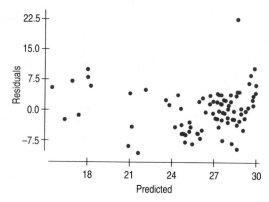

a) State what you want to know, identify the variables, and give the appropriate hypotheses.
b) Check the assumptions and conditions.
c) If the conditions are met, complete the analysis.

T 35. Education and mortality. The software output below is based on the mortality rate (deaths per 100,000 people) and the education level (average number of years in school) for 58 U.S. cities.

Variable	Count	Mean	StdDev
Mortality	58	942.501	61.8490
Education	58	11.0328	0.793480

Dependent variable is: Mortality
R-squared = 41.0%
s = 47.92 with 58 − 2 = 56 degrees of freedom

Variable	Coefficient	SE(Coeff)
Intercept	1493.26	88.48
Education	−49.9202	8.000

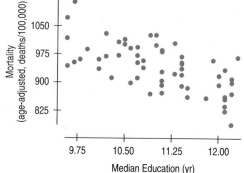

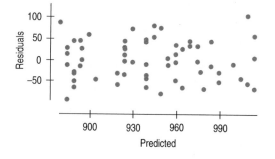

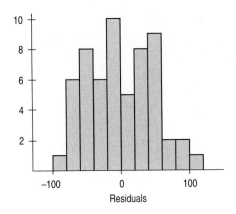

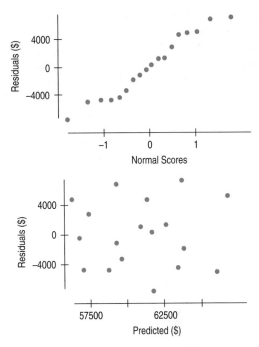

a) Comment on the assumptions for inference.

b) Is there evidence of a strong association between the level of *education* in a city and the *mortality* rate? Test an appropriate hypothesis and state your conclusion.

c) Can we conclude that getting more education is likely (on average) to prolong your life? Why or why not?

d) Find a 95% confidence interval for the slope of the true relationship.

e) Explain what your interval means.

f) Find a 95% confidence interval for the average *mortality* rate in cities where the adult population completed an average of 12 years of school.

🝔 **36. Property assessments.** The software outputs below provide information about the size (in square feet) of 18 homes in Ithaca, NY, and the city's assessed value of those homes.

Variable	Count	Mean	StdDev	Range
SqFt	18	2003.39	264.727	890
Asse$$	18	60946.7	5527.62	19710

Dependent variable is: Asse$$
R-squared = 32.5%
s = 4682 with 18 − 2 = 16 degrees of freedom

Variable	Coefficient	SE(Coeff)
Intercept	37108.8	8664
SqFt	11.8987	4.290

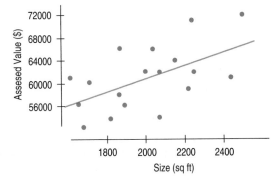

a) Explain why inference for linear regression is appropriate with these data.

b) Is there a significant association between the *size* of a home and its *assessed value*? Test an appropriate hypothesis and state your conclusion.

c) What percentage of the variability in assessed value is explained by this regression?

d) Give a 90% confidence interval for the slope of the true regression line, and explain its meaning in the proper context.

e) From this analysis, can we conclude that adding a room to your house will increase its assessed value? Why or why not?

f) The owner of a home measuring 2100 square feet files an appeal, claiming that the $70,200 assessed value is too high. Do you agree? Explain your reasoning.

just checking

Answers

1. A high *t*-ratio of 3.27 indicates that the slope is different from zero—that is, that there is a linear relationship between height and mouth size. The small P-value says that a slope this large would be very unlikely to occur by chance if, in fact, there was no linear relationship between the variables.

2. Not really. The R^2 for this regression is only 15.3%, so height doesn't account for very much of the variability in mouth size.

3. The value of *s* tells the standard deviation of the residuals. Mouth sizes have a mean of 60.3 cubic centimeters. A standard deviation of 15.7 in the residuals indicates that the errors made by this regression model can be quite large relative to what we are estimating. Errors of 15 to 30 cubic centimeters would be common.

QUICK REVIEW

With these last two chapters you have added important analytical tools to your ways of looking at data. Here's a brief summary of those key concepts and skills, as well as an overview of statistical inference:

▶ Inferences about distributions of counts use chi-square models.

- To see if an observed distribution is consistent with a proposed model, use a goodness-of-fit test.

- To see if two or more observed distributions could have arisen from populations with the same model use a test of homogeneity.

▶ Inference about association between two variables tests the hypothesis that it is plausible to consider the variables independent.

- If the variables are categorical, display the data in a contingency table and use a chi-square test of independence.

- If the variables are quantitative, display them with a scatterplot. You may use a linear regression *t*-test if there appears to be a linear association for which the residuals are random, consistent in terms of spread, and approximately Normal.

▶ You can now use statistical inference to answer questions about means, proportions, distributions, and associations.

- No inference procedure is valid unless the underlying assumptions are true. Always check the conditions before proceeding. Many of those checks should be made by examining a graph.

- You can make inferences about a single proportion or the difference of two proportions using Normal models.

- You can make inferences about one mean, the difference of two independent means, or the mean of paired differences using *t*-models.

- You can make inferences about distributions using chi-square models.

- You can make inferences about association between categorical variables using chi-square models.

- You can make inferences about linear association between quantitative variables using *t*-models.

If you look back at where we've been in this book, you'll see that statistical inference relies on almost everything we've seen. In Chapters 12 and 13 we learned techniques of collecting data using randomization—that's what makes inference possible at all. In Chapters 3, 4, and 7 we learned to plot our data and to look for the patterns and relationships we use to check the conditions that allow inference. In Chapters 3, 5, and 8 we learned about the summary statistics we use to do the mechanics of inference. We use our knowledge of randomness and probability from Chapters 11, 14, and 15 to help us think clearly about uncertainty, and the probability models of Chapters 6, 16, and 17 to measure our uncertainty precisely. Ultimately, the Central Limit Theorem of Chapter 18 makes all of inference possible.

Remember (have we said this often enough yet?): Never use any inference procedure without first checking the assumptions and conditions. On the next page we summarize the new types of inference procedures, the corresponding formulas, and the assumptions and conditions. You'll find complete summaries for all our inference procedures inside the back cover of the book. Have a look. Then you'll be ready for more opportunities to practice using these concepts and skills . . . .

Quick Guide to Inference

Think			Show				Tell?
Inference about?	One group or two?	Procedure	Model	Parameter	Estimate	SE	Chapter
Distributions (one categorical variable)	One sample	Goodness-of-Fit	χ^2 $df = cells - 1$			$\sum \dfrac{(\text{obs} - \text{exp})^2}{\text{exp}}$	26
	Many independent groups	Homogeneity χ^2 Test	χ^2 $df = (r-1)(c-1)$				
Independence (two categorical variables)	One sample	Independence χ^2 Test					
Association (two quantitative variables)	One sample	Linear Regression t-Test or Confidence Interval for β	t $df = n - 2$	β_1	b_1	$\dfrac{s_e}{s_x \sqrt{n-1}}$ (compute with technology)	27
		Confidence Interval for μ_ν		μ_ν	$\hat{y}_\nu$	$\sqrt{SE^2(b_1) \cdot (x_\nu - \bar{x})^2 + \dfrac{s_e^2}{n}}$	
		Prediction Interval for y_ν		y_ν	$\hat{y}_\nu$	$\sqrt{SE^2(b_1) \cdot (x_\nu - \bar{x})^2 + \dfrac{s_e^2}{n} + s_e^2}$	

Assumptions for Inference   And the Conditions That Support or Override Them

Distributions/Association (χ^2)

- **Goodness-of-fit** (df = # of cells − 1; one variable, one sample compared with population model)
  1. Data are counts.
  2. Data in sample are independent.
  3. Sample is sufficiently large.

  1. (Are they?)
  2. SRS and $n < 10\%$ of the population.
  3. All expected counts ≥ 5.

- **Homogeneity** [df = $(r-1)(c-1)$; samples from many populations compared on one variable]
  1. Data are counts.
  2. Data in groups are independent.
  3. Groups are sufficiently large.

  1. (Are they?)
  2. SRSs and $n < 10\%$ OR random allocation.
  3. All expected counts ≥ 5.

- **Independence** [df = $(r-1)(c-1)$; sample from one population classified on two variables]
  1. Data are counts.
  2. Data are independent.
  3. Sample is sufficiently large.

  1. (Are they?)
  2. SRSs and $n < 10\%$ of the population.
  3. All expected counts ≥ 5.

Regression (t, df $= n - 2$)

- **Association** between two quantitative variables ($\beta = 0$?)
  1. Form of relationship is linear.
  2. Errors are independent.
  3. Variability of errors is constant.
  4. Errors have a Normal model.

  1. Scatterplot looks approximately linear.
  2. No apparent pattern in residuals plot.
  3. Residuals plot has consistent spread.
  4. Histogram of residuals is approximately unimodal and symmetric or Normal probability plot reasonably straight.*

(*less critical as n increases)

REVIEW EXERCISES

1. Genetics. Two human traits controlled by a single gene are the ability to roll one's tongue and whether one's ear lobes are free or attached to the neck. Genetic theory says that people will have neither, one, or both of these traits in the ratio 1:3:3:9 (1 attached, noncurling; 3 attached, curling; 3 free, noncurling; 9 free, curling). An Introductory Biology class of 122 students collected the data shown. Are they consistent with the genetic theory? Test an appropriate hypothesis and state your conclusion.

	Trait			
	Attached, noncurling	Attached, curling	Free, noncurling	Free, curling
Count	10	22	31	59

2. Tableware. Nambe Mills manufactures plates, bowls, and other tableware made from an alloy of several metals. Each item must go through several steps, including polishing. To better understand the production process and its impact on pricing, the company checked the polishing time (in minutes) and the retail price (in US$) of these items. The regression analysis is shown below. The scatterplot showed a linear pattern, and residuals were deemed suitable for inference.

Dependent variable is: Price
R-squared = 84.5%
s = 20.50 with 59 − 2 = 57 degrees of freedom

Variable	Coefficient	SE(Coeff)
Intercept	−2.89054	5.730
Time	2.49244	0.1416

a) How many different products were included in this analysis?
b) What fraction of the variation in retail price is explained by the polishing time?
c) Create a 95% confidence interval for the slope of this relationship.
d) Interpret your interval in this context.

3. Hard water. In an investigation of environmental causes of disease, data were collected on the annual mortality rate (deaths per 100,000) for males in 61 large towns in England and Wales. In addition, the water hardness was recorded as the calcium concentration (parts per million, or ppm) in the drinking water. Here are the scatterplot and regression analysis of the relationship between mortality and calcium concentration.

Dependent variable is: mortality
R-squared = 43%
s = 143.0 with 61 − 2 = 59 degrees of freedom

Variable	Coefficient	SE(Coeff)
Intercept	1676	29.30
calcium	−3.23	0.48

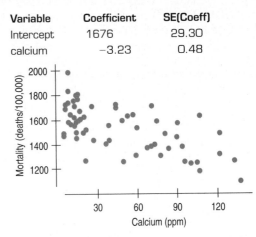

a) Is there an association between the hardness of the water and the mortality rate? Write the appropriate hypothesis.
b) Assuming the assumptions for regression inference are met, what do you conclude?
c) Create a 95% confidence interval for the slope of the true line relating calcium concentration and mortality.
d) Interpret your interval in context.

4. Mutual funds. In March 2002, *Consumer Reports* listed the rate of return for several large cap mutual funds over the previous 3-year and 5-year periods. ("Large cap" refers to companies worth over $10 billion.)
a) Create a 95% confidence interval for the difference in rate of return for the 3- and 5-year periods covered by these data. Clearly explain what your interval means.
b) It's common for advertisements to carry the disclaimer that "past returns may not be indicative of future performance," but do these data indicate that there was an association between 3-year and 5-year rates of return?

	Annualized Returns (%)	
Fund Name	**3-year**	**5-year**
Ameristock	7.9	17.1
Clipper	14.1	18.2
Credit Suisse Strategic Value	5.5	11.5
Dodge & Cox Stock	15.2	15.7
Excelsior Value	13.1	16.4
Harbor Large Cap Value	6.3	11.5
ICAP Discretionary Equity	6.6	11.4
ICAP Equity	7.6	12.4
Neuberger Berman Focus	9.8	13.2
PBHG Large Cap Value	10.7	18.1
Pelican	7.7	12.1
Price Equity Income	6.1	10.9
USAA Cornerstone Strategy	2.5	4.9
Vanguard Equity Income	3.5	11.3
Vanguard Windsor	11.0	11.0

5. Resume fraud. In 2002 the Veritas Software company found out that its chief financial officer did not actually have the MBA he had listed on his resume. They fired him, and the value of the company's stock dropped 19%. Kroll, Inc., a firm that specializes in investigating such matters, said that they believe as many as 25% of background checks might reveal false information. How many such random checks would they have to do to estimate the true percentage of people who misrepresent their backgrounds to within ±5% with 98% confidence?

T 6. Paper airplanes. In preparation for a regional paper airplane competition, a student tried out her latest design. The distances her plane traveled (in feet) in 11 trial flights are given here. (The world record is an astounding 193.01 feet!) The data were 62, 52, 68, 23, 34, 45, 27, 42, 83, 56, and 40 feet.

Here are some summaries:

Count	11
Mean	48.3636
Median	45
StdDev	18.0846
StdErr	5.45273
IntQRange	25
25th %tile	35.5000
75th %tile	60.5000

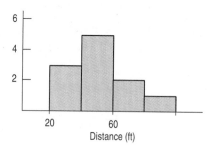

a) Construct a 95% confidence interval for the true distance.
b) Based on your confidence interval, is it plausible that the mean distance is 40 ft? Explain.
c) How would a 99% confidence interval for the true distance differ from your answer in part a? Explain briefly, without actually calculating a new interval.
d) How large a sample size would the student need to get a confidence interval half as wide as the one you got in part a, at the same confidence level?

T 7. Back to Montana. The respondents to the Montana poll described in Exercise 17 in Chapter 26 were also classified by income level: low (under $20,000), middle ($20,000–$35,000), or high (over $35,000). Is there any evidence that party enrollment there is associated with income? Test an appropriate hypothesis about this table, and state your conclusions.

	Democrat	Republican	Independent
Low	30	16	12
Middle	28	24	22
High	26	38	6

T 8. Wild horses. Large herds of wild horses can become a problem on some federal lands in the West. Researchers hoping to improve the management of these herds collected data to see if they could predict the number of foals that would be born based on the size of the current herd. Their attempt to model this herd growth is summarized in the output shown.

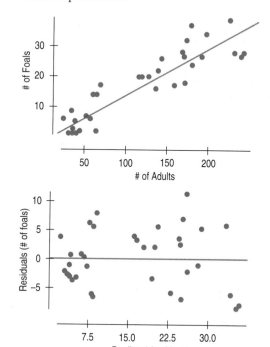

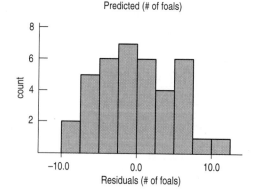

Variable	Count	Mean	StdDev
Adults	38	110.237	71.1809
Foals	38	15.3947	11.9945

Dependent variable is: Foals
R-squared = 83.5%
s = 4.941 with 38 − 2 = 36 degrees of freedom

Variable	Coefficient	SE(Coeff)	t-ratio	P-value
Constant	−1.57835	1.492	−1.06	0.2970
Adults	0.153969	0.0114	13.5	≤ 0.0001

a) How many herds of wild horses were studied?
b) Are the conditions necessary for inference satisfied? Explain.
c) Create a 95% confidence interval for the slope of this relationship.
d) Explain in this context what that slope means.
e) Suppose that a new herd with 80 adult horses is located. Estimate, with a 90% prediction interval, the number of foals that may be born.

9. Lefties and music. In an experiment to see if left- and right-handed people have different abilities in music, subjects heard a tone and were then asked to identify which of several other tones matched the first. Of 76 right-handed subjects, 38 were successful in completing this test, compared with 33 of 53 lefties. Is this strong evidence of a difference in musical abilities based on handedness?

T 10. AP Statistics scores. In 2001, more than 41,000 Statistics students nationwide took the Advanced Placement Examination in Statistics. The national distribution of scores and the results at Ithaca High School are shown in the table.

		Ithaca High School	
Score	National Distribution	Number of boys	Number of girls
5	11.5%	13	13
4	23.4%	21	15
3	24.9%	6	13
2	19.1%	7	3
1	21.1%	4	2

a) Is the distribution of scores at this high school significantly different from the national results?
b) Was there a significant different between the performances of boys and girls at this school?

T 11. Polling. How accurate are pollsters in predicting the outcomes of Congressional elections? The table shows the actual number of Democrat seats in the House of Representatives and the number predicted by the Gallup organization for nonpresidential election years between World War II and 1998.
a) Is there a significant difference between the number of seats predicted for the Democrats and the number they actually held? Test an appropriate hypothesis and state your conclusions.

Democrat Congressmen		
Year	Predicted	Actual
1946	190	188
1950	235	234
1954	232	232
1958	272	283
1962	259	258
1966	247	248
1970	260	255
1974	292	291
1978	277	277
1982	275	269
1986	264	258
1990	260	267
1994	201	204
1998	211	211

b) Is there a strong association between the pollsters' predictions and the outcomes of the elections? Test an appropriate hypothesis and state your conclusions.

T 12. Twins. In 2000 The *Journal of the American Medical Association* published a study that examined a sample of pregnancies that resulted in the birth of twins. Births were classified as preterm with intervention (induced labor or cesarean), preterm without such procedures, or term or postterm. Researchers also classified the pregnancies by the level of prenatal medical care the mother received (inadequate, adequate, or intensive). The data, from the years 1995–1997, are summarized in the table below. Figures are in thousands of births. (*JAMA* 284 [2000]: 335–341)

		Twin Births 1995–1997 (in thousands)			
		Preterm (induced or Cesarean)	Preterm (without procedures)	Term or postterm	Total
Level of Prenatal Care	Intensive	18	15	28	61
	Adequate	46	43	65	154
	Inadequate	12	13	38	63
	Total	76	71	131	278

Is there evidence of an association between the duration of the pregnancy and the level of care received by the mother?

T 13. Twins, again. After reading of the *JAMA* study in Exercise 12, a large city hospital examined their records of twin births for several years, and found the data summarized in the table below. Is there evidence that the way the hospital deals with pregnancies involving twins may have changed?

		1990	1995	2000
Outcome of Pregnancy	Preterm (induced or cesarean)	11	13	19
	Preterm (without procedures)	13	14	18
	Term or postterm	27	26	32

14. Preemies. Do the effects of being born prematurely linger into adulthood? Researchers examined 242 Cleveland area children born prematurely between 1977 and 1979, and compared them with 233 children of normal birth weight; 24 of the "preemies" and 12 of the other children were described as being of "subnormal height" as adults. Is this evidence that babies born with a very low birth weight are more likely to be smaller than normal adults? ("Outcomes in Young Adulthood for Very-Low-Birth-Weight Infants," *New England Journal of Medicine*, 346, no. 3 [January 2002])

T 15. LA rainfall. The Los Angeles Almanac Web site reports recent annual rainfall (in inches), as shown in the table.
a) Create a 90% confidence interval for the mean annual rainfall in LA.
b) If you wanted to estimate the mean annual rainfall with a margin of error of only 2 inches, how many years' data would you need?
c) Do these data suggest any change in annual rainfall as time passes? Check for an association between rainfall and year.

Year	Rain (in.)	Year	Rain (in.)
1980	8.96	1991	21.00
1981	10.71	1992	27.36
1982	31.28	1993	8.14
1983	10.43	1994	24.35
1984	12.82	1995	12.46
1985	17.86	1996	12.40
1986	7.66	1997	31.01
1987	12.48	1998	9.09
1988	8.08	1999	11.57
1989	7.35	2000	17.94
1990	11.99	2001	4.42

T 16. Age and party. The Gallup Poll conducted a representative telephone survey during the first quarter of 1999. Among the reported results was the following table concerning the preferred political party affiliation of respon-

dents and their ages. Is there evidence of age-based differences in party affiliation in the United States?

Age	Republican	Democratic	Independent	Total
18–29	241	351	409	1001
30–49	299	330	370	999
50–64	282	341	375	998
65+	279	382	343	1004
Total	**1101**	**1404**	**1497**	**4002**

a) Will you conduct a test of homogeneity or independence? Why?
b) Test an appropriate hypothesis.
c) State your conclusion, including an analysis of differences you find (if any).

T 17. Eye and hair color. A survey of 1021 school-age children was conducted by randomly selecting children from several large urban elementary schools. Two of the questions concerned eye and hair color. In the survey, the following codes were used:

Hair Color	Eye Color
1 = Blond	1 = Blue
2 = Brown	2 = Green
3 = Black	3 = Brown
4 = Red	4 = Grey
5 = Other	5 = Other

The Statistics students analyzing the data were asked to study the relationship between eye and hair color.
a) One group of students produced the output shown below. What kind of analysis is this? What are the null and alternative hypotheses? Is the analysis appropriate? If so, summarize the findings, being sure to include any assumptions you've made and/or limitations to the analysis. If it's not an appropriate analysis, explicitly state why not.

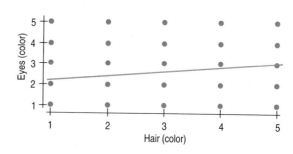

Dependent variable is: Eyes
R-squared = 3.7%
s = 1.112 with 1021 − 2 = 1019 degrees of freedom

Variable	Coefficient	SE(Coeff)	t-ratio	P-value
Intercept	1.99541	0.08346	23.9	≤0.0001
Hair	0.211809	0.03372	0.28	≤0.0001

b) A second group of students used the same data to produce the output shown below. The table displays counts and standardized residuals in each cell. What kind of analysis is this? What are the null and alternative hypotheses? Is the analysis appropriate? If so, summarize the findings, being sure to include any assumptions you've made and/or limitations to the analysis. If it's not an appropriate analysis, explicitly state why not.

Eye Color

		1	2	3	4	5
Hair Color	1	143 7.67540	30 0.41799	58 −5.88169	15 −0.63925	12 −0.31451
	2	90 −2.57141	45 0.29019	215 1.72235	30 0.49189	20 −0.08246
	3	28 −5.39425	15 −2.34780	190 6.28154	10 −1.76376	10 −0.80382
	4	30 2.06116	15 2.71589	10 −4.05540	10 2.37402	5 0.75993
	5	10 −0.52195	5 0.33262	15 −0.94192	5 1.36326	5 2.07578

$$\sum \frac{(Observed - Expected)^2}{Expected} = 223.6 \quad \text{P-value} < 0.00001$$

18. **Depression and the Internet.** The September 1998 issue of the *American Psychologist* published an article reporting on an experiment examining "the social and psychological impact of the Internet on 169 people in 73 households during their first 1 to 2 years online." In the experiment, a sample of households was offered free Internet access for one or two years in return for allowing their time and activity online to be tracked. The members of the households who participated in the study were also given a battery of tests at the beginning and again at the end of the study. One of the tests measured the subjects' levels of depression on a 4-point scale, with higher numbers meaning the person was more depressed. Internet usage was measured in average number of hours per week. The regression analysis examines the association between the subjects' depression levels and the amounts of Internet use. The conditions for inference were satisfied.

Dependent variable is: Depression After
R-squared = 4.6%
s = 0.4563 with 162 − 2 = 160 degrees of freedom

Variable	Coefficient	SE(coeff)	t-ratio	Prob
Constant	0.565485	0.0399	14.2	≤0.0001
Intr_use	0.019948	0.0072	2.76	0.0064

a) Do these data indicate that there is an association between Internet use and depression? Test an appropriate hypothesis and state your conclusion clearly.
b) One conclusion of the study was that those who spent more time online tended to be more depressed at the end of the experiment. News headlines said that too much time on the Internet can lead to depression. Does the study support this conclusion? Explain.
c) As noted, the subjects' depression levels were tested at both the beginning and the end of this study; higher scores indicated the person was more depressed. Results are summarized in the table. Is there evidence that the depression level of the subjects changed during this study?

Depression Level
162 subjects

Variable	Mean	StdDev
DeprBfore	0.730370	0.487817
DeprAfter	0.611914	0.461932
Difference	−0.118457	0.552417

19. **Pregnancy.** In 1998 a San Diego reproductive clinic reported 42 live births to 157 women under the age of 38, but only 7 successes for 89 clients aged 38 and older. Is this evidence of a difference in the effectiveness of the clinic's methods for older women?

a) Test the appropriate hypotheses using the two-proportion z-procedure.
b) Repeat the analysis using an appropriate chi-square procedure.
c) Explain how the two results are equivalent.

20. **Family planning.** A 1954 study of 1438 pregnant women examined the association between the woman's education level and the occurrence of unplanned pregnancies, producing these data:

	Education Level		
	<3 yr HS	3+ yr HS	Some college
Number of pregnancies	591	608	239
% unplanned	66.2%	55.4%	42.7%

Do these data provide evidence of an association between family planning and education level? (*Fertility Planning and Fertility Rates by Socio-Economic Status*, Social and Psychological Factors Affecting Fertility, 1954)

T 21. Old Faithful. As you saw in an earlier chapter, Old Faithful isn't all that faithful. Eruptions do not occur at uniform intervals, and may vary greatly. Can we improve our chances of predicting the time of the next eruption if we know how long the previous eruption lasted?
a) Describe what you see in this scatterplot.

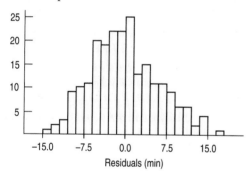

b) Write an appropriate hypothesis.
c) Here are a histogram of the residuals and the residuals plot. Do you think the assumptions for inference are met? Explain.

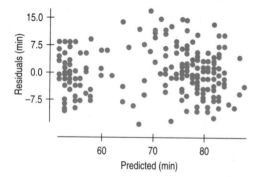

d) State a conclusion based on this regression analysis:

Dependent variable is: Interval
R-squared = 77.0%
s = 6.159 with 222 − 2 = 220 degrees of freedom

Variable	Coefficient	SE(Coeff)	t-ratio	P-value
Constant	33.9668	1.428	23.8	≤0.0001
Duration	10.3582	0.3822	27.1	≤0.0001

Variable	Mean	StdDev
Duration	3.57613	1.08395
Interval	71.0090	12.7992

e) The second table shows the summary statistics for the two variables. Create a 95% confidence interval for the mean length of time that will elapse following a 2-minute eruption.
f) You arrive at Old Faithful just as eruption ends. Witnesses say it lasted 4 minutes. Create a 95% prediction interval for the length of time you will wait to see the next eruption.

22. Togetherness. Are good grades in high school associated with family togetherness? A simple random sample of 142 high-school students was asked how many meals per week their families ate together. Their responses produced a mean of 3.78 meals per week, with a standard deviation of 2.2. Researchers then matched these responses against the students' grade point averages. The scatterplot appeared to be reasonably linear, so they went ahead with the regression analysis, seen below. No apparent pattern emerged in the residuals plot.

Dependent variable: GPA
R-squared = 11.0%
s = 0.6682 with 142 − 2 = 140 df

Variable	Coefficient	SE(Coeff)
Intercept	2.7288	0.1148
Meals/wk	0.1093	0.0263

a) Is there evidence of an association? Test an appropriate hypothesis and state your conclusion.
b) Do you think this association would be useful in predicting a student's grade point average? Explain.
c) Are your answers to parts a and b contradictory? Explain.

23. Learning math. Developers of a new math curriculum called "Accelerated Math" compared performances of students taught by their system with control groups of students in the same schools who were taught using traditional instructional methods and materials. Statistics about pretest and posttest scores are shown in the table. (J. Ysseldyke and S. Tardrew, *Differentiating Math Instruction*, Renaissance Learning, 2002)
a) Did the groups differ in average math score at the start of this study?
b) Did the group taught using the Accelerated Math program show a significant improvement in test scores?
c) Did the control group show significant improvement in test scores?

d) Were gains significantly higher for the Accelerated Math group than for the control group?

		Instructional Method	
		Acc. math	**Control**
Number of students		231	245
Pretest	**Mean**	560.01	549.65
	St. Dev	84.29	74.68
Post-test	**Mean**	637.55	588.76
	St. Dev	82.9	83.24
Individual gain	**Mean**	77.53	39.11
	St. Dev.	78.01	66.25

24. Pesticides. A study published in 2002 in the journal *Environmental Health Perspectives* examined the gender ratios of children born to workers exposed to dioxin in Russian pesticide factories. The data covered the years 1961 to 1988 in the city of Ufa, Bashkortostan, Russia. Of 227 children born to workers exposed to dioxin, only 40% were male. Overall in the city of Ufa, the proportion of males was 51.2%. Is this evidence that human exposure to dioxin may result in the birth of more girls? (An interesting note: It appeared that paternal exposure was most critical; 51% of babies born to mothers exposed to the chemical were boys.)

25. Dairy sales. Peninsula Creameries sells both cottage cheese and ice cream. The CEO recently noticed that in months when the company sells more cottage cheese, it seems to sell more ice cream as well. Two of his aides were assigned to test whether this is true or not. The first aide's plot and analysis of sales data for the past 12 months (in millions of pounds for cottage cheese and for ice cream) appears below.

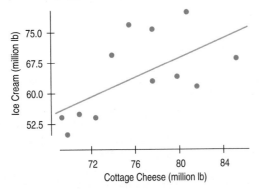

Dependent variable is: Ice cream
R-squared = 36.9%
s = 8.320 with 12 − 2 = 10 degrees of freedom

Variable	Coefficient	SE(Coeff)	t-ratio	P-value
Constant	−26.5306	37.68	−0.704	0.4975
Cottage C ...	1.19334	0.4936	2.42	0.0362

The other aide looked at the differences in sales of ice cream and cottage cheese for each month, and created the following output:

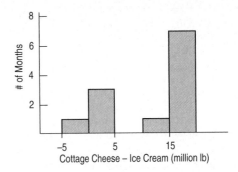

Cottage Cheese-Ice Cream
Count   12
Mean   11.8000
Median   15.3500
StdDev   7.99386
IntQRange   14.3000
25th %tile   3.20000
75th %tile   17.5000

Test HO: $\mu(CC - IC) = 0$ vs Ha: $\mu(CC - IC) \neq 0$
Sample Mean = 11.800000 t-Statistic = 5.113 w/11 df
Prob = 0.0003
Lower 95% bound = 6.7209429
Upper 95% bound = 16.879057

a) Which analysis would you use to answer the CEO's question? Why?
b) What would you tell the CEO?
c) Which analysis would you use to test whether the company sells more cottage cheese or ice cream in a typical year? Why?
d) What would you tell the CEO about this other result?
e) What assumptions are you making in the analysis you chose in part a? What assumptions are you making in the analysis in part c?
f) Next month's cottage cheese sales are 82 million pounds. Ice cream sales are not yet available. How much ice cream do you predict Peninsula Creameries will sell?
g) Give a 95% confidence interval for the true slope of the regression equation of ice cream sales by cottage cheese sales.
h) Explain what your interval means.

26. Infliximab. In an article appearing in the journal *Lancet* in 2002, medical researchers reported on the experimental use of the arthritis drug infliximab in treating Crohn's disease. In a trial, 573 patients were given initial 5-mg injections of the drug. Two weeks later, 335 had responded positively. These patients were then randomly assigned to three groups. Group I received continued injections of a placebo, Group II continued with 5 mg of infliximab, and Group III received 10 mg of the drug. After 30 weeks, 23 of 110 Group I patients were in remission, compared with 44 of 113 Group II and 50 of 112 Group III patients. Do these data indicate that continued treatment with infliximab is of value for Crohn's disease patients who exhibit a positive initial response to the drug?

T 27. Weight loss. A weight loss clinic advertises that its program of diet and exercise will allow clients to lose 10 pounds in one month. A local reporter investigating weight reduction gets permission to interview a randomly selected sample of clients who report the given weight losses during their first month in this program. Create a confidence interval to test the clinic's claim that typical weight loss is 10 pounds.

Pounds Lost	
9.5	9.5
13	9
9	8
10	7.5
11	10
9	7
5	8
9	10.5
12.5	10.5
6	9

28. Education vs. income. The information below examines the median income and education level (years in school) for several U.S. cities.

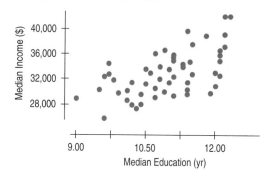

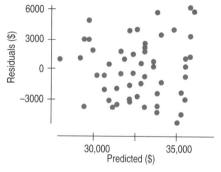

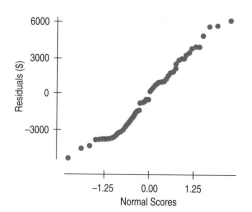

Variable	Count	Mean	StdDev
Education	57	10.9509	0.848344
Income	57	32742.6	3618.01

Dependent variable is: Income
R-squared = 32.9%
s = 2991 with 57 − 2 = 55 degrees of freedom

Variable	Coefficient	SE(Coeff)	t-ratio	P-value
Intercept	5970.05	5175	1.15	0.2537
Education	2444.79	471.2	5.19	≤0.0001

a) Do you think the assumptions for inference are met? Explain.

b) Does there appear to be an association between education and income levels in these cities?

c) Would this association appear to be weaker, stronger, or the same if data were plotted for individual people rather than for cities in aggregate? Explain.

d) Create and interpret a 95% confidence interval for the slope of the true line that describes the association between income and education.

e) Predict the median income for cities where residents spent an average of 11 years in school. Describe your estimate with a 90% confidence interval, and interpret that result.

T 29. Diet. Thirteen overweight women volunteered for a study to determine whether eating specially prepared crackers before a meal could help them lose weight. The subjects were randomly assigned to eat crackers with different types of fiber (bran fiber, gum fiber, both, and a control cracker). Unfortunately, some of the women developed uncomfortable bloating and upset stomachs. Researchers suspected that some of the crackers might be at fault. The contingency table of "Cracker" versus "Bloat" shows the relationship between the four different types of crackers and the reported bloating. The study was paid for by the manufacturers of the gum fiber. What would you recommend to them about the prospects for marketing their new diet cracker?

		Bloat	
		Little/None	Moderate/Severe
Cracker	Bran	11	2
	Gum	4	9
	Combo	7	6
	Control	8	4

T 30. Cramming. Students in two basic Spanish classes were required to learn 50 new vocabulary words. One group of 45 students received the list on Monday and studied the words all week. Statistics summarizing this group's scores on Friday's quiz are given. The other group of 28 students did not get the vocabulary list until Thursday. They also took the quiz on Friday after "cramming" Thursday night. Then, when they returned to class the following Monday they were retested—without advance warning. Both sets of test scores for these students are shown.

Group 1

Fri.
Number of students = 45
Mean = 43.2 (of 50)
StDev = 3.4
Students passing (score $\geq$ 40) = 33%

a) Did the week-long study group have a mean score significantly higher than that of the overnight crammers?
b) Was there a significant difference in the percentages of students who passed the quiz on Friday?
c) Is there any evidence that when students cram for a test their "learning" does not last for 3 days?
d) Use a 95% confidence interval to estimate the mean number of words that might be forgotten by crammers.
e) Is there any evidence that how much students forget depends on how much they "learned" to begin with?

Group 2

Fri.	Mon.	Fri.	Mon.
42	36	50	47
44	44	34	34
45	46	38	31
48	38	43	40
44	40	39	41
43	38	46	32
41	37	37	36
35	31	40	31
43	32	41	32
48	37	48	39
43	41	37	31
45	32	36	41
47	44		

APPENDIXES

$Range = Max - Min$

$IQR = Q3 - Q1$

Outlier Rule-of-Thumb: $y < Q1 - 1.5 \times IQR$ or $y > Q3 + 1.5 \times IQR$

$$\bar{y} = \frac{\sum y}{n}$$

$$s = \sqrt{\frac{\sum (y - \bar{y})^2}{n - 1}}$$

$$z = \frac{y - \mu}{\sigma} \text{ (model based)}$$

$$z = \frac{y - \bar{y}}{s} \text{ (data based)}$$

$$r = \frac{\sum z_x z_y}{n - 1}$$

$$\hat{y} = b_0 + b_1 x \qquad \text{where } b_1 = r\frac{s_y}{s_x} \text{ and } b_0 = \bar{y} - b_1\bar{x}$$

$P(\mathbf{A}) = 1 - P(\mathbf{A}^C)$

$P(\mathbf{A} \text{ or } \mathbf{B}) = P(\mathbf{A}) + P(\mathbf{B}) - P(\mathbf{A} \text{ and } \mathbf{B})$

$P(\mathbf{A} \text{ and } \mathbf{B}) = P(\mathbf{A}) \times P(\mathbf{B}|\mathbf{A})$

$$P(\mathbf{B}|\mathbf{A}) = \frac{P(\mathbf{A} \text{ and } \mathbf{B})}{P(\mathbf{A})}$$

If $\mathbf{A}$ and $\mathbf{B}$ are independent, $P(\mathbf{B}|\mathbf{A}) = P(\mathbf{B})$

$E(X) = \mu = \sum x \cdot P(x)$	$Var(X) = \sigma^2 = \sum (x - \mu)^2 P(x)$
$E(X \pm c) = E(X) \pm c$	$Var(X \pm c) = Var(X)$
$E(aX) = aE(X)$	$Var(aX) = a^2 Var(X)$
$E(X \pm Y) = E(X) \pm E(Y)$	$Var(X \pm Y) = Var(X) + Var(Y)$ if X and Y are independent

Geometric: $P(x) = q^{x-1}p$ $\mu = \dfrac{1}{p}$ $\sigma = \sqrt{\dfrac{q}{p^2}}$

Binomial: $P(x) = {}_nC_x p^x q^{n-x}$ $\mu = np$ $\sigma = \sqrt{npq}$

$\hat{p} = \dfrac{x}{n}$ $\mu(\hat{p}) = p$ $SD(\hat{p}) = \sqrt{\dfrac{pq}{n}}$

Sampling distribution of $\bar{y}$:
(CLT) As n grows, the sampling distribution approaches the Normal model with

$$\mu(\bar{y}) = \mu_y \qquad SD(\bar{y}) = \frac{\sigma}{\sqrt{n}}$$

Inference:

Confidence interval for parameter = ***statistic ± critical value × SD(statistic)***

$$\text{Test statistic} = \frac{Statistic - Parameter}{SD(statistic)}$$

Parameter	Statistic	SD(statistic)	SE(statistic)
p	$\hat{p}$	$\sqrt{\dfrac{pq}{n}}$	$\sqrt{\dfrac{\hat{p}\hat{q}}{n}}$
$p_1 - p_2$	$\hat{p}_1 - \hat{p}_2$	$\sqrt{\dfrac{p_1 q_1}{n_1} + \dfrac{p_2 q_2}{n_2}}$	$\sqrt{\dfrac{\hat{p}_1 \hat{q}_1}{n_1} + \dfrac{\hat{p}_2 \hat{q}_2}{n_2}}$
μ	$\bar{y}$	$\dfrac{\sigma}{\sqrt{n}}$	$\dfrac{s}{\sqrt{n}}$
$\mu_1 - \mu_2$	$\bar{y}_1 - \bar{y}_2$	$\sqrt{\dfrac{\sigma_1^2}{n_1} + \dfrac{\sigma_2^2}{n_2}}$	$\sqrt{\dfrac{s_1^2}{n_1} + \dfrac{s_2^2}{n_2}}$
μ_d	$\bar{d}$	$\dfrac{\sigma_d}{\sqrt{n}}$	$\dfrac{s_d}{\sqrt{n}}$
σ_ε	$s_e = \sqrt{\dfrac{\sum(y - \hat{y})^2}{n - 2}}$		
β_1	b_1		$\dfrac{s_e}{s_x \sqrt{n - 1}}$
μ_ν	$\hat{y}_\nu$		$\sqrt{SE^2(b_1) \cdot (x_\nu - \bar{x})^2 + \dfrac{s_e^2}{n}}$
y_ν	$\hat{y}_\nu$		$\sqrt{SE^2(b_1) \cdot (x_\nu - \bar{x})^2 + \dfrac{s_e^2}{n} + s_e^2}$

Pooling: For testing difference between proportions: $\hat{p}_{pooled} = \dfrac{y_1 + y_2}{n_1 + n_2}$

For testing difference between means: $s_p = \sqrt{\dfrac{(n_1 - 1)s_1^2 + (n_2 - 1)s_2^2}{n_1 + n_2 - 2}}$

Substitute these pooled estimates in the respective SE formulas for both groups when assumptions and conditions are met.

Chi-square: $\chi^2 = \sum \dfrac{(obs - exp)^2}{exp}$

Here are the "answers" to the exercises for the chapters and the unit reviews. As we said in Chapter 1, the answers are outlines of the complete solution. Your solution should follow the model of the Step-by-Step examples, where appropriate. You should explain the context, show your reasoning and calculations, and draw conclusions. For some problems, what you decide to include in an argument may differ somewhat from the answers here. But, of course, the numerical part of your answer should match the numbers in the answers shown.

CHAPTER 2

1. Answers will vary.
3. *Who*—50 recent oil spills
    Cases—Each of the 50 recent oil spills is a case.
  What—Spillage amount and cause of puncture
  When—Reported in 1995
  Where—Worldwide
  Why—To use in designing new tankers
  How—Not specified
  Variable—Spillage amount
    Type—Quantitative
    Units—Not specified
  Variable—Cause of puncture
    Type—Categorical
5. *Who*—54 bears
    Cases—Each bear is a case.
  What—Weight, neck size, length, and sex
  When—Not specified
  Where—Not specified
  Why—To estimate weight from easier-to-measure variables
  How—Researchers collected data on 54 bears they were able to catch
  Variable—Weight
    Type—Quantitative
    Units—Not specified
  Variable—Neck size
    Type —Quantitative
    Units—Not specified
  Variable—Length
    Type—Quantitative
    Units—Not specified
  Variable—Sex
    Type—Categorical
7. *Who*—Arby's sandwiches
    Cases—Each sandwich is a case.
  What—Type of meat, number of calories, and serving size
  When—Not specified
  Where—Arby's restaurants
  Why—To assess nutritional value of sandwiches
  How—Report by Arby's restaurants
  Variable—Type of meat
    Type—Categorical
  Variable—Number of calories
    Type—Quantitative
    Units—Calories
  Variable—Serving size

    Type—Quantitative
    Units—Ounces
9. *Who*—882 births
    Cases—Each of the 882 births is a case.
  What—Mother's age, length of pregnancy, type of birth, level of prenatal care, birth weight of baby, gender of baby, and baby's health problems
  When—1998–2000
  Where—Large city hospital
  Why—Researchers were investigating the impact of prenatal care on newborn health.
  How—Not specified exactly, but probably from hospital records
  Variable—Mother's age
    Type—Quantitative
    Units—Probably years
  Variable—Length of pregnancy
    Type—Quantitative
    Units—Weeks
  Variable—Birth weight of baby
    Type—Quantitative
    Units—Not specified, probably pounds and ounces
  Variable—Type of birth
    Type—Categorical
  Variable—Level of prenatal care
    Type—Categorical
  Variable—Sex
    Type—Categorical
  Variable—Baby's health problems
    Type—Categorical
11. *Who*—25,892 men aged 30 to 87
    Cases—Each of the 25,892 men was an individual.
  What—Fitness level and cause of death
  When—Over a 10-year period prior to the article being published in May 2002
  Where—Not specified
  Why—To look for an association between fitness level and death from cancer
  How—Researchers tracked the group of men over a 10-year period.
  Variable—Fitness level
    Type—Categorical
  Variable—Death from cancer
    Type—Categorical (yes, no)
13. *Who*—Experiment subjects
    Cases—Each subject is an individual.
  What—Treatment (herbal cold remedy or sugar solution) and cold severity
  When—Not specified
  Where—Not specified

Why—To test efficacy of herbal remedy on common cold
How—The scientists set up an experiment
Variable—Treatment
　Type—Categorical
Variable—Cold severity rating
　Type—Quantitative (perhaps ordinal categorical)
　Units—Scale from 0 to 5
Concerns—The severity of a cold seems subjective and difficult to quantify. Scientists may feel pressure to report negative findings of herbal product.

15. *Who*—Automobiles in student and staff lots
　Cases—Each car is a case.
What—Make, country of origin, type of vehicle, and age of vehicle
When—Not specified
Where—A large university
Why—Not specified
How—A survey was taken in campus parking lots
Variable—Make
　Type—Categorical
Variable—Country of origin
　Type—Categorical
Variable—Type of vehicle
　Type—Categorical
Variable—Age of vehicle
　Type—Quantitative
　Units—Years

17. *Who*—Streams
　Cases—Each stream is a case.
What—Name of stream, substrate of the stream, acidity of the water, temperature, BCI
When—Not specified
Where—Upstate New York
Why—To study ecology of streams
How—Not specified
Variable—Stream name
　Type—Identifier
Variable—Substrate
　Type—Categorical
Variable—Acidity of water
　Type—Quantitative
　Units—pH
Variable—Temperature
　Type—Quantitative
　Units—Degrees Celsius
Variable—BCI
　Type—Quantitative
　Units—Not specified

19. *Who*—All airline flights in the United States
　Cases—Each flight is a case.
What—Type of aircraft, number of passengers, whether departures and arrivals were on schedule, mechanical problems
When—Current
Where—United States
Why—Required by the Federal Aviation Administration
How—Data are collected from airline flight information.
Variable—Type of aircraft
　Type—Categorical
Variable—Number of passengers
　Type—Quantitative
　Units—Count

Variable—Departed on time?
　Type—Categorical
Variable—Arrived on time?
　Type—Categorical
Variable—Mechanical problems?
　Type— Categorical

21. *Who*—41 refrigerator models
　Cases—Each of the 41 refrigerator models is a case.
What—Brand, cost, size, type, estimated annual energy cost, overall rating, and repair history
When—2002
Where—United States
Why—To provide information to the readers of *Consumer Reports*
How—Not specified
Variable—Brand
　Type—Categorical
Variable—Cost
　Type—Quantitative
　Units—Not specified (dollars)
Variable—Size
　Type—Quantitative
　Units—Cubic feet
Variable—Type
　Type—Categorical
Variable—Estimated annual energy cost
　Type—Quantitative
　Units—Not specified (dollars)
Variable—Overall rating
　Type—Categorical (ordinal)
Variable—Percent requiring repair in last 5 years
　Type—Quantitative
　Units—Percent

23. *Who*—Days
What—Sleep, wake early, TV, hours standing, mood
When—2001
Where—At home (?)
Why—Analyze sleep patterns
How—Daily recording
Variable—Sleep
　Type—Quantitative
　Units—Hours
Variable—Wake early
　Type—Categorical
Variable—TV
　Type—Categorical
Variable—Hours standing
　Type—Quantitative
　Units—Hours
Variable—Mood
　Type—Quantitative
　Units—Arbitrary scale

25. *Who*—Kentucky Derby races
What—Date, winner, margin, jockey, net proceed to winner, duration, track condition
When—1875 to 2004
Where—Churchill Downs, Louisville, Kentucky
Why—To see trends in horse racing?
How—Official statistics collected at race
Variable—Year
　Date—Quantitative

Units—Day and year
Variable—Winner
   Type—Identifier
Variable—Margin
   Type—Quantitative
   Units—Horse lengths
Variable—Jockey
   Type—Categorical
Variable—Net proceeds to winner
   Type—Quantitative
   Units—Dollars
Variable—Duration
   Type—Quantitative
   Units—Minutes and seconds
Variable—Track condition
   Type—Categorical

CHAPTER 3

1. Answers will vary.   **3.** Answers will vary.

5. 1,755 students applied for admission to the magnet schools program. 53% were accepted, 17% were wait-listed, and the other 30% were turned away.

7. a) Yes. We can add because these categories do not overlap (each person is assigned only one cause of death).
   b) $100 - (30.3 + 23.0 + 8.4 + 7.9 + 4.1) = 26.3\%$
   c) Either a bar chart or pie chart with "other" added would be appropriate. A bar chart is shown.

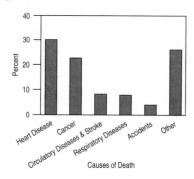

Causes of Death

9. a) Individuals must have been able to select more than one phenomenon. Because of this, it is not reasonable to add those percents.
   b) No, there may be people who believe in more than one of these, or none. We can't tell from these percentages.
   c) A bar chart displays the percentages well:

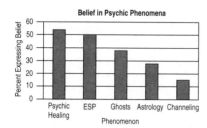

11. The bar chart shows that grounding is the most frequent cause of oil spills, although the other three sources—collision, fire, and

hull failure—are nearly as common. Very few have unknown causes. A pie chart seems appropriate as well. One can see that the four types of causes are nearly evenly split.

13. a) $67\% - 46\% = 21\%$ (approximately)
   b) No, the chart makes it look like about 2 to 3 times as much.
   c) To maintain equivalent intervals, the vertical axis should begin at 0, not near 30%.
   d)

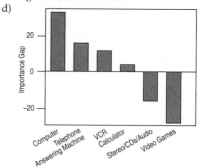

   e) Teens tend to think that some technologies are critically important to own, although they don't use them every day. On the other hand, they use some technologies that are less important very often. The importance gap highlights this difference.

15. a)

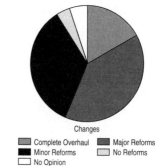

Changes

Complete Overhaul ▪ Major Reforms
Minor Reforms ▫ No Reforms
No Opinion

   b) The bar chart makes comparisons of percentages easier and keeps the order of the response.
   c) The vast majority favor some type of reform. While only 17% of the American adults surveyed recommended a complete overhaul, 74% of the respondents recommended either major or minor reforms in the way that corporations are audited.

17. a) A bar chart would be appropriate. A pie chart would not because these are counts and not fractions of a whole.
   b) The *Who* is athletic trainers. Reporting counts of complications could be misleading. A trainer who treated many patients with cryotherapy would be more likely to have seen a complication than one who used cryotherapy rarely. We'd prefer a study in which the *Who* was patients, so we could assess the risks of each complication.

19. a) 16.6%   b) 11.8%   c) 37.7%   d) 53.0%

21. 1,755 students applied for admission to the magnet schools program. 53% were accepted, 17% were wait-listed, and the other 30% were turned away. While the overall acceptance rate was 53%, 93.8% of Blacks and Hispanics were accepted, compared to only 16.6% of Asians and 35.5% of Whites. Overall, 29.5% of applicants were Black or Hispanic, but only 6% of those turned away were. Asians accounted for 16.6% of all applicants, but 24.9% of those turned away were. Whites were 54% of the applicants and 68% of those who were turned away. It appears that the admissions decisions were not independent of the applicant's ethnicity.

23. a) 82.5%   b) 12.9%   c) 11.1 %
   d) 13.4%   e) 85.7%
   f) Many charts are possible. Here is a side-by-side bar chart.

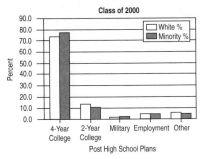

g) The white and minority students' plans are very similar. The small differences should be interpreted with caution because the total number of minority students is small. There is little evidence of an association between race and plans.

25. a) 67.1 %   b) 31.3 %   c) 93.9%   d) 74.1 %
   e) No, the percent of Canadian citizens overall speaking French is 31.3%, whereas the percent of citizens of Quebec speaking French is 93.9%. These percentages are very different and should be the same if language knowledge and province of residence are independent.

27. a) 9.3%   b) 24.7%   c) 80.8%
   d) Yes, there is a strong association between weather and ability to forecast weather. On days he forecast rain, he was correct 30% of the time. When he forecast no rain, he was correct 97.5% of the time.

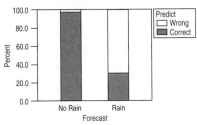

29. The change in attitude in 10 years was fairly small. There was a slight decrease in the response "One works full time, other part time," and there seems to be a similar increase in the response "One works, other works at home."

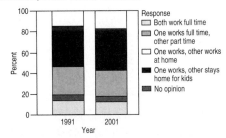

31. a)

	Total	Percent
Blood Pressure Low	95	20.0%
Normal	232	48.9%
High	147	31.0%

b)

	Under 30	Percent
Blood Pressure Low	27	27.6%
Normal	48	49.0%
High	23	23.5%

	30–49	Percent
Blood Pressure Low	37	20.7%
Normal	91	50.8%
High	51	28.5%

	Over 50	Percent
Blood Pressure Low	31	15.7%
Normal	93	47.2%
High	73	37.1%

c)

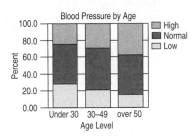

d) As age increases, the percent of adults with high blood pressure increases. On the other hand, the percent of adults with low blood pressure decreases.

e) No, but it gives an indication that it might. There might be additional reasons that explain the differences in blood pressures.

33. No, it could be related to other socioeconomic variables. Perhaps women who became mothers were less likely to continue their educations.

35. a) 34.1%   b) 3.8%
   c) In Europe, a greater percentage of toys are sold in toy chains, toy, hobby, and game retailers, as well as department stores. Those distribution channels account for 75% of the sales in Europe. In North America, general merchandise and food, drug, and miscellaneous outlets and toy chains have the highest percentages, totaling 86% of the toy sales.

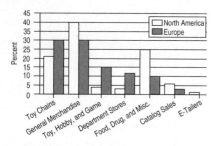

d) The distribution channel varies by region, although general merchandise stores are either the most important or a close second in all regions. Toy chains are especially important in Europe, and toy and hobby shops are significant in Asia. Only in North America are a sizable percentage of toys sold in general food and drug stores. E-tailing of toys was relatively unimportant in the year 1999.

37. a) 160 of 1300, or 12.3%
b) Yes. Major surgery: 15.3% vs. minor surgery: 6.7%
c) Large hospital: 13%, small hospital: 10%
d) Large hospital: Major 15% vs. minor 5%
Small hospital: Major 20% vs. minor 8%
e) No. Smaller hospitals have a higher rate for both kinds of surgery, even though it's lower "overall."
f) The small hospital has a larger percentage of minor surgeries (83.3%) than the large hospital (20%). Minor surgeries have a lower delay rate, so the small hospital looks better "overall."

39. a) 42.6%
b) A higher percentage of males than females were admitted: Males: 47.2% to females: 30.9%.
c) Program 1: Males 61.9%, females 82.4%
Program 2: Males 62.9%, females 68.0%
Program 3: Males 33.7%, females 35.2%
Program 4: Males 5.9%, females 7.0%
d) The comparisons in c) show that males have a *lower* admittance rate in every program, even though the overall rate shows males with a higher rate of admittance. This is an example of Simpson's paradox.

CHAPTER 4

1. Answers will vary.
3. a) Unimodal (near 0) and skewed. Many seniors will have 0 or 1 speeding tickets. Some may have several, and a few may have more than that.
b) Probably unimodal and slightly skewed to the right. It is easier to score 15 strokes over the mean than 15 strokes under the mean.
c) Probably unimodal and symmetric. Weights may be equally likely to be over or under the average.
d) Probably bimodal. Men's and women's distributions may have different modes. It may also be skewed to the right, since it is possible to have very long hair, but hair length can't be negative.
5. Skewed to right. Bimodal. One mode near 1, another near 8. All data between 1 and 37. Some outliers on the high end.
7. a) Bimodal. Looks like two groups. Modes are near 6% and 46%. No real outliers.
b) Looks like two groups of cereals, a low-sugar and a high-sugar group.

9. a) 78%
b) Skewed to the right with at least one high outlier. Most of the wineries are less than 90 acres with a few high ones. The mode is between 0 and 30 acres.

11. a)

Stem	Leaf
2.2	65
2.2	
2.1	68
2.1	12
2.0	768797
2.0	2343

2.1|7 = $2.179

b) The distribution of gas prices is skewed to the right, centered around $2.10 per gallon, with most stations charging between $2.05 and $2.13. The lowest and highest prices were $2.03 and $2.27.
c) There's a gap; no stations charge between $2.19 and $2.25.

13. Skewed to the right, mode in low 30s. Three low outliers, then a gap from 9 to 22.

15. a) This is not a histogram. The horizontal axis should split the number of home runs hit in each year into bins. The vertical axis should show the number of years in each bin.
b)

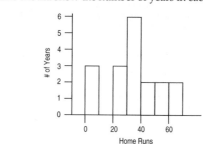

17. This distribution is nearly uniform. Values range from 65 to 155 horsepower. The center is near 105.

Stem	Leaf
15	05
14	2
13	0358
12	0559
11	00555
10	359
9	00577
8	0058
7	01158
6	55889

6|5 = 65 horsepower

19. a)

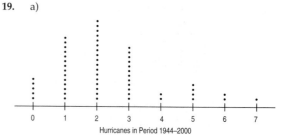

b) Slightly skewed to the right. Unimodal, mode near 2. Possibly a second mode near 5. No outliers.

21. Skewed to the right, possibly bimodal with one fairly symmetric group near 4.4, another at 5.6. Two outliers in middle seem not to belong to either group.

```
Stem │ Leaf
  57 │ 8
  56 │ 27
  55 │ 1
  54 │
  53 │
  52 │ 9
  51 │
  50 │ 8
  49 │
  48 │ 2
  47 │ 3
  46 │ 034
  45 │ 267
  44 │ 015
  43 │ 0199
  42 │ 669
  41 │ 22
```

4│12 = 4.12 pH

23. a) They should be put on the same scale, from 0 to 20 days.

b) Men have a mode at 1 day then tapering off from there. Women have a mode near 5 days with a sharp drop afterward.

c) A possible reason is childbirth.

25. Histogram bin widths are too wide to be useful.

27. a) Histogram bins are too narrow to be useful.

b) Skewed to the left, mode near 170, several outliers below 100. Fairly tightly clustered except for outliers.

29. Neither appropriate nor useful. Zip codes are categorical data, not quantitative. But they do contain *some* information. The leading digit gives a rough East to West placement in the United States. So we see that they have almost no customers in the Northeast, but a bar chart by leading digit would be more appropriate.

31. What is the *x*-axis? What units does productivity have here?

33. a)

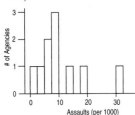
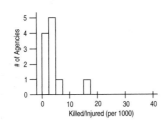

b) Assaults have a much more spread-out distribution. Both are skewed to the right. Killed/Injured has a center near 4, assaults near 10. There are high outliers in both distributions.

c) Outliers are BATF and National Park Service in assaults and National Park Service in number killed/injured.

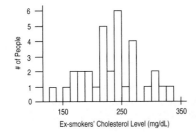

35. a) Rounding the data values and splitting each stem into five groups gives: (other answers are possible)

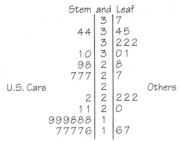

1│16 means 16 mpg

b) Data for U.S. car mileage have a lower center and are skewed to the right with a mode near 19 mpg. Perhaps bimodal with a second mode near 28 mpg. Others have more of the cars at a higher mode near 32 and only a few with lower mode near 22.

37. a)

```
Stem │ Leaf
   8 │ 0148
   7 │ 9
   6 │ 02  3
   5 │ 6
   4 │
   3 │ 25
   2 │ 8
```

2│8 = 28 thousand $ per mW

b) Left skewed with some gaps.

c)

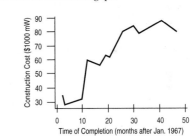

d) Costs are generally increasing over time.

39. a) Most of the data are found in the far left of this histogram. The distribution is very skewed to the right.

b) Re-expressing the data by, for example, logs or square roots might help make the distribution more nearly symmetric.

41. a) The logarithm makes the histogram more unimodal and symmetric. It is easy to see that the center is around 3.5 in log assets.

b) That has a value of around 2,500 million dollars.

c) That has a value of around 1,000 million dollars.

CHAPTER 5

1. Answers will vary.

3. a) $1001.50   b) 1025, 850, 1200   c) 835, 350

5. a) Median will probably be unaffected. The mean will be larger.

b) The range and standard deviation will increase; IQR will be unaffected.

7. a) Mean $525, median $450
  b) 2 employees earn more than the mean.
  c) The median because of the outlier.
  d) The IQR will be least sensitive to the outlier of $1200, so it would be the best to report.

9. a) The standard deviation will be larger for set 2, since the values are more spread out. SD(set 1) = 2.2, SD(set 2) = 3.2.
  b) The standard deviation will be larger for set 2, since 11 and 19 are farther from 15 than are 14 and 16. Other numbers are the same. SD(set 1) = 3.6, SD(set 2) = 4.5.
  c) The standard deviation will be the same for both sets, since the values in the second data set are just the values in the first data set + 80. The spread has not changed. SD(set 1) = 4.2. SD(set 2) = 4.2.

11. a) Since these data are strongly skewed to the right, the median and IQR are the best statistics to report.
  b) The mean will be larger than the median because the data are skewed to the right.
  c) Median 4 million. The IQR is 4.5 million (Q_3 = 6 million, Q_1 = 1.5 million).
  d) The distribution of populations of the states and Washington, D.C., is unimodal and skewed to the right. The median population is 4 million. One state is an outlier, with a population of 34 million.

13. According to the fence rule, 61 is not technically an outlier. The median is 24.5. The two quartiles are 14 and 33, so the IQR is 19. That means 1.5 IQRs is 28.5. Adding 28.5 to the upper quartile of 33 gives a fence of 61.5. So, 61 is just within the fence and not an outlier by that definition. But it certainly looks unusual compared with the other 9 seasons!

15. The southern and western states appear to have significantly higher growth rates between 1990 and 2000 than the northeastern and midwestern states.

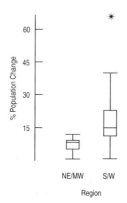

17. a) The histogram of height is most nearly symmetric and shows no outliers. That makes it the best candidate for summarizing with a mean.
  b) The histogram of sip size shows a high outlier. The standard deviation is sensitive to outliers, so we'd prefer to use the IQR for this one.

19. a) Both girls have a median score of about 17 points per game, but Scyrine is much more consistent. Her IQR is about 2 points, while Alexandra's is over 10.
  b) If the coach wants a consistent performer, she should take Scyrine. She'll almost certainly deliver somewhere between

15 and 20 points. But, if she wants to take a chance and needs a "big game," she should take Alexandra. Alex scores over 24 points about a quarter of the time. (On the other hand, she scores under 11 points as often.)

21. Women appear to marry about 3 years younger than men, but the two distributions are very similar in shape and spread.

23. a)   Seneca   b) Seneca   c) Keuka
  d) The Cayuga and Seneca wineries have about the same average price; the boxplots show similar medians and similar IQRs, even though Seneca's range is larger. Keuka has consistently higher prices except for one low outlier, and a more consistent pricing as shown by the smaller IQR. Distributions for all three appear to be roughly symmetric around their centers.

25. a) Because the mean is larger than the median, we might suppose that the distribution is right skewed. An outlier or outliers might explain why the SD is so much larger than the IQR.
  b) The differences in the statistics could be caused by skewness or by outliers on the high end.
  c) The distribution of times in the men's 100-m swim in Sydney is unimodal, with most times between 48 and 53 seconds. There is a slight skew to the right, with three times between 58 and 63 and one very extreme outlier at about 113 seconds.

27. a) Class 3
  b) Class 3
  c) Class 3 because it is skewed. Median would be higher than the mean because it is skewed to the left.
  d) Class 1
  e) Class 1

29. a) Essentially symmetric, very slightly skewed to the right with two high outliers at 36 and 48. Most victims are between the ages of 16 and 24.
  b) The slight increase between ages 22 and 24 is apparent in the histogram but not in the boxplot. It may be a second mode.
  c) The median would be the most appropriate measure of center because of the slight skew and the extreme outliers.
  d) The IQR would be the most appropriate measure of spread because of the slight skew and the extreme outliers.

31. a) Probably slightly left skewed. The mean is slightly below the median and the 25th percentile is farther from the median than the 75th percentile.
  b) No, all data are within the fences.
  c)

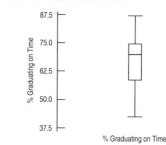

  d) The 48 universities graduate, on average, about 68% of freshman "on time," with percents ranging from 43% to 87%. The middle 50% of these universities graduate between 59% and 75% of their freshman in 4 years.

33. a) *Who:* Student volunteers
  What: Memory test
  Where, when: Not specified

How: Students took memory test 2 hours after drinking caffeine-free, half-dose caffeine, or high-caffeine soda. *Why:* To see if caffeine makes you more alert and aids memory retention.

b) Drink: categorical; Test score: quantitative.

c)

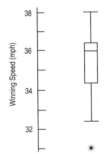

d) The participants scored about the same with no caffeine and low caffeine. The medians for both were 21 points, with slightly more variation for the low-caffeine group. The high-caffeine group generally scored lower than the other two groups on all measures of the 5-number summary: min, lower quartile, median, upper quartile, and max. It is not clear if the differences are significant.

35. a) About 36 mph
    b) Q_1 about 34.5 mph and Q_3 about 36.5 mph
    c) The range appears to be 7 mph from about 31 to 38 mph. The IQR is about 2 mph.
    d) We can't know exactly, but the boxplot may look something like this:

e) The median winning speed has been about 36 mph, with a max of about 38 and a min of about 31 mph. The middle 50% of the speeds appeared to range from 34.5 to 36.5 mph, for an IQR of 2 mph. The middle 20% of the data appear to be tightly clustered around 36. Only about 5% of the Kentucky Derby winners raced at speeds of more than 37 mph.

37. a) Boys   b) Boys   c) Girls
    d) The boys appeared to have more skew, as their scores were less symmetric between quartiles. The girls' quartiles are the same distance from the median, although the left tail stretches a bit farther to the left.
    e) Girls. Their median and upper quartiles are larger. The lower quartile is slightly lower, but close.
    f) $[14(4.2) + 11(4.6)]/25 = 4.38$

39. a)

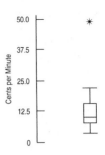

b) The mean cost per minute for Net2Phone rates is 13.7 cents. The median cost per minute is 9.9 cents per minute. Because of the outliers, the median is the more appropriate measure of center to use.
c) The IQR for the cost per minute of the phone calls is 7.6 cents. The standard deviation is 11.8 cents per minute. Again, because of the outliers, the better choice for measure of spread is the IQR.
d) Yes, both India and Pakistan are outliers. They each lie more than 1.5 times the IQR above Q_3.
e) The distribution of the long-distance rates for Net2Phone is unimodal and symmetric except for two outliers, India and Pakistan, at 49 cents per minute. The median cost is 9.9 cents per minute. The middle 50% of costs per minute are between 7.9 and 15.5 cents per minute for an IQR of 7.6 cents per minute.
f) These data represent only 24 of the 250 countries that Net2Phone offers service to. These might not be representative of the entire group.

41. a) 5-number summary: 275, 448, 499, 531, 604
    IQR: 83
    Mean: 487.2
    SD: 72.3

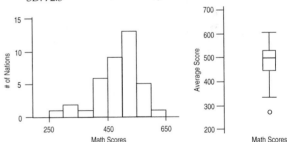

b) The distribution is unimodal and skewed to the left. As shown in the boxplot, there is one outlier, an average score of 275. The median was 499, while the mean was 487.2, slightly below the median score. The middle 50% of the nations scored between 448 and 531 for an IQR of 83 points.

43. In the year 2000, per capita gasoline use by state in the United States averaged around 500 gallons per person (mean 488, median 502). States varied in their consumption with a standard deviation of 62 gallons, ranging from a min of 297 (NC) to a max of 587 (TN). The two low outliers were NC and HI. The IQR of

82 gallons reflects the fact that 50% of the states varied from 453 to 533 gallons per capita.

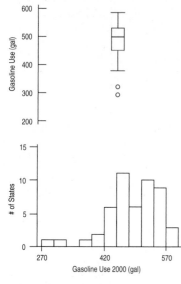

45. a) Although numeric codes have been assigned to the different titles, these data are categorical, not quantitative. The mean of 54.41 is meaningless.
b) The typical reasons are skewness and/or outliers.
c) No. Here the numbers are just codes. Most of the people probably had titles of Mr. or Mrs., making the "median" 1, but these summary statistics are meaningless.
47. No, boxplots are for quantitative data, and these are categorical, although coded as numbers. The numbers used for hair color and eye color are arbitrary, so the boxplot and any accompanying statistics for eye color make no sense.
49. The analysis would probably be improved by using the log-transformed data. The distributions are more symmetric, and it's easier to compare the groups. The outliers are eliminated.

CHAPTER 6

1. a) Skewed to the right; mean is higher than median.
b) $350 and $950.
c) Minimum $350. Mean $750. Median $550. Range $1200. IQR $600. Q1 $400. SD $400.
d) Minimum $330. Mean $770. Median $550. Range $1320. IQR $660. Q1 $385. SD $440.
3. Lowest score = 910. Mean = 1230. SD = 120. $Q_3 = 1350$. Median = 1270. IQR = 240.
5. In January, a high of 55 is not quite 2 standard deviations above the mean, whereas in July a high of 55 is more than 2 standard deviations lower than the mean. So it's less likely to happen in July.
7. a) Megan   **b)** Anna
9. a) About 1.81 standard deviations below the mean
b) 1000 ($z = -1.81$) is more unusual than 1250 ($z = 1.17$).
11. a) Mean = $1152 - 1000 = 152$ pounds; SD is unchanged at 84 pounds.
b) Mean = $0.40(1152) = \$460.80$; SD = $0.40(84) = \$33.60$.
13. Min = $0.40(980) - 20 = \$372$; median = $0.40(1140) - 20 = \$436$; SD = $0.40(84) = \$33.60$; IQR = $0.40(102) = \$40.80$.

15. College professors can have between 0 and maybe 40 (or possibly 50) years experience. A standard deviation of 1/2 year is impossible, because many professors would be 10 or 20 SDs away from the mean, whatever it is. A SD of 16 years would mean that 2 SDs on either side of the mean is plus or minus 32, for a range of 64 years. That's too high. So, the SD must be 6 years.
17. a)

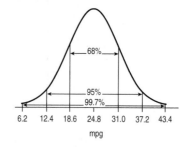

b) 18.6 to 31.0 mpg   **c)** 16%
d) 13.5%   **e)** less than 12.4 mpg
19. Any weight more than 2 standard deviations below the mean, or less than $1152 - 2(84) = 984$ pounds, is unusually low. We expect to see a steer below $1152 - 3(84) = 900$ pounds only rarely.
21. a) 16%   **b)** 5.7%
c) Because the Normal model doesn't fit well.
d) Distribution is skewed to the right with an outlier on the high side.
23. a)

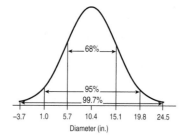

b) Between 1.0 and 19.8 inches
c) 2.5%   **d)** 34%   **e)** 16%
25. Since the histogram is not unimodal and symmetric, it is not wise to have faith in numbers from the Normal model.
27. a) 16%
b) 16% of the students should be watching less than -1.27 hours a week, but no one can watch less than 0 hours, so the model doesn't fit well.
c) Data are strongly skewed to the right, not symmetric.
29. a) 6.68%   **b)** 98.78%   **c)** 71.63%   **d)** 61.71%
31. a) 0.842   **b)** -0.675 **c)** -1.881 **d)** -1.645 to 1.645
33. a) 12.2%   **b)** 71.6%   **c)** 23.3%
35. a) 1259.7 lb   **b)** 1081.3 lb   **c)** 1108 lb to 1196 lb
37. a) 1130.7 lb   **b)** 1347.4 lb   **c)** 113.3 lb
39. a) 79.55   **b)** 18.50   **c)** 95.79   **d)** -2.77
41. a)

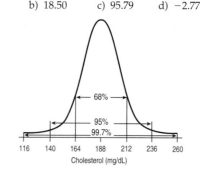

b) 30.85%   c) 17.00%   d) 32 points   e) 212.9 points
43. a) 11.1%   b) (35.9, 40.5) inches   c) 40.5 inches
45. The average Dutch man has a *z*-score of 2.31 in Athens. About 1% of Greek men are this tall or taller.
47. Mean 12.1 months, SD 1.3 months
49. a) 5.3 grams   b) 6.4 grams
   c) Younger because SD is smaller.
   d) Mean 62.7 grams, SD 6.2 grams

PART I REVIEW

1. a)

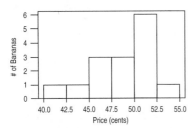

   b) Median 49 cents, IQR 6 cents
   c) The distribution is unimodal and left skewed. The center is near 50 cents; values range from 42 cents to 53 cents.
3. a) If enough sopranos have a height of 65 inches, this can happen.
   b) The distribution of heights for each voice part is roughly symmetric. The basses are slightly taller than the tenors. The sopranos and altos have about the same median height. Heights of basses and sopranos are more consistent than those of altos and tenors.
5. a) It means their heights are also more variable.
   b) The *z*-score for women to qualify is 2.40, compared with 1.75 for men, so it is harder for women to qualify.
7. a) *Who*—People who live near Watsamatta University
   What—Age, attended college? Favorable opinion of Watsamatta?
   When—Not stated
   Where—Region around Watsamatta U.
   Why—To report to the university's directors
   How—Sampled and phoned 850 local residents
   b) Age—Quantitative (years); attended college?—categorical; favorable opinion?—categorical.
   c) The fact that the respondents know they are being interviewed by the university's staff may influence answers.
9. a) These are categorical data, so mean and standard deviation are meaningless.
   b) Not appropriate. Even if it fits well, the Normal model is meaningless for categorical data.
11. a)

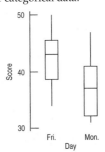

   b) The scores on Friday were higher by about 5 points. This is a drop of more than 10% of the average score and shows that

students fared worse on Monday after preparing for the test on Friday. The spreads are about the same, but the scores on Monday are a bit skewed to the right.
   c)

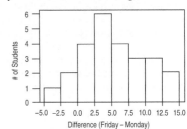

   d) The changes (Friday − Monday) are unimodal and centered near 4 points, with a spread of about 5 (SD). They are fairly symmetric, but slightly skewed to the right. Only 3 students did better on Monday (had a negative difference).
13. a) Categorical
   b) Go fish. All you need to do is match the denomination. The denominations are not ordered. (Answers will vary.)
   c) Gin rummy. All cards are worth their value in points (face cards are 10 points). (Answers will vary.)
15. a) Annual mortality rate for males (quantitative) in deaths per 100,000 and water hardness (quantitative) in parts per million.
   b) Calcium is skewed right, possibly bimodal. There looks to be a mode down near 12 ppm that is the center of a fairly tight symmetric distribution and another mode near 62.5 ppm that is the center of a much more spread out, symmetric (almost uniform) distribution. Mortality, however, appears unimodal and symmetric with the mode near 1500 deaths per 100,000.
17. a) They are on different scales.
   b) January's values are lower and more spread out.
   c) Roughly symmetric but slightly skewed to the left. There are more low outliers than high ones. Center is around 40 degrees with an IQR of around 7.5 degrees.
19. a) Bimodal with modes near 2 and 4.5 minutes. Fairly symmetric around each mode.
   b) Because there are two modes, which probably correspond to two different groups of eruptions, an average might not make sense.
   c) The intervals between eruptions are longer for long eruptions. There is very little overlap. More than 75% of the short eruptions had intervals less than about an hour (62.5 minutes), while more than 75% of the long eruptions had intervals longer than about 75 minutes. Perhaps the interval could even be used to predict whether the next eruption will be long or short.
21. a)

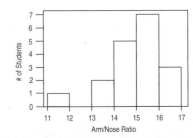

   The distribution is left skewed with a center of about 15. It has an outlier between 11 and 12.
   b) Even though the distribution is somewhat skewed, the mean and median are close. The mean is 15.0 and the SD is 1.25.

c) Yes. 11.8 is already an outlier. 9.3 is more than 4.5 SDs below the mean. It is a *very* low outlier.

23. If we look only at the overall statistics, it appears that the follow-up group is insured at a much lower rate than those not traced (11.1% of the time as compared with 16.6%). But most of the follow-up group were black, who have a lower rate of being insured. When broken down by race, the follow-up group actually has a higher rate of being insured for both blacks and whites. So the overall statistic is misleading and is attributable to the difference in race makeup of the two groups.

25. a)

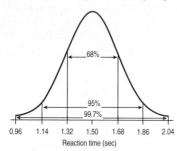

b) According to the model, reaction times are symmetric with center at 1.5 seconds. About 95% of all reaction times are between 1.14 and 1.86 seconds.

c) 8.2%

d) 24.1%

e) Quartiles are 1.38 and 1.62 seconds, so the IQR is 0.24 seconds.

f) The slowest 1/3 of all drivers have reaction times of 1.58 seconds or more.

27. a)

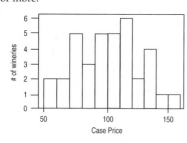

b) Mean $100.25, SD $25.54 Case Price.

c) The distribution is somewhat symmetric and unimodal, but the center is rather flat, almost uniform.

d) 64%. The Normal model seems to work reasonably well, since it predicts 68%.

29. a) 38 cars

b) Possibly because the distribution is skewed to the right.

c) Center—median is 148.5 cubic inches. Spread—IQR is 126 cubic inches.

d) No. It's bigger than average, but smaller than more than 25% of cars. The upper quartile is at 231 inches.

e) No. 1.5 IQR is 189, and 105 − 189 is negative, so there can't be any low outliers. 231 + 189 = 420. There aren't any cars with engines bigger than this, since the maximum has to be at most 105 (the lower quartile) + 275 (the range) = 380.

f) Because the distribution is skewed to the right, this is probably not a good approximation.

g) Mean, median, range, quartiles, IQR, and SD all get multiplied by 16.4.

31. a) *Who*—100 health food store customers

What—Have you taken a cold remedy?, and Effectiveness (scale 1 to 10)

When—Not stated

Where—Not stated

Why—Promotion of herbal medicine

How—In-person interviews

b) Have you taken a cold remedy?—categorical. Effectiveness—categorical or ordinal.

c) No. Customers are not necessarily representative, and the council had an interest in promoting the herbal remedy.

33. a) 27.5%

b) If this were a random sample of all voters, yes.

c) 50.1%   d) 10.2%   e) 27.3%   f) 40.9%

35. a) Republican—1101, Democrat—1404, Independent—1497; or Republican—27.5%, Democrat—35.1%, Independent—37.4%

b)

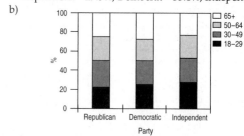

c) Age distributions are similar along party lines, with Independents having slightly more young voters and Republicans having slightly fewer. Age groups are fairly evenly distributed, with each party having about 25% of each age group.

d) Nearly independent with exceptions noted in part c.

37. a) *Who*—Years from 1991 to 2000

What—Bicycle injuries reported

When—1991 to 2000

Where—Massachusetts

Why—To study trends in bicycle safety

How—Safety bureau reports

b)

Stem	Leaf
17	6
16	
15	2
14	5
13	1478
12	8
11	2
10	3

10|3 = 1030 injuries

c)

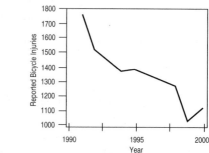

d) In the stemplot, it is clear that some years may be unusual.

e) The downward trend is visible in the timeplot.

f) In the decade 1991 to 2000, reported bicycle injuries decreased steadily from about 1800 per year to around 1100 per year.

39. a) Mean 1.29 hours, SD 0.43 hours
   b) 1.4 hours
   c) 0.89 hours (or 53.4 minutes)
   d) Survey results vary, and the mean and the SD may have changed.

CHAPTER 7

1. a) Weight in ounces: explanatory; Weight in grams: response. (Could be other way around.) To predict the weight in grams based on ounces. Scatterplot: positive, straight, strong (perfectly linear relationship).
   b) Circumference: explanatory. Weight: response. To predict the weight based on the circumference. Scatterplot: positive, linear, moderately strong.
   c) Shoe size: explanatory; GPA: response. To try to predict GPA from shoe size. Scatterplot: no direction, no form, very weak.
   d) Miles driven: explanatory; Gallons remaining: response. To predict the gallons remaining in the tank based on the miles driven since filling up. Scatterplot: negative, straight, moderate.

3. a) Altitude: explanatory; Temperature: response. (Other way around possible as well.) To predict the temperature based on the altitude. Scatterplot: negative, possibly straight, weak to moderate.
   b) Ice cream cone sales: explanatory. Air conditioner sales: response—although the other direction would work as well. To predict one from the other. Scatterplot: positive, straight, moderate.
   c) Age: explanatory; Grip strength: response. To predict the grip strength based on age. Scatterplot: curved down, moderate. Very young and elderly would have grip strength less than that of adults.
   d) Reaction time: explanatory; Blood alcohol level: response. To predict blood alcohol level from reaction time test. (Other way around is possible.) Scatterplot: positive, nonlinear, moderately strong.

5. a) None   b) 3 and 4   c) 2, 3, and 4
   d) 1 and 2   e) 3 and possibly 1

7. There seems to be a very weak—or possibly no—relation between brain size and performance IQ.

9. a)

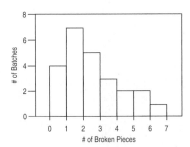

   b) Unimodal, skewed to the right. The skew.
   c) The positive, somewhat linear relation between batch number and broken pieces.

11. a) 0.006   b) 0.777 c) −0.923 d) −0.487

13. a) Yes. It shows a linear form and no outliers.

b) There is a strong, positive, linear association between drop and speed; the greater the coaster's initial drop, the higher the top speed.

15. a) −0.649
   b) The correlation will not change. Correlation does not change with change of units.
   c) There is a moderately strong negative linear association between number of calories consumed and time at the table for toddlers. Longer times are associated with fewer calories.
   d) There appears to be a relation between time at the table and calories consumed, but there is no reason to believe increased time *causes* decreased calorie consumption. There may be a lurking variable.

17. a)

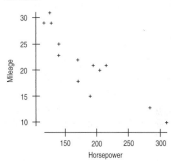

   b) Negative, linear, strong.
   c) −0.878
   d) There is a strong linear relation in a negative direction between horsepower and mpg. As the horsepower increases, the fuel economy decreases.

19.

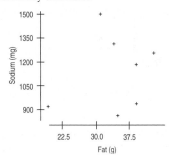

(Plot could have explanatory and predictor variables swapped.) Correlation is 0.199. There does not appear to be a relation between sodium and fat content in burgers, especially without the low-fat low-sodium item. The correlation of 0.199 shows a weak relationship, even with the outlier included.

21. a) Yes, the scatterplot appears to be somewhat linear.
   b) As the number of runs increases, the attendance also increases.
   c) There is a positive association, but it does not *prove* that more fans will come if the number of runs increases. Association does not indicate causality.

23. There may be an association but not a correlation unless the variables are quantitative. There could be a correlation between average number of hours of TV watched per week per person and the number of crimes committed per year. Even if there is a relationship, it doesn't mean one causes the other.

25. a) Actually, yes, taller children will tend to have higher reading scores, but this doesn't imply causation.

b) Older children are generally both taller and are better readers. Age is the lurking variable.

27. The scatterplot is not linear; correlation is not appropriate.

29. a) Assuming the relation is linear, a correlation of −0.772 shows a strong relation in a negative direction.
b) Continent is a categorical variable. Correlation does not apply.

31. This is categorical data even though it is represented by numbers. The correlation is meaningless.

33. A scatterplot shows a generally straight scattered pattern with no outliers. The correlation between *drop* and *duration* is 0.35, indicating that rides on coasters with greater initial drops generally last somewhat longer, but the association is weak.

35. a)

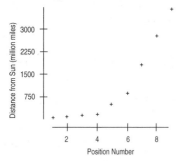

The relation between position and distance is nonlinear, with a positive direction. There is very little scatter from the trend.
b) The relation is not linear.
c)

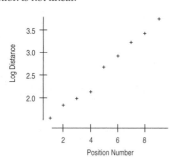

The relation between position number and log of distance appears to be roughly linear.

CHAPTER 8

1.

$\bar{x}$	s_x	$\bar{y}$	s_y	r	$\hat{y} = b_0 + b_1x$
a) 10	2	20	3	0.5	$\hat{y} = 12.5 + 0.75x$
b) 2	0.06	7.2	1.2	−0.4	$\hat{y} = 23.2 − 8x$
c) 12	6	152	30	−0.8	$\hat{y} = 200 − 4x$
d) 2.5	1.2	25	100	0.6	$\hat{y} = −100 + 50x$

3. a) Model is appropriate.
b) Model is not appropriate. Relationship is nonlinear.
c) Model may not be appropriate. Spread is changing.

5. The line $\hat{y} = 7.0 + 1.1x$ minimizes the sum of squared vertical distances from the points to the line.

7. a) Price (in thousands of dollars) is y and size (in square feet) is x.
b) Slope is thousands of $ per square foot.
c) Positive. Larger homes should cost more.

9. Size differences explain 71.4% of the variation in home prices.

11. a)   0.845
b) Price should be 0.845 SDs above the mean in price.
c) Price should be 1.690 SDs below the mean in price.

13. a) Every square foot increase in size increases the average price $0.061 × 1000, or $61.00.
b) 230.82 thousand, or $230,820.
c) $115,020; $6000 is the residual.

15. a) Probably. The residuals show some initially low points, but there is no clear curvature.
b) Tar content explains 92.4% of the variability in nicotine.

17. a) $r = 0.961$
b) Nicotine should be 1.922 SDs below average.
c) Tar should be 0.961 SDs above average.

19. a) $\widehat{nicotine} = 0.15403 + 0.065052\ tar$
b) 0.414 mg
c) Every extra milligram of tar is associated with an increase of 0.065 mg of nicotine.
d) We'd expect a cigarette with no tar to have 0.154 mg of nicotine.
e) 0.1094 mg

21. 300 pounds/foot. If a "typical" car is 15 feet long, all of 3, 30, and 3000 would give ridiculous weights.

23. a) R^2 does not tell whether the model is appropriate, but measures the strength of the linear relationship. High R^2 could also be due to an outlier.
b) Predictions based on a regression line are for average values of y for a given x. The actual wingspan will vary around the prediction.

25. a) Probably not. Your score is better than about 97.5% of people, assuming scores follow the Normal model. Your next score is likely to be closer to the mean.
b) The friend probably should retake the test. His score is better than only about 16% of people. His score is likely to be closer to the mean.

27. a) Moderately strong, fairly straight, and positive. Possibly some outliers (higher than expected math scores).
b) The student with 500 verbal and 800 math.
c) Positive, fairly strong linear relationship. 46.9% of variation in math scores is explained by verbal scores.
d) *Predicted math* = 217.7 + 0.662 × *verbal*.
e) Every point of verbal score adds an average 0.662 points to the predicted math score.
f) 548.5 points.
g) 53.0 points.

29. a) 0.685
b) *Predicted verbal* = 162.1 + 0.71 × *math*.
c) The observed verbal score is higher than predicted from the math score.
d) 516.7 points.
e) 559.6 points.
f) Regression to the mean. Someone whose math score is below average is predicted to have a verbal score below average, but not as far (in SDs). So if we use *that* verbal score to predict math, they will be even closer to the mean in predicted math score than their observed math score. If we kept cycling back

and forth, eventually we would predict the mean of each and stay there.

31. a)

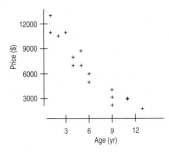

b) Negative, linear, strong.
c) Yes.
d) −0.946
e) Age accounts for 89.4% of the variation in advertised prices.
f) Other factors contribute—options, condition, mileage, etc.

33. a) *Predicted price* = 12319.6 − 924 × *years*.
b) Every extra year of age decreases average value by $924.
c) The average new Corolla costs $12,319.60.
d) $5851.60
e) Negative residual. Its price is below the predicted value for its age.
f) −$1579.60
g) No. After age 13, the model predicts negative prices. The relationship is no longer linear.

35. a)

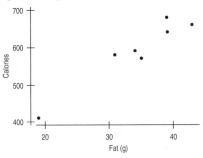

b) 92.3% of the variation in calories can be accounted for by the fat content.
c) *Predicted calories* = 211.0 + 11.06 × *calories/fat gram*.
d)

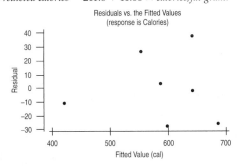

Residuals show no clear pattern, so the model seems appropriate.
e) Could say a fat-free burger still has 211.0 calories, but this is extrapolation (no data close to 0).

f) Every gram of fat adds 11.06 calories, on average.
g) 553.5 calories.

37. a) The regression was for predicting calories from fat, not the other way around.
b) *Predicted fat grams* = −15.0 + 0.083 *grams/calories*.
Predict 34.8 grams of fat.

39. a) 82.9 minutes.
b) The linear relationship is strong, but without seeing the data we don't know if the model is appropriate.
c) Every extra minute of digital time adds 0.44 minutes of analog time, on average.
d) It lasts longer in analog mode than predicted for its number of digital minutes.

41. a) 0.578
b) CO_2 levels account for 33.4% of the variation in mean temperature.
c) *Predicted mean temperature* = 15.3066 + 0.004 × CO_2.
d) Each part per million of CO_2 adds 0.004 degrees to mean temperature (Celsius).
e) One *could* say that with no CO_2 in the atmosphere, there would be a temperature of 15.3066 degrees Celsius, but this is extrapolation.
f) No.
g) Predicted 16.7626 degrees C.

43. a) *Predicted % body fat* = −27.4 + 0.25 × *weight*.
b) Residuals look randomly scattered around 0, so conditions are satisfied.
c) For each pound of weight, average body fat increases 0.25 percent.
d) Reliable is relative. R^2 is 48.5%, but residuals have a standard deviation of 7%, so variation around the line is large.
e) 0.9 percent.

45. a) $\widehat{highjump}$ = 2.643 − 0.00656 *time*. Every additional second in the 800-m run is associated with a decrease of 0.0066 meters (on average) in the high jump.
b) *R-squared* is 24.6%.
c) Yes, the slope is negative. Faster runners tend to jump higher as well.
d) The residuals plot is fairly patternless, but there is one point for an extremely slow runner.
e) Not especially. The residual standard deviation is 0.058 meters, which is not much smaller than the SD of all high jumps (0.066 meters). I would be skeptical of the accuracy of the predictions.

47. a) As calcium levels increase, mortality rate decreases. Relationship is fairly strong, negative, and linear.
b) *Predicted mortality* = 1676.0 − 3.23 × *calcium*.
c) Mortality decreases 3.23 deaths per 100,000, on average, for each part per million of calcium. The intercept indicates a baseline mortality of 1676 deaths per 100,000 with no calcium, but this is extrapolation.
d) Exeter has 348.6 fewer deaths per 100,000 than the model predicts.
e) 1353 deaths per 100,000.
f) Calcium concentration explains 43.0% of the variation in death rate per 100,000 people.

49.

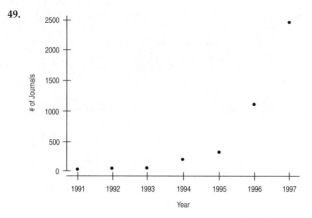

The growth is not linear. Need to re-express y or x first.

CHAPTER 9

1. a) The trend appears to be somewhat linear up to about 1940, but from 1940 to about 1970 the trend appears to be nonlinear. From 1975 or so, to present, the trend appears to be linear.
b) Relatively strong for certain time periods.
c) No, as a whole, the graph is clearly nonlinear. Within certain time periods the correlation is high (1975 to the present).
d) Overall, no. You could fit a linear model to the time period 1975 to 1995, but why? You don't need to interpolate, since every year is reported and extrapolation seems very dangerous.

3. a) Answers may vary. Using the data for 1955–1995, the scatterplot is relatively linear with some curvature. The residual plot shows a definite trend, indicating that the data are not linear. If you used the line, for 2005, the predicted age is 25.3 years.
b) Not much since, the data are not truly linear, and 2005 is 10 years from the last data point (extrapolating is risky).
c) No, that extrapolation of more than 50 years would be absurd. There's no reason to believe the trend from 1955 to 1995 will continue.

5. a) No. We need to see the scatterplot first to see if the conditions are satisfied and models are always wrong.
b) No, the linear model might not fit the data at all.

7. a) The graph shows that on average, students progress at about one reading level per year. This graph shows averages for each grade. The linear trend has been increased by using averages.
b) Very close to 1.
c) The individual data points would show much more scatter, and the correlation would be lower.
d) A slope of 1 would indicate that for each 1-year grade level increase, the average reading level is increasing by 1 year.

9. a) As the avg. monthly temperature increases by 1°F, the cost decreases by $2.13. So warmer temperatures indicate lower costs.
b) For an avg. monthly temperature of 0°F, the cost is predicted to be $133.
c) Too high; the residuals (observed − predicted) around 32°F are negative, showing that the model overestimates the costs.
d) $111.70
e) About $105.70

f) No, the residuals show a definite curved pattern. The data are probably not linear.
g) No, there would be no difference. The relationship does not depend on the units.

11. a) 1) High leverage, small residual.
2) No, not influential for the slope.
3) Correlation would decrease because outlier has large z_x and z_y, increasing correlation.
4) Slope wouldn't change much because the outlier is in line with other points.
b) 1) High leverage, probably small residual.
2) Yes, influential.
3) Correlation would weaken and become less negative.
4) Slope would increase toward 0, since outlier makes it negative.
c) 1) Some leverage, large residual.
2) Yes, somewhat influential.
3) Correlation would increase, since scatter would decrease.
4) Slope would increase slightly.
d) 1) Little leverage, large residual.
2) No, not influential.
3) Correlation would become stronger and become more negative because scatter would decrease.
4) Slope would increase and get a little less negative.

13. 1) e   **2)** d   **3)** c   **4)** b   **5)** a

15. a) Stronger. Both slope and correlation would increase.
b) Restricting the study to nonhuman animals would justify it.
c) Moderately strong.
d) For every year increase in life expectancy, the gestation period increases by about 15.5 days, on average.
e) About 270.5 days.

17. Perhaps high blood pressure causes high body fat, high body fat causes high blood pressure, or both could be caused by a lurking variable such as a genetic or lifestyle issue.

19. No. There is one high-leverage outlier, National Park Service. With that point, R-squared is 46.2%; without that point, R-squared is 0.

21. Answers may vary. There seem to be two time periods. From 1980 to 1993 or so, there was a consistent increase of about 1% a year. A regression of those years shows:
Predicted illegitimate percentage = −1988.94 + 1.014 *year*
with an R^2 of 98.9%. The years from 1994 to 1998 are nearly flat. A model of those years shows:
Predicted illegitimate percentage = −87.28 + 0.060 *year*
with an R^2 of 17.3%.
How to use the data depends on the purpose of the study.

23. a) The scatterplot is clearly nonlinear; however, the last few years, say from 1972 on, do appear to be linear.
b) Using the data from 1972 to 2000 gives $r = 0.998$ and $\widehat{CPI}$ = −9247.62 + 4.710 *year*. Predicted CPI in 2010 = 219.72.

25. a) The scatterplot appears to be nearly a perfectly straight line.
b) $\widehat{ChiCPI}$ = 0.899 + 0.9457 *NYCPI*. Chicago's CPI goes up on average about 0.95 points for every point that New York's CPI increases. The R^2 value of 99.73% indicates that this is a very strong relationship. The variables are quantitative and there are no outliers.
c) Yes, the residuals are not random. There are periods in which the inflation in Chicago tends to go faster or slower than the linear relationship with New York's CPI would indicate. In the late 70s, inflation in Chicago was increasing faster than NY's. In the 80s that relationship was reversed.

27. The stock market crash and World War II are clearly visible in their impact on CPI.

CHAPTER 10

1. a) 2.8
  b) 16.44
  c) 7.84
  d) 0.36
  e) 2.09
3. a) Fairly linear, negative, strong.
  b) Gas mileage decreases an average 7.652 mpg for each thousand pounds of weight.
  c) No. Residuals show a curved pattern.
5. a) Residuals are more randomly spread around 0, with some low outliers.
  b) *Predicted gallons per 100 miles = 0.625 + 1.178 × Weight.*
  c) For each additional 1000 pounds of weight, an additional 1.178 gallons will be needed to drive 100 miles.
  d) 21.06 miles per gallon.
7. About 31.59 dactyls.
9. a)

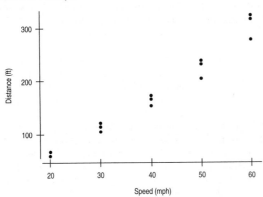

Predicted distance = −65.9 + 5.98 speed.
But residuals have a curved shape, so linear model is not appropriate.

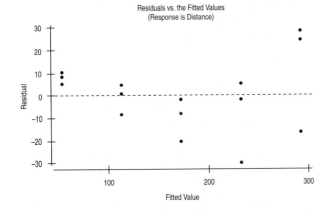

b)

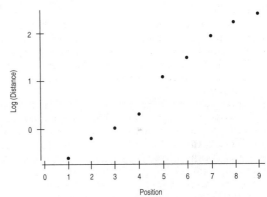

$\sqrt{Distance}$ linearizes the plot.
  c) *Predicted* $\sqrt{Distance}$ = 3.30 + 0.235 × *Speed.*
  d) 263.4 feet.
  e) 390.2 feet.
  f) Fairly confident, since R^2 = 98.4%, and s is small.
11. a) Log(*Salary*) works well.
  b) Predicted Log(*Salary*) = −113.47 + 0.057 × Year.
  c) 31.18 million dollars.
13. a)

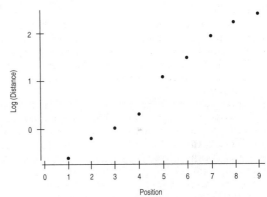

Log(*Distance*) against position works pretty well.
Predicted log(*Distance*) = 1.245 + 0.271 × *Position number.*
  b) Pluto's residual is not especially larger in the log scale. However, a model without Pluto predicts the 9th planet should be 5741 million miles. Pluto, at "only" 3666 million miles doesn't fit very well, giving support to the argument that Pluto doesn't behave like a planet.
15. a) *Predicted log(Distance) = 3.635.* Pluto's log(*Distance*) is 3.564, and Quaoar's is 3.602. Quaoar is better predicted.
  b) *Predicted log(Distance) = 1.299 + 0.235 New Position.* The R^2 has gone up to 99.5%. It's a slightly better fit.

17. a)

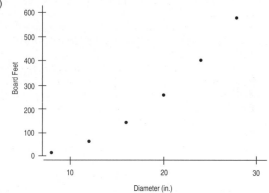

$$\sqrt{\widehat{bdft}} = -4 + diam$$
The model is exact.
b) 36 board feet.
c) 1024 board feet.

19.

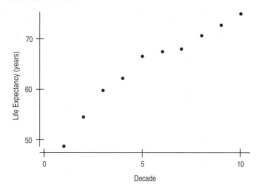

$$\log \widehat{life} = 3.879 + 0.18497 \log decade$$

21. The relationship cannot be made straight by the methods of this chapter.

23. a) $\sqrt{\widehat{left}} = 8.39 - 0.070(age)$
b) 68.8 years

25. a) $\log(\widehat{journals}) = -686.76 + 0.346(year)$
b) 21497.04 journals.
c) Seems too high. Internet growth has exploded, but the model is unlikely to be very accurate.

PART II REVIEW

1. % over 50, 0.69.
% under 20, −0.71.
% Graduating on time, −0.51.
% Full-time Faculty, 0.09.

3. a) There does not appear to be a linear relationship.
b) Nothing, there is no reason to believe that the results for the Finger Lakes region are representative of the vineyards of the world.
c) *Predicted Case Price = 92.77 + 0.567 × Years.*
d) Only 2.7% of the variation in case price is accounted for by the ages of vineyards. Most of that is caused by two outliers. We are better off using the mean price rather than this model.

5. a) *Predicted TwinBirths = −4316980 + 2214.19 × Year.*
b) For each 1-year increase in time, the number of twins born in a year increases by approximately 2214.2.
c) 115,835.2 births. The scatterplot appears to be somewhat linear, but there is some curvature in the pattern. There is no reason to believe that the increase will continue to be linear 5 years out from the data.
d) The residuals plot shows a definite curved pattern, so the relation is not linear.

7. a) −0.520
b) Negative, not strong, somewhat linear, but with more variation as pH increases.
c) The BCI would also be average.
d) The predicted BCI will be 1.56 SDs of BCI below the mean BCI.

9. a) *Predicted Manatee Deaths = −45.9 + 0.132 × Powerboat Registrations* (in 1000s).
b) For each increase of 10,000 motorboat registrations, the number of manatees killed increases by 1.32.
c) If there were 0 motorboat registrations, the number of manatee deaths would be −45.9. This is obviously a silly extrapolation.
d) The predicted number is 67.4 deaths. The actual number of deaths was 81. The residual is 81 − 67.4 = 13.6. The model underestimated the number of deaths by 13.6.
e) Negative residuals would suggest that the actual number of deaths was lower than the predicted number.
f) Over time the number of motorboat registrations has increased, and the number of manatee kills has increased. The trend may continue.

11. a) −0.984   b) 96.9%   c) 32.95 mph   d) 1.66 mph
e) Slope will increase.
f) Correlation will weaken (become less negative).
g) Correlation is the same, regardless of units.

13. a) Weight (but unable to verify linearity).
b) As weight increases, mileage decreases.
c) Weight accounts for 81.5% of the variation in fuel efficiency.

15. a) *Predicted Horsepower = 3.50 + 34.314 × Weight.*
b) Thousands. For the equation to have predicted values between 60 and 160, the X values would have to be in thousands of pounds.
c) Yes. The residual plot does not show any pattern.
d) 115.0 horsepower.

17. a) The scatterplot shows a fairly strong linear relation in a positive direction. There seem to be two distinct clusters of data.
b) *Predicted Interval = 33.967 + 10.358 × Duration.*
c) As the duration of the previous eruption increases by 1 minute, the time between eruptions increases by about 10.4 minutes on average.
d) Since 77% of the variation in interval is accounted for by duration, and the error standard deviation is 6.16 minutes, the prediction will be relatively accurate.
e) 75.4 minutes.
f) A residual is the observed value minus the predicted value. So the residual = 79 − 75.4 = 3.6 minutes, indicating that the model underestimated the interval in this case.

19. a) r = 0.888. Although r is high, you must look at the scatterplot and verify that the relation is linear in form.

b)

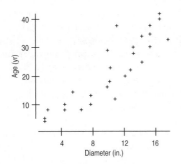

The association between diameter and age appears to be strong, somewhat linear, and positive.
c) *Predicted Age* = −0.97 + 2.21 × *Diameter*.
d)

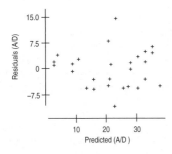

The residuals show a curved pattern (and two outliers).
e) The residuals for five of the seven largest trees (15 in. or larger) are positive, indicating that the predicted values underestimate the age.

21. Most houses have areas between 1000 and 5000 square feet. Increasing 1000 square feet would result in either 1000(.008) = 8 thousand dollars, 1000(.08) = 80 thousand dollars, 1000(.8) = 800 thousand dollars, or 1000(8) = 8000 thousand dollars. Only $80,000 is reasonable, so the slope must be 0.08.

23. a) The model predicts % smoking from year, not the other way around.
b) *Predicted Year* = 2009.9 − 1.130 × *% Smoking*.
c) The smallest % smoking given is 10.4, and an extrapolation for *x* = 0 is probably too far from the given data. The prediction is not very reliable in spite of the strong correlation.

25. The relation shows a negative direction, with a somewhat linear form, but perhaps with some slight curvature. There are several model outliers.

27. a) 71.9%.
b) As latitude increases, the January temperature decreases.
c) *Predicted January Temperature* = 108.80 − 2.111 × *Latitude*.
d) As the latitude increases by 1 degree, the average January temperature drops by about 2.11 degrees on average.
e) The *y*-intercept would indicate that the average January temperature is 108.8 when the latitude is 0. However, this is extrapolation and may not be meaningful.
f) 24.4 degrees.
g) The equation underestimates the average January temperature.

29. a) The scatterplot shows a strong, linear, positive association.
b) There is an association, but it is likely that training and

technique have increased over time and affected both jump performances.
c) Neither; the change in units does not affect the correlation.
d) The long jumper would jump 0.92 SDs above the mean long jump on average.

31. a) No relation; the correlation would probably be close to 0.
b) The relation would have a positive direction and the correlation would be strong, assuming students were studying French in each grade level. Otherwise, no correlation.
c) No relation; correlation close to 0.
d) The relation would have a positive direction and the correlation would be strong, since vocabulary would increase with each grade level.

33. *Predicted Calories* = 560.7 − 3.08 × *Time*.
Each minute extra at the table results in 3.08 fewer calories being consumed on average. Perhaps the hungry children eat fast and eat more.

35. There seems to be a strong, positive, linear relationship with one high-leverage point (Northern Ireland) that makes the overall R^2 quite low. Without that point, the R^2 increases to 61.5%. Of course, these data are averaged across thousands of households, and so the correlation appears to be higher than individuals would be. Any conclusions about individuals would be suspect.

37. a) 3.842   b) 501.187   c) 4.0

39. a) 30,818 pounds.
b) 1302 pounds.
c) 31,187.6 pounds.
d) I would be concerned about using this relation if we needed accuracy closer than 1000 pounds or so, as the residuals are more than ± 1000 pounds.
e) Negative residuals will be more of a problem, as the predicted weight would overestimate the weight of the truck; trucking companies might be inclined to take the ticket to court.

41. The original data are nonlinear, with a significant curvature. Using reciprocal square root of diameter gave a scatterplot that is nearly linear:

$$1/\sqrt{drain\ time} = 0.0024 + 0.219\ diameter.$$

43. The predicted values are (12, 774), (24, 738), (36, 702), (48, 666).
The residuals are (12, 26), (24, −58), (36, 38), (48, −6).
The squared residuals are (12, 676), (24, 3364), (36, 1444), (48, 36).

The "least squares" regression equation minimizes the sum of the squares of the residuals.

CHAPTER 11

1. Yes. You cannot predict the outcome beforehand.

3. A machine pops up numbered balls. If it were truly random, the outcome could not be predicted and the outcomes would be equally likely. It is random only if the balls generate numbers in equal frequencies.

5. a) The outcomes are not equally likely; for example, tossing 5 heads does not have the same probability as tossing 0 or 9 heads, but the simulation assumes they are equally likely.
b) The even-odd assignment assumes that the player is equally likely to score or miss the shot. In reality, the likelihood of making the shot depends on the player's skill.
c) Suppose a hand has 4 aces. This might be represented by 1, 1, 1, 1, and any other number. The likelihood for the first ace in

the hand is not the same as for the second or third or fourth. But with this simulation, the likelihood is the same for each.

7. The conclusion should indicate that the simulation *suggests* that the average length of the line would be 3.2 people. Future results might not match the simulated results exactly.

9. a) The component is one voter voting. An outcome is a vote for our candidate or not. Use two random digits, giving 00–54 a vote for your candidate and 55–99 for the underdog.
   b) A trial is 100 votes. Examine 100 two-digit random numbers and count how many people voted for each candidate. Whoever gets the majority of votes wins that trial.
   c) The response variable is whether the underdog wins or not.

11. Answers will vary, but average answer will be about 51.5%.

13. Answers will vary, but average answer will be about 26%.

15. a) Answers will vary, but you should win about 10% of the time.
    b) You should win at the same rate with any number.

17. Answers will vary, but you should win about 10% of the time.

19. Answers will vary, but average answer will be about 1.9 tests.

21. Answers will vary, but average answer will be about 1.24 points.

23. Do the simulation in two steps. First simulate the payoffs. Then count until $500 is reached. Answers will vary, but average should be near 10.2 customers.

25. Answers will vary, but average answer will be about 3 children.

27. Answers will vary, but average answer will be about 7.5 rolls.

29. No, it will happen about 40% of the time.

31. Answers will vary, but average answer will be about 37.5%.

33. 3 women will be selected about 7.8% of the time.

35. Answers will vary, but average answer will be about 28 envelopes.

CHAPTER 12

1. a) Population—Human resources directors of Fortune 500 companies.
   b) Parameter—Proportion who don't feel surveys intruded on their work day.
   c) Sampling Frame—list of HR directors at Fortune 500 companies.
   d) Sample—23% who responded.
   e) Method—Questionnaire mailed to all (nonrandom).
   f) Bias—Nonresponse. Hard to generalize because who responds is related to the question itself.

3. a) Population—All U.S. adults.
   b) Parameter—Proportion who have used and benefited from alternative medicine.
   c) Sampling Frame—all Consumer's Union subscribers.
   d) Sample—Those who responded (random).
   e) Method—Questionnaire to all (nonrandom).
   f) Bias—Nonresponse. Those who respond may have strong feelings one way or another.

5. a) Population—Adults.
   b) Parameter—Proportion who think drinking and driving is a serious problem.
   c) Sampling Frame—Bar patrons.
   d) Sample—Every 10th person leaving the bar.
   e) Method—Systematic sampling.
   f) Bias—Those interviewed had just left a bar. They probably think drinking and driving is less of a problem than do adults in general.

7. a) Population—Soil around a former waste dump.
   b) Parameter—Concentrations of toxic chemicals.
   c) Sampling Frame—Accessible soil around the dump.
   d) Sample—16 soil samples.

   e) Method—Not clear.
   f) Bias—Don't know if soil samples were randomly chosen. If not, may be biased toward more or less polluted soil.

9. a) Population—Snack food bags.
   b) Parameter—Weight of bags, proportion passing inspection.
   c) Sampling Frame—All bags produced each day.
   d) Sample—Bags in 10 randomly selected cases, 1 bag from each case for inspection
   e) Method—Multistage sampling.
   f) Bias—Should be unbiased.

11. a) Voluntary response. Only those who both see the ad *and* feel stongly enough will respond.
    b) Cluster sampling. One school may not be typical of all.
    c) Attempted census. Will have nonresponse bias.
    d) Stratified sampling with follow-up. Should be unbiased.

13. a) This is a multistage design, with a cluster sample at the first stage and a simple random sample for each cluster.
    b) If any of the three churches you pick at random is not representative of all churches, then you'll introduce sampling error by the choice of that church.

15. a) This is a systematic sample.
    b) It is likely to be representative of those waiting for the roller coaster. Indeed, it may do quite well if those at the front of the line respond differently (after their long wait) than those at the back of the line.
    c) The sampling frame is patrons willing to wait for the roller coaster on that day at that time. It should be representative of the people in line, but not of all people at the amusement park.

17. a) Answers will definitely differ. Question 1 will probably get many "No" answers, while Question 2 will get many "Yes" answers. This is wording bias.
    b) "Do you think standardized tests are appropriate for deciding whether a student should be promoted to the next grade?" (Other answers will vary.)

19. Only those who think it worth the wait are likely to be in line. Those who don't like roller coasters are unlikely to be in the sampling frame, so the poll won't get a fair picture of whether park patrons overall would favor still more roller coasters.

21. a) Biased toward yes because of "pollute." "Should companies be responsible for any costs of environmental clean-up?"
    b) Biased towards no because of "old enough to serve in the military." "Do you think the drinking age should be lowered from 21?"

23. a) Not everyone has an equal chance. People with unlisted numbers, people without phones, and those at work cannot be reached.
    b) Generate random numbers and call at random times.
    c) Under the original plan, those families in which one person stays home are more likely to be included. Under the second plan, many more are included. People without phones are still excluded.
    d) It improves the chance of selected households being included.
    e) This takes care of phone numbers. Time of day may be an issue. People without phones are still excluded.

25. a) Answers will vary.
    b) Your own arm length. Parameter is your own arm length; population is all possible measurements of it.
    c) Population is now the arm lengths of your friends. The average estimates the mean of these lengths.
    d) Probably not. Friends are likely to be of the same age, and not very diverse, or representative of the larger population.

27. a) Assign numbers 001 to 120 to each order. Use random numbers to select 10 transactions to examine.

b) Sample proportionately within each type. (Do a stratified random sample.)

29. a) Select three cases at random, then select one jar randomly from each case.

b) Use random numbers to choose 3 cases from numbers 61 through 80; then use random numbers between 1 and 12 to select the jar from each case.

c) No. Multistage sampling.

31. a) Depends on the yellow page listings used. If from regular (line) listings, this is fair if all doctors are listed. If from ads, probably not, as those doctors may not be typical.

b) Not appropriate. This cluster sample will probably contain listings for only one or two business types.

CHAPTER 13

1. a) Experiment.

b) Bipolar disorder patients.

c) Omega-3 fats from fish oil, two levels.

d) 2 treatments.

e) Improvement (fewer symptoms?).

f) Design not specified.

g) Blind (due to placebo), unknown if double-blind.

h) Individuals with bipolar disease improve with high-dose omega-3 fats from fish oil.

3. a) Observational study.

b) Prospective.

c) Men and women with moderately high blood pressure and normal blood pressure, unknown selection process.

d) Memory and reaction time.

e) As there is no random assignment, there is no way to know that high blood pressure *caused* subjects to do worse on memory and reaction time tests. A lurking variable may also be the cause.

5. a) Observational study.

b) Retrospective.

c) Swedish men, unknown selection process.

d) Risk of kidney cancer.

e) As there is no random assignment, there is no way to know that the overweight or high blood pressure caused the higher risk for kidney cancer.

7. a) Experiment.

b) Postmenopausal women.

c) Alcohol—2 levels; blocking variable—estrogen supplements (2 levels).

d) 2 treatments × 2 blocking levels.

e) Increase in estrogen levels.

f) Blocked.

g) Not blind.

h) Indicates that alcohol comsumption *for those taking estrogen supplements* may increase estrogen levels.

9. a) Experiment.

b) Locations in a garden.

c) 1 factor: traps (2 levels).

d) 2 treatments.

e) Number of bugs in the trap.

f) Blocked by location.

g) Not blind.

h) One type of trap is more effective than the other.

11. a) Observational study.

b) Retrospective.

c) Women in Finland, unknown selection process with data from church records.

d) Women's lifespans.

e) As there is no random assignment, there is no way to know that having sons or daughters shortens or lengthens the lifespan of mothers.

13. a) Observational.

b) Prospective.

c) People with or without depression, unknown selection process.

d) Frequency of crying in response to sad situations.

e) There is no apparent difference in crying response (to sad movies) for depressed and nondepressed groups.

15. a) Experiment.

b) Rats.

c) 1 factor: sleep deprivation; four levels.

d) 4 treatments.

e) Glycogen content in the brain.

f) No discussion of randomness.

g) Blinding is not discussed.

h) The conclusion could be that rats deprived of sleep have significantly lower glycogen levels and may need sleep to restore that brain energy fuel. Extrapolating to humans would be very speculative.

17. a) Experiment.

b) People experiencing migraines.

c) 2 factors (pain reliever and water temperature), 2 levels each.

d) 4 treatments.

e) Level of pain relief.

f) Completely randomized over 2 factors.

g) Blind, as subjects did not know if they received the pain medication or the placebo, but not blind, as the subjects will know if they are drinking regular or ice water.

h) It may indicate whether pain reliever alone or in combination with ice water gives pain relief, but patients are not blinded to ice water, so placebo effect may also be the cause of any relief seen caused by ice water.

19. a) Experiment.

b) Athletes with hamstring injuries.

c) 1 factor: type of exercise program (2 levels).

d) 2 treatments.

e) Time to return to sports.

f) Completely randomized.

g) No blinding—subjects must know what kind of exercise they do.

h) Can determine which of the two exercise programs is more effective.

21. Answers may vary. Use a random number generator to randomly select 24 numbers from 01 to 24 without replication. Assign the first 8 numbers to the first group, the second 8 numbers to the second group, and the third group of 8 numbers to the third group.

23. a) First, they are using athletes who have a vested interest in the success of the shoe by virtue of their sponsorship. They should choose other athletes. Second, they should randomize the order of the runs, not run all the races with their shoes second. They should blind the athletes by disguising the shoes if possible, so they don't know which is which. The timers shouldn't know which athletes are running with which shoes, either. Finally, they should replicate several times since times will vary under both shoe conditions.

b) Because of the problems in (a), the results they obtain may be biased in favor of their shoes. In addition, the results obtained for Olympic athletes may not be the same as for the general runner.

25. a) Allowing athletes to self-select treatments could confound the results. Other issues such as severity of injury, diet, age, etc., could also affect time to heal, and randomization should equalize the two treatment groups with respect to any such variables.

b) A control group could have revealed whether either exercise program was better (or worse) than just letting the injury heal.

c) Doctors who evaluated the athletes to approve their return to sports should not know which treatment the subject had engaged in.

d) It's hard to tell. The difference of 15 days seems large, but the standard deviations indicate that there was a great deal of variability in the times.

27. a) The differences among the Mozart and quiet groups were more than would have been expected from ordinary sampling variation.

b)

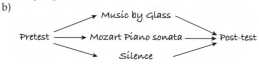

c) The Mozart group seems to have the smallest median difference and thus the *least* improvement, but there does not appear to be a significant difference.

d) No, if anything there is less improvement, but the difference does not seem significant compared with the usual variation.

29. a) Observational. Randomly select a group of children, ages 10 to 13, have them taste the cereal, and ask if they like the cereal.

b) Answers may vary. Get volunteers ages 10 to 13. Each volunteer will taste both cereals, randomizing the order in which they taste them. Compare the percentage of favorable ratings for each cereal.

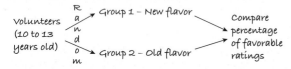

c) Answers may vary. From the volunteers, identify the children who watch Frump and identify the children who do not watch Frump. Use a blocked design to reduce variation in cereal preference that may be associated with watching the Frump cartoon.

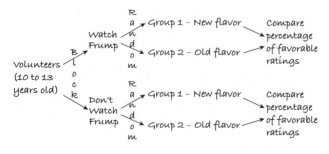

31. a) Observational, prospective study.

b) The supposed relation between health and wine consumption might be explained by the confounding variables of income and education.

c) None of these. While the variables have a relation, there is no causality indicated for the relation.

33. a) Arrange the 20 containers in 20 separate locations. Use a random number generator to identify the 10 containers that should be filled with water.

b) Guessing, the dowser should be correct about 50% of the time. A record of 60% (12 out of 20) does not appear to be significantly different.

c) Answers may vary. You would need to see a high level of success—say, 90% to 100%, that is, 18 to 20 correct.

35. Randomly assign half the reading teachers in the district to use each method. Students should be randomly assigned to teachers as well. Make sure to block both by school and grade (or control grade by using only one grade). Construct an appropriate reading test to be used at the end of the year and compare scores.

37. a) They mean that the difference is higher than they would expect from normal sampling variability.

b) An observational study.

c) No. Perhaps the differences are attributable to some confounding variable (like people are more likely to engage in riskier behaviors on the weekend) rather than the day of admission.

d) Perhaps people have more serious accidents and traumas on weekends and are thus more likely to die as a result.

39. Answers may vary. This experiment has 1 factor (pesticide), at 3 levels (pesticide A, pesticide B, no pesticide), resulting in 3 treatments. The response variable is the number of beetle larvae found on each plant. Randomly select a third of the plots to be sprayed with pesticide A, a third with pesticide B, and a third to be sprayed with no pesticide (since the researcher also wants to know whether the pesticides even work at all). To control the experiment, the plots of land should be as similar as possible, with regard to amount of sunlight, water, proximity to other plants, etc. If not, plots with similar characteristics should be blocked together. If possible, use some inert substance as a placebo pesticide on the control group, and do not tell the counters of the beetle larvae which plants have been treated with pesticides. After a given period of time, count the number of beetle larvae on each plant and compare the results.

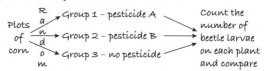

41. Answers may vary. Find a group of volunteers. Each volunteer will be required to shut off the machine with his or her left hand and the right hand. Randomly assign the left or right hand to be used first. Complete the first attempt for the whole group. Now repeat the experiment with the alternate hand. Check the differences in time for the left and right hands.

43. a)   Jumping with or without a parachute.

b) Volunteer skydivers (the dim-witted ones).

c) A parachute that looks real but doesn't work.

d) A good parachute and a placebo parachute.

e) Whether parachutist survives the jump (or extent of injuries).

f) All should jump from the same altitude in similar weather conditions and land on similar surfaces.

g) Randomly assign people the parachutes.

h) The skydivers (and the people involved in distributing the parachute packs) shouldn't know who got a working chute. And the people evaluating the subjects after the jumps should not be told who had a real parachute either!

PART III REVIEW

1. Observational prospective study. Indications of behavior differences can be seen in the two groups. May show a link between premature birth and behavior, but there may be lurking variables involved.

3. Experiment, matched by gender and weight, randomization within blocks of two pups of same gender and weight. Factor is type of diet. Treatments are low-calorie diet and allowing the dog to eat all it wants. Response variable is length of life. Can conclude that, on average, dogs with a lower-calorie diet live longer.

5. Observational prospective study. Indicates folate *may* help in reducing colon cancer for those with family histories of the disease.

7. Sampling. Probably a simple random sample, although may be stratified by type of firework. Population is all fireworks produced each day. Parameter is proportion of duds. Can determine if the day's production is ready for sale.

9. Observational retrospective study. Living near strong electromagnetic fields may be associated with more leukemia than normal. Lurking variables, such as socioeconomic level, may be involved.

11. Experiment. Blocked by sex of rat. Randomization is not specified. Factor is type of hormone given. Treatments are leptin and insulin. Response variable is lost weight. Can conclude that hormones can help suppress appetites in rats, and the type of hormone varies by gender.

13. Experiment. Factor is gene therapy. Hamsters were randomized to treatments. Treatments were gene therapy or not. Response variable is heart muscle condition. Can conclude that gene therapy is beneficial (at least in hamsters).

15. Sampling. Population is all oranges on the truck. Parameter is proportion of unsuitable oranges. Procedure is probably simple random sampling. Can conclude whether or not to accept the truckload.

17. Observational prospective study. Physically fit men may have a lower risk of death from cancer.

19. Answers will vary. This is a simulation problem. Using a random digits table or software, call 0–4 a loss, and 5–9 a win for the gambler on a game. Use blocks of 5 digits to simulate a week's pick.

21. Answers will vary.

23. a) Experiment. Actively manipulated candy giving, diners were randomly assigned treatments, control group was those with no candy, lots of dining parties.
    b) It depends on when the decision was made. If early in the meal, the server may give better treatment to those who will receive candy—biasing the results.
    c) A difference in response so large it cannot be attributed to natural sampling variability.

25. a) Voluntary response. Only those who feel strongly will pay for the 900 phone call.
    b) "If it would help future generations live a longer, healthier life, would you be in favor of human cloning?"

27. a) Simulation results will vary. Average will be around 5.8 points.

b) Simulation results will vary. Average will also be around 5.8 points.
   c) Answers will vary.

29. a) Yes.
    b) No. Residences without phones are excluded. Residences with more than one phone had a higher chance.
    c) No. People who respond to the survey may be of age but not be registered voters.
    d) No. Households who answered the phone may be more likely to have someone at home when the phone call was generated. These may not be representative of all households.

31. a) Does not prove it. There may be other confounding variables. Only way to prove this would be to do a controlled experiment.
    b) Alzheimer's usually shows up late in life. Perhaps smokers have died of other causes before Alzheimer's can be seen.
    c) An experiment would be unethical. One could design a prospective study in which groups of smokers and nonsmokers are followed for many years and the incidence of Alzheimer's is tracked.

33.

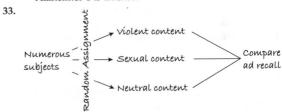

Numerous subjects will be randomly assigned to see shows with violent, sexual, or neutral content. They will see the same commercials. After the show, they will be interviewed for their recall of brand names in the commercials.

35. a) May have been a simple random sample, but given the relative equality in age groups, may have been stratified.
    b) 35.1%.
    c) We don't know. Perhaps cell phones or unlisted numbers were excluded, and Democrats have more (or fewer) of those. Probably OK, though.
    d) Do party affiliations differ for different age groups?

37. The factor in the experiment will be type of bird control. I will have three treatments: scarecrow, netting, and no control. I will randomly assign several different areas in the vineyard to one of the treatments, taking care that there is sufficient separation that the possible effect of the scarecrow will not be confounded. At the end of the season, the response variable will be the proportion of bird-damaged grapes.

39. a) We want all subjects treated as alike as possible. If there were no "placebo surgery," subjects would know this and perhaps behave differently.
    b) The experiment was intended to see if there was a difference in the effectiveness of the two treatments. (If we wanted to generalize, we would need to assume that the results for these volunteers is the same as on all patients who might need this operation.)
    c) Statistically significant means a difference in results so large it cannot be explained by natural sampling variability.

41. a) Use stratified sampling to select 2 first-class passengers and 12 from coach.
    b) Number passengers alphabetically, 01 = Bergman to 20 = Testut. Read in blocks of two, ignoring any numbers more

than 20. This gives 65, 43, 67, 11 (selects Fontana), 27, 04 (selects Castillo).

c) Number passengers alphabetically 001 to 120. Use the random number table to find three-digit numbers in this range until 12 have been selected.

43. Simulation results will vary.

(Use integers 00 to 99 as a basis. Use integers 0 to 69 to represent a tee shot on the fairway. If on the fairway, use digits 00 to 79 to represent on the green. If off the fairway, use 00 to 39 to represent getting on the green. If not on the green, use digits 00 to 89 to represent landing on the green. For the first putt, use digits 00 to 19 to represent making the shot. For subsequent putts, use digits 00 to 89 to represent making the shot.)

CHAPTER 14

1. In this context "truly random" should mean that every number is equally likely to occur and knowing one outcome will not affect the next.

3. If events are independent, there is no "Law of Averages." She would be wrong to think they are "due" for a harsh winter.

5. There is no "Law of Averages" for independent events. His chance for a hit does not change based on recent successes or failures.

7. a) There is a small chance you would have to pay out much more than the $300.
   b) Many customers pay for insurance. The small risk for any one customer is spread among all.

9. a) Legitimate.
   b) Legitimate.
   c) Not legitimate (sum more than 1).
   d) Legitimate.
   e) Not legitimate (can't have negatives or values more than 1).

11. a) 0.72   b) 0.89   c) 0.28

13. a) 0.5184   b) 0.0784   c) 0.4816

15. a) Repair needs for the two cars must be independent.
    b) Maybe not. An owner may treat the two cars similarly, taking good (or poor) care of both. This may decrease (or increase) the likelihood that each needs to be repaired.

17. a) 332/1005 = 0.330.
    b) 80/1005 + 30/1005 = 110/1005 = 0.109.

19. a) 0.176
    b) 0.779
    c) Responses are independent.
    d) People were polled at random.

21. a) 0.4712
    b) 0.7112
    c) $(1 - 0.76) + 0.76(1 - 0.38)$ or $1 - (0.76)(0.38)$

23. a) 1. 0.30   2. 0.30   3. 0.90   4. 0.0
    b) 1. 0.027   2. 0.128   3. 0.512   4. 0.271

25. a) Disjoint (can't be both red and orange).
    b) Independent.
    c) No. Once you know that one of a pair of disjoint events has occurred, the other is impossible.

27. a) 0.0046   b) 0.125   c) 0.296   d) 0.421   e) 0.995

29. a) 0.027   b) 0.063   c) 0.973   d) 0.014

31. a) 0.024   b) 0.250   c) 0.543

33. 0.078

35. a) For any day with a valid three-digit date, the chance is 0.001, or 1 in 1000. For many dates in October through December,

the probability is 0. (No three digits will make 10/15, for example.)

b) There are 65 days when the chance to match is 0. (Oct 10–31, Nov. 10–30, and Dec. 10–31.) The chance for no matches on the remaining 300 days is 0.741

c) 0.259
d) 0.049

CHAPTER 15

1. a) S = {HH, HT, TH, TT}, equally likely.
   b) S = {0, 1, 2, 3}, not equally likely.
   c) S = {H, TH, TTH, TTT}, not equally likely.
   d) S = {1, 2, 3, 4, 5, 6}, not equally likely.

3. a) 0.68   b) 0.32   c) 0.04

5. a) 0.31   b) 0.48   c) 0.31

7. a) 0.515   b) 0.630   c) 0.292   c) 0.773

9. a) 0.50   b) 1.00   c) 0.077   d) 0.333

11. a) 0.11   b) 0.27   c) 0.407   d) 0.344

13. a) 0.134   b) 0.622   c) 0.309   d) 0.178   e) 0.551

15. 0.21

17. a) 0.145   b) 0.118   c) 0.414   d) 0.217

19. a) 0.318   b) 0.955   c) 0.071   d) 0.009

21. a) 32%
    b) 0.135
    c) No, 7% of juniors have taken both.
    d) No, the probability that a junior has taken a computer course is 0.23. The probability that a junior has taken a computer course *given* he or she has taken a Statistics course is 0.135.

23. a) 0.266
    b) No, 26.6 % of homes with garages have pools; 21% of homes overall have pools.
    c) No, 17% of homes have both.

25. Yes, $P(\text{Ace}) = 1/13$. $P(\text{Ace}|\text{Spade}) = 1/13$.

27. a) Yes; no one is both under 30 and over 65.
    b) No. Almost 22% of the respondents are under 30, but (of course) no one over 65 is.
    c) No. 65 of the 1005 respondents were over 65 and chose Clinton.
    d) No. Over 43% of all respondents chose Clinton, but only 39% of those over 65 did.

29. No, 28.8% of men with OK blood pressure have high cholesterol, but 40.7% of those with high blood pressure do.

31. a) 95.4%
    b) Probably. 95.4% of people with cell phones had land lines, and 95.6% of all people did.

33. No. Only 34% of men were Democrats, but over 41% of all voters were.

35. a) No, the probability that the luggage arrives on time depends on whether the flight is on time. The probability is 95% if the flight is on time and only 65% if not.
    b) 0.695

37. 0.975

39. a) No, the probability of missing work for day-shift employees is 0.01. It is 0.02 for night-shift employees. The probability depends on whether they work day or night shift.
    b) 1.4%

41. 57.1%

43. a) 0.20   b) 0.272   c) 0.353   d) 0.033

45. 0.563

CHAPTER 16

1. a) 19   b) 4.2

3. a)

Amount won	$0	$5	$10	$30
P(Amount won)	$\frac{26}{52}$	$\frac{13}{52}$	$\frac{12}{52}$	$\frac{1}{52}$

b) $4.13
c) $4 or less (answers may vary)

5. a)

Children	1	2	3
P(Children)	0.5	0.25	0.25

b) 1.75 children
c) 0.87 boys

Boys	0	1	2	3
P(Boys)	0.5	0.25	0.125	0.125

7. $27,000
9. a) 7   b) 1.89
11. $5.44
13. 0.83
15. a) 1.7   b) 0.9
17. $\mu = 0.64, \sigma = 0.93$
19. a) No. The probability of winning the second depends on the outcome of the first.
b) 0.42
c) 0.08
d)

Games won	0	1	2
P(Games won)	0.42	0.50	0.08

e) $\mu = 0.66, \sigma = 0.62$

21. a)

Number good	0	1	2
P(Number good)	0.067	0.467	0.467

b) 1.40
c) 0.61
23. a) $\mu = 30, \sigma = 6$ b) $\mu = 26, \sigma = 5$
c) $\mu = 30, \sigma = 5.39$ d) $\mu = -10, \sigma = 5.39$
25. a) $\mu = 240, \sigma = 12.80$ b) $\mu = 140, \sigma = 24$
c) $\mu = 720, \sigma = 34.18$ d) $\mu = 60, \sigma = 39.40$
27. a) 1.8
b) 0.87
c) Cartons are independent of each other.
29. $\mu = 13.6, \sigma = 2.55$ (assuming the hours are independent of each other).
31. a) There will be many gains of $150 with a few large losses.
b) $\mu = \$300, \sigma = \$8485.28$
c) $\mu = \$1,500,000, \sigma = \$600,000$
d) Yes. $0 is 2.5 SDs below the mean for 10,000 policies.
e) Losses are independent of each other. A major catastrophe with many policies in an area would violate the assumption.
33. a) 1 oz   b) 0.5 oz   c) 0.023
d) $\mu = 4$ oz, $\sigma = 0.5$ oz
e) 0.159
f) $\mu = 12.3$ oz, $\sigma = 0.54$ oz

35. a) 12.2 oz   b) 0.51 oz   c) 0.058
37. a) $\mu = 200.57$ sec, $\sigma = 0.46$ sec
b) No, $Z = \dfrac{199.48 - 200.57}{0.461} = -2.36$. There is only 0.009 probability of swimming that fast or faster.
39. a) A = price of a pound of apples; P = price of a pound of potatoes; Profit = $100A + 50P - 2$
b) $63.00
c) $20.62
d) Mean—no; SD—yes (independent sales prices).

CHAPTER 17

1. a) No. More than two outcomes are possible.
b) Yes, assuming the people are unrelated to each other.
c) No. The chance of a heart changes as cards are dealt.
d) No, 500 is more than 10% of 3000.
e) If packages in a case are independent of each other, yes; otherwise, no.
3. a) Use single random digits. Let 0, 1 = Tiger. Count the number of random numbers until a 0 or 1 occurs.
c) Results will vary.
d)

x	1	2	3	4	5	6	7	8	≥ 9
$P(x)$	0.2	0.16	0.128	0.102	0.082	0.066	0.052	0.042	0.168

5. a) Use single random digits. Let 0, 1 = Tiger. Examine random digits in groups of five, counting the number of 0's and 1's.
c) Results will vary.
d)

x	0	1	2	3	4	5
$P(x)$	0.33	0.41	0.20	0.05	0.01	0.0

7. a) 0.0819   b) 0.0064   c) 0.992
9. 5
11. a) 25   b) 0.185   c) 0.217   d) 0.693
13. a) 0.0745   b) 0.502   c) 0.211
d) 0.0166   e) 0.0179   f) 0.9987
15. a) 0.65   b) 0.75   c) 7.69 picks
17. a) $\mu = 10.44, \sigma = 1.16$
b) 0.812   c) 0.475   d) 0.00193   e) 0.998
19. a) 0.118   b) 0.324   c) 0.744   d) 0.580
21. a) $\mu = 56, \sigma = 4.10$
b) Yes, $np = 56 \geq 10$, $nq = 24 \geq 10$.
c) In a match with 80 serves, approximately 68% of the time she will have between 51.9 and 60.1 good serves, approximately 95% of the time she will have between 47.8 and 64.2 good serves, and approximately 99.7% of the time she will have between 43.7 and 68.3 good serves.
d) Normal, approx.: 0.014; Binomial, exact: 0.016
23. a) $\mu = 18.75, \sigma = 4.05$
b) Yes, $np = 18.75 \geq 10$, $nq = 131.25 \geq 10$.
c) No, 22 is only 0.8 SD above the mean; this is likely to happen by natural sampling variability.
25. Normal, approx.: 0.053; Binomial, exact: 0.061
27. The mean number of sales should be 24 with SD 4.60. Ten sales is more than 3.0 SDs below the mean. He was probably misled.

29. a) 0.0953
b) 0.0427
31. a) 4 cases
b) 0.9817
33. a) 5   b) 0.066   c) 0.107
d) $\mu = 24, \sigma = 2.19$
e) Normal, approx.: 0.819; Binomial, exact: 0.848
35. $\mu = 20, \sigma = 4$. I'd want *at least* 32 (3 SDs above the mean). (Answers will vary.)
37. Probably not. There's a more than 9% chance that he could hit 4 shots in a row, so he can expect this to happen nearly once in every 10 sets of 4 shots he takes. That does not seem unusual.
39. Yes. We'd expect him to make 22 shots with a standard deviation of 3.15 shots. 32 shots is more than 3 standard deviations above the expected value, an unusually high rate of success.

PART IV REVIEW

1. a) 0.34
b) 0.27
c) 0.069
d) No, 2% of cars have both types of defects.
e) Of all cars with cosmetic defects, 6.9% have functional defects. Of cars without cosmetic defects, 7.0% have functional defects. The probabilities here are estimates, so these are probably close enough to say the defects are independent.
3. a) C = Price to China; F = Price to France; Total = $3C + 5F$
b) $\mu = \$5500, \sigma = \$672.68$
c) $\mu = \$500, \sigma = \$180.28$
d) Means—no. Standard deviations—yes; ticket prices must be independent of each other for different countries, but all tickets to the same country are at the same price.
5. a) $\mu = -\$0.20, \sigma = \$1.89$
b) $\mu = -\$0.40, \sigma = \$2.67$
7. a) 0.106
b) 0.651
c) 0.442
9. a) 0.590
b) 0.328
c) 0.00856
11. a) $\mu = 15.2, \sigma = 3.70$
b) Yes. $np \geq 10$ and $nq \geq 10$
c) Normal, approx.: 0.080; Binomial, exact: 0.097
13. a) 0.0173
b) 0.591
c) Left: 960; right: 120; both: 120
d) $\mu = 120, \sigma = 10.39$
e) About 68% chance of between 109 and 130; about 95% between 99 and 141; about 99.7% between 89 and 151.
15. a) Men's heights are more variable than women's.
b) Men
c) M = Man's height; W = Woman's height; $M - W$ is how much taller the man is.
d) 5.1"
e) 3.75"
f) 0.913
g) If independent, it should be about 91.3%. We are told 92%. This difference seems small and may be due to natural sampling variability.

17. a) No. The chance is 1.6×10^{-7}.
b) 0.952
c) 0.063
19. $-\$2080.00$
21. a) 0.717   b) 0.588
23. a) $\mu = 100, \sigma = 8$
b) $\mu = 1000, \sigma = 60$
c) $\mu = 100, \sigma = 8.54$
d) $\mu = -50, \sigma = 10$
25. a) Many do both, so the two categories can total more than 100%.
b) No. They can't be disjoint. If they were, the total would be 100% or less.
c) No. Probabilities are different for boys and girls.
d) 0.0524
27. a) 21 days
b) 1649.73 soms
c) 3300 soms extra. About 157 soms "cushion" each day.
29. No, you'd expect 541.2 homeowners, with an SD of 13.56. 523 is 1.34 SDs below the mean; not unusual.
31. a) 0.018   b) 0.300   c) 0.26
33. a) 6   b) 15   c) 0.402
35. a) 34%
b) 35%
c) 31.4%
d) 31.4% of classes that used calculators used computer assignments, while in classes that didn't use calculators, 30.6% used computer assignments. These are close enough to think the choice is probably independent.
37. a) A Poisson model
b) 0.1 failures
c) 0.090
d) 0.095
39. a) 1/11   b) 7/22   c) 5/11   d) 0   e) 19/66
41. a) Expected number of stars with planets.
b) Expected number of planets with intelligent life.
c) Probability of a planet with a suitable environment having intelligent life.
d) f_l: If a planet has a suitable environment, the probability that life develops.
f_i: If a planet develops life, the probability that the life evolves intelligence.
f_c: If a planet has intelligent life, the probability that it develops radio communication.
43. 0.991

CHAPTER 18

1. a) Symmetric, because probability of heads and tails is equal.
b) 0.5
c) 0.125
d) $np = 8 < 10$
3. a) About 68% should have proportions between 0.4 and 0.6, about 95% between 0.3 and 0.7, and about 99.7% between 0.2 and 0.8.
b) $np = 12.5, nq = 12.5$; both are ≥ 10.

c)

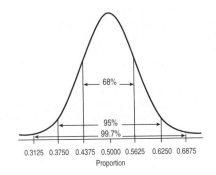

$np = nq = 32$; both are ≥ 10.
d) Becomes narrower (less spread around 0.5).

5. This is a fairly unusual result: about 2.26 SDs below the mean. The probability of that is about 0.012. So, in a class of 100 this is certainly a reasonable possibility.

7. a)

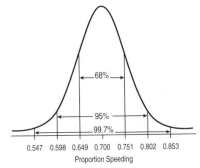

b) Both $np = 56$ and $nq = 24 \geq 10$. Drivers *may* be independent of each other, but if flow of traffic is very fast, they may not be. Or weather conditions may affect all drivers. In these cases they may get more or fewer speeders than they expect.

9. a) $\mu = 7\%, \sigma = 1.8\%$
   b) Assume that clients pay independently of each other that we have a random sample of all possible clients, and that these represent less than 10% of all possible clients. $np = 14$ and $nq = 186$ are both at least 10.
   c) 0.048

11.

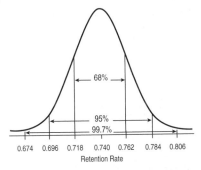

These are not random samples, and not all colleges may be typical (representative). $np = 296, nq = 104$ are large enough.

13. Yes; if their students were typical, a retention rate of $522/603 = 86.6\%$ would be over 7 standard deviations above the expected rate of 74%.

15. 0.212. Reasonable that those polled are independent of each other and represent less than 10% of all potential voters. We assume the sample was selected at random. Success/Failure Condition met: $np = 208, nq = 192$. Both ≥ 10.

17. 0.088 using $N(0.08, 0.022)$, model.

19. Answers will vary. Using $\mu + 3\sigma$ for "very sure," the restaurant should have 89 nonsmoking seats. Assumes customers at any time are independent of each other, a random sample, and represent less than 10% of all potential customers. $np = 72, nq = 48$, so Normal model is reasonable ($\mu = 0.60, \sigma = 0.045$).

21. a) Normal, center at μ, standard deviation $\sigma/\sqrt{n}$.
   b) Standard deviation will be smaller. Center will remain the same.

23.

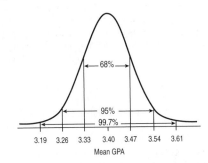

Normal, $\mu = 3.4, \sigma = 0.07$. We assume that the students are randomly assigned to the seminars and represent less than 10% of all possible students, and that individual's GPAs are independent of one another.

25. a) 21.1%
   b) 276.8 days or more
   c) $N(266, 2.07)$
   d) 0.002

27. a) There are more premature births than very long pregnancies. Modern practice of medicine stops pregnancies at about 2 weeks past normal due date.
   b) Parts (a) and (b)—yes—we can't use Normal model if it's very skewed. Part (c)—no—CLT guarantees a Normal model for this large sample size.

29. a) $\mu = \$2.00, \sigma = \$3.61$
   b) $\mu = \$4.00, \sigma = \$5.10$
   c) 0.191. Model is $N(80, 22.83)$.

31. a) $\mu = 2.869, \sigma = 1.312$
   b) No. The score distribution in the sample should resemble that in the population, somewhat uniform for scores 1–4 and about half as many 5's.
   c) Approximately $N\left(2.869, \dfrac{1.312}{\sqrt{40}}\right)$.

33. About 7.6%, based on $N(2.76, 0.168)$.

35. a) $N(2.9, 0.045)$
   b) 0.0131
   c) 2.97 gm/mi

37. a) Can't use a Normal model to estimate probabilities. The distribution is skewed right—not Normal.
   b) 4 is probably not a large enough sample to say the average follows the Normal model.
   c) No. This is 3.16 SDs above the mean.

39. a) 0.0003. Model is $N(384, 34.15)$.
   b) $\$427.77$ or more

41. a) 0.734
   b) 0.652. Model is $N(10, 12.81)$.
   c) 0.193. Model is $N(120, 5.774)$.
   d) 0.751. Model is $N(10, 7.394)$.

CHAPTER 19

1. He believes the true proportion is within 4% of his estimate, with some (probably 95%) degree of confidence.
3. a) Population—all cars; sample—those actually stopped at the checkpoint; p—proportion of all cars with safety problems; $\hat{p}$—proportion actually seen with safety problems (10.4%); if sample (a cluster sample) is representative, then the methods of this chapter will apply.
   b) Population—general public; sample—those who logged onto the Web site; p—population proportion of those who favor prayer in school; $\hat{p}$—proportion of those who voted in the poll who favored prayer in school (81.1%); can't use methods of this chapter—sample is biased and nonrandom.
   c) Population—parents at the school; sample—those who returned the questionnaire; p—proportion of all parents who favor uniforms; $\hat{p}$—proportion of respondents who favor uniforms (60%); have to use caution with methods of this chapter, since nonresponse may bias results.
   d) Population—students at the college; sample—the 1632 students who entered that year; p—proportion of all students who will graduate on time; $\hat{p}$—proportion of that year's students who graduate on time (85.0%); can use methods of this chapter if that year's students (a cluster sample) are viewed as a representative sample of all possible students at the school.
5. a) Not correct. This implies certainty.
   b) Not correct. Different samples will give different results. Many fewer than 95% will have 88% on-time orders.
   c) Not correct. The interval is about the population proportion, not the sample proportion in different samples.
   d) Not correct. In this sample, we *know* 88% arrived on time.
   e) Not correct. The interval is about the parameter, not about the days.
7. a) False   b) True   c) True   d) False
9. On the basis of this sample, we are 90% confident that the proportion of Japanese cars is between 29.9% and 47.0%.
11. a) 2.5%
   b) The pollsters are 90% confident the true proportion of adults who believe in ghosts is within 2.5% of the estimated 38%.
   c) Larger. More confidence for a given sample requires a wider interval.
   d) 3.9%
   e) A smaller margin of error will imply less confidence.
13. a) (12.7%, 18.6%)
   b) We are 95% confident, based on this sample, that the proportion of all auto accidents that involve teenage drivers is between 12.7% and 18.6%.
   c) About 95% of all random samples will produce confidence intervals that contain the true population proportion.
   d) Contradicts. The interval is completely below 20%.
15. Probably nothing. Those who bothered to fill out the survey may be a biased sample.
17. a) (39.9%, 48.2%)
   b) No
19. a) Wording
   b) (45.5%, 51.5%)
   c) Smaller—the sample size was larger.
21. a) (18.2%, 21.8%)
   b) We are 98% confident, based on the sample, that between 18.2% and 21.8% of English children are deficient in vitamin D.
   c) About 98% of all random samples will produce a confidence interval that contains the true proportion.
23. a) The students' birth orders are likely to be independent. The sample was random and consisted of less than 10% of the population of students. There were 20 successes and 206 failures.
   b) (5.1%, 12.6%)
   c) We're 95% confident that between 5.1% and 12.6% of all college students are "only" children.
   d) If we were to select repeated samples like this we'd expect about 95% of the confidence intervals we created to contain the true proportion of all college students who are "only" children.
25. a) (48.9%, 55.1%)
   b) For Clinton, the 95% confidence interval is 39.9% to 46.1%. It seems unlikely that half of the entire population thinks Hillary fits the bill.
27. a) Wider. The sample size is about one-fourth the sample size for all adults. This will make the confidence interval about twice as wide.
   b) (56.0%, 68.0%)
29. a) (15.5%, 26.3%)   b) 612
   c) Sample may not be random or representative. Deer that may legally be hunted may not be representative of all sexes and ages.
31. a) 141   b) 318   c) 564
33. 1801
35. 384 total, using $p = 0.15$.
37. 90%

CHAPTER 20

1. a) $H_0: p = 0.30$; $H_A: p < 0.30$
   b) $H_0: p = 0.50$; $H_A: p \neq 0.50$
   c) $H_0: p = 0.20$; $H_A: p > 0.20$
3. Statement d is correct.
5. No, we can say only that there is a 27% chance of seeing the observed effectiveness just from natural sampling variation. There is no *evidence* that the new formula is more effective, but we can't conclude that they are equally effective.
7. a) No. There's a 25% chance of losing twice in a row. That's not unusual.
   b) 0.125
   c) No, we expect that to happen 1 time in 8.
   d) Maybe 5? The chance of 5 losses in a row is only 1 in 32, which seems unusual.
9. 1) Use p, not $\hat{p}$, in hypotheses.
   2) The question was about failing to meet the goal, so H_A should be $p < 0.96$.
   3) Did not check $0.04(200) = 8$. Since $nq < 10$, the Success/Failure Condition is violated.
   4) $188/200 = 0.94$: $SD(\hat{p}) = \sqrt{\dfrac{(0.96)(0.04)}{200}} = 0.014$
   5) z is incorrect; should be $z = \dfrac{0.94 - 0.96}{0.014} = -1.43$.
   6) $P = P(z < -1.43) = 0.076$
   7) There is only weak evidence that the new system does not work.
11. a) $H_0: p = 0.30$; $H_A: p > 0.30$
   b) Possibly an SRS; we don't know if the sample is less than 10% of his customers but could be viewed as less than 10% of all

possible customers; $(0.3)(80) \geq 10$ and $(0.7)(80) \geq 10$. Wells are independent only if customers don't have farms on the same underground springs.

c) $z = 0.73$; P-value $= 0.232$

d) If his dowsing is no different from standard methods, there is more than a 23% chance of seeing results as good as those of the dowser's, or better, by natural sampling variation.

e) These data provide no evidence that the dowser's chance of finding water is any better than normal drilling.

13. a) H_0: $p_{2000} = 0.34$; H_A: $p_{2000} < 0.34$

b) Students were randomly sampled and should be independent. 34% and 66% of 8302 are greater than 10. 8302 students are less than 10% of the entire student population of the United States.

c) P $= 0.029$

d) With such a small P-value, I reject H_0. There has been a statistically significant decrease in the proportion of students who have no absences.

e) No. A difference this small, although statistically significant, is not meaningful. We might look at new data in a couple of years.

15. a) Based on these data, we are 95% confident the proportion of adults in 1995 who had never smoked cigarettes is between 48.7% and 55.3%.

b) H_0: $p = 0.44$; H_A: $p \neq 0.44$. Since 44% is not in the confidence interval, we will reject H_0 and conclude the proportion of adults in 1995 who had never smoked was more than in the 1960s.

17. H_0: $p = 0.20$; H_A: $p > 0.20$. SRS (not clear from information provided); 22 is more than 10% of the population of 150; $(0.20)(22) < 10$. Do not proceed with a test.

19. H_0: $p = 0.03$; H_A: $p < 0.03$. $\hat{p} = 0.015$. One mother having twins will not impact another, so observations are independent; not an SRS; sample is less than 10% of all births. However, the mothers at this hospital may not be representative of all teenagers; $(0.03)(469) = 14.07 \geq 10$; $(0.97)(469) \geq 10$. $z = -1.91$; P-value $= 0.0278$. With a P-value this low, reject H_0. These data show evidence that the rate of twins born to teenage girls at this hospital is less than the national rate of 3%. It is not clear whether this can be generalized to all teenagers.

21. H_0: $p = 0.25$; H_A: $p > 0.25$. SRS; sample is less than 10% of all potential subscribers; $(0.25)(500) \geq 10$; $(0.75)(500) \geq 10$. $z = 1.24$; P-value $= 0.1076$. The P-value is high, so do not reject H_0. These data do not show that more than 25% of current readers would subscribe; the company should not go ahead with the WebZine on the basis of these data.

23. H_0: $p = 0.40$; H_A: $p < 0.40$. Data are for all executives in this company and may not be able to be generalized to all companies; $(0.40)(43) \geq 10$; $(0.60)(43) \geq 10$. $z = -1.31$; P-value $= 0.0955$. Because the P-value is high, we fail to reject H_0. These data do not show that the proportion of women executives is less than the 40% women in the company in general.

25. H_0: $p = 0.109$; H_A: $p > 0.109$. $\hat{p} = 0.118$; $z = 1.198$; P-value $= 0.115$. Because the P-value is high, we fail to reject H_0. These data do not show the dropout rate has increased from 10.9%.

27. H_0: $p = 0.90$; H_A: $p < 0.90$. $\hat{p} = 0.844$; $z = -2.05$; P-value $= 0.0201$. Because the P-value is so low, we reject H_0. There is strong evidence that the actual rate at which passengers with lost luggage are reunited with it within 24 hours is less than the 90% claimed by the airline.

29. a) Yes; assuming this sample to be a typical group of people, P $= 0.0008$. This cancer rate is very unusual.

b) No, this group of people may be atypical for reasons that have nothing to do with the radiation.

CHAPTER 21

1. If there is no difference in effectiveness, the chance of seeing an observed difference this large or larger is 4.7% by natural sampling variation.

3. $\alpha = 0.10$: Yes. The P-value is less than 0.05, so it's less than 0.10. But to reject H_0 at $\alpha = 0.01$, the P-value must be below 0.01, which isn't necessarily the case.

5. a) There is only a 1.1% chance of seeing a sample proportion as low as 89.4% vaccinated by natural sampling variation if 90% have really been vaccinated.

b) We conclude that p is below 0.9, but a 95% confidence interval would show that the true proportion is between (0.889, 0.899). Most likely, a decrease from 90% to 89.9% would not be considered important. On the other hand, with 1,000,000 children a year vaccinated, even 0.1% represents about 1000 kids—so this may very well be important.

7. a) (1.9%, 4.1%)

b) Because 5% is not in the interval, there is strong evidence that fewer than 5% of all men use work as their primary measure of success.

c) $\alpha = 0.01$; it's a lower-tail test based on a 98% confidence interval.

9. a) (0.459, 0.521)

b) Since 0.50 is in the confidence interval, we can't reject the hypothesis that $p = 0.50$.

c) Democrats could claim that Bush's approval rating might have sunk to 46%, while Republicans could argue that it was possible that a majority (52%) still approved of the President.

11. a) The null is that the level of home ownership remains the same. The alternative is that it rises.

b) The city concludes that home ownership is on the rise, but in fact the tax breaks don't help.

c) The city abandons the tax breaks, but they were helping.

d) A Type I error causes the city to forego tax revenue, while a Type II error withdraws help from those who might have otherwise been able to buy a home.

e) The power of the test is the city's ability to detect an actual increase in home ownership.

13. a) It is decided the shop is not meeting standards when it is.

b) The shop is certified as meeting standards when it is not.

c) Type I

d) Type II

15. a) The probability of detecting the shop is not meeting standards when it is not.

b) 40 cars. Larger n.

c) 10%. More chance to reject H_0.

d) A lot. Larger differences are easier to detect.

17. a) One-tailed. The company wouldn't be sued if "too many" minorities were hired.

b) Deciding the company is discriminating when it is not.

c) Deciding the company is not discriminating when it is.

d) The probability of correctly detecting discrimination when it exists.

e) Increases power.

f) Lower, since n is smaller.

19. a) One-tailed. Software is supposed to decrease the dropout rate.
b) $H_0: p = 0.13$; $H_A: p < 0.13$
c) He buys the software when it doesn't help students.
d) He doesn't buy the software when it does help students.
e) The probability of correctly deciding the software is helpful.

21. a) $z = -3.21$, $p = 0.0007$. The change is statistically significant. A 95% confidence interval is (2.3%, 8.5%). This is clearly lower than 13%. If the cost of the software justifies it, the professor should consider buying the software.
b) The chance of observing 11 or fewer dropouts in a class of 203 is only 0.07% if the dropout rate is really 13%.

23. a) 0.0464
b) Type I
c) 37.6%
d) Increase the number of shots. Or keep the number of shots at 10 but increase alpha by declaring that 8, 9, or 10 will be deemed as having improved.

25. Laborers only.

27. a) (21.8%, 30.2%)
b) No, 20% is below the low end of the interval.

CHAPTER 22

1. a) Stratified
b) 6% higher among males
c) 4%
d)

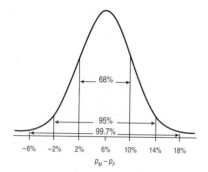

e) Yes; a poll result showing little difference is only 1–2 standard deviations below the expected outcome.

3. a) Yes. Random sample; less than 10% of the population; samples are independent; more than 10 successes and failures in each sample.
b) (0.055, 0.140)
c) We are 95% confident, based on these samples, that the proportion of American women age 65 and older who suffer from arthritis is between 5.5% and 14.0% more than the proportion of American men of the same age who suffer from arthritis.
d) Yes; the entire interval lies above 0.

5. a) 0.035
b) (0.356, 0.495)
c) We are 95% confident, based on these data, that the proportion of pets with a malignant lymphoma in homes where herbicides are used is between 35.6% and 49.5% higher than the proportion of pets with lymphoma in homes where no pesticides are used.

7. a) Experiment. Men were randomly assigned to have surgery or not.
b) (0.006, 0.080)

c) Since the entire interval lies above 0, there is evidence that surgery may be effective in preventing death from prostate cancer.

9. a) Differences have a larger standard error than single samples.
b) No. The confidence interval 2% ± 4% includes 0.

11. a) Prospective study
b) $H_0: p_1 - p_2 = 0$; $H_A: p_1 - p_2 \neq 0$ where p_1 is the proportion of students whose parents disapproved of smoking who became smokers and p_2 is the proportion of students whose parents are lenient about smoking who became smokers.
c) Yes. We assume the students were randomly selected; they are less than 10% of the population; samples are independent; at least 10 successes and failures in each sample.
d) $z = -1.17$, P-value = 0.2422. These samples do not show evidence that parental attitudes influence teens' decisions to smoke.
e) If there is no difference in the proportions, there is about a 24% chance of seeing the observed difference or larger by natural sampling variation.
f) Type II

13. a) (−0.065, 0.221)
b) We are 95% confident that the proportion of teens whose parents disapprove of smoking who will eventually smoke is between 6.5% less and 22.1% more than for teens with parents who are lenient about smoking.
c) 95% of all random samples will produce intervals that contain the true difference.

15. a) $H_0: p_1 - p_2 = 0$; $H_A: p_1 - p_2 \neq 0$. $z = 3.56$, P-value = 0.0004. With a P-value this low, we reject H_0. There is a significant difference in the clinic's effectiveness. Younger mothers have a higher birth rate than older mothers. Note that the Success/Failure Condition is met based on the pooled estimate of p.
b) We are 95% confident, based on these data, that the proportion of successful live births at the clinic is between 10.0% and 27.8% higher for mothers under 28 than in those 38 and older. However, the Success/Failure Condition is not met for the older women, since #successes < 10. We should be cautious in interpreting the level of the confidence interval.

17. $H_0: p_1 - p_2 = 0$; $H_A: p_1 - p_2 > 0$. $z = 1.18$, P-value = 0.118. With a P-value this high, we fail to reject H_0. These data do not show evidence of a decrease in the voter support for the candidate.

19. a) $H_0: p_1 - p_2 = 0$; $H_A: p_1 - p_2 \neq 0$. $z = -0.39$, P-value = 0.6951. With a P-value this high, we fail to reject H_0. There is no evidence of racial differences in the likelihood of multiple births, based on these data.
b) Type II

21. a) $H_0: p_1 - p_2 = 0$; $H_A: p_1 - p_2 > 0$. $z = 2.17$, P-value = 0.0148. With a P-value this low, we reject H_0. These data do suggest that mammograms may reduce breast cancer deaths.
b) Type I

23. a) We are 95% confident, based on this study, that between 67.0% and 83.0% of patients with joint pain will find medication A effective.
b) We are 95% confident, based on this study, that between 51.9% and 70.3% of patients with joint pain will find medication B effective.
c) Yes, they overlap. This might indicate no difference in the effectiveness of the medications, although this is not a proper test.

d) We are 95% confident that the proportion of patients with joint pain who will find medication A effective is between 1.7% and 26.1% higher than the proportion who will find medication B effective.

e) No. There is a difference in the effectiveness of the medications.

f) To estimate the variability in the difference of proportions, we must add variances. The two one-sample intervals do not. The two-sample method is the correct approach.

25. A 95% confidence interval is $(-1.6\%, 5.6\%)$; 0%—or no bounce—is a plausible value. They should have said that there was no evidence of a bounce from their poll, however, since they can't prove there was none at all.

27. The conditions are satisfied to test $H_0: p_{young} = p_{old}$ against $H_A: p_{young} > p_{old}$. The one-sided P-value is 0.0619, so we may reject the null hypothesis. Although the evidence is not strong, *Time* may be justified in saying that younger men are more comfortable discussing personal problems.

29. We assume these innings are typical of results we anticipate in future games, and test a two-sided hypothesis that the Giants are equally likely to score whether the other team pitches to Bonds or walks him. The P-value of 0.075 provides some evidence that the Giants will score in a higher percentage of innings if Bonds is intentionally walked.

PART V REVIEW

1. H_0: There is no difference in cancer rates, $p_1 - p_2 = 0$.
H_A: The cancer rate in those who use the herb is higher, $p_1 - p_2 > 0$.

3. a) 10.29

b) Not really. The z-score is -1.11. Not any evidence to suggest that the proportion for Monday is low.

c) Yes. The z-score is 2.26 with a P-value of 0.024 (two-sided).

d) Some births are scheduled for the convenience of the doctor and/or the mother.

5. a) $H_0: p = 0.40$; $H_A: p < 0.40$

b) Random sample; less than 10% of all California gas stations, $0.4(27) = 10.8, 0.6(27) = 16.2$. Assumptions and conditions are met.

c) $z = -1.49$, P-value $= 0.0677$

d) With a P-value this high, we fail to reject H_0. These data do not provide evidence that the proportion of leaking gas tanks is less than 40% (or that the new program is effective in decreasing the proportion).

e) Yes, Type II.

f) Increase α, increase the sample size.

g) Increasing α—increases power, lowers chance of Type II error, but increases chance of Type I error.
Increasing sample size—increases power, costs more time and money.

7. a) The researcher believes that the true proportion of "A's" is within 10% of the estimated 54%, namely, between 44% and 64%.

b) Small sample

c) No, 63% is contained in the interval.

9. a) Pew believes that the true proportion is within 3% of the 33% from the sample; that is, between 30% and 36%.

b) Larger, since it's a smaller sample.

c) We are 95% confident that the proportion of active traders who rely on the Internet for investment information is between 38.7% and 51.3%, based on this sample.

d) Larger, since it's a smaller sample.

11. a) Bimodal!

b) μ, the population mean. Sample size does not matter.

c) $\sigma/\sqrt{n}$; sample size does matter.

d) It becomes closer to a Normal model and narrower as the sample size increases.

13. a) $\mu = 0.80$ $\sigma = 0.028$

b) Yes. $0.8(200) = 160, 0.2(200) = 40$. Both ≥ 10.

c)

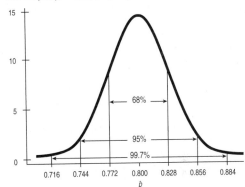

d) 0.039

15. H_0: There is no difference, $p_1 - p_2 = 0$ H_A: Early births have increased, $p_1 - p_2 < 0$. $z = -0.729$, P-value $= 0.2329$. Because the P-value is so high, we do not reject H_0. These data do not show an increase in the incidence of early birth of twins.

17. a) H_0: There is no difference, $p_1 - p_2 < 0$. H_A: Treatment prevents deaths from eclampsia, $p_1 - p_2 < 0$.

b) Samples are random and independent; less than 10% of all pregnancies (or eclampsia cases); more than 10 successes and failures in each group.

c) 0.8008

d) There is not sufficient evidence to conclude that magnesium sulfide is effective in preventing eclampsia deaths.

e) Type II

f) Increase the sample size, increase α.

g) Increasing sample size: decreases variation in the sampling distribution, is costly. Increasing α: Increases likelihood of rejecting H_0, increases chance of Type I error.

19. a) It is not clear What the pollster asked. Otherwise they did fine.

b) Stratified sampling.

c) 4%

d) 95%

e) Smaller sample size.

f) Wording and order of questions (response bias).

21. a) H_0: There is no difference, $p = 0.143$. H_A: The fatal accident rate is lower in girls, $p < 0.143$. $z = -1.67$, P-value $= 0.0479$. Because the P-value is low, we reject H_0. These data give some evidence that the fatal accident rate is lower for girls than for teens in general.

b) If the proportion is really 14.3%, we will see the observed proportion (11.3%) or lower 4.8% of the time by sampling variation.

23. a) One would expect many small fish, with a few large ones.

b) We don't know the exact distribution, but we know it's not Normal.

c) Probably not. With a skewed distribution, a sample size of five is not a large enough sample to say the mean is approximately Normal.

d) 0.961

25. a) Yes. 0.8(60) = 48, 0.2(60) = 12. Both are ≥ 10.

b) 0.834

c) Higher. Bigger sample means smaller standard deviation for $\hat{p}$.

d) Answers will vary. A sample size of 500 makes the probability 0.997.

27. a) 54.4 to 62.5%

b) Based on this study, with 95% confidence the proportion of Crohn's disease patients who will respond favorable to infliximab is between 54.4% and 62.5%.

c) 95% of all such random samples will produce confidence intervals that contain the true proportion of patients who respond favorably.

29. At least 423, assuming that p is near 50%.

31. a) Random sample (?); certainly less than 10% of all preemies and normal babies; more than 10 failures and successes in each group. 1.7% to 16.3% greater for normal-birth-weight children.

b) Since 0 is not in the interval, there is evidence that preemies have a lower high school graduation rate than children of normal birth weight.

c) Type I, since we rejected the null hypothesis.

33. a) H_0: The computer is undamaged. H_A: The computer is damaged.

b) 20% of good PCs will be classified as damaged (bad), while all damaged PCs will be detected (good).

c) 3 or more.

d) 20%

e) By switching to two or more as the rejection criterion, 7% of the good PCs will be misclassified, but only 10% of the bad ones will, increasing the power from 20% to 90%.

35. The null hypothesis is that Bush's support in the population is 45% (the Carter benchmark). The one-tailed P-value is 0.0055, strong evidence that Bush's approval rating in July 2004 was higher than 45%.

CHAPTER 23

1. a) 1.74   b) 2.37   c) 0.0524   d) 0.0889

3. Shape and center do not change; spread becomes narrower.

5. a) The confidence interval is for the population mean, not the individual cows in the study.

b) The confidence interval is not for individual cows.

c) We *know* the average gain in this study was 56 pounds!

d) The average weight gain of all cows does not vary. It's what we're trying to estimate.

e) No. There is not a 95% chance for another sample to have an average weight gain between 45 and 67 pounds. There is a 95% chance that another sample will have its average weight gain within two standard errors of the true mean.

7. a) No. A confidence interval is not about individuals in the population.

b) No. It's not about individuals in the sample, either.

c) No. We know the mean cost for students in the sample was $850.

d) No. A confidence interval is not about other sample means.

e) Yes. A confidence interval estimates a population parameter.

9. a) Based on this sample, we can say, with 95% confidence, that the mean pulse rate of adults is between 70.9 and 74.5 beats per minute.

b) 1.8 beats per minute

c) Larger

11. a) Yes. Randomly selected group; less than 10% of the population; the histogram is not unimodal and symmetric, but it is not highly skewed and there are no outliers, so with a sample size of 52, the CLT says $\bar{y}$ is approximately Normal.

b) (98.06, 98.51) degrees F

c) We are 98% confident, based on the data, that the average body temperature for an adult is between 98.05°F and 98.51°F.

d) 98% of all such random samples will produce intervals containing the true mean temperature.

e) These data suggest that the true normal temperature is somewhat less than 98.6°F.

*f) The sign test rejects that the median is 98.6 with a two-sided P-value of 0.0183. The z-value is −2.3591. (Exact binomial probabilities give a two-sided P-value of 0.0259.) Same conclusion as in part e.

13. a) Narrower. We need fewer values to be less confident.

b) Advantage: more chance of including the true value. Disadvantage: wider interval.

c) Narrower; due to the larger sample, the SE will be smaller.

d) About 252

15. a) (298.5, 321.5)

b) Normal population, random sample. These seem reasonable here—but we would need to check the Nearly Normal Condition.

c) With 95% confidence, the mean sodium content in these "reduced sodium" hot dogs is between 298.5 mg and 321.5 mg, based on this sample.

17. a) Yes. Larger sample, smaller SE.

b) 4.1 mg sodium

c) With 95% confidence, the mean sodium content of all hot dogs is between 309.7 mg and 326.3 mg, based on this sample.

d) 0.7(465) = 325.5. Since the confidence interval extends above this, it is possible these should not be labeled as "reduced sodium."

19. If in fact the mean cholesterol of pizza eaters does not indicate a health risk, then only 7 of every 100 samples would have mean cholesterol levels as high (or higher) as observed in this sample.

21. a) Upper-tail. We want to show it will hold 500 pounds (or more) easily.

b) They will decide the stands are safe, when they're not.

c) They will decide the stands are unsafe, when they are in fact safe.

23. a) Decrease α. This means a smaller chance of declaring the stands safe, if they are not.

b) The probability of correctly detecting that the stands are capable of holding more than 500 pounds.

c) Decrease the standard deviation—probably costly. Increase the sample size—takes more time for testing and is costly. Increase α—more Type I errors. Increase the "design load" to be well above 500 pounds—again, costly.

25. a) H_0: $\mu = 23.3$; H_A: $\mu > 23.3$

b) We have a random sample of the population. Population may not be normally distributed, as it would be easier to have a few much older men at their first marriage than some very young men. However, with a sample size of 40, $\bar{y}$ should be approximately Normal. We should check the histogram for severity of skewness and possible outliers.

c) $(\bar{y} - 23.3)/(s/\sqrt{40}) \sim t_{39}$

d) 0.1447

e) If the average age at first marriage is still 23.3 years, there is a 14.5% chance of getting a sample mean of 24.2 years or older simply from natural sampling variation.

f) We have not shown that the average age at first marriage has increased from the mean of 23.3 years.

27. a) Probably a representative sample; the Nearly Normal Condition seems reasonable from a Normal probability plot. The histogram is nearly uniform, with no outliers or skewness.

b) $\bar{y} = 28.78$, $s = 0.40$

c) (28.36, 29.21) grams

d) Based on this sample, we are 95% confident the average weight of the content of Ruffles bags is between 28.36 and 29.21 grams.

e) The company is erring on the safe side, as it appears that, on average, it is putting in slightly more chips than stated.

29. a) Type I; he mistakenly rejected the null hypothesis that $p = 0.10$ (or worse).

b) Yes. These are a random sample of bags and the Nearly Normal Condition is met; $t = -2.51$ with 7 df for a one-sided P-value of 0.0203.

31. No. Based on this large random sample, $t = -1.45$; P-value = 0.1495. Because the P-value is high, we fail to reject H_0. These data do not provide evidence to support the claim that the average weight of all cars is not 3000 pounds. We retain the null hypothesis that the mean is 3000 pounds.

33. a) Random sample; the Nearly Normal Condition seems reasonable from a Normal probability plot. The histogram is roughly unimodal and symmetric with no outliers.

b) (1187.9, 1288.4) chips

c) Based on this sample, the mean number of chips in an 18-ounce bag is between 1187.9 and 1288.4, with 95% confidence. The *mean* number of chips is clearly greater than 1000. However, if the claim is about individual bags, then it's not necessarily true. If the mean is 1188 and the SD deviation is near 94, then 2.5% of the bags will have fewer than 1000 chips, using the Normal model. If in fact the mean is 1288, the proportion below 1000 will be less than 0.1%, but the claim is still false.

35. a) The Normal probability plot is relatively straight, with one outlier at 93.8 sec. Without the outlier, the conditions seem to be met. The histogram is roughly unimodal and symmetric with no other outliers.

b) $t = -2.63$, P-value = 0.0160. With the outlier included, we might conclude that the mean completion time for the maze is not 60 seconds; in fact, it is less.

c) $t = -4.46$, P-value = 0.0003. Because the P-value is so small, we reject H_0. Without the outlier, we conclude that the average completion time for the maze is not 60 seconds; in fact, it is less. The outlier here did not change the conclusion.

d) The maze does not meet the "one-minute average" requirement. Both tests rejected a null hypothesis of a mean of 60 seconds.

*e) Sign test gives $z = 2.400$ with a one-sided P-value of 0.0082, rejecting that the median is 60 seconds. (Exact binomial probability gives 0.0133.) Same conclusion as in part d, even with the outlier included.

CHAPTER 24

1. a) 2.927

b) Larger

c) Based on this sample, we are 95% confident that students who learn Math using the CPMP method will score on average between 5.57 and 11.43 points better on a test solving applied Algebra problems with a calculator than students who learn by traditional methods.

d) Yes; 0 is not in the interval.

3. a) $H_0: \mu_C - \mu_T = 0$ vs. $H_A: \mu_C - \mu_T \neq 0$

b) Yes. Groups are independent, though we don't know if students were randomly assigned to the programs. Sample sizes are large, so CLT applies.

c) If the means for the two programs are really equal, there is less than a 1 in 10,000 chance of seeing a difference as large as or larger than the observed difference just from natural sampling variation.

d) On average, students who learn with the CPMP method do significantly worse on Algebra tests that do not allow them to use calculators than students who learn by traditional methods.

5. a) (1.36, 4.64)

b) No; 5 minutes is beyond the high end of the interval.

7.

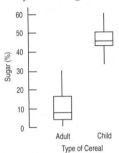

Random sample—questionable, but probably representative, independent samples, less than 10% of all cereals; boxplot shows no outliers—not exactly symmetric, but these are reasonable sample sizes. Based on these samples, with 95% confidence, children's cereals average between 32.49% and 40.80% more sugar content than adult's cereals.

9. $H_0: \mu_N - \mu_C = 0$ vs. $H_A: \mu_N - \mu_C > 0$; $t = 2.076$; P-value = 0.0228; df = 33.4. Because of the small P-value, we reject H_0. These data do suggest that new activities are better. The mean reading comprehension score for the group with new activities is significantly (at $\alpha = 0.05$) higher than the mean score for the control group.

(*Can't use a Tukey test because one group has the lowest and highest score.)

11.

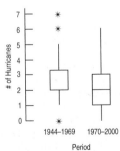

There are several concerns here. First, we don't have a random sample. We have to assume that the actual number of hurricanes in a given year is a random sample of the hurricanes that might occur under similar weather conditions. Also, the data for 1944–1969 are not symmetric and have three outliers. The outliers will tend to make the average for the period 1944–1969 larger. These data are not appropriate for inference. The boxplots provide little evidence of a change in the mean number of hurricanes in the two periods.

13. a)

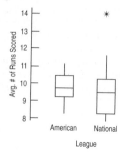

Both are reasonably symmetric, except for the outlier in the National League.

b) Based on these data, the average number of runs in an American League stadium is between 9.34 and 10.27, with 95% confidence.

c) There is an outlier. It could be deleted, and mentioned separately.

d) Yes. The boxplot indicates it is an outlier.

e) We want to work directly with the average difference. The two separate confidence intervals do not answer questions about the difference. The difference has a different standard deviation.

15. a) $(-0.69, 1.14)$

b) Based on these data, with 95% confidence, American League stadiums average between –0.69 and 1.14 more runs than National League stadiums.

c) No; 0 is in the interval.

d) No. The interval is now $(-0.19, 1.23)$. It still contains 0.

17. These are not two independent samples. These are before and after scores for the same individuals.

19. a) These data do not provide evidence of a difference in ad recall between shows with sexual content and violent content.

b) $H_0: \mu_S - \mu_N = 0$ vs. $H_A: \mu_S - \mu_N \neq 0$. $t = -6.08$, df = 213.99, P-value = 5.5×10^{-9}. Because the P-value is low, we reject H_0. These data show that ad recall between shows with sexual and neutral content is different; those who saw shows with neutral content had higher average recall.

21. a) $H_0: \mu_V - \mu_N = 0$ vs. $H_A: \mu_V - \mu_N \neq 0$. $t = -7.21$, df = 201.96, P-value = 1.1×10^{-11}. Because of the very small P-value, we reject H_0. There is a significant difference in mean ad recall between shows with violent content and neutral content; viewers of shows with neutral content remember more brand names, on average.

b) With 95% confidence, the average number of brand names remembered 24 hours later is between 1.45 and 2.41 higher for viewers of neutral content shows than for viewers of sexual content shows, based on these data.

23. a) The 95% confidence interval for the difference is $(0.61, 5.39)$. 0 is not in the interval, so scores in 1996 were significantly higher. (Or the t, with more than 7500 df, is 2.459 for a P-value of 0.0070.)

b) Since both samples were very large, there shouldn't be a difference in how certain you are, assuming conditions are met.

25. a) We can be 95% confident that the interval 74.8 ± 178.05 minutes includes the true difference in mean crossing times between men and women. Because the interval includes zero, we cannot be confident that there is any difference at all.

b) Independence Assumption: There is no reason to believe that the swims are not independent or that the two groups are not independent of each other.

Randomization Condition: The swimmers are not a random sample from any identifiable population, but they may be representative of swimmers who tackle challenges such as this.

Nearly Normal Condition: the boxplots show no outliers. The histograms are unimodal; the histogram for men is somewhat skewed to the right.

27. Independent Groups Assumption: The runners are different women, so the groups are independent. The Randomization Condition is satisfied since the runners are selected at random for these heats.

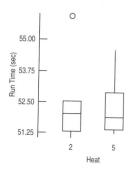

Nearly Normal Condition: The boxplots show an outlier, but we will proceed and then redo the analysis with the outlier deleted. When we include the outlier, $t = 0.035$ with a two-sided P-value of 0.97. With the outlier deleted, $t = -1.14$, with P = 0.2837. Either P-value is so large that we fail to reject the null hypothesis of equal means and conclude that there is no evidence of a difference in the mean times for runners in unseeded heats.

29. With $t = -4.57$ and a very low P-value of 0.0013, we reject the null hypothesis of equal mean velocities. There is strong evidence that golf balls hit off Stinger tees will have a higher mean initial velocity.

31. The 90% confidence interval for the difference in mean number of months to publication between *Applied Statistics* and *The American Statistician* is $(8.43, 11.57)$ months. We assume that the articles are a random sample of the kinds of articles typically submitted to these two journals, and that they represent fewer than 10% of all articles that could have been submitted. Both samples are large.

33. a) $H_0: \mu_R - \mu_N = 0$ vs. $H_A: \mu_R - \mu_N < 0$. $t = -1.36$, df = 20.00, P-value = 0.0945. Because the P-value is large, we fail to reject H_0. These data show no evidence of a difference in mean number of objects recalled between listening to rap or no music at all.

b) Didn't conclude a difference.

CHAPTER 25

1. a) Randomly assign 50 hens to each of the two kinds of feed. Compare production at the end of the month.

b) Give all 100 hens the new feed for 2 weeks and the old food for 2 weeks, randomly selecting which feed the hens get first. Analyze the differences in production for all 100 hens.

c) Matched pairs. Because hens vary in egg production, the matched-pairs design will control for that.

3. a) Show the same people ads with and without sexual images and record how many products they remember in each group. Randomly decide which ads a person sees first. Examine the differences for each person.

b) Randomly divide volunteers into two groups. Show one group ads with sexual images and the other group ads without. Compare how many products each group remembers.

5. a) Matched pairs—same cities in different time periods.

b) There is a significant difference (P-value = 0.0244) in the labor force participation rate for women in these cities; women's participation increased between 1968 and 1972.

7. a) The paired t-test is appropriate since we have pairs of Fridays in 5 different months. Data from adjacent Fridays within a month may be more similar than randomly chosen Fridays.

b) We conclude that there is evidence (P-value 0.0212) that the mean number of cars found on the M25 motorway on Friday the 13th is less than on the previous Friday.

c) We don't know if these Friday pairs were selected at random. Obviously, if these are the Fridays with the largest differences, this will affect our conclusion. The Nearly Normal Condition appears to be met by the differences, but the sample size is small.

9. a) Even if the individual times show a trend of improving speed over time, the differences may well be independent of each other. They are subject to random year-to-year fluctuations, and we may believe these data to be representative of similar races. We don't have any information with which to check the Nearly Normal Condition.

b) With 95% confidence we can say the mean time difference is between -14.54 minutes (men are faster) and $+14.72$ minutes (women are faster).

c) The interval contains 0, so we would not reject the hypothesis of no mean difference at $\alpha = 0.05$. We can't discern a difference between the female wheelchair times and the male running times.

11. a) Same cows before and after injection; the cows should be representative of others of their breed; cows are independent of each other; less than 10% of all cows; don't know about nearly Normal differences.

b) (12.66, 15.34)

c) Based on this sample, with 95% confidence, the average increase in milk production for Ayrshire cows given BST is between 12.66 and 15.34 pounds per day.

d) 0.25(47) = 11.75. The average increase is much more than this, so we would recommend he go to the extra expense.

13.

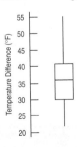

Data are paired for each city; cities are independent of each other; boxplot shows the temperature differences are symmetric, with no outliers. This is probably not a random sample, so we might be wary of inferring that this difference applies to all European cities. Based on these data, we are 90% confident that the average temperature in European cities in July is between 32.3°F and 41.3°F higher than in January.

15. Based on these data we are 90% confident that boys, on average, can do between 1.6 and 13.0 more push-ups than girls. (independent samples—not paired).

17. a) Paired sample test. Data are before/after for the same workers; workers randomly selected; assume less than 10% of all this company's workers; boxplot of differences shows them to be symmetric, with no outliers.

b) $H_0: \mu_D = 0$ vs. $H_A: \mu_D > 0$. $t = 3.60$, P-value = 0.0029. Because P < 0.01, reject H_0. These data show that average job satisfaction has increased after implementation of the exercise program.

*c) 8 of 10 people's satisfaction increased, giving a one-sided P-value (from binomial probability) of 0.0549. Not quite significant. Using the sign test, we do not have quite enough evidence to suggest at $\alpha = 0.05$ that there has been an increase.

19. a) $H_0: \mu_D = 0$ vs. $H_A: \mu_D > 0$

b) 0.0025

c) If there is no gain of additional hours of sleep with the herb, the chance of seeing a mean difference as large or larger than the one observed is about one-quarter percent.

d) The data provide evidence that the herb is helpful in gaining additional sleep.

e) Type I

21. $H_0: \mu_D = 0$ vs. $H_A: \mu_D \neq 0$. Data are paired by brand; brands are independent of each other; less than 10% of all yogurts (questionable); boxplot of differences shows an outlier (100) for Great Value:

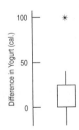

With the outlier included, the mean difference (Strawberry − Vanilla) is 12.5 calories with a t-stat of 1.332, with 11 df, for a P-value of 0.2098. Deleting the outlier, the difference is even smaller, 4.55 calories with a t-stat of only 0.833 and a P-value of 0.4241. With P-values so large, we do not reject H_0. We conclude that the data do not provide evidence of a difference in mean calories.

23. a) Not a simple random sample, but most likely representative; stops most likely independent of each other; boxplot is symmetric with no outliers.

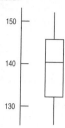

Based on these data, with 95% confidence, the average braking distance for these tires on dry pavement is between 133.6 and 145.2 feet.

b) Not a simple random sample, but most likely representative; stops most likely independent of each other; less than 10% of all possible wet stops; a Normal probability plot is relatively straight.

Based on these data, with 95% confidence, the average increase in distance for these tires on wet pavement is between 51.4 and 74.6 feet.

25. a) Cars were probably not a simple random sample, but may be representative in terms of stopping distance; boxplot does not show outliers, but does indicate right skewness. A 95% confidence interval for the mean stopping distance on dry pavement is (131.8, 145.6) feet.

b) Data are paired by car; cars were probably not randomly chosen, but representative; boxplot shows an outlier (car 4) with a difference of 12. With deletion of that car, a Normal probability plot of the differences is relatively straight.

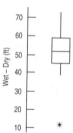

Having deleted an outlier difference of 12 feet, we estimate with 95% confidence that the average braking distance is between 47.2 and 62.8 feet more on wet pavement than on dry, based on this sample. (With the outlier, the confidence interval is 38.8 to 62.6 feet.)

27. Using a *t*-test for paired differences, $t = -0.86$ and two-tailed $P = 0.396$. With a P-value so high, we fail to reject the null hypothesis of no mean difference. There is no evidence that sexual images in ads affects people's ability to remember the product being advertised.

29. a) 60% is 30 strikes; $H_0: \mu = 30$ vs. $H_A: \mu > 30$. $t = 6.07$, P-value $= 3.92 \times 10^{-6}$. With a very small P-value, we reject H_0. There is very strong evidence that players can throw more than 60% strikes after training, based on this sample.

b) $H_0: \mu_D = 0$ vs. $H_A: \mu_D > 0$. $t = 0.135$, P-value $= 0.4472$. With such a high P-value, we do not reject H_0. These data provide no evidence that the program has improved pitching in these Little League players.

*c) Of the 20 pitchers, 9 increased their number of strikes, 9 decreased, and 2 stayed the same. This results in a one-sided P-value of 0.50 (using exact or z-test) and provides no evidence that the program increases strikes. Same conclusion as in part (b).

PART VI REVIEW

1. a) $H_0: \mu_{Jan} - \mu_{Jul} = 0$; $H_A: \mu_{Jan} - \mu_{Jul} \neq 0$. $t = -1.94$, df = 43.68, P-value = 0.0590. Since P < 0.10, reject the null. These data show a significant difference in mean age to crawl between January and July babies.

b) $H_0: \mu_{Apr} - \mu_{Oct} = 0$; $H_A: \mu_{Apr} - \mu_{Oct} \neq 0$. $t = -0.92$; df = 59.40; P-value = 0.3610. Since P > 0.10, do not reject the null; these data do not show a significant difference between April and October with regard to the mean age at which crawling begins.

c) These results are not consistent with the claim.

3. $H_0: p = 0.26$; $H_A: p \neq 0.26$. $z = 0.946$; P-value = 0.3443. Because the P-value is high, we do not reject H_0. These data do not show that the Denver area is different from the national rate in the proportion of businesses with women owners.

5. Based on these data we are 95% confident that the mean difference in aluminum oxide content is between –3.37 and 1.65. The means in aluminum oxide content of the pottery made at the two sites could reasonably be the same.

7. a) $H_0: p_{ALS} - p_{Other} = 0$; $H_A: p_{ALS} - p_{Other} > 0$. $z = 2.52$; P-value = 0.0058. With such a low P-value, we reject H_0. This is strong evidence that there is a higher proportion of varsity athletes among ALS patients than those with other disorders.

b) Observational retrospective study. To make the inference, one must assume the patients studied are representative.

9. $H_0: \mu = 7.41$; $H_A: \mu \neq 7.41$. $t = 2.18$; df = 111; P-value = 0.0313. With such a low P-value, we reject H_0. Assuming that Missouri babies fairly represent the United States, these data show that American babies are different from Australian babies in birth weight; American babies are heavier, on average.

11. a) If there is no difference in the average fish sizes, the chance of seeing an observed difference this large just by natural sampling variation is 0.1%.

b) If cost justified, feed them a natural diet.

c) Type I

13. a) Assuming the conditions are met, from these data we are 95% confident that patients with cardiac disease average between 3.39 and 5.01 years older than those without cardiac disease.

b) Older patients are at greater risk from a variety of other health issues, and perhaps more depressed.

15. a) Stratified sample survey.

b) We are 95% confident that boys are between 7.0% and 17.0% more likely to play computer games than girls, based on the survey.

c) Yes. The entire interval lies above 0.

17. Based on the data, we are 95% confident that the mean difference in words misunderstood is between −3.76 and 3.10. Because 0 is in the confidence interval, we would conclude that the two tapes are not equivalent.

19. a)

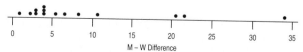

The countries that appear to be outliers are Spain, Italy, and Portugal. They are all Mediterranean countries.

b) $H_0: \mu_D = 0; H_A: \mu_D > 0$.
$t = 5.56; df = 10;$ P-value = 0.0001. With such a low P-value, we reject H_0. These data show that European men are more likely than women to read newspapers.

21. Based on the survey, we are 95% confident that the proportion of the American adults who would agree with the statement is between 57.0% and 63.0%.

23. Data are matched pairs (before and after for the same rooms); less than 10% of all rooms in a large hotel; uncertain how these rooms were selected (are they representative?). Histogram shows that differences are roughly unimodal and symmetric, with no outliers. A 95% confidence interval for the difference, before − after is (0.58, 2.65) counts. Since the entire interval is above 0, these data show that the new air-conditioning system was effective in reducing average bacteria counts.

25. a) We are 95% confident that between 19.77% and 38.66% of children with bipolar symptoms will be helped with medication and psychotherapy, based on this study.
    b) 221 children

27. a) From this histogram, about 115 loaves or more. (Not Normal.) This assumes the last 100 days are typical.
    b) Large sample size; CLT says $\bar{y}$ will be approximately Normal.
    c) From the data, we are 95% confident that the bakery will sell between 101.2 and 104.8 loaves of bread on an average day.
    d) 25
    e) Yes, 100 loaves per day is too low—the entire confidence interval is above that.

29. a) $H_0: p_{High} - p_{Low} = 0; H_A: p_{High} - p_{Low} \neq 0. z = -3.57;$ P-value = 0.0004. Because the P-value is so low, we reject H_0. These data show the IRS risk is different in the two groups; people who consume dairy products often have a lower risk on average.
    b) Doesn't prove it; this is not an experiment.

31. Based on these data, we are 95% confident that seeded clouds will produce an average of between −4.76 and 559.56 more acre-feet of rain than unseeded clouds. Since the interval contains negative values, it may be that seeding is unproductive.

33. a) Randomizing order of the tasks helps avoid bias and memory effects. Randomizing the cards helps avoid bias as well.
    b) $H_0: \mu_D = 0; H_A: \mu_D > 0$
    c)

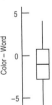

Boxplot of the differences looks symmetric with no outliers.
    d) $t = -1.70;$ P-value = 0.0999; do not reject H_0 because P > 0.05.
    e) The data do not provide evidence that the color or written word dominates.

35. a) Different samples give different means; this is a fairly small sample. The difference may be due to natural sampling variation.
    b) $H_0: \mu = 100; H_A: \mu < 100$
    c) Batteries selected are a SRS (representative); less than 10% of the company's batteries; lifetimes are approximately Normal.

d) $t = -1.0;$ P-value = 0.1666; do not reject H_0. This sample does not show that the average life of the batteries is significantly less than 100 hours.
e) Type II.

CHAPTER 26

1. a) Chi-square test of independence. We have one sample and two variables. We want to see if the variable *account type* is independent of the variable *trade type*.
   b) Other test. The variable *account size* is quantitative, not counts.
   c) Chi-square test of homogeneity. We want to see if the distribution of one variable, *courses*, is the same for two groups (resident and nonresident students).

3. a) 10
   b) Goodness-of-fit
   c) H_0: The die is fair (all faces have $p = 1/6$).
      H_A: The die is not fair.
   d) Count data; rolls are random and independent; expected frequencies all bigger than 5.
   e) 5
   f) $\chi^2 = 5.600$, P-value = 0.3471
   g) Because the P-value is high, do not reject H_0. The data show no evidence that the die is unfair.

5. a) Weights are quantitative, not counts.
   b) Count the number of each kind of nut, assuming the company's percentages are based on counts rather than weights.

7. H_0: The police force represents the population (29.2% white, 28.2% black, etc.). H_A: The police force is not representative of the population. $\chi^2 = 16516.88$, df = 4, P-value = 0.0000. Because the P-value is so low, we reject H_0. These data show that the police force is not representative of the population. In particular, there are too many white officers in relationship to their membership in the community.

9. a) $\chi^2 = 5.671$, df = 3, P-value = 0.1288. With a P-value this high, we fail to reject H_0. Yes, these data are consistent with those predicted by genetic theory.
   b) $\chi^2 = 11.342$, df = 3, P-value = 0.0100. Because of the low P-value, we reject H_0. These data provide evidence that the distribution is not as specified by genetic theory.
   c) With small samples, many more data sets will be consistent with the null hypothesis. With larger samples, small discrepancies will show evidence against the null hypothesis.

11. a) 40.2%   b) 8.1%
    c) 62.2%   d) 285.48
    e) H_0: Survival was independent of status on the ship.
       H_A: Survival depended on the status.
    f) 3
    g) We reject the null hypothesis. Survival depended on status. We can see that first-class passengers were more likely to survive than any other class.

13. a) Independence.
    b) H_0: Choice of college is independent of birth order.
       H_A: There is an association between birth order and choice of college.
    c) Counted Data Condition—these are counts of students; Randomization Condition—a class is not a random sample, but there's little reason to think this particular group of students isn't representative; Expected Cell Frequency Condition—the expected frequencies are low for both the Social Science and

Professional Colleges for third and fourth or higher birth orders (check residuals later).
d) 9
e) With a P-value this low, we reject the null hypothesis. There is some evidence of an association between birth order and choice of college.
f) Unfortunately, 3 of the 4 largest standardized residuals are in cells with expected counts less than 5. We should be very wary of drawing conclusions from this chi-square test.

15. a) Experiment—actively imposed treatments (different drinks)
b) Homogeneity
c) H_0: The rate of urinary tract infection is the same for all three groups. H_A: The rate of urinary tract infection is different among the groups.
d) Count data; random assignment to treatments; all expected frequencies larger than 5.
e) 2
f) $\chi^2 = 7.776$, P-value $= 0.020$.
g) With a P-value this low, we reject H_0. These data provide reasonably strong evidence there is a difference in urinary tract infection rates between cranberry juice drinkers, lactobacillus drinkers, and the control group.
h) The standardized residuals are

	Cranberry	Lactobacillus	Control
Infection	-1.87276	1.19176	0.68100
No Infection	1.24550	-0.79259	-0.45291

From the standardized residuals (and the sign of the residuals), it appears those who drank cranberry juice were less likely to develop urinary tract infections; those who drank lactobacillus were more likely to have infections.

17. a) Independence
b) H_0: *Political affiliation* is independent of *gender*.
H_A: There is a relationship between *political affiliation* and *gender*.
c) Count data; probably a random sample, but can't extend results to other states; all expected frequencies greater than 5.
d) $\chi^2 = 4.851$, df $= 2$, P-value $= 0.0884$.
e) Because of the high P-value, we do not reject H_0. These data do not provide evidence of a relationship between political affiliation and gender.

19. H_0: *Political affiliation* is independent of *region*. H_A: There is a relationship between political affiliation and region. $\chi^2 = 13.849$, df $= 4$, P-value $= 0.0078$. With a P-value this low, we reject H_0. *Political affiliation* and *region* are related. Examination of the residuals shows that those in the West are more likely to be Democrat than Republican; those in the Northeast are more likely to be Republican than Democrat.

21. a) Homogeneity
b) H_0: The grade distribution is the same for both professors.
H_A: The grade distributions are different.
c)

	Dr. Alpha	Dr. Beta
A	6.667	5.333
B	12.778	10.222
C	12.222	9.778
D	6.111	4.889
F	2.222	1.778

Three cells have expected frequencies less than 5.

23. a)

	Dr. Alpha	Dr. Beta
A	6.667	5.333
B	12.778	10.222
C	12.222	9.778
Below C	8.333	6.667

All expected frequencies are now larger than 5.
b) Decreased from 4 to 3.
c) $\chi^2 = 9.306$, P-value $= 0.0255$. Because the P-value is so low, we reject H_0. The grade distributions for the two professors are different. Dr. Alpha gives fewer A's and more grades below C than Dr. Beta.

25. $\chi^2 = 14.058$, df $= 1$, P-value $= 0.0002$. With a P-value this low, we reject H_0. There is evidence of racial steering. Blacks are much less likely to rent in Section A than Section B.

27. a) $z = 3.74936$, $z^2 = 14.058$.
b) P-value $(z) = 0.0002$ (same as in Exercise 25).

29. $\chi^2 = 2815.968$, df $= 9$, P-value <0.0001. Because the P-value is so low, we reject H_0. There are definite differences in education levels attained by the groups. The largest component indicates that Hispanics are more likely to not have high-school diplomas. The next largest residual indicates that fewer whites than expected do not have a high-school diploma.

31. $\chi^2 = 0.870$, df $= 2$, P-value $= 0.6471$. Because the P-value is so large, we do not reject H_0. These data show no evidence of a difference in higher education opportunities between blacks and Hispanics.

33. A goodness-of-fit test with 2 df yields $\chi^2 = 12.26$, with P $= 0.002$. Compared to the top 500 universities, the top 100 seem to be more heavily concentrated in the Americas and sparser in the Asia, Africa, and Pacific regions.

CHAPTER 27

1. a) $\widehat{\%Yes} = -5.583 + 0.999 \times Year$. The percentage who would vote for a woman has been increasing about 1% per year.
b) H_0: There has been no change in percentage of voters who would vote for a woman candidate for president, $\beta_1 = 0$.
H_A: There has been a change, $\beta_1 > 0$.
c) $t = 15.1$, P-value <0.0001. With such a low P-value, we reject H_0. There has been a significant change. Voters are increasingly willing to vote for a woman for president.
d) 94.2% of the variation in percentage willing to vote for a woman is accounted for by the regression on year.

3. a) H_0: There has been no change in the percentage expressing no opinion about voting for a woman president, $\beta_1 = 0$.
H_A: There has been a change, $\beta_1 \neq 0$.
b) $t = -1.21$, P-value $= 0.2458$. Because the P-value is so large, we do not reject H_0. These data do not indicate the percentage of "no opinion" responses has changed.
c) The plot indicates no real trend. The high value in 1945 and low value in 1990 are influential. True slope does not appear to be negative after discounting influential points at the ends. Answer to the last question remains the same because we did not reject H_0 in part b.

5. a) H_0: Difference in age at first marriage has not been decreasing, $\beta_1 = 0$. H_A: Difference in age at first marriage has been decreasing, $\beta_1 < 0$.

b) Residual plot shows no obvious pattern; histogram is not particularly Normal, but shows no obvious skewness or outliers.

c) $t = -4.35$, P-value $= 0.00015$. With such a low P-value, we reject H_0. These data show evidence that difference in age at first marriage is decreasing.

7. Based on these data, we are 95% confident that the average difference in age at first marriage is changing at a rate between -0.035 and -0.013 years per year.

9. a) H_0: *Fuel economy* and *weight* are not (linearly) related, $\beta_1 = 0$. H_A: *Fuel economy* changes with *weight*, $\beta_1 \neq 0$. P-value < 0.0001, indicating strong evidence of an association.

b) Yes, the conditions seem satisfied. Histogram of residuals is unimodal and symmetric; residual plot looks OK, but some "thickening" of the plot with increasing values.

c) $t = -12.2$, P-value < 0.0001. These data show evidence that gas mileage decreases with the weight of the car.

11. a) $(-9.57, -6.86)$ mpg per 1000 pounds.

b) Based on these data, we are 95% confident that *fuel efficiency* decreases between 6.86 and 9.57 miles per gallon on average for each additional 1000 pounds of *weight*.

13. a) Based on this regression, we are 95% confident that 2500-pound cars will average between 27.44 and 28.97 miles per gallon.

b) Based on the regression, a 3450-pound car will get between 16.00 and 24.81 miles per gallon, with 95% confidence.

15. a) Yes. $t = 2.73$, P-value $= 0.0079$. With a P-value so low, we reject H_0. There is a positive relationship between calories and sodium content.

b) No. $R^2 = 9\%$ and s appears to be large, although without seeing the data it is a bit hard to tell.

17. Plot of *calories* against *fiber* does not look linear; this is borne out in the residuals plot, which also shows increasing variance as predicted values get large. The histogram of residuals is right skewed.

19. a) H_0: No (linear) relationship between *BCI* and *pH*, $\beta_1 = 0$. H_A: There is a relationship, $\beta_1 \neq 0$.

b) $t = -7.73$ with 161 df; P-value < 0.0001

c) There is a negative relationship; *BCI* decreases as *pH* increases at an average of 197.7 *BCI* units per increase of 1 *pH*.

21. a) H_0: No linear relationship between *population* and *ozone*, $\beta_1 = 0$. H_A: *Ozone* increases with *population*, $\beta_1 > 0$. $t = 3.48$, P-value $= 0.0018$. With a P-value so low, we reject H_0. These data show evidence that ozone does increase with population.

b) Yes; population explains 84% of the variability on ozone level, and s is just over 5 parts per million.

23. a) Based on this regression, each additional million residents is expected to increase average ozone level by between 3.29 and 10.01 ppm, with 90% confidence.

b) Based on the regression, the mean ozone level for cities with 600,000 people is between 18.47 and 27.29 ppm, with 90% confidence.

25. a) 33 batteries.

b) Data plot is roughly linear with lots of scatter; plot of residuals vs. predicted values shows no overt patterns; Normal probability plot of residuals is reasonably straight.

c) H_0: No linear relationship between *cost* and *cranking amps*, $\beta_1 = 0$. H_A: *Cranking amps* increase with cost, $\beta_1 > 0$. $t = 3.23$; P-value $= 0.0029$. With a P-value so low, we reject H_0. These data provide evidence that more expensive batteries do have more cranking amps.

d) No. $R^2 = 25.2\%$ and $s = 116$ amps. Considering that the range of amperage is only about 400 amps, an s of 116 is not very useful.

e) $\widehat{Cranking\ amps} = 384.59 + 4.15 \times Cost$.

f) $(1.97, 6.32)$ cold cranking amps per dollar.

g) Cranking amps increase, on average, between 1.97 and 6.32 with every dollar of battery cost increase, with 90% confidence.

27.

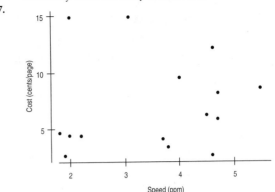

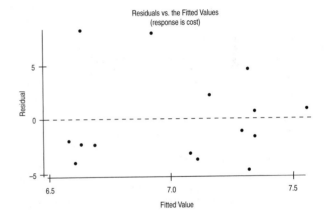

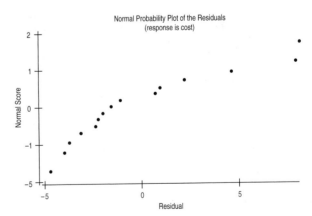

Data plot does not show any indication of a linear relationship. There may be a curved relationship. Normal probability plot of the residuals is not linear (residuals not Normal). Inference for linear regression is not valid here.

29. a) H_0: No linear relationship between *waist size* and *%body fat*, $\beta_1 = 0$. H_A: *%Body fat* increases with *waist size*, $\beta_1 > 0$. $t = 8.14$; P-value < 0.0001. *%Body fat* seems to increase with *waist size*.

b) With 95% confidence, mean percent body fat for people with 40-inch waists is between 23.58 and 29.02, based on this regression.

31. a) The regression model is $\widehat{Midterm2} = 12.005 + 0.721 Midterm1$ with output:

	Estimate	Std Error	t-ratio	P-value
Intercept	12.00543	15.9553	0.752442	0.454633
Slope	0.72099	0.183716	3.924477	0.000221
RSquare			0.198982	
s			16.78107	
n			64	

b) The scatterplot shows a weak somewhat linear positive relationship. There are several outlying points, but removing them only makes the relationship slightly stronger. There is no obvious pattern in the residual plot. The regression model appears appropriate. The small P-value for the slope shows that the slope is statistically distinguishable from 0 even though the R^2 value of 0.199 suggests that the overall relationship is weak.

c) No. The R^2 value is only 0.199 and the value of s of 16.8 points indicates that she would not be able to predict performance on *Midterm2* very accurately.

33. a) We want to know whether the times to swim across Lake Ontario have changed over time. We know the time of each swim in minutes and the year of the swim. $H_0: \beta_1 = 0$ (no linear relationship); $H_A: \beta_1 \neq 0$.

b) The Straight Enough Condition fails because there is an outlier.

c) We cannot continue this analysis.

d) With the outlier removed, the relationship looks straight enough. Swims are separate, but some swimmers performed the feat more than once, so they may not be entirely independent. There is no thickening of the plot. The normal probability plot of the residuals looks straight. The conditions appear to be met.

e) The analysis shows that swim times have been increasing at about 14.9 minutes per year. The P-value of 0.0158 is small enough to reject the null hypothesis, so this looks like a real increase.

f) The SE depends on three things: the standard deviation of the residuals around the line, σ_e, the standard deviation of x, and the number of data values. Only the first of these was changed substantially by removing the outlier. Indeed, we can see that s_e was 447.3 with the outlier present and only 277.2 when it was removed. This accounts for most of the change in the SE and therefore in the P-value. Because the outlier was at a value of x near the middle of the x range, it didn't affect the slope very much.

35. a) Data plot looks linear; no overt pattern in residuals; histogram of residuals roughly symmetric and unimodal.

b) H_0: No linear relationship between *education* and *mortality*, $\beta_1 = 0$. $H_A: \beta_1 < 0$. $t = -6.24$; P-value <0.001. There is evidence that cities in which the mean education level is higher also tend to have a lower mortality rate.

c) No. Data are on cities, not individuals. Also, these are observational data. We cannot predict causal consequences from them.

d) $(-65.95, -33.89)$ deaths per 100,000 people.

e) Deaths per 100,000 people decrease on average between 33.89 and 65.95 for each year of average education, based on the regression.

f) Based on the regression, the average mortality rate for cities with an average of 12 years of school will be between 874.239 and 914.196 deaths per 100,000 people.

PART VII REVIEW

1. H_0: The proportions are as specified by the ratio 1:3:3:9; H_A: The proportions are not as stated. $\chi^2 = 5.01$; df = 3; P-value = 0.1711. Since P > 0.05, we fail to reject H_0. These data do not provide evidence to indicate that the proportions are other than 1:3:3:9.

3. a) H_0: *Mortality* and *calcium concentration* in water are not linearly related, $\beta_1 = 0$; H_A: They are related, $\beta_1 \neq 0$.

b) $t = -6.73$; P-value <0.0001. There is a significant negative relationship between calcium in drinking water and mortality.

c) $(-4.19, -2.27)$ deaths per 100,000 for each ppm calcium.

d) Based on the regression, we are 95% confident that mortality (deaths per 100,000) decreases on average between 2.27 and 4.19 for each part per million of calcium in drinking water.

5. 404 checks

7. H_0: *Income* and *party* are independent. H_A: *Income* and *party* are not independent. $\chi^2 = 17.19$; P-value = 0.0018. With such a small P-value, we reject H_0. These data show evidence that income level and party are not independent. Examination of components shows Democrats are most likely to have low incomes; Independents are most likely to have middle incomes, and Republicans most likely to have high incomes.

9. $H_0: p_L - p_R = 0$; $H_A: p_L - p_R \neq 0$. $z = 1.38$; P-value = 0.1683. Since P > 0.05, we do not reject H_0. These data do not provide evidence of a difference in musical abilities between right- and left-handed people.

11. a) $H_0: \mu_D = 0$; $H_A: \mu_D \neq 0$.
Boxplot of the differences indicates a strong outlier (1958). With the outlier kept in, the *t*-stat is 0, with a P-value of 1.00 (two sided). There is no evidence of a difference (on average) of actual and that predicted by Gallup. With the outlier taken out, the *t*-stat is still only -0.8525 with a P-value of 0.4106, so the conclusion is the same.

b) H_0: There is no (linear) relationship between predicted and actual number of Democratic seats won ($\beta_1 = 0$). H_A: There is a relationship ($\beta_1 \neq 0$). The relationship is very strong, with an R^2 of 97.7%. The *t*-stat is 22.56. Even with only 12 df, this is clearly significant (P-value <0.0001). There is an outlying residual (1958), but without it, the regression is even stronger.

13. $\chi^2 = 0.69$; P-value = 0.9526. Since P > 0.05, we do not reject H_0. These data provide no evidence to show that the way the hospital deals with twin pregnancies has changed.

15. a) Based on these data, the average annual rainfall in LA is between 11.65 and 17.39 inches, with 90% confidence.

b) About 46 years

c) No. The regression equation is $\widehat{Rain} = -51.684 + 0.033 \times Year$. $R^2 = 0.1\%$. The *t*-stat for the slope is 0.12 with a P-value of 0.9029.

17. a) This is a linear regression that is meaningless—the data are categorical.

b) This is a two-way table, that is appropriate. H_0: *eye* and *hair* color are independent. H_A: *Eye* and *hair* color are not independent. However, four cells have expected counts less than 5, so the χ^2 analysis is not valid unless cells are merged. However with a χ^2 value of 223.6 with 16 df and a P-value <0.0001, the results are not likely to change if we merge appropriate eye colors.

19. a) $H_0: p_Y - p_O = 0$; $H_A: p_Y - p_O \neq 0$. $z = 3.56$; P-value $= 0.0004$. With such a small P-value, we reject H_0. We conclude there is evidence of a difference in effectiveness; it appears the methods are not as good for older women.

b) $\chi^2 = 12.70$; P-value $= 0.0004$. Same conclusion.

c) The P-values are the same; $z^2 = (3.563944)^2 = 12.70 = \chi^2$.

21. a) Positive direction, generally linear trend; moderate scatter.

b) H_0: There is no linear relationship between *interval* and *duration*. $\beta_1 = 0$. H_A: There is a linear relationship $\beta_1 \neq 0$.

c) Yes; histogram is unimodal and roughly symmetric; residuals plot shows random scatter.

d) $t = 27.1$; P-value ≤ 0.001. With such a small P-value, we reject H_0. There is a significant positive linear relationship between duration and time to next eruption of Old Faithful.

e) Based on this regression, the average time to next eruption after a 2-minute eruption is between 53.24 and 56.12 minutes, with 95% confidence.

f) Based on this regression, we will have to wait between 63.23 and 87.57 minutes after a 4-minute eruption, with 95% confidence.

23. a) $t = 1.42$, P-value $= 0.1574$. Since $P > 0.05$, we do not reject H_0. The two groups did not differ in ability at the start of the study.

b) $t = 15.11$; P-value < 0.0001. The group taught using the accelerated Math program showed a significant improvement.

c) $t = 9.24$; P-value < 0.0001. The control group showed a significant improvement in test scores.

d) $t = 5.78$; P-value < 0.0001. The Accelerated Math group had significantly higher gains than the control group.

25. a) The regression—he wanted to know about association.

b) There is a moderate relationship between cottage cheese and ice cream sales; for every million pounds of cottage cheese, 1.19 million pounds of ice cream are sold on average.

c) Testing if the mean difference is 0.(matched *t*-test). Regression won't answer this question.

d) The company sells more cottage cheese than ice cream, on average.

e) part (a)—linear relationship; residuals have a Normal distribution; residuals are independent with equal variation about the line. (c)—Observations are independent; differences are approximately Normal; less than 10% of all possible months' data.

f) About 71.32 million pounds.

g) (0.09, 2.29)

h) From this regression, every million pounds of cottage cheese sold is associated with an increase in ice cream sales of between 0.09 and 2.29 million pounds.

27. Based on these data, the average weight loss for the clinic is between 8.24 and 10.06 pounds, with 95% confidence. The clinic's claim is plausible.

29. $\chi^2 = 8.23$; P-value $= 0.0414$. There is evidence of an association between *cracker type* and *bloating*. Standardized residuals for the gum cracker are -1.32 and 1.58. Prospects for marketing this cracker are not good.

Note: Page numbers in **boldface** denote a definition; *n* indicates a footnote.

TABLE OF RANDOM DIGITS

Row										
1	96299	07196	98642	20639	23185	56282	69929	14125	38872	94168
2	71622	35940	81807	59225	18192	08710	80777	84395	69563	86280
3	03272	41230	81739	74797	70406	18564	69273	72532	78340	36699
4	46376	58596	14365	63685	56555	42974	72944	96463	63533	24152
5	47352	42853	42903	97504	56655	70355	88606	61406	38757	70657
6	20064	04266	74017	79319	70170	96572	08523	56025	89077	57678
7	73184	95907	05179	51002	83374	52297	07769	99792	78365	93487
8	72753	36216	07230	35793	71907	65571	66784	25548	91861	15725
9	03939	30763	06138	80062	02537	23561	93136	61260	77935	93159
10	75998	37203	07959	38264	78120	77525	86481	54986	33042	70648
11	94435	97441	90998	25104	49761	14967	70724	67030	53887	81293
12	04362	40989	69167	38894	00172	02999	97377	33305	60782	29810
13	89059	43528	10547	40115	82234	86902	04121	83889	76208	31076
14	87736	04666	75145	49175	76754	07884	92564	80793	22573	67902
15	76488	88899	15860	07370	13431	84041	69202	18912	83173	11983
16	36460	53772	66634	25045	79007	78518	73580	14191	50353	32064
17	13205	69237	21820	20952	16635	58867	97650	82983	64865	93298
18	51242	12215	90739	36812	00436	31609	80333	96606	30430	31803
19	67819	00354	91439	91073	49258	15992	41277	75111	67496	68430
20	09875	08990	27656	15871	23637	00952	97818	64234	50199	05715
21	18192	95308	72975	01191	29958	09275	89141	19558	50524	32041
22	02763	33701	66188	50226	35813	72951	11638	01876	93664	37001
23	13349	46328	01856	29935	80563	03742	49470	67749	08578	21956
24	69238	92878	80067	80807	45096	22936	64325	19265	37755	69794
25	92207	63527	59398	29818	24789	94309	88380	57000	50171	17891
26	66679	99100	37072	30593	29665	84286	44458	60180	81451	58273
27	31087	42430	60322	34765	15757	53300	97392	98035	05228	68970
28	84432	04916	52949	78533	31666	62350	20584	56367	19701	60584
29	72042	12287	21081	48426	44321	58765	41760	43304	13399	02043
30	94534	73559	82135	70260	87936	85162	11937	18263	54138	69564
31	63971	97198	40974	45301	60177	35604	21580	68107	25184	42810
32	11227	58474	17272	37619	69517	62964	67962	34510	12607	52255
33	28541	02029	08068	96656	17795	21484	57722	76511	27849	61738
34	11282	43632	49531	78981	81980	08530	08629	32279	29478	50228
35	42907	15137	21918	13248	39129	49559	94540	24070	88151	36782
36	47119	76651	21732	32364	58545	50277	57558	30390	18771	72703
37	11232	99884	05087	76839	65142	19994	91397	29350	83852	04905
38	64725	06719	86262	53356	57999	50193	79936	97230	52073	94467
39	77007	26962	55466	12521	48125	12280	54985	26239	76044	54398
40	18375	19310	59796	89832	59417	18553	17238	05474	33259	50595

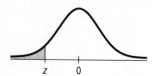

Table Z

Areas under the standard normal curve

Second decimal place in z										
0.09	0.08	0.07	0.06	0.05	0.04	0.03	0.02	0.01	0.00	z
0.0001	0.0001	0.0001	0.0001	0.0001	0.0001	0.0001	0.0001	0.0001	0.0001	−3.8
0.0001	0.0001	0.0001	0.0001	0.0001	0.0001	0.0001	0.0001	0.0001	0.0001	−3.7
0.0001	0.0001	0.0001	0.0001	0.0001	0.0001	0.0001	0.0001	0.0002	0.0002	−3.6
0.0002	0.0002	0.0002	0.0002	0.0002	0.0002	0.0002	0.0002	0.0002	0.0002	−3.5
0.0002	0.0003	0.0003	0.0003	0.0003	0.0003	0.0003	0.0003	0.0003	0.0003	−3.4
0.0003	0.0004	0.0004	0.0004	0.0004	0.0004	0.0004	0.0005	0.0005	0.0005	−3.3
0.0005	0.0005	0.0005	0.0006	0.0006	0.0006	0.0006	0.0006	0.0007	0.0007	−3.2
0.0007	0.0007	0.0008	0.0008	0.0008	0.0008	0.0009	0.0009	0.0009	0.0010	−3.1
0.0010	0.0010	0.0011	0.0011	0.0011	0.0012	0.0012	0.0013	0.0013	0.0013	−3.0
0.0014	0.0014	0.0015	0.0015	0.0016	0.0016	0.0017	0.0018	0.0018	0.0019	−2.9
0.0019	0.0020	0.0021	0.0021	0.0022	0.0023	0.0023	0.0024	0.0025	0.0026	−2.8
0.0026	0.0027	0.0028	0.0029	0.0030	0.0031	0.0032	0.0033	0.0034	0.0035	−2.7
0.0036	0.0037	0.0038	0.0039	0.0040	0.0041	0.0043	0.0044	0.0045	0.0047	−2.6
0.0048	0.0049	0.0051	0.0052	0.0054	0.0055	0.0057	0.0059	0.0060	0.0062	−2.5
0.0064	0.0066	0.0068	0.0069	0.0071	0.0073	0.0075	0.0078	0.0080	0.0082	−2.4
0.0084	0.0087	0.0089	0.0091	0.0094	0.0096	0.0099	0.0102	0.0104	0.0107	−2.3
0.0110	0.0113	0.0116	0.0119	0.0122	0.0125	0.0129	0.0132	0.0136	0.0139	−2.2
0.0143	0.0146	0.0150	0.0154	0.0158	0.0162	0.0166	0.0170	0.0174	0.0179	−2.1
0.0183	0.0188	0.0192	0.0197	0.0202	0.0207	0.0212	0.0217	0.0222	0.0228	−2.0
0.0233	0.0239	0.0244	0.0250	0.0256	0.0262	0.0268	0.0274	0.0281	0.0287	−1.9
0.0294	0.0301	0.0307	0.0314	0.0322	0.0329	0.0336	0.0344	0.0351	0.0359	−1.8
0.0367	0.0375	0.0384	0.0392	0.0401	0.0409	0.0418	0.0427	0.0436	0.0446	−1.7
0.0455	0.0465	0.0475	0.0485	0.0495	0.0505	0.0516	0.0526	0.0537	0.0548	−1.6
0.0559	0.0571	0.0582	0.0594	0.0606	0.0618	0.0630	0.0643	0.0655	0.0668	−1.5
0.0681	0.0694	0.0708	0.0721	0.0735	0.0749	0.0764	0.0778	0.0793	0.0808	−1.4
0.0823	0.0838	0.0853	0.0869	0.0885	0.0901	0.0918	0.0934	0.0951	0.0968	−1.3
0.0985	0.1003	0.1020	0.1038	0.1056	0.1075	0.1093	0.1112	0.1131	0.1151	−1.2
0.1170	0.1190	0.1210	0.1230	0.1251	0.1271	0.1292	0.1314	0.1335	0.1357	−1.1
0.1379	0.1401	0.1423	0.1446	0.1469	0.1492	0.1515	0.1539	0.1562	0.1587	−1.0
0.1611	0.1635	0.1660	0.1685	0.1711	0.1736	0.1762	0.1788	0.1814	0.1841	−0.9
0.1867	0.1894	0.1922	0.1949	0.1977	0.2005	0.2033	0.2061	0.2090	0.2119	−0.8
0.2148	0.2177	0.2206	0.2236	0.2266	0.2296	0.2327	0.2358	0.2389	0.2420	−0.7
0.2451	0.2483	0.2514	0.2546	0.2578	0.2611	0.2643	0.2676	0.2709	0.2743	−0.6
0.2776	0.2810	0.2843	0.2877	0.2912	0.2946	0.2981	0.3015	0.3050	0.3085	−0.5
0.3121	0.3156	0.3192	0.3228	0.3264	0.3300	0.3336	0.3372	0.3409	0.3446	−0.4
0.3483	0.3520	0.3557	0.3594	0.3632	0.3669	0.3707	0.3745	0.3783	0.3821	−0.3
0.3859	0.3897	0.3936	0.3974	0.4013	0.4052	0.4090	0.4129	0.4168	0.4207	−0.2
0.4247	0.4286	0.4325	0.4364	0.4404	0.4443	0.4483	0.4522	0.4562	0.4602	−0.1
0.4641	0.4681	0.4721	0.4761	0.4801	0.4840	0.4880	0.4920	0.4960	0.5000	−0.0

For $z \leq -3.90$, the areas are 0.0000 to four decimal places.

Table Z (cont.)
Areas under the
standard normal curve

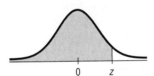

z	0.00	0.01	0.02	0.03	0.04	0.05	0.06	0.07	0.08	0.09
					Second decimal place in z					
0.0	0.5000	0.5040	0.5080	0.5120	0.5160	0.5199	0.5239	0.5279	0.5319	0.5359
0.1	0.5398	0.5438	0.5478	0.5517	0.5557	0.5596	0.5636	0.5675	0.5714	0.5753
0.2	0.5793	0.5832	0.5871	0.5910	0.5948	0.5987	0.6026	0.6064	0.6103	0.6141
0.3	0.6179	0.6217	0.6255	0.6293	0.6331	0.6368	0.6406	0.6443	0.6480	0.6517
0.4	0.6554	0.6591	0.6628	0.6664	0.6700	0.6736	0.6772	0.6808	0.6844	0.6879
0.5	0.6915	0.6950	0.6985	0.7019	0.7054	0.7088	0.7123	0.7157	0.7190	0.7224
0.6	0.7257	0.7291	0.7324	0.7357	0.7389	0.7422	0.7454	0.7486	0.7517	0.7549
0.7	0.7580	0.7611	0.7642	0.7673	0.7704	0.7734	0.7764	0.7794	0.7823	0.7852
0.8	0.7881	0.7910	0.7939	0.7967	0.7995	0.8023	0.8051	0.8078	0.8106	0.8133
0.9	0.8159	0.8186	0.8212	0.8238	0.8264	0.8289	0.8315	0.8340	0.8365	0.8389
1.0	0.8413	0.8438	0.8461	0.8485	0.8508	0.8531	0.8554	0.8577	0.8599	0.8621
1.1	0.8643	0.8665	0.8686	0.8708	0.8729	0.8749	0.8770	0.8790	0.8810	0.8830
1.2	0.8849	0.8869	0.8888	0.8907	0.8925	0.8944	0.8962	0.8980	0.8997	0.9015
1.3	0.9032	0.9049	0.9066	0.9082	0.9099	0.9115	0.9131	0.9147	0.9162	0.9177
1.4	0.9192	0.9207	0.9222	0.9236	0.9251	0.9265	0.9279	0.9292	0.9306	0.9319
1.5	0.9332	0.9345	0.9357	0.9370	0.9382	0.9394	0.9406	0.9418	0.9429	0.9441
1.6	0.9452	0.9463	0.9474	0.9484	0.9495	0.9505	0.9515	0.9525	0.9535	0.9545
1.7	0.9554	0.9564	0.9573	0.9582	0.9591	0.9599	0.9608	0.9616	0.9625	0.9633
1.8	0.9641	0.9649	0.9656	0.9664	0.9671	0.9678	0.9686	0.9693	0.9699	0.9706
1.9	0.9713	0.9719	0.9726	0.9732	0.9738	0.9744	0.9750	0.9756	0.9761	0.9767
2.0	0.9772	0.9778	0.9783	0.9788	0.9793	0.9798	0.9803	0.9808	0.9812	0.9817
2.1	0.9821	0.9826	0.9830	0.9834	0.9838	0.9842	0.9846	0.9850	0.9854	0.9857
2.2	0.9861	0.9864	0.9868	0.9871	0.9875	0.9878	0.9881	0.9884	0.9887	0.9890
2.3	0.9893	0.9896	0.9898	0.9901	0.9904	0.9906	0.9909	0.9911	0.9913	0.9916
2.4	0.9918	0.9920	0.9922	0.9925	0.9927	0.9929	0.9931	0.9932	0.9934	0.9936
2.5	0.9938	0.9940	0.9941	0.9943	0.9945	0.9946	0.9948	0.9949	0.9951	0.9952
2.6	0.9953	0.9955	0.9956	0.9957	0.9959	0.9960	0.9961	0.9962	0.9963	0.9964
2.7	0.9965	0.9966	0.9967	0.9968	0.9969	0.9970	0.9971	0.9972	0.9973	0.9974
2.8	0.9974	0.9975	0.9976	0.9977	0.9977	0.9978	0.9979	0.9979	0.9980	0.9981
2.9	0.9981	0.9982	0.9982	0.9983	0.9984	0.9984	0.9985	0.9985	0.9986	0.9986
3.0	0.9987	0.9987	0.9987	0.9988	0.9988	0.9989	0.9989	0.9989	0.9990	0.9990
3.1	0.9990	0.9991	0.9991	0.9991	0.9992	0.9992	0.9992	0.9992	0.9993	0.9993
3.2	0.9993	0.9993	0.9994	0.9994	0.9994	0.9994	0.9994	0.9995	0.9995	0.9995
3.3	0.9995	0.9995	0.9995	0.9996	0.9996	0.9996	0.9996	0.9996	0.9996	0.9997
3.4	0.9997	0.9997	0.9997	0.9997	0.9997	0.9997	0.9997	0.9997	0.9997	0.9998
3.5	0.9998	0.9998	0.9998	0.9998	0.9998	0.9998	0.9998	0.9998	0.9998	0.9998
3.6	0.9998	0.9998	0.9999	0.9999	0.9999	0.9999	0.9999	0.9999	0.9999	0.9999
3.7	0.9999	0.9999	0.9999	0.9999	0.9999	0.9999	0.9999	0.9999	0.9999	0.9999
3.8	0.9999	0.9999	0.9999	0.9999	0.9999	0.9999	0.9999	0.9999	0.9999	0.9999

For $z \geq 3.90$, the areas are 1.0000 to four decimal places.

Two tail probability	0.20	0.10	0.05	0.02	0.01	
One tail probability	0.10	0.05	0.025	0.01	0.005	

Table T	df	0.20	0.10	0.05	0.02	0.01	df
Values of t_α	1	3.078	6.314	12.706	31.821	63.657	1
	2	1.886	2.920	4.303	6.965	9.925	2
	3	1.638	2.353	3.182	4.541	5.841	3
	4	1.533	2.132	2.776	3.747	4.604	4
	5	1.476	2.015	2.571	3.365	4.032	5
	6	1.440	1.943	2.447	3.143	3.707	6
	7	1.415	1.895	2.365	2.998	3.499	7
	8	1.397	1.860	2.306	2.896	3.355	8
	9	1.383	1.833	2.262	2.821	3.250	9
	10	1.372	1.812	2.228	2.764	3.169	10
	11	1.363	1.796	2.201	2.718	3.106	11
	12	1.356	1.782	2.179	2.681	3.055	12
	13	1.350	1.771	2.160	2.650	3.012	13
	14	1.345	1.761	2.145	2.624	2.977	14
	15	1.341	1.753	2.131	2.602	2.947	15
	16	1.337	1.746	2.120	2.583	2.921	16
	17	1.333	1.740	2.110	2.567	2.898	17
	18	1.330	1.734	2.101	2.552	2.878	18
	19	1.328	1.729	2.093	2.539	2.861	19
	20	1.325	1.725	2.086	2.528	2.845	20
	21	1.323	1.721	2.080	2.518	2.831	21
	22	1.321	1.717	2.074	2.508	2.819	22
	23	1.319	1.714	2.069	2.500	2.807	23
	24	1.318	1.711	2.064	2.492	2.797	24
	25	1.316	1.708	2.060	2.485	2.787	25
	26	1.315	1.706	2.056	2.479	2.779	26
	27	1.314	1.703	2.052	2.473	2.771	27
	28	1.313	1.701	2.048	2.467	2.763	28
	29	1.311	1.699	2.045	2.462	2.756	29
	30	1.310	1.697	2.042	2.457	2.750	30
	32	1.309	1.694	2.037	2.449	2.738	32
	35	1.306	1.690	2.030	2.438	2.725	35
	40	1.303	1.684	2.021	2.423	2.704	40
	45	1.301	1.679	2.014	2.412	2.690	45
	50	1.299	1.676	2.009	2.403	2.678	50
	60	1.296	1.671	2.000	2.390	2.660	60
	75	1.293	1.665	1.992	2.377	2.643	75
	100	1.290	1.660	1.984	2.364	2.626	100
	120	1.289	1.658	1.980	2.358	2.617	120
	140	1.288	1.656	1.977	2.353	2.611	140
	180	1.286	1.653	1.973	2.347	2.603	180
	250	1.285	1.651	1.969	2.341	2.596	250
	400	1.284	1.649	1.966	2.336	2.588	400
	1000	1.282	1.646	1.962	2.330	2.581	1000
	∞	1.282	1.645	1.960	2.326	2.576	∞
Confidence levels		80%	90%	95%	98%	99%	

Two tails: $-t_{\alpha/2}$ 0 $t_{\alpha/2}$, with areas $\frac{\alpha}{2}$ in each tail.

One tail: 0 t_α, with area α in the tail.

Right tail probability		0.10	0.05	0.025	0.01	0.005
Table χ	df					
	1	2.706	3.841	5.024	6.635	7.879
Values of χ^2_α	2	4.605	5.991	7.378	9.210	10.597
	3	6.251	7.815	9.348	11.345	12.838
	4	7.779	9.488	11.143	13.277	14.860
	5	9.236	11.070	12.833	15.086	16.750
	6	10.645	12.592	14.449	16.812	18.548
	7	12.017	14.067	16.013	18.475	20.278
	8	13.362	15.507	17.535	20.090	21.955
	9	14.684	16.919	19.023	21.666	23.589
	10	15.987	18.307	20.483	23.209	25.188
	11	17.275	19.675	21.920	24.725	26.757
	12	18.549	21.026	23.337	26.217	28.300
	13	19.812	22.362	24.736	27.688	29.819
	14	21.064	23.685	26.119	29.141	31.319
	15	22.307	24.996	27.488	30.578	32.801
	16	23.542	26.296	28.845	32.000	34.267
	17	24.769	27.587	30.191	33.409	35.718
	18	25.989	28.869	31.526	34.805	37.156
	19	27.204	30.143	32.852	36.191	38.582
	20	28.412	31.410	34.170	37.566	39.997
	21	29.615	32.671	35.479	38.932	41.401
	22	30.813	33.924	36.781	40.290	42.796
	23	32.007	35.172	38.076	41.638	44.181
	24	33.196	36.415	39.364	42.980	45.559
	25	34.382	37.653	40.647	44.314	46.928
	26	35.563	38.885	41.923	45.642	48.290
	27	36.741	40.113	43.195	46.963	49.645
	28	37.916	41.337	44.461	48.278	50.994
	29	39.087	42.557	45.722	59.588	52.336
	30	40.256	43.773	46.979	50.892	53.672
	40	51.805	55.759	59.342	63.691	66.767
	50	63.167	67.505	71.420	76.154	79.490
	60	74.397	79.082	83.298	88.381	91.955
	70	85.527	90.531	95.023	100.424	104.213
	80	96.578	101.879	106.628	112.328	116.320
	90	107.565	113.145	118.135	124.115	128.296
	100	118.499	124.343	129.563	135.811	140.177

Assumptions for Inference   And the Conditions That Support or Override Them

Proportions (z)

- **One sample**
  1. Individuals are independent.
  2. Sample is sufficiently large.

  1. SRS and $n < 10\%$ of the population.
  2. Successes and failures each ≥ 10.

- **Two groups**
  1. Groups are independent.
  2. Data in each group are independent.

  3. Both groups are sufficiently large.

  1. (Think about how the data were collected.)
  2. Both are SRSs and $n < 10\%$ of populations OR random allocation.
  3. Successes and failures each ≥ 10 for both groups.

Means (t)

- **One Sample** (df $= n - 1$)
  1. Individuals are independent.
  2. Population has a Normal model.

  1. SRS and $n < 10\%$ of the population.
  2. Histogram is unimodal and symmetric.*

- **Matched pairs** (df $= n - 1$)
  1. Data are matched.
  2. Individuals are independent.
  3. Population of differences is Normal.

  1. (Think about the design.)
  2. SRS and $n < 10\%$ OR random allocation.
  3. Histogram of differences is unimodal and symmetric.*

- **Two independent groups** (df from technology)
  1. Groups are independent.
  2. Data in each group are independent.
  3. Both populations are Normal.

  1. (Think about the design.)
  2. SRSs and $n < 10\%$ OR random allocation.
  3. Both histograms are unimodal and symmetric.*

Distributions/Association (χ^2)

- **Goodness of fit** (df $=$ # of cells $- 1$; one variable, one sample compared with population model)
  1. Data are counts.
  2. Data in sample are independent.
  3. Sample is sufficiently large.

  1. (Are they?)
  2. SRS and $n < 10\%$ of the population.
  3. All expected counts ≥ 5.

- **Homogeneity** [df $= (r - 1)(c - 1)$; many groups compared on one variable]
  1. Data are counts.
  2. Data in groups are independent.
  3. Groups are sufficiently large.

  1. (Are they?)
  2. SRSs and $n < 10\%$ OR random allocation.
  3. All expected counts ≥ 5.

- **Independence** [df $= (r - 1)(c - 1)$; sample from one population classified on two variables]
  1. Data are counts.
  2. Data are independent.
  3. Sample is sufficiently large.

  1. (Are they?)
  2. SRSs and $n < 10\%$ of the population.
  3. All expected counts ≥ 5.

Regression (t, df $= n - 2$)

- **Association** between two quantitative variables ($\beta = 0$?)
  1. Form of relationship is linear.
  2. Errors are independent.
  3. Variability of errors is constant.
  4. Errors have a Normal model.

  1. Scatterplot looks approximately linear.
  2. No apparent pattern in residuals plot.
  3. Residuals plot has consistent spread.
  4. Histogram of residuals is approximately unimodal and symmetric, or normal probability plot reasonably straight.*

(*less critical as n increases)

Think			Show				Tell?
Inference about?	One group or two?	Procedure	Model	Parameter	Estimate	SE	Chapter
Proportions	One sample	1-Proportion z-Interval	z	p	$\hat{p}$	$\sqrt{\dfrac{\hat{p}\hat{q}}{n}}$	19
		1-Proportion z-Test				$\sqrt{\dfrac{p_0 q_0}{n}}$	20, 21
	Two independent groups	2-Proportion z-Interval	z	$p_1 - p_2$	$\hat{p}_1 - \hat{p}_2$	$\sqrt{\dfrac{\hat{p}_1 \hat{q}_1}{n_1} + \dfrac{\hat{p}_2 \hat{q}_2}{n_2}}$	22
		2-Proportion z-Test				$\sqrt{\dfrac{\hat{p}\hat{q}}{n_1} + \dfrac{\hat{p}\hat{q}}{n_2}},\ \hat{p} = \dfrac{y_1 + y_2}{n_1 + n_2}$	22
Means	One sample	t-Interval / t-Test	t df $= n - 1$	μ	$\bar{y}$	$\dfrac{s}{\sqrt{n}}$	23
	Two independent groups	2-Sample t-Test / 2-Sample t-Interval	t df from technology	$\mu_1 - \mu_2$	$\bar{y}_1 - \bar{y}_2$	$\sqrt{\dfrac{s_1^2}{n_1} + \dfrac{s_2^2}{n_2}}$	24
	Matched pairs	Paired t-Test / Paired t-Interval	t df $= n - 1$	μ_d	$\bar{d}$	$\dfrac{s_d}{\sqrt{n}}$	25
Distributions (one categorical variable)	One sample	Goodness-of-Fit	χ^2 df $=$ cells $- 1$			$\sum \dfrac{(Obs - Exp)^2}{Exp}$	26
	Many independent groups	Homogeneity χ^2 Test					
Independence (two categorical variables)	One sample	Independence χ^2 Test	χ^2 df $= (r - 1)(c - 1)$				
Association (two quantitative variables)	One sample	Linear Regression t-Test or Confidence Interval for β	t df $= n - 2$	β_1	b_1	$\dfrac{s_e}{s_x \sqrt{n - 1}}$ (compute with technology)	27
		Confidence Interval for μ_ν		μ_ν	$\hat{y}_\nu$	$\sqrt{SE^2(b_1) \cdot (x_\nu - \bar{x})^2 + \dfrac{s_e^2}{n}}$	
		Prediction Interval for y_ν		y_ν	$\hat{y}_\nu$	$\sqrt{SE^2(b_1) \cdot (x_\nu - \bar{x})^2 + \dfrac{s_e^2}{n} + s_e^2}$	
Inference about?	One group or two?	Procedure	Model	Parameter	Estimate	SE	Chapter